Basic Analysis III

Basic Analysis III: Mappings on Infinite Dimensional Spaces

The cephalopods were beginning to teach themselves in the open ocean.

James K. Peterson
Department of Mathematical Sciences
Clemson University

CRC Press
Taylor & Francis Group
Boca Raton London New York

CRC Press is an imprint of the
Taylor & Francis Group, an **informa** business
A CHAPMAN & HALL BOOK

First edition published 2020
by CRC Press
6000 Broken Sound Parkway NW, Suite 300, Boca Raton, FL 33487-2742

and by CRC Press
2 Park Square, Milton Park, Abingdon, Oxon, OX14 4RN

CRC Press is an imprint of Taylor & Francis Group

ISBN: 978-1-138-05508-7 (hbk)
ISBN: 978-1-315-16620-9 (ebk)
LCCN: 2019059882

Dedication We dedicate this work to all of our students who have been learning these ideas of analysis through our courses. We have learned as much from them as we hope they have from us. We are a firm believer that all our students are capable of excellence and that the only path to excellence is through discipline and study. We have always been proud of our students for doing so well on this journey. We hope these notes in turn make you proud of our efforts.

Abstract This book introduces you to more about analysis by looking at spaces of various types. Spaces are collections of objects with certain properties and here we will focus on three types: metric, normed and inner product spaces. Our emphasis will be on the new things we find when the underlying vector spaces are not finite dimensional. Hence, this text continues your training in the *abstract* way of looking at the world. As always, we feel that is a most important skill to have when your life's work will involve quantitative modeling to gain insight into the real world.

Acknowledgments I want to acknowledge the great debt I have to my wife, Pauli, for her patience in dealing with the long hours spent in typing and thinking. You are the love of my life.

The cover for this book is an original painting by us done in July 2017. It shows the cephalopods have now learned to teach themselves abstract mathematics in the open sea.

Table of Contents

Part I

Introduction

Chapter 1

Introduction

We believe that all students who are seriously interested in mathematics at the master's and doctoral level should have a passion for analysis even if it is not the primary focus of their own research interests. So you should all understand that our own passion for the subject will shine though in the notes that follow! And, it goes without saying that we assume that you are all mature mathematically and eager and interested in the material! Now, the present course focuses on ideas from linear analysis and linear functional analysis which is the next step after you have learned the basic concepts of analysis covered in (Peterson (18) 2020) and (Peterson (20) 2020). Also, for those of you who are preparing to take a qualifying or preliminary examination in analysis, this material is part of the material you will be tested on. The senior year in analysis and this course generally comprise material from fairly standard textbooks. Of course, in our notes we put our own spin on it, but we encourage you to read widely.

1.1 The Analysis Courses

In outline form, the basic analysis courses should cover the following material using textbooks equivalent to the ones listed below:

(A): Undergraduate Analysis, text **Advanced Calculus: An Introduction to Analysis**, by Fulks (Fulks (5) 1978). Our take on this material is seen in (Peterson (18) 2020). Additional material not covered in courses now but very useful is discussed in (Peterson (20) 2020).

(B): Introduction to Abstract Spaces, text **Introduction to Functional Analysis and Applications**, by Kreyszig (Kreyszig (11) 1989). Our take on this material is found in (Peterson (19) 2020).

(C): Measure Theory and Abstract Integration, texts **General Theory of Functions and Integration**, by Taylor (Taylor (24) 1985) and **Real Analysis** by Royden (Royden (22) 2017). Our take on this material is developed in (Peterson (16) 2019).

(D): Functional Analysis, texts such as **Introduction to Functional Analysis and Applications**, by Ervin Kreyszig and **Introduction to Topology and Modern Analysis** by Simmons (Simmons (23) 1963) are very nice introductions. Essentially, at this point we wish to blend topology, analysis and algebra together. Our take on this material is seen in (Peterson (17) 2020).

In addition, a nice book that organizes the many interesting examples and counterexamples in this area is good to have on your shelf. We recommend the text **Counterexamples in Analysis** (Gelbaum and Olmstead (6) 2003). There are thus essentially five courses required to teach you enough of the concepts of mathematical analysis to enable you to read technical literature (such as engineering, control, physics, mathematics, statistics and so forth) at the beginning research level. Here are some more details about these courses.

1.1.1 Senior Level Analysis

Typically, this is a full two semester sequence that discusses thoroughly what we would call the analysis of functions of a real variable. This is the material we cover in (Peterson (18) 2020). This two semester sequence covers the following:

Advanced Calculus I: This course studies sequences and functions whose domain is simply the real line. There are, of course, many complicated ideas, but everything we do here involves things that act on real numbers to produce real numbers. If we call these things that act on other things, **OPERATORS**, we see that this course is really about real–valued operators on real numbers. This course invests a lot of time in learning how to be precise with the notion of convergence of sequences of objects, which happen to be real numbers, to other numbers.

1. Basic Logic, Inequalities for Real Numbers, Functions
2. Sequences of Real Numbers, Convergence of Sequences
3. Subsequences and the Bolzano - Weierstrass Theorem
4. Cauchy Sequences
5. Continuity of Functions
6. Basic Convex Analysis
7. Consequences of Continuity
8. Uniform Continuity
9. Differentiability of Functions
10. Consequences of Differentiability
11. Taylor Series Approximations
12. Calculus in $\Re^2$

Advanced Calculus II: In this course, we rapidly become more abstract. First, we develop carefully the concept of the Riemann Integral. We show that although differentiation is intellectually quite a different type of limit process, it is intimately connected with the Riemann integral. Also, for the first time, we begin to explore the idea that we could have sequences of objects other than real numbers. We study carefully their convergence properties. We learn about two fundamental concepts: pointwise and uniform convergence of sequences of objects called functions. We are beginning to see the need to think about sets of objects, such as functions, and how to define the notions of convergence and so forth in this setting.

1. The Riemann Integral
2. Sequences of Functions
3. Uniform Convergence of Sequence of Functions
4. Series of Functions
5. Fourier Series
6. Power Series and Applications to ODEs

There is also much material from functions of n variables such as the inverse and implicit funtion theorem, operators on finite dimensional vector spaces and much more that we simply do not cover anymore in courses. Still this material needs to be mastered. We assume serious students will want to learn this on their own and this is the content of (Peterson (20) 2020).

1.1.2 The Graduate Analysis Courses

There are three basic courses at most universities. First, linear analysis, then measure and integration and finally, functional analysis. Linear analysis is the core analysis course and all master's students usually take it. Also, the linear analysis and measure theory courses are the two courses which we test prospective Ph.D. students on as part of the analysis preliminary examination. The content of these courses must also fit within a web of other responsibilities. Many students are typically weak in abstraction coming in, so if we teach the material too fast, we lose them. Now if 20 students take linear analysis, usually 15 or 75% are already committed to an M.S. program which emphasizes Operations Research, Statistics, Algebra/ Combinatorics or Computation in addition to Applied Analysis. Hence, currently, there are only about 5 students in linear analysis who might be interested in an M.S. specialization in analysis. The other students typically either don't like analysis at all and are only there because they have to be or they like analysis but it is part of their studies in number theory, partial differential equations for the Computation area and so forth. Either way, the students will not continue to study analysis for a degree specialization. However, we think it is important for all students to both know and appreciate this material. Traditionally, there are several ways to go.

The Cynical Approach: Nothing you can do will make students who don't like analysis change their mind. So teach the material hard and fast and target the $2-3$ students who can benefit. The rest will come along for the ride and leave the course convinced that analysis is just like they thought – too hard and too complicated. If you do this approach, you can pick about any book you like. Most books for our students are too abstract and so are very hard for them to read. But the $2-3$ students who can benefit from material at this level will be happy with the book. *We admit this is not our style although some think it is a good way to find the really bright analysis students.* We prefer the alternate Enthusiastic *"maybe I can get them interested anyway"* Approach: The instructor scours the available literature in order to make up notes to lead the students *"gently"* into the required abstract way of thinking. We haven't had much luck finding a published book for this so as is our preferred plan of action: we type up notes such as the ones you have in your hand. These notes start out handwritten and slowly mature into the typed versions. We believe it is important to actively try to get all the students interested but, of course, this is never completely successful. However, we still think there is great value in this approach and it is the one we have been trying for many years. The course in Linear Analysis is what we cover (and more) in this text, our third course on basic analysis.

Linear Analysis: Our constraints here are to choose content so the students are adequately exposed to a more abstract way of thinking about the world. This is the material we cover in the notes you are presently starting to read. We would like to cover

- Metric Spaces.
- Vector Spaces with a Norm.
- Vector spaces with an Inner Product.
- Linear Operators
- Basic Linear Functional Analysis such as Hahn - Banach Theorems and so Forth.
- The Open Mapping and Closed Graph Theorem.

It doesn't sound like much but there is a lot of material in here the students haven't seen. For example, we typically focus a lot on how we are really talking about sets of objects with some additional structure. A set plus a way to measure distance between objects gives a metric space; if we can add and scale objects, we get a vector space; if we have a vector space and add the structure that allows us to project a vector to a subspace, we get an inner product space. We also mention we could have a set of objects and define one operation to obtain a group or if we define a special collection of sets we call open, we get a topological space and so forth. If we work hard, we can help open their minds to the fact that each of the many sub disciplines in the Mathematical Sciences focuses on special structure we add to a set to help us solve problems in that arena.

There are lots of ways to cover the important material in these topic areas and even many ways to decide on exactly what is important from metric, normed and inner product spaces. So there is that kind of freedom, but not so much freedom that you can decide to drop say, inner product spaces. For example, we could use Stürm - Liouville systems as an example when the discussion turns to eigenvalues of operators. It is nice to use projection theorems in an inner product setting as a big finishing application, but remember the students are weak in background, e.g., their knowledge of ordinary differential equations and Calculus in $\Re^n$ is normally weak. So we are limited in our coverage of the completeness of an orthonormal sequence in an inner product space in many respects. In fact, since the students don't know measure theory (which is in (Peterson (16) 2019)) the discussion of Hilbert Spaces of functions is inherently weak. However, we decided to address this by constructing the reals from scratch and then showing how to build a set of functions $\mathbb{L}_2([a, b])$ which is a Hilbert Space without having to resort to measure theory. Of course, this is all equivalent to what you would get with a measure theory approach, but it has the advantage of allowing the students to see how all the details work without brushing stuff under the rug, so to speak! We always prefer that approach as we think everyone at this level wants to have a really good understanding of stuff and in general, they don't like to be told to just accept results. If you look carefully at that material, you need to cover Hahn - Banach results and dual spaces to to do it right. However, we run out of time to cover such advanced topics in one semester. But we can do it all in a book treatment! Further, we believe this text is about teaching the students the abstract way of thinking about problems and hence, we feel there is great value in teaching very, very carefully the basics of this material.

Also, this material is a nice prerequisite for partial differential equations and ordinary differential equations as well as statistics and probability courses. This course takes a huge amount of time for lecture preparation and student interaction in your office, so when we teach this material, we slow down in our research output!

In more detail, in this book, we begin to rephrase all of our knowledge about convergence of sequence of objects in a much more general setting.

1. Metric Spaces: A set of objects and a way of measuring distance between objects which satisfies certain special properties. This function is called a **metric** and its properties were chosen to mimic the properties that the absolute value function has on the real line. We learn to understand convergence of objects in a general metric space. It is really important to note that there is NO additional structure imposed on this set of objects; no linear structure (i.e. vector space structure), no notion of a special set of elements called a basis which we can use to represent arbitrary elements of the set. The **metric** in a sense generalizes the notion of distance between numbers. We can't really measure the size of an object by itself, so we do not yet have a way of generalizing the idea of size or length.

   A fundamentally important concept now emerges: the notion of completeness and how it is related to our choice of metric on a set of objects. We learn a clever way of constructing an abstract representation of the completion of any metric space, but at this time, we have no practical way of seeing this representation.

2. Normed Spaces: We add linear structure to the set of objects and a way of measuring the magnitude of an object; that is, there is now an operation we think of as addition and another operation which allows us to scale objects and a special function called a **norm** whose value for a given object can be thought of as the object's magnitude. We

then develop what we mean by convergence in this setting. Since we have a vector space structure, we can now begin to talk about a special subset of objects called a **basis** which can be used to find a useful way of representing an arbitrary object in the space.

Another most important concept now emerges: the cardinality of this basis may be finite or infinite. We begin to explore the consequences of a space being finite versus infinite dimensional.

3. Inner Product Spaces: To a set of objects with vector space structure, we add a function called an **inner product** which generalizes the notion of dot product of vectors. This has the extremely important consequence of allowing the inner product of two objects to zero even though the objects are not the same. Hence, we can develop an abstract notion of the orthogonality of two objects. This leads to the idea of a basis for the set of objects in which all the elements are mutually orthogonal. We then finally can learn how to build representations of arbitrary objects efficiently.
4. Completions: We learn how to complete an arbitrary metric, normed or inner product space in an abstract way, but we know very little about the practical representations of such completions.
5. Linear Operators: We study a little about functions whose domain is one set of objects and whose range is another. These functions are typically called operators. We learn a little about them here.
6. Linear Functionals: We begin to learn the special role that real-valued functions acting on objects play in analysis. These types of functions are called linear functionals and learning how to characterize them is the first step in learning how to use them. We just barely begin to learn about this here.

Measure Theory This is covered in (Peterson (16) 2019). This course generalizes the notion of integration to a very abstract setting. Roughly speaking, we first realize that the Riemann integral is a linear mapping from the space of bounded real-valued functions on a compact interval into the reals which has a number of interesting properties. We then study how we can generalize such mappings so that they can be applied to arbitrary sets X, a special collection of subsets of X called a sigma-algebra and a new type of mapping called a measure which on $\Re$ generalizes our usual notion of the length of an interval. We discuss the following:

1. The Riemann Integral
2. Measures on a Sigma-algebra S in the set X and Integration with respect to the Measure.
3. Measures Specialized to Sigma-algebras on the set $\Re^n$ and Integrations with respect to These Measures. The canonical example of this is Lebesgue measure on $\Re^n$.
4. Differentiation and Integration in These Abstract Settings and Their Connections.

1.1.3 More Advanced Courses

It is also recommended that students consider taking a course in what is called Functional Analysis. Our version is (Peterson (17) 2020) which tries hard to put topology, analysis and algebra ideas together into one pot. It is carefully designed for self study as this course is just not being offered much as it should be. While not part of the qualifying examination, in this course, we can finally develop in a careful way the necessary tools to work with linear operators, weak convergence and so forth. This is a huge area of mathematics, so there are many possible ways to design an introductory course. Our notes cover

1. Topology

2. Topological Vector Spaces
3. Differential Geometry
4. Constructing Topologies for Graph Based Computational Models
5. Basic Degree Theory

We use Octave (Eaton et al. (2) 2020), which is an open source GPL licensed (Free Software Foundation (4) 2020) clone of MATLAB®, as a computational engine and we are not afraid to use it as an adjunct to all the theory we go over. Of course, you can use MATLAB® (MATLAB (14) 2018 - 2020) also if your university or workplace has a site license or if you have a personal license. Get used to it: theory, computation and science go hand in hand! Well, buckle up and let's get started!

1.2 Table of Contents

This text is based on quite a few years of teaching first year graduate courses in analysis and began from handwritten notes starting roughly in the late 1990's but with material being added all the time. Along the way, these notes have been helped by the students we have taught and also by our own research interests as research informs teaching and teaching informs research in return.

In this text, we have chosen to go over the following blocks of material.

Part One These are our beginning remarks you are reading now which are in Chapter 1.

Part Two: Metric Spaces Here we are concerned with metric spaces and how we construct their completion.

- In Chapter 2 we spend a lot of time proving something most of you have taken for granted for a long time. This is the existence of the set $\Re$ as a complete metric space where Cauchy sequences always converge to an element of $\Re$. You should go through this carefully at this point in your education so you understand it really is a construction process and the thing we call $\Re$ is what we choose to identify with a space of equivalence classes of the right things. We also talk a lot about general metric spaces and the familiar friends we call the sequence spaces. But now we very explicitly explore everything more abstractly.
- Chapter 3 is a long chapter on how we complete metric spaces using the fact that we have the complete metric space $\Re$ to build on. We put a lot of effort into completing various space of integrable functions for our use later. You should know this sort of completion is also done using integration based on measures but that is discussed in (Peterson (16) 2019) and so that point of view is inaccessible for now. But please keep this in mind as the proper way to proceed is to know both approaches.

Part Three: Normed Linear Spaces We discuss the idea of normed linear spaces carefully.

- In Chapter 4 we go back to covering the ideas of vector spaces again. In the first two books (Peterson (18) 2020) and (Peterson (20) 2020) we have gone over these ideas quite a bit and tried to give you extensive examples of how this stuff is used. But here, we want to be more abstract such as proving all vector spaces have a basis by using a non constructive approach involving Zorn's Lemma. So this is uncharted territory.
- Chapter 5 looks at normed linear spaces again. We have discussed these a fair bit in the first two books which are mostly modeled on $\Re^n$ with occasional side trips to function spaces. But now we want to introduce norms properly and discuss important ideas such as Schauder bases for various spaces. We also want to revisit compactness again and stress how our vision of what that means changes in infinite dimensional spaces.

- Chapter 6 is about linear operators between normed linear spaces and we no longer focus mostly on finite dimensional spaces. Hence matrix representations are not as useful. We also talk about the extension of the eigenvalue - eigenvector problem to this setting.

Part Four: Inner Product Spaces In this part, we go over many ideas loosely built around the idea of inner product spaces.

- Chapter 7 introduces the inner product on a vector space and works out the eigenvalue and eigenvector structure of a Hermitian matrix using matrix norm ideas. We show such matrices always have an orthonormal basis of eigenvectors. This is almost the same as the proof we used for symmetric matrices in (Peterson (20) 2020). It is also almost the same proof we use later for the case of a self-adjoint linear operator in Chapter 14 which we do later. Please pay attention to how these proofs need to be modified to make the transition to the infinite dimensional setting. We also introduce Hilbert Spaces and spend a lot of time completing the integrable functions with the least squares norm so we can characterize the Hilbert Space of square integrable functions better. We finish with some properties of inner product space.
- Chapter 8 works out the detail of complete orthonormal sequences in a Hilbert Space and projections.
- Chapter 9 is a first basic discussion of dual spaces and the double dual of a space. We find the duals of a few sequence spaces too. We also introduce the idea of weak convergence.
- In Chapter 10 we discuss the Hahn - Banach theorem and its extensions. We use Zorn's Lemma here again.
- Chapter 11 We now have better tools at our disposal and we can talk about reflexive spaces intelligently. We also work out the dual of the set of continuous functions which is a non trivial result which requires we introduce Riemann - Stieltjes integration in a fairly quick way. We cover that topic more fully in (Peterson (16) 2019) so if you want you can look ahead at that discussion. We finish with sesquilinear forms and adjoints.
- Chapter 12 presents some important theorems from linear functional analysis called the Open Mapping and Closed Graph Theorem. To do that properly, we also have to discuss a very abstract idea called first and second category metric spaces.

Part Five: Operators Here, we want to work carefully the theory of self-adjoint linear operators and one example of them: the Stürm - Liouville Operators.

- Chapter 13 works out the details of the Stürm - Liouville Operators. We reintroduce a few ideas from ODEs too to help you remember some of those details. We look carefully at the eigenvalue - eigenfunction behavior here.
- In Chapter 14 we look carefully at the general details of the self-adjoint operators and work out their eigenvalue - eigenfunction properties just like we did for symmetric and Hermitian matrices earlier. The tools we use are a bit different as the spaces are infinite dimensional.

Part Six: Topics in Applied Modeling We want to finish with a non standard application of these ideas to game theory.

- Chapter 15 discusses bounded charges on rings and fields of subsets.
- Chapter 16 is an introduction to games of transferable utility. In our discussion, we even have to look at a Riemann Integral extension, so great fun!

Part Seven: Summing It All Up In Chapter 17 we talk about the things you have learned here and where you should go next to continue learning basic analysis.

There is much more we could have done, but these topics are a nice introduction into the further use of abstraction in your way of thinking about models and should prepare you well for the next step which is measure theory and functional analysis and topology.

Jim Peterson
School of Mathematical and Statistical Sciences
Clemson University

Part II

Metric Spaces

Chapter 2

Metric Spaces

How do we measure the distance between two objects? In fact, what do we mean by the term **distance**? This chapter will explore this idea carefully.

2.1 The Construction of the Reals

First, it is time we construct the real numbers. We have always assumed the real numbers satisfy the completeness axiom and that Cauchy sequences of real numbers converge to a real number. Now let's show this in more detail. Recall

Axiom 1 The Completeness Axiom

Let S be a set of real numbers which is nonempty and bounded above. Then the supremum of S exists and is finite.

Let S be a set of real numbers which is nonempty and bounded below. Then the infimum of S exists and is finite.

Of course, this just assumes we have this set called the real numbers. Let's see if we can make it less mysterious. Consider the set of all rational numbers and let's measure the distance between two rationals x and y by the usual absolute value function, $d(x, y) = |x - y|$. It is easy to see

M1: The number $d(x, y)$ is always defined and non-negative. Note this number is a rational number too.

M2: $d(x, y) = 0 \Longleftrightarrow x = y$.

M3: $d(x, y) = d(y, x)$; i.e. the order in which the rationals occurs in the distance calculation does not matter.

M4: Since $|x+y| \leq |x+z|+|z+y|$ by the triangle inequality, we know $d(x, y) \leq d(x, z)+d(z, y)$ also and so d satisfies the triangle inequality.

More formally, $d : \mathbb{Q} \times \mathbb{Q} \rightarrow \mathbb{Q}$.

A set F is a **field** if there are operations $\oplus$ and $\odot$ on F that satisfy

F1: $x \oplus (y \oplus z) = (x \oplus y) \oplus z$ and $x \odot (y \odot z) = (x \odot y) \odot z$: associativity of $\oplus$ and $\odot$.

F2: $x \oplus y = y \oplus x$ and $x \odot y = y \odot x$: commutativity of $\oplus$ and $\odot$.

F3: $x \odot (y \oplus z) = x \odot y \oplus x \odot z$: distributivity of $\odot$ over $\oplus$.

F4: $x \oplus 0 = x$: existence of the additive identity.

F5: $x \odot 1 = x$: existence of the multiplicative identity.

F6: $x \oplus -x = 0$: existence of the additive inverse.

F7: If $x \neq 0$, then there is y so that $x \odot y = 1$.

Homework

Exercise 2.1.1 *Assuming the real numbers are a field (we will be constructing them in a bit) convince yourself the complex numbers are a field.*

Exercise 2.1.2 *Look up the quaternions. Do they form a field? They actually form a division ring which has all the properties of a field except $\odot$ is not commutative. Look up division ring (or algebra).*

Exercise 2.1.3 *Look up the octonions and understand why they are not a field. They are a non associative, non commutative division ring (or algebra).*

For $\mathbb{Q}$, $\oplus$ and $\odot$ are just the usual addition and multiplication of rational numbers and it is straightforward to show $\mathbb{Q}$ is a field.

2.1.1 Totally Ordering of the Rationals

A **totally ordered set** is a set X allow with a binary relation $\preceq$ so that

TO1: $x \preceq x$, reflexivity.

TO2: $x \preceq y$ and $y \preceq x$ implies $x = y$, antisymmetry.

TO3: $x \preceq y$ and $y \preceq z$ implies $x \preceq y$, transitivity.

TO4: $x \preceq y$ or $y \preceq x$, totality, i.e. any two elements can be compared.

For $\mathbb{Q}$, we use $\preceq = \leq$ for this ordering. Hence $\mathbb{Q}$ is a totally ordered field.

If the field elements are compatible with the ordering $\preceq$, we must have

Comp1: $x \preceq y$ implies $(x \oplus z) \preceq (y \oplus z)$, compatibility of $\preceq$ and $\oplus$.

Comp2: $0 \preceq x$ and $0 \preceq y$ then $0 \preceq x \odot y$, preservation of $\preceq$ under $\odot$.

Note the order $\leq$ and the operations $+$ and $\cdot$ for $\mathbb{Q}$ are comparable. Hence $\mathbb{Q}$ is a totally ordered field whose ordering is compatible with its operations. For any totally ordered field F whose ordering is compatible with its operations, we can define the idea of a Cauchy sequence of rationals as follows: the sequence $(x_n) \in \mathbb{Q}$ is said to be a Cauchy sequence over F if for all $\epsilon \in F > 0$ there is N so that $n, m > N$ implies $|x_n - x_m| < \epsilon$. In particular, we can choose $F = \mathbb{Q}$ itself. Then the sequence $(x_n) \in \mathbb{Q}$ is said to be a Cauchy sequence over $\mathbb{Q}$ if for all $\epsilon \in \mathbb{Q} > 0$ there is N so that $n, m > N$ implies $|x_n - x_m| < \epsilon$.

The mapping $|\cdot| : \mathbb{Q} \times \mathbb{Q} \to \mathbb{Q}$ that satisfies $M1$, $M2$, $M3$ and $M4$ is called a **metric** over the field $\mathbb{Q}$ and we call the pair $(\mathbb{Q}, |\cdot|)$ a metric space.

2.1.2 The Construction of the Reals

Now let's get on with the construction of the real numbers. To construct the real number system, we can use various approaches, but we will use one built on equivalence classes of Cauchy sequences of rational numbers.

Step 1: Let

$$S = \{(x_n) \mid x_n \in \mathbb{Q}, (x_n) \text{ is a Cauchy sequence in } (\mathbb{Q}, |\cdot|)\}$$

Step 2: Define an equivalence relation on S by

$$(x_n) \sim (y_n) \iff \lim_{n\to\infty} |x_n - y_n| = 0$$

Note convergence here is in the field $\mathbb{Q}$.

Step 3: Let $\widetilde{\mathbb{Q}} = S/\sim$, the set of all equivalence classes in S under $\sim$. Let the elements of $\widetilde{\mathbb{Q}}$ be denoted by $[(x_n)]$.

Note we don't know that Cauchy sequences of rational numbers converge, so we cannot use that in any of our arguments. We are going to identify the set $\widetilde{\mathbb{Q}}$ with the reals by showing it is a field which is totally ordered whose operations are compatible with the ordering and finally by showing it is a complete field: i.e. every non empty subset of $\widetilde{\mathbb{Q}}$ which is bounded above has a least upper bound. This property is what we called the Completeness Axiom earlier.

Property One: $\widetilde{\mathbb{Q}}$ is a field. In what follows, the equivalence class associated with the constant rational sequence $(r, r, r, \ldots)$ will be denoted $[(\bar{r})]$ for convenience. There are many things to check here and we will leave most of them to you as they are tedious. For the set $\widetilde{\mathbb{Q}}$ define operations $\oplus$ and $\odot$ by

$$[(x_n)] \oplus [(y_n)] = [(x_n + y_n)], \quad [(x_n)] \odot [(y_n)] = [(x_n\, y_n)]$$

It is easy to see these are well-defined. It is straightforward to verify

F1: $[(x_n)] \oplus ([(y_n)] \oplus [(z_n)]) = ([(x_n)] \oplus [(y_n)]) \oplus [(z_n)]$ and $[(x_n)] \odot ([(y_n)] \odot [(z_n)]) = ([(x_n)] \odot [(y_n)]) \odot [(z_n)]$: associativity of $\oplus$ and $\odot$.

F2: $[(x_n)] \oplus [(y_n)] = [(y_n)] \oplus [(x_n)]$ and $[(x_n)] \odot [(y_n)] = [(y_n)] \odot [(x_n)]$: commutativity of $\oplus$ and $\odot$.

F3: $[(x_n)] \odot ([(y_n)] \oplus [(z_n)]) = [(x_n)] \odot [(y_n)] \oplus [(x_n)] \odot [(z_n)]$: distributivity of $\odot$ over $\oplus$.

F4: $[(x_n)] \oplus [(\bar{0})] = [(x_n)]$: existence of the additive identity.

F5: $[(x_n)] \odot [(\bar{1})] = [(x_n)]$: existence of the multiplicative identity.

F6: $[(x_n)] \oplus [(-x_n)] = [(\bar{0})]$: existence of the additive inverse.

F7: If $[(x_n)] \neq [(\bar{0})]$, then there is $[(y_n)]$ so that $[(x_n)]\,[(y_n)] = [(\bar{1})]$. This one is a bit tricky. Assume $[(x_n)] \neq [(\bar{0})]$. Remember we don't know that $\lim_n x_n$ exists at all so we have to argue differently. The sequence (x_n) is equivalent to the sequence $(\hat{x}_n = x_n + sign(x_n)\,\frac{1}{n})$. If $x_n = -\frac{1}{n}$, $\hat{x}_n = \frac{2}{n}$ which is not zero and if $x_n = \frac{1}{n}$, $\hat{x}_n = \frac{2}{n}$ which is also not zero. Hence $(\hat{x}_n)$ is not $[(\bar{0})]$ and is equivalent to $[(x_n)]$. Let $(\hat{y}_n = \frac{1}{\hat{x}_n})$. Then $\hat{x}_n\,\hat{y}_n = 1$ always. Hence, the multiplicative inverse of $[(x_n)]$ is $[(\hat{y}_n)]$.

This shows $\widetilde{\mathbb{Q}}$ is a field.

2.1.2.1 Homework

Exercise 2.1.4 *Let X be a set of objects and let the binary relationship be set containment, $\subset$. Why is X with $\subset$ not a totally ordered set?*

Exercise 2.1.5 `mergesort` *is a useful algorithm for ordering a list of objects which requires that the operation $A \leq B$ make sense. Can you define* `mergesort` *on complex numbers?*

Exercise 2.1.6 *Are the expansions of a $\sqrt{2}$ with respect to base 3 and base 5 in the same equivalence class as defined above? Show why or why not this is true in detail.*

Exercise 2.1.7 *Is the expansion of $e = \lim_n (1 + \frac{1}{n})^n$ with respect to base 10 in the same equivalence as the number we call e? Show why or why not this is true in detail.*

2.1.3 The Equivalence Classes of Q are Totally Ordered

Property Two: $\tilde{\mathbb{Q}}$ is a totally ordered set. We define the ordering $\preceq$ by

$$(x_n) \preceq (y_n) \iff (x_n) \sim (y_n) \text{ or } \exists\, N\; n > N \Longrightarrow x_n \leq y_n$$

We need to understand some of the details of this ordering.

Lemma 2.1.1 If (x_n^1) is a Subsequence of (x_n), then $(x_n^1) \sim (x_n)$

> *Let (x_n) be a Cauchy sequence of rationals. Then if (x_{n_k}) is a subsequence of (x_n), $(x_{n_k}) \sim (x_n)$.*

Proof 2.1.1
Let $y_k = x_{n_k} - x_k$. Then since $(x_{n_k} - x_k)$ is a Cauchy sequence, given $\epsilon > 0$, there is N so that $p, q > N$ imply

$$|(x_p - x_q)| < \epsilon \Longrightarrow |x_{n_k} - x_k| < \epsilon$$

when $k, n_k > N$. This tells us $|y_k| < \epsilon$ when k is sufficiently large. Hence, $\lim_k |x_{n_k} - x_k| = 0$ and so $(x_{n_k}) \sim (x_n)$. ■

Now define $[(x_n)] \preceq [(y_n)]$ to mean any representative (u_n) of $[(x_n)]$ and (v_n) of $[(y_n)]$ also satisfies $(u_n) \preceq (v_n)$.

Lemma 2.1.2 $[(x_n)] \preceq [(y_n)]$ if and only if $(x_n) \preceq (y_n)$

> *$[(x_n)] \preceq [(y_n)]$ if and only if $(x_n) \preceq (y_n)$.*

Proof 2.1.2
Let (x_n) and (y_n) be Cauchy sequences of rationals. Then if $(x_n) \preceq (y_n)$, there is N so that $x_n \leq y_n$ for $n > N$ or the sequences are similar. The case where they are similar is easy, so we will only argue the case where they are different. First, we show there is M so that $u_n \leq y_n$ for $n \geq M$. Suppose this is not true. Then for all $k > M$, there are $n_k > k$ so that $u_{n_k} > y_{n_k}$. Now since $(u_n) - (x_n)$ is a Cauchy sequence, given $\epsilon > 0$, there is P so that $n_k > P$ implies

$$-\epsilon < u_{n_k} - x_{n_k} < \epsilon$$

So $n_k > P$ gives

$$0 < u_{n_k} - y_{n_k} = u_{n_k} - x_{n_k} + x_{n_k} - y_{n_k} \leq u_{n_k} - x_{n_k} < \epsilon$$

Thus, $n_k > P$

$$|u_{n_k} - y_{n_k}| < \epsilon$$

This says $(u_{n_k}) \sim (y_{n_k}) \sim (y_n)$ *which is not possible as* $(u_{n_k}) \sim (x_n)$ *and we assume* (x_n) *and* (y_n) *are not similar. Hence, we must be able to find a N so that* $u_n \leq y_n$ *for* $n > N$*. The argument for an arbitrary* (v_n) *representative of* $[(y_n)]$ *is almost the same.*

If we assume $[(x_n)] \preceq [(y_n)]$*, then by definition, we have* $(x_n) \preceq (y_n)$*.* ■

Let's show $\preceq$ is a total ordering on $\widetilde{\mathbb{Q}}$.

TO1: $[(x_n)] \preceq [(x_n)]$, reflexivity.

TO2: $[(x_n)] \preceq [(y_n)]$ and $[(y_n)] \preceq [(x_n)]$ implies $[(x_n)] = [(y_n)]$, antisymmetry. If all both are similar, this is true. Otherwise, there is N_1 and N_2 so that if $n > \max\{N_1, N_2\}$, then $n > N$ implies $x_n \leq y_n$ and $y_n \leq x_n$. But then $\lim_n |x_n - y_n| = 0$ and they are the same equivalence class.

TO3: $[(x_n)] \preceq [(y_n)]$ and $[(y_n)] \preceq [(z_n)]$ implies $[(x_n)] \preceq [(z_n)]$, transitivity.

TO4: $[(x_n)] \preceq [(y_n)]$ or $[(y_n)] \preceq [(x_n)]$, totality. This property is a little harder to see. Let's assume(x_n) and (y_n) are not similar. Let's also assume for all m, there is m_k and m_{k+1} so that $m < m_k < m_{k+1}$ and

$$x_{m_k} \quad \geq \quad y_{m_k} \text{ and } x_{m_{k+1}} \leq y_{m_{k+1}}$$

Let $\epsilon > 0$ be given, then since $(x_n - y_n)$ is a Cauchy sequence, there is N so that

$$n, m > N \quad \Longrightarrow \quad -\epsilon/2 < (x_n - y_n) - (x_m - y_m) < \epsilon/2$$

or for $m > N$

$$(x_m - y_m) - \epsilon/2 < (x_{m_k} - y_{m_k}) < (x_m - y_m) + \epsilon/2$$
$$(x_m - y_m) - \epsilon/2 < (x_{m_{k+1}} - y_{m_{k+1}}) < (x_m - y_m) + \epsilon/2$$

or for $m > N$

$$(x_m - y_m) - \epsilon/2 \leq 0 \text{ and } 0 \geq (x_m - y_m) + \epsilon/2 \quad \Longrightarrow \quad -\epsilon/2 \leq (x_m - y_m) \leq \epsilon/2$$

This implies $\lim_m |x_m - y_m| = 0$ or $(x_n) \sim (y_n)$ which we have assumed is not true. This argument is for the sign alteration $+$, $-$ etc., but it is easy to see the argument works just as well if we start with a $-$ and then alternate. Hence, there must be N so that either $x_n \geq y_n$ or $y_n \geq x_n$ if $n > N$. Thus, either $(x_n) \leq (y_n)$ or $(y_n) \leq (x_n)$. Therefore the ordering is total; i.e. any two elements can be compared.

This shows $\widetilde{\mathbb{Q}}$ is a totally ordered field.

2.1.3.1 Homework

Exercise 2.1.8 *Let* (x_n) *be the decimal expansion of* $\sqrt{2}$ *in base* 10 *and* (y_n) *be the decimal expansion of* $\sqrt{3}$ *in base* 2*. Then* $[(x_n)]$ *and* $[(y_n)]$ *are the equivalence classes for* $\sqrt{3}$ *and* $\sqrt{2}$ *in* $\widetilde{\mathbb{Q}}$*. Show explicitly why* $[(x_n)] \preceq [(y_n)]$*.*

Exercise 2.1.9 *Let (x_n) be the decimal expansion of e in base 10 and (y_n) be the decimal expansion of $\sqrt{5}$ in base 8. Then $[(x_n)]$ and $[(y_n)]$ are the equivalence classes for e and $\sqrt{5}$ in $\widetilde{\mathbb{Q}}$. Show explicitly why $[(y_n)] \preceq [(x_n)]$.*

Exercise 2.1.10 *How do you interpret $3^{\sqrt{2}}$ in $\widetilde{\mathbb{Q}}$?*

Exercise 2.1.11 *How do you interpret $(\sqrt{3})^{\sqrt{2}}$ in $\widetilde{\mathbb{Q}}$?*

2.1.4 The Ordering and the Operations are Compatible

Property Three: The ordering $\preceq$ and the operations $\oplus$ and $\odot$ are compatible

Comp1: $[(x_n)] \preceq [(y_n)]$ implies $([(x_n)] \oplus [(z_n)]) \preceq ([(y_n)] \oplus [(z_n)])$, compatibility of $\preceq$ and $\oplus$.

Comp2: if $[(\bar{0})] \preceq [(x_n)]$ and $[(\bar{0})] \preceq [(y_n)]$ then $[(\bar{0})] \preceq [(x_n)] \odot [(y_n)]$, compatibility of $\preceq$ and $\odot$, preservation of $\preceq$ under $\odot$.

It is easy to see the ordering $\preceq$ we have defined for $\widetilde{\mathbb{Q}}$ is compatible with the operations $\oplus$ and $\odot$. This shows $\widetilde{\mathbb{Q}}$ is a totally ordered field whose ordering is compatible with its operations.

$\widetilde{\mathbb{Q}}$ satisfies an additional property that $\mathbb{Q}$ does not: its order $\preceq$ is **complete**. This means every nonempty subset of $\mathbb{Q}$ that is bounded above, has a least upper bound. We define these terms in the usual way as we did in (Peterson (18) 2020). This is what we have called the **Completeness Axiom**. To prove this, suppose $U \subset \widetilde{\mathbb{Q}}$ is bounded above. This means there is a $[(w_n)]$ so that $[(x_n)] \preceq [(w_n)]$ for all $[(x_n)] \in U$. Since a Cauchy sequence must be bounded, we can replace the bound $[(w_n)]$ by a larger constant rational sequence $[(M)]$. Since U is not empty, choose a $[(z_n)] \in U$. Since it is a bounded set of rational numbers, choose a rational number L so that $[(L)] \preceq [(z_n)]$. Next, we construct sequences (b_n) and (t_n).

Set $b_0 = L$ and $t_0 = M$.

For $n = 0$, consider the number $m_0 = (b_0 + t_0)/2$. If $[(m_0)]$ is an upper bound for U set $t_1 = m_0$ and $b_1 = b_0$. Otherwise, set $b_1 = m_0$ and $t_1 = t_0$.

For $n = 1$, consider the number $m_1 = (b_1 + t_1)/2$. If $[(m_1)]$ is an upper bound for U set $t_2 = m_1$ and $b_2 = b_1$. Otherwise, set $b_2 = m_1$ and $t_2 = t_1$.

If you draw a simple picture of this behavior, you can see we are constructing via induction two Cauchy sequences of rationals (t_n) and (b_n) with $(t_n, t_n, t_n, \ldots)$ an upper bound for U for all n and $(b_n, b_n, b_n, \ldots)$ never an upper bound for U. Hence, the equivalence class $[(t_n)]$ is an upper bound for U. From our construction, you can see $\lim_n |t_n - b_n| = 0$ so $(b_n) \sim (t_n)$. Hence, $[(b_n)] = [(t_n)]$. Now let $[(v_n)]$ be an upper bound for U with $[(v_n)] \prec [(t_n)] = [(b_n)]$. Hence, there is an N so that $v_n < b_n$ if $n > N$. Since each constant sequence $(b_k, b_k, \ldots)$ is not an upper bound, there are sequences (x_n^k) in U so that $(b_k, b_k, \ldots) \preceq (x_1^k, x_2^k, \ldots)$ implying there is N_k so that $b_k < x_j^k$ for $j > N_k$.

Also, we have $(v_k) \preceq (b_k)$ which implies there is P so $k > P$ implies $v_k \leq b_k$.

Since (v_n) is an upper bound for U, $(v_j) \succeq (x_j^k)$ for all k which tells us there is M_k so that $v_j \geq x_j^k$ if $j > M_k$. So if $j > \max\{N_k, M_k, P\}$,

$$x_j^k \leq v_j \quad \leq \quad \leq b_j \leq b_k < x_j^k$$

because the sequence $(b)j)$ is non-decreasing. This says $(v_n) \prec (x_n^k)$ and $(v_n) \succeq (x_n^k)$ which is not possible. Hence, our assumption (v_n) is an upper bound below (t_n) is not possible and (t_n) is the least upper bound for U.

Thus, we have shown $\tilde{\mathbb{Q}}$ is a totally ordered field which is complete: i.e. satisfies the completeness axiom (also called the least upper bound axiom). We identify $\tilde{\mathbb{Q}}$ with what we normally call the real numbers $\Re$.

2.1.4.1 Homework

Exercise 2.1.12 *Prove $\leq$ and $+$ and $*$ are compatible in $\mathbb{Q}$. This is easy, of course, but it shows you what is involved in this discussion.*

Exercise 2.1.13 *What goes wrong when you try to find a good ordering for $\mathbb{C}$?*

Exercise 2.1.14 *Prove* **Comp1** *for our ordering.*

Exercise 2.1.15 *Prove* **Comp2** *for our ordering.*

2.1.5 Cauchy Sequences of Equivalence Classes Converge

Now let's show Cauchy sequences in $\tilde{\mathbb{Q}}$ converge to an element of $\tilde{\mathbb{Q}}$. To do this, we have to have a way of measuring distance between elements of $\tilde{\mathbb{Q}}$.

Definition 2.1.1 A Metric for the Completion of the Rationals

Let $\tilde{d} : \tilde{\mathbb{Q}} \times \tilde{\mathbb{Q}} \rightarrow \tilde{\mathbb{Q}}$ be defined by $\tilde{d}([(x_n)], [(y_n)]) = [(|x_n - y_n|)]$

We can then prove $(\tilde{\mathbb{Q}}, \tilde{d})$ is a metric space. Note the metric here has range in the totally ordered complete field $\tilde{\mathbb{Q}}$ instead of the field $\mathbb{Q}$.

Theorem 2.1.3 The Completed Rationals is a Metric Space

$\tilde{d}$ is a metric on $\tilde{\mathbb{Q}}$ and $(\tilde{\mathbb{Q}}, \tilde{d})$ is a metric space.

Proof 2.1.3
We want $\tilde{d}$ to satisfy the usual properties we would expect for a way of measuring distance.

M1: *It is easy to see $\tilde{d}([(x_n)], [(y_n)]) = [(|x_n - y_n|)]$ is well-defined. It is non negative because $[(|x_n - y_n|)] \succeq [(\bar{0})]$.*

M2: *$\tilde{d}([(x_n)], [(y_n)]) = [(\bar{0})]$ implies $[(|x_n - y_n|)] = [(\bar{0})]$. That implies $[(x_n)] \sim [(y_n)]$.*

M3: *$\tilde{d}([(x_n)], [(y_n)]) = \tilde{d}([(y_n)], [(x_n)])$ follows from the definition.*

M4: *We know $|x_n - y_n| \leq |x_n - z_n| + |z_n - y_n|$. Thus, $[(|x_n - y_n)] \preceq [(|x_n - z_n|)] + [(|z_n - y_n|)]$ or $\tilde{d}([(x_n)], [(y_n)]) \preceq \tilde{d}([(x_n)], [(z_n)]) + \tilde{d}([(z_n)], [(y_n)])$.*

■

We can now show $\mathbb{Q}$ is **dense** in $(\tilde{\mathbb{Q}}, \tilde{d})$ with the proper embedding.

Theorem 2.1.4 The Rationals are Dense in the Completed Rationals

Embed $\mathbb{Q}$ into $(\tilde{\mathbb{Q}}, \tilde{d})$ by $q \rightarrow [(\bar{q})]$. Then this embedding of $\mathbb{Q}$ is dense in $(\tilde{\mathbb{Q}}, \tilde{d})$.

Proof 2.1.4
Let $[(\bar{\epsilon})] > 0$. It is easy to see that there is a rational ϵ with $\epsilon > 0$ so that for sufficiently large n, the representative $(x_n($for be $[(\bar{\epsilon})]$ satisfies $x_n > \epsilon$. Then for this ϵ there is N so that $|x_n - x_m| < \epsilon$ if $n, m > N$. Choose the constant rational sequence $[(\bar{x_{N+1}})]$. Then $\tilde{d}([(\bar{x_{N+1}})], [(x_n)]) = [(x_{N+1} - x_n)]$. But $|x_{N+1} - x_n| < \epsilon$ for $n > N$ which implies $[(|x_{N+1} - x_n|)] \prec [(\bar{\epsilon})]$. This shows the embedding of $\mathbb{Q}$ is dense in $(\widetilde{\mathbb{Q}}, \tilde{d})$. ∎

We are finally ready to show Cauchy sequences of elements in $(\widetilde{\mathbb{Q}}, \tilde{d})$ converge to an element of $(\widetilde{\mathbb{Q}}, \tilde{d})$.

Theorem 2.1.5 The Constructed Reals are Complete

Let $([(x_n^p)])$ be a Cauchy sequence in $(\widetilde{\mathbb{Q}}, \tilde{d})$. Then there is an $[(x_n)]$ in $(\widetilde{\mathbb{Q}}, \tilde{d})$ so that $\tilde{d}([(x_n^p)], [(x_n)]) \rightarrow [\mathbf{0}]$.

Proof 2.1.5
For each n, let $[(\bar{r_p})]$ be the embedded constant rational sequence which satisfies $\tilde{d}([(x_n^p)], [(\overline{r_p})]) \prec [(\overline{1/p})]$. by our arguments about the denseness of $\mathbb{Q}$, we see we may assume without loss of generality, that we are just dealing with the rational $1/p$. Then let $[(x_p)]$ be the sequence defined by $x_p = r_p$.

$$\tilde{d}([(x_n^p)], [(\overline{x_p})]) = [(|x_n^p - r_p|)]$$

Now, there is P so that $\frac{1}{P} < \epsilon$ and so if $p > P$,

$$\tilde{d}([(x_n^p)], [(\overline{r_p})]) \prec [(\overline{1/p})] \prec [(\bar{\epsilon})]$$

which tells us $[(x_n^p)] \rightarrow [(x_p)]$ in $\tilde{d}$ and so $(\widetilde{\mathbb{Q}}, \tilde{d})$ is complete. ∎

To summarize, we have constructed a totally ordered complete field from $\mathbb{Q}$ which we call $\widetilde{\mathbb{Q}}$ for which the least upper bound axiom holds and in which Cauchy sequences converge. This is the field we usually call $\Re$. We identify real numbers with Cauchy sequences of rational numbers where we measure distance using the usual absolute value function. Since $\tilde{d}([(x_n)], [(y_n)]) = [(|x_n - y_n|)]$, if we let $\alpha = [[(x_n)]]$ and $\beta = [(y_n)]$ and we define $|\alpha - \beta| = [(|x_n - y_n|)]$, we get our usual absolute value function defined on the elements of $\widetilde{\mathbb{Q}}$ as $\alpha - \beta|$. We can also define the value of $\lim_n |x_n - y_n| = |\alpha - \beta|$ as well. Hence, the value of $\lim_n |x_n - y_n|$ divides the set of Cauchy sequences of rationals into the equivalence classes of S we have seen. From now on, we will ignore all of this complication and simply treat the completed rationals as the field we know and love called $\Re$!
With all this said, we can mimic quite a bit of this to complete an arbitrary metric space.

2.1.5.1 Homework

Exercise 2.1.16 *For the equivalence class $\sqrt{2} \equiv [(x_n)]$ For a given $\epsilon > 0$ find an equivalence class $r \equiv [(\bar{r})]$ which satisfies $\tilde{d}(r, \sqrt{2}) < \epsilon$.*

Exercise 2.1.17 *For the equivalence class $2^{\sqrt{2}} \equiv [(x_n)]$ For a given $\epsilon > 0$ find an equivalence class $r \equiv [(\bar{r})]$ which satisfies $\tilde{d}(r, 2^{\sqrt{2}}) < \epsilon$.*

Exercise 2.1.18 *Look up the Archimedian property for $\Re$ and show it is satisfied.*

Exercise 2.1.19 *Look back at the infimum and supremum tolerance lemma as discussed in (Peterson (18) 2020) and think about how it can be phrased more carefully in the language of $\widetilde{\mathbb{Q}}$.*

2.2 The Idea of a Metric

To measure the distance between two objects, we lay down a ruler and mark the position of the first object at one point on the ruler, mark the position of the second object at another point on the ruler and subtract to get the distance. If the distance of the ruler a and the other distance is b, then the distance between the two objects is $b - a$. It doesn't matter which object plays the role of the first one and it should be clear the distance between the two objects can be written $|a - b|$. At this point, we just assume there is some sort of coordinate system imposed on the set of all possible objects, so that the idea of using a **ruler** makes sense. Understanding how that coordinate system can be built is a bit harder and we will not talk about that here. But at any rate, we can say distance measurements should not depend on which object is measured first. It is also pretty clear the distance between two objects is not negative and if the distance between two objects is zero, the objects should be the same.

Now let $\boldsymbol{X}$ be the set of objects. We have gone through the ideas here for the rationals and for their completion into the reals. But now, we won't bias ourselves by saying what these objects are. The object in $\boldsymbol{X}$ could be an integer, a real number, a sequence, a function etc. We will also only use the field $\Re$ for what we will call metrics. Consider a mapping $d : \boldsymbol{X} \times \boldsymbol{X} \rightarrow \Re$ so that

M1: $d(\boldsymbol{x}, \boldsymbol{y}) \geq 0$ for all objects $\boldsymbol{x}$ and $\boldsymbol{y}$ in $\boldsymbol{X}$. This corresponds to our understanding that the physical distance between objects is a non-negative number.

M2: $d(\boldsymbol{x}, \boldsymbol{y}) = 0$ if and only if $\boldsymbol{x} = \boldsymbol{y}$. This corresponds to our understanding two objects are the same if they are zero distance apart.

M3: $d(\boldsymbol{x}, \boldsymbol{y}) = d(\boldsymbol{y}, \boldsymbol{x})$ for all objects $\boldsymbol{x}$ and $\boldsymbol{y}$ in $\boldsymbol{X}$. This corresponds to our understanding that in physical distance measurements, it does not matter what order we use for our measurements.

This new mapping d seems to act like what we would call a distance measurement on the set of objects $\boldsymbol{X}$. But there is one other important property that we need. We know for three real numbers, the triangle inequality holds: i.e. $|a - b| \leq |a - c| + |c - b|$ for any numbers a, b and c. The absolute value function is a way of measuring distance between numbers which satisfies **M1**, **M2** and **M3**. Hence $|\cdot - \cdot|$ is an example of the mapping d acting on the set of objects $\Re$. So if you want the mapping d to act like a distance measurement on any arbitrary set of objects, we also want d to satisfy

M4: $d(\boldsymbol{x}, \boldsymbol{y}) \leq d(\boldsymbol{x}, \boldsymbol{z}) + d(\boldsymbol{z}, \boldsymbol{y})$ for all objects $\boldsymbol{x}$, $\boldsymbol{y}$ and $\boldsymbol{z}$ in $\boldsymbol{X}$. This corresponds to our understanding that we should have a triangle inequality for any such mapping d.

Such a mapping d on a set of objects $\boldsymbol{X}$ defines what is called a **metric space**.

Definition 2.2.1 Metric Space

Let $\boldsymbol{X}$ be a set of objects. Consider a mapping $d : \boldsymbol{X} \times \boldsymbol{X} \rightarrow \Re$ so that

M1: $d(\boldsymbol{x}, \boldsymbol{y}) \geq 0$ *for all objects $\boldsymbol{x}$ and $\boldsymbol{y}$ in $\boldsymbol{X}$.*

M2: $d(\boldsymbol{x}, \boldsymbol{y}) = 0$ *if and only if $\boldsymbol{x} = \boldsymbol{y}$.*

M3: $d(\boldsymbol{x}, \boldsymbol{y}) = d(\boldsymbol{y}, \boldsymbol{x})$ *for all objects $\boldsymbol{x}$ and $\boldsymbol{y}$ in $\boldsymbol{X}$.*

M4: $d(\boldsymbol{x}, \boldsymbol{y}) \leq d(\boldsymbol{x}, \boldsymbol{z}) + d(\boldsymbol{z}, \boldsymbol{y})$ *for all objects $\boldsymbol{x}$, $\boldsymbol{y}$ and $\boldsymbol{z}$ in $\boldsymbol{X}$.*

The mapping d is called a **metric** *and the pair $(\boldsymbol{X}, d)$ is called a* **metric space**.

There are many consequences here.

- The metric d induces a **topology** of open and closed sets on $\boldsymbol{X}$. We have

$$B(\boldsymbol{x}, r) \quad = \quad \{\boldsymbol{y} \in \boldsymbol{X} \mid d(\boldsymbol{x}, \boldsymbol{y}) < r\}$$

which is the **open ball** of radius r about $\boldsymbol{x}$.

$$\overline{B}(\boldsymbol{x}, r) = \{\boldsymbol{y} \in \boldsymbol{X} \mid d(\boldsymbol{x}, \boldsymbol{y}) \leq r\}$$

which is the **closed ball** of radius r about $\boldsymbol{x}$.

$$\hat{B}(\boldsymbol{x}, r) = \{\boldsymbol{y} \in \boldsymbol{X} \mid 0 < d(\boldsymbol{x}, \boldsymbol{y}) < r\}$$

which is the **punctured ball** of radius r about $\boldsymbol{x}$ which does not contain $\boldsymbol{x}$ itself.

- **Convergence**: A sequence $(\boldsymbol{x_n})$ in $\boldsymbol{X}$ is said to **converge** to $\boldsymbol{x}$ in $\boldsymbol{X}$ if

  $$\forall \epsilon > 0, \exists N \ni d(\boldsymbol{x_n}, \boldsymbol{x}) < \epsilon \text{ if } n > N$$

  We use the notation $\boldsymbol{x_n} \rightarrow \boldsymbol{x}$.

- **Cauchy Sequence**: A sequence $(\boldsymbol{x_n})$ in $\boldsymbol{X}$ is said to be a **Cauchy Sequence** if

  $$\forall \epsilon > 0, \exists N \ni d(\boldsymbol{x_n}, \boldsymbol{x_m}) < \epsilon \text{ if } n, m > N$$

  From our earlier work, it is clear convergence sequences are Cauchy sequences as this proof is an easy modification of arguments in the first two volumes. If every Cauchy sequence in $\boldsymbol{X}$ converges to some object $\boldsymbol{x}$ in $\boldsymbol{X}$, we say the metric space $(\boldsymbol{X}, d)$ is a **Complete Metric Space**.

- **Continuity**: If $f : \boldsymbol{X} \rightarrow \Re$ is a mapping from the metric space $(\boldsymbol{X}, d)$ to the real numbers, we say f is continuous at $\boldsymbol{x_0}$ if

  $$\forall \epsilon > 0, \exists \delta > 0 \ni |f(\boldsymbol{x}) - f(\boldsymbol{x_0})| < \epsilon \text{ if } d(\boldsymbol{x}, \boldsymbol{x_0}) < \delta$$

  If $f : (\boldsymbol{X}, d_X) \rightarrow (\boldsymbol{Y}, d_Y)$ is a mapping from the metric space $(\boldsymbol{X}, d_X)$ to the metric space $(\boldsymbol{Y}, d_Y)$, we say f is continuous at $\boldsymbol{x_0}$ if

  $$\forall \epsilon > 0, \exists \delta > 0 \ni d_Y(f(\boldsymbol{x}), f(\boldsymbol{x_0})) < \epsilon \text{ if } d_X(\boldsymbol{x}, \boldsymbol{x_0}) < \delta$$

  We have used these ideas for the particular metric spaces such as $\Re^n$ with various norms and spaces of matrices.

There are many more which are easy proofs which will come up now and then in our future discussions. Most of these we will leave to you to prove.

Homework

Exercise 2.2.1 *Prove if (x_n) is a sequence in the metric space (X, d) which converges to x, then it must be a Cauchy sequence.*

Exercise 2.2.2 *Prove if (x_n) is a sequence in the metric space (X, d) which converges to x, then any subsequence of it converges to the same limit x.*

Exercise 2.2.3 *Prove the limit of a convergence sequence in the metric space (X, d) is unique.*

Exercise 2.2.4 *Prove the modification of the backwards triangle inequality for $|\cdot|$ for a metric d.*

2.3 Examples

There are many examples of metric spaces which are also vector spaces as many spaces of objects also have a well-defined way of **adding** two objects together to create a new object.

The Discrete Metric Space: For the space of objects $\boldsymbol{X}$, define $d_{dis}(\boldsymbol{x}, \boldsymbol{y}) = 1$ when $\boldsymbol{x} \neq \boldsymbol{y}$ and 0 if they match. Note any mapping $f : (\boldsymbol{X}, d_{dis}) \rightarrow \Re$ satisfies

$$d_{dis}(\boldsymbol{x}, \boldsymbol{y}) \quad < \quad 1 \Longrightarrow \boldsymbol{x} = \boldsymbol{y} \Longrightarrow |f(\boldsymbol{x}) - f(\boldsymbol{y})| = 0$$

Thus, for any $\epsilon > 0$, if we choose $\delta = 1/2$, then $|f(\boldsymbol{x}) - f(\boldsymbol{y})| = 0 < \epsilon$. Therefore all real-valued mappings f are continuous in this metric space.

$\mathbb{Z}$: The integers which the absolute value function are a metric space. You can easily verify the properties of $d(n, m) = |n \ - m|$. What is the topology here? Note

$$B(n, r) \quad = \quad \begin{cases} \{n\} & 0 < r < 1 \\ \{n-1, n, n+1\} & 1 \leq 1 < 2 \\ \{n-2, n-1, n, n+1, n+2\} & 2 \leq 1 < 3 \vdots \end{cases}$$

and so forth. Note each subset of $\mathbb{Z}$ which contains integers is open as each integer n in the set is an interior point. The complement of each subset is itself a set of integers and is therefore also open. Hence each subset is both open and closed. An alternate way to interpret the mapping $f : \mathbb{Z} \rightarrow \Re$ is that it is the sequence $(f(n))_{n=-\infty}^{\infty}$. If $\mathscr{O}$ is an open subset of $\Re$, the inverse image of $\mathscr{O}$ is

$$f^{-1}(\mathscr{O}) \quad = \quad \{n | f(n) \in \mathscr{O}\}$$

which is an open set. Thus the real mappings f are continuous.

$\mathbb{Z}_p$: The integers modulo p also form a metric space. Here $\mathbb{Z}_p = \{\boldsymbol{1}, \ldots, \boldsymbol{p}\}$. For $\boldsymbol{i}$ and $\boldsymbol{j}$ in $\mathbb{Z}_p$, we define $d(\boldsymbol{i}, \boldsymbol{j}) = |\boldsymbol{i} - \boldsymbol{j}|$. It is easy to see $M1$ to $M4$ are satisfied. The same comments about topology and continuity apply here too.

$\sum^p$: Let $S_i = \mathbb{Z}_p$ for $p \geq 2$ and $\sum^p = \prod_{i=-\infty}^{\infty} S_i$. This is called the bi-infinity tuple of elements of $\mathbb{Z}_p$. They are also called **symbol sequences**. Of course, this is also a sequence whose elements are members of $\mathbb{Z}_p$. Thus if $\boldsymbol{s} \in \sum_p$, we have

$$\boldsymbol{s} \quad = \quad \{\ldots, s_{-n}, s_{n-1}, \ldots, s_{-1}, s_0, s_1, \ldots, s_{n-1}, s_n, \ldots\}$$

where each $s_i \in \mathbb{Z}_p$. We often abbreviate this as

$$\boldsymbol{s} \quad = \quad \{\cdots s_{-n}\ s_{n-1} \cdots s_{-1}\ s_0\ s_1 \cdots s_{n-1}, s_n, \cdots\}$$

or simply

$$\boldsymbol{s} \quad = \quad \cdots s_{-n}\ s_{n-1} \cdots s_{-1}\ s_0\ s_1 \cdots s_{n-1}, s_n, \cdots$$

A metric d_{Σ_p} on $\sum_p$ is defined as follows

$$d_{\Sigma_p}(\boldsymbol{s}, \boldsymbol{t}) \quad = \quad \sum_{i=-\infty}^{\infty} \frac{1}{2^{|i|}} \frac{|s_i - t_i|}{1 + |s_i - t_i|}$$

Note the partial sums satisfy

$$
\begin{aligned}
T_{-n,m} = \frac{1}{2^{M+1}} &\leq \sum_{i=-n}^{m} \frac{1}{2^{|i|}} = \sum_{i=-n}^{0} \frac{1}{2^{|i|}} + \sum_{i=1}^{m} \frac{1}{2^i} \\
&= \sum_{i=0}^{n} \frac{1}{2^i} + \sum_{i=1}^{m} \frac{1}{2^i} \leq 4
\end{aligned}
$$

Hence, the set of numbers $\{T_{-n,m} \mid n, m$ non-negative integers $\}$ is bounded above and so by the completeness axiom has a finite supremum α. Then, for $\epsilon = \frac{1}{k}$, there are indices $(-n_k, m_k)$ with n_k increasing and unbounded and m_k increasing and unbounded so that

$$
\alpha - \frac{1}{k} < T_{-n_k,m_k} \leq T_{-n,n} < \alpha + \frac{1}{k}
$$

This immediately tells us the sequence $(T_{-n,m})$ converges to α. Hence, the usual shorthand letting

$$
\alpha = \sum_{i=-\infty}^{\infty} \frac{1}{2^{|i|}} \frac{|s_i - t_i|}{1 + |s_i - t_i|}
$$

can be used. We conclude d_{Σ_p} is a well-defined map on $\sum_p \times \sum_p$. Let $\phi(u) = \frac{u}{1+u}$ for all $u \geq 0$. Then, we could write

$$
d_{\Sigma_p}(\boldsymbol{s}, \boldsymbol{t}) = \sum_{i=-\infty}^{\infty} \frac{1}{2^{|i|}} \phi(|s_i - t_i|)
$$

which is a useful in showing $M4$ is satisfied. Note for non-negative u and v, we have

$$
\begin{aligned}
\phi(u+v) &= \frac{u+v}{1+(u+v)} = \frac{u}{1+(u+v)} + \frac{v}{1+(u+v)} \leq \frac{u}{1+u} + \frac{v}{1+v} \\
&= \phi(u) + \phi(v)
\end{aligned}
$$

Thus, for $\boldsymbol{s}$, $\boldsymbol{t}$ and $\boldsymbol{w}$ in $\sum_p$

$$
\begin{aligned}
\sum_{i=-n}^{m} \frac{1}{2^{|i|}} \phi(((s_i - w_i) + (w_i - t_i)|) &\leq \sum_{i=-n}^{m} \frac{1}{2^{|i|}} \phi(|s_i - w_i|) + \sum_{i=-n}^{m} \frac{1}{2^{|i|}} \phi(|w_i - t_i|) \\
&\leq d_{\Sigma_p}(\boldsymbol{s}, \boldsymbol{w}) + d_{\Sigma_p}(\boldsymbol{w}, \boldsymbol{t})
\end{aligned}
$$

This immediately implies property $M4$ holds. Properties $M1$, $M2$ and $M3$ are easy to see. Hence, d_{Σ_p} is a metric on $\sum_p$.

2.3.1 Homework

Exercise 2.3.1 *Define the metric on* $\Re^2$ *by* $d(\boldsymbol{x}, \boldsymbol{y}) = \max_{i=1,2} |x_i = y_i|$. *First, prove this is a metric on* $\Re^2$ *and second describe the topology of open sets. Tell us what the sets* $B_r(\boldsymbol{x}, r)$ *look like.*

Exercise 2.3.2 *Define the metric on the set of continuous functions on the finite interval* $[a, b]$ *by* $d(\boldsymbol{x}, \boldsymbol{y}) = \max_{a \leq t \leq b} |x(t) - y(t)|$. *First, prove this is a metric on* $\Re^2$ *and second describe the topology of open sets. Tell us what the sets* $B_r(\boldsymbol{x}, r)$ *look like.*

Exercise 2.3.3 *If we define the mapping d by $d(\boldsymbol{x}, \boldsymbol{y}) = \int_a^b x(t)y(t)dt$ for any continuous functions $\boldsymbol{x}$ and $\boldsymbol{y}$ on the finite interval $[a, b]$, is d a metric?*

Exercise 2.3.4 *If we define the mapping d by $d(\boldsymbol{x}, \boldsymbol{y}) = \int_a^b x(t)y(t)dt$ for any Riemann Integrable $\boldsymbol{x}$ and $\boldsymbol{y}$ on the finite interval $[a, b]$, is d a metric?*

2.4 More on Symbol Sequences

Let's explore the topology of $\sum_p$ a bit. When are two symbol sequences **close**?

Lemma 2.4.1 Distance Estimates for Symbol Sequences

For all $\boldsymbol{s}$ and $\boldsymbol{t}$ in $\sum_p$

$$
\begin{aligned}
d_{\Sigma_p}(\boldsymbol{s}, \boldsymbol{t}) &< \frac{1}{2^{M+1}} \Longrightarrow s_i = t_i \text{ if } |i| \leq M \\
s_i = t_i \text{ if } |i| \leq M &\Longrightarrow d_{\Sigma_p}(\boldsymbol{s}, \boldsymbol{t}) \leq \frac{1}{2^{M-1}}
\end{aligned}
$$

Proof 2.4.1
We prove the first assertion by contradiction. Assume $d_{\Sigma_p}(\boldsymbol{s}, \boldsymbol{t}) < \frac{1}{2^{M+1}}$ and there is an index j, $|j| \leq M$ with $s_j \neq t_j$. Then the term in the metric d_{Σ_p} for index j satisfies

$$
\frac{1}{2^{|j|}} \frac{|s_j - t_j|}{1 + |s_j - t_j|} \geq \frac{1}{2^{|j|}} \frac{1}{2}
$$

To see this, note $\phi(u) = \frac{u}{1+u}$ is monotonically increasing for all non-negative u as $\phi'(u) = \frac{1}{(1+u)^2} > 0$. Since $|s_j - t_j| \geq 1$ by assumption, $\phi(|s_j - t_j|) \geq \frac{1}{2}$. Now this tells us

$$
d_{\Sigma_p}(\boldsymbol{s}, \boldsymbol{t}) \geq \frac{1}{2^{|j|}} \frac{|s_j - t_j|}{1 + |s_j - t_j|} \geq \frac{1}{2^{|j|+1}} \geq \frac{1}{2^{M+1}}
$$

which contradicts the assumption that $d_{\Sigma_p}(\boldsymbol{s}, \boldsymbol{t}) < \frac{1}{2^{M+1}}$. Hence, we cannot find such an index j and we conclude $s_i = t_i$ when $|i| \leq M$.

To prove the remaining assertion, if $s_i = t_i$ for $|i| \leq M$, then for sufficiently $n > M$,

$$
\begin{aligned}
\sum_{i=-n}^{n} \frac{1}{2^{|i|}} \phi(|s_i - t + i|) &= \sum_{i=-n}^{-(M+1)} \frac{1}{2^{|i|}} \phi(|s_i - t + i|) + \sum_{i=-M+1}^{n} \frac{1}{2^{|i|}} \phi(|s_i - t + i|) \\
&\leq \sum_{i=-n}^{-(M+1)} \frac{1}{2^{|i|}} + \sum_{i=-M+1}^{n} \frac{1}{2^{|i|}} = 2 \sum_{i=M+1}^{n} \frac{1}{2^{|i|}} \\
&= 2 \left(\frac{1 - \frac{1}{2^{n+1}}}{1 - \frac{1}{2}} - \frac{1 - \frac{1}{2^{M+1}}}{1 - \frac{1}{2}} \right) = 4 \left(\frac{1}{2^{M+1}} - \frac{1}{2^{n+1}} \right) \\
&= \frac{1}{2^{M-1}}
\end{aligned}
$$

This implies $d_{\Sigma_p}(\boldsymbol{s}, \boldsymbol{t}) \leq \frac{1}{2^{M-1}}$. ■

So what does this mean? It means two symbol sequences are close as long as they agree **on a long central block**. Next, what does $B(\boldsymbol{s}, \epsilon)$ look like? Well, given $\epsilon > 0$ there is an m with $\frac{1}{2^{M+1}} < \epsilon$. By Lemma 2.4.1, we then know $d_{\Sigma_p}(\boldsymbol{s}, \boldsymbol{t}) < \frac{1}{2^{M+1}} < \epsilon$ implies $s_i = t_i$ for $|i| \leq M$. Thus,

$$B(\boldsymbol{s}, \epsilon) \quad \subset \quad \{\boldsymbol{t} \in \sum_p \mid t_i = s_i, \ |i| \leq M\}$$

Call the set on the right hand side Q_M for convenience. Thus, $B(\boldsymbol{s}, \epsilon) \subset Q_M$. Also, if $\boldsymbol{t} \in Q_{M+2}$, by the second assertion of Lemma 2.4.1, we have

$$d_{\Sigma_p}(\boldsymbol{s}, \boldsymbol{t}) \quad \leq \quad \frac{1}{2^{M+2-1}} < \epsilon$$

Thus, we have shown $Q_{M+2} \subset B(\boldsymbol{s}, \epsilon) \subset Q_M$. Also, it is clear $Q_{M+1} \subset \overline{B}(\boldsymbol{s}, \epsilon)$. For example, if $\epsilon = 0.13$, then $M = 2$ and we have $Q_{2+j} \subset B(\boldsymbol{s}, 0.13)$ for all $j \geq 1$, $Q_{2+1} \subset \overline{B}(\boldsymbol{s}, 0.13)$ and $B(\boldsymbol{s}, 0.13) \subset Q_2$. Now Q_M consists of symbol sequences that match $\{s_{-M}, \ldots s_0.\ldots, s_M\}$. Hence, $B(\boldsymbol{s}, 0.13)$ contains all symbol sequences that match $\boldsymbol{s}$ on the central block from $-(2+j)$ to $2+j$ for all $j \geq 1$. It could contain other symbol sequences, but it definitely contains these.

Finally, what about convergence in this metric? Let $\boldsymbol{s_n} \to \boldsymbol{s}$ in d_{Σ_p}. Then given $\epsilon > 0$, there is an N so that $d_{\Sigma_p}(\boldsymbol{s_n}, \boldsymbol{s}) < \epsilon$ if $n > N$. By part one of our Lemma, since there is an M so that $\frac{1}{2^{M+1}} < \epsilon$, we have $s_{ni} = s_i$ for all $|i| \leq M$, for all $n > M$.

Now consider the **shift** map, $\boldsymbol{\sigma} : \sum_p \to \sum_p$ which is defined by $\boldsymbol{\sigma}(\boldsymbol{s})$ is the symbol sequence with terms $(\ \boldsymbol{\sigma}(\boldsymbol{s})\ )_i = s_{i+1}$.

Lemma 2.4.2 Properties of the Symbol Shift Map

$\boldsymbol{\sigma} : \sum_p \to \sum_p$ is continuous and onto.

Proof 2.4.2

Pick $\boldsymbol{s}$ and pick an $\epsilon > 0$. Choose M so that $\frac{1}{2^{M-2}} < \epsilon$. Then if $\delta = \frac{1}{2^{M+1}}$, by Lemma 2.4.1, $d_{\Sigma_p}(\boldsymbol{s}, \boldsymbol{t}) < \delta$ implies $s_i = t_i$ for $|i| \leq M$. Thus, $(\ \boldsymbol{\sigma}(\boldsymbol{s})\ )_i = (\ \boldsymbol{\sigma}(\boldsymbol{t})\ )_i$ $|i| \leq M - 1$. Then, by the second assertion of the Lemma, $d_{\Sigma_p}(\boldsymbol{\sigma}(\boldsymbol{s}), \boldsymbol{\sigma}(\boldsymbol{t})) < \frac{1}{2^{M-2}} < \epsilon$. Thus, $\boldsymbol{\sigma}$ is continuous at $\boldsymbol{s}$. Since $\boldsymbol{s}$ is arbitrary, $\boldsymbol{\sigma}$ is continuous on $\sum_p$.

To show $\boldsymbol{\sigma}$ is an onto map, given any symbol sequence $\boldsymbol{s}$, the symbol sequence $\mathbf{t}$ defined by $t_i = s_{i-1}$ is mapped by $\boldsymbol{\sigma}$ to $\boldsymbol{s}$. Hence, the map is surjective. ■

2.4.1 Homework

Exercise 2.4.1 *Analyze the function $\phi(t) = \frac{t}{1+t}$ for all $t \neq 0$.*

Exercise 2.4.2 *Does our analysis here change if we scale by 3^{-i} instead of 2^{-1}?*

Exercise 2.4.3 *Work out all the details for Σ_3.*

Exercise 2.4.4 *Work out all the details for Σ_7.*

Exercise 2.4.5 *If $\epsilon = 0.03$, find what M is and then determine the block.*

2.5 Symbol Spaces over Two Symbols

Let's specialize to the symbol space over two symbols $\sum_2$ or binary symbol sequences and we will start the indexing at $i = 0$ for a change to let you see how to do this sort of analysis in this context. Hence,

$$\Sigma_2 \quad = \quad \{s_0, s_1, s_2, \ldots, s_n, \ldots \mid s_j \in \mathbb{Z}_2\}$$

Hence, the choices for each s_i are just 0 or 1. As usual, we identify

$$\boldsymbol{s} = (s_i)_{i \geq 0} \quad = \quad s_0 s_1 s_2 \cdots$$

We will use a different metric now

$$d_{\Sigma_2}(\boldsymbol{s}, \boldsymbol{t}) \quad = \quad \sum_{i=0}^{\infty} \frac{1}{2^i} |s_i - t_i|$$

which is easily seen to be well-defined by the comparison test. We don't need to use the ϕ function here because $|s_i - t_i| \leq 1$ always. It is clear $M1$, $M2$ and $M3$ are satisfied. For the triangle inequality, note for symbol sequences $\boldsymbol{s}$, $\boldsymbol{t}$ and $\boldsymbol{w}$, we have

$$|s_i - t_i| \quad \leq \quad |s_i - w_i| + |w_i - t_i|$$

from which $M4$ follows after a straightforward argument.

Lemma 2.5.1 Distance Estimates for Binary Symbol Sequences

For all $\boldsymbol{s}$ and $\boldsymbol{t}$ in $\sum_2$

$$d_{\Sigma_2}(\boldsymbol{s}, \boldsymbol{t}) < \frac{1}{2^n} \quad \Longrightarrow \quad s_i = t_i \text{ if } i \leq n$$

$$s_i = t_i \text{ if } i \leq n \quad \Longrightarrow \quad d_{\Sigma_2}(\boldsymbol{s}, \boldsymbol{t}) \leq \frac{1}{2^n}$$

Proof 2.5.1

If $d_{\Sigma_2}(\boldsymbol{s}, \boldsymbol{t}) < \frac{1}{2^n}$, then

$$\sum_{i=0}^{\infty} \frac{1}{2^i} |s_i - t_i| \quad < \quad \frac{1}{2^n}$$

If $s_k = t_k$ for an index $0 \leq k \leq n$, then

$$\sum_{i=0}^{\infty} \frac{1}{2^i} |s_i - t_i| \quad \geq \quad \frac{|s_k - t_k|}{2^k} = \frac{1}{2^k} \geq \frac{1}{2^n}$$

But this contradicts our assumption that $d_{\Sigma_2}(\boldsymbol{s}, \boldsymbol{t}) < \frac{1}{2^n}$. Hence, we can find now such index and the first assertion is true.

If $s_i = t_i$ for $i \leq n$, then

$$d_{\Sigma_2}(\boldsymbol{s}, \boldsymbol{t}) \quad = \quad \sum_{i=n+1}^{\infty} \frac{1}{2^i} |s_i - t_i| \leq \sum_{i=n+1}^{\infty} \frac{1}{2^i} = \frac{1}{2^n}$$

This proves the second assertion. ■

Now let's look at the shift map on the binary symbol space. We have $\boldsymbol{\sigma} : \Sigma_2 \rightarrow \Sigma_2$ defined by

$$\boldsymbol{\sigma}(s_0 s_1 s_2 \cdots) = s_1 s_2 \cdots$$

Hence, $\boldsymbol{\sigma}$ simply removes the leading entry of the binary symbol sequence.

Lemma 2.5.2 The Binary Symbol Shift Map is Continuous

$\boldsymbol{\sigma} : \Sigma_2 \rightarrow \Sigma_2$ is continuous.

Proof 2.5.2
Choose $\boldsymbol{s} \in \Sigma_2$. Pick $\epsilon > 0$ arbitrarily. Now

$$\begin{aligned} d_{\Sigma_2}(\boldsymbol{\sigma}(\boldsymbol{s}), \boldsymbol{\sigma}(\boldsymbol{t})) &= \sum_{i=0}^{\infty} \frac{1}{2^i} |s_{i+1} - t_{i+1}| = \sum_{i=1}^{\infty} \frac{1}{2^{i-1}} |s_i - t_i| \\ &= 2 \sum_{i=1}^{\infty} \frac{1}{2^i} |s_i - t_i| = 2\, d_{\Sigma_2}(\boldsymbol{s}, \boldsymbol{t}) \end{aligned}$$

Thus, if we choose $\delta = \epsilon/2$, then

$$d_{\Sigma_2}(\boldsymbol{\sigma}(\boldsymbol{s}), \boldsymbol{\sigma}(\boldsymbol{t})) < \epsilon$$

if $d_{\Sigma_2}(\boldsymbol{s}, \boldsymbol{t}) < \delta$. This shows $\boldsymbol{\sigma}$ is continuous on Σ_2. ■

2.5.1 Homework

Exercise 2.5.1 *Define $\Sigma_3 = s_0 s_1 s_2 \cdots$ for $s_i \in \mathbb{Z}_3$. Define $d : \Sigma_3 \times \Sigma_3 \rightarrow \Re$ by*

$$d(\boldsymbol{s}, \boldsymbol{t}) = \sum_{i=0}^{\infty} \frac{|s_i - t_i|}{3^i}$$

Prove d is a metric.

Exercise 2.5.2 *Define $\Sigma_3 = s_0 s_1 s_2 \cdots$ for $s_i \in \mathbb{Z}_3$. Define $\hat{d} : \Sigma_3 \times \Sigma_3 \rightarrow \Re$ by*

$$\hat{d}(\boldsymbol{s}, \boldsymbol{t}) = \sum_{i=0}^{\infty} \frac{|s_i - t_i|}{2^i}$$

Prove $\hat{d}$ is a metric.

Exercise 2.5.3 *Prove the map $\sigma : (\Sigma_2, d) \rightarrow (\Sigma_2, \hat{d})$ is continuous.*

Exercise 2.5.4 *Prove the map $\sigma : (\Sigma_3, d) \rightarrow (\Sigma_3, \hat{d})$ is continuous.*

Exercise 2.5.5 *Prove the map $\sigma : (\Sigma_5, d) \rightarrow (\Sigma_5, \hat{d})$ is continuous.*

2.6 Periodic Points of a Map

A point is a periodic point or fixed point of a map f from a space into itself if $f(\boldsymbol{x}) = \boldsymbol{x}$. We can get some insight into this idea by looking for the fixed points of the shift map $\boldsymbol{\sigma}$ on Σ_2. If $\boldsymbol{s}$ is a fixed point of $\boldsymbol{\sigma}$ then for $\boldsymbol{s} = s_0 s_1 s_2 \cdots$,

$$\boldsymbol{\sigma}(\boldsymbol{s}) = \boldsymbol{s} \quad \Longrightarrow s_1 s_2 \cdots = s_0 s_1 s_2 \cdots$$

Thus, any constant symbol sequence $s_0 s_0 \cdots$ is a fixed point. Now let $\boldsymbol{t} = s_0 s_1 : s_0 s_1 : \cdots$ be the symbol sequence which repeats the block $s_0 s_1$. Then note

$$\boldsymbol{\sigma}(\,\boldsymbol{\sigma}(\boldsymbol{t})) = \boldsymbol{\sigma}(s_1 : s_0 s_1 : \cdots) = \boldsymbol{t}$$

and hence $\boldsymbol{t}$ is a periodic or fixed point of $\boldsymbol{\sigma}(\,\boldsymbol{\sigma}) = \boldsymbol{\sigma}^2$. We call $\boldsymbol{t}$ a periodic point of period 2. Since the underlying space is $\mathbb{Z}_2$, there are 2 periodic points of period 1: $0000\cdots$ and $1111\cdots$ and 4 periodic points of period two which are built from the blocks

$$(s_0 s_1) \quad \in \quad \{(00), (10), (01), (11)\}$$

In general, there are 2^n blocks of the form $s_0 s_1 \cdots s_{n-1}$ which give us the periodic points of period n. That is

$$\boldsymbol{\sigma}^n(s_0 s_1 \cdots s_{n-1} : s_0 s_1 \cdots s_{n-1} : \cdots) \quad = \quad s_0 s_1 \cdots s_{n-1} : s_0 s_1 \cdots s_{n-1} : \cdots$$

The set of all periodic points of order n is a special subset of Σ_2 called a **dense** subset.

Definition 2.6.1 Dense Subsets of a Metric Space

Let (X, d) be a metric space and let $S \subset X$. We say S is **dense** *in X if given x in X, for all $\epsilon > 0$, there is a $\boldsymbol{y_\epsilon} \in S$ so that $d(\boldsymbol{x}, \boldsymbol{y_\epsilon}) < \epsilon$. This implies given $\boldsymbol{x} \in X$, there is a sequence $\boldsymbol{y_n}$ so that $\boldsymbol{y_n} \to \boldsymbol{x}$ in the d metric.*

An even more special subset is a dense subset that is countable. We have discussed countable sets in the earlier volumes, so let's just do a quick review. A subset $S \subset X$ of a metric space is countable if there is a one to one and onto map f from S to $\mathbb{N}$. In this case, we usually identify the elements of S with a sequence by associating each $\boldsymbol{x} \in S$ with its unique integer $f(\boldsymbol{x})$ which is called f_n. Thus

$$S \quad = \quad f_1 f_2 \cdots$$

Let

$$Per(\sigma) \quad = \quad \{\boldsymbol{t} \in \Sigma_2 \mid \boldsymbol{t} \text{ is periodic of period } n \text{ for some positive integer } n\}$$

Lemma 2.6.1 $Per(\sigma)$ **is dense in** Σ_2 **with Metric** d_{Σ_2}

$Per(\boldsymbol{\sigma})$ is dense in Σ_2 with metric d_{Σ_2}.

Proof 2.6.1

Let $\boldsymbol{s} = s_0 s_1 \cdots s_n s_{n+1} \cdots$ in Σ_2. Let $\boldsymbol{t_n} = s_0 s_1 \cdots s_n : s_0 s_1 \cdots s_n : \cdots$ in Σ_2. Since $t_{ni} = s_i$ for $i \leq n$, we know $d_{\Sigma_2}(\boldsymbol{s}, \boldsymbol{t_n}) \leq \frac{1}{2^n}$. This shows we can find a periodic point of order n arbitrarily close to $\boldsymbol{s}$ by choosing n large enough. Thus, $Per(\boldsymbol{\sigma})$ is dense in Σ_2 with metric d_{Σ_2}. ■

We can find more structure if we keep looking. Let

$$\begin{aligned} \boldsymbol{s}^* &= \{0 : 1!00 : 01 : 10 : 11!000 : 001 : 010 : \cdots : 111! \cdots\} \\ &= \{\text{ sequences of length 1! sequences of length 2! sequences of length 3!} \cdots\} \end{aligned}$$

Lemma 2.6.2 Orbits are Dense in Σ_2

Let $S = \{\boldsymbol{\sigma}^n(\boldsymbol{s}^*) \mid n \in \mathbb{N}\}$. This is called the **orbit** *of $\boldsymbol{s}^*$. Then S is dense in Σ_2.*

Proof 2.6.2
Let $\boldsymbol{s} \in \Sigma_2$ and let $\epsilon > 0$ be chosen. The block $s_0 s_1 \cdots s_n$ is of length $n+1$. $\boldsymbol{\sigma}^2$ applied to $\boldsymbol{s}^*$ strips off the 2 sequences of length 1. Then $\boldsymbol{\sigma}^6$ applied to $\boldsymbol{s}^*$ strips off the 2 sequences of length 1 and length 2. Let $M_n = 2^1 + 2^2 + \ldots + 2^n$. Then $\boldsymbol{\sigma}^{M_n}$ applied to $\boldsymbol{s}^*$ strips off all sequences up to and including length n. Hence, $\boldsymbol{\sigma}^{M_n}(\boldsymbol{s}^*)$ and $\boldsymbol{s}$ both start with $s_0 s_1 \cdots s_{n+1}$. Hence, we know

$$d_{\Sigma_2}(\boldsymbol{\sigma}^{M_n}(\boldsymbol{s}^*), \boldsymbol{s}) \leq \frac{1}{2^n}$$

So given $\epsilon > 0$, there is an M so that $\frac{1}{2^M} < \epsilon$. We conclude there is an M_n so that $d_{\Sigma_2}(\boldsymbol{\sigma}^{M_n}(\boldsymbol{s}^*), \boldsymbol{s})$ $\leq \epsilon$ for all $n > M$. This shows S is dense in Σ_2. ■

Let $T \subset \Sigma_2$ be defined by

$$T = \{\boldsymbol{t} \mid t_j = 0 \Longrightarrow t_{j+1} = 1\}$$

It is easy to see if $\boldsymbol{t} \in T$, then so is $\boldsymbol{\sigma}(\boldsymbol{t})$.

Lemma 2.6.3 The Orbit of s^* is Dense in T

The orbit of $\boldsymbol{s}^*$, $Per(\boldsymbol{s}^*)$, is dense in T.

Proof 2.6.3
Let $\boldsymbol{t} = t_0 t_1 t_2 \cdots$ be in T. Define $\boldsymbol{t_n}$ in T by

- *If $t_n = 0$, we know $t_{n+1} = 1$. So in this case let*

$$\boldsymbol{t_n^a} = t_0 t_1 \cdots t_n t_{n+1} : t_0 t_1 \cdots t_n t_{n+1} : \cdots = t_0 t_1 \cdots 01 : t_0 t_1 \cdots 01 : \cdots$$

  We see $\boldsymbol{t_n}$ is in T.

- *If $t_n = 1$, use*

$$\boldsymbol{t_n^b} = t_0 t_1 \cdots t_n : t_0 t_1 \cdots t_n : \cdots = t_0 t_1 \cdots 1 : t_0 t_1 \cdots 1 : \cdots$$

  which is also in T. In both cases, $\boldsymbol{s}$ and $\boldsymbol{t_n^a}$andt$\boldsymbol{t_n^b}$ match for indices $0 \leq i \leq n$ and so

$$d_{\Sigma_2}(\boldsymbol{t}, \boldsymbol{t_n^a}) \leq \frac{1}{2^n}, \quad d_{\Sigma_2}(\boldsymbol{t}, \boldsymbol{t_n^b}) \leq \frac{1}{2^n}$$

Also, note $\boldsymbol{t_n^a}$ is periodic of order $n+2$ and $\boldsymbol{t_n^b}$ is periodic of order $n+1$.

Given $\epsilon > 0$, there is an M so that $\frac{1}{2^M} < \epsilon$. Hence, choosing $n > M$, we can find an element of T (either $\boldsymbol{t_n^a}$ or $\boldsymbol{t_n^b}$) satisfying $d_{\Sigma_2}(\boldsymbol{t}, \boldsymbol{t_n^a}) < \epsilon$ or $d_{\Sigma_2}(\boldsymbol{t}, \boldsymbol{t_n^b}) < \epsilon$. This shows the orbit of $\boldsymbol{s}^*$ is dense in T. ■

Homework

Exercise 2.6.1 *Are the rational numbers dense in the reals?*

Exercise 2.6.2 *Is $\mathbb{Q} \times \mathbb{Q}$ dense in $\Re^2$?*

Exercise 2.6.3 *Let $0.s_0 s_1 s_2 \cdots$ be the binary expansion of the number $\frac{1}{\sqrt{2}}$. Map this to $\boldsymbol{s}$ in Σ_2. Compute $d_{\Sigma_2}(\boldsymbol{\sigma}^n(\boldsymbol{s}^*), \boldsymbol{s})$ for $n \leq 10$.*

Exercise 2.6.4 *Let $0.s_0 s_1 s_2 \cdots$ be the binary expansion of the number $\frac{1}{e}$. Map this to $\boldsymbol{s}$ in Σ_2. Compute $d_{\Sigma_2}(\boldsymbol{\sigma}^n(\boldsymbol{s}^*), \boldsymbol{s})$ for $n \leq 20$.*

Exercise 2.6.5 *Prove the polynomials with rational coefficients are dense in the set of continuous functions on $[a, b]$ using the metric $\|f - g\| = \max_{a \leq t \leq b} |f(t) - g(t)|$ for any continuous function f and g on $[a, b]$.*

2.7 Completeness

Another important topic is that of Cauchy sequences in a metric space and whether or not a metric space is complete.

Definition 2.7.1 Cauchy Sequence in a Metric Space

A sequence $(\boldsymbol{x_n})$ in a metric space (X, d) is a Cauchy sequence if

$$\forall \epsilon > 0 \, \exists N \ni n, m > N \implies d(\boldsymbol{x_n}, \boldsymbol{x_m}) < \epsilon$$

A Cauchy sequence need not converge to an object in the metric space but if a sequence in the metric space converges, then it must be a Cauchy sequence.

Theorem 2.7.1 Convergent Sequences in a Metric Space are Cauchy Sequences

Let $(\boldsymbol{x_n})$ be a sequence in the metric space (X, d) which converges to $\boldsymbol{x}$. Then $(\boldsymbol{x_n})$ is a Cauchy sequence.

Proof 2.7.1
Let $\epsilon > 0$ be given. Since the sequence converges, there is N so that $d(\boldsymbol{x_n}, \boldsymbol{x}) < \epsilon/2$ if $n > N$. Thus, if $n, m > N$

$$d(\boldsymbol{x_n}, \boldsymbol{x_m}) \leq d(\boldsymbol{x_n}, \boldsymbol{x}) + d(\boldsymbol{x}, \boldsymbol{x_m}) < \epsilon/2 + \epsilon/2 = \epsilon$$

which shows the sequence is a Cauchy sequence. ■

The rational numbers $\mathbb{Q}$ with the metric given by the absolute value function forms a metric space $(\mathbb{Q}, d = |\cdot|)$. There is a procedure we will discuss very carefully later which lets us build a new metric space $(\hat{\mathbb{Q}}, \hat{d})$ from $(\mathbb{Q}, d) = |\cdot|$ which is always complete. The space $\hat{\mathbb{Q}}$ is what we call $\Re$ and any real number is really a member of $\hat{\mathbb{Q}}$. Since $\Re$ is complete, we can prove other metric spaces are compete easily.

Theorem 2.7.2 The Completeness of $(\Re^n, d_\infty)$

For any $\boldsymbol{x}$ and $\boldsymbol{y}$ in $\Re^n$, define $d_\infty(\boldsymbol{x}, \boldsymbol{y}) = \max\{|x_1 - y_1|, \ldots, |x_n - y_n|\}$. Then d_∞ is a metric on $\Re^n$ and the metric space $(\Re^n, d_\infty)$ is complete; i.e. if $(\boldsymbol{x_k})$ is a Cauchy sequence in this metric space, there is an $\boldsymbol{x}$ in $(\Re^n, d_\infty)$ so that $\boldsymbol{x_k}$ converges in d_∞ to $\boldsymbol{x}$.

Proof 2.7.2
We leave the proof that the d_∞ is a metric to you. We will actually work this out later and you have

seen this in (Peterson (18) 2020) anyway. Assume $(\boldsymbol{x_k})$ is a Cauchy sequence. The element $\boldsymbol{x_k}$ is a sequence itself which we denote by $(x_{k,j})$ for $1 \leq j \leq n$. Then for a given $\epsilon > 0$, there is an N so that

$$\max_{1 \leq j \leq n} |x_{k,j} - x_{m,j}| \quad < \quad \epsilon/2 \text{ when } m > k > N_\epsilon$$

So for each fixed index j, we have

$$|x_{k,j} - x_{m,j}| \quad < \quad \epsilon/2 \text{ when } m > k > N_\epsilon$$

This says for fixed j, the sequence $(x_{k,j})$ is a Cauchy sequence of real numbers and hence must converge to a real number we will call a_j. This defines a new vector $\boldsymbol{a} \in \Re^n$. Does $\boldsymbol{x_k} \to \boldsymbol{a}$ in the d_∞ metric? We use the continuity of the function $|\cdot|$ to see for any $k > N_\epsilon$, we have

$$\lim_{m \to \infty} |x_{k,j} - x_{m,j}| \quad \leq \quad \epsilon/2 \Longrightarrow |x_{k,j} - \lim_{m \to \infty} x_{m,j}| \leq \epsilon/2$$

This argument works for all j and so $|x_{k,j} - a_j| \leq \epsilon/2$ when $k > N_\epsilon$ for all j which implies $\max_{1 \leq j \leq n} |x_{k,j} - a_j| \leq \epsilon/2 < \epsilon$ or $\|\boldsymbol{x_k} - \boldsymbol{a}\|_\infty < \epsilon$ when $k > N_\epsilon$. So $\boldsymbol{x_k} \to \boldsymbol{a}$ in d_∞. ■

If $\Re$ is complete so is $\mathbb{C}$. If $(\boldsymbol{z_n})$ is a Cauchy sequence of complex numbers, then $\boldsymbol{z_n} = x_n + \boldsymbol{i}y_n$ and

$$\forall \epsilon > 0 \, \exists N \ni n, m > N \quad \Longrightarrow \quad \sqrt{x_n^2 + y_n^2} < \epsilon$$

This tells us (x_n) and (y_n) are both Cauchy sequences of real numbers and so converge to $\boldsymbol{x}$ and $\boldsymbol{y}$ respectively. Thus, $\boldsymbol{z_n} \to \boldsymbol{z} = x + \boldsymbol{i}y$. It is also easy to see

Theorem 2.7.3 The Completeness of $(\mathbb{C}^n, d_\infty)$

$(\mathbb{C}^n, d_\infty)$ is complete; i.e. if $(\boldsymbol{z_k})$ is a Cauchy sequence in this space there is an $\boldsymbol{z}$ in $(\mathbb{C}^n, d_\infty)$ so that $\boldsymbol{z_k}$ converges in d_∞ to $\boldsymbol{z}$.

Proof 2.7.3

Assume $(\boldsymbol{z_k})$ is a Cauchy sequence. The element $\boldsymbol{z_k}$ is a sequence itself which we denote by $(z_{k,j})$ for $1 \leq j \leq n$ where $z_{k,j} = x_{k,j} + \boldsymbol{i}y_{k,j}$. Then for a given $\epsilon > 0$, there is an N so that

$$\max_{1 \leq j \leq n} \sqrt{(x_{k,j} - x_{m,j})^2 + (y_{k,j} - y_{m,j})^2} \quad < \quad \epsilon/2 \text{ when } m > k > N_\epsilon$$

So for each fixed index j, we have

$$|x_{k,j} - x_{m,j}| \quad < \quad \epsilon/2 \text{ when } m > k > N_\epsilon, \quad |y_{k,j} - y_{m,j}| < \epsilon/2 \text{ when } m > k > N_\epsilon$$

This says for fixed j, the sequences $(x_{k,j})$ and $(y_{k,j})$ are Cauchy sequences of real numbers and hence must converge to a real numbers a_j and b_j. This defines vectors $\boldsymbol{z} = a + \boldsymbol{i}b$ in $\mathbb{C}^n$. Does $\boldsymbol{z_k} \to \boldsymbol{z}$ in the $\|\cdot\|_\infty$ norm? We use the continuity of the function $|\cdot|$ to see for any $k > N_\epsilon$, we have

$$\lim_{m \to \infty} |x_{k,j} - x_{m,j}| \quad \leq \quad \epsilon/2 \Longrightarrow |x_{k,j} - \lim_{m \to \infty} x_{m,j}| \leq \epsilon/2$$

This argument works for all j and so $|x_{k,j} - a_j| \leq \epsilon/2$ when $k > N_\epsilon$ for all j which implies $\max_{1 \leq j \leq n} |x_{k,j} - a_j| \leq \epsilon/2 < \epsilon$ or $\|\boldsymbol{x_k} - \boldsymbol{a}\|_\infty < \epsilon$ when $k > N_\epsilon$. So $\boldsymbol{x_k} \to \boldsymbol{a}$ in $\|\cdot\|_\infty$. A similar argument shows $\boldsymbol{y_k} \to \boldsymbol{b}$ in $\|\cdot\|_\infty$. Hence, $\boldsymbol{z_k} \to \boldsymbol{z}$ in the d_∞. ■

Homework

Exercise 2.7.1 *Let X be a set of objects finite or infinite in number. Define the metric $d(x, y)$ on X by*

$$d(x,y) = \begin{cases} 1, & if x = y \\ 0, & if x \neq y \end{cases}$$

This is called the discrete metric on X.

- *Prove d is a metric on X.*
- *Characterize the Cauchy sequences of (X, d).*

Exercise 2.7.2 *Let A be a symmetric 3×3 matrix of non-negative real numbers. Let X be the finite set labeled $\{x_1, x_2, x_3\}$ and define the function d on $X \times X$ by $d(x_i.x_j) = A_{ij}$.*

- *Let $A = \begin{bmatrix} 0 & a & b \\ a & 0 & c \\ b & c & 0 \end{bmatrix}$. Prove if $a = b + c$, d defines a metric.*
- *For a given A which generates a metric, characterize the Cauchy sequences of (X, d).*

Exercise 2.7.3 *Let $A = \begin{bmatrix} 0 & 2 & 1 \\ 2 & 0 & 3 \\ 1 & 3 & 0 \end{bmatrix}$. Let X be a set of three elements labeled by $\{X_1, X_2, X_3\}$. Define $d : X \times X \to \Re$ by $d(X_i, X_j) = A_{ij}$.*

- *Prove d is a metric on X.*
- *Characterize the Cauchy sequences of (X, d).*

Exercise 2.7.4 *Let $A = \begin{bmatrix} 0 & 2 & a \\ 2 & 0 & 3 \\ a & 3 & 0 \end{bmatrix}$. Let X be a set of three elements labeled by $\{X_1, X_2, X_3\}$. Define $d : X \times X \to \Re$ by $d(X_i, X_j) = A_{ij}$.*

- *Characterize the values of a that allow d to be a metric on X.*
- *Characterize the Cauchy sequences of (X, d).*

Next, let's look at vector convergence using another way to measure the distance between vectors: the d_p metric for $p \geq 1$.

Theorem 2.7.4 $(\Re^n, d_p)$ **is Complete**

For any $\boldsymbol{x}$ and $\boldsymbol{y}$ in $\Re^n$, define $d_p(\boldsymbol{x}, \boldsymbol{y}) = \left(\sum_{i=1}^{n} |x_i - y_i|^p\right)^{1/p}$ for $p \geq 1$. We have discussed d_p in (Peterson (18) 2020) already and we will discuss it later here as well, but it is straightforward to show d_p is a metric with the triangle inequality established using Minkowski"s inequality. So d_∞ is a metric on $\Re^n$. Then, $(\Re^n, d_p)$ is complete; i.e. if $(\boldsymbol{x_k})$ is a Cauchy sequence in this space, there is an $\boldsymbol{x}$ in $(\Re^n, d_p)$ so that $\boldsymbol{x_k}$ converges in d_p to $\boldsymbol{x}$.

Proof 2.7.4

Let $(\boldsymbol{x_k})$ be a Cauchy sequence in $\|\cdot\|_p$. Then given $\epsilon > 0$, there is an N_ϵ so that $m > k > N_\epsilon$ implies

$$\|\boldsymbol{x_k} - \boldsymbol{x_m}\|_p = \left(\sum_{j=1}^{n} |x_{k,j} - x_{m,j}|^p\right)^{\frac{1}{p}} \quad < \quad \epsilon/2.$$

Thus, if $m > k > N_\epsilon$, $\sum_{j=1}^n |x_{k,j} - x_{m,j}|^p < (\epsilon/2)^p$. Since this is a sum on non-negative terms, each term must be less than $(\epsilon/2)^p$. So we must have $|x_{k,j} - x_{m,j}|^p < (\epsilon/2)^p$ or $|x_{k,j} - x_{m,j}| < (\epsilon/2)$ when $m > k > N_\epsilon$. This tells us immediately the sequence of real numbers $(x_{k,j})$ is a Cauchy sequence of real numbers and so must converge to a number we will call a_j. This defines the vector $\boldsymbol{a} \in \Re^n$. Since $|\cdot|^p$ is continuous, we can say

$$\begin{aligned}(\epsilon/2)^p \quad &\geq \quad \lim_{m\to\infty}\left(\sum_{j=1}^{n} |x_{k,j} - x_{m,j}|^p\right) = \left(\sum_{j=1}^{n} |x_{k,j} - \lim_{m\to\infty} x_{m,j}|^p\right) \\ &= \quad \sum_{j=1}^{n} |x_{k,j} - a_j|^p\end{aligned}$$

This says immediately that $\|\boldsymbol{x_k} - \boldsymbol{a}\|_p < \epsilon$ when $k > N_\epsilon$ so $\boldsymbol{x_k} \to \boldsymbol{a}$ in d_p. ■

A similar argument shows

Theorem 2.7.5 $(\mathbb{C}^n, d_p)$ **is Complete**

$(\mathbb{C}^n, d_p)$ is complete; i.e. if $(\boldsymbol{z_k})$ is a Cauchy sequence in this space, there is an $\boldsymbol{z}$ in $(\mathbb{C}^n, d_p)$ so that $\boldsymbol{z_k}$ converges in d_p to $\boldsymbol{z}$.

Proof 2.7.5

We leave the proof of this to you. ■

Homework

Exercise 2.7.5 $(\mathbb{C}^n, d_p)$ *is complete for $n > 1$.*

Exercise 2.7.6 *For $\boldsymbol{x} = \begin{bmatrix} x_1 \\ x_2 \end{bmatrix}$ in $\Re^2$ and $D = \begin{bmatrix} 2 & 0 \\ 0 & 4 \end{bmatrix}$, define $\rho^2(\boldsymbol{x}) = \boldsymbol{x}^T D \boldsymbol{x} = 2x_1^2 + 4x_2^2$.*

- *Using Minkowski's Inequality for $\Re^2$, first prove $2x_1y_1 + 4x_2y_2 = \boldsymbol{y}^T D\boldsymbol{x} \leq \rho(\boldsymbol{x})\rho(\boldsymbol{y})$. Do this by looking at the inner product of $\begin{bmatrix} \sqrt{2}x_1 \\ 2x_2 \end{bmatrix}$ with itself.*
- *Then prove $\rho^2(\boldsymbol{x} + \boldsymbol{y}) \leq (\rho(\boldsymbol{x}) + \rho(\boldsymbol{y}))^2$.*
- *Define $d(\boldsymbol{x}, \boldsymbol{y}) = \rho(\boldsymbol{x} - \boldsymbol{y})$. Prove d is a metric.*

Exercise 2.7.7 *For $\boldsymbol{x} = \begin{bmatrix} x_1 \\ x_2 \end{bmatrix}$ in $\Re^2$ and $D = \begin{bmatrix} 3 & 0 \\ 0 & 8 \end{bmatrix}$, define $\rho^2(\boldsymbol{x}) = \boldsymbol{x}^T D \boldsymbol{x} = 3x_1^2 + 8x_2^2$.*

- *Using Minkowski's Inequality for $\Re^2$, first prove $3x_1y_1 + 8x_2y_2 = \boldsymbol{y}^T D\boldsymbol{x} \leq \rho(\boldsymbol{x})\rho(\boldsymbol{y})$. Do this by looking at the inner product of $\begin{bmatrix} \sqrt{3}x_1 \\ \sqrt{8}x_2 \end{bmatrix}$ with itself.*
- *Then prove $\rho^2(\boldsymbol{x} + \boldsymbol{y}) \leq (\rho(\boldsymbol{x}) + \rho(\boldsymbol{y}))^2$.*

- *Define* $d(\boldsymbol{x}, \boldsymbol{y}) = \rho(\boldsymbol{x} - \boldsymbol{y})$*. Prove* d *is a metric.*

Exercise 2.7.8 *For* $\boldsymbol{x} = \begin{bmatrix} x_1 \\ x_2 \end{bmatrix}$ *in* $\Re^2$ *and* $D = \begin{bmatrix} \lambda_1^2 & 0 \\ 0 & \lambda_2^2 \end{bmatrix}$ *where* λ_1 *and* λ_2 *are both positive. Define* $\rho^2(\boldsymbol{x}) = \boldsymbol{x}^T D\boldsymbol{x} = \lambda_1^2 x_1^2 + \lambda_2 x_2^2$.

- *Using Minkowski's Inequality for* $\Re^2$*, first prove* $\lambda_1^2 x_1 y_1 + \lambda_2^2 x_2 y_2 = \boldsymbol{y}^T D\boldsymbol{x} \leq \rho(\boldsymbol{x})\rho(\boldsymbol{y})$*. Do this by looking at the inner product of* $\begin{bmatrix} \lambda_1 x_1 \\ \lambda_2 x_2 \end{bmatrix}$ *with itself.*
- *Then prove* $\rho^2(\boldsymbol{x} + \boldsymbol{y}) \leq (\rho(\boldsymbol{x}) + \rho(\boldsymbol{y}))^2$.
- *Define* $d(\boldsymbol{x}, \boldsymbol{y}) = \rho(\boldsymbol{x} - \boldsymbol{y})$*. Prove* d *is a metric.*

Exercise 2.7.9 *For* $\boldsymbol{x} = \begin{bmatrix} x_1 \\ x_2 \\ \vdots \\ x_n \end{bmatrix}$ *in* $\Re^n$ *and* D *an* $n \times n$ *diagonal matrix whose diagonal entries are* λ_i^2 *with each* λ_i *positive. Define* $\rho^2(\boldsymbol{x}) = \boldsymbol{x}^T D\boldsymbol{x} = \sum_{i=1}^n \lambda_i^2 x_i^2$.

- *Using Minkowski's Inequality for* $\Re^n$*, first prove* $\lambda_1^2 x_1 y_1 + \ldots + \lambda_n^2 x_n y_n = \boldsymbol{y}^T D\boldsymbol{x} \leq \rho(\boldsymbol{x})\rho(\boldsymbol{y})$*. Do this by looking at the inner product of* $\begin{bmatrix} \lambda_1 x_1 \\ \vdots \\ \lambda_n x_n \end{bmatrix}$ *with itself.*
- *Then prove* $\rho^2(\boldsymbol{x} + \boldsymbol{y}) \leq (\rho(\boldsymbol{x}) + \rho(\boldsymbol{y}))^2$.
- *Define* $d(\boldsymbol{x}, \boldsymbol{y}) = \rho(\boldsymbol{x} - \boldsymbol{y})$*. Prove* d *is a metric.*

Exercise 2.7.10 *Let* $\boldsymbol{x}$ *and* $\boldsymbol{y}$ *be in* $\Re^n$ *and let* A *be an* $n \times n$ *positive definite symmetric matrix. Define* $\rho^2(\boldsymbol{x}) = \boldsymbol{x}^T A\boldsymbol{x}$*. Define* $d(\boldsymbol{x}, \boldsymbol{y}) = \rho(\boldsymbol{x} - \boldsymbol{y})$*. Prove* d *is a metric. This proof uses the decomposition* $A = P^T DP$ *where* D *is a diagonal matrix as we have used in the previous exercise. Look at the explanations in (Peterson (20) 2020) for more details.*

One more idea here is very important: the idea of **separability** of a space.

Definition 2.7.2 Separable Subsets in a Metric Space

We say S *is a separable subset of the metric space* (X, d) *is* S *is countable and dense in* X.

This will be an important tool later.

2.7.1 Homework

Exercise 2.7.11 *Prove the rational numbers are dense in* $\Re$ *and the complex numbers with rational real and imaginary parts are dense in* $\mathbb{C}$.

Exercise 2.7.12 *Prove the vectors with rational components are dense in* $\Re^n$ *and the complex vectors with rational components are dense in* $\mathbb{C}^n$.

Exercise 2.7.13 *Prove the polynomials with rational coefficients are dense in the set of continuous functions on* $[a, b]$ *with the infinity norm. We know we can approximate any continuous function to within a given* $\epsilon > 0$ *in the infinity norm using a Bernstein polynomial. The proof thus requires you to prove you can approximate any polynomial with a polynomial with rational coefficients to any accuracy desired.*

Exercise 2.7.14 *Repeat the exercise above for the metric* $d_1(x,y) = \int_a^b |x(t) - y(t)|dt$.

Exercise 2.7.15 *Repeat the exercise above for the metric* $d_2(x,y) = \sqrt{\int_a^b |x(t) - y(t)|^2 dt}$.

2.8 Function Metric Spaces

Let's explore various spaces of function spaces. For most sets of objects, there are multiple useful ways to define a metric.

$C(\Omega)$**:** For the Ω the finite interval $[a,b]$, we let

$$C([a,b]) \quad = \quad \{\boldsymbol{x} : [a,b] \to \Re \mid \boldsymbol{x} \text{ is continuous on } [a,b]\}$$

and define $d_\infty(\boldsymbol{x},\boldsymbol{y}) = \sup_{a \leq t \leq b} |x(t) - y(t)|$. Since $\boldsymbol{x}$ and $\boldsymbol{y}$ are continuous on $[a,b]$, so is $|\boldsymbol{x} - \boldsymbol{y}|$ and therefore a maximum is achieved. We could therefore write $d_\infty(\boldsymbol{x},\boldsymbol{y}) = \max_{a \leq t \leq b} |x(t) - y(t)|$ instead. We note d_∞ is a metric.

M1: The maximum is always finite and non-negative.

M2: If $d_\infty(\boldsymbol{x},\boldsymbol{y}) = 0$, then $\max_{a \leq t \leq b} |x(t) - y(t)| = 0$ implying $|\boldsymbol{x}(t) - \boldsymbol{y}(t)| = 0$ for all t. Thus, $\boldsymbol{x}(t) = \boldsymbol{y}(t)$ for all t or $\boldsymbol{x} = \boldsymbol{y}$. On the other hand, if $\boldsymbol{x} = \boldsymbol{y}$, we see $d_\infty(\boldsymbol{x},\boldsymbol{y}) = 0$.

M3: It is clear $d_\infty(\boldsymbol{x},\boldsymbol{y}) = d_\infty(\boldsymbol{y},\boldsymbol{x})$.

M4: For any t, given any $\boldsymbol{z} \in C([a,b])$,

$$\begin{aligned} |\boldsymbol{x}(t) - \boldsymbol{y}(t)| &\leq |\boldsymbol{x}(t) - \boldsymbol{z}(t)| + |\boldsymbol{z}(t) - \boldsymbol{y}(t)| \\ &\leq d_\infty(\boldsymbol{x},\boldsymbol{z}) + d_\infty(\boldsymbol{z},\boldsymbol{y}) \end{aligned}$$

This shows

$$d_\infty(\boldsymbol{x},\boldsymbol{y}) \quad \leq \quad d_\infty(\boldsymbol{x},\boldsymbol{z}) + d_\infty(\boldsymbol{z},\boldsymbol{y})$$

$B(\,(a,b)\,)$**:** Here $\Omega = (a,b)$ for a finite interval. The definition is

$$B(\,(a,b)\,) \quad = \quad \{\boldsymbol{x} : (a,b) \to \Re \mid \boldsymbol{x} \text{ is bounded on } (a,b)\}$$

and define $d_\infty(\boldsymbol{x},\boldsymbol{y}) = \sup_{a \leq t \leq b} |x(t) - y(t)|$. It is not possible to replace the sup by max here as the maximum need not be attained. The definition of $B([a,b])$ is similar. Also, note for

$$C(\,(a,b)\,) \quad = \quad \{\boldsymbol{x} : (a,b) \to \Re \mid \boldsymbol{x} \text{ is continuous on } (a,b)\}$$

$d_\infty(\boldsymbol{x},\boldsymbol{y})$ need not be finite so it cannot serve as a metric. Proving d_∞ is a metric on $B((a,b))$ is essentially the argument we used in the case of $C([a,b])$ so we won't repeat it.

$RI(\Omega)$**:** For $\Omega = [a,b]$ where $[a,b]$ is a finite interval, we define

$$RI(\,(a,b)\,) \quad = \quad \{\boldsymbol{x} \in B([a,b]) \mid \boldsymbol{x} \text{ is Riemann Integrable on } [a,b]\}$$

and define $d_1(\boldsymbol{x},\boldsymbol{y}) = \int_a^b |\boldsymbol{x} - \boldsymbol{y}(t)|dt$ which is well-defined as $\boldsymbol{x}$ and $\boldsymbol{y}$ are Riemann Integrable.

M1: The integral of a non-negative function in non-negative.

M2: If $\boldsymbol{x} = \boldsymbol{y}$, we see $d_1(\boldsymbol{x}, \boldsymbol{y}) = 0$. But if $d_1(\boldsymbol{x}, \boldsymbol{y}) = 0$, then we can't say the two functions match as they can differ on a set of content zero. Hence $M2$ fails here.

M3: It is clear $d_1(\boldsymbol{x}, \boldsymbol{y}) = d_1(\boldsymbol{y}, \boldsymbol{x})$.

M4: For any t, given any $\boldsymbol{z} \in RI([a, b])$,

$$|\boldsymbol{x}(t) - \boldsymbol{y}(t)| \leq |\boldsymbol{x}(t) - \boldsymbol{z}(t)| + |\boldsymbol{z}(t) - \boldsymbol{y}(t)|$$

This shows

$$d_1(\boldsymbol{x}, \boldsymbol{y}) \leq d_1(\boldsymbol{x}, \boldsymbol{z}) + d_1(\boldsymbol{z}, \boldsymbol{y})$$

Thus $(RI([a, b]), d_1)$ is **not** a metric space. However, $(C([a, b]), d_1)$ **is** a metric space.

M1: The integral of a non-negative function in non-negative.

M2: If $\boldsymbol{x} = \boldsymbol{y}$, we see $d_1(\boldsymbol{x}, \boldsymbol{y}) = 0$. But if $d_1(\boldsymbol{x}, \boldsymbol{y}) = 0$, let's assume there is a point t_0 where $\boldsymbol{x}(t_0) \neq \boldsymbol{y}(t_0)$. Hence, $|\boldsymbol{x}(t_0) - \boldsymbol{y}(t_0)| = r > 0$. Thus, because of continuity, there is a $\delta > 0$ so that $r/2 < |\boldsymbol{x}(t) - \boldsymbol{y}(t)|$ for all $t_0 - \delta < t < t_0 + \delta$ with δ small enough so that $(t_0 - \delta, t_0 + \delta) \subset [a, b]$. We then have

$$\begin{aligned} 0 &= \int_a^{t_0-\delta} |\boldsymbol{x}(t) - \boldsymbol{y}(t)| dt + \int_{t_0-\delta}^{t_0+\delta} |\boldsymbol{x}(t) - \boldsymbol{y}(t)| dt + \int_{t_0+\delta}^{b} |\boldsymbol{x}(t) - \boldsymbol{y}(t)| dt \\ &\geq 0 + 2\delta r > 0 \end{aligned}$$

This is a contradiction and so no such point t_0 can exist. Thus $\boldsymbol{x} = \boldsymbol{y}$ and $M2$ is satisfied.

M3: It is clear $d_1(\boldsymbol{x}, \boldsymbol{y}) = d_1(\boldsymbol{y}, \boldsymbol{x})$.

M4: For any t, given any $\boldsymbol{z} \in RI([a, b])$,

$$|\boldsymbol{x}(t) - \boldsymbol{y}(t)| \leq |\boldsymbol{x}(t) - \boldsymbol{z}(t)| + |\boldsymbol{z}(t) - \boldsymbol{y}(t)|$$

This shows

$$d_1(\boldsymbol{x}, \boldsymbol{y}) \leq d_1(\boldsymbol{x}, \boldsymbol{z}) + d_1(\boldsymbol{z}, \boldsymbol{y})$$

Thus $(C([a, b]), d_1)$ is a metric space.

Homework

Exercise 2.8.1 *Prove* $(B([a, b]), d_\infty)$ *is a metric space.*

Exercise 2.8.2 *Let* $[a, b]$ *be a finite interval*

$$C^1([a, b]) = \{\boldsymbol{x} \in C([a, b]) \textit{ and } \boldsymbol{x}' \in C([a, b])\}$$

and define $d(\boldsymbol{x}, \boldsymbol{y}) = d_\infty(\boldsymbol{x}, \boldsymbol{y}) + d_\infty(\boldsymbol{x}', \boldsymbol{y}')$. *Prove* $(C^1([a, b]), d)$ *is a metric space.*

Exercise 2.8.3 *Let* $[a, b]$ *be a finite interval*

$$C^1([a, b]) = \{\boldsymbol{x} \in C([a, b]) \textit{ and } \boldsymbol{x}' \in C([a, b])\}$$

and define $d(\boldsymbol{x}, \boldsymbol{y}) = \max\{d_\infty(\boldsymbol{x}, \boldsymbol{y}), d_\infty(\boldsymbol{x}', \boldsymbol{y}')\}$. *Prove* $(C^1([a, b]), d)$ *is a metric space.*

Exercise 2.8.4 *Let* $[a, b]$ *be a finite interval*

$$C^1([a, b]) = \{\boldsymbol{x} \in C([a, b]) \textit{ and } \boldsymbol{x}' \in C([a, b])\}$$

and define $d(\boldsymbol{x}, \boldsymbol{y}) = \sqrt{(d_\infty(\boldsymbol{x}, \boldsymbol{y}))^2 + (d_\infty(\boldsymbol{x}', \boldsymbol{y}'))^2}$*. Prove* $(C^1([a,b]), d)$ *is a metric space.*

Exercise 2.8.5 *Is* $B(\mathbf{0}, 1) in (C([a,b]), d_\infty) \subset B(\mathbf{0}, 1) in (C([a,b]), d_1)$*?*

Exercise 2.8.6 *Is* $B(\mathbf{0}, 1) in (C([a,b]), d_1) \subset B(\mathbf{0}, 1) in (C([a,b]), d_\infty)$*?*

Exercise 2.8.7 *Let* $[a,b]$ *be a finite interval*

$$X([a,b]) = \{\boldsymbol{x} \in C([a,b]) \text{ and } \boldsymbol{x}' \text{ exists on } [a,b]\}$$

and define $d(\boldsymbol{x}, \boldsymbol{y}) = d_\infty(\boldsymbol{x}, \boldsymbol{y}) + d_\infty(\boldsymbol{x}', \boldsymbol{y}')$*. Is* $(X([a,b]), d)$ *is a metric space?*

2.9 Sequences and Series of Complex Numbers

We are now going to discuss sequences and series of special interest to us. In (Peterson (18) 2020), we have carefully gone over the idea of sequences and series of real numbers. Now we want to extend these ideas to sequences and series of complex numbers.

Given any sequence of complex numbers, (a_n) for $n \geq 1$, the convergence of such a sequence is defined as usual.

Definition 2.9.1 Complex Sequence Convergence and Divergence

Let $(a_n)_{n \geq k}$ *be a sequence of complex numbers. Let* a *be a complex number. We say the sequence* a_n *converges to* a *if*

$$\forall \epsilon > 0, \exists N \ni n > N \Longrightarrow |a_n - a| < \epsilon$$

where $\cdot|$ *here is the complex magnitude; i.e. if* $z = a + ib$*, then* $|z| = \sqrt{a^2 + b^2}$*. We usually just write* $a_n \to a$ *as* $n \to \infty$ *to indicate this convergence. We call* a *the limit of the sequence* $(a_n)_{n \geq k}$*. All of our usual theorems about convergence of real sequences essentially follow through but, of course, any results involving comparison will fail as we cannot compare complex numbers, only their magnitudes.*

The sequence $(a_n)_{n \geq k}$ **does not converge** *to a complex number* a *if we can find a positive number* ϵ *so that no matter what* N *we choose, there is always at least one* $n > N$ *with* $|a_n - a| > \epsilon$*. We write this using mathematical language as*

$$\exists \epsilon > 0, \ni \forall N \exists n > N \text{ with } |a_n - a| \geq \epsilon$$

If a sequence does not converge, we say it **diverges.**

We can construct from a complex sequence a new sequence, called the sequence of **Partial Sums**, as follows:

$$\begin{aligned} S_1 &= a_1 \\ S_2 &= a_1 + a_2 \\ S_3 &= a_1 + a_2 + a_3 \\ &\vdots \\ S_n &= a_1 + a_2 + a_3 + \ldots + a_n \end{aligned}$$

Now we are interested in whether or not the sequence of partial sums of a complex sequence converges. We need more notation.

Definition 2.9.2 The Sum of a Complex Series or Convergence of a Complex Series

Let $(a_n)_{n\geq 1}$ be any complex sequence and let $(S_n)_{n\geq 1}$ be its associated sequence of partial sums.
(a) If $\lim_{n\to\infty} S_n$ exists, we denote the value of this limit by S. Since this is the same as $\lim_{n\to\infty} \sum_{i=1}^{n} a_i = S$, we use the symbol $\sum_{i=1}^{\infty} a_i$ to denote S. But remember it is a **symbol** *for a limiting process. Again, note the choice of summation variable i is immaterial. We also say $\sum_{i=1}^{\infty} a_i$ is the* **infinite series** *associated with the sequence $(a_n)_{n\geq 1}$.*
(b) If the $\lim_{n\to\infty} S_n$ does not exist, we say the **series** $\sum_{i=1}^{\infty} a_i$ **diverges.**

We can get a lot of information about a complex series by looking at the series we get by summing the complex magnitudes of the terms of the base sequence used to construct the partial sums.

Definition 2.9.3 Complex Absolute Convergence and Conditional Convergence

Let $(a_n)_{n\geq 1}$ be any complex sequence and let $\sum_{i=1}^{\infty} |a_i|$ be the complex series we construct from (a_n) by taking the complex magnitude of each term a_i in the sequence.
(a) if $\sum_{i=1}^{\infty} |a_i|$ **converges** *we say the series* **converges absolutely.**
(b) (a) if $\sum_{i=1}^{\infty} |a_i|$ **diverges** *but $\sum_{i=1}^{\infty} a_i$* **converges**, *we say the series* **converges conditionally.**

Since the series of complex magnitudes is a standard series of real numbers, all of our usual tools about series of non-negative terms apply. The first in the nth term test whose proof is virtually identical to the one for real sequences.

Theorem 2.9.1 The Complex n^{th} Term Test

Assume $\sum_{n=1}^{\infty} a_n$ converges, then $a_n \to 0$. Note this says if a_n does **not** *converge to 0 or fails to converge at all, the original series must* **diverge.**

Proof 2.9.1
Since $\sum_{n=1}^{\infty} a_n$ converges, we know $S_n \to S$ for some S. Thus, given $\epsilon > 0$, there is an N so that

$$n > N \Longrightarrow |S_n - S| < \epsilon/2$$

Now pick any $\hat{n} > N+1$. Then $\hat{n}-1$ and $\hat{n}$ are both greater than N. Thus, since $a_{\hat{n}}$ is the difference between two successive terms in (S_n), we have

$$|a_{\hat{n}}| = |S_{\hat{n}} - S + S - S_{\hat{n}-1}| \leq |S_{\hat{n}} - S| + |S - S_{\hat{n}-1}| < \epsilon/2 + \epsilon/2 = \epsilon$$

Since the choice of $\hat{n} > N$ was arbitrary, we see we have shown $|a_n| < \epsilon$ for $n > N$ with the choice of $\epsilon > 0$ arbitrary. Hence, $\lim_{n\to\infty} a_n = 0$. ■

If the sequence (S_n) converges to S, we also know the sequence is a Cauchy sequence. Since $\mathbb{C}$ is **complete** if (S_n) is a Cauchy sequence, it must converge. So we can say

$$(S_n) \text{ converges} \Longleftrightarrow (S_n) \text{ is a Cauchy Sequence}$$

Now (S_n) is a Cauchy sequence means given $\epsilon > 0$ there is an N so that

$$n, m > N \implies |S_n - S_m| < \epsilon$$

For the moment, assume $n > m$. Then

$$S_n - S_m = \left(\sum_{i=1}^{m} a_i + \sum_{i=m+1}^{n} a_i \right) - \sum_{i=1}^{m} a_i = \sum_{i=m+1}^{n} a_i$$

and for $m < n$, we would get

$$S_m - S_n = \left(\sum_{i=1}^{n} a_i + \sum_{i=n+1}^{m} a_i \right) - \sum_{i=1}^{n} a_i = \sum_{i=n+1}^{m} a_i$$

We can state this as a theorem!

Theorem 2.9.2 The Cauchy Criterion for Complex Series

$$\sum_{n=1}^{\infty} a_n \text{ converges} \iff \left(\forall \epsilon > 0 \, \exists N \ni \left| \sum_{i=m+1}^{n} a_i \right| < \epsilon \text{ if } n > m > N \right)$$

Now use this Cauchy criterion, we can say more about the consequences of absolute convergence.

Theorem 2.9.3 A Complex Series that Converges Absolutely also Converges

If $\sum_{n=1}^{\infty} a_n$ converges absolutely then it also converges.

Proof 2.9.2

We know $\sum_{n=1}^{\infty} |a_n|$ converges. Let $\epsilon > 0$ be given. By the Cauchy Criterion for series, there is an N so that

$$\big| |a_{m+1}| + \ldots + |a_n| \big| < \epsilon \text{ if } n \geq m > N$$

Thus

$$|a_{m+1}| + \ldots + |a_n| < \epsilon \text{ if } n > m > N$$

The triangle and reversed triangle inequality for complex numbers also hold, so

$$|a_{m+1} + \ldots + a_m| \leq |a_{m+1}| + \ldots + |a_n| < \epsilon \text{ if } n > m > N$$

Thus $| \sum_{i=m+1}^{n} a_i | < \epsilon$ when $n > m > N$. Hence, the sequence of partial sums is a Cauchy sequence which tells us the $\sum_{n=1}^{\infty} a_n$ converges. ■

Homework

Exercise 2.9.1 *Redo the idea of a geometric series $\sum_{n=0}^{\infty} c^n$ where c is a complex number. The proof that this series converges to $\frac{1}{1-c}$ if $|c| < 1$ is virtually identical.*

Exercise 2.9.2 *Prove the ratio test for complex series.*

Exercise 2.9.3 *Consider the set of all functions f defined on $[a, b] \subset \Re$ with complex values. Prove f is continuous on $[a, b]$ as a complex function if and only if its real and imaginary parts are continuous on $[a, b]$.*

Exercise 2.9.4 *For complex-valued functions on $[a, b] \subset \Re$ with the infinity norm prove a sequence of complex-valued functions which converges uniformly with respect to this norm has a continuous limit function. You will have to figure out how to redefine uniform convergence here. Note we have not discussed norms officially yet but you should all know what norm we are talking about here.*

Exercise 2.9.5 *For the series $\sum_{n=0}^{\infty}((6n^3 + 3n + 20) + (n^2 + 5)\boldsymbol{i})c^n$ where c is a complex number determine the interval of convergence. and if the series converges at the endpoints.*

2.10 Sequence Metric Spaces

Given a sequence of complex numbers (a_n), which we assume has indexing starting at $n = 1$ for convenience, the set of all complex sequences has a variety of interesting subsets which might even be subspaces. It is clear the set of all sequences is a vector space over the reals but whether or not a subset is a subspace depends on whether the subset is closed under scalar multiplication and vector addition.

In the context of complex sequences, a sequence is bounded refers to its bound with respect to the complex magnitude. Clearly, the set of all sequences contains sequences which may not be bounded and may not converge. Let S be the set of all sequences of complex numbers and assume for convenience their indexing begins at 1. For two sequences $\boldsymbol{x}$ and $\boldsymbol{y}$, define

$$d(\boldsymbol{x}, \boldsymbol{y}) = \sum_{n=1}^{\infty} \frac{1}{2^i} \phi(|x_i - y_i|)$$

where $\phi(u) = \frac{u}{1+u}$ for non-negative u is a function we have seen before in our discussion of symbol sequences. From those discussions, we see d is a well-defined metric on S. Notice d could be defined as

$$d(\boldsymbol{x}, \boldsymbol{y}) = \sum_{n=1}^{\infty} \frac{1}{3^i} \phi(|x_i - y_i|)$$

and so forth for other base exponents different from 2^i and 3^i. As long as the comparison test applied to the series of non-negative numbers we get using the complex magnitudes gives us a geometric series all is good.

The set of all bounded sequences S becomes a metric space too. For two complex sequences $\boldsymbol{x}$ and $\boldsymbol{y}$,

$$d(\boldsymbol{x}, \boldsymbol{y}) = \sup_i |x_i - y_i|$$

M1: The supremum here always exists and is non-negative.

M2: If $\boldsymbol{x} = \boldsymbol{y}$, we see $d(\boldsymbol{x}, \boldsymbol{y}) = 0$. If $d(\boldsymbol{x}, \boldsymbol{y}) = 0$, then $\sup_i |x_i - y_i| = 0$ or $x_i = y_i$ for all i telling us $\boldsymbol{x} = \boldsymbol{y}$.

M3: It is clear $d(\boldsymbol{x}, \boldsymbol{y}) = d(\boldsymbol{y}, \boldsymbol{x})$.

M4: Given any bounded sequence $\boldsymbol{z}$,

$$\begin{aligned} |x_i - y_i| &\leq |x_i - z_i| + |z_i - y_i| \\ &\leq \sup_i |x_i - z_i| + \sup_i |z_i - y_i| = d(\boldsymbol{x}, \boldsymbol{z}) + d(\boldsymbol{z}, \boldsymbol{y}) \end{aligned}$$

This shows

$$d(\boldsymbol{x}, \boldsymbol{y}) \quad \leq \quad d(\boldsymbol{x}, \boldsymbol{z}) + d(\boldsymbol{z}, \boldsymbol{y})$$

Thus (S, d) is a metric space. This sequence space is called ℓ^∞ and this metric is called d_∞. We also use these names for the corresponding sequences of real numbers (S, d) and we know which one we are referring to from context. In a rare circumstance, we might have to tell you this. In this section, we will occasionally remind you of this just to make sure it is clear.

Another common subset of the space of all sequence is

$$S \quad = \quad \{\boldsymbol{x} = (x_n) \mid \sum_{n=1}^{\infty} |x_n| \text{ converges } \}$$

S becomes a metric space too. For two sequences $\boldsymbol{x}$ and $\boldsymbol{y}$,

$$d(\boldsymbol{x}, \boldsymbol{y}) \quad = \quad \sum_{i=1}^{\infty} |x_i - y_i|$$

M1: Since

$$\sum_{i=1}^{n} |x_i - y_i| \quad \leq \quad \sum_{i=1}^{n} |x_i| + \sum_{i=1}^{n} |y_i| \leq \sum_{i=1}^{\infty} |x_i| + \sum_{i=1}^{\infty} |y_i|$$

we see $\sum_{i=1}^{\infty} |x_i - y_i|$ converges and in non-negative.

M2: If $\boldsymbol{x} = \boldsymbol{y}$, we see $d(\boldsymbol{x}, \boldsymbol{y}) = 0$. If $d(\boldsymbol{x}, \boldsymbol{y}) = 0$, then $|x_i - y_i| = 0$ or $x_i = y_i$ for all i telling us $\boldsymbol{x} = \boldsymbol{y}$.

M3: It is clear $d(\boldsymbol{x}, \boldsymbol{y}) = d(\boldsymbol{y}, \boldsymbol{x})$.

M4: Given any other sequence $\boldsymbol{z}$ in S,

$$\begin{aligned} \sum_{i=1}^{n} |x_i - y_i| \quad &\leq \quad \sum_{i=1}^{n} |x_i - z_i| + \sum_{i=1}^{n} |z_i - y_i| \\ &\leq \quad d(\boldsymbol{x}, \boldsymbol{z}) + d(\boldsymbol{z}, \boldsymbol{y}) \end{aligned}$$

This shows

$$d(\boldsymbol{x}, \boldsymbol{y}) \quad \leq \quad d(\boldsymbol{x}, \boldsymbol{z}) + d(\boldsymbol{z}, \boldsymbol{y})$$

Thus (S, d) is a metric space. This sequence space is called ℓ^1 and this metric is called d_1.

Homework

Exercise 2.10.1 *Let $\boldsymbol{x}, \boldsymbol{y} \in \Re^n$ and let A be an $n \times n$ positive definite symmetric matrix. Prove this version of Hölder's Inequality: $\boldsymbol{x}^T A\boldsymbol{y} \leq \sqrt{\boldsymbol{x}^T \boldsymbol{x}}\ \sqrt{\boldsymbol{y}^T A^T A\boldsymbol{y}}$. To do this, let $\boldsymbol{w} = A\boldsymbol{y}$ and use the original Hölder's inequality.*

Exercise 2.10.2 *Let $\boldsymbol{x}, \boldsymbol{y} \in \Re^n$ and let A be an $n \times n$ positive definite matrix. Prove this version of Hölder's Inequality: $\boldsymbol{x}^T A\boldsymbol{y} \leq \sum_{i=1}^{n} |x_i| \ \max_{1 \leq i \leq n} |A\boldsymbol{y}|_i$.*

Exercise 2.10.3 *Let $\boldsymbol{x}, \boldsymbol{y} \in \ell^2$ and let $A : \ell^2 \rightarrow \ell^2$ satisfy $\sum_{i=1}^{\infty} x_i (A\boldsymbol{y})_i > 0$ for all $\boldsymbol{x}$ and $\boldsymbol{y}$ in ℓ^2 that are not zero: this is a positive definite condition in this setting. Can you prove a version of Hölder's Inequality here?*

Exercise 2.10.4 *Let $\boldsymbol{x}$ be any sequence. This means all x_i are real numbers.*

- *Define $\phi(t) = \frac{t}{1+t}$ for all $t \geq 0$. Prove $\phi(s+t) \leq \phi(s) + \phi(t)$ for all such s and t.*
- *Prove the series $\sum_{i=1}^{\infty} \frac{1}{2^i} \phi(|x_i|)$ converges for all sequences $\boldsymbol{x}$.*
- *Define $d(\boldsymbol{x}, \boldsymbol{y}) = \sum_{i=1}^{\infty} \frac{1}{2^i} \phi(|x_i - y_i|)$ for all sequences $\boldsymbol{x}$ and $\boldsymbol{y}$. Prove d is a metric on the set of all sequences. Does it matter that we use $\frac{1}{2^i}$ as the nonlinear scaling factor here?*

Exercise 2.10.5 *Repeat the previous exercise with the nonlinear scaling factor $\frac{1}{3^i}$.*

There are many other sequence spaces here. We can classify some of them using conjugate exponents.

Definition 2.10.1 Conjugate Exponents

we say the positive numbers p and q are **conjugate exponents** *if $p > 1$ and $1/p + 1/q = 1$. If $p = 1$, we define its conjugate exponent to be $q = \infty$.*

Comment 2.10.1 *Conjugate exponents satisfy some fundamental identities. Clearly, if $p > 1$, $\frac{1}{p} + \frac{1}{q} \Longrightarrow 1 = \frac{p+q}{p\,q}$ and also $pq = p + q$ and $(p-1)\,(q-1) = 1$. We will use these identities quite a bit.*

The are additional new sequence spaces we want to discuss called the ℓ^p spaces where $p \geq 1$ will have a conjugate exponent q. We have just shown that (ℓ^∞, d_∞) and (ℓ^1, d_1) are metric spaces. To discuss these spaces carefully, we need a standard result we call the $\boldsymbol{\alpha} - \boldsymbol{\beta}$ lemma whose proof is in many texts such as (Peterson (18) 2020).

Lemma 2.10.1 The $\alpha - \beta$ Lemma

Let α and β be positive real numbers and p and q be conjugate exponents. Then $\alpha\,\beta \leq \frac{\alpha^p}{p} + \frac{\beta^q}{q}$.

Proof 2.10.1
You should look at the proof in (Peterson (18) 2020) to refresh your memory of how this is done. ■

The new sequence spaces we want to define are

Definition 2.10.2 The ℓ^p Sequence Space

Let $p \geq 1$. The collection of all complex sequences, $(a_n)_{n=1}^{\infty}$ for which $\sum_{n=1}^{\infty} |a_n|^p$ converges is denoted by the symbol ℓ^p.
(1) $\ell^1 = \{(a_n)_{n=1}^{\infty} : \sum_{n=1}^{\infty} |a_n|$ converges.$\}$
(2) $\ell^2 = \{(a_n)_{n=1}^{\infty} : \sum_{n=1}^{\infty} |a_n|^2$ converges.$\}$
We also define $\ell^\infty = \{(a_n)_{n=1}^{\infty} : \sup_{n \geq 1} |a_n| < \infty\}$.

There is a fundamental inequality connecting sequences in ℓ^p and ℓ^q when p and q are conjugate exponents called **Hölder's Inequality**. Its proof is straightforward but has a few tricks. We have proven this in (Peterson (18) 2020) and also in (Peterson (20) 2020) but the review is good and we will try to add some additional insights. Note these arguments which originally use the absolute value function as they were argued for the case of real values, do not change at all really using complex magnitudes.

Theorem 2.10.2 Complex Hölder's Inequality

Let $p > 1$ and p and q be conjugate exponents. If $x \in \ell^p$ and $y \in \ell^q$, then

$$\sum_{n=1}^{\infty} |x_n \, y_n| \;\leq\; \left(\sum_{n=1}^{\infty} |x_n|^p\right)^{1/p} \left(\sum_{n=1}^{\infty} |y_n|^q\right)^{1/q}$$

where $x = (x_n)$ and $y = (y_n)$.

Proof 2.10.2

This inequality is clearly true if either of the two sequences x and y are the zero sequence. So we can assume both x and y have some nonzero terms in them. Then $x \in \ell^p$, we know

$$0 < u \;=\; \left(\sum_{n=1}^{\infty} |x_n|^p\right)^{1/p} < \infty, \quad 0 < v = \left(\sum_{n=1}^{\infty} |y_n|^q\right)^{1/q} < \infty$$

Now define new sequences, $\hat{x}$ and $\hat{y}$ by $\hat{x}_n = x_n/u$ and $\hat{y}_n = y_n/v$. Then, we have

$$\sum_{n=1}^{\infty} |\hat{x}_n|^p \;=\; 1, \quad \sum_{n=1}^{\infty} |\hat{y}_n|^q = 1.$$

Now apply the $\alpha - \beta$ Lemma to $\alpha = |\hat{x_n}|$ and $\beta = |\hat{y_n}|$ for any nonzero terms $\hat{x}_n$ and $\hat{y}_n$. Then $|\hat{x_n} \, \hat{y_n}| \leq |\hat{x_n}|^p/p + |\hat{y_n}|^q/q$. This is also true, of course, if either $\hat{x}_n$ or $\hat{y}_n$ are zero although the $\alpha - \beta$ lemma does not apply. Now sum over N terms to get

$$\sum_{n=1}^{N} |\hat{x_n} \, \hat{y_n}| \;\leq\; \frac{1}{p} \sum_{n=1}^{N} |\hat{x_n}|^p + \frac{1}{q} \sum_{n=1}^{N} |\hat{y_n}|^q \leq \frac{1}{p} \sum_{n=1}^{\infty} |\hat{x_n}|^p + \frac{1}{q} \sum_{n=1}^{\infty} |\hat{y_n}|^q = 1$$

So we have $\sum_{n=1}^{\infty} |\hat{x_n} \, \hat{y_n}| \leq 1$ which implies

$$\sum_{n=1}^{\infty} |x_n \, y_n| \leq u \, v \;=\; \left(\sum_{n=1}^{\infty} |x_n|^p\right)^{1/p} \left(\sum_{n=1}^{\infty} |y_n|^q\right)^{1/q}$$

■

We can also do this inequality for the case $p = 1$ and $q = \infty$.

Theorem 2.10.3 Complex Hölder's Theorem for $p = 1$ and $q = \infty$

If $x \in \ell^1$ and $y \in \ell^\infty$, then $\sum_{n=1}^{\infty} |x_n y_n| \leq \left(\sum_{n=1}^{\infty} |x_n|\right) \sup_{n \geq 1} |y_n|$.

Proof 2.10.3

We know since $y \in \ell^\infty$, $|y_n| \leq \sup_{k \geq 1} |y_k|$. Thus, $\sum_{n=1}^{N} |x_n y_n| \leq \left(\sum_{n=1}^{\infty} |x_n|\right) \sup_{k \geq 1} |y_k|$.

We see the sequence of partial sums $\sum_{n=1}^{N} |x_n y_n|$ is bounded above by $\left(\sum_{n=1}^{\infty} |x_n|\right) \sup_{k \geq 1} |y_k|$. This gives us our result. ■

There is also an associated inequality called Minkowski's Inequality. The general version is

Theorem 2.10.4 Complex Minkowski's Inequality

Let $p \geq 1$ and let x and y be in ℓ^p, Then, $\boldsymbol{x} + \boldsymbol{y}$ is in ℓ^p also and

$$\left(\sum_{n=1}^{\infty} |x_n + y_n|^p\right)^{\frac{1}{p}} \leq \left(\sum_{n=1}^{\infty} |x_n|^p\right)^{\frac{1}{p}} + \left(\sum_{n=1}^{\infty} |y_n|^p\right)^{\frac{1}{p}}$$

and for x and y in ℓ^∞,

$$\sup_{n\geq 1} |x_n + y_n| \leq \sup_{n\geq 1} |x_n| + \sup_{n\geq 1} |y_n|$$

Proof 2.10.4
(1): $p = \infty$
We know $|x_n + y_n| \leq |x_n| + |y_n|$ by the triangle inequality. So we have $|x_n + y_n| \leq \sup_{n\geq 1} |x_n| + \sup_{n\geq 1} |y_n|$. Thus, the right hand side is an upper bound for all the terms of the left side. We then can say $\sup_{n\geq 1} |x_n + y_n| \leq \sup_{n\geq 1} |x_n| + \sup_{n\geq 1} |y_n|$ which is the result for $p = \infty$.

(2): $p = 1$
Again, we know $|x_n + y_n| \leq |x_n| + |y_n|$ by the triangle inequality. Sum the first N terms on both sides to get

$$\sum_{n=1}^{N} |x_n + y_n| \leq \sum_{n=1}^{N} |x_n| + \sum_{n=1}^{N} |y_n| \leq \sum_{n=1}^{\infty} |x_n| + \sum_{n=1}^{\infty} |y_n|$$

The right hand side is an upper bound for the partial sums on the left. Hence, we have

$$\sum_{n=1}^{\infty} |x_n + y_n| \leq \sum_{n=1}^{\infty} |x_n| + \sum_{n=1}^{\infty} |y_n|$$

(3) $1 < p < \infty$
We have

$$\begin{aligned} |x_n + y_n|^p &= |x_n + y_n|\, |x_n + y_n|^{p-1} \leq |x_n|\, |x_n + y_n|^{p-1} + |y_n|\, |x_n + y_n|^{p-1} \\ \sum_{n=1}^{N} |x_n + y_n|^p &\leq \sum_{n=1}^{N} |x_n|\, |x_n + y_n|^{p-1} + \sum_{n=1}^{N} |y_n|\, |x_n + y_n|^{p-1} \end{aligned}$$

Let $a_n = |x_n|$, $b_n = |x_n + y_n|^{p-1}$, $c_n = |y_n|$ and $d_n = |x_n + y_n|^{p-1}$. Hölder's Inequality applies just fine to finite sequences: i.e. sequences in $\Re^N$. So we have

$$\sum_{n=1}^{N} a_n b_n \leq \left(\sum_{n=1}^{N} a_n^p\right)^{\frac{1}{p}} \left(\sum_{n=1}^{N} b_n^q\right)^{\frac{1}{q}}$$

But $b_n^q = |x_n + y_n|^{q(p-1)} = |x_n + y_n|^p$ using the conjugate exponents identities we established. So we have found

$$\sum_{n=1}^{N} |x_n|\, |x_n + y_n|^{p-1} \leq \left(\sum_{n=1}^{N} |x_n|^p\right)^{\frac{1}{p}} \left(\sum_{n=1}^{N} |x_n + y_n|^p\right)^{\frac{1}{q}}$$

We can apply the same reasoning to the terms c_n and d_n to find

$$\sum_{n=1}^{N} |y_n|\,|x_n + y_n|^{p-1} \quad \leq \quad \left(\sum_{n=1}^{N} |y_n|^p\right)^{\frac{1}{p}} \left(\sum_{n=1}^{N} |x_n + y_n|^p\right)^{\frac{1}{q}}$$

We can use the inequalities to get

$$\left(\sum_{n=1}^{N} |x_n + y_n|^p\right)^{1-\frac{1}{q}} \leq \left(\sum_{n=1}^{N} |x_n|^p\right)^{\frac{1}{p}} + \left(\sum_{n=1}^{N} |y_n|^p\right)^{\frac{1}{p}}$$

But $1 - 1/q = 1/p$, so we have $\left(\sum_{n=1}^{N} |x_n + y_n|^p\right)^{\frac{1}{p}} \leq \left(\sum_{n=1}^{N} |x_n|^p\right)^{\frac{1}{p}} + \left(\sum_{n=1}^{N} |y_n|^p\right)^{\frac{1}{p}}$.

Now apply the final estimate to find $\left(\sum_{n=1}^{N} |x_n+y_n|^p\right)^{\frac{1}{p}} \leq \left(\sum_{n=1}^{\infty} |x_n|^p\right)^{\frac{1}{p}} + \left(\sum_{n=1}^{\infty} |y_n|^p\right)^{\frac{1}{p}}$.
This says the right hand side is an upper bound for the partial sums on the left side. Hence, we know

$$\left(\sum_{n=1}^{\infty} |x_n + y_n|^p\right)^{\frac{1}{p}} \quad \leq \quad \left(\sum_{n=1}^{\infty} |x_n|^p\right)^{\frac{1}{p}} + \left(\sum_{n=1}^{\infty} |y_n|^p\right)^{\frac{1}{p}}$$

■

Homework

Exercise 2.10.6 *If $x \in \ell^p > 1$ and $p' > p$, prove $\boldsymbol{x} \in \ell^{p'}$. Hence, $\ell^{p'} \subset \ell^p$ if $p' > p > 1$.*

Exercise 2.10.7 *If $x \in \ell^p \leq 1$ and $p' < p$, prove $\boldsymbol{x} \in \ell^{p'}$. Hence, $\ell^{p'} \subset \ell^p$ if $p' < p \leq 1$.*

Exercise 2.10.8 *Let $\boldsymbol{x} = (x_i)_{i\geq 1} \in \ell^2$. Define the right* shift by one *operator $A : \ell^2 \to \ell^2$ by $A(\boldsymbol{x})$ is the new ℓ^2 sequence defined by $A(x_1, x_2, x_3, \ldots) = (x_2, x_3, \ldots)$. Work out Hölder's and Minkowski's Inequalities applied to $\boldsymbol{x}$ and $\boldsymbol{y} = A(\boldsymbol{x})$.*

Exercise 2.10.9 *Let $\boldsymbol{x} = (x_i)_{i\geq 1} \in \ell^2$. Define the right* shift by two *operator $A : \ell^2 \to \ell^2$ by $A(\boldsymbol{x})$ is the new ℓ^2 sequence defined by $A(x_1, x_2, x_3, \ldots) = (x_3, x_4, \ldots)$. Work out Hölder's and Minkowski's Inequalities applied to $\boldsymbol{x}$ and $\boldsymbol{y} = A(\boldsymbol{x})$.*

Exercise 2.10.10 *Let $\boldsymbol{x} = \left(\frac{1}{2^i}\right)_{i\geq 2}$ and $\boldsymbol{y} = \left(\frac{1}{3^i}\right)_{i\geq 2}$. Work out Hölder's and Minkowski's Inequalities applied to $\boldsymbol{x}$ and $\boldsymbol{y}$.*

Exercise 2.10.11 *Let $\boldsymbol{x} = \left(\frac{1}{2^i}\right)_{i\geq 1}$ and $\boldsymbol{y} = \left(\frac{1}{3^i}\right)_{i\geq 4}$. Work out Hölder's and Minkowski's Inequalities applied to $\boldsymbol{x}$ and $\boldsymbol{y}$.*

When we specialize to $\mathbb{C}^n$ we obtain

Theorem 2.10.5 Complex Hölder's Inequality in $\mathbb{C}^n$ or $\Re^n$

Let $p > 1$ and p and q be conjugate exponents. If $\boldsymbol{x} \in \mathbb{C}^n$, then the associated sequence $(x_j) = \{x_1, \ldots, x_n, 0, \ldots\}$ is in ℓ^p for all p. We have for any two such sequences (x_j) and (y_j)

$$\sum_{i=1}^{n} |x_j\, y_j| \;\leq\; \left(\sum_{j=1}^{n} |x_j|^p\right)^{1/p} \left(\sum_{j=1}^{n} |y_j|^q\right)^{1/q}$$

and $\sum_{j=1}^n |x_j y_j| \leq \left(\sum_{j=1}^n |x_j|\right) \max_{1 \leq j \leq n} |y_j|$.

and

Theorem 2.10.6 Minkowski's Inequality in $\mathbb{C}^n$

Let $p \geq 1$ and let $\boldsymbol{x}$ and $\boldsymbol{y}$ be in $\mathbb{C}^n$ with associated sequences (x_j) and (y_j) Then,

$$\left(\sum_{j=1}^{n} |x_j + y_j|^p\right)^{\frac{1}{p}} \;\leq\; \left(\sum_{j=1}^{n} |x_j|^p\right)^{\frac{1}{p}} + \left(\sum_{j=1}^{n} |y_j|^p\right)^{\frac{1}{p}}$$

and for the case $p = \infty$, we have

$$\sup_{1 \leq j \leq n} |x_j + y_j| \;\leq\; \sup_{1 \leq j \leq n} |x_j| + \sup_{1 \leq j \leq n} |y_j|$$

In $\mathbb{C}^n$, define

1. $\|\boldsymbol{x}\|_1 = \sum_{j=1}^n |x_j|$

2. $\|\boldsymbol{x}\|_2 = \sqrt{\sum_{j=1}^n |x_j|^2}$

3. $\|\boldsymbol{x}\|_\infty = \max_{1 \leq j \leq n} |x_j$

and in general $\|\boldsymbol{x}\|_p = (\sum_{j=1}^n |x_j|^p)^{1/p}$ for any $p \geq 1$. Minkowski's Inequality tells in all cases

$$\|\boldsymbol{x} + \boldsymbol{y}\|_p \;\leq\; \|\boldsymbol{x}\|_p + \|\boldsymbol{y}\|_p$$

Homework

Exercise 2.10.12 *Let $\boldsymbol{x}$ and $\boldsymbol{y}$ be in $\mathbb{C}^n$ and let A be an $n \times n$ Hermitian matrix (look this up if you don't remember it). Let $\boldsymbol{v} = A(\boldsymbol{x})$ and $\boldsymbol{w} = A(\boldsymbol{y})$. Work out Hölder's and Minkowski's Inequalities applied to $\boldsymbol{v}$ and $\boldsymbol{w}$ expressing the results in terms of $\boldsymbol{x}$ and $\boldsymbol{y}$.*

Exercise 2.10.13 *Let* $\boldsymbol{x} = \begin{bmatrix} 1 + i2 \\ -2 + i4 \end{bmatrix}$ *and* $\boldsymbol{y} = \begin{bmatrix} -5 + i6 \\ 9 - i7 \end{bmatrix}$ *and let* $A = \begin{bmatrix} 1 & -1 + i1 \\ 3 - i2 & 4 - i6 \end{bmatrix}$. *Let $\boldsymbol{v} = A(\boldsymbol{x})$ and $\boldsymbol{w} = A(\boldsymbol{y})$. Work out Hölder's and Minkowski's Inequalities applied to $\boldsymbol{v}$ and $\boldsymbol{w}$ for all choices of p and its conjugate exponent q.*

Exercise 2.10.14 *Let* $\boldsymbol{x} = \begin{bmatrix} 1 + i2 \\ -2 + i4 \\ -6 + i2 \end{bmatrix}$ *and* $\boldsymbol{y} = \begin{bmatrix} -5 + i6 \\ 9 + i7 \\ 20 - i10 \end{bmatrix}$ *and let* $A = \begin{bmatrix} 1 & -1 + i1 & 4 - i2 \\ 3 & -4 - i2 & 3 - i4 \\ 3 - i2 & 4 + i6 & 10 \end{bmatrix}$. *Let $\boldsymbol{v} = A(\boldsymbol{x})$ and $\boldsymbol{w} = A(\boldsymbol{y})$. Work out Hölder's and Minkowski's Inequalities applied to $\boldsymbol{v}$ and $\boldsymbol{w}$ for all choices of p and its conjugate exponent q.*

2.10.1 Sequence Space Metrics

We define the mapping $d_p : \ell^p \times \ell^p \rightarrow \Re$ by

$$d_p(\boldsymbol{x}, \boldsymbol{y}) = \begin{cases} \left(\sum_{i=1}^{\infty} |x_i - y_i|^p\right)^{\frac{1}{p}}, & p > 1 \\ \sup_i |x_i - y_i|, & p = \infty \end{cases}$$

Theorem 2.10.7 (ℓ^p, d_p) **is a Metric Space**

The sequence spaces ℓ^p is a metric space under the metric d_p.

Proof 2.10.5
Let $\boldsymbol{x}$ and $\boldsymbol{y}$ be in ℓ^p. ($p > 1$):

M1: *It is clear $\sum_{i=1}^{\infty} |x_i - y_i|^p$ is finite and non-negative.*

M2: *If $\boldsymbol{x} = \boldsymbol{y}$, we see $d_p(\boldsymbol{x}, \boldsymbol{y}) = 0$. If $d_p(\boldsymbol{x}, \boldsymbol{y}) = 0$, then $|x_i - y_i|^p = 0$ or $x_i = y_i$ for all i telling us $\boldsymbol{x} = \boldsymbol{y}$.*

M3: *It is clear $d_p(\boldsymbol{x}, \boldsymbol{y}) = d_p(\boldsymbol{y}, \boldsymbol{x})$.*

M4: *Given any other sequence $\boldsymbol{z}$ in ℓ^p, by Minkowski's Inequality*

$$\left(\sum_{i=1}^{\infty} |x_i - y_i|^p\right)^{\frac{1}{p}} \leq \left(\sum_{i=1}^{\infty} |x_i - z_i|^p\right)^{\frac{1}{p}} + \left(\sum_{i=1}^{\infty} |z_i - y_i|^p\right)^{\frac{1}{p}}$$

This shows

$$d_p(\boldsymbol{x}, \boldsymbol{y}) \leq d_p(\boldsymbol{x}, \boldsymbol{z}) + d_p(\boldsymbol{z}, \boldsymbol{y})$$

So (ℓ^p, d_p) is a metric space.

($p = \infty$):
In this case it is easy to verify $M1$ through $M4$ and we leave it to you. In fact, we have done this in the earlier volumes.

■

Homework

Exercise 2.10.15 *Prove the $p = \infty$ case.*

Exercise 2.10.16 *Let $\boldsymbol{x} = \left(\frac{1}{2^i}\right)_{i \geq 1}$ and $\boldsymbol{y} = \left(\frac{1}{3^i}\right)_{i \geq 1}$. Calculate $d_p(\boldsymbol{x}, \boldsymbol{y})$ for all relevant p.*

Exercise 2.10.17 *Let $\boldsymbol{x} = \left(\frac{1}{2^i}\right)_{i \geq 2}$ and $\boldsymbol{y} = \left(\frac{1}{3^i}\right)_{i \geq 4}$. Calculate $d_p(\boldsymbol{x}, \boldsymbol{y})$ for all relevant p.*

Exercise 2.10.18 *Let $\boldsymbol{x} = \left(\frac{(-1)^i}{5^i}\right)_{i \geq 2}$ and $\boldsymbol{y} = \left(\frac{1}{7^i}\right)_{i \geq 1}$. Calculate $d_p(\boldsymbol{x}, \boldsymbol{y})$ for all relevant p.*

Exercise 2.10.19 *Let $\boldsymbol{x} = \left(\frac{1}{6^i}\right)_{i \geq 4}$ and $\boldsymbol{y} = \left(\frac{(-1)^i}{3^i}\right)_{i \geq 5}$. Calculate $d_p(\boldsymbol{x}, \boldsymbol{y})$ for all relevant p.*

2.11 Hölder's and Minkowski's Inequality in Function Spaces

We can also prove these theorems in other spaces. Let $C([a,b])$ be the set of all real-valued continuous functions on the finite interval $[a,b]$. We know $\boldsymbol{x} \in RI([a,b])$ and so $g \circ x$ is Riemann Integrable for any continuous g. For example, let $p = \sqrt{2}$. Then we could define $|x(t)| = e^{\ln(|x(t)|)}$ and $|x(t)|^{\sqrt{2}} = e^{\sqrt{2}\ln(|x(t)|)}$. As long as the domains line up, $\ln(|x(t)|)$ is therefore integrable and so $e^{\sqrt{2}\ln(|x(t)|)}$ is also integrable. We can see $|x|^p$ is integrable for $p \geq 1$.

Theorem 2.11.1 Hölder's Inequality for Continuous Functions

Let $p > 1$ and p and q be conjugate exponents. Let $\boldsymbol{x}$ and $\boldsymbol{y}$ be in $C([a,b])$. Then

$$\int_a^b |x(t)y(t)| \leq \left(\int_a^b |x(t)|^p dt\right)^{\frac{1}{p}} \left(\int_a^b |y(t)|^q dt\right)^{\frac{1}{q}}$$

Proof 2.11.1
This inequality is clearly true if either of the two functions $\boldsymbol{x}$ and $\boldsymbol{y}$ are the zero function. So we can assume both $\boldsymbol{x}$ and $\boldsymbol{y}$ are both nonzero. Since they are continuous, this means $\int_a^b |x(t)|^p dt$ and $\int_a^b |y(t)|^q dt$ are not zero. So

$$0 < u = \left(\int_a^b |x(t)|^p dt\right)^{\frac{1}{p}}, \quad 0 < v = \left(\int_a^b |y(t)|^q dt\right)^{\frac{1}{q}}$$

Now define new functions, $\hat{\boldsymbol{x}}$ and $\hat{\boldsymbol{y}}$ by $\hat{\boldsymbol{x}} = \boldsymbol{x}/u$ and $\hat{\boldsymbol{y}} = \boldsymbol{y}/v$. Then, we have

$$\int_a^b |\hat{x}(t)|^p dt = \frac{1}{u^p}\int_a^b |x(t)|^p dt = \frac{u^p}{u^p} = 1$$
$$\int_a^b |\hat{y}(t)|^q dt = \frac{1}{v^q}\int_a^b |y(t)|^q dt = \frac{v^q}{v^q} = 1$$

Now apply the $\alpha - \beta$ Lemma to $\alpha = |\hat{x(t)}|$ and $\beta = |\hat{y(t)}|$. Then $|\hat{x(t)}\,\hat{y(t)}| \leq |\hat{x(t)}|^p/p + |\hat{y(t)}|^q/q$. This is true even if $\hat{x}_n$ or $\hat{y}_n$ are zero and in the case they are not, the $\alpha - \beta$ lemma applies. Then

$$\int_a^b |\hat{x}(t)\,\hat{y}(t)|dt \leq \frac{1}{p}\int_a^b |\hat{x}(t)|^p dt + \frac{1}{q}\int_a^b |\hat{y}(t)|^q dt$$
$$= \frac{1}{p}\frac{1}{u^p}\,u^p + \frac{1}{q}\frac{1}{v^q}\,v^q = \frac{1}{p} + \frac{1}{q} = 1$$

We conclude

$$\int_a^b |\hat{x}(t)\,\hat{y}(t)|dt = \frac{1}{uv}\int_a^b |x(t)\,y(t)|dt \leq 1$$

which implies the result:

$$\int_a^b |x(t)\,y(t)|dt \leq uv$$

■

We can also do this inequality for the case $p = 1$ and $q = \infty$.

Theorem 2.11.2 Hölder's Theorem for Continuous Functions $p = 1$ and $q = \infty$

If $\boldsymbol{x}$ and $\boldsymbol{y}$ are continuous functions on $[a, b]$, then

$$\int_a^b |x(t)y(t)|dt \leq \left(\sup_{a \leq t \leq b} |y(t)|\right) \int_a^b |x(t)|dt$$

Proof 2.11.2
We know

$$|x(t)y(t)| \leq \left(\sup_{a \leq t \leq b} |y(t)|\right) |x(t)|$$

and the result then follows immediately. ■

Minkowski's Inequality for continuous functions is then

Theorem 2.11.3 Minkowski's Inequality for Continuous Functions

Let $p \geq 1$. Then, if $\boldsymbol{x}$ and $\boldsymbol{y}$ are continuous on $[a, b]$,

$$\left(\int_a^b |x(t) + y(t)|^p\right)^{\frac{1}{p}} \leq \left(\int_a^b |x(t)|^p\right)^{\frac{1}{p}} + \left(\int_a^b |y(t)|^p\right)^{\frac{1}{p}}$$

$$\max_{a \leq t \leq b} |x(t) + y(t)| \leq \max_{a \leq t \leq b} |x(t)| + \max_{a \leq t \leq b} |y(t)|$$

Proof 2.11.3
(1): $p = \infty$
We know $|x(t)+y(t)| \leq |x(t)|+|y(t)|$ so we have $|x(t)+y(t)| \leq \sup_{a \leq t \leq b} |x(t)|+\sup_{a \leq t \leq b} |y(t)|$. Thus, the right hand side is an upper bound for all the terms of the left side. We then can say $\sup_{a \leq t \leq b} |x(t) + y(t)| \leq \sup_{a \leq t \leq b} |x(t)| + \sup_{a \leq t \leq b} |y(t)|$ which is the result for $p = \infty$.

(2): $p = 1$
Again, we know $|x(t) + y(t)| \leq |x(t)| + |y(t)|$. Thus,

$$\int_a^b |x(t) + y(t)|dt \leq \int_a^b |x(t)|dt + \int_a^b |y(t)|dt$$

(3) $1 < p < \infty$
We have

$$|x(t) + y(t)|^p = |x(t) + y(t)|\,|x(t) + y(t)|^{p-1} \leq |x(t)|\,|x(t) + y(t)|^{p-1} + |y(t)|\,|x(t) + y(t)|^{p-1}$$

Therefore

$$\int_a^b |x(t) + y(t)|^p\,dt \leq \int_a^b |x(t)|\,|x(t) + y(t)|^{p-1}\,dt + \int_a^b |y(t)|\,|x(t) + y(t)|^{p-1}\,dt$$

By the Hölder's inequality

$$\int_a^b |x(t)|\,|x(t)+y(t)|^{p-1}\,dt \;\le\; \left(\int_a^b |x(t)|^p\,dt\right)^{\frac{1}{p}} \left(\int_a^b |x(t)+y(t)|^{q(p-1)}\,dt\right)^{\frac{1}{q}}$$

$$\int_a^b |y(t)|\,|x(t)+y(t)|^{p-1}\,dt \;\le\; \left(\int_a^b |y(t)|^p\,dt\right)^{\frac{1}{p}} \left(\int_a^b |x(t)+y(t)|^{q(p-1)}\,dt\right)^{\frac{1}{q}}$$

Since $q(p-1)=p$, we have

$$\int_a^b |x(t)|\,|x(t)+y(t)|^{p-1}\,dt \;\le\; \left(\int_a^b |x(t)|^p\,dt\right)^{\frac{1}{p}} \left(\int_a^b |x(t)+y(t)|^{p}\,dt\right)^{\frac{1}{q}}$$

$$\int_a^b |y(t)|\,|x(t)+y(t)|^{p-1}\,dt \;\le\; \left(\int_a^b |y(t)|^p\,dt\right)^{\frac{1}{p}} \left(\int_a^b |x(t)+y(t)|^{p}\,dt\right)^{\frac{1}{q}}$$

Thus,

$$\begin{aligned}\int_a^b |x(t)+y(t)|^p\,dt \;\le\; & \left(\int_a^b |x(t)|^p\,dt\right)^{\frac{1}{p}} \left(\int_a^b |x(t)+y(t)|^{p}\,dt\right)^{\frac{1}{q}} \\ & + \left(\int_a^b |y(t)|^p\,dt\right)^{\frac{1}{p}} \left(\int_a^b |x(t)+y(t)|^{p}\,dt\right)^{\frac{1}{q}}\end{aligned}$$

or

$$\left(\int_a^b |x(t)+y(t)|^p\,dt\right)^{1-\frac{1}{q}} \;\le\; \left(\int_a^b |x(t)|^p\,dt\right)^{\frac{1}{p}} + \left(\int_a^b |y(t)|^p\,dt\right)^{\frac{1}{p}}$$

which gives the result. ■

Homework

Exercise 2.11.1 *This is a familiar exercise. Prove if f is continuous on $[a,b]$ with $f \ge 0$, then $\int_a^b f(s)ds = 0$ implies $f = 0$ on $[a,b]$.*

Exercise 2.11.2 *Again, this is a familiar exercise. Prove if f is continuous and nonzero on $[a,b]$ then $\int_a^b f^2(s)ds = 0$ implies $f = 0$ on $[a,b]$.*

Exercise 2.11.3 *Let $f(t) = t^2$ and $g(t) = 2t+3$ on $[-1,2]$. Work out the Hölder's and Minkowski's Inequalities for all relevant choices of p.*

Exercise 2.11.4 *Let $f(t) = \sin(t)$ and $g(t) = \cos(t)$ on $[-3,4]$. Work out the Hölder's and Minkowski's Inequalities for all relevant choices of p.*

What happens if we drop continuity? The same comments about integrability of p powers hold. Since $\boldsymbol{x} \in RI([a,b])$, $g \circ x$ is Riemann Integrable for any continuous g. We can see $|x|^p$ is integrable for $p \ge 1$.

Theorem 2.11.4 Hölder's Inequality for Riemann Integrable Functions

Let $p > 1$ and p and q be conjugate exponents. Let $\boldsymbol{x}$ and $\boldsymbol{y}$ be in $RI([a,b])$. Then

$$\int_a^b |x(t)y(t)| \leq \left(\int_a^b |x(t)|^p dt\right)^{\frac{1}{p}} \left(\int_a^b |y(t)|^q dt\right)^{\frac{1}{q}}$$

Proof 2.11.4

This inequality is clearly true if either of the two functions $\boldsymbol{x}$ and $\boldsymbol{y}$ are the zero function. This time, let's assume both $\int_a^b |x(t)|^p dt$ and $\int_a^b |y(t)|^q dt$ are both nonzero. So

$$0 < u = \left(\int_a^b |x(t)|^p dt\right)^{\frac{1}{p}}, \quad 0 < v = \left(\int_a^b |y(t)|^q dt\right)^{\frac{1}{q}}$$

Now define new functions, $\hat{\boldsymbol{x}}$ and $\hat{\boldsymbol{y}}$ by $\hat{\boldsymbol{x}} = \boldsymbol{x}/u$ and $\hat{\boldsymbol{y}} = \boldsymbol{y}/v$. The rest of the argument is then the same. ■

We can also do this inequality for the case $p = 1$ and $q = \infty$.

Theorem 2.11.5 Hölder's Theorem for Riemann Integrable Functions $p = 1$ and $q = \infty$

If $\boldsymbol{x}$ and $\boldsymbol{y}$ are Riemann Integrable functions on $[a,b]$, then

$$\int_a^b |x(t)y(t)|dt \leq \left(\sup_{a \leq t \leq b} |y(t)|\right) \int_a^b |x(t)|dt$$

Proof 2.11.5

If $\boldsymbol{x}$ and $\boldsymbol{y}$ are Riemann Integrable functions on $[a,b]$, then they are bounded and

$$|x(t)y(t)| \leq \left(\sup_{a \leq t \leq b} |y(t)|\right) |x(t)|$$

The result then follows immediately. ■

Minkowski's Inequality for Riemann Integrable functions is then

Theorem 2.11.6 Minkowski's Inequality for Riemann Integrable Functions

Let $p \geq 1$. Then, if $\boldsymbol{x}$ and $\boldsymbol{y}$ are Riemann Integrable on $[a,b]$,

$$\left(\int_a^b |x(t) + y(t)|^p\right)^{\frac{1}{p}} \leq \left(\int_a^b |x(t)|^p\right)^{\frac{1}{p}} + \left(\int_a^b |y(t)|^p\right)^{\frac{1}{p}}$$

$$\max_{a \leq t \leq b} |x(t) + y(t)| \leq \sup_{a \leq t \leq b} |x(t)| + \max_{a \leq t \leq b} |y(t)|$$

Proof 2.11.6

(1): $p = \infty$

We know both $\boldsymbol{x}$ and $\boldsymbol{y}$ are bounded and $|x(t) + y(t)| \leq |x(t)| + |y(t)|$ so we have $|x(t) + y(t)| \leq$

$\sup_{a \le t \le b} |x(t)| + \sup_{a \le t \le b} |y(t)|$. *This shows the result.*

(2): $p = 1$
Again, we know $|x(t) + y(t)| \le |x(t)| + |y(t)|$. *Thus,*

$$\int_a^b |x(t) + y(t)| dt \le \int_a^b |x(t)| dt + \int_a^b |y(t)| dt$$

(3) $1 < p < \infty$
This is the same argument as before. ■

Homework

Exercise 2.11.5 *Let*

$$f(t) = \begin{cases} t^2, & 0 \le t < 1 \\ 3t + 5, & 1 \le t \le 2 \end{cases} \quad g(t) = \begin{cases} -2t^2, & 0 \le t < 1 \\ 4t + 1, & 1 \le t \le 2 \end{cases}$$

Work out the Hölder's and Minkowski's Inequalities for all relevant choices of p.

Exercise 2.11.6 *Let*

$$f(t) = \begin{cases} t^2, & 0 \le t < 1 \\ 3t + 5, & 1 \le t \le 2 \end{cases} \quad g(t) = \begin{cases} -2t^2, & 0 \le t < 0.5 \\ 4t + 1, & 0.5 \le t \le 2 \end{cases}$$

Work out the Hölder's and Minkowski's Inequalities for all relevant choices of p.

Exercise 2.11.7 *Let*

$$f(t) = \begin{cases} 2t^2, & 1 \le t \le 4 \\ t + 5, & 4 < t \le 8 \end{cases} \quad g(t) = \begin{cases} -3t^2, & 1 \le t \le 4 \\ 6t + 1, & 4 < t \le 8 \end{cases}$$

Work out the Hölder's and Minkowski's Inequalities for all relevant choices of p.

Exercise 2.11.8 *Let*

$$f(t) = \begin{cases} 2t^2, & 1 \le t < 4 \\ t + 5, & 4 \le t \le 8 \end{cases} \quad g(t) = \begin{cases} -3t^2, & 1 \le t < 2 \\ 6t + 1, & 2 \le t \le 8 \end{cases}$$

Work out the Hölder's and Minkowski's Inequalities for all relevant choices of p.

2.11.1 Function Space Metrics

For the finite interval $[a, b]$, we define the mapping $d_p : C([a, b]) \times C([a, b]) \to \Re$ by

$$d_p(\boldsymbol{x}, \boldsymbol{y}) = \begin{cases} \left(\int_a^b |x(t) - y(t)|^p \right)^{\frac{1}{p}}, & p > 1 \\ \max_{a \le t \le b} |x(t) - y(t)|, & p = \infty \end{cases}$$

Theorem 2.11.7 $(C([a, b]), d_p)$ **is a Metric Space**

The function space $C([a, b])$ *is a metric space under the metric* d_p.

Proof 2.11.7
Let $\boldsymbol{x}$ and $\boldsymbol{y}$ be in $C([a,b])$. ($p > 1$):

M1: *It is clear $\int_a^b |x(t) - y(t)|^p dt$ is finite and non-negative.*

M2: *If $\boldsymbol{x} = \boldsymbol{y}$, we see $d_p(\boldsymbol{x}, \boldsymbol{y}) = 0$. If $d_p(\boldsymbol{x}, \boldsymbol{y}) = 0$, then by the same arguments we used for d_1 earlier for continuous functions, we know the non-negative continuous function $|x(t) - y(t)|^p$ cannot be nonzero at any point in $[a, b]$. Hence, $x(t) = y(t)$ always and so $\boldsymbol{x} = \boldsymbol{y}$.*

M3: *It is clear $d_p(\boldsymbol{x}, \boldsymbol{y}) = d + p(\boldsymbol{y}, \boldsymbol{x})$.*

M4: *Given any other continuous $\boldsymbol{z}$ on $[a, b]$ by Minkowski's Inequality*

$$\left(\int_a^b |x(t) - y(t)|^p\right)^{\frac{1}{p}} \leq \left(\int_a^b |x(t) - z(t)|^p\right)^{\frac{1}{p}} + \left(\int_a^b |z(t) - y_(t)|^p\right)^{\frac{1}{p}}$$

This shows

$$d_p(\boldsymbol{x}, \boldsymbol{y}) \leq d_p(\boldsymbol{x}, \boldsymbol{z}) + d + p(\boldsymbol{z}, \boldsymbol{y})$$

So $(C([a,b]), d_p)$ is a metric space.

($p = \infty$):
In this case it is also easy to verify $M1$ through $M4$ and we leave it to you. ∎

Theorem 2.11.8 $(RI([a,b]), d_p)$ **is not a Metric Space**

For $p > 1$, function space $RI([a, b])$ is **not** *metric space under the metric d_p. However, for $p = \infty$, it is a metric space.*

Proof 2.11.8
Let $\boldsymbol{x}$ and $\boldsymbol{y}$ be in $RI([a,b])$. ($p > 1$):
It is easy to see $M1$, $M3$ and $M4$ are satisfied, however $M2$ is not. If $d_p(\boldsymbol{x}, \boldsymbol{y}) = 0$ this does not imply $\boldsymbol{x} = \boldsymbol{y}$. So $(RI([a, b]), d_p)$ is not a metric space.

($p = \infty$):
In this case, all we have to do is verify $M2$. If $d_\infty(\boldsymbol{x}, \boldsymbol{y}) = 0$, then $\sup_{a \leq t \leq b} |x(t) - y(t)| = 0$. Thus, $|x(t) - y(t)| = 0$ for all t implying $\boldsymbol{x} = \boldsymbol{y}$. The converse is easy. Since $M1$, $M3$ and $M4$ are also satisfied, we see $(RI([a, b]), d_\infty)$ is a metric space. ∎

Homework

Exercise 2.11.9 *Let $C^1([a, b])$ be the set of functions with a continuous first derivative on $[a, b]$. For any $\boldsymbol{x}$ and $\boldsymbol{y}$ in $C^1([a, b])$, define $d(\boldsymbol{x}, \boldsymbol{y}) = \sqrt{d_2(\boldsymbol{x}, \boldsymbol{y})^2 + d_2(\boldsymbol{x}', \boldsymbol{y}')^2}$. Prove d is a metric on $C^1([a, b])$.*

Exercise 2.11.10 *Let $C^1([a, b])$ be the set of functions with a continuous first derivative on $[a, b]$. For any $\boldsymbol{x}$ and $\boldsymbol{y}$ in $C^1([a, b])$, define $d(\boldsymbol{x}, \boldsymbol{y}) = d_\infty(\boldsymbol{x}, \boldsymbol{y}) + d_\infty(\boldsymbol{x}', \boldsymbol{y}')$. Prove d is a metric on $C^1([a, b])$.*

Exercise 2.11.11 *Let M be the set of all 2×2 matrices whose components are continuous functions on $[a, b]$. Hence, if $A \in M$, $A_{ij} \in C([a, b])$ for all appropriate indices i and j. For any A and B in M, define $d(A, B) = \sqrt{\sum_{i=1}^{2} \sum_{j=1}^{2} d_\infty(A_{ij}, B_{ij})}$. Prove d is a metric on M.*

Exercise 2.11.12 *Let M be the set of all 2×2 matrices whose components are continuous functions on $[a, b]$. Hence, if $A \in M$, $A_{ij} \in C([a, b])$ for all appropriate indices i and j. For any A and B in M, define $d(A, B) = \sqrt{\sum_{i=1}^{2} \sum_{j=1}^{2} d_1(A_{ij}, B_{ij})}$. Prove d is a metric on M.*

Exercise 2.11.13 *Let M be the set of all 2×2 matrices whose components are continuous functions on $[a, b]$. Hence, if $A \in M$, $A_{ij} \in C([a, b])$ for all appropriate indices i and j. For any A and B in M, define $d(A, B) = \sqrt{\sum_{i=1}^{2} \sum_{j=1}^{2} d_2(A_{ij}, B_{ij})}$. Prove d is a metric on M.*

Exercise 2.11.14 *Let M be the set of all 2×2 matrices whose components are functions with a continuous derivative on $[a, b]$. Hence, if $A \in M$, $A_{ij} \in C^1([a, b])$ for all appropriate indices i and j. For any A and B in M, define $d(A, B) = \sqrt{\sum_{i=1}^{2} \sum_{j=1}^{2} d(A_{ij}, B_{ij})}$ where $d(A_{ij}, B_{ij}) = d_\infty(A_{ij}, B_{ij}) + d_\infty(A'_{ij}, B'_{ij})$. Prove d is a metric on M.*

Exercise 2.11.15 *Let M be the set of all 2×2 matrices whose components are functions with a continuous derivative on $[a, b]$. Hence, if $A \in M$, $A_{ij} \in C^1([a, b])$ for all appropriate indices i and j. For any A and B in M, define $d(A, B) = \sqrt{\sum_{i=1}^{2} \sum_{j=1}^{2} d(A_{ij}, B_{ij})}$ where $d(A_{ij}, B_{ij}) = \sqrt{d_2^2(A_{ij}, B_{ij}) + d_2^2(A'_{ij}, B'_{ij})}$. Prove d is a metric on M.*

2.12 More Completeness Results

Let's look at whether or not some metric spaces are complete.

Theorem 2.12.1 **$(C([0, 1]), d_1)$ is not Complete**

> *The metric space $(C([0, 1]), d_1)$ is not complete.*

Proof 2.12.1
Consider the metric space $(C([0, 1]), d_1)$ and look at the sequence of functions (x_n) given by

$$x_n(t) = \begin{cases} 0, & 0 \leq t \leq \frac{1}{2} \\ n(t - \frac{1}{2}), & \frac{1}{2} < t < \frac{1}{2} + \frac{1}{n} \\ 1, & \frac{1}{2} + \frac{1}{n} \leq t \leq 1 \end{cases}$$

Pick any $t > \frac{1}{2}$. Then there is N so that $n > N$ implies $\frac{1}{2} + \frac{1}{n} < t$ and so $x_n(t) = 1$ for all $n > N$. Thus $x_n(t) \to 1$ when $t > \frac{1}{2}$. It is also easy to see $x_n(t) = 0$ when $0 \leq t \leq \frac{1}{2}$ for all n. We have shown (x_n) converges pointwise to x where

$$x(t) = \begin{cases} 0, 0 \leq t \leq \frac{1}{2} \\ 1, \frac{1}{2} < t \leq 1 \end{cases}$$

It is also straightforward to see this sequence is a Cauchy sequence. Note for $n > m$

$$\begin{aligned} \int_0^1 |x_n(t) - x_m(t)| &= \int_0^{\frac{1}{2}} \mathbf{0}\, dt + \int_{\frac{1}{2}}^{\frac{1}{2}+\frac{1}{m}} (x_m(t) - x_n(t))\, dt + \int_{\frac{1}{2}+\frac{1}{m}}^1 \mathbf{0}\, dt \\ &= \frac{n - m}{m\, n} \leq \frac{1}{m} \end{aligned}$$

Thus, given $\epsilon > 0$, there is N so that $n, m > N$ so that $d_1(x_n, x_m) < \epsilon$. Hence, this sequence is a Cauchy sequence in $(C([0,1]), d_1)$.

Is there a function $x \in C([0,1])$ so that $d_1(x_n, x) \to 0$? Let $\epsilon > 0$ and assume

$$d_1(x_n, x) < \epsilon \quad \longrightarrow \quad \int_0^{\frac{1}{2}} |0 - x(t)|\, dt + \int_{\frac{1}{2}}^{\frac{1}{2}+\frac{1}{n}} |x_n(t) - x(t)|\, dt + \int_{\frac{1}{2}+\frac{1}{n}}^{1} |1 - x(t)|\, dt < \epsilon$$

for all $n > N$. This would imply all three pieces are less than ϵ for $n > N$. Thus, for all $n > N$

$$\int_0^{\frac{1}{2}} |x(t)|\, dt \quad < \quad \epsilon$$

which would tell us $\int_0^{\frac{1}{2}} |x(t)|\, dt = 0$. Since we assume x is continuous on $[0,1]$ this will force $x(t) = 0$ on $[0, \frac{1}{2}]$. The third piece would give

$$\int_{\frac{1}{2}+\frac{1}{n}}^{1} |1 - x(t)|\, dt < \epsilon$$

Now, since x is continuous,

$$\int_{\frac{1}{2}}^{\frac{1}{2}+\frac{1}{n}} |1 - x(t)| dt \quad \leq \quad \|1 - x\|_\infty \, \frac{1}{n}$$

and so $\int_0^{\frac{1}{2}+\frac{1}{n}} |1 - x(t)| dt \to 0$. Combining, we have

$$\lim_{n\to\infty} \int_{\frac{1}{2}}^{1} |1 - x(t)| dt \quad = \quad \lim_{n\to\infty} \int_{\frac{1}{2}}^{\frac{1}{2}+\frac{1}{n}} |1 - x(t)| dt + \lim_{n\to\infty} \int_{\frac{1}{2}+\frac{1}{n}}^{1} |1 - x(t)| dt = 0$$

Since $1 - x$ is continuous, this forces $x(t) = 1$ on $[\frac{1}{2}, 1]$. This is not possible as this means $x(\frac{1}{2}) = 0$ and 1. Hence, our assumption such an x exists is wrong. We conclude that $(C([0,1]), d_1)$ is not complete as we have found a Cauchy sequence that does not converge to an element of the space. ■

Homework

Exercise 2.12.1 *For the example above, if you redefine the limit function for the sequence at a finite number of points is this redefined function still a limit function for the sequence using the d_1 metric?*

Exercise 2.12.2 *For the example above, if you redefine the limit function for the sequence on any set of content zero is this redefined function still a limit function for the sequence using the d_1 metric?*

Exercise 2.12.3 *Design a similar sequence of continuous functions on $[0,2]$ which is a Cauchy sequence but does not converge in d_1 to a continuous function. This shows $C([0,2], d_1)$ is not complete.*

Exercise 2.12.4 *Design a similar sequence of continuous functions on $[-1,2]$ which is a Cauchy sequence but does not converge in d_1 to a continuous function. This shows $C([-1,2], d_1)$ is not complete.*

Theorem 2.12.2 $(C([0,1]), d_2)$ **is not Complete**

The metric space $(C([0,1]), d_2)$ is not complete.

Proof 2.12.2
Next, consider the metric space $(C([0,1]), d_2)$ and look at the same sequence of functions (x_n) given by

$$x_n(t) = \begin{cases} 0, & 0 \le t \le \frac{1}{2} \\ n(t - \frac{1}{2}), \frac{1}{2} < t < \frac{1}{2} + \frac{1}{n} & \\ 1, & \frac{1}{2} + \frac{1}{n} \le t \le 1 \end{cases}$$

As usual, (x_n) converges pointwise to x where

$$x(t) = \begin{cases} 0, 0 \le t \le \frac{1}{2} \\ 1, \frac{1}{2} < t \le 1 \end{cases}$$

It is also straightforward to see this sequence is a Cauchy sequence. Note for $n > m$

$$\begin{aligned} (d_2(x_n, x_m))^2 = \int_0^1 |x_n(t) - x_m(t)|^2 &= \int_0^{\frac{1}{2}} \mathbf{0}\, dt + \int_{\frac{1}{2}}^{\frac{1}{2}+\frac{1}{m}} |x_m(t) - x_n(t)|^2\, dt + \int_{\frac{1}{2}+\frac{1}{m}}^1 \mathbf{0}\, dt \\ &\le \int_{\frac{1}{2}}^{\frac{1}{2}+\frac{1}{m}} |x_m(t) - x_n(t)|\, |x_m(t) + x_n(t)|\, dt \\ &\le 2 \int_{\frac{1}{2}}^{\frac{1}{2}+\frac{1}{m}} |x_m(t) - x_n(t)|\, dt \\ &= 2 \frac{n - m}{m\, n} \le \frac{2}{m} \end{aligned}$$

Thus, given $\epsilon > 0$, there is N so that $n, m > N$ so that $d_2(x_n, x_m) = \sqrt{\frac{2}{m}} < \epsilon$. Hence, this sequence is a Cauchy sequence in $(C([0,1]), d_2)$.

Is there a function $x \in C([0,1])$ so that $d_2(x_n, x) \to 0$? Let $\epsilon > 0$ and assume

$$(d_2(x_n, x))^2 < \epsilon \longrightarrow \int_0^{\frac{1}{2}} |0 - x(t)|^2\, dt + \int_{\frac{1}{2}}^{\frac{1}{2}+\frac{1}{n}} |x_n(t) - x(t)|^2\, dt + \int_{\frac{1}{2}+\frac{1}{n}}^1 |1 - x(t)|^2\, dt < \epsilon$$

for all $n > N$. This would imply all three pieces are less than ϵ for $n > N$. Thus, for all $n > N$

$$\int_0^{\frac{1}{2}} |x(t)|^2\, dt < \epsilon$$

which would tell us $\int_0^{\frac{1}{2}} |x(t)|^2\, dt = 0$. Since we assume x is continuous on $[0,1]$ this will force $x(t) = 0$ on $[0, \frac{1}{2}]$. The third piece would give

$$\int_{\frac{1}{2}+\frac{1}{n}}^1 |1 - x(t)|^2\, dt < \epsilon$$

Now, since x is continuous,

$$\int_{\frac{1}{2}}^{\frac{1}{2}+\frac{1}{n}} |1 - x(t)|^2 dt \le \|1 - x\|_\infty^2 \frac{1}{n}$$

and so $\int_0^{\frac{1}{2}+\frac{1}{n}} |1-x(t)|^2 dt \to 0$. *Combining, we have*

$$\lim_{n\to\infty}\int_{\frac{1}{2}}^{1} |1-x(t)|^2 dt = \lim_{n\to\infty}\int_{\frac{1}{2}}^{\frac{1}{2}+\frac{1}{n}} |1-x(t)|^2 dt + \lim_{n\to\infty}\int_{\frac{1}{2}+\frac{1}{n}}^{1} |1-x(t)|^2 dt = 0$$

Since $1-x$ *is continuous, this forces* $x(t)=1$ *on* $[\frac{1}{2},1]$. *This is not possible as this means* $x(\frac{1}{2})=0$ *and* 1. *Hence, our assumption such an* x *exists is wrong. We conclude that* $(C([0,1]),d_2)$ *is not complete either as we have found a Cauchy sequence that does not converge to an element of the space.* ■

Homework

Exercise 2.12.5 *Design a similar sequence of continuous functions on* $[0,2]$ *which is a Cauchy sequence but does not converge in* d_2 *to a continuous function. This shows* $C([0,2],d_2)$ *is not complete.*

Exercise 2.12.6 *Design a similar sequence of continuous functions on* $[-3,2]$ *which is a Cauchy sequence but does not converge in* d_1 *to a continuous function. This shows* $C([-3,2],d_2)$ *is not complete.*

Exercise 2.12.7 *For the example we have looked at, if you redefine the limit function for the sequence at a finite number of points is this redefined function still a limit function for the sequence using the* d_2 *metric?*

Exercise 2.12.8 *For the examples we have looked at, if you redefine the limit function for the sequence on any set of content zero is this redefined function still a limit function for the sequence using the* d_2 *metric?*

Exercise 2.12.9 *Consider the metric space* $(C([0,1]),d_3)$ *and look again at the sequence of functions* (x_n) *given by*

$$x_n(t) = \begin{cases} 0, & 0 \le t \le \frac{1}{2} \\ n(t-\frac{1}{2}), \frac{1}{2} < t < \frac{1}{2}+\frac{1}{n} & \\ 1, & \frac{1}{2}+\frac{1}{n} \le t \le 1 \end{cases}$$

Prove this sequence in a Cauchy sequence in $C([0,1])$ *with the* d_3 *metric and hence* $(C([0,1]),d_3)$ *is not complete.*

If you redefine the limit function for this sequence at a finite number of points is this redefined function still a limit function for the sequence using the d_3 *metric?*

Exercise 2.12.10 *Consider the metric space* $(C([0,1]),d_4)$ *and look again at the sequence of functions* (x_n) *given by*

$$x_n(t) = \begin{cases} 0, & 0 \le t \le \frac{1}{2} \\ n(t-\frac{1}{2}), \frac{1}{2} < t < \frac{1}{2}+\frac{1}{n} & \\ 1, & \frac{1}{2}+\frac{1}{n} \le t \le 1 \end{cases}$$

Prove this sequence in a Cauchy sequence in $C([0,1])$ *with the* d_4 *metric and hence* $(C([0,1]),d_4)$ *is not complete.*

If you redefine the limit function for this sequence at a finite number of points is this redefined function still a limit function for the sequence using the d_4 *metric?*

Another way to look at these results is in terms of closure. The sequence (x_n) above does converge to a Riemann Integrable function x. So the subset $C([0,1]) \subset RI([0,1])$ cannot be a closed subset with respect to d_1 or d_2 convergence. This is because $x_n \to x$ in both d_1 and d_2 yet $x \notin C([0,1])$.

With respect to the supremum metric d_∞, $(C([a,b]), d_\infty)$ is a complete metric space. We have proven this result in (Peterson (18) 2020) but let's go through it again so we can compare ideas.

Theorem 2.12.3 $(C([a,b]), d_\infty)$ **is Complete**

$(C([a,b]), d_\infty)$ is a complete metric space.

Proof 2.12.3

Let (x_n) be a Cauchy sequence in $(C([a,b]), d_\infty)$. Then for all $\epsilon > 0$, there is N so that $n > m > M$ implies $d_\infty(x_n, x_m) < \epsilon/2$. In particular, for a fixed t, this says $(x_n(t))$ is a Cauchy sequence in $\Re$. Since $\Re$ is complete, we know there is a value a_t so that $x_n(t) \to a_t$. Define the limit function x by $x(t) = a_t$. Hence, for $n > m > N$

$$\lim_{n\to\infty} \max_{a\le t\le b} |x_n(t) - x_m(t)| \le \epsilon/2 \quad \longrightarrow \quad \max_{a\le t\le b} |x(t) - x_m(t)| < \epsilon$$

for $m > N$. This says $x_m \to x$ in d_∞ metric. To show x is continuous, pick s. Then since there is N so that $|x_n(t) - x(t)| < \epsilon/3$ for all $a \le t \le b$, we have

$$\begin{aligned} |x(t) - x(s)| &\le |x(t) - x_{N+1}(t)| + |x_{N+1}(t) - x_{N+1}(s)| + |x_{N+1}(s) - x(s)| \\ &< 2\epsilon/3 + |x_{N+1}(t) - x_{N+1}(s)| \end{aligned}$$

But x_{N+1} is a continuous function, so there is a $\delta > 0$ so that $|x_{N+1}(t) - x_{N+1}(s)| < \epsilon/3$ if $|t - s| < \delta$. We conclude $|x(t) - x(s)| < \epsilon$ if $|t - s| < \delta$. This shows x is continuous and that $(C([a,b]), d_\infty)$ is complete. ■

Theorem 2.12.4 (ℓ^∞, d_∞) **is Complete**

(ℓ^∞, d_∞) is a complete metric space.

Proof 2.12.4

Let $(\boldsymbol{x_n})$ be a Cauchy sequence. Then given $\epsilon > 0$, there is N so that

$$n, m > N \implies \sup_i |x_{n,i} - x_{m,i}| < \frac{\epsilon}{2}$$

where each $\boldsymbol{x_n} = (x_{n,i}) \in \ell^\infty$. Thus, if $n, m > N$, $|x_{n,i} - x_{m,i}| < \frac{\epsilon}{2}$. It follows that for fixed i the sequence $(x_{n,i})$ is a Cauchy sequence of real numbers. Since $\Re$ is complete, there is a real number we will label as x_i so that $x_{n,i} \to x_i$. This defines a sequence $\boldsymbol{x} = (x_i)$.

For our ϵ, we can say for $n > m > N$,

$$\lim_{n\to\infty} |x_{n,i} - x_{m,i}| = |\lim_{n\to\infty} x_{n,i} - x_{m,i}| = |x_i - x_{m.i}| \le \frac{\epsilon}{2}$$

This shows $\sup_i |x_{m,i} - x_i| < \epsilon$ when $m > N$. Thus $\boldsymbol{x_m} \to \boldsymbol{x}$ in d_∞ metric.

Finally, we also know for any chosen $m > N$*, we can say*

$$|x_i| \;\leq\; |x_{m,i} - x_i| + |x_{m.i}| < \epsilon/2 + \sup_i |x_{m.i}|$$

But $\boldsymbol{x_m} \in \ell^\infty$*, so* $\sup_i |x_{m.i}| < \infty$*. Hence* $\boldsymbol{x}$ *is bounded too and so is in* ℓ^∞*. We conclude* (ℓ^∞, d_∞) *is complete.* ■

Theorem 2.12.5 (ℓ^p, d_p) **is Complete for** $p \geq 1$

(ℓ^p, d_p) is a complete metric space for $p \geq 1$.

Proof 2.12.5

Let $p \geq 1$ be chosen and let $(\boldsymbol{x_n})$ be a Cauchy sequence in (ℓ^p, d_p). Then given $\epsilon > 0$, there is N so that

$$n, m > N \;\implies\; \left(\sum_{j=1}^{\infty} |x_{n,i} - x_{m,i}|^p\right)^{\frac{1}{p}} < \frac{\epsilon}{2}$$

where each $\boldsymbol{x_n} = (x_{n,i}) \in \ell^p$. Hence, for a fixed i and for $n, m > N$,

$$|x_{n,i} - x_{m,i}|^p \;<\; \left(\frac{\epsilon}{2}\right)^p$$

or $|x_{n,i} - x_{m,i}| < \frac{\epsilon}{2}$ for fixed i and $n, m > N$. This implies $(x_{n,i})$ for fixed i is a Cauchy sequence of real numbers. Since $\Re$ is complete, there is a number x_i so that $x_{n,i} \to x_i$. This defines the sequence $\boldsymbol{x} = (x_i)$.

We also know if $n, m > N$, then

$$\sum_{j=1}^{k} |x_{n,i} - x_{m,i}|^p \;<\; \left(\frac{\epsilon}{2}\right)^p$$

and so

$$\lim_{n \to \infty} \sum_{j-1}^{k} |x_{n,i} - x_{m,i}|^p \;\leq\; \left(\frac{\epsilon}{2}\right)^p$$

or if $m > N$,

$$\sum_{j-1}^{k} |x_i - x_{m,i}|^p \;\leq\; \left(\frac{\epsilon}{2}\right)^p$$

which implies

$$\sum_{j-1}^{\infty} |x_i - x_{m,i}|^p \;\leq\; \left(\frac{\epsilon}{2}\right)^p$$

or $\boldsymbol{x_m} \rightarrow \boldsymbol{x}$ in d_p metric.
Pick any $m > M$, then

$$\sum_{j-1}^{\infty} |x_i - x_{m,i}|^p \leq \left(\frac{\epsilon}{2}\right)^p \implies \boldsymbol{x_m} - \boldsymbol{x} \in \ell^p.$$

Now

$$\begin{aligned}\left(\sum_{j=1}^{\infty} |x_j|^p\right)^{\frac{1}{p}} &= \left(\sum_{j=1}^{\infty} |(x_j - x_{m,j}) + x_{m,j}|^p\right)^{\frac{1}{p}} \\ &\leq \left(\sum_{j=1}^{\infty} |x_j - x_{m,j}|^p\right)^{\frac{1}{p}} + \left(\sum_{j=1}^{\infty} |x_{m,j}|^p\right)^{\frac{1}{p}}\end{aligned}$$

by Minkowski's Inequality. We conclude $\left(\sum_{j=1}^{\infty} |x_j|^p\right)^{\frac{1}{p}} < \infty$ and so $\boldsymbol{x} \in \ell^p$. This shows (ℓ^p, d_p) is complete for $p \geq 1$. ■

Homework

Exercise 2.12.11 *Let M_1 be the set of all 3×3 matrices whose components are sequences in ℓ^1. For any A and b in M_1, define $d_{M_1}(A, B) = \sum_{i=1}^{3} \sum_{j=1}^{3} d_1(A_{ij}, B_{ij})$. Prove d_{M_1} is a metric on M_1 and M_1 is complete with respect to d_{M_1}.*

Exercise 2.12.12 *Let M_2 be the set of all 4×4 matrices whose components are sequences in ℓ^2. For any A and b in M_2, define $d_{M_2}(A, B) = \sum_{i=1}^{4} \sum_{j=1}^{4} d_2(A_{ij}, B_{ij})$. Prove d_{M_2} is a metric on M_2 and M_2 is complete with respect to d_{M_2}.*

Exercise 2.12.13 *Let M_∞ be the set of all 5×5 matrices whose components are sequences in ℓ^∞. For any A and b in M_∞, define $d_{M_\infty}(A, B) = \sum_{i=1}^{4} \sum_{j=1}^{4} d_\infty(A_{ij}, B_{ij})$. Prove d_{M_∞} is a metric on M_∞ and M_∞ is complete with respect to d_{M_∞}.*

2.13 More on Separability

Let's look at separability now.

Theorem 2.13.1 (ℓ^∞, d_∞) **is not Separable**

(ℓ^∞, d_∞) is not a separable space.

Proof 2.13.1
Let $\boldsymbol{x} = (x_{n,i})$ be a sequence with $x_{n,i} \in \{0, 1\}$. This sequence $\boldsymbol{x}$ is associated with the real number $\hat{x} = \sum_{i=1}^{\infty} \frac{x_{n,i}}{2^i}$; i.e. this is the binary expansion of the real number $\hat{x}$. The number of points in $[0, 1]$ is not countable as we discussed in (Peterson (18) 2020). Each point $\hat{y}$ in $[0, 1]$ has a unique binary representation under the following assumptions. We note the series

$$\sum_{i=m+1}^{\infty} \frac{1}{2^i} = \sum_{i=0}^{\infty} \frac{1}{2^i} - \sum_{i=0}^{m} \frac{1}{2^i} = \frac{1}{1 - \frac{1}{2}} - \frac{1 - \left(\frac{1}{2}\right)^{m+1}}{1 - \frac{1}{2}}$$

$$= 2\left(1 - 1 + \left(\frac{1}{2}\right)^{m+1}\right) = \frac{1}{2^m}$$

Hence, the binary expansion $\underbrace{0\cdots0}_{m\ zeros}\ \underbrace{111\cdots}_{rest\ 1}$ *equals the binary expansion* $\underbrace{0\cdots0}_{m-1\ zeros}\ \underbrace{1\,000\cdots}_{rest\ 0}$*. Hence, some special real numbers in* $[0,1]$ *have two binary representations. We can avoid this by always choosing expansions that do not terminate in an infinite number of* 1*'s.*

Thus, it is possible for each $\hat{y}$ *in* $[0,1]$ *to have a unique binary representation. However, what we need for this proof is that different* $\hat{y}$*'s in* $[0,1]$ *have different binary expansions and this is always true whether or not we restrict the form of the binary representations to force uniqueness for each expansion. We conclude from the above there are uncountably many sequences of zeros and ones in* ℓ^∞ *of the from* $\boldsymbol{x}$ *given at the start of this proof.*

Take two such sequences $\boldsymbol{x_1}$ *and* $\boldsymbol{x_2}$*. If* $\boldsymbol{x_1}$ *and* $\boldsymbol{x_2}$ *are not equal, they differ in at least one slot* i*. Thus* $\sup_i |x_{1,i} - x_{2,i}| = 1$*. If* $\boldsymbol{x}$ *is any sequence of zeros and ones, then*

$$B(\boldsymbol{x}, \frac{1}{3}) = \{\boldsymbol{y} \in \ell^\infty \mid \sup_i |x_i = y_i| < \frac{1}{3}\}$$

cannot contain any other sequence of zeros and ones. as they are clearly 1 *apart.*

Now assume $M \subset \ell^\infty$ *is dense. Let* $\boldsymbol{x}$ *be a binary sequence. Then there is a* $\boldsymbol{w}$ *in* M *so that* $d_\infty(\boldsymbol{x}, \boldsymbol{w}) < \frac{1}{3}$ *because* M *is dense in* ℓ^∞*. Since each* $B(\boldsymbol{x}, \frac{1}{3})$ *can only contain one binary sequence. This means* M *must contain all binary sequences which means it is not countable. Thus* (ℓ^∞, d_∞) *is not separable.* ■

Theorem 2.13.2 (ℓ^p, d_p) **is Separable for** $p \geq 1$

> *(ℓ^p, d_p) is a separable space for $p \geq 1$.*

Proof 2.13.2
Let $p \geq 1$ *be chosen and let* $M \subset \ell^p$ *be defined by*

$$M = \{\boldsymbol{x} \in \ell^p \mid \boldsymbol{x} = (r_1, \ldots, r_n, 0, 0, 0, \cdots),\ r_i \in \mathbb{Q}, n \in \mathbb{N}\}$$

In words, M *is the set of all sequences with only a finite number of nonzero terms and each term is a rational number. Note for fixed* n

$$M_n = \{\boldsymbol{x} \in \ell^p \mid \boldsymbol{x} = (r_1, \ldots, r_n, 0, 0, 0, \cdots),\ r_i \mathbb{Q}\}$$

is equivalent to $\mathbb{Q}^n$ *which is countable. Since* $M = \cup_n M_n$ *which is the countable union of countable sets, we know* M *is countable also. Note a sequence of the form* $\boldsymbol{x} = (r_1, r_2, \ldots, r_n, r_{n+1}, \ldots)$ *which consists of an infinite number of nonzero rationals is not in* M*.*

Let $\boldsymbol{x} \in \ell^p$ *be arbitrarily chosen. Thus, for given* $\epsilon > 0$*, there is* N *so that* $m > N$ *implies* $|\sum_{j=N+1}^{m} |x_j|^p| < \frac{\epsilon^p}{4}$*. Since* $\mathbb{Q}$ *is dense in* $\Re$*, to each* x_j *there is a rational* r_j *so that* $|x_j - r_j|^p < \frac{\epsilon^p}{2^{j+2}}$*. Hence,*

$$\sum_{j=1}^{N} |x_j - r_j|^p < \sum_{j=1}^{N} \frac{\epsilon^p}{2^{j+3}} = \frac{\epsilon^p}{8} \sum_{j=1}^{N} \frac{1}{2^j} = \frac{\epsilon^p}{8} \frac{1 - \left(\frac{1}{2}\right)^{N+1}}{1 - \frac{1}{2}} < \frac{\epsilon^p}{4}$$

Let $\boldsymbol{y} = (r_1, r_2, \ldots, r_N, 0, 0, \cdots) \in M$. *Then for* $m > N$,

$$\sum_{j=1}^{m} |x_j - y_j|^p \quad = \quad \sum_{j=1}^{N} |x_j - r_j|^p + \sum_{j=N+1}^{m} |x_j|^p < \frac{\epsilon^p}{4} + \frac{\epsilon^p}{4} = \frac{\epsilon^p}{2}$$

We conclude

$$\sum_{j=1}^{m} |x_j - y_j|^p \quad \leq \quad \frac{\epsilon^p}{2} < \epsilon^p$$

Thus, $\sum_{j=1}^{m} |x_j - y_j|^p < \epsilon^p$ *for all* $m > N$ *which tells us* $d_p(\boldsymbol{x}, \boldsymbol{y}) < \epsilon$. *We conclude* M *is countable and dense and so* (ℓ^p, d_p) *is separable.* ∎

We will use these ideas later.

Homework

Exercise 2.13.1 *Let* M_1 *be the set of all* 3×3 *matrices whose components are sequences in* ℓ^1. *For any* A *and* b *in* M_1, *define* $d_{M_1}(A, B) = \sum_{i=1}^{3} \sum_{j=1}^{3} d_1(A_{ij}, B_{ij})$. *Prove* M_1 *is separable with respect to this metric.*

Exercise 2.13.2 *Let* M_2 *be the set of all* 4×4 *matrices whose components are sequences in* ℓ^2. *For any* A *and* b *in* M_2, *define* $d_{M_2}(A, B) = \sum_{i=1}^{4} \sum_{j=1}^{4} d_2(A_{ij}, B_{ij})$. *Prove* M_1 *is separable with respect to this metric.*

Exercise 2.13.3 *Let* M_∞ *be the set of all* 5×5 *matrices whose components are sequences in* ℓ^∞. *For any* A *and* b *in* M_∞, *define* $d_{M_\infty}(A, B) = \sum_{i=1}^{4} \sum_{j=1}^{4} d_\infty(A_{ij}, B_{ij})$. *Prove* M_1 *is separable with respect to this metric.*

Chapter 3

Completing a Metric Space

Now that we have completed a construction of the real numbers with all its familiar properties and seen a large variety of examples and basic results, let's go through how we can modify the procedure we used to complete $\mathbb{Q}$ to complete an arbitrary metric space (X, d).

3.1 The Completion of a Metric Space

It is easy to see that $\mathbb{Q}$ together with the standard absolute value function gives the metric space $(\mathbb{Q}, |\cdot|)$.

Step 1: Let

$$S \quad = \quad \{(x_n) \mid x_n \in X, (x_n) \text{ is a Cauchy sequence in } (X, d)\}$$

Step 2: Define an equivalence relation on S by

$$(x_n) \sim (y_n) \quad \Longleftrightarrow \quad \lim_{n\to\infty} d(x_n, y_n) = 0$$

Note convergence here is in the field $\Re$ which we have already constructed from the rationals.

Step 3: Let $\widetilde{X} = S/\sim$, the set of all equivalence classes in S under $\sim$. Let the elements of $\widetilde{X}$ be denoted by $[(x_n)]$.

Step 4: Extend the metric d on X to the metric $\tilde{d}$ on $\widetilde{X}$ by defining

$$\tilde{d}(\tilde{x}, \tilde{y}) \quad = \quad \lim_{n\to\infty} d(x_n, y_n)$$

for any equivalence classes $\tilde{x}$ and $\tilde{y}$ in $\widetilde{X}$ and any choice of representatives $(x_n) \in \tilde{x}$ and $(y_n) \in \tilde{y}$. To show $\tilde{d}$ is a metric on $\widetilde{\mathbb{Q}}$, we need to show it satisfies the properties of a metric.

M1: First, we show the definition of $\tilde{d}$ makes sense. By the triangle inequality

$$\begin{aligned} d(x_n, y_n) \quad &\leq \quad d(x_n, x_m) + d(x_m, y_m) + d(y_m, y_n) \\ &\Longrightarrow \quad d(x_n, y_n) - d(x_m, y_m) \leq d(x_n, x_m) + d(y_n, y_m) \end{aligned}$$

Since (x_n) and (y_n) are Cauchy sequences, given $\epsilon > 0$, there is N so that $n, m > N$ implies $d(x_n, x_m) + d(y_n, y_m) < \epsilon$. Thus, if $n, m > N$ then $d(x_n, y_n) - d(x_m, y_m) < \epsilon$.

By the same technique, we can also say

$$\begin{aligned} d(x_m, y_m) &\leq d(x_m, x_n) + d(x_n, y_n) + d(y_n, y_m) \\ &\Longrightarrow d(x_m, y_m) - d(x_n, y_n) \leq d(x_n, x_m) + d(y_n, y_m) \end{aligned}$$

Again, there is N so that $n, m > N$ implies $d(x_n, x_m) + d(y_n, y_m) < \epsilon$. Thus, if $n, m > N$ then $d(x_m, y_m) - d(x_n, y_n|) < \epsilon$. Combining we have $|\, d(x_m, y_m) - d(x_n, y_n) \,| < \epsilon$ when $n, m > N$. This shows the sequence $(d(x_n, y_n))$ is a Cauchy sequence in $\Re$ and so must converge.

If (x_n) and (x'_n) are in $[(x_n)]$ and (y_n) and (y'_n) are in $[(y_n)]$, then

$$\begin{aligned} d(x_n, y_n) &\leq d(x_n, x'_n) + d(x'_n, y'_n) + d(y'_n, y_n) \\ \Longrightarrow d(x_n, y_n) - d(x'_n, y'_n) &\leq d(x_n, x'_n) + +d(y'_n, y_n) \end{aligned}$$

Since $(x_n) \sim (x'_n)$ and $(y_n) \sim (y'_n)$, we know $\lim_n d(x_n, x'_n) = lim_n d(y_n, y'_n) = 0$. So given $\epsilon > 0$, there is N so that $d(x_n, x'_n) + d(y_n, y'_n) < \epsilon$ if $n > N$. We conclude

$$n > N \Longrightarrow d(x_n, y_n) - d(x'_n, y'_n) \quad < \quad \epsilon$$

A similar argument shows

$$n > N \Longrightarrow d(x'_n, y'_n) - d(x_n, y_n) \quad < \quad \epsilon$$

Combining, we have $|d(x_n, y_n) - d(x'_n, y'_n)| < \epsilon$ if $n > N$. This tells us the value of $\tilde{d}$ is the same for both pairs of representatives.

Hence, d is well-defined and always non-negative.

M2: If $\tilde{d}([(x_n)], [(y_n)]) = 0$, then $\lim_n d(x_n, y_n) = 0$. This says $(x_n) \sim (y_n)$ and so $[(x_n)] = [(y_n)]$. It is also easy to see that if $[(x_n)] = [(y_n)]$, then $\tilde{d}([(x_n)], [(y_n)]) = 0$.

M3: It is clear $\tilde{d}([(x_n)], [(y_n)]) = \tilde{d}([(y_n)], [(x_n)])$.

M4: We know for $[(x_n)], [(y_n)]$ and $[(z_n)]$, that

$$\begin{aligned} d(x_n, y_n) &\leq d(x_n, y_n) + d(z_n, y_n) \\ &\Longrightarrow \tilde{d}([(x_n)], [(y_n)]) \leq \tilde{d}([(x_n)], [(z_n)]) + \tilde{d}([(z_n)], [(y_n)]) \end{aligned}$$

Therefore, $\tilde{d}$ is a metric on $\widetilde{X}$.

Step 5: Define the mapping $T : X \to \widetilde{X}$ by $T(x) = [(\bar{x})]$ where $\bar{x} = (x, x, x, \ldots)$ is the constant sequence. Then T is $1 - 1$, onto and **isometric**; i.e. $\tilde{d}([(\bar{x})], [(\bar{y})]) = d(x, y)$.

1-1: If $[(\bar{x})] = [(\bar{y})]$, then $x = y$.

Onto: If $[(\bar{x})]$ is chosen, clearly $T(x) = [(\bar{x})]$.

Isometric: $\tilde{d}([(\bar{x})], [(\bar{y})]) = \lim_n d(x_n = x, y_n = y) = d(x, y)$.

Step 6: $T(X)$ is dense in $(\widetilde{X}, \tilde{d})$. Let $\epsilon > 0$ be given and choose $[(x_n)]$. Since (x_n) is a Cauchy sequence, for given $\epsilon > 0$, there is N so that $n, m > N$ implies $d(x_n, x_m) < \epsilon/2$. Thus, for all $n > N$, $d(x_{N+1}, x_n) < \epsilon/2$. Note

$$\tilde{d}([(x_n)], [(\overline{x_{N+1}})]) \quad = \quad \lim_n d(x_{N+1}, x_n) \leq \epsilon/2 < \epsilon$$

This says $\tilde{d}([(x_n)], T(x_{N+1})) < \epsilon$ and shows $T(X)$ is dense in $(\widetilde{X}, \tilde{d})$.

Step 7: $(\widetilde{X}, \tilde{d})$ is complete. To show this, let $([(x_n^p)])$ be a Cauchy sequence in $(\widetilde{X}, \tilde{d})$. Since $T(X)$ is dense, there is a sequence $(T(c_p))$ so that $\tilde{d}([(x_n^p)], T(c_p)) < \frac{1}{p}$ for all p. Now

$$\begin{aligned} \tilde{d}(T(c_p), T(c_q)) &\leq \tilde{d}(T(c_p), [(x_n^p)]) + \tilde{d}([(x_n^p)], [(x_n^q)]) + \tilde{d}(T(c_p), [(x_n^p)]) \\ &\leq \frac{1}{p} + \tilde{d}([(x_n^p)], [(x_n^q)]) + \frac{1}{q} \end{aligned}$$

Since $([(x_n^p)])$ is a Cauchy sequence, for $\epsilon > 0$, there is N_1 so that $p, q > N_1$ implies $\tilde{d}([(x_n^p)], [(x_n^q)]) < \epsilon/3$. Also, there is N_2 so that $1/k < \epsilon/3$ if $k > N_2$. Hence, if $p, q > \max\{N_1, N_2\}$, $\tilde{d}(T(c_p), T(c_q)) = d(c_p, c_q) < \epsilon$. We see (c_n) is a Cauchy sequence in X and so defines an equivalence $[(c_n)]$.

Now

$$\begin{aligned} \tilde{d}([(x_n^p)], [(c_n)]) &\leq \tilde{d}([(x_n^p)], T([(\bar{c}_p)])) + \tilde{d}(T([(\bar{c}_p)]), [(c_n)]) \\ &< \frac{1}{p} + \tilde{d}(T([(\bar{c}_p)]), [(c_n)]) = \frac{1}{p} + \lim_n d(c_p, c_n) \end{aligned}$$

Since (c_n) is a Cauchy sequence, there is N_1 so that $p, n > N_1$ implies $d(c_p, c_n) < \epsilon/2$. Further, there is N_2 so that $p > N_2$ implies $1/p < \epsilon/2$. We conclude $p > \max\{N_1, N_2\}$ gives

$$\tilde{d}([(x_n^p)], [(c_p)]) \leq \epsilon/2 + \epsilon/2 < \epsilon$$

So $[(x_n^p)] \to [(c_n)]$ in $\tilde{d}$ and we see $(\widetilde{X}, \tilde{d})$ is complete.

Homework

Exercise 3.1.1 *Describe the equivalence class* $[(\overline{\frac{1}{3}})]$.

Exercise 3.1.2 *What is the metric distance between the equivalence classes* $[(\overline{\frac{1}{3}})]$. *and* $[(\overline{\frac{1}{5}})]$.

Exercise 3.1.3 *Consider what we typically call the real number* $\sqrt{2}$. *This is really an identification of a Cauchy sequence in the completed space we discuss here. Using the base* 10 *expansion of* $\sqrt{2}$, *find an explicit representative of the equivalence class we call* $\sqrt{2}$. *Further, find a Cauchy sequence of equivalence classes which converge to the equivalence class identified with* $\sqrt{2}$.

Exercise 3.1.4 *Consider what we typically call the real number* e. *This is really an identification of a Cauchy sequence in the completed space we discuss here. Using the base* 10 *expansion of* e, *find an explicit representative of the equivalence class we call* e. *Further, find a Cauchy sequence of equivalence classes which converge to the equivalence class identified with* e.

Exercise 3.1.5 *Consider what we typically call the real number* e *again. Using the sequence definition of* e, *find an explicit representative of the equivalence class called* e. *Further, find a Cauchy sequence of equivalence classes which converge to the equivalence class identified with* e.

Note this exercise and the one before it both give a representative for the equivalence class identified with e.

Exercise 3.1.6 *Given two Cauchy sequences of equivalence classes that converge to the equivalence classes identified with* e *and* $\sqrt{2}$, *show how we find the distance between these equivalence classes.*

Exercise 3.1.7 *Given two Cauchy sequences of equivalence classes that converge to the equivalence classes identified with* e^2 *and* $\sqrt{3}$*, show how we find the distance between these equivalence classes.*

3.2 Completing the Integrable Functions

Let's first look at continuous functions which are a subset of all Riemann Integrable functions. We have already seen that Cauchy sequences of continuous functions need not converge to a continuous function. It depends on the choice of metric d. Since $(C([a,b]), d_\infty)$ is known to be complete, there is no need to do the construction process above to generate $(\widetilde{C}([a,b]), \widetilde{d_\infty})$. However, we know $(C([a,b]), d_1)$ and $(C([a,b]), d_2)$ are not complete. Let's go through the steps of constructing $(\widetilde{C}([a,b]), \widetilde{d_1})$.

Step 1: Define

$$S = \{(x_n) \mid x_n \in C([a,b]), (x_n) \text{ is a Cauchy sequence in } (C([a,b]), d_1)\}$$

Step 2: Define an equivalence relation on S by

$$(x_n) \sim (y_n) \iff \lim_{n\to\infty} d_1(x_n, y_n) = 0$$

Step 3: Let $\widetilde{C}([a,b]) = S/\sim$, the set of all equivalence classes in S under $\sim$. Let the elements of $\widetilde{C}([a,b])$ be denoted by $[(x_n)]$.

Step 4: Extend the metric d_1 on $C([a,b])$ to the metric $\tilde{d_1}$ on $\widetilde{C}([a,b])$ by defining

$$\tilde{d_1}(\tilde{x}, \tilde{y}) = \lim_{n\to\infty} d_1(x_n, y_n)$$

for any equivalence classes $\tilde{x}$ and $\tilde{y}$ in $\widetilde{C}([a,b])$ and any choice of representatives $(x_n) \in \tilde{x}$ and $(y_n) \in \tilde{y}$. The proof the $\tilde{d_1}$ is a metric is just like the general case.

Step 5: Define the mapping $T : C([a,b]) \to \widetilde{C}([a,b])$ by $T(x) = [(\bar{x})]$ where $\bar{x} = (x, x, x, \ldots)$ is the constant sequence. Note each entry x is actually a continuous function on $[a,b]$. Then T is 1-1, onto and **isometric**.

Step 6: $T(C([a,b]))$ is dense in $(\widetilde{C}([a,b]), \tilde{d_1})$.

Step 7: $(\widetilde{C}([a,b]), \tilde{d_1})$ is complete.

So going back to a nice example in Chapter 2.12, the sequence in $C([0,1])$ with metric d_1 given by (x_n) given by

$$x_n(t) = \begin{cases} 0, & 0 \le t \le \frac{1}{2} \\ n(t - \frac{1}{2}), \frac{1}{2} < t < \frac{1}{2} + \frac{1}{n} & \\ 1, & \frac{1}{2} + \frac{1}{n} \le t \le 1 \end{cases}$$

is a Cauchy sequence which does not converge. It does converge pointwise to the discontinuous function

$$x(t) = \begin{cases} 0, 0 \le t \le \frac{1}{2} \\ 1, \frac{1}{2} < t \le 1 \end{cases}$$

This Cauchy sequence determines an equivalence in $(\widetilde{C}([0,1]), \tilde{d}_1)$ called $[(\bar{x_n})]$ and we know $\tilde{d}_1([(\bar{x_n})], [(\bar{x})]) \rightarrow 0$. Thus, from a certain point of view, the pointwise limit function x is a member of this equivalence class. Of course, if we look at (y_n) given by

$$y_n(t) = \begin{cases} 0, & 0 \leq t < \frac{1}{2} \\ n(t - \frac{1}{2}), \frac{1}{2} \leq t < \frac{1}{2} + \frac{1}{n} \\ 1, & \frac{1}{2} + \frac{1}{n} \leq t \leq 1 \end{cases}$$

this is a Cauchy sequence which also does not converge. It does converge pointwise to the discontinuous function

$$y(t) = \begin{cases} 0, 0 \leq t < \frac{1}{2} \\ 1, \frac{1}{2} \leq t \leq 1 \end{cases}$$

Clearly (y_n) is in $[(x_n)]$ as $\lim_n d_1(x_n, y_n) = 0$. This relabeling of the equivalence class is similar to what we do with real numbers. The real number $\sqrt{3}$ can be represented by the sequences of rational numbers given by the expansion in terms of base 3, base 10 and so on. All are in the equivalence class $[(x_n)]$ where (x_n) is one choice of such an expansion. So the functions x and y are examples of discontinuous functions which are in $[(x_n)]$ in this case.

Let's look at Riemann Integrable functions using the d_1 metric. As we have discussed, d_1 is not a metric really as there are Riemann Integrable functions with $d_1(x, y) = 0$ even though $x \neq y$. The functions x and y above are examples of that. However, we can think of d_1 as a metric here if we switch to equivalence classes. Let $RI([a, b])$ be the set of all Riemann Integrable functions on $[a, b]$. Define the equivalence relation $\sim$ by $x \sim y$ if $d_1(x, y) = 0$. Let $\mathbb{RI}([a, b]) = RI([a, b])/\sim$ and denote an equivalence class by $[x]$ as usual for any Riemann Integrable function x on $[a, b]$. Extend d_1 to equivalence classes by

$$\mathbb{D}_1([x], [y]) = d_1(x, y)$$

where x and y are any two representatives of these equivalence classes. This is always a non-negative number and if x' and y' are two other representatives, then

$$\begin{aligned} \int_a^b |x'(t) - y'(t)|dt &\leq \int_a^b |x'(t) - x(t)|dt + \int_a^b |x(t) - y(t)|dt + \int_a^b |y(t) - y'(t)|dt \\ &= \int_a^b |x(t) - y(t)|dt \end{aligned}$$

A similar argument shows $\int_a^b |x(t) - y(t)|dt \leq \int_a^b |x'(t) - y'(t)|dt$. It follows immediately that $d_1(x, y) = d_1(x', y')$. Hence, this value is independent of the choice of representatives and so $\mathbb{D}_1$ is well-defined. This shows property $M1$ for a metric. For property $M2$, note

$$\mathbb{D}_1([x], [y]) = d_1(x, y) = 0 \Longrightarrow |x - y| = 0 \ a.e$$

which tells us $x - y$ is $[0]$. The other properties $M3$ and $M4$ are easily established. Thus, $(\mathbb{RI}([a, b]), \mathbb{D}_1)$ is a metric space. Now we complete this metric space.

Let's go through the steps of constructing $(\mathbb{L}_1([a, b], \widetilde{\mathbb{D}}_1)) \equiv (\widetilde{\mathbb{RI}}([a, b]), \widetilde{\mathbb{D}}_1)$.

Step 1: Define

$$\mathbb{S} = \{([x_n]) \mid [x_n] \in \mathbb{RI}([a, b]), ([x_n]) \text{ is a Cauchy sequence in } (\mathbb{RI}([a, b]), \mathbb{D}_1)\}$$

Step 2: Define an equivalence relation on $\mathbb{S}$ by

$$([x_n]) \sim ([y_n]) \iff \mathbb{D}_1([x_n],[y_n]) = 0$$

Step 3: Let $\mathbb{L}_1([a,b]) = \mathbb{S}/\sim$, the set of all equivalence classes in $\mathbb{S}$ under $\sim$. Let the elements of $\mathbb{L}_1([a,b])$ be denoted by $[\,[(x_n)]\,]$. This is a messy notation, so let's use $\mathscr{X} = [\,[(x_n)]\,]$.

Step 4: Extend the metric $\mathbb{D}_1$ on $\mathbb{RI}([a,b])$ to the metric $\widetilde{\mathbb{D}}_1$ on $\mathbb{L}_1([a,b])$ by defining

$$\widetilde{\mathbb{D}}_1(\mathscr{X},\mathscr{Y}) = \lim_{n\to\infty} \mathbb{D}_1([(x_n)],[(y_n)]) = \lim_{n\to\infty} d_1|x_n,y_n|$$

for any equivalence classes $\mathscr{X}$ and $\mathscr{Y}$ in $\mathbb{L}_1([a,b])$ and any choice of representatives $([x_n]) \in \mathscr{X}$ and $([y_n]) \in \mathscr{Y}$. The proof $\widetilde{\mathbb{D}}_1$ is a metric is just like the general case.

Step 5: Define the mapping $T : \mathbb{RI}([a,b]) \to \mathbb{L}_1([a,b])$ by

$$T([x]) = [\,[\bar{x}]\,] = \bar{\mathscr{X}}$$

where $[\bar{x}] = ([x],[x],[x],\ldots)$ is the constant sequence. Note each entry $[x]$ is actually an equivalence class in $\mathbb{RI}([a,b])$. The T is 1-1, onto and **isometric**.

Step 6: $T(\mathbb{RI}([a,b]))$ is dense in $(\mathbb{L}_1([a,b]),\widetilde{\mathbb{D}}_1)$.

Step 7: $(\mathbb{L}_1([a,b]),\widetilde{\mathbb{D}}_1)$ is complete.

So the discontinuous function x given by

$$x(t) = \begin{cases} 0, 0 \le t \le \frac{1}{2} \\ 1, \frac{1}{2} < t \le 1 \end{cases}$$

generates the equivalence class $[x]$ in $\mathbb{RI}([0,1])$. This in turn is embedded in $\mathbb{L}_1([0,1])$ as the constant sequence $T([x]) = [\,[\bar{x}]\,] = \bar{\mathscr{X}}$. The sequence of equivalence classes $[x_n]$ where

$$x_n(t) = \begin{cases} 0, & 0 \le t \le \frac{1}{2} \\ n(t-\frac{1}{2}), \frac{1}{2} < t < \frac{1}{2}+\frac{1}{n} & \\ 1, & \frac{1}{2}+\frac{1}{n} \le t \le 1 \end{cases}$$

is a Cauchy sequence of equivalence classes in $\mathbb{RI}([0,1])$ which are embedded as the sequence $(T([x_n]))$ into $\mathbb{L}_1([0,1])$ as constant sequences. This Cauchy sequence of equivalence classes in $\mathbb{L}_1([0,1])$ converges to $T([x])$. Note

$$\widetilde{\mathbb{D}}_1(T([x_n]),T([x])) = \lim_n \mathbb{D}_1([x_n],[x]) = \lim_n d_1(x_n,x) = \int_0^1 |x_n(t)-x(t)|dt = 0$$

Thus, the function which the Cauchy sequence (x_n) in $(RI([0,1]),d_1)$ converges to is the equivalence class $[T([x])]$ in $\mathbb{L}_1([0,1])$.

Let y be another function with $[T(y)]$ in $[T(x)]$. Note $\mathscr{Y} = [[y_n]]$ is in $[T([x])]$ if $\mathbb{D}_1([[y_n]],[[\bar{y}]]) = 0$. This says $\lim_n d_1(y_n,y) = 0$ and $d_1(x,y) = 0$. So any Riemann Integrable function y which is similar to x under d_1 is similar to the Cauchy sequence $(T([x_n]))$. Hence the limiting equivalence class here is the set of all Riemann Integrable functions y with $d_1(x,y) = 0$.

Homework

Exercise 3.2.1 *Let*

$$f(t) = \begin{cases} 1, & 1 \le t < 2 \\ 2, & 2 \le t \le 4 \end{cases}$$

- *Find a representative for the equivalence class* $[\bar{f}]$ *using* d_1.
- *Find a representative for the equivalence class* $[[\bar{f}]]$ *using* $\mathbb{D}_1$,

Exercise 3.2.2 *Let*

$$f(t) = \begin{cases} -2, & -4 \le t \le 4 \\ 5, & 4 < t \le 10 \end{cases}$$

- *Find a representative for the equivalence class* $[\bar{f}]$ *using* d_1.
- *Find a representative for the equivalence class* $[[\bar{f}]]$ *using* $\mathbb{D}_1$,

Exercise 3.2.3 *Let*

$$f(t) = \begin{cases} 1, & 1 \le t < 2 \\ 2, & 2 \le t \le 4 \end{cases} \quad g(t) = \begin{cases} -2, & 1 \le t < 2 \\ 5, & 2 \le t \le 4 \end{cases}$$

Show explicitly how we find the distance between $[[\bar{f}]]$ *and* $[[\bar{g}]]$.

Exercise 3.2.4 *Let*

$$f(t) = \begin{cases} -1, & -2 \le t < 4 \\ 12, & 4 \le t \le 10 \end{cases} \quad g(t) = \begin{cases} 2, & -2 \le t < 6 \\ 8, & 6 \le t \le 10 \end{cases}$$

Show explicitly how we find the distance between $[[\bar{f}]]$ *and* $[[\bar{g}]]$.

Exercise 3.2.5 *Let*

$$f(t) = \begin{cases} -1, & -2 \le t < 2 \\ 3, & 2 \le t < 6 \\ -5, & 6 \le t < 10 \end{cases} \quad g(t) = \begin{cases} 20, & -2 \le t < 0 \\ 10, & 0 \le t \le 8 \\ 3, & 8 < t \le 10 \end{cases}$$

Show explicitly how we find the distance between $[[\bar{f}]]$ *and* $[[\bar{g}]]$.

Part III

Normed Linear Spaces

Chapter 4

Vector Spaces

We are now going to explore vector spaces in some generality. If we have a set of objects $\boldsymbol{u}$ with a way to add them to create new objects in the set and a way to *scale* them to make new objects, this is formally called a **Vector Space** with the set denoted by $\boldsymbol{V}$.

4.1 Vector Spaces over a Field

Definition 4.1.1 Abstract Vector Space

Let $\boldsymbol{V}$ be a set of objects $\boldsymbol{u}$ with an additive operation $\oplus$ and a scaling method $\odot$. Formally, this means

VS1: *Given any $\boldsymbol{u}$ and $\boldsymbol{v}$, the operation of adding them together is written $\boldsymbol{u} \oplus \boldsymbol{v}$ and results in the creation of a new object in the vector space. This operation is* commutative *which means the order of the operation is not important; so $\boldsymbol{u} \oplus \boldsymbol{v}$ and $\boldsymbol{v} \oplus \boldsymbol{u}$ give the same result. Also, this operation is associative as we can group any two objects together first, perform this addition $\oplus$ and then do the others and the order of the grouping does not matter.*

VS2: *Given any $\boldsymbol{u}$ and any number c (either real or complex, depending on the type of vector space we have), the operation $c \odot \boldsymbol{u}$ creates a new object. We call such numbers* scalars.

VS3: *The scaling and additive operations are nicely compatible in the sense that order and grouping is not important. These are called the* distributive *laws for scaling and addition. They are*

$$\begin{aligned} c \odot (\boldsymbol{u} \oplus \boldsymbol{v}) &= (c \odot \boldsymbol{u}) \oplus (c \odot \boldsymbol{v}) \\ (c + d) \odot \boldsymbol{u} &= (c \odot \boldsymbol{u}) \oplus (d \odot \boldsymbol{u}). \end{aligned}$$

VS4: *There is a special object called $\boldsymbol{o}$ which functions as a* zero *so we always have $\boldsymbol{o} \oplus \boldsymbol{u} = \boldsymbol{u} \oplus \boldsymbol{o} = \boldsymbol{u}$.*

VS5: *There are* additive inverses *which means to each $\boldsymbol{u}$ there is a unique object $\boldsymbol{u}^\dagger$ so that $\boldsymbol{u} \oplus \boldsymbol{u}^\dagger = \boldsymbol{o}$.*

Comment 4.1.1 *These laws imply*

$$(0+0) \odot \boldsymbol{u} \quad = \quad (0 \odot \boldsymbol{u}) \oplus (0 \odot \boldsymbol{u})$$

which tells us $0 \odot \boldsymbol{u} = 0$. *A little further thought then tells us that since*

$$\begin{aligned} \mathbf{0} &= (1-1) \odot \boldsymbol{u} \\ &= (1 \odot \boldsymbol{u}) \oplus (-1 \odot \boldsymbol{u}) \end{aligned}$$

we have the additive inverse $\boldsymbol{u}^{\dagger} = -1 \odot \boldsymbol{u}$.

Comment 4.1.2 *We usually say this simpler. The set of objects* $\boldsymbol{V}$ *is a vector space over its scalar field if there are two operations which we denote by* $\boldsymbol{u} + \boldsymbol{v}$ *and* $c\boldsymbol{u}$ *which generate new objects in the vector space for any* $\boldsymbol{u}$, $\boldsymbol{v}$ *and scalar* c. *We then just add that these operations satisfy the usual commutative, associative and distributive laws and there are unique additive inverses.*

Comment 4.1.3 *The objects are often called* vectors *and sometimes we denote them by* $\boldsymbol{u}$ *although this notation is often too cumbersome.*

Vector spaces have two other important ideas associated with them. We have already talked about linearly independent objects in $\Re^2$ or $\Re^3$. In general, for any vector space, we would use the same definitions with just a few changes:

Definition 4.1.2 Two Linearly Independent Objects in a Vector Space

Let $\boldsymbol{E}$ *and* $\boldsymbol{F}$ *be two objects in a vector space* $\boldsymbol{V}$. *We say* $\boldsymbol{E}$ *and* $\boldsymbol{F}$ *are* **linearly dependent** *if we can find nonzero scalars* α *and* β *so that*

$$\alpha \odot \boldsymbol{E} + \beta \odot \boldsymbol{F} \quad = \quad \mathbf{0}.$$

Otherwise, we say they are **linearly independent**.

Homework

Exercise 4.1.1 *What does it mean for two real* 2×2 *matrices to be linearly dependent objects?*

Exercise 4.1.2 *What does it mean for two real two dimensional vectors to be linear dependent objects?*

Exercise 4.1.3 *What does it mean for two functions on* $[a, b]$ *to be linearly dependent objects?*

Exercise 4.1.4 *Let* $f(t) = t^2$ *and* $g(t) = 2t + 8$ *on* $[-1, 4]$. *Are* f *and* g *linearly dependent or independent on* $[-1, 4]$*?*

Exercise 4.1.5 *When are two polynomials independent on* $[a, b]$*?*

Exercise 4.1.6 *If* f *and* g *are in* $C([a, b])$, *prove the functions* $F(x) = \int_a^x f(s)ds$ *and* $G(x) = \int_a^x g(s)ds$ *are also linearly independent on* $[a, b]$.

Exercise 4.1.7 *If* f *and* g *are in* $C^1([a, b])$ *and are linearly independent, is it always true* f' *and* g' *are linearly independent as well?*

Exercise 4.1.8 *Prove the set* $\{1, t, t^2, t^3\}$ *are linearly independent in* $C([0, 1])$.

Exercise 4.1.9 *Prove the set* $\{t^2, t, t^4, t^6\}$ *are linearly independent in* $C([0, 1])$.

We can then easily extend this idea to any finite collection of such objects as follows.

Definition 4.1.3 Finitely Many Linearly Independent Objects

Let $\{\boldsymbol{E}_i : 1 \leq i \leq N\}$ be N objects in a vector $\boldsymbol{V}$. We say the set $\{\boldsymbol{E}_i : 1 \leq i \leq N\}$ is a **linearly dependent** *set if we can find nonzero constants α_1 to α_N, not all 0, so that*

$$\alpha_1 \odot \boldsymbol{E}_1 + \ldots + \alpha_N \odot \boldsymbol{E}_N = \boldsymbol{0}.$$

Note we have changed the way we define the constants. When there are more than two objects involved, we can't say, in general, that all *of the constants must be nonzero. Otherwise, we say the set is a* **linearly independent** *set.*

The next concept is that of the **span** of a collection of vectors.

4.1.1 The Span of a Set of Vectors

Definition 4.1.4 The Span of a Set of Vectors

Given a finite set of vectors in a vector space $\boldsymbol{V}$, $\mathscr{W} = \{\boldsymbol{u}_1, \ldots, \boldsymbol{u}_N\}$ for some positive integer N, the span of $\mathscr{W}$ is the collection of all new vectors of the form $\sum_{i=1}^{N} c_i \boldsymbol{u}_i$ for any choices of scalars $c_1, \ldots, c_N$. It is easy to see $\mathscr{W}$ is a vector space itself and since it is a subset of $\mathscr{V}$, we call it a vector subspace. *The span of the set $\mathscr{W}$ is denoted by $\boldsymbol{Sp}\mathbf{W}$. If the set of vectors $\mathscr{W}$ is not finite, the definition is similar but we say the span of $\mathscr{W}$ is the set of all vectors which can be written as $\sum_{i=1}^{N} c_i \boldsymbol{u}_i$ for some finite set of vectors $\boldsymbol{u}_1, \ldots \boldsymbol{u}_N$ from $\mathscr{W}$.*

The reason we define span differently for a non finite set of vectors is because to do that we really need a metric. If we had a set $\mathscr{W}$ which was countably infinite, we could label it as the sequence (W_i). If an element in the span of $\mathscr{W}$ could be written as

$$S = \sum_{i=1}^{\infty} \alpha_i W_i$$

That would imply we have elements S_n defined by

$$\begin{aligned} S_n &= \sum_{i=1}^{n} \alpha_i W_i, \text{ this is in the vector space} \\ S &= \lim_{n \to \infty} S_n \end{aligned}$$

We know for each n, the finite sums are in the vector space but at this point, with no way of defining convergence in the vector space, we do not know how to interpret this **limit**. If the vector space had a metric d we would understand this limit as there is an object in the vector space S so that for all $\epsilon > 0$, there is N so that $d(S_n, S) < \epsilon$ if $n > N$. But without a metric, this limit notion is impossible to define. It is possible to define an idea of limit without a metric, but to do so we have to define what the open sets are in the vector space to generate what is called a topological space. But since we are just in a vector space now with no additional structure, we cannot do that. Hence, we define spanning for an infinite set of objects in the way that we do above.

Homework

Exercise 4.1.10 *Let $f(t) = t^2$ and $g(t) = 2t + 8$ on $[-1, 4]$. Describe the span of f and g.*

Exercise 4.1.11 *Describe the span of the functions 1, t and t^2 on $[a, b]$. The set of polynomials of degree n can then be described as the span of the powers $\{1, t, t^2, \ldots, t^n\}$.*

Exercise 4.1.12 *If f and g are in $C([a,b])$, prove the functions $F(x) = \int_a^x f(s)ds$ and $G(x) = \int_a^x g(s)ds$ are also linearly independent on $[a,b]$.*

Exercise 4.1.13 *If two vectors in $\Re^2$ are linearly independent describe their span.*

Exercise 4.1.14 *If two vectors in $\Re^3$ are linearly independent describe their span.*

Exercise 4.1.15 *Define the 2×2 matrices A_i by*

$$A_1 = \begin{bmatrix} 1 & 0 \\ 0 & 0 \end{bmatrix}, \quad A_2 = \begin{bmatrix} 0 & 1 \\ 0 & 0 \end{bmatrix}, \quad A_3 = \begin{bmatrix} 0 & 0 \\ 1 & 0 \end{bmatrix}, \quad A_4 = \begin{bmatrix} 0 & 0 \\ 0 & 1 \end{bmatrix},$$

Describe the span of A_i.

Exercise 4.1.16 *Define the 3×3 matrices A_i similar to what you did in the previous problem. Describe the span of A_i.*

The comments above lead to the notation of a *basis* for a vector space.

Definition 4.1.5 A Basis for a Finite Dimensional Vector Space

If the set of vectors $\{\boldsymbol{V}_1, \ldots, \boldsymbol{V}_n\}$ is a linearly independent spanning set for the vector space $\mathscr{V}$, we say this set is a **basis** *for $\mathscr{V}$ and that the* **dimension** *of $\mathscr{V}$ is n.*

Comment 4.1.4 *From our comments, we know the maximum number of linear independent objects in a vector space of dimension n is n.*

We can modify this idea for vectors spaces that do not have a finite basis. First, we extend the idea of linear independence to sets that are not necessarily finite.

Definition 4.1.6 Linear Independence for Non Finite Sets

Given a set of vectors in a vector space $\mathscr{V}$, $\mathscr{W}$, we say $\mathscr{W}$ is a linearly independent subset if every finite set of vectors from $\mathscr{W}$ is linearly independent in the usual manner.

We have already defined the span of a set of vectors that is not finite. So as a final matter, we want to define what we mean by a **basis** for a vector space which does not have a finite basis.

Definition 4.1.7 A Basis for an Infinite Dimensional Vector Space

Given a set of vectors in an infinite dimensional vector space $\mathscr{V}$, $\mathscr{W}$, we say $\mathscr{W}$ is a basis *for $\mathscr{V}$ if the span of $\mathscr{W}$ is all of $\mathscr{V}$ and if the vectors in $\mathscr{W}$ are linearly independent. Hence, a basis is a linearly independent spanning set for $\mathscr{V}$. Recall, this means any vector in $\mathscr{V}$ can be expressed as a finite linear combination of elements from $\mathscr{W}$ and any finite collection of objects from $\mathscr{W}$ is linearly independent in the usual sense for a finite dimensional vector space.*

The vector space $C([0,1])$ does not have a finite basis. The set $\mathscr{W} = \{x_0(t) = 1, x_1 1(t) = t^1, x_2(t) = t^2, \ldots, x_n(t) = t^n, \ldots\}$ is a countable infinite set. Let $\{x_{n1}, x_{n2}, \ldots, x_{np}\}$ be any finite subset of $\mathscr{W}$ with $n_i < n_{i+1}$ for all i. The linearly independent object equation is

$$\sum_{i=1}^{p} \alpha_{n_i} x_{n_i}(t) = 0, \; 0 \leq t \leq 1$$

To make it easier to see what is going on, add all the other powers to t from $i = 0$ to $i = n_p$ and get the equation

$$\sum_{i=1}^{n_p} \alpha_i x_i(t) = 0,\ 0 \leq t \leq 1$$

This equation is still true after n_p derivatives giving $\alpha_{n_p} = 0$. Now back substitute through the n_p derivative equations to prove $\alpha_i = 0$ for all other i. Hence, $\mathscr{W}$ is a linearly independent set in $C([0,1])$. Since $\sqrt{t}$ is continuous on $[0,1]$ and cannot be written as a finite combination of powers of t; i.e. $\sqrt{t}$ is not a polynomial in t, we see the span of $\mathscr{W}$ is not $C([0,1])$. Thus, $\mathscr{W}$ is not a basis. However, since $C([0,1])$ contains a countably infinite linearly independent set, $C([0,1])$ is an example of an infinite dimensional vector space.

Homework

Exercise 4.1.17 *As usual, define the* 2×2 *matrices* A_i *by*

$$A_1 = \begin{bmatrix} 1 & 0 \\ 0 & 0 \end{bmatrix},\quad A_2 = \begin{bmatrix} 0 & 1 \\ 0 & 0 \end{bmatrix},\quad A_3 = \begin{bmatrix} 0 & 0 \\ 1 & 0 \end{bmatrix},\quad A_4 = \begin{bmatrix} 0 & 0 \\ 0 & 1 \end{bmatrix},$$

Is $\{A_1, A_2, A_2, A_4\}$ *a basis for the set of all* 2×2 *real-valued matrices?*

Exercise 4.1.18 *Define the* 3×3 *matrices* A_i *similar to what you did in the previous problem. Is* $\{A_i | 1 \leq i \leq 9\}$ *a basis for the set of all* 3×3 *real-valued matrices?*

Exercise 4.1.19 *Let* $\{\sin(\pi t), \sin(2\pi t), \sin(3\pi t)\}$ *be three functions on* $[0,1]$. *Prove this is a linearly independent set and is a basis for the span of* $\{\sin(\pi t), \sin(2\pi t), \sin(3\pi t)\}$. *Do this the hard way by writing down the linear dependence equation and showing all the constants must be zero.*

Exercise 4.1.20 *Let* $\{\cos(\pi t), \cos(2\pi t), \cos(3\pi t)\}$ *be three functions on* $[0,1]$. *Prove this is a linearly independent set and is a basis for the span of* $\{\cos(\pi t), \cos(2\pi t), \cos(3\pi t)\}$. *Do this the hard way by writing down the linear dependence equation and showing all the constants must be zero.*

Exercise 4.1.21 *Let* $\{\sin(\pi t), \sin(2\pi t), \sin(3\pi t)\}$ *be three functions on* $[0,1]$. *Prove this is a linearly independent set and is a basis for the span of* $\{\sin(\pi t), \sin(2\pi t), \sin(3\pi t)\}$. *Do this the easy way now:*

- *Write down the linear dependence equation* $\sum_{i=1}^{3} a_i \sin(i\pi t) = 0$ *on* $[0,1]$ *and multiply by* $\sin(j\pi t)$.
- *Integrate both sides from* 0 *to* 1 *and show this forces* $a_j = 0$.
- *From this show the set is linearly independent and so is a basis for the span of these functions.*

Exercise 4.1.22 *Let* $\{\cos(\pi t), \cos(2\pi t), \cos(3\pi t)\}$ *be three functions on* $[0,1]$. *Prove this is a linearly independent set and is a basis for the span of* $\{\cos(\pi t), \cos(2\pi t), \cos(3\pi t)\}$.

- *Write down the linear dependence equation* $\sum_{i=1}^{3} a_i \cos(i\pi t) = 0$ *on* $[0,1]$ *and multiply by* $\cos(j\pi t)$.
- *Integrate both sides from* 0 *to* 1 *and show this forces* $a_j = 0$.
- *From this show the set is linearly independent and so is a basis for the span of these functions.*

Exercise 4.1.23 *Use the integration idea of the last two exercises so show* $\{\sin(i\pi t) | 1 \leq 1 \leq n\}$ *is a linearly independent set for any* n.

Exercise 4.1.24 *Use the integration idea of the last two exercises so show* $\{\cos(i\pi t) | 0 \leq 1 \leq n\}$ *is a linearly independent set for any* n.

Exercise 4.1.25 *Show the last two exercises tell us that the sets* $\{\sin(i\pi t) | i \geq 1\}$ *and* $\{\cos(i\pi t) | i \geq 0\}$ *are two countably infinite linearly independent sets in* $C([0,1])$. *Again, this shows* $C([0,1])$ *cannot have a finite basis.*

4.2 Every Vector Space Has a Basis

We can prove every vector has a basis which is called a **Hamel Basis**. To do this, we must use a powerful axiom called **Zorn's Lemma**. This means our proof will be **non constructive**. To understand the problem we have in trying to find a basis for a vector space, let's do a thought experiment. Let X be a nonempty vector space. Since it is nonempty, it has at least one object in it. Call this $\boldsymbol{x_1}$. Let $< \boldsymbol{x_1} >$ denote the span of this object. Now consider the set

$$X \backslash < \boldsymbol{x_1} > \quad = \quad \{\boldsymbol{x} \mid \boldsymbol{x} \notin < \boldsymbol{x_1} >\}$$

If $X \backslash < \boldsymbol{x_1} >$ is not empty, there is an element $\boldsymbol{x_2}$ in there. The linearly independent equation then is

$$\alpha_1 \boldsymbol{x_1} + \alpha_2 \boldsymbol{x_2} = \boldsymbol{0}$$

Now if there were nonzero scalars for which this equation was true that would mean $\boldsymbol{x_2}$ is in the span of $\boldsymbol{x_1}$ which is not possible. So these two elements are linear independent.

Now look at $X \backslash < \boldsymbol{x_1}, \boldsymbol{x_2} >$ where $< \boldsymbol{x_1}, \boldsymbol{x_2} >$ is the span of these two objects. If there is an $\boldsymbol{x_3}$ in $X \backslash < \boldsymbol{x_1}, \boldsymbol{x_2} >$, it is easy to see $\{\boldsymbol{x_1}, \boldsymbol{x_2}, \boldsymbol{x_3}\}$ is a linearly independent set. Clearly, we can continue this process leading to an algorithm

$$\exists\, \boldsymbol{x_{i+1}} \in X \backslash < \boldsymbol{x_1}, \ldots, \boldsymbol{x_i} > \text{ or } X = < \boldsymbol{x_1}, \ldots, \boldsymbol{x_i} >$$

If at the ith step, the answer to whether or not the span of $\{\boldsymbol{x_1}, \ldots, \boldsymbol{x_i}\}$ is all of X is **yes**, we have $\{\boldsymbol{x_1}, \ldots, \boldsymbol{x_i}\}$ is a linearly independent spanning set and so is a basis for X. On the other hand, if this process does not terminate after a finite number of steps we find two possibilities

A countably infinite basis: There is a linearly independent sequence $(\boldsymbol{x_i})$ in X which is a basis. We could say $X = < \boldsymbol{x_1}, \ldots, \boldsymbol{x_i}, \ldots >$

There is no countable basis: In this case, our algorithm completes without finding a linearly independent spanning set.

To prove any nonempty vector space has a basis, we must turn to new tools. This new tool is called **Zorn's Lemma**.

Definition 4.2.1 Partial Ordering on a Set

Let M be a nonempty set. A partial ordering on M is a binary relation $R \subset M \times M$ satisfying

PO1: *$(a, a) \in R$. This is read as aRa or $a \preceq a$ and is called reflexivity.*

PO2: *$(a, b) \in R$ and $(b, a) \in R$ implies $a = b$. This is read as aRb and bRa implies $a = b$ or $a \preceq b$ and $b \preceq a$ implies $a = b$. This is called antisymmetry.*

PO3: *$(a, b) \in R, (b, c) \in R$ implies $(a, c) \in R$. This is read as aRb and bRc implies aRc or $a \preceq b$ and $b \preceq c$ implies $a \preceq c$. This is called transitivity.*

The pair $(M, \preceq)$ is called a partially ordered set or **poset**. *We often just say M is a poset and leave out mention of the ordering $\preceq$.*

Comment 4.2.1 *These are the same properties as $TO1$, $TO2$ and $TO3$ of a total ordering which we used when we were completing $\mathbb{Q}$. A partial ordering $\preceq$ does not say that any two objects can be compared though which is property $TO4$.*

Comment 4.2.2 *The set of all subsets of a nonempty set M can be partially ordered by set containment $\subset$. Since two subsets can be disjoint, not all subsets can be compared. Hence, this is not a total ordering.*

Comment 4.2.3 *We say $a, b \in M$ are comparable, if either $a \preceq b$ or $b \preceq a$ or both.*

Homework

Exercise 4.2.1 *Show $\mathbb{C}$ is partially ordered by $\leq$ where $a + bi \leq c + di$ if and only if $a \leq c$ and $b \leq d$ but it is not a total order.*

Exercise 4.2.2 *Show the set of two dimensional vectors with non-negative components is partially ordered by $\leq$ where $A \leq B$ if and only if $A_1 \leq B_1$ and $A_2 \leq B_2$ but it is not a total order.*

Exercise 4.2.3 *Show the set of three dimensional vectors with non-negative components is partially ordered by $\leq$ where $A \leq B$ if and only if $A_1 \leq B_1$ and $A_2 \leq B_2$ and $A_3 \leq B_3$ but it is not a total order.*

Now let's define **chains**.

Definition 4.2.2 Totally Ordered Sets or Chains

Let $(M, \preceq)$ be a collection of subsets of a nonempty poset. Then if any two elements in M are comparable, we say M is a totally ordered set or chain.

Comment 4.2.4 *Hence, if $\preceq$ satisfies $TO4$, M is a chain.*

Definition 4.2.3 Upper Bounds and Maximal Objects of Posets

An upper bound of a subset W in a poset M is an object $U \in M$ so that $x \preceq U$ for all $x \in W$.

A maximal object of M is $m \in M$ so that $x \in M$ with $m \preceq x$ implies $m = x$.

Comment 4.2.5 *Note U need not be in W, of course, but it is in M. And such U need not exist.*

We can now state **Zorn's Lemma**.

Lemma 4.2.1 Zorn's Lemma

Let M be a nonempty poset. Assume every chain $C \subset M$ has an upper bound. Then M has at least one maximal object.

We will use this to prove every vector space has a basis.

Theorem 4.2.2 Every Nonempty Vector Space Has a Basis

Every nonempty Vector Space has a Basis.

Proof 4.2.1
Let X be a nonempty vector space and M be the set of all linearly independent subsets of X. M is nonempty as we assume X is nonempty. We make M into a poset by using set inclusion as a partial ordering. Let $C \subset M$ be a chain. Then C consists of objects of M which we will call c_α where α is taken from some index set Λ. For example, if C was countable, we could write $C = \{c_n \, n \geq 1\} = (c_n)$ but the index set need not be countable. Hence, in general, we write $C = (c_\alpha)$, $\alpha \in \Lambda$. Let $U = \cup_{\alpha \in \Lambda} c_\alpha$. We see $c_\alpha \in U$ for all α so $c_\alpha \preceq U$ for all indices.

Is $U \in M$? Take any finite subset of U and label this subset as $\{x_1, \ldots, x_p\}$ for some positive integer p. Clearly, $x_i = c_{\alpha_i}$ for some $\alpha_i \in \Lambda$. Since there are only a finite number of subsets here, there is Q so that $\{x_1, \ldots, x_p\} \subset \cup_{i=1}^{Q} c_i$ where each c_i is a linearly independent subset of X. Since the sets c_i are in the chain C, they must be comparable. Hence, for each pair i, j, $c_i \preceq c_j$, $c_j \preceq c_i$ or both. Since there are only a finite number of subsets here, there will be an index q_0 with each x_i in $\{x_1, \ldots, x_p\}$ a subset of c_{q_0}. This says $\{x_1, \ldots, x_p\}$ is a subset of X which is linearly independent. Since our choice of a finite number of elements of U was arbitrary, this says U is a linearly independent subset of X and so is in M.

Now apply Zorn's Lemma to conclude that M has at least one maximal object B. Let $Y = span(B)$, the span of B. It is clear Y is a subspace of X. Assume $Y \neq X$. Then, there is a $z \in X \setminus Y$ which is nonzero. Let $D = B \cup \{z\}$. This is a linearly independent set and so must be in M. Also $B \preceq D$. But B is maximal, so $B = D$. However, this is impossible. Hence, our assumption that $Y \neq X$ is wrong and we have $X = span(B)$. This shows B is a basis for X. ■

Comment 4.2.6 *This basis is called the Hamel Basis S. So given $x \in X$ there are a finite number of elements b_i from S and associated scalars α_i so that $x = \sum_{i=1}^{p} \alpha_i b_i$ for some positive integer p. However, there is no algorithm or procedure to calculate these coefficients and this is a significant problem.*

Homework

Exercise 4.2.4 *Look up the* **Well Ordering Theorem**, **The Axiom of Choice**, **The Hausdorff Maximum Principle** *and* **Tukey's Lemma**. *These are all equivalent to* **Zorn's Lemma** *and the proof that they are is quite involved and not something we go over in the early analysis courses now. However, it is well worth your effort to do so. It is very abstract and so encourages your growth in thinking in that manner. See (Hewitt and Stromberg (8) 1975) for a long discussion of this.*

Exercise 4.2.5 *Let X be the set of all sets in $\Re$ partially ordered by set inclusion. Let M be the poset $\{(0, \frac{1}{1}, \ldots, \frac{1}{i}, \ldots) | i \geq 1\}$ where i is an integer. Is M a chain?*

Exercise 4.2.6 *Let X be the set of all polynomials partially ordered by set inclusion. Let P_n be the set of all polynomials of degree $\leq n$ for the integer $n \geq 0$. Let $M = \{P_0, P_1, P_2, P_3, \ldots\}$. Is M a chain?*

Exercise 4.2.7 *Let X be the set of all polynomials partially ordered by set inclusion. Let P_n be the set of all polynomials of degree $\leq n$ for the integer $n \geq 0$. Let $M = \cup_{n \geq 0} P_n$. Is M a chain?*

Chapter 5

Normed Linear Spaces

Many vector spaces also have a way to measure the magnitude or size of an object in the vector space which is called a **norm**.

5.1 Norms

Given a basis $\boldsymbol{E}$ of $\Re^n$, $\boldsymbol{E} = \{\boldsymbol{e}_1, \ldots, \boldsymbol{e}_n\}$, we know any vector $\boldsymbol{x}$ in $\Re^n$ has a decomposition $\left[\boldsymbol{x}\right]_{\boldsymbol{E}}$ where the components of $\boldsymbol{X}$ with respect to the basis $\boldsymbol{E}$ are x_i with

$$\boldsymbol{x} = x_1\boldsymbol{e}_1 + \ldots + x_n\boldsymbol{e}_n$$

We often wish to **measure** the magnitude or size of a vector $\boldsymbol{x}$. We have talked about doing this before and have used the term **norm** to denote the mapping that does the job. Let's be specific now.

Definition 5.1.1 Norm on a Vector Space

Let $\boldsymbol{V}$ be a vector space over the reals and let $\boldsymbol{\rho} : \boldsymbol{V} \rightarrow \Re$ satisfy

N1: $\boldsymbol{\rho}(\boldsymbol{x}) \geq 0$ *for all* $\boldsymbol{x} \in \boldsymbol{V}$.

N2: $\boldsymbol{\rho}(\boldsymbol{x}) = 0$ *if and only if* $\boldsymbol{x} = \boldsymbol{0}$

N3: $\boldsymbol{\rho}(\alpha\boldsymbol{x}) = |\alpha|\boldsymbol{\rho}(\boldsymbol{x})$ *for all real numbers* α *and* $\boldsymbol{x} \in \boldsymbol{V}$

N4: $\boldsymbol{\rho}(\boldsymbol{x} + \boldsymbol{y}) \leq \boldsymbol{\rho}(\boldsymbol{x}) + \boldsymbol{\rho}(\boldsymbol{y})$ *for all* $\boldsymbol{x}, \boldsymbol{y} \in \boldsymbol{V}$

Note given a norm, we can use it to define a metric: define $d(\boldsymbol{x}, \boldsymbol{y}) = \boldsymbol{\rho}(\boldsymbol{x} - \boldsymbol{y})$.

Homework

Exercise 5.1.1 *Given a norm $\boldsymbol{\rho}$ on the vector space X, prove $d(\boldsymbol{x}, \boldsymbol{y}) = \boldsymbol{\rho}(\boldsymbol{x} - \boldsymbol{y})$ defines a metric.*

Exercise 5.1.2 *The easiest norm to look at is $\rho(x) = |x|$ for $x \in \Re$. Prove this is a norm on $\Re$.*

Exercise 5.1.3 *The next easiest norm to look at is $\rho(\boldsymbol{x}) = \sqrt{x_1^2 + x_2^2}$ for $\boldsymbol{x} = \begin{bmatrix} x_1 \\ x_2 \end{bmatrix}$ in $\Re^2$. Prove this is a norm on $\Re^2$ which we call $\| \cdot \|_2$, We know vectors are special cases of sequences of real numbers so to prove $N4$ we can use Minkowski's Inequality.*

Exercise 5.1.4 *Another norm is* $\rho(\boldsymbol{x}) = |x_1| + |x_2|$ *for* $\boldsymbol{x} = \begin{bmatrix} x_1 \\ x_2 \end{bmatrix}$ *in* $\Re^2$. *Prove this is a norm on* $\Re^2$ *which we call* $\|\cdot\|_1$,

Exercise 5.1.5 *Another norm is* $\rho(\boldsymbol{x}) = \max(|x_1|, ||x_2|)$ *for* $\boldsymbol{x} = \begin{bmatrix} x_1 \\ x_2 \end{bmatrix}$ *in* $\Re^2$. *Prove this is a norm on* $\Re^2$ *which we call* $\|\cdot\|_\infty$,

Exercise 5.1.6 *Define* $\|\cdot\|_p$ *on* $\Re^n$ *for* $p \geq 1$ *including the case* $p = \infty$ *and prove these are norms.*

Exercise 5.1.7 *For* $\boldsymbol{x} = \begin{bmatrix} x_1 \\ x_2 \end{bmatrix}$ *in* $\Re^2$ *and* $D = \begin{bmatrix} 2 & 0 \\ 0 & 4 \end{bmatrix}$, *define* $\rho^2(\boldsymbol{x}) = \boldsymbol{x}^T D\boldsymbol{x} = 2x_1^2 + 4x_2^2$.

- *Using Minkowski's Inequality for* $\Re^2$, *first prove* $2x_1y_1 + 4x_2y_2 = \boldsymbol{y}^T D\boldsymbol{x} \leq \rho(\boldsymbol{x})\rho(\boldsymbol{y})$. *Do this by looking at the inner product of* $\begin{bmatrix} \sqrt{2}x_1 \\ 2x_2 \end{bmatrix}$ *with itself.*
- *Then prove* $\rho^2(\boldsymbol{x} + \boldsymbol{y}) \leq (\rho(\boldsymbol{x}) + \rho(\boldsymbol{y}))^2$. *This will show property* $N4$.

Exercise 5.1.8 *For* $\boldsymbol{x} = \begin{bmatrix} x_1 \\ x_2 \end{bmatrix}$ *in* $\Re^2$ *and* $D = \begin{bmatrix} 3 & 0 \\ 0 & 8 \end{bmatrix}$, *define* $\rho^2(\boldsymbol{x}) = \boldsymbol{x}^T D\boldsymbol{x} = 3x_1^2 + 8x_2^2$.

- *Using Minkowski's Inequality for* $\Re^2$, *first prove* $3x_1y_1 + 8x_2y_2 = \boldsymbol{y}^T D\boldsymbol{x} \leq \rho(\boldsymbol{x})\rho(\boldsymbol{y})$. *Do this by looking at the inner product of* $\begin{bmatrix} \sqrt{3}x_1 \\ \sqrt{8}x_2 \end{bmatrix}$ *with itself.*
- *Then prove* $\rho^2(\boldsymbol{x} + \boldsymbol{y}) \leq (\rho(\boldsymbol{x}) + \rho(\boldsymbol{y}))^2$. *This will show property* $N4$.

Exercise 5.1.9 *For* $\boldsymbol{x} = \begin{bmatrix} x_1 \\ x_2 \end{bmatrix}$ *in* $\Re^2$ *and* $D = \begin{bmatrix} \lambda_1^2 & 0 \\ 0 & \lambda_2^2 \end{bmatrix}$ *where* λ_1 *and* λ_2 *are both positive. Define* $\rho^2(\boldsymbol{x}) = \boldsymbol{x}^T D\boldsymbol{x} = \lambda_1^2 x_1^2 + \lambda_2 x_2^2$.

- *Using Minkowski's Inequality for* $\Re^2$, *first prove* $\lambda_1^2 x_1y_1 + \lambda_2^2 x_2y_2 = \boldsymbol{y}^T D\boldsymbol{x} \leq \rho(\boldsymbol{x})\rho(\boldsymbol{y})$. *Do this by looking at the inner product of* $\begin{bmatrix} \lambda_1 x_1 \\ \lambda_2 x_2 \end{bmatrix}$ *with itself.*
- *Then prove* $\rho^2(\boldsymbol{x} + \boldsymbol{y}) \leq (\rho(\boldsymbol{x}) + \rho(\boldsymbol{y}))^2$. *This will show property* $N4$.

Exercise 5.1.10 *For* $\boldsymbol{x} = \begin{bmatrix} x_1 \\ x_2 \\ \vdots \\ x_n \end{bmatrix}$ *in* $\Re^n$ *and* D *a* $n \times n$ *diagonal matrix whose diagonal entries are* λ_i^2 *with each* λ_i *positive. Define* $\rho^2(\boldsymbol{x}) = \boldsymbol{x}^T D\boldsymbol{x} = \sum_{i=1}^n \lambda_i^2 x_i^2$.

- *Using Minkowski's Inequality for* $\Re^n$, *first prove* $\lambda_1^2 x_1y_1 + \ldots + \lambda_n^2 x_ny_n = \boldsymbol{y}^T D\boldsymbol{x} \leq \rho(\boldsymbol{x})\rho(\boldsymbol{y})$. *Do this by looking at the inner product of* $\begin{bmatrix} \lambda_1 x_1 \\ \vdots \\ \lambda_n x_n \end{bmatrix}$ *with itself.*
- *Then prove* $\rho^2(\boldsymbol{x} + \boldsymbol{y}) \leq (\rho(\boldsymbol{x}) + \rho(\boldsymbol{y}))^2$. *This will show property* $N4$.

Exercise 5.1.11 *Let* $\boldsymbol{x}$ *and* $\boldsymbol{y}$ *be in* $\Re^n$ *and let* A *be an* $n \times n$ *positive definite symmetric matrix. Define* $\rho^2(\boldsymbol{x}) = \boldsymbol{x}^T A\boldsymbol{x}$. *Prove* ρ *is a norm on* $\Re^n$. *This proof uses the decomposition* $A = P^T DP$ *where* D *is a diagonal matrix as we have used in the previous exercise. Look at the explanations in (Peterson (20) 2020) for more details.*

Comment 5.1.1 *We usually denote the norm* $\boldsymbol{\rho}$ *by the symbol* $\|\cdot\|$ *and sometimes add a subscript to remind us where the norm comes from. We will see many examples of this soon.*

5.1.1 Sequence Space Norms

The set of either real or complex-valued sequences and its subsets give us many interesting examples of normed linear spaces.

Bounded Sequences: The set of all bounded sequences S becomes a normed space too. We will look at this for complex-valued sequences, but the ideas hold for real-valued sequences also. For the complex sequence $\boldsymbol{x}$, define

$$\|\boldsymbol{x}\| = \sup_i |x_i|$$

N1: The supremum here always exists and is non-negative.

N2: If $\boldsymbol{x} = \boldsymbol{0}$, we see $\|\boldsymbol{x}\| = 0$. If $\|\boldsymbol{x}\| = 0$, then $\sup_i |x_i| = 0$ or $x_i = 0$ for all i telling us $\boldsymbol{x} = \boldsymbol{0}$.

N3: It is clear $\|\alpha \boldsymbol{x}\| = |\alpha| \|\boldsymbol{y}\|$.

N4: Since

$$\begin{aligned} |x_i + y_i| &\leq |x_i| + |y_i| \\ &\leq \sup_i |x_i| + \sup_i |y_i| = \|\boldsymbol{x}\| + \|\boldsymbol{z}\| \end{aligned}$$

we have the result.

Thus $(S, \|\cdot\|)$ is a normed space. This sequence space is called ℓ^∞ and this norm is called $\|\cdot\|_\infty$. We also use these names for the corresponding sequences of real numbers and we know which one we are referring to from context.

Sequences whose series are absolutely convergent: Another common subset of the space of all sequence is

$$S = \{\boldsymbol{x} = (x_n) \sum_{n=1}^{\infty} |x_n| \text{ converges } \}$$

S becomes a normed space too. For the sequence $\boldsymbol{x}$, define $\|\cdot\|$ by

$$\|\boldsymbol{x}\| = \sum_{i=1}^{\infty} |x_i|$$

N1: This is clear.

N2: If $\boldsymbol{x} = \boldsymbol{0}$, we see $\|\boldsymbol{x}\| = 0$. If $\|\boldsymbol{x}\| = 0$, then $\sum_{i=1}^{\infty} |x_i| = 0$ or $x_i = 0$ for all i telling us $\boldsymbol{x} = \boldsymbol{0}$.

N3: It is clear $\|\alpha \boldsymbol{x}\| = |\alpha|| : \|\boldsymbol{y}\|$.

N4: Since

$$\begin{aligned} \sum_{i=1}^{n} |x_i + y_i| &\leq \sum_{i=1}^{n} |x_i| + \sum_{i=1}^{n} |y_i| \\ &\leq \|\boldsymbol{x}\| + \|\boldsymbol{y}\| \end{aligned}$$

the result follows immediately.

Thus $(S, \|\cdot\|)$ is a normed space. This sequence space is called ℓ^1 and this metric is called $\|\cdot\|_1$.

The $1 \leq p < \infty$ spaces : We have already defined the sequence spaces ℓ^p for $p > 1$ We define the mapping $\|\cdot\|_p : \ell^p \rightarrow \Re$ by

$$\|\boldsymbol{x}\|_p = \left(\sum_{i=1}^{\infty} |x_i|^p\right)^{\frac{1}{p}}$$

Let $\boldsymbol{x}$ be in ℓ^p.

N1: This is clear $\sum_{i=1}^{\infty} |x_i|^p$ is finite and non-negative.

N2: If $\boldsymbol{x} = \boldsymbol{y}$, we see $\|\boldsymbol{x}\|_p = 0$. If $\|\boldsymbol{x}\|_p = 0$, then $|x_i|^p = 0$ or $x_i = 0$ for all i telling us $\boldsymbol{x} = \boldsymbol{0}$.

N3: This is easy to see.

N4: By Minkowski's Inequality

$$\left(\sum_{i=1}^{\infty} |x_i + y_i|^p\right)^{\frac{1}{p}} \leq \left(\sum_{i=1}^{\infty} |x_i|^p\right)^{\frac{1}{p}} + \left(\sum_{i=1}^{\infty} |y_i|^p\right)^{\frac{1}{p}}$$

This shows the result.

So $(\ell^p, \|\cdot\|_p)$ is a normed space.

We can also prove a result that shows the infinity norm is obtained through a limit on p.

Theorem 5.1.1 The Infinity Norm is the Limit on p of the p Norm

If $\boldsymbol{x} \in \ell^p$ for $p > 1$, then $\|\boldsymbol{x}\|_\infty = \lim_{p \rightarrow \infty} \|\boldsymbol{x}\|_p$.

Proof 5.1.1
You will prove this in the exercises below. ∎

Homework

Exercise 5.1.12 *If $\boldsymbol{x} \in \ell^p$, p finite and $p > 1$, prove $\boldsymbol{x}$ is bounded; i.e. $\boldsymbol{x} \in \ell^\infty$.*

Exercise 5.1.13 *You were asked to prove this earlier in another exercise, but it is good to make sure you get it by asking you to do it again. If $\boldsymbol{x} \in \ell^p > 1$ and $p' > p$, prove $\boldsymbol{x} \in \ell^{p'}$. Hence, $\ell^{p'} \subset \ell^p$ if $p' > p > 1$.*

Exercise 5.1.14 *Let $\boldsymbol{x} \neq \boldsymbol{0}$ be in $\Re^n$. Assume there is one component whose absolute value is the biggest. Let the index where this occurs be L. Prove*

- *Since $\boldsymbol{x} = x_L\left[\frac{x_i}{X_L}, \ldots, \frac{x_n}{x_L}\right]$, $\|\boldsymbol{x}\|_p^p = |x_L|^p \sum_{i \neq L}\left(\frac{|x_i|}{|X_L|}\right)^p + |x_L|^p$.*
- *If $|c| < 1$, prove $\lim_{p' \rightarrow \infty} |c|^{1/p'} = 1$.*
- *The terms $\frac{|x_i|}{|x_L|}$ for $i \neq L$ are all less than one. Prove there is M so if $p' > M$,*

$$\left(\sum_{i \neq L}\left(\frac{|x_i|}{|X_L|}\right)^{p'}\right)^{1/p'} < 1$$

- *Prove* $\lim_{p\infty\infty} \|\boldsymbol{x}\|_p = \|\boldsymbol{x}\|_\infty = |x_L|$.

Exercise 5.1.15 *Let* $\boldsymbol{x} \neq \boldsymbol{0}$ *be in* $\Re^n$. *Assume there are two components whose absolute value is the biggest. Let the indices where this occurs be* L_1 *and* L_2. *Mimic the proof in the last exercise to prove* $\lim_{p\infty} \|\boldsymbol{x}\|_p = \|\boldsymbol{x}\|_\infty$.

Exercise 5.1.16 *Let* $\boldsymbol{x} \neq \boldsymbol{0}$ *be in* $\Re^n$. *Mimic the last two proofs into one general proof that shows* $\lim_{p\infty\infty} \|\boldsymbol{x}\|_p = \|\boldsymbol{x}\|_\infty$.

Exercise 5.1.17 *Let* $\boldsymbol{x} \in \ell^p$. *Then* $\boldsymbol{x} \in \ell^\infty$ *also. We will assume no individual* $|x_i| = \|\boldsymbol{x}\|_\infty$. *You might think how the argument below would change if this was not true.*

- *Prove for a given* $\epsilon > 0$, *there is a* N *so that* $\sum_{i=N+1}^{\infty} |x_i|^{p'} < \left(\frac{\epsilon}{2}\right)^p$ *for all* $p' > p$. *Note since* $\boldsymbol{x} \in \ell^p$, *there is an* M *so that* $|x_i|^p < 1$ *if* $i > M$. *Thus,* $|x_i|^{p'} < |x_i|^p < 1$ *and so* $\sum_{i=N+1}^{\infty} |x_i|^{p'} < \sum_{i=N+1}^{\infty} |x_i|^p < \left(\frac{\epsilon}{2}\right)^p$.

- *Again, we will use the fact that if* $|c| < 1$, *prove* $\lim_{p'\to\infty} |c|^{1/p'} = 1$.

- *Using Minkowski's Inequality and the remarks above*

$$\begin{aligned}
\left| \left(\sum_{i=1}^{\infty} |x_i|^{p'} \right)^{1/p'} - \|\boldsymbol{x}\|_\infty \right| &\leq \|\boldsymbol{x}\|_\infty \left| \sum_{i=1}^{N} \left(\frac{|x_i|}{\|\boldsymbol{x}\|_\infty} \right)^{p'} - 1 \right| + \left(\sum_{i=N+1}^{\infty} |x_i|^{p'} \right)^{1/p'} \\
&\leq \|\boldsymbol{x}\|_\infty^p \left| \sum_{i=1}^{N} \left(\frac{|x_i|}{\|\boldsymbol{x}\|_\infty} \right)^{p'} - 1 \right| + \frac{\epsilon}{2}
\end{aligned}$$

- *Prove there is* M *so* $p' > M$ *implies*

$$\sum_{i=1}^{N} \left(\frac{|x_i|}{\|\boldsymbol{x}\|_\infty} \right)^{p'} < 1$$

Thus, $\lim_{p'\to\infty} \sum_{i=1}^{N} \left(\frac{|x_i|}{\|\boldsymbol{x}\|_\infty} \right)^{p'} = 1$.

- *Using all these pieces, prove* $\lim_{p\infty\infty} \|\boldsymbol{x}\|_p = \|\boldsymbol{x}\|_\infty$.

5.1.2 Function Space Norms

$C(\Omega)$**:** For the Ω the finite interval $[a, b]$, we let

$$C([a,b]) = \{\boldsymbol{x} : [a,b] \to \Re \mid \boldsymbol{x} \text{ is continuous on } [a,b]\}$$

and define $\|\boldsymbol{x}\|_\infty(\boldsymbol{x}) = \sup_{a \leq t \leq b} |x(t)|$. Since $\boldsymbol{x}$ is continuous on $[a,b]$, so is $|\boldsymbol{x}|$ and therefore a maximum is achieved. We could therefore write $\|\boldsymbol{x}\| = \max_{a \leq t \leq b} |x(t)|$ instead. We note $\|\cdot\|$ is a norm.

N1: The maximum is always finite and non-negative.

N2: If $\|\boldsymbol{x}\| = 0$, then $\max_{a \leq t \leq b} |x(t)| = 0$ implying $|\boldsymbol{x}(t)| = 0$ for all t. Thus, $\boldsymbol{x}(t) = 0$ for all t. On the other hand, if $\boldsymbol{x} = \boldsymbol{0}$, we see $\|\boldsymbol{x}\|_\infty = 0$.

M3: It is clear $\|\alpha\boldsymbol{x}\|_\infty = |\alpha| \|\boldsymbol{x}\|_\infty$.

M4: For any t, given any $\boldsymbol{z} \in C([a,b])$,

$$\begin{aligned} |\boldsymbol{x}(t) + \boldsymbol{y}(t)| &\leq |\boldsymbol{x}(t)| + |\boldsymbol{y}(t)| \\ &\leq \|\boldsymbol{x}\|_\infty + \|\boldsymbol{y}\| \end{aligned}$$

This implies $N4$ is satisfied.

$B(\ (a,b)\ )$**:** Here $\Omega = (a,b)$ for a finite interval. The definition is

$$B(\ (a,b)\ ) \quad = \quad \{\boldsymbol{x} : (a,b) \to \Re \mid \boldsymbol{x} \text{ is bounded on } (a,b)\}$$

and define $\|\boldsymbol{x}\|_\infty = \sup_{a \leq t \leq b} |x(t)|$. It is not possible to replace the sup by max here as the maximum need not be attained. The definition of $B([a,b])$ is similar. Also, note for

$$C(\ (a,b)\ ) \quad = \quad \{\boldsymbol{x} : (a,b) \to \Re \mid \boldsymbol{x} \text{ is continuous on } (a,b)\}$$

$\|\boldsymbol{x}\|_\infty$ need not be finite so it cannot serve as a norm. Proving $\|\cdot\|_\infty$ is a norm on $B((a,b))$ is essentially the argument we used in the case of $C([a,b])$ so we won't repeat it.

$RI(\Omega)$**:** For $\Omega = [a,b]$ where $[a,b]$ is a finite interval, we define

$$RI(\ (a,b)\ ) \quad = \quad \{\boldsymbol{x} \in B([a,b]) \mid \boldsymbol{x} \text{ is Riemann Integrable on } [a,b]\}$$

and define $\|\boldsymbol{x}\|_1 = \int_a^b |\boldsymbol{x}| dt$ which is well-defined as $\boldsymbol{x}$ is Riemann Integrable.

N1: The integral of a non-negative function is non-negative.

N2: If $\boldsymbol{x} = \boldsymbol{y}$, we see $\|\boldsymbol{x}\|_1 = 0$. But if $\|\boldsymbol{x}\|_1 = 0$, then we can't say the two functions match as they can differ on a set of content zero. See our discussion of the completion of Riemann Integrable functions in Chapter 7.4 that explains more about this case. Hence $N2$ fails here.

N3: It is clear $\|\alpha\boldsymbol{x}\|_1 = |\alpha|\ \|\boldsymbol{x}\|_1$.

N4: For any t, $|\boldsymbol{x}(t) + \boldsymbol{y}(t)| \leq |\boldsymbol{x}(t)| + |\boldsymbol{y}(t)|$. This shows $\|\boldsymbol{x} + \boldsymbol{y}\|_1 \leq \|\boldsymbol{x}\|_1 + \|\boldsymbol{y}\|_1$.

Thus $(RI([a,b]), d_1)$ is **not** a normed space. However, $(C([a,b]), d_1)$ **is** a normed space as property $N2$ is satisfied.

N1: The integral of a non-negative function is non-negative.

N2: If $\boldsymbol{x} = \boldsymbol{0}$, we see $\|\boldsymbol{x}\|_1 = 0$. If $\|\boldsymbol{x}\|_1 = 0$, by the same argument we used in showing d_1 is a metric on $C([a,b])$, we have $\boldsymbol{x} = \boldsymbol{0}$.

N3: It is clear $\|\boldsymbol{\alpha x}\|_1 = |\alpha|\ \|\boldsymbol{x}\|_1$.

M4: The usual arguments show this is true also.

Thus $(C([a,b])([a,b]), d_1)$ is a normed space.

Homework

Exercise 5.1.18 *Prove* $(B([a,b]), \|\cdot\|_\infty)$ *is a normed space.*

Exercise 5.1.19 *Let* $[a,b]$ *be a finite interval*

$$C^1([a,b]) \quad = \quad \{\boldsymbol{x} \in C([a,b]) \textit{ and } \boldsymbol{x}' \in C([a,b])\}$$

and define $\|\boldsymbol{x}\| = \|\boldsymbol{x}\|_\infty + \|\boldsymbol{x}'\|_\infty$. *Prove* $(C^1([a,b]), \|\cdot\|)$ *is a normed space.*

Exercise 5.1.20 *Let $[a,b]$ be a finite interval*

$$C^1([a,b]) = \{\boldsymbol{x} \in C([a,b]) \text{ and } \boldsymbol{x}' \in C([a,b])\}$$

and define $\|\boldsymbol{x}\| = \max\{\|\boldsymbol{x}\|_\infty, \|\boldsymbol{x}'\|_\infty\}$. Prove $(C^1([a,b]), \|\cdot\|)$ is a normed space.

Exercise 5.1.21 *Let $[a,b]$ be a finite interval*

$$C^1([a,b]) = \{\boldsymbol{x} \in C([a,b]) \text{ and } \boldsymbol{x}' \in C([a,b])\}$$

and define $\|\boldsymbol{x}\| - 2 = \sqrt{(\|\boldsymbol{x}\|_\infty^2 + \|\boldsymbol{x}'\|_\infty)^2}$. Prove $(C^1([a,b]), \|\cdot\|)$ is a normed space.

Exercise 5.1.22 *Let $[a,b]$ be a finite interval*

$$X([a,b]) = \{\boldsymbol{x} \in C([a,b]) \text{ and } \boldsymbol{x}' \text{ exists on } [a,b]\}$$

and define $\|\boldsymbol{x}\| = \|\boldsymbol{x}\|_\infty + \|\boldsymbol{x}'\|_\infty$. Is $(X([a,b]), \|\cdot\|)$ is a normed space?

Exercise 5.1.23 *Note all of the work above can be done with virtually no change if the sequences are complex-valued. Simply use the complex modulus instead of the absolute value.*

A metric does not have to be induced by a norm. Let S be the set of all real or complex-valued sequences. Then d defined by

$$d(\boldsymbol{x}, \boldsymbol{y}) = \sum_{i=1}^{\infty} \frac{1}{2^i} \frac{|x_i - y_i|}{1 + |x_i - y_i|}$$

is a metric on S. We have discussed this metric in the context of symbol spaces and we won't repeat these arguments here. Note this metric works by non linearly scaling the contributions for each component of the sequences. If a metric is induced by a norm, we can prove this result.

Theorem 5.1.2 Metrics Induced by a Norm are Translation Invariant and Scale Linearly

Let $(X, \|\cdot\|)$ be a normed linear space and let (X, d) be the induced metric space. Then d satisfies a translation invariance and linear scaling property.

Translation Invariance: $d(x+z, y+z) = d(x,y)$

Linear Scaling: $d(\alpha x, \alpha y) = |\alpha|\, d(x,y)$

Proof 5.1.2

Translation Invariance:

$$d(x+z, y+z) = \|(x+z) - (y+z)\| = \|x-y\| = d(x,y)$$

Linear Scaling:

$$d(\alpha x, \alpha y) = \|\alpha x - \alpha y\| = |\alpha|\, \|x-y\| = d(x,y)$$

■

In our example, let $\boldsymbol{x} = (1,1,0,0,\ldots)$ and $\boldsymbol{y} = (2,2,0,0,\ldots)$. Then

$$\begin{aligned} d(\boldsymbol{x},\boldsymbol{0}) &= \frac{1}{2}\frac{1}{1+1} + \frac{1}{4}\frac{1}{1+1} = \frac{3}{8} \\ d(2\boldsymbol{x},2\boldsymbol{0}) &= d(\boldsymbol{y},\boldsymbol{0}) = \frac{1}{2}\frac{2}{1+2} + \frac{1}{4}\frac{2}{1+2} = \frac{1}{6} + \frac{1}{3} = \frac{1}{2} \\ 2d(\boldsymbol{x},\boldsymbol{0}) &= 2\frac{3}{8} \neq d(2\boldsymbol{x},2\boldsymbol{0}) \end{aligned}$$

So this metric cannot be induced by any norm.

Homework

Exercise 5.1.24 *For any $\boldsymbol{x}$ which is a real-valued sequence $d(\boldsymbol{x},\boldsymbol{y}) = \sum_{i=1}^{\infty} \frac{1}{4^i}\frac{|x_i - y_i|}{1+|x_i-y_i|}$ find an example which shows this metric cannot be induced by a norm.*

Exercise 5.1.25 *For any $\boldsymbol{x}$ which is a real-valued sequence $d(\boldsymbol{x},\boldsymbol{y}) = \sum_{i=1}^{\infty} \frac{1}{8^i}\frac{|x_i - y_i|}{1+|x_i-y_i|}$ find an example which shows this metric cannot be induced by a norm.*

Exercise 5.1.26 *Let f and g be any two bounded functions on the interval $[a,b]$. Define $d(f,g) = \frac{\|f-g\|_\infty}{1+\|f-g\|_\infty}$. Prove d is a metric on the space of bounded functions $B([a,b])$ which is not induced by a norm.*

Exercise 5.1.27 *Let f and g be any differentiable functions on $[a,b]$. We do not assume the derivative is continuous but we do assume it exists everywhere so each derivative is a bounded function. Define $d(f,g) = \frac{\|f-g\|_\infty}{1+\|f-g\|_\infty} + \frac{\|f'-g'\|_\infty}{1+\|f'-g'\|_\infty}$. Prove d is a metric on this space which is not induced by a norm.*

5.2 The Schauder Basis

Although all vector spaces X possess a Hamel Basis B, the Hamel basis is not very useful. Given $x \in X$, it is not usually possible to construct the appropriate scalars α_i in the **known to exist** representation $x = \sum_{i=1}^{n} \alpha_i x_i$ for a finite set $\{x_1,\ldots,x_n\}$ from the Hamel Basis B. We can come up with a better notion of a special set like a **basis** if we have additional structure on the vector space X. If we have a norm, we can talk about the existence of countable dense subsets. Suppose M in X was a dense subset. Then given $x \in X$ and $\epsilon > 0$ there is $m \in M$ so that $\|m - x\| < \epsilon$ where we use the induced metric d from $\|\cdot\|$ to compute the distance. So for the sequence of tolerances $\epsilon = \frac{1}{n}$, we could construct a sequence (m_n) in M so that $\|m_n - x\| < \frac{1}{n}$ which clearly tells us $\|m_n - x\| \to 0$. However, we still cannot calculate the elements m_n in general. Also, M need not be countable.

What if M was countable and dense? This is the same as saying X is a separable normed space with the norm $\|\cdot\|$. Then we have $M = (m_n)_{n=1}^{\infty}$ and given $x \in X$, there is a sequence $(m_{n_k})_{k=1}^{\infty}$ so that $\|m_{n_k} - x\| \to x$. But we still do not know how to find the elements m_{n_k}. Let's change our attack. Take a countable set $M = (m_n)_{n=1}^{\infty}$ which is linearly independent so that E, the collection of all finite linear combinations of elements from M, is dense in X. Then, given $x \in X$, there is a sequence $(e_n)_{n=1}^{\infty}$ in E with $\|e_n - x\| < \frac{1}{n}$. But $e_n = \sum_{j=1}^{p_n} \alpha_j^{n_j} m_{n_j}$ for some positive integer p_n, scalars $\{\alpha_1^{n_j},\ldots,\alpha_{p_n}^{n_{p_n}}\}$ and elements $\{m_{n_1},\ldots,m_{n_{p_n}}\}$ in M. Clearly, $\|e_n - x\| \to 0$. Note by adding zero scalars as needed, we can rewrite e_n as $e_n = \sum_{j=1}^{p_n} \alpha_j^n m_j$ and although

$$\lim_{n\to\infty} e_n = \lim_{n\to} \sum_{j=1}^{p_n} \alpha_j^n m_j$$

converges in norm, this **is not** the same as the usual partial sum notation for a series:

$$x = \lim_{n \to \infty} \sum_{j=1}^{n} \beta_j \, m_j$$

because the scalars α_j^n depend on n too.

However, we would still like to find a special set $M = (m_n)_{n=1}^{\infty}$ so that given $x \in X$, there is a unique sequence of scalars $(\beta_j)_{j=1|^{\infty}}$ with $x = \lim_{n \to \infty} \sum_{j=1}^{n} \beta_j \, m_j$. It is now clear that just having a countable dense subset M of X is not enough. The special set we are looking for is called a **Schauder Basis** for $(X, \|\cdot\|)$.

Definition 5.2.1 A Schauder Basis for a normed linear space

> *Let $(X, \|\cdot\|)$ be a non empty normed linear space. Let $M = (m_n)_{n=1}^{\infty}$ be a countable subset of X so that given $x \in X$, there is a unique sequence of scalars $(\alpha_n)_{n=1}^{\infty}$ so that $x = \lim_{n \to \infty} \sum_{j=1}^{n} \alpha_j \, m_j$. This says $x = \sum_{j=1}^{\infty} \alpha_j m_j$ is an infinite series in X whose partial sums converge in norm to x. This set M is called a Schauder Basis for X.*

Comment 5.2.1 *Given a finite subset of M, we can assume this is of the form $\{m_1, \ldots, m_p\}$ for some p. The linear independence equation gives*

$$c_1 m_1 + \ldots c_p m_p = 0$$

If $\{m_1, \ldots, m_p\}$ was a linearly dependent set, then there would be some m_q in the span of the other elements. But $m_q = 1 m_q$ is already the expansion of the element m_q in X with respect to the Schauder basis. So linear dependence would say m_1 has a non unique expansion which is not true. So M must be a linearly independent set.

Comment 5.2.2 *If E is the set of all finite linear combinations of the Schauder Basis M with rational coefficients then E is countable and dense in X. So if X has a Schauder Basis, it must be separable. You will prove this in the exercises.*

Comment 5.2.3 *Since ℓ^{∞} is* **not** *separable, it cannot have a Schauder Basis even though it does have a Hamel Basis.*

Homework

Exercise 5.2.1 *If the normed linear space $(X, \|\cdot\|)$ has a Schauder Basis it must be separable.*

Exercise 5.2.2 *Let $\mathbf{E}$ be any basis for $\Re^n$. Is $\mathbf{E}$ a Schauder Basis?*

Exercise 5.2.3 *Let $\mathbf{E}$ be any basis for the set of $n \times n$ real-valued matrices. Is $\mathbf{E}$ a Schauder Basis?*

Exercise 5.2.4 *Let M be the set of all powers of t on $[0, 1]$ starting with t^0. Is M a Schauder Basis for $C([0, 1])$?*

Exercise 5.2.5 *This uses ideas about Fourier Series from (Peterson (18) 2020). If M is the set of all functions of the form $\sin(n\pi t)$ on $[0, 1]$ for $n \geq 1$ an integer, is M a Schauder Basis for $C([0, 1])$?*

Exercise 5.2.6 *What is the difference between a Schauder Basis and a basis for the normed linear space $(X, \|\cdot\|)$?*

5.2.1 Schauder Basis Examples

The p spaces for $1 \leq p < \infty$: Let $E = (\boldsymbol{e_n})_{n=1}^{\infty}$ where

$$\boldsymbol{e_n} = (0, \ldots, 0, \underbrace{1}_{\text{slot } n}, 0, \ldots)$$

Then $\boldsymbol{e_n} \in \ell^p$ for all n. Given $\boldsymbol{x} = (x_n)_{n=1}^{\infty}$ in ℓ^p, we know $\sum_{j=1}^{\infty} |x_j|^p < \infty$. Let $\boldsymbol{S_n} = \sum_{j=1}^{n} x_j \boldsymbol{e_j}$ Then

$$\begin{aligned} \|\boldsymbol{S_n} - \boldsymbol{x}\|_p^p &= \sum_{j=1}^{\infty} |S_{nj} - x_j|^p = \sum_{j=1}^{n} |x_j - x_j|^p + \sum_{j=n+1}^{\infty} |0 - x_j|^p \\ &= \sum_{j=n+1}^{\infty} |x_j|^p \end{aligned}$$

Since $\boldsymbol{x} \in \ell^p$, given $\epsilon > 0$, there is N so that $n > N$ implies $\sum_{j=n+1}^{\infty} |x_j|^p < \left(\frac{\epsilon}{2}\right)^p$. In particular, we see for any $n > N$,

$$\|\boldsymbol{S_n} - \boldsymbol{x}\|_p^p = \sum_{j=n+1}^{\infty} |x_j|^p < \left(\frac{\epsilon}{2}\right)^p$$

which tells us $\|\boldsymbol{S_n} - \boldsymbol{x}\|_p < \frac{\epsilon}{2}$ when $n > N$. Hence, $\boldsymbol{S_n} \to \boldsymbol{x}$ in $\|\cdot\|_p$ norm. Therefore, there is a sequence of scalars, $(\alpha_j)_{j=1}^{\infty}$ with $\alpha_j = x_j$ so that $\sum_{j=1}^{\infty} \alpha_j \boldsymbol{e_j} = \boldsymbol{x}$.

Is this sequence of scalars unique? Let $\boldsymbol{T_n} = \sum_{j=1}^{n} \beta_j \boldsymbol{e_j}$ and $\boldsymbol{S_n} = \sum_{j=1}^{n} \alpha_j \boldsymbol{e_j}$ and assume $\boldsymbol{S_n} \to \boldsymbol{x}$ and $\boldsymbol{T_n} \to \boldsymbol{x}$ in $\|\cdot\|_p$ norm. Since $\boldsymbol{T_n} \to \boldsymbol{x}$, given $\epsilon > 0$, there is N so that $n > N$ implies

$$\|\boldsymbol{T_n} - \boldsymbol{x}\|_p^p < \epsilon^p \Longrightarrow \sum_{j=1}^{n} |\beta_j - x_j|^p + \sum_{j=n+1}^{\infty} |x_j|^p < \epsilon^p$$

Thus, if $n > N$

$$\sum_{j=1}^{n} |\beta_j - x_j|^p < \epsilon^p \Longrightarrow |\beta_j - x_j|^p < \epsilon^p, \; 1 \leq j \leq n$$

We see then that $|\beta_j - x_j| < \epsilon$ for all j implying $\beta_j = x_i$ always. Hence, the sequence of scalars is unique and E is a Schauder Basis for $(\ell^p, \|\cdot\|_p)$ and every $\boldsymbol{x} \in \ell^p$ has representation $\boldsymbol{x} = \sum_{j=1}^{\infty} x_j \boldsymbol{e_j}$ where $\boldsymbol{x} = (x_j)$.

The set of all sequences that converge: Let $\boldsymbol{c}$ denote the set of all real or complex-valued sequences that converge with the norm $\|\boldsymbol{x}\| = \sup_j |x_j|$ for $\boldsymbol{x} \in \boldsymbol{c}$ with $\boldsymbol{x} = (x_j)$. Note we know (x_j) converges so it is bounded and it is easy to see this is a norm for this space. let x_∞ denote the limit value of this sequence. Let $E = (\boldsymbol{e_n})_{n=0}^{\infty}$ where

$$\begin{aligned} \boldsymbol{e_0} &= (1, 1, 1, \ldots), \text{ all ones} \\ \boldsymbol{e_n} &= (0, \ldots, 0, \underbrace{1}_{\text{slot } n}, 0, \ldots) \end{aligned}$$

We will show E is a Schauder Basis for $\boldsymbol{c}$ with this norm and any $\boldsymbol{x} = (x_j) \in c$ has representation $\boldsymbol{x} = x_\infty \boldsymbol{e_0} + \sum_{j=1}^{\infty}(x_n - x_\infty)\boldsymbol{e_n}$. Let $\boldsymbol{S_n} = x_\infty \boldsymbol{e_0} + \sum_{j=1}^{n} x_j \boldsymbol{e_j}$ Then

$$\begin{aligned}
\|\boldsymbol{S_n} - \boldsymbol{x}\|_\infty &= \sup_j |S_{nj} - x_j| \\
&= \sup_j |((x_\infty + x_1 - x_\infty), (x_\infty + x_2 - x_\infty), \ldots, (x_\infty + x_n - x_\infty), x_\infty, \ldots)) \\
&\quad - (x_1, x_2, \ldots, x_n, \ldots))| \\
&= \sup_j((x_1 - x_1), (x_2 - x_2), \ldots, (x_n - x_n), x_\infty - x_{N+1}, x_\infty - x_{N+2}, \ldots)| \\
&= \sup_j |(\underbrace{0, \ldots, 0}_{n \text{ slots}}, x_\infty - x_{N+1}, \ldots)| = \sup_{j \geq N} |x_\infty - x_j|
\end{aligned}$$

But $x_n \to x_\infty$, so given $\epsilon > 0$, there is N so if $n > N$, $|x_\infty - x_j| < \frac{\epsilon}{2}$. Hence, we have $\|\boldsymbol{S_n} - \boldsymbol{x}\|_\infty \leq \frac{\epsilon}{2} < \epsilon$ when $n > N$. Thus, $\boldsymbol{S_n} \to \boldsymbol{x}$ in this norm and the representation of $\boldsymbol{x} = x_\infty \boldsymbol{e_0} + \sum_{j=1}^{\infty}(x_n - x_\infty)\boldsymbol{e_n}$.

Is the representation unique? Assume

$$\boldsymbol{x} \quad = \quad x_\infty \boldsymbol{e_0} + \sum_{j=1}^{\infty}(x_n - x_\infty)\boldsymbol{e_n} = \alpha_0 \boldsymbol{e_0} + \sum_{j=1}^{\infty} \alpha_n \boldsymbol{e_n}$$

Now let $\boldsymbol{T_n} = \alpha_0 \boldsymbol{e_0} + \sum_{j=1}^{n} \alpha_n \boldsymbol{e_n}$. By assumption $\boldsymbol{S_n} - \boldsymbol{T_n} \to 0$ and

$$\begin{aligned}
\boldsymbol{S_n} - \boldsymbol{T_n} &= ((x_\infty + x_1 - x_\infty), (x_\infty + x_2 - x_\infty), \ldots, (x_\infty + x_n - x_\infty), x_\infty, \ldots)) \\
&\quad -(\alpha_0 + \alpha_1, \ldots, \alpha_0 + \alpha_n, \alpha_0, \ldots) \\
&= (x_1 - \alpha_0 - \alpha_1, \ldots, x_n - \alpha_0 - \alpha_n, x_\infty - \alpha_0, \ldots)
\end{aligned}$$

and so

$$\|\boldsymbol{S_n} - \boldsymbol{T_n}\|_\infty \quad = \quad \sup_j(\sup_{1 \leq j \leq n} |x_j - \alpha_0 - \alpha_j|, |x_\infty - \alpha_0|)$$

Since $\boldsymbol{S_n} - \boldsymbol{T_n} \to 0$, given $\epsilon > 0$, there is N so that $\|\boldsymbol{S_n} - \boldsymbol{T_n}\|_\infty < \epsilon$ if $n > N$. Therefore

$$\sup_j(\sup_{1 \leq j \leq n} |x_j - \alpha_0 - \alpha_j|, |x_\infty - \alpha_0|) \quad < \quad \epsilon$$

when $n > N$. This tells us $|x_\infty - \alpha_0| < \epsilon$ and since ϵ is arbitrary, we have $x_\infty = \alpha_0$. Using this we have

$$\sup_{1 \leq j \leq n} |x_j - x_\infty - \alpha_j| < \epsilon$$

for $n > N$. Hence, $|x_j - x_\infty - \alpha_j| < \epsilon$ for all j implying $\alpha_j = x_j - x_\infty$ always. Hence the scalars are unique and E is a Schauder Basis for $\boldsymbol{c}$ with the $\| \cdot \|_\infty$ norm.

Homework

Exercise 5.2.7 *Find a Schauder Basis for the set of all sequences that converge to zero. Note the* $\boldsymbol{e_0}$ *element is no longer needed. This space is called* c_0.

Exercise 5.2.8 *Find a Schauder Basis for* $\ell^p \times \ell^p$ *for* $1 \leq p < \infty$.

Exercise 5.2.9 *Find a Schauder Basis for* $c \times c$.

Exercise 5.2.10 *Find a Schauder Basis for $c_0 \times c_0$.*

5.3 The Linear Combination Theorem

In a normed linear space $(X, \|\cdot\|)$ there is much more structure to work with. A consequence of the underlying vector space structure is the following theorem.

Theorem The Linear Combination Theorem (LCT)

Let $(X, \|\cdot\|)$ be a normed linear space and let $\{\boldsymbol{x}_1, \ldots, \boldsymbol{x}_n\}$ be a finite linearly independent subset of X. Then there is a positive constant $\mathbf{c}$ so that

$$\left\|\sum_{i=1}^{n} \alpha_i \boldsymbol{x}_i\right\| \geq \mathbf{c} \sum_{i=1}^{n} |\alpha_i|$$

for all choices of scalars $\{\alpha_1, \ldots, \alpha_n\}$. This results holds where the underlying field is real or the complex.

This theorem has **profound** consequences. First, all finite dimensional normed linear spaces are like $\Re^n$: they are complete.

Theorem Finite Dimensional Normed Linear Spaces Must Be Complete

Let $(X, \|\cdot\|)$ be a finite dimensional normed linear space. Then X must be complete with respect to the metric induced by the norm.

Comment *If $(X, \|\cdot\|)$ is any normed linear space, perhaps infinite dimensional, let Y be a finite dimensional subspace of X. We can make Y into a normed linear space too by simply using the norm $\|\cdot\|$ for Y. Hence, Y must be complete.*

The comment above leads to the next theorem.

Theorem Finite Dimensional Subspaces of a Normed Linear Space Must Be Closed

Let $(X, \|\cdot\|)$ be a normed linear space and let Y be a finite dimensional subspace. Then $(Y, \|\cdot\|)$ is complete with respect to the metric induced by the norm and is therefore a closed subspace. That is, if $(\boldsymbol{y}_n)$ is a norm convergent sequence in Y, then the limit point $\boldsymbol{y}$ is in Y which tells us Y is closed.

This all leads to the fundamental result that it does not matter which norm you choose to use in a finite dimensional normed linear space: different norms generate equivalent topologies and convergence.

Theorem All Norms on a Finite Dimensional Normed Linear Space are Equivalent

Let $(X, \|\cdot\|_1)$ and $(X, \|\cdot\|_2)$ be two finite dimensional normed linear spaces that share the underlying vector space X. Then $\|\cdot\|_1$ and $\|\cdot\|_2$ are equivalent: i.e. there are positive constants α and β so that for all $\boldsymbol{x} \in X$,

$$\alpha \|\boldsymbol{x}\|_1 \leq \|\boldsymbol{x}\|_2, \quad \beta \|\boldsymbol{x}\|_2 \leq \|\boldsymbol{x}\|_1$$

Comment *This theorem says convergent sequences with respect to one norm also converge with respect to the other norm. Also, open balls in one norm are scaled versions of open balls in another norm.*

Homework

Exercise 5.3.1 *Let $\overline{B_1^1(\mathbf{0})}$ be the closed unit ball in $(C([0,1]), \|\cdot\|_1)$ and $\overline{B_1^\infty(\mathbf{0})}$ be the closed unit ball in $(C([0,1]), \|\cdot\|_\infty)$.*

- *If $\boldsymbol{f} \in \overline{B_1^1(\mathbf{0})}$ explain why $\boldsymbol{f}$ can have arbitrarily large infinity norm. Hence there is no radius R so that $\overline{B_1^1(\mathbf{0})} \subset \overline{B_R^1(\mathbf{0})}$ where $\overline{B_R^1(\mathbf{0})}$ denotes the closed ball of radius R. It is easy to see this if you look at the appropriate C^∞ bumps as discussed in (Peterson (18) 2020).*
- *Prove $\boldsymbol{f} \in \overline{B_1^\infty(\mathbf{0})} \subset \overline{B_1^1(\mathbf{0})}$.*

Exercise 5.3.2 *If $\boldsymbol{x} \in \Re^n$, use Hölder's Inequality to show $\|\boldsymbol{x}\|_1 \leq \sqrt{n}\|\boldsymbol{x}\|_2$.*

Exercise 5.3.3 *If $\boldsymbol{x} \in \Re^n$, use Hölder's Inequality to show $\|\boldsymbol{x}\|_1 \leq n\|\boldsymbol{x}\|_\infty$.*

Exercise 5.3.4 *If $\boldsymbol{x} \in C([0,1])$, use Hölder's Inequality to show $\|\boldsymbol{x}\|_1 \leq \|\boldsymbol{x}\|_2$.*

Exercise 5.3.5 *Consider the normed linear space $(C([0,1]), \|\cdot\|_\infty)$ and the sequence of functions $(\boldsymbol{x_n})$ given by*

$$x_n(t) = \begin{cases} 0, & 0 \leq t \leq \frac{1}{2} \\ n(t-\frac{1}{2}), \frac{1}{2} < t < \frac{1}{2}+\frac{1}{n} & \\ 1, & \frac{1}{2}+\frac{1}{n} \leq t \leq 1 \end{cases}$$

The sequence $(\boldsymbol{x_n})$ is contained in the closed unit ball in $C([0,1])$ with respect to the norm $\|\cdot\|_\infty$.

- *Prove the limit of this sequence with respect to this norm does not exist.*
- *Prove the limit of this sequence with respect to the norm $\|\cdot\|_1$ does exist as a Riemann Integrable function but the limit is not continuous.*
- *Prove that $\overline{B_1^\infty(\mathbf{0})}$ is a closed subset in $(C([0,1]), \|\cdot\|_\infty)$.*

Exercise 5.3.6 *Consider the normed linear space $(RI([0,1]), \|\cdot\|_1)$ and the sequence of functions $(\boldsymbol{x_n})$ given by*

$$x_n(t) = \begin{cases} 0, & 0 \leq t \leq \frac{1}{2} \\ n(t-\frac{1}{2}), \frac{1}{2} < t < \frac{1}{2}+\frac{1}{n} & \\ 1, & \frac{1}{2}+\frac{1}{n} \leq t \leq 1 \end{cases}$$

Prove the limit of this sequence with respect to the $\|\cdot\|_1$ norm exists but it is not in $\overline{B_1^\infty(\mathbf{0})}$. Hence, the set $\overline{B_1^\infty(\mathbf{0})}$ in the normed linear space $(C([0,1]), \|\cdot\|_1)$ is not closed with respect to the $\|\cdot\|_1$ norm.

More succinctly, letting **FDNLS** stand for **Finite Dimensional Normed Linear Space**, we have

$$\textbf{Linear Combination Theorem} \quad \Longrightarrow \quad \begin{cases} \textbf{FDNLS complete} \\ \textbf{Finite Dimensional Subspaces of a NLS are complete} \\ \textbf{Finite Dimensional Subspaces of a NLS are closed} \\ \textbf{All Norms on FDNLS are equivalent} \end{cases}$$

Let's assume the LCT is true. Using it we can complete the proofs of the other conjectures.

Theorem 5.3.1 Finite Dimensional Normed Linear Spaces Must Be Complete

Let $(X, \|\cdot\|)$ be a finite dimensional normed linear space. Then X must be complete with respect to the metric induced by the norm.

Proof 5.3.1

Since X is finite dimensional, there is a basis $\boldsymbol{E} = \{\boldsymbol{e_1}, \ldots, \boldsymbol{e_n}\}$ for some positive integer n. Let $(\boldsymbol{x_p})$ be a Cauchy sequence in X. Then each $\boldsymbol{x_p}$ has representation

$$\boldsymbol{x_p} = \sum_{i=1}^{n} \alpha_{pi}\boldsymbol{e_i}$$

that is unique. Apply the LCT to this linearly independent set. This implies there is a $\mathbf{c} > 0$ so that for each p

$$||\boldsymbol{x_p}|| = \left\|\sum_{i=1}^{n} \alpha_{pi}\boldsymbol{e_i}\right\| \geq \mathbf{c}\sum_{i=1}^{n} |\alpha_{pi}|$$

Since $(\boldsymbol{x_p})$ is a Cauchy sequence, given $\epsilon > 0$, there is N so that $\|\boldsymbol{x_p} - \boldsymbol{x_q}\| < \epsilon\mathbf{c}$ when $p, q > N$. Thus,

$$\begin{aligned} \epsilon\mathbf{c} &> \|\boldsymbol{x_p} - \boldsymbol{x_q}\| = \left\|\sum_{i=1}^{n} \alpha_{pi}\boldsymbol{e_i} - \sum_{i=1}^{n} \alpha_{qi}\boldsymbol{e_i}\right\| \\ &= \left\|\sum_{i=1}^{n}(\alpha_{pi} - \alpha_{qi})\boldsymbol{e_i}\right\| \geq \mathbf{c}\sum_{i=1}^{n} |\alpha_{pi} - \alpha_{qi}| \end{aligned}$$

This implies $\sum_{i=1}^{n} |\alpha_{pi} - \alpha_{qi}| < \epsilon$ when $p, q > N$. It then follows that $|\alpha_{pi} - \alpha_{qi}| < \epsilon$ when $p, q > N$. This tells us the sequences (α_{pi}) for fixed i are Cauchy sequences of real numbers. Since $\Re$ is complete, there are numbers α_i so that $\alpha_{pi} \rightarrow \alpha_i$. Define $\boldsymbol{x} = \sum_{i=1}^{n} \alpha_i\boldsymbol{e_i} \in X$. Then

$$\|\boldsymbol{x_p} - \boldsymbol{x}\| = \left\|\sum_{i=1}^{n}(\alpha_{pi} - \alpha_i)\boldsymbol{e_i}\right\| \leq \sum_{i=1}^{n} |\alpha_{pi} - \alpha_i|\ \|\boldsymbol{e_i}\|$$

Let $B = \max|\{\|\boldsymbol{e_1}\|, \ldots, \boldsymbol{e_n}\|\}$. Then $\|\boldsymbol{x_p} - \boldsymbol{x}\| \leq B\sum_{i=1}^{n} |\alpha_{pi} - \alpha_i|$. Since $\alpha_{pi} \rightarrow \alpha_i$, for $\epsilon > 0$, there are N_i so that $|\alpha_{pi} - \alpha_i| < \frac{\epsilon}{nB}$ for $p > N_i$. So for $n > N = \max\{N_1, \ldots, N_n\}$, all these estimates hold and

$$\|\boldsymbol{x_p} - \boldsymbol{x}\| \leq B\sum_{i=1}^{n} \frac{\epsilon}{nB} = \epsilon$$

for $n > N$. We see $\boldsymbol{x_p} \rightarrow \boldsymbol{x}$ and X is complete. ■

This result allows us to prove

Theorem 5.3.2 Finite Dimensional Subspaces of a Normed Linear Space Must be Closed

Let $(X, \|\cdot\|)$ be a normed linear space and let Y be a finite dimensional subspace. Then $(Y, \|\cdot\|)$ is complete with respect to the metric induced by the norm and is therefore a closed subspace. That is, if $(\boldsymbol{y_n})$ is a norm convergent sequence in Y, then the limit point $\boldsymbol{y}$ is in Y which tells us Y is closed.

Proof 5.3.2

Let $(\boldsymbol{y_n})$ be a convergent sequence in Y with respect to the norm. Then there is $\boldsymbol{x} \in X$ so that $\boldsymbol{y_n} \rightarrow \boldsymbol{x}$. But a convergent sequence is a Cauchy sequence and since Y is finite dimensional, $(Y, \|\cdot\|)$ is a complete normed linear space. Thus, there is $\boldsymbol{y} \in Y$ so that $\boldsymbol{x_n} \rightarrow \boldsymbol{y}$. But limits must *be unique, so $\boldsymbol{x} = \boldsymbol{y}$. We conclude the limit is in Y and Y is closed.* ∎

There are many examples of subspaces that are not closed.

- If Y is the set of sequences in ℓ^∞ where $\boldsymbol{y}$ has only a finite number of nonzero entries, it is easy to see Y is a vector subspace of ℓ^∞. Let $\boldsymbol{y_n} = (1, \frac{1}{2}, \frac{1}{3}, \ldots, \frac{1}{n}, 0, 0, \ldots)$. Each $\boldsymbol{y_n} \in Y$. Let $\boldsymbol{y} = (1/n) \in \ell^\infty$. Then $\|\boldsymbol{y_n} - \boldsymbol{y}\|_\infty = \sup_{i>n} \left|\frac{1}{i}\right| = \frac{1}{n+1}$. Clearly, this implies $\boldsymbol{y_n} \to \boldsymbol{y}$ yet $\boldsymbol{y} \notin Y$ as it does not have a finite number of nonzero terms. So Y is not closed.

- Let Y be the sequences in ℓ^2 which only have a finite number of nonzero terms. Then Y is a subspace of ℓ^2. Let $\boldsymbol{y_n} = (1, \frac{1}{2}, \frac{1}{3}, \ldots, \frac{1}{n}, 0, 0, \ldots)$. Each $\boldsymbol{y_n} \in Y$. Let $\boldsymbol{y} = (1/n) \in \ell^2$. Then

$$\|\boldsymbol{y_n} - \boldsymbol{y}\|_2^2 = \sum_{i=1}^{\infty} |y_{ni} - y_i|^2 = \sum_{i=n+1}^{\infty} \frac{1}{i^2}$$

  Clearly, this implies $\boldsymbol{y_n} \to \boldsymbol{y}$ yet $\boldsymbol{y} \notin Y$ as it does not have a finite number of nonzero terms. So Y is not closed.

- Let Y be the set of all polynomials in $C([-1, 1])$. Then Y is a subspace. Let $y_n(t) = 1 + t + \frac{t^2}{2!} + \cdots + \frac{t^n}{n!}$. Then y_n is a partial sum of the Taylor series expansion for e^t on $[-1, 1]$ which converges everywhere. So $y_n(t) \to e^t$ pointwise on $[-1, 1]$. Rewrite this in terms of the remainder to get

$$\begin{aligned} \|y_n - y\|_\infty &= \sup_{-1 \le t \le 1} \left| \sum_{i=n+1}^{\infty} \frac{t^i}{i!} \right| \\ &= \sup_{-1 \le t \le 1} \frac{e^{d_t}}{(n+1)!} |t|^{n+1} \end{aligned}$$

  for some d_t between -1 and 1. The maximum here is always achieved at e^1. So

$$\|y_n - y\|_\infty \le \frac{e}{(n+1)!} \sup_{-1 \le t \le 1} |t|^{n+1} = \frac{e}{(n+1)!}$$

  This tells us $y_n \to y$ in $\|\cdot\|_\infty$ norm. But $y(t) = e^t$ is not a polynomial and so it is not in Y. Hence, Y is not closed.

Now we show all norms on a finite dimensional normed linear space are equivalent.

Theorem 5.3.3 All Norms on a Finite Dimensional Normed Linear Space are Equivalent

Let $(X, \|\cdot\|_1)$ and $(X, \|\cdot\|_2)$ be two finite dimensional normed linear spaces that share the underlying vector space X. Then $\|\cdot\|_1$ and $\|\cdot\|_2$ are equivalent: i.e. there are positive constants α and β so that for all $\boldsymbol{x} \in X$,

$$\alpha \|\boldsymbol{x}\|_1 \le \|\boldsymbol{x}\|_2, \quad \beta \|\boldsymbol{x}\|_2 \le \|\boldsymbol{x}\|_1$$

Proof 5.3.3
Let $\|\cdot\|_1$ and $\|\cdot\|_2$ be two norms on X. Let $\boldsymbol{E} = \{\boldsymbol{e_1}, \ldots, \boldsymbol{e_n}\}$ be a basis for X where n is some fixed positive integer. Then $\boldsymbol{x} \in X$ has a representation $\boldsymbol{x} = \sum_{i=1}^n \alpha_i \boldsymbol{e_i}$ and by the LCT there is a

constant $\boldsymbol{c_2}$ *so that*

$$\|\boldsymbol{x}\|_2 = \left\|\sum_{i=1}^{n} \alpha_i \boldsymbol{e_i}\right\|_2 \geq \boldsymbol{c_2} \sum_{i=1}^{n} |\alpha_i|$$

Let $B_1 = \max_i \|\boldsymbol{e_i}\|_1$. *Then*

$$\|\boldsymbol{x}\|_1 = \left\|\sum_{i=1}^{n} \alpha_i \boldsymbol{e_i}\right\|_1 \leq \sum_{i=1}^{n} |\alpha_i| \, \|\boldsymbol{e_i}\|_1 \leq B_1 \sum_{i=1}^{n} |\alpha_i| \leq \frac{B_1}{\boldsymbol{c_2}} \|\boldsymbol{x}\|_2$$

and so $\frac{\boldsymbol{c_2}}{B_1} \|\boldsymbol{x}\|_1 \leq |\boldsymbol{x}\|_2$ *which tells us* $\alpha = \frac{\boldsymbol{c_2}}{B_1}$.

A similar argument tells us what β *is here. Let* $B_2 = \max_i \|\boldsymbol{e_i}\|_2$. *Then*

$$\|\boldsymbol{x}\|_2 = \left\|\sum_{i=1}^{n} \alpha_i \boldsymbol{e_i}\right\|_2 \leq \sum_{i=1}^{n} |\alpha_i| \, \|\boldsymbol{e_i}\|_2 \leq B_2 \sum_{i=1}^{n} |\alpha_i|$$

Then by the LCT again, there is $\boldsymbol{c_1}$ *so that*

$$\|\boldsymbol{x}\|_1 = \left\|\sum_{i=1}^{n} \alpha_i \boldsymbol{e_i}\right\|_1 \geq \boldsymbol{c_1} \sum_{i=1}^{n} |\alpha_i|$$

Using this, we find

$$\|\boldsymbol{x}\|_2 \leq B_2 \sum_{i=1}^{n} |\alpha_i| \leq \frac{B_2}{\boldsymbol{c_1}} \|\boldsymbol{x}\|_1$$

and so $\frac{\boldsymbol{c_1}}{B_2} \|\boldsymbol{x}\|_2 \leq |\boldsymbol{x}\|_1$ *which tells us* $\beta = \frac{\boldsymbol{c_1}}{B_2}$. ■

Homework

Exercise 5.3.7 *Prove the set of solutions of* $u'' + 5u' + 6u = 0$ *is a two dimensional vector subspace in the set of all continuous functions* $C(\Re)$.

- *Prove it is a closed and complete subspace of* $C([0,T])$ *for any* $T > 0$ *with norm* $\|\cdot\|_\infty$.
- *Prove it is a closed and complete subspace of* $C([0,T])$ *for any* $T > 0$ *with norm* $\|\cdot\|_1$.
- *Prove it is a closed and complete subspace of* $C([0,T])$ *for any* $T > 0$ *with norm* $\|\cdot\|_2$.
- *Note the norms we use above will not work on the vector space* $C(\Re)$*!*

Exercise 5.3.8 *Find the general solution to* $u'' + 2u' + u = 0$ *and show these solutions form a two dimensional vector subspace in the set of all continuous functions* $C(\Re)$.

- *Prove it is a closed and complete subspace of* $C([0,T])$ *for any* $T > 0$ *with norm* $\|\cdot\|_\infty$.
- *Prove it is a closed and complete subspace of* $C([0,T])$ *for any* $T > 0$ *with norm* $\|\cdot\|_1$.
- *Prove it is a closed and complete subspace of* $C([0,T])$ *for any* $T > 0$ *with norm* $\|\cdot\|_2$.

Exercise 5.3.9

- *Find the general complex solution to* $u'' + 6u' + 25u = 0$ *and show these solutions form a two dimensional vector subspace over the set of all complex-valued continuous functions.*

- *Find the general real solution to* $u'' + 6u' + 25u = 0$ *and show these solutions form a two dimensional vector subspace over the set of all continuous functions.*
- *For the complex case, the subspace of solutions is a closed and complete subspace of the space all complex-valued continuous functions on* $[0, T]$ *for any* $T > 0$ *with the* $\|\cdot\|_\infty$ *norm.*
- *For the real case, the subspace of solutions is a closed and complete subspace of the space all real-valued continuous functions on* $[0, T]$ *for any* $T > 0$ *with the* $\|\cdot\|_\infty$ *norm.*
- *For the real case, the subspace of solutions is a closed and complete subspace of the space all real-valued continuous functions on* $[0, T]$ *for any* $T > 0$ *with the* $\|\cdot\|_1$ *norm.*
- *For the real case, the subspace of solutions is a closed and complete subspace of the space all real-valued continuous functions on* $[0, T]$ *for any* $T > 0$ *with the* $\|\cdot\|_2$ *norm.*

Exercise 5.3.10 *Find the general solution to*

$$\begin{aligned} x'(t) &= -4\,x(t) + \;y(t) \\ y'(t) &= -5\,x(t) + 2\,y(t) \end{aligned}$$

and show it forms a two dimensional vector subspace in the set of all continuous two dimensional vector functions in $\Re$.

- *show the subspace of solutions is a closed and complete subspace of the space all real-valued continuous two dimensional vector functions on* $[0, T]$ *for any* $T > 0$ *with the* $\|\cdot\|_\infty$ *norm.*
- *show the subspace of solutions is a closed and complete subspace of the space all real-valued continuous two dimensional vector functions on* $[0, T]$ *for any* $T > 0$ *with the* $\|\cdot\|_1$ *norm.*
- *show the subspace of solutions is a closed and complete subspace of the space all real-valued continuous two dimensional vector functions on* $[0, T]$ *for any* $T > 0$ *with the* $\|\cdot\|_2$ *norm.*

Exercise 5.3.11 *Find the general solution to*

$$\begin{aligned} x'(t) &= 5\,x(t) + \;y(t) \\ y'(t) &= -7\,x(t) - 3\,y(t) \end{aligned}$$

and show it forms a two dimensional vector subspace in the set of all continuous two dimensional vector functions in $\Re$.

- *show the subspace of solutions is a closed and complete subspace of the space all real-valued continuous two dimensional vector functions on* $[0, T]$ *for any* $T > 0$ *with the* $\|\cdot\|_\infty$ *norm.*
- *show the subspace of solutions is a closed and complete subspace of the space all real-valued continuous two dimensional vector functions on* $[0, T]$ *for any* $T > 0$ *with the* $\|\cdot\|_1$ *norm.*
- *show the subspace of solutions is a closed and complete subspace of the space all real-valued continuous two dimensional vector functions on* $[0, T]$ *for any* $T > 0$ *with the* $\|\cdot\|_2$ *norm.*

Exercise 5.3.12 *Find the general real and complex vector-valued general solution to*

$$\begin{aligned} x'(t) &= 2\,x(t) + 5\,y(t) \\ y'(t) &= -x(t) + 4\,y(t) \\ x(0) &= 6 \end{aligned}$$

$$y(0) \quad = \quad -1$$

and show they forms a two dimensional vector subspace in the set of all continuous two dimensional complex-valued two dimensional or real-valued continuous two dimensional vector functions.

- *In the complex case, show the subspace of solutions is a closed and complete subspace of the space all complex-valued continuous two dimensional vector functions on $[0, T]$ for any $T > 0$ with the $\|\cdot\|_\infty$ norm.*

- *In the real case, show the subspace of solutions is a closed and complete subspace of the space all real-valued continuous two dimensional vector functions on $[0, T]$ for any $T > 0$ with the $\|\cdot\|_\infty$ norm.*

- *In the real case, show the subspace of solutions is a closed and complete subspace of the space all real-valued continuous two dimensional vector functions on $[0, T]$ for any $T > 0$ with the $\|\cdot\|_1$ norm.*

- *In the real case, show the subspace of solutions is a closed and complete subspace of the space all real-valued continuous two dimensional vector functions on $[0, T]$ for any $T > 0$ with the $\|\cdot\|_2$ norm.*

We are now ready to prove the LCT.

Theorem 5.3.4 The Linear Combination Theorem (LCT)

Let $(X, \|\cdot\|)$ be a normed linear space and let $\{\boldsymbol{x_1}, \ldots, \boldsymbol{x_n}\}$ be a finite linearly independent subset of X. Then there is a positive constant $\mathbf{c}$ so that

$$\left\|\sum_{i=1}^{n} \alpha_i \boldsymbol{x_i}\right\| \quad \geq \quad \mathbf{c} \sum_{i=1}^{n} |\alpha_i|$$

for all choices of scalars $\{\alpha_1, \ldots, \alpha_n\}$. This results holds where the underlying field is real or the complex.

Proof 5.3.4

First note any $\mathbf{c} > 0$ will work if all the scalars are zero. So we can restrict our attention to the sets of scalars with $s = \sum_{i=1}^{n} |\alpha_i| > 0$. Hence, we are looking for a $\mathbf{c} > 0$ so that $\frac{1}{s} \|\sum_{i=1}^{n} \alpha_i \boldsymbol{x_i}\| \geq \mathbf{c}$. Note $-1 \leq \frac{\alpha_i}{s} = \frac{\alpha_i}{\sum_{i=1}^{n} |\alpha_i|} \leq 1$ and so $\sum_{i-1}^{n} \frac{|\alpha_i|}{s} = 1$. Let $\beta_i = \frac{\alpha_i}{s}$. Then, using the scaling principle of the norm, this can be rewritten as

$$\left\|\sum_{i=1}^{n} \frac{\alpha_i}{s} \boldsymbol{x_i}\right\| \geq \mathbf{c} \Longrightarrow \|\textstyle\sum_{i=1}^{n} \beta_i \, \boldsymbol{x_i}\| \geq \mathbf{c} \tag{5.1}$$

for all β_i with $\beta_1 \in [-1, 1]$, $\sum_{i=1}^{n} \beta_i \in [-1, 1]$ and $\sum_{i=1}^{n} |\beta_i| = 1$. Hence, proving the LCT means proving the inequality Equation 5.1 holds subject to the restrictions on β_i.
We will do this by contradiction. We assume this is not true. Hence for all positive integers p, there is a set of scalars $\{\beta_{p1}, \ldots, \beta_{pn}\}$ so that

$$\sum_{i=1}^{n} \beta_{pi} \in [-1, 1], \;\; \beta_{pi} \in [-1, 1], \;\; \sum_{i=1}^{n} |\beta_{pi}| = 1, \;\; \left\|\sum_{i=1}^{n} \beta_{pi} \boldsymbol{x_i}\right\| < \frac{1}{p} \tag{5.2}$$

Let $\boldsymbol{y_p} = \sum_{i=1}^{n} \beta_{pi}\boldsymbol{x_i}$. Note $\boldsymbol{y_p} \to 0$ in $\|\cdot\|$ as given $\epsilon > 0$, there is P so that $p > P$ implies

$$\|\boldsymbol{y_p} - \boldsymbol{0}\| < \frac{1}{p} < \epsilon$$

We will show $(\boldsymbol{y_p})$ has a subsequence $(\boldsymbol{y_{p_k}})$ with $\boldsymbol{y_{p_k}} \to \boldsymbol{x} = \sum_{i=1}^{n} \beta_i\boldsymbol{x_i}$ with not all scalars β_i zero; i.e. $\sum_{i=1}^{n} |\beta_i| > 0$. Since we already know $\boldsymbol{y_p} \to \boldsymbol{0}$, the subsequential limit $\boldsymbol{x}$ must be the same. So we know $\boldsymbol{x} = \boldsymbol{0}$. Hence, $\boldsymbol{0} = \sum_{i=1}^{n} \beta_i\boldsymbol{x_i}$ with not all scalars β_i zero. This says the set $\{\boldsymbol{x_1}, \ldots, \boldsymbol{x_n}\}$ is linearly dependent which is a contradiction. Hence, if we can find this subsequence, we will have a contradiction and so our assumption that Equation 5.1 is false is wrong.

Note for each p

$$\boldsymbol{y_p} \quad = \quad \beta_{p1}\boldsymbol{x_1} + \beta_{p2}\boldsymbol{x_2} + \ldots + \beta_{pn}\boldsymbol{x_n}$$

So the sequence $(\beta_{p1})_{p=1}^{\infty}$ is bounded as it is contained in $[-1, 1]$. By the Bolzano - Weierstrass Theorem, there is a subsequence (β_{p1}^1) which must converge to a scalar $\beta_1 \in [-1, 1]$; i.e. $\beta_{p1}^1 \to \beta_1$.

This subsequence determines a subsequence of (β_{p2}) we will denote by (β_{p2}^1). Now apply the same sort of argument. The sequence $(\beta_{p2}^1)_{p=1}^{\infty}$ is bounded as it is contained in $[-1, 1]$. By the Bolzano - Weierstrass Theorem, there is a subsequence β_{p2}^2 which must converge to a scalar $\beta_2 \in [-1, 1]$; i.e. $\beta_{p2}^2 \to \beta_2$. Note since subsequences converge to the original limit value, we also have the subsequence $\beta_{p1}^2 \to \beta_1$.

Now look at the subsequences (β_{p1}^2), (β_{p2}^2) to (β_{pn}^2). By the same sort of argument, there is a subsequence β_{p3}^3 which converges to $\beta_3 \in [-1, 1]$. Then we have the subsequences (β_{p1}^3) and (β_{p2}^3) still converge to β_1 and β_3.

This process terminates after n steps. We find subsequences

$$\begin{aligned} \beta_{p1}^n &\to \beta_1 \in [-1, 1] \\ &\vdots \\ \beta_{pn}^n &\to \beta_n \in [-1, 1] \end{aligned}$$

Define $\boldsymbol{y_p^n} = \sum_{i=1}^{n} \beta_{pi}^n\, \boldsymbol{x_i}$ and $\boldsymbol{x} = \sum_{i=1}^{n} \beta_i\, \boldsymbol{x_i}$. Let $B = \max\{\|\boldsymbol{x_1}\|, \ldots, \|\boldsymbol{x_n}\|\}$ and let $\epsilon > 0$ be chosen. Since $\beta_{pi}^n \to \beta_i$, there is N_i so $n > N_i$ implies $|\beta_{pi}^n - \beta_i| < \frac{\epsilon}{nB}$. So for $p > N = \max\{N_1, \ldots, N_n\}$, we have all these estimates hold and

$$\|\boldsymbol{y_p^n} - \boldsymbol{x}\| \quad = \quad \left\| \sum_{i=1}^{n} (\beta_{pi}^n - \beta_i)\, \boldsymbol{x_i} \right\| \leq \sum_{i=1}^{n} |\beta_{pi}^n - \beta_i|\, \|\boldsymbol{x_i}\| < \sum_{i=1}^{n} \frac{\epsilon}{nB} B = \epsilon$$

This tells us $\boldsymbol{y_p^n} \to \boldsymbol{x}$. But we know $\sum_{i=1}^{n} |\beta_{pi}^n| = 1$ and $\sum_{i=1}^{n} |\beta_{pi}^n| \to \sum_{i=1}^{n} |\beta_i|$. So $\sum_{i=1}^{n} |\beta_i| = 1$. This tells us not all the scalars β_i are zero which shows the contradiction. Hence, we have established the LCT holds. ■

This is a nice example of how much we can accomplish with abstraction. It is really hard to find the constants c that must exist. To make sure you get this point, we will guide you through two such calculations.

Homework

Exercise 5.3.13 *Let $f(t) = t$ and $g(t) = t^2$ on $[0,1]$. We want to find a constant $c > 0$ so that for all α and β the following inequality holds:*

$$\int_0^1 |\alpha t + \beta t^2| dt \quad \geq c(|\alpha| + |\beta|)$$

This is a lot harder than it looks. It is enough to look at the cases where both α and β are nonzero.

- *If both α and β are positive or the case where they are both negative are easy. Show we have*

$$\int_0^1 ||\alpha| t + |\beta| t^2| dt \quad = \quad \frac{|\alpha|}{2} + \frac{|\beta|}{3} \geq \frac{1}{6}(|\alpha| + |\beta|)$$

- *Show the only other case is when α and β have opposite signs in which case it is enough to study $\int_0^1 ||\alpha|t - |\beta|t^2|dt$. Show there are two cases to consider:*

  1. *$\frac{|\alpha|}{|\beta|} \geq 1$ which implies the integrand is always positive and we find*

  $$\int_0^1 ||\alpha| t - |\beta| t^2| dt \quad = \quad \frac{|\alpha|}{2} - \frac{|\beta|}{3} \geq \frac{1}{12}(|\alpha| + |\beta|)$$

  because $|\alpha| > |\beta|$ here.

  2. *$\frac{|\alpha|}{|\beta|} < 1$ so that we write*

  $$\begin{aligned}\int_0^1 ||\alpha| t - |\beta| t^2| dt \quad &= \quad \int_0^{\frac{|\alpha|}{|\beta|}} (|\alpha| t - |\beta| t^2) dt + \int_{\frac{|\alpha|}{|\beta|}}^1 (|\beta| t^2 - |\alpha| t) dt \\ &= \quad \frac{|\alpha|}{3}\left(\frac{|\alpha|}{|\beta|}\right)^2 + \frac{1}{3}|\beta| - \frac{1}{2}|\alpha|\end{aligned}$$

  Show if $u = \frac{|\alpha|}{|\beta|}$ we have

  $$\frac{\frac{|\alpha|}{3}\left(\frac{|\alpha|}{|\beta|}\right)^2 + \frac{1}{3}|\beta| - \frac{1}{2}|\alpha|}{|\alpha| + |\beta|} \quad = \quad f(u) = \frac{\frac{1}{3} - \frac{1}{2}u + \frac{1}{3}u^3}{1+u}$$

  We only need to consider this function on $(0,1)$. Show using standard calculus that f has a positive absolute minimum on $(0,1)$ which furnishes another value of c.

- *The smallest of all these c values is the one we seek.*

Exercise 5.3.14 *Let $f(t) = t$ and $g(t) = t^2$ on $[0,1]$. We want to find a constant $c > 0$ so that for all α and β the following inequality holds:*

$$\sqrt{\int_0^1 |\alpha t + \beta t^2|^2 dt} \quad \geq c(|\alpha| + |\beta|)$$

This is a again a lot harder than it looks. It is enough to look at the cases where both α and β are nonzero.

- *If both α and β are positive or the case where they are both negative is not so bad. Show we have*

$$\int_0^1 (|\alpha| t + |\beta| t^2|)^2 dt \quad = \quad \int_0^1 (|\alpha|^2 t^2 + 2|\alpha||\beta| t^3 + |\beta|^2 t^4) dt$$

$$\begin{aligned} &= \frac{|\alpha|^2}{3} + 2\frac{|\alpha||\beta|}{4} + \frac{|\beta|^2}{5} \geq \frac{|\alpha|^2}{5} + 2\frac{|\alpha||\beta|}{5} + \frac{|\beta|^2}{5} \\ &= \frac{(|\alpha| + |\beta|)^2}{5} \Longrightarrow \sqrt{\int_0^1 (|\alpha|t + |\beta|t^2|)^2 dt} \geq \frac{1}{\sqrt{5}}(|\alpha| + |\beta|) \end{aligned}$$

- *If α and β are different signs, then the work is more difficult. We only need to consider*

$$\begin{aligned} \int_0^1 (|\alpha|t - |\beta|t^2|)^2 dt &= \int_0^1 (|\alpha|^2 t^2 - 2|\alpha||\beta|t^3 + |\beta|^2 t^4) dt \\ &= \frac{|\alpha|^2}{3} - \frac{|\alpha||\beta|}{2} + \frac{|\beta|^2}{5} \end{aligned}$$

We want

$$\sqrt{\frac{|\alpha|^2}{3} - \frac{|\alpha||\beta|}{2} + \frac{|\beta|^2}{5}} \geq c(|\alpha| + |\beta|) \Longrightarrow \frac{\frac{|\alpha|^2}{3} - \frac{|\alpha||\beta|}{2} + \frac{|\beta|^2}{5}}{(|\alpha| + |\beta|)^2} \geq c^2$$

Let $u = \frac{|\alpha|}{|\beta}$ and show

$$\frac{\frac{|\alpha|^2}{3} - \frac{|\alpha||\beta|}{2} + \frac{|\beta|^2}{5}}{(|\alpha| + |\beta|)^2} = f(u) = \frac{\frac{1}{3}u^2 - \frac{1}{2}u + \frac{1}{5}}{(1+u)^2}$$

We need to consider this function on $(0, \infty)$. Show using standard calculus that f has a positive absolute minimum on $(0, 1)$ which furnishes another value of c.

- *The smallest of all these c values is the one we seek.*

Exercise 5.3.15 *Let $f(t) = t$ and $g(t) = t^3$ on $[0, 1]$. We want to find a constant $c > 0$ so that for all α and β the following inequality holds:*

$$\int_0^1 |\alpha t + \beta t^3| dt \geq c(|\alpha| + |\beta|)$$

Oh boy, this is ugly!

Exercise 5.3.16 *Let $f(t) = t$ and $g(t) = t^3$ on $[0, 1]$. We want to find a constant $c > 0$ so that for all α and β the following inequality holds:*

$$\sqrt{\int_0^1 |\alpha t + \beta t^3|^2 dt} \geq c(|\alpha| + |\beta|)$$

This is even uglier but look at it this way: you always wanted to show the world you know your calculus right?

5.4 Compactness

Let's look at the notion of compact subsets of a normed linear space. We know subsets of $\Re^n$ are topologically compact if and only if the subset is bounded if and only if the subset is sequentially compact. The space $\Re^n$ is an example of a finite dimensional normed linear space. What is true in an arbitrary normed linear space? Let's redefine these types of compactness in the normed setting.

Definition 5.4.1 Sequentially Compact Subsets in a Normed Linear Space

Let $(X, \|\cdot\|)$ be a normed linear space. We say $S \subset X$ is sequentially compact if any sequence $(\boldsymbol{x_n}) \subset S$ contains at least one subsequence $(\boldsymbol{x_n^1})$ which converges to $\boldsymbol{x} \in S$.

The definition is almost the same in a metric space.

Definition 5.4.2 Sequentially Compact Subsets in a Metric Space

Let (X, d) be a metric space. We say $S \subset X$ is sequentially compact if any sequence $(\boldsymbol{x_n}) \subset S$ contains at least one subsequence $(\boldsymbol{x_n^1})$ which converges to $\boldsymbol{x} \in S$.

We often use **compact** as a short hand for **sequentially compact** just to save space.
The first theorem we can prove is familiar to us from our settings in $\Re$ and $\Re^n$ which we discussed in the earlier volumes.

Theorem 5.4.1 Sequentially Compact in a Normed Linear or Metric Space implies Closed and Bounded

Whether X is a normed linear space with norm $\|\cdot\|$ or a metric space with metric d, if $S \subset X$ is sequentially compact, then it is closed and bounded.

Proof 5.4.1
S **is closed**: *Let $(\boldsymbol{x_n})$ be a convergent sequence in S which is a subset of the normed linear space $(X, \|\cdot\|)$. Then, there is $\boldsymbol{x} \in X$ so that $\boldsymbol{x_n} \to \boldsymbol{x}$ in norm. Because S is sequentially compact, there is a subsequence $(\boldsymbol{x_n^1})$ which converges to $\boldsymbol{y} \in S$. But the subsequential limit must be the same as the limit of a convergent sequence. Hence, $\boldsymbol{y} = \boldsymbol{x}$ and we see S contains its limit points. So S is closed.*

If $(\boldsymbol{x_n})$ be a convergent sequence in S which is a subset of the metric space (X, d). Then, there is $\boldsymbol{x} \in X$ so that $d(\boldsymbol{x_n}, \boldsymbol{x}) \to 0$. Because S is sequentially compact, there is a subsequence $(\boldsymbol{x_n^1})$ which converges to $\boldsymbol{y} \in S$. But the subsequential limit must be the same as the limit of a convergent sequence. Hence, $\boldsymbol{y} = \boldsymbol{x}$ and we see S contains its limit points. So S is closed. Note the argument for the norm and the metric are essentially the same.

S **is bounded**: *Let $(\boldsymbol{x_n})$ be a convergent sequence in S which is a subset of the normed linear space $(X, \|\cdot\|)$. Assume S is not bounded. Then for all n, there is $\boldsymbol{x_n} \in S$ so that $\|\boldsymbol{x_n}\| > n$. But $(\boldsymbol{x_n})$ is a sequence in S and so there is a subsequence $(\boldsymbol{x_{n_k}})$ which converges to $\boldsymbol{y} \in S$. Then for $\epsilon = 1$ there is N so $n_k > N$ implies $\|\boldsymbol{x_{n_k}} - \boldsymbol{x}\| < 1$. The backwards triangle inequality for the norm gives $n_K > N$ implies*

$$\|\boldsymbol{x_{n_k}}\| \leq \|\boldsymbol{x}\| + 1$$

and so

$$\|\boldsymbol{x_{n_k}}\| \leq \max\{\max_{1 \leq n_k \leq N} \|\boldsymbol{x_{n_k}}\|, \|\boldsymbol{x}\| + 1\}$$

which says the subsequence is bounded in norm. But by construction, it is not. Hence our assumption the set is not bounded in norm in wrong.

In the metric space setting, we have to decide what it means for a set to be bounded. A reasonable definition is that S is bounded if there is $R > 0$ so that $d(\boldsymbol{x}, \boldsymbol{y}) < R$ for all $\boldsymbol{x}$ and $\boldsymbol{y}$ in S. Another way to look at this is to fix $\boldsymbol{x_0}$ in S and then boundedness of S means $d(\boldsymbol{x}, \boldsymbol{x_0}) < R$ for all $\boldsymbol{x} \in S$. Let's assume S is not bounded. Then for all n, there is $\boldsymbol{x_n}$ so that $d(\boldsymbol{x_n}, \boldsymbol{x_0}) > n$. But $(\boldsymbol{x_n}) \subset S$

and so it has a convergence subsequence $(\boldsymbol{x}_{n_k})$ *which converges to* $\boldsymbol{x} \in S$. *Thus for* $\epsilon = 1$, *there is* N *so that* $n_K > N$ *implies* $d(\boldsymbol{x}_{n_k}, \boldsymbol{x}) < 1$. *Now apply the backwards triangle inequality to find*

$$d(\boldsymbol{x}_{n_k}, \boldsymbol{x}_0) \quad < \quad d(\boldsymbol{x}, \boldsymbol{x}_0) + 1$$

We conclude

$$d(\boldsymbol{x}_{n_k}, \boldsymbol{x}_0) \quad \leq \quad \max\{ \max_{1 \leq n_k \leq N} d(\boldsymbol{x}_{n_k}, \boldsymbol{x}),\ d(\boldsymbol{x}, \boldsymbol{x}_0) + 1\}$$

which says the subsequence $d(\boldsymbol{x}_{n_k}, \boldsymbol{x}_0)$ *is bounded. However, we know it is not. This contradiction means our assumption* S *is not bounded is wrong.* ■

There are big differences between compactness in the metric space and normed space setting. We explore this in the exercises below.

Homework

Exercise 5.4.1 *Let* X *be the set of all sequences and let the metric here be* $d(\boldsymbol{x}, \boldsymbol{y}) = \sum_{i=1}^{\infty} \frac{1}{2^i} \frac{|x_i - y_i|}{1 + |x_i - y_i|}$ *for any sequences* $\boldsymbol{x}$ *and* $\boldsymbol{y}$. *Let* $S = \{\boldsymbol{e_n}\}$ *where* $\boldsymbol{e_n}$ *is the usual sequence with a* 1 *in the* n^{th} *slot and* 0*'s elsewhere. Prove* S *is compact in this metric space. Note* X *is also a vector space which is not finite dimensional which is not a normed linear space.*

Exercise 5.4.2 *Let* X *be the set of all sequences and let the metric here be* $d(\boldsymbol{x}, \boldsymbol{y}) = \sum_{i=1}^{\infty} \frac{1}{3^i} \frac{|x_i - y_i|}{1 + |x_i - y_i|}$ *for any sequences* $\boldsymbol{x}$ *and* $\boldsymbol{y}$. *Let* $S = \{\boldsymbol{f_n}\}$ *where* $\boldsymbol{f_n}$ *is the sequence with a* 1 *in the first* n *slots and* 0*'s elsewhere. Prove* S *is compact in this metric space. Note* X *is also a vector space which is not finite dimensional which is not a normed linear space.*

Exercise 5.4.3 *Let* X *be the set of all sequences and let the metric here be* $d(\boldsymbol{x}, \boldsymbol{y}) = \sum_{i=1}^{\infty} \frac{1}{4^i} \frac{|x_i - y_i|}{1 + |x_i - y_i|}$ *for any sequences* $\boldsymbol{x}$ *and* $\boldsymbol{y}$. *Let* $S = \{\boldsymbol{g_n}\}$ *where* $\boldsymbol{g_n}$ *is the sequence with a* 0 *in the first* n *slots and* 1*'s elsewhere. Prove* S *is compact in this metric space. Note* X *is also a vector space which is not finite dimensional which is not a normed linear space.*

Exercise 5.4.4 *Let* X *be the set of all sequences and let the metric here be* $d(\boldsymbol{x}, \boldsymbol{y}) = \sum_{i=1}^{\infty} \frac{1}{5^i} \frac{|x_i - y_i|}{1 + |x_i - y_i|}$ *for any sequences* $\boldsymbol{x}$ *and* $\boldsymbol{y}$. *Let* $S = X$ *itself. Let* $(\boldsymbol{x_n})$ *be any sequence in* S. *Let* $\boldsymbol{y_n}$ *be the sequence which is* x_{nn} *in the* n^{th} *slot and* 0*s else and let* $\boldsymbol{y} = (x_{nn})_{n \geq 1}$. *Prove* $\boldsymbol{y_n}$ *converges to* $\boldsymbol{y}$ *in* d *showing* X *is compact. Again note* X *is also a vector space which is not finite dimensional which is not a normed linear space.*

Now we need to focus on normed linear spaces as we need the vector space structure for the next result.

Theorem 5.4.2 In Finite Dimensional Normed Linear Spaces, Sets are Sequentially Compact if and only if They are Closed and Bounded

Let $(X, \|\cdot\|)$ *be a finite dimensional normed linear space. Then* $S \subset X$ *is sequentially compact if and only if it is closed and bounded.*

Proof 5.4.2
By the previous theorem, we know if S *is sequentially compact, it is closed and bounded. So we must show the other direction. Let* S *be closed and bounded and assume the dimension of* X *is* n. *Since* S *is bounded, there is* $B > 0$ *so that* $\|\boldsymbol{x}\| < B$ *for all* $\boldsymbol{x} \in S$. *Let* $(\boldsymbol{x_p})$ *be a sequence in* S. *Then each* $\boldsymbol{x_p}$ *as a representation with respect to a basis* $\boldsymbol{E} = \{\boldsymbol{E_1}, \ldots, \boldsymbol{E_n}\}$ *given by* $\boldsymbol{x_p} = \sum_{i=1}^{p} \alpha_{pi} \boldsymbol{E_i}$. *By*

the LCT, there is a $\boldsymbol{c} > 0$ *so that*

$$B \quad > \quad \left\| \sum_{i=1}^{p} \alpha_{pi} \boldsymbol{E_i} \right\| \geq \boldsymbol{c} \sum_{i=1}^{p} |\alpha_{pi}|$$

This implies $\sum_{i=1}^{p} |\alpha_{pi}| < \frac{B}{\boldsymbol{c}}$*, for* $1 \leq i \leq n$ *and for all* p*. Applying the Bolzano - Weierstrass Theorem* n *times as we did in the LCT proof, we find subsequences*

$$(\alpha_{p1}^n),\ (\alpha_{p2}^n), \ldots, (\alpha_{pn}^n), \quad \subset \left[-\frac{B}{\boldsymbol{c}}, \frac{B}{\boldsymbol{c}} \right]$$

and scalars $\alpha_1, \ldots, \alpha_n$ *so that* $\alpha_{pi}^n \rightarrow \alpha_i$ *with* $\alpha_i \in \left[-\frac{B}{\boldsymbol{c}}, \frac{B}{\boldsymbol{c}} \right]$*. Define the subsequence* $\boldsymbol{x_p^n} = \sum_{i=1}^{n} \alpha_{pi}^n \boldsymbol{E_i}$*. This is a subsequence of* $(\boldsymbol{x_p})$ *which converges in norm to* $\boldsymbol{x} = \sum_{i=1}^{n} \alpha_i \boldsymbol{E_i}$*. Since* S *is closed by assumption, this tells us* $\boldsymbol{x} \in S$*. This shows* S *is sequentially compact.* ■

Now look at the previous exercises within the context of a norm. We must change the metric we use, of course.

Homework

Exercise 5.4.5 *Prove if* $S \subset X$ *where* X *is a finite dimensional vector space, then if* S *is compact in the normed linear space* $(X, \boldsymbol{\rho_1})$ *for norm* $\boldsymbol{\rho_1}$ *it is also compact in normed linear space* $(X, \boldsymbol{\rho_2})$ *for norm* $\boldsymbol{\rho_2}$

Exercise 5.4.6 *Prove if* $S \subset X$ *where* X *is a finite dimensional vector space, then if* S *is closed in the normed linear space* $(X, \boldsymbol{\rho_1})$ *for norm* $\boldsymbol{\rho_1}$ *it is also closed in normed linear space* $(X, \boldsymbol{\rho_2})$ *for norm* $\boldsymbol{\rho_2}$

Exercise 5.4.7 *Prove if* $S \subset X$ *where* X *is a finite dimensional vector space, then if* S *is bounded in the normed linear space* $(X, \boldsymbol{\rho_1})$ *for norm* $\boldsymbol{\rho_1}$ *it is also bounded in normed linear space* $(X, \boldsymbol{\rho_2})$ *for norm* $\boldsymbol{\rho_2}$

Let's look at some examples of closed and bounded subsets of infinite dimensional normed linear spaces which are not sequentially compact.

A subset in ℓ^2**:** Let $S \subset \ell^2$ be the $S = (\boldsymbol{e_i})$ where $\boldsymbol{e_i}$ is the sequence of all zeroes except for a 1 in the nth slot. Note $\|\boldsymbol{e_i}\|_2 = 1$ for all i and $\|\boldsymbol{e_i} - \boldsymbol{e_j}\|_2 = \sqrt{1^2 + 1^2} = \sqrt{2}$ for all $i \neq j$. Clearly S is bounded. Now if $(\boldsymbol{e_{p_k}})$ was a sequence from S, we know if it converged it would be a Cauchy sequence. However, we also know $\|\boldsymbol{e_{p_n}} - \boldsymbol{e_{p_m}}\|_2 = \sqrt{2}$ is $p_n \neq p_m$ and is 0 when they match. Hence, no sequence in S can converge in norm unless it is a constant sequence for indices sufficiently large. All such sequences converge in norm to one of the $\boldsymbol{e_n}$. Hence S contains its limit points and so it is closed. We have shown then that S is closed and bounded. But our earlier argument shows there are sequences of S which cannot have a convergence subsequence because the distance between terms is always $\sqrt{2}$. Hence, S is not sequentially compact.

A subset in ℓ^∞**:** Let $S \subset \ell^\infty$ be the set $S = (\boldsymbol{e_i})$ where $\boldsymbol{x_i}$ is the sequence which has a 1 in the first n slots and 0 from slot $n+1$ on. The norm here is the sup norm so we have $\|\boldsymbol{x_i}\|_\infty = 1$ for all i. The set S is therefore bounded. Also, $\|\boldsymbol{x_i} - \boldsymbol{x_j}\|_\infty = 1$ when $i \neq j$ so the sequence $(\boldsymbol{x_{n_k}})$ can never converge as it is not a Cauchy sequence. Thus S is not sequentially compact. However, any sequence from S which is a constant sequence $\boldsymbol{x_p}$ for indices sufficiently large will converge to $\boldsymbol{x_p}$. Hence, each point in S is a limit point and so S is closed.

Homework

Exercise 5.4.8 *Let $S \subset \ell^2$ be $S = (\boldsymbol{f_i})$ where $\boldsymbol{f_i}$ is the sequence which is 0 for the first n slots and 1 else. Prove S is not compact.*

Exercise 5.4.9 *Let $S \subset \ell^2$ be $S = (\boldsymbol{f_i})$ where $\boldsymbol{g_i}$ is the sequence which is 1 for the first n slots and 0 else. Prove S is not compact.*

Exercise 5.4.10 *Let $S \subset \ell^\infty$ be $S = (\boldsymbol{f_i})$ where $\boldsymbol{f_i}$ is the sequence which is 0 for the first n slots and 1 else. Prove S is not compact.*

Exercise 5.4.11 *Let $S \subset \ell^\infty$ be the the $S = (\boldsymbol{f_i})$ where $\boldsymbol{g_i}$ is the sequence which is 1 for the first n slots and 0 else. Prove S is not compact.*

It seems that closed and bounded subsets of a normed linear space need not be sequentially compact. We can prove a more precise result once we have the following result.

Lemma 5.4.3 Riesz's Lemma

Let $(X, \|\cdot\|)$ be a normed linear space and let Y and Z be subspaces of X with Y closed and $Y \subset Z$ a proper subset. Then given $\lambda \in (0,1)$, there is $\boldsymbol{z_\lambda} \in Z$, $\|\boldsymbol{z_\lambda}\| = 1$ so that $\inf_{\boldsymbol{y} \in Y} \|\boldsymbol{z_\lambda} - \boldsymbol{y}\| \geq \lambda$.

Proof 5.4.3

Since $Y \subset Z$ is a proper subset, there is $\boldsymbol{v} \in Z \setminus Y$. Let $\boldsymbol{a} = \inf_{\boldsymbol{y} \in Y} \|\boldsymbol{v} - \boldsymbol{y}\|$. From the Infimum Tolerance Lemma, given $\epsilon > 0$, there is $\boldsymbol{y_0} \in Y$ so that $\boldsymbol{a} \leq \|\boldsymbol{v} - \boldsymbol{y_0}\| < \boldsymbol{a} + \epsilon$. In particular for $\epsilon = \frac{1}{n}$, there is $\boldsymbol{y_n} \in Y$ so that $\boldsymbol{a} \leq \|\boldsymbol{v} - \boldsymbol{y_n}\| < \boldsymbol{a} + \frac{1}{n}$. This defines a sequence $(\boldsymbol{y_n})$ in Y and clearly $\|\boldsymbol{v} - \boldsymbol{y_n}\| \to \boldsymbol{a}$ as $n \to \infty$.

$(\boldsymbol{a} > 0)$: If $\boldsymbol{a} = 0$, we would have $\|\boldsymbol{v} - \boldsymbol{y_n}\| \to 0$ as $n \to \infty$ or $\boldsymbol{y_n} \to \boldsymbol{v}$. But Y is closed and so $\boldsymbol{v}$ would have to be in Y; this is not possible as $\boldsymbol{v} \in Z \setminus Y$.

(Find $\boldsymbol{z_\lambda}$): Since $\boldsymbol{a} > 0$, $\frac{\boldsymbol{a}}{\lambda} > 0$ as $0 < \lambda < 1$. Now use $\frac{\boldsymbol{a}}{\lambda} - \boldsymbol{a}$ as the ϵ in the use of the Infimum Tolerance Lemma. There is then $\boldsymbol{y_0} \in Y$ so $0 < \boldsymbol{a} \leq \|\boldsymbol{v} - \boldsymbol{y_0}\| < \boldsymbol{a} + \epsilon = \frac{\boldsymbol{a}}{\lambda}$. Let $\boldsymbol{z_\lambda} = \frac{\boldsymbol{v} - \boldsymbol{y_0}}{\|\boldsymbol{v} - \boldsymbol{y_0}\|}$. Since this is a linear combination of elements in Z, it is in Z also. Then $\|\boldsymbol{z_\lambda}\| = 1$ and for all $\boldsymbol{y} \in T$,

$$\begin{aligned} \|\boldsymbol{z_\lambda} - \boldsymbol{y}\| &= \left\| \frac{\boldsymbol{v} - \boldsymbol{y_0}}{\|\boldsymbol{v} - \boldsymbol{y_0}\|} - \boldsymbol{y} \right\| = \frac{1}{\|\boldsymbol{v} - \boldsymbol{y_0}\|} \left\| \boldsymbol{v} - \boldsymbol{y_0} - \boldsymbol{y}\|\boldsymbol{v} - \boldsymbol{y_0}\| \right\| \\ &= \frac{1}{\|\boldsymbol{v} - \boldsymbol{y_0}\|} \left\| \boldsymbol{v} - (\boldsymbol{y_0} + \boldsymbol{y}\|\boldsymbol{v} + \boldsymbol{y_0}\|) \right\| \end{aligned}$$

But $\boldsymbol{y_0} + \boldsymbol{y}\|\boldsymbol{v} + \boldsymbol{y_0}\|$ is a linear combination of two elements in the subspace Y. From the definition of the infimum value $\boldsymbol{a}$, we then have $\left\| \boldsymbol{v} - (\boldsymbol{y_0} + \boldsymbol{y}\|\boldsymbol{v} + \boldsymbol{y_0}\|) \right\| \geq \boldsymbol{a}$.

This tells us $\|\boldsymbol{z_\lambda} - \boldsymbol{y}\| \geq \frac{\boldsymbol{a}}{\|\boldsymbol{v} - \boldsymbol{y_0}\|}$. However, we know $\|\boldsymbol{v} - \boldsymbol{y_0}\| < \frac{\boldsymbol{a}}{\lambda}$ which implies $\frac{1}{\|\boldsymbol{v} - \boldsymbol{y_0}\|} > \frac{\lambda}{\boldsymbol{a}}$. We conclude

$$\|\boldsymbol{z_\lambda} - \boldsymbol{y}\| \geq \frac{\lambda}{\boldsymbol{a}} \boldsymbol{a} = \boldsymbol{\lambda}$$

with $\|\boldsymbol{z_\lambda}\| = 1$. This is what we wanted to prove. ■

A huge consequence of this is the following.

Theorem 5.4.4 The Closed Ball of Radius 1 is Sequentially Compact in a Normed Linear Space Implies the Space is Finite Dimensional

Let $(X, \|\cdot\|)$ be a normed linear space. Then $\overline{B(\mathbf{0}, 1)}$ is sequentially compact implies X is finite dimensional.

Proof 5.4.4

We assume X is infinite dimensional and show it leads to a contradiction. Since X is not finite dimensional, $X \neq \emptyset$. Pick $\boldsymbol{x_1} \in X$ not zero. Let $\mathbf{e_1} = \frac{\boldsymbol{x_1}}{\|\boldsymbol{x_1}\|}$. Then $\|\mathbf{e_1}\| = 1$ and so $\mathbf{e_1} \in \overline{B(\mathbf{0}, 1)}$.

Let $Z = X$ and $Y =< \mathbf{e_1} >$, the span of $\mathbf{e_1}$. Since X is not one dimensional, Y is a proper subset of Z. Also since Y is finite dimensional, Y is closed. Apply Riesz's Lemma to conclude there is $\mathbf{e_2} \in Z \setminus Y$ with $\|\mathbf{e_2}\| = 1$ with $\|\mathbf{e_2} - \boldsymbol{y}\| \geq \frac{1}{2}$ for all $\boldsymbol{y} \in Y =< \mathbf{e_1} >$. Note also $\mathbf{e_2} \in \overline{B(\mathbf{0}, 1)}$. Hence, $\|\mathbf{e_2} - \mathbf{e_1}\| \geq \frac{1}{2}$.

Let $Z = X$ and $Y =< \mathbf{e_1}, \mathbf{e_2} >$. Since X is not one dimensional, Y is a proper subset of Z. Also since Y is finite dimensional, Y is closed. Apply Riesz's Lemma to conclude there is $\mathbf{e_3} \in Z \setminus Y$ with $\|\mathbf{e_3}\| = 1$ with $\|\mathbf{e_3} - \boldsymbol{y}\| \geq \frac{1}{2}$ for all $\boldsymbol{y} \in Y =< \mathbf{e_1}, \mathbf{e_2} >$. Note also $\mathbf{e_3} \in \overline{B(\mathbf{0}, 1)}$. Hence, $\|\mathbf{e_3} - \mathbf{e_1}\| \geq \frac{1}{2}$ and $\|\mathbf{e_3} - \mathbf{e_2}\| \geq \frac{1}{2}$.

Continuing by induction, we obtain a sequence $(\boldsymbol{e_n}) \subset \overline{B(\mathbf{0}, 1)}$, $\|\boldsymbol{e_n}\| = 1$ and $\|\boldsymbol{e_i} - \boldsymbol{e_j}\| \geq \frac{1}{2}$ when $i \neq j$. It is clear this sequence cannot converge as if it did it would be a Cauchy sequence and it is clearly not that. Hence, this sequence can have no convergent subsequences which tells us $\overline{B(\mathbf{0}, 1)}$ is **not** *sequentially compact. This contradiction means our assumption X is infinite dimensional is wrong and so X must be finite dimensional.* ■

Homework

Exercise 5.4.12 *Look at Riesz's Lemma in $\Re^2$ with $\|\cdot\|_2$. Let Y be the x axis which is a closed and proper subspace of $\Re^2$ and pick $\lambda = \frac{1}{2}$. Convince yourself this is a good way to interpret Riesz's Lemma here. Look at the quadrant one only. Draw the horizontal line $y = \lambda$ for $0 < \lambda < 1$. Consider the point $\boldsymbol{z}_\lambda = \begin{bmatrix} \sqrt{1-\lambda^2} \\ \lambda \end{bmatrix}$. The distance from $\boldsymbol{z_\lambda}$ to the x axis satisfies*

$$\min_{y \in \Re} \left\| \begin{bmatrix} \sqrt{1-\lambda^2} \\ \lambda \end{bmatrix} - \begin{bmatrix} \sqrt{1-y^2} \\ 0 \end{bmatrix} \right\|_2 \geq \lambda$$

which is the what Riesz's Lemma says. Make sure you draw the pictures here.

Exercise 5.4.13 *For the exercise above, draw the pictures explaining Riesz's Lemma for Quadrant 2, 3 and 4 as well.*

Exercise 5.4.14 *Look at Riesz's Lemma in $\Re^2$ with $\|\cdot\|_1$. Make sure you draw the pictures here for all four quadrants.*

Exercise 5.4.15 *Look at Riesz's Lemma in $\Re^2$ with $\|\cdot\|_\infty$. Make sure you draw the pictures here for all four quadrants.*

Exercise 5.4.16 *Draw pictures to explain Riesz's Lemma in $\Re^3$ with $\|\cdot\|_2$ using Y is the $x - y$ plane.*

Exercise 5.4.17 *Draw pictures to explain Riesz's Lemma in $\Re^3$ with $\|\cdot\|_1$ using Y is the $x - y$ plane.*

Exercise 5.4.18 *Draw pictures to explain Riesz's Lemma in $\Re^3$ with $\|\cdot\|_\infty$ using Y is the $x - y$ plane..*

Exercise 5.4.19 *Let X be the set of all sequences which are bounded with the infinity norm. Note X is also a vector space which is not finite dimensional which is normed linear space. Prove the closure of the ball $B(\mathbf{0}; r)$ is not compact.*

Exercise 5.4.20 *Let X be the set of all sequences which converge with the infinity norm. Note X is also a vector space which is not finite dimensional which is normed linear space. Prove the closure of the ball $B(\mathbf{0}; r)$ is not compact for any $r > 0$.*

Exercise 5.4.21 *If X is the set of all continuously differentiable functions on $[0, 1]$ with the supremum norm, prove the closure of the ball $B(\mathbf{0}; r)$ is not compact for any $r > 0$.*

Let's look at compact subsets of $(C([a, b]), \| \cdot \|)$ where the norm can be $\| \cdot \|_\infty$, $\| \cdot \|_1$ or $\| \cdot \|_2$ or perhaps others. We need to define a few new things.

Definition 5.4.3 Nowhere Dense Sets

Let $(X, \| \cdot \|)$ be a normed linear space and let $S \subset X$. We say $\mathbf{p}$ is an interior point of S if there is a radius $r > 0$ so that $B(\mathbf{p}; r) \subset S$ and S is open if all of its points are interior points. We say a set S is nowhere dense if the closure of the set has no interior points.

One of the projects we typically assign in the second semester or part of the material in (Peterson (18) 2020) is about the construction of generalized Cantor Sets which are actually discussed very carefully in the context of Lebesgue measure which we go over in (Peterson (16) 2019). Cantor sets are closed sets in $[0, 1]$ and do not have interior points. Hence, there are nowhere dense sets in $[0, 1]$. A Cantor Set is defined as follows: Let (a_n) for $n \geq 0$ be a fixed sequence of real numbers which satisfy

$$a_0 = 1, \quad 0 < 2a_n < a_{n-1}$$

There is a lot we can do to analyze the structure of this set of points and the classic Cantor Set is the one we get by using $a_n = \frac{1}{3^n}$ for $n \geq 0$ which clearly satisfies the defining equation. These sets are very important in analysis as we use them later when we talk about spaces which are second category. Important examples of nowhere dense sets are the compact sets of an infinite dimensional normed linear space.

Theorem 5.4.5 Compact Subsets of an Infinite Dimensional Normed Linear Space are Nowhere Dense

Let $(X, \| \cdot \|)$ be an infinite dimensional normed linear space. Then any compact subset of X is nowhere dense.

Proof 5.4.5
Let $S \subset X$ be compact. Then S is closed and bounded. If S had an interior point $\mathbf{p}$, then there is $r > 0$ so that $B(\mathbf{p}; r) \subset S$ which implies $\overline{B(\mathbf{p}; \frac{r}{2})} \subset S$. In general, closed subsets T of a compact set S are also compact. Here is the short argument: Since S is bounded, so is T. If $(\boldsymbol{x_n})$ is a sequence in T, it is also a sequence in S and so it has a convergent subsequence so that $\boldsymbol{x^1_n} \rightarrow \boldsymbol{x}$ in norm where $\boldsymbol{x} \in S$. But T is closed and so $\boldsymbol{x}$ must be in T. Thus, T is sequentially compact.

Thus, $\overline{B(\mathbf{p}; \frac{r}{2})}$ is a compact subset of X which is infinite dimensional. Note this implies the translate $-\mathbf{p} + \overline{B(\mathbf{p}; \frac{r}{2})} = \overline{B(\mathbf{0}; \frac{r}{2})}$ is compact and the scaled set $\frac{2}{r}(-\mathbf{p} + \overline{B(\mathbf{p}; \frac{r}{2})}\,) = \overline{B(\mathbf{0}; 1)}$ is compact. But this is not possible if X is infinite dimensional. So our assumption S has an interior point is wrong. We conclude any compact subset of X is nowhere dense if X is infinite dimensional. ■

So what does a compact subset of an infinite dimensional set look like? We can give you one nice example using the idea of an **equicontinuous family** of continuous functions and use the **Arzela - Ascoli Theorem** to prove such a family is a compact set.

Definition 5.4.4 Equicontinuous Families of Continuous Functions

Let $\mathscr{F}$ be a family of functions in the space $C([a,b])$ for a finite interval $[a,b]$. The family is equicontinuous at t_0 in $[a,b]$ if for all $\epsilon > 0$, there is a $\delta > 0$ so that for all $f \in \mathscr{F}$, $|f(t_0) - f(t)| < \epsilon$ when $|t - t_0| < \delta$.

The family is uniformly equicontinuous, if for all $\epsilon > 0$, there is a $\delta > 0$ so that for all $f \in \mathscr{F}$, $|f(s) - f(t)| < \epsilon$ when $|t - s| < \delta$.

We also need to know about an idea called **uniform boundedness** of a family of functions.

Definition 5.4.5 Uniform Boundedness of a family of functions

Let $\mathscr{F}$ be a family of functions in the space $C([a,b])$ for a finite interval $[a,b]$. We say the family is uniformly bounded if there is $B > 0$ so that $|f(t)| \leq B$ for all $f \in \mathscr{F}$ and all $t \in [a,b]$. This is same as saying $\|f\|_\infty \leq B$ for all $f \in \mathscr{F}$.

There is a famous result about such families called the Arzela - Ascoli Theorem.

Theorem 5.4.6 Arzela - Ascoli Theorem

Let $\mathscr{F} \subset C([a,b])$ for a finite $[a,b]$ be an infinite family of functions. Assume $\mathscr{F}$ is uniformly bounded and equicontinuous. Then there exists a sequence $(x_n) \subset \mathscr{F}$ that converges uniformly to a continuous function x on $[a,b]$.

Proof 5.4.6
Let's identify $\mathbb{Q}$ with the enumeration (r_n). Since $\mathscr{F}$ is uniformly bounded by a positive number B, the set of points $\{(f(r_1)) | f \in \mathscr{F}\}$ is bounded by B also. By the Bolzano - Weierstrass Theorem, there is then a sequence of distinct functions (f_n^1) from $\mathscr{F}$ so that $(f_n^1(r_1))$ converges to a real number α_1.

Now repeat the same argument for the set of points $\{(f_n^1(r_2)) | f \in \mathscr{F}\}$. There is a sequence of distinct functions $(f_n^2) \subset (f_n^1)$ from $\mathscr{F}$ so that $(f_n^2(r_2))$ converges to a real number α_2. Note since $(f_n^2) \subset (f_n^1)$, we know $(f_n^2(r_1))$ also converges to α_1.

By induction, we find a chain of subsequences

$$(f_n^1) \supset (f_n^2) \supset (f_n^3) \supset \ldots \supset (f_n^p) \supset \ldots$$

so that for each k, (f_n^k) converges at $r_1, \ldots, r_k$ to $\alpha_1, \ldots, \alpha_k$. The sequences look like this:

$$(f_n^k) = \left((f_n^k)_1, (f_n^k)_2, (f_n^k)_3, (f_n^k)_4, \ldots, (f_n^k)_k, \ldots \right)$$

where $(f_n^k)_j$ is the jth element in the sequence. So consider this table

$$\begin{array}{cccccccc}
\boxed{(f_n^1)_1} & (f_n^1)_2 & (f_n^1)_3 & (f_n^1)_4 & \cdots & (f_n^1)_k & \cdots \\
(f_n^2)_1 & \boxed{(f_n^2)_2} & (f_n^2)_3 & (f_n^2)_4 & \cdots & (f_n^2)_k & \cdots \\
(f_n^3)_1 & (f_n^3)_2 & \boxed{(f_n^3)_3} & (f_n^3)_4 & \cdots & (f_n^3)_k & \cdots \\
\vdots & & & & & & \\
(f_n^k)_1 & (f_n^k)_2 & (f_n^k)_3 & (f_n^k)_4 & \cdots & \boxed{(f_n^k)_k} & \cdots \\
\vdots & & & & & &
\end{array}$$

Look at the diagonal sequence shown in the boxes, $(f_n^k)_k$. Fix the rational number r_p and consider the sequence $((f_n^k)_k)(r_p)$. This is a subsequence of $(f_n^p)(r_p)$ when $k > p$ and so converges to α_p. So this **diagonal** *sequence converges on $\mathbb{Q}$. For a given $\epsilon > 0$, there is a $N(r_k)$ so that $p, q > N(r_k)$ imply*

$$|(f_n^p)_p(r_k) - (f_n^q)_q(r_k)| < \frac{\epsilon}{3}$$

Since the family is equicontinuous, for each $x \in [a,b]$, there is a $\delta_x > 0$ so that $|f(x) - f(t)| < \frac{\epsilon}{3}$ for all $f \in \mathscr{F}$ and $|x - t| < \delta_x$. This can easily be rephrased as

$$t \in B(x;\delta_x) \quad \Longrightarrow \quad |f(x) - f(t)| < \frac{\epsilon}{3}$$

But the collection $\{B(x,\delta_x)\}$ is an open cover of the compact set $[a,b]$. So there is a finite subcover, $\{B(x_1,\delta_{x_1}),\ldots,B(x_P,\delta_{x_P})\}$. Pick any $x \in [a,b]$. Then $x \in B(x_i,\delta_{x_i})$ for some $1 \leq i \leq P$ and there is a rational number r_ℓ in there too. Thus

$$\begin{aligned}
|(f_n^p)_p(x) - (f_n^q)_q(x)| \quad &\leq \quad |(f_n^p)_p(x) - (f_n^p)_p(r_\ell)| + |(f_n^p)_p(r_\ell) - (f_n^q)_q(r_\ell)| \\
&+ \quad |(f_n^q)_p(r_\ell) - (f_n^p)_p(x)|
\end{aligned}$$

Now $x, r_\ell \in B(x_i,\delta_{x_i})$ and so $|(f_n^p)_p(x) - (f_n^p)_p(r_\ell)| < \frac{\epsilon}{3}$ and $|(f_n^q)_q(x) - (f_n^q)_q(r_\ell)| < \frac{\epsilon}{3}$. Thus,

$$|(f_n^p)_p(x) - (f_n^q)_q(x)| \quad \leq \quad \frac{2}{3}\epsilon + |(f_n^p)_p(r_\ell) - (f_n^q)_q(r_\ell)|$$

But, we also know $p, q > N(r_\ell)$ implies

$$|(f_n^p)_p(r_\ell) - (f_n^q)_q(r_\ell)| < \frac{\epsilon}{3}$$

Thus, if $p, q > N(r_\ell)$ $|(f_n^p)_p(x) - (f_n^q)_q(x)| < \epsilon$. We can do this for any x in $[a,b]$. Thus, since each will need a potentially different r_ℓ from the finite number of balls in the open cover, we collect these into the set of rationals $\{r_{\ell_1},\ldots,r_{\ell_P}\}$ and letting $N = \max\{N_{r_{\ell_1}},\ldots,N_{r_{\ell_P}}\}$, we see $p, q > N$ implies $|(f_n^p)_p(x) - (f_n^q)_q(x)| < \epsilon$. This diagonal sequence therefore satisfies the Cauchy Criterion for uniform convergence on $[a,b]$ and so there is a continuous function x on $[a,b]$ so that $(f_n^k)_k \to x$ uniformly on $[a,b]$. ■

Homework

Exercise 5.4.22 *Let $X = C([0,1])$ with the supremum norm. Define the set S by*

$$S = \{\boldsymbol{y} \in X \mid y(t) = \int_0^t f(x(s))\, ds, \text{ for some } \boldsymbol{x} \in X \text{ with } \|\boldsymbol{x}\|_\infty \leq 1\}$$

where $f(u) = u^2$.

- *Note if $\boldsymbol{y} \in S$, then $\|\boldsymbol{y}\|_\infty \leq 1$ so that S is a uniformly bounded subset of X.*
- *Note for any $\boldsymbol{y_1}$ and $\boldsymbol{y_2}$ in S, there are $\boldsymbol{x_1}$ and $\boldsymbol{x_2}$ so that*

$$|y_1(t) - y_2(t)| \leq 2\|\boldsymbol{x_1} - \boldsymbol{x_2}\|_\infty$$

- *Note if $\boldsymbol{y} \in S$, then*

$$|y(t_1) - y(t_2)| \leq \|\boldsymbol{x}\|_\infty^2 \, |t_1 - t_2| < |t_1 - t_2|$$

Show this says S is a uniformly equicontinuous family in $C([0,1])$.

- *Use the Arzela - Ascoli Theorem to prove S is compact.*
- *Prove S is therefore nowhere dense in $C([0,1])$.*

Exercise 5.4.23 *Let $X = C([0,1])$ with the supremum norm. Define the set S by*

$$S = \{\boldsymbol{y} \in X \mid y(t) = \int_0^t f(x(s))\, ds, \text{ for some } \boldsymbol{x} \in X \text{ with } \|\boldsymbol{x}\|_\infty \leq 1\}$$

where f is continuous and globally Lipschitz on $\Re$; i.e. there is a $K > 0$ so that $|f(u) - f(v)| < K|u - v|$ for all u and v is $\Re$. Note this says $|f(x_1(t)) - f(x_2(t))| < K|x_1(t) - x_2(t)|$.

- *Note if $\boldsymbol{y} \in S$, then there is a $B > 0$ so that $\|\boldsymbol{y}\|_\infty \leq B$. Hence, S is a uniformly bounded subset of X.*
- *Note if $\boldsymbol{y} \in S$, then for any $\boldsymbol{y_1}$ and $\boldsymbol{y_2}$ in S*

$$|y_1(t) - y_2(t)| \leq K\|\boldsymbol{x_1} - \boldsymbol{x_2}\|_\infty$$

- *Note if $\boldsymbol{y} \in S$, then*

$$|y(t_1) - y(t_2)| \leq \max_{0 \leq u \leq 1} |f(u)| \, |t_1 - t_2|$$

Show this says S is a uniformly equicontinuous family in $C([0,1])$.

- *Use the Arzela - Ascoli Theorem to prove S is compact.*
- *Prove S is therefore nowhere dense in $C([0,1])$.*

Chapter 6

Linear Operators on Normed Spaces

6.1 Linear Transformations between Normed Linear Spaces

Let's start by defining linear transformations between vector spaces carefully.

Definition 6.1.1 Linear Transformations between Vector Spaces

Let $\boldsymbol{X}$ and $\boldsymbol{Y}$ be two vector spaces over $\Re$. We say $\boldsymbol{T} : \boldsymbol{X} \rightarrow \boldsymbol{Y}$ is a linear transformation if

$$\boldsymbol{T}(\alpha \boldsymbol{x} + \beta \boldsymbol{y}) = \alpha \boldsymbol{T}(\boldsymbol{x}) + \beta \boldsymbol{T}(\boldsymbol{y})$$

Before we go on, let's remind ourselves of what a matrix is.

Definition 6.1.2 Real Matrices

Let $\boldsymbol{M}$ be a collection of real numbers organized in a table of m rows and n columns. The number M_{ij} refers to the entry in the ith row and j^{th} column of this table. The table is organized like this

$$\begin{matrix} M_{11} & \dots & M_{1n} \\ M_{21} & \dots & M_{2n} \\ \vdots & \vdots & \vdots \\ M_{m1} & \dots & M_{mn} \end{matrix}$$

We identify the matrix $\boldsymbol{M}$ with this table and write

$$\boldsymbol{M} = \begin{bmatrix} M_{11} & \dots & M_{1n} \\ \vdots & \vdots & \vdots \\ M_{m1} & \dots & M_{mn} \end{bmatrix}$$

The set of all matrices with m rows and n columns is denoted by $\boldsymbol{M}_{mn}$. We also say $\boldsymbol{M}$ is an $m \times n$ matrix.

Comment 6.1.1 *The n columns of $\boldsymbol{M}$ are clearly vectors in $\Re^m$ and the m rows of $\boldsymbol{M}$ are vectors in $\Re^n$. Thus, we have some additional identifications:*

$$\boldsymbol{M} = \begin{bmatrix} M_{11} & \dots & M_{1n} \\ \vdots & \vdots & \vdots \\ M_{m1} & \dots & M_{mn} \end{bmatrix} = \begin{bmatrix} \boldsymbol{C_1} & \dots & \boldsymbol{C_n} \end{bmatrix} = \begin{bmatrix} \boldsymbol{R_1} \\ \vdots \\ \boldsymbol{R_m} \end{bmatrix}$$

where column i is denoted by $\boldsymbol{C_j} \in \Re^m$ and row j is denoted by $\boldsymbol{R_j} \in \Re^n$. Note the rows are vectors displayed as the transpose of the usual column vector notation we use.

6.1.1 Basic Properties

Consider a matrix $\boldsymbol{M}$ in $\boldsymbol{M_{mn}}$. We define the action of $\boldsymbol{M}$ on the vector space $\Re^n$ by

$$\boldsymbol{M}(\boldsymbol{x}) = \left(\begin{bmatrix} \boldsymbol{R_1} & \dots & \boldsymbol{R_m} \end{bmatrix}\right)(\boldsymbol{x}) = \begin{bmatrix} < \boldsymbol{R_1}, \boldsymbol{x} > \\ \vdots \\ < \boldsymbol{R_m}, \boldsymbol{x} > \end{bmatrix}$$

We generally just write $\boldsymbol{Mx} = \boldsymbol{M}(\boldsymbol{x})$ to save a few parentheses. It is easy to see this definition of the action of $\boldsymbol{M}$ on $\Re^n$ defines a linear transformation from $\Re^n$ to $\Re^m$.

The set of $\boldsymbol{x}$ in $\Re^n$ with $\boldsymbol{Mx} = \boldsymbol{0} \in \Re^m$ is called the **kernel** or **nullspace** of $\boldsymbol{M}$. The span of the columns of $\boldsymbol{M}$ is called the column space of $\boldsymbol{M}$. We can prove a fundamental result.

Theorem 6.1.1 If $\boldsymbol{M} \in \boldsymbol{M_{mn}}$, then dim(kernel) + dim(column space) $= n$

If $\boldsymbol{M}$ is a $m \times n$ matrix, if $p = dim(ker(\boldsymbol{M}))$ and q is the dimension of the span of the columns of $\boldsymbol{M}$, then $p + q = n$.

Proof 6.1.1
It is easy to see that $ker(\boldsymbol{M})$ is a subspace of $\Re^n$ with dimension $p \leq n$. Let $\{\boldsymbol{K_1}, \dots, \boldsymbol{K_p}\}$ be an orthonormal basis of $ker(\boldsymbol{M})$. We can use GSO to construct an additional $n - p$ vectors in $\Re^n$, $\{\boldsymbol{L_1}, \dots, \boldsymbol{L_{n-p}}\}$ that are all orthogonal to $\ker(\boldsymbol{M})$ with length one. The subspace spanned by $\{\boldsymbol{L_1}, \dots, \boldsymbol{L_{n-p}}\}$ is perpendicular to the subspace $ker(\boldsymbol{M})$ and is called $(\ker(\boldsymbol{M}))^\perp$ to denote it is what is called an orthogonal complement. We note $\Re^n = ker(\boldsymbol{M}) \oplus span\{\boldsymbol{L_1}, \dots, \boldsymbol{L_{n-p}}\}$.

Thus, $\{\boldsymbol{K_1}, \dots, \boldsymbol{K_p}, \boldsymbol{L_1}, \dots, \boldsymbol{L_{n-p}}\}$ is an orthonormal basis for $\Re^n$ and hence $\boldsymbol{x} = a_1\boldsymbol{K_1} + \dots + a_p\boldsymbol{K_p} + b_1\boldsymbol{L_1} + \dots + b_{n-p}\boldsymbol{L_{n-p}}$. It follows using the linearity of $\boldsymbol{M}$ that

$$\boldsymbol{Mx} = b_1\boldsymbol{ML_1} + \dots b_{n-p}\boldsymbol{ML_{n-p}}$$

Now if $b_1\boldsymbol{ML_1} + \dots b_{n-p}\boldsymbol{ML_{n-p}} = 0$, this means $\boldsymbol{M}(b_1\boldsymbol{L_1} + \dots + b_{n-p}\boldsymbol{L_{n-p}}) = 0$ too. This says $b_1\boldsymbol{L_1} + \dots + b_{n-p}\boldsymbol{L_{n-p}}$ is in both $ker(\boldsymbol{M})$ and the $span\{\boldsymbol{L_1}, \dots, \boldsymbol{L_{n-p}}\}$ which forces $b_1\boldsymbol{L_1} + \dots + b_{n-p}\boldsymbol{L_{n-p}} = 0$. But $\{\boldsymbol{L_1}, \dots, \boldsymbol{L_{n-p}}\}$ is a linearly independent set and so all coefficients $b_i = 0$. This says the set $\{\boldsymbol{ML_1}, \dots, \boldsymbol{ML_{n-p}}\}$ is a linearly independent set in $\Re^m$.

Finally, note the column space of $\boldsymbol{M}$ is defined to be the $span\{\boldsymbol{C_1}, \dots, \boldsymbol{C_n}\}$. A vector in this span has the look $\boldsymbol{y} = \sum_{i=1}^m a_i\boldsymbol{C_i}$. This is the same as $\boldsymbol{Ma}$ where $a \in \Re^n$. However, we know $\Re^n$ has the orthonormal basis $\{\boldsymbol{K_1}, \dots, \boldsymbol{K_p}, \boldsymbol{L_1}, \dots, \boldsymbol{L_{n-p}}\}$ and so $\boldsymbol{a} = \sum_{i=1}^p b_i\boldsymbol{K_i} + \sum_{i=1}^{n-p} c_i\boldsymbol{L_i}$. Applying $\boldsymbol{M}$ we find

$$\boldsymbol{Ma} = c_1\boldsymbol{ML_1} + \dots + c_{n-p}\boldsymbol{ML_{n-p}}$$

Since $\{\boldsymbol{ML_1}, \ldots, \boldsymbol{ML_{n-p}}\}$ is a linearly independent, we see the range of $\boldsymbol{M}$, the column space of $\boldsymbol{M}$ has dimension $n-p$. Thus, we have shown

$$dim(ker(\boldsymbol{M})) + dim(span(\{\boldsymbol{C_1}, \ldots, \boldsymbol{C_n}\})) = n$$

where, of course, we also have $n - p \leq m$. ■

Comment 6.1.2 *Thus, if $n-p=m$, the kernel of $\boldsymbol{M}$ must be $n-m$ dimensional.*

6.2 Input - Output Ratios

The **size** of a matrix can be measured in terms of its rows and columns, but a better way is to think about how it acts on vectors. We can think of a matrix as an **engine** which transforms inputs into outputs and it is a natural thing to ask about the ratio of output to input. Such a ratio gives a measure of how the matrix alters the data. Since we can't divide by vectors, we typically use a measure of vector size for the output and input sides. Recall the **Euclidean norm** of a vector $\boldsymbol{x}$ is $\|\boldsymbol{x}\| = \sqrt{x_1^2 + \cdots + x_n^2}$ where x_1 through x_n are the components of $\boldsymbol{x}$ in $\Re^n$ with respect to the standard basis in $\Re^n$. This is also the $\|\cdot\|_2$ norm we have discussed before. Let $\boldsymbol{A}$ be an $n \times n$ matrix. Then $\boldsymbol{A}$ transforms vectors in $\Re^n$ to new vectors in $\Re^n$. We measure the ratio of output to input by calculating $\left\| \frac{\boldsymbol{A}(\boldsymbol{x})}{\|\boldsymbol{x}\|_2} \right\|_2$; of course, this doesn't make sense if $\boldsymbol{x} = \boldsymbol{0}$. We want to see how big this ratio can get, so we are interested in the maximum value of $\left\| \frac{\boldsymbol{A}(\boldsymbol{x})}{\|\boldsymbol{x}\|_2} \right\|_2$; Now the matrix $\boldsymbol{A}$ is linear; i.e. $\boldsymbol{A}(c\boldsymbol{x}) = c\boldsymbol{A}(\boldsymbol{x})$. So in particular, if $y \neq 0$, we can say $\boldsymbol{A}(\boldsymbol{y}) = \|\boldsymbol{y}\|_2 \, \boldsymbol{A}\left(\frac{\boldsymbol{y}}{\|\boldsymbol{y}\|_2}\right)$. Thus,

$$\frac{\|\boldsymbol{A}(\boldsymbol{y})\|_2}{\|\boldsymbol{y}\|_2} = \frac{\|\boldsymbol{y}\|_2 \, \boldsymbol{A}\left(\frac{\boldsymbol{y}}{\|\boldsymbol{y}\|_2}\right)}{\|\boldsymbol{y}\|_2} = \frac{\boldsymbol{A}\left(\frac{\boldsymbol{y}}{\|\boldsymbol{y}\|_2}\right)}{\frac{\boldsymbol{y}}{\|\boldsymbol{y}\|_2}}$$

Now let $\boldsymbol{x} = \frac{\boldsymbol{y}}{\|\boldsymbol{y}\|_2}$ and we have $\frac{\|\boldsymbol{A}(\boldsymbol{y})\|_2}{\|\boldsymbol{y}\|_2} = \frac{\|\boldsymbol{A}(\boldsymbol{x})\|_2}{\|\boldsymbol{x}\|_2}$ where $\boldsymbol{x}$ has norm 1. So

$$\max_{\|\boldsymbol{y}\| \neq 0} \left(\frac{\|\boldsymbol{A}(\boldsymbol{y}\|)}{\|\boldsymbol{y}\|} \right) = \max_{\|\boldsymbol{x}\|=1} \|\boldsymbol{A}(\boldsymbol{x})\|$$

In order to understand this, first note that the numerator $\|\boldsymbol{A}(\boldsymbol{x})\|$ is a continuous function of its arguments x_1 through x_n. We can't use just any arguments either; here we can only use those for which $x_1^2 + \ldots + x_n^2 = 1$. The set of such x_i's in $\Re^n$ is a bounded set which contains all of its boundary points. For example, in $\Re^2$, this set is the unit circle $x_1^2 + x_2^2 = 1$ and in $\Re^3$, it is the surface of the sphere $x_1^2 + x_2^2 + x_3^2 = 1$. These sets are bounded and contain their boundary points and are therefore compact subsets. We know any continuous function must have a global maximum and a global minimum over a compact domain. Hence, the problem of finding the maximum of $\|\boldsymbol{A}(\boldsymbol{x})\|$ over the closed and bounded set $\|\boldsymbol{x}\| = 1$ has a solution. We define this maximum ratio to be the **norm** of the matrix $\boldsymbol{A}$. We denote the matrix norm as usual by $\|\boldsymbol{A}\|$ and so we have

Definition 6.2.1 The Norm of a Matrix

Let $\boldsymbol{A}$ be an $n \times n$ matrix. The **norm** *of $\boldsymbol{A}$ is*

$$\|\boldsymbol{A}\| = \max_{\|\boldsymbol{x}\|=1} \left(\frac{\|\boldsymbol{A}(\boldsymbol{x})\|}{\|\boldsymbol{x}\|} \right)$$

Note we do not say the output of the linear mapping is bounded in norm; i.e. we only say the ratio of output to input may be bounded.

Homework:

Exercise 6.2.1 *Let* $T : \ell^2 \rightarrow \ell^2$ *be the right shift operator; i.e. if* $\boldsymbol{x} = \{x_1, x_2, \ldots\}$ *then* $T(\boldsymbol{x}) = \{x_2, x_3, \ldots\}$. *Show* $\sup_{\|\boldsymbol{x}\|_2=1} \|T(\boldsymbol{x})\|_2 \leq 1$.

Exercise 6.2.2 *Let* A *be an* $n \times n$ *matrix with* n *linearly independent eigenvectors* $\boldsymbol{E_i}$ *with real eigenvalues* λ_i. *Let* $\| \cdot \|$ *denote any norm used on* $\Re^n$ *for both the domain and the range. Prove*

$$\sup_{\|\boldsymbol{x}\|=1} \|A(\boldsymbol{x})\| \quad \leq \quad \frac{EL \sum_{i=1}^n |x_i|}{c \sum_{i=1}^n |x_i|} = \frac{EL}{c}$$

where the components of any $\boldsymbol{x}$ *are denoted by* x_i *as usual,* $L = \max_{1 \leq i\ eqn} |\lambda_i|$, $E = \max_{1 \leq i \leq n} \|\boldsymbol{E_i}\|$ *and c is the constant we obtain by invoking the Linear Combination Theorem.*

Exercise 6.2.3 *We have studied symmetric matrices and their decompositions in (Peterson (20) 2020) extensively. Let's apply what we know to the study of the Hessian of a nice two dimensional map. Let* $f(x, y) = 2x^2 + 5xy + 6y^2 + 7x + 9y + 10$.

- *Find the Hessian for this map and its eigenvalues and eigenvectors. Note we can choose the eigenvectors to be an orthonormal basis for* $\Re^2$.
- *Use the* $\| \cdot \|_2$ *norm on* $\Re^2$ *here and prove*

$$\|H\| = \sup_{\|\boldsymbol{x}=1} \|H(\boldsymbol{x})\| \quad \leq \quad \sqrt{|\lambda_1|^2 |x_1|^2 + |\lambda_2|^2 |x_2|^2} \leq \max\{|\lambda_1|, |\lambda_2|\}$$

- *Prove* $\|H\| \geq \max\{|\lambda_1|, |\lambda_2|\}$
- *Combining, this proves* $\|H\| = \max\{|\lambda_1|, |\lambda_2|\}$.

Note make sure you fill in the numerical values here also.

Exercise 6.2.4 *Using the approach of the last exercise, let* $f(x, y) = 5x^2 + 15xy + 8y^2 + 70x + 90y + 30$.

- *Find the Hessian for this map and its eigenvalues and eigenvectors. Note we can choose the eigenvectors to be an orthonormal basis for* $\Re^2$.
- *Use the* $\| \cdot \|_2$ *norm on* $\Re^2$ *here and prove*

$$\|H\| = \sup_{\|\boldsymbol{x}=1} \|H(\boldsymbol{x})\| \quad \leq \quad \sqrt{|\lambda_1|^2 |x_1|^2 + |\lambda_2|^2 |x_2|^2} \leq \max\{|\lambda_1|, |\lambda_2|\}$$

- *Prove* $\|H\| \geq \max\{|\lambda_1|, |\lambda_2|\}$
- *Combining, this proves* $\|H\| = \max\{|\lambda_1|, |\lambda_2|\}$.

Note make sure you fill in the numerical values here also.

6.3 Linear Operators between Normed Linear Spaces

We now study function T whose domain is a subspace of a normed linear space $(X, \| \cdot \|_X)$ and whose range is contained in another normed linear space $(Y, \| \cdot \|_Y)$. We have already thought about this when X and Y are $\Re^n$ and $\Re^m$ and when X and Y are both finite dimensional normed spaces and so isomorphic to some $\Re^n$ and $\Re^m$. We can use a variety of norms on these finite dimensional spaces and so far we have typically used $\|\cdot\|_2$. However, we can be more general and certainly the normed spaces can be infinite dimensional. Let's be clear about what we mean by a **linear operator** or **linear map** between normed spaces.

Definition 6.3.1 Linear Operators on Normed Linear Spaces

Let $(X, \|\cdot\|_X)$ and $(Y, \|\cdot\|_Y)$ be two normed linear spaces. Let $T : dom(T) \subset X \to Y$ be a mapping and assume $dom(T)$ is a subspace of X and $T(\alpha\boldsymbol{x} + \beta\boldsymbol{y}) = \alpha T(\boldsymbol{x}) + \beta T(\boldsymbol{y})$ for all α, β in the scalar field F of the vector space and for all $\boldsymbol{x}, \boldsymbol{y}$ in X. We say T is a linear operator from X to Y.

If $dom(T) = X$ and $Y = \Re$, we say T is a real linear functional on X. If $Y = \mathbb{C}$, we say T is a complex linear functional on X.

We can also have a general mapping between X and Y without any assumption of linearity. In this case the domain need not be a subspace of X. Formally,

Definition 6.3.2 Nonlinear Operators on Normed Linear Spaces

Let $(X, \|\cdot\|_X)$ and $(Y, \|\cdot\|_Y)$ be two normed linear spaces. Let $T : \Omega \subset X \to Y$ be a mapping. We say T is an operator on X and there is no requirement that Ω be a subspace of X and no requirement T be linear.

If $dom(T) = X$ and $Y = \Re$, T is a real functional on X and if $Y = \mathbb{C}$, T is a complex functional on X.

We need to look at some examples.
(**Linear Functionals on Continuous Functions**):
$T : (C([0,1]), \|\cdot\|_\infty) \to \Re$ by $T(\boldsymbol{x}) = \int_0^1 x(t)dt$. Clearly, T is a linear functional.

(**Linear Operators on Differentiable Functions**):
$T : (C^1([0,1]), \|\cdot\|) \to (C([0,1]), \|\cdot\|_\infty)$ by $T(\boldsymbol{x}) = \boldsymbol{x}'$. Recall

$$C^1([0,1]) \quad = \quad \{\boldsymbol{x} : [0,1] \to \Re \mid \boldsymbol{x}' \in C([0,1])\}$$

and here $\|\boldsymbol{x}\| = \|\boldsymbol{x}\|_\infty + \|\boldsymbol{x}'\|_\infty$. Clearly T is linear. We see $dom(T) = C^1([0,1])$ which is a subspace of $C([0,1])$. Also note the kernel of T is $\Re$ and so is one dimensional.

(**Linear Operators on Finite Dimensional Normed Linear Spaces**):
Let's consider linear operators on finite dimensional spaces more carefully. Let $(X, \|\cdot\|_X)$ and $(Y, \|\cdot\|_Y)$ be two finite dimensional normed spaces with $dim(X)- = n$ and $dim(Y) = m$. Let $T : X \to Y$ be any linear mapping. As we have discussed for a basis $\boldsymbol{E} = \{\boldsymbol{E}_1, \ldots, \boldsymbol{E}_n\}$ of X and a basis $\boldsymbol{F} = \{\boldsymbol{F}_1, \ldots, \boldsymbol{F}_n\}$ of Y. Since X and Y are finite dimensional vector spaces, we can prove

Theorem 6.3.1 Linear Operators between Finite Dimensional Vector Spaces

Let $\boldsymbol{X}$ and $\boldsymbol{Y}$ be two finite dimensional vector spaces over $\Re$. Assume $\boldsymbol{X}$ has dimension n and $\boldsymbol{Y}$ has dimension m. Let $\boldsymbol{T}$ be a linear transformation between the spaces. Given a basis $\boldsymbol{E}$ for $\boldsymbol{X}$ and a basis $\boldsymbol{F}$ for $\boldsymbol{Y}$, we can identify $\Re^n$ with the span of $\boldsymbol{E}$ and $\Re^m$ with the span of $\boldsymbol{F}$. Then here is an $n \times m$ matrix $\boldsymbol{T_{EF}} : \Re^n \to \Re^m$ so that

$$\left[\boldsymbol{Tx}\right]_{\boldsymbol{F}} = \begin{bmatrix} T_{11} & \ldots & T_{1n} \\ \vdots & \vdots & \vdots \\ T_{m1} & \ldots & T_{mn} \end{bmatrix}_{\boldsymbol{EF}} \left[\boldsymbol{x}\right]_{\boldsymbol{E}}$$

Proof 6.3.1
Given $\boldsymbol{x}$ in $\boldsymbol{X}$, we can write

$$\boldsymbol{x} = x_1 \boldsymbol{E_1} + \ldots + x_n \boldsymbol{E_n}$$

By the linearity of $\boldsymbol{T}$, we then have

$$\boldsymbol{T}(\boldsymbol{x}) = x_1 \boldsymbol{T}(\boldsymbol{E_1}) + \ldots + x_n \boldsymbol{T}(\boldsymbol{E_n})$$

But each $\boldsymbol{T}(\boldsymbol{E_i})$ is in $\boldsymbol{F}$, so we have the expansions

$$\begin{aligned} \boldsymbol{T}(\boldsymbol{E_1}) &= \beta_{11}\boldsymbol{F_1} + \ldots + \beta_{m1}\boldsymbol{F_m} \\ \boldsymbol{T}(\boldsymbol{E_2}) &= \beta_{12}\boldsymbol{F_1} + \ldots + \beta_{m2}\boldsymbol{F_m} \\ \vdots &= \vdots \\ \boldsymbol{T}(\boldsymbol{E_n}) &= \beta_{1n}\boldsymbol{F_1} + \ldots + \beta_{mn}\boldsymbol{F_m} \end{aligned}$$

which implies since $\boldsymbol{Tx}$ is also in $\boldsymbol{Y}$ that

$$\boldsymbol{T}(\boldsymbol{x}) = x_1 \left(\sum_{j=1}^{m} \beta_{j1} \boldsymbol{F_j} \right) + \ldots + x_n \left(\sum_{j=1}^{m} \beta_{jn} \boldsymbol{F_j} \right) = \sum_{j=1}^{m} \gamma_j \boldsymbol{F_j}$$

Now reorganize these sums to obtain

$$\sum_{j=1}^{m} \left(\sum_{i=1}^{n} \beta_{ji} x_i \right) \boldsymbol{F_j} = \sum_{j=1}^{m} \gamma_j \boldsymbol{F_j}$$

But expansions with respect to the $\boldsymbol{F}$ basis are unique, so we have

$$\begin{bmatrix} \gamma_1 \\ \vdots \\ \gamma_m \end{bmatrix}_{\boldsymbol{F}} = \begin{bmatrix} \beta_{11} & \beta_{12} & \beta_{13} & \ldots & \beta_{1n} \\ \beta_{21} & \beta_{22} & \beta_{23} & \ldots & \beta_{2n} \\ \vdots & \vdots & \vdots & \vdots & \vdots \\ \beta_{m1} & \beta_{m2} & \beta_{m3} & \ldots & \beta_{mn} \end{bmatrix}_{\boldsymbol{EF}} \begin{bmatrix} x_1 \\ \vdots \\ x_n \end{bmatrix}_{\boldsymbol{E}}$$

Let $T_{ij} = \beta_{ij}$ and we have found the matrix $\boldsymbol{T_{EF}}$ so that

$$\left[\boldsymbol{T}(\boldsymbol{x})\right]_{\boldsymbol{F}} = \boldsymbol{T_{EF}} \left[\boldsymbol{x}\right]_{\boldsymbol{E}}$$

■

Homework

Exercise 6.3.1 *Define T on the set of two dimensional continuously differentiable vector functions on the interval $[a, b]$ by*

$$T\left(\begin{bmatrix} \boldsymbol{x} \\ \boldsymbol{y} \end{bmatrix}\right) = \begin{bmatrix} \boldsymbol{x}' \\ \boldsymbol{y}' \end{bmatrix}$$

for any such functions $\boldsymbol{x}$ and $\boldsymbol{y}$.

- *Prove T is a linear map between vector spaces whether or not we impose a norm structure.*
- *Use the norm*

$$\left\| \begin{bmatrix} \boldsymbol{x} \\ \boldsymbol{y} \end{bmatrix} \right\| = \|\boldsymbol{x}\|_\infty + \|\boldsymbol{y}\|_\infty$$

on the domain and the range of T. Prove $\|T\|$ is not bounded.

Exercise 6.3.2 *Define T on the set of two dimensional continuously differentiable vector functions on the interval $[a, b]$ by*

$$T\left(\begin{bmatrix} \boldsymbol{x} \\ \boldsymbol{y} \end{bmatrix}\right) = \begin{bmatrix} \boldsymbol{x}' \\ \boldsymbol{y}' \end{bmatrix}$$

for any such functions $\boldsymbol{x}$ and $\boldsymbol{y}$.

- *Prove T is a linear map between vector spaces whether or not we impose a norm structure.*
- *Use the norm*

$$\left\| \begin{bmatrix} \boldsymbol{x} \\ \boldsymbol{y} \end{bmatrix} \right\| = \|\boldsymbol{x}\|_\infty + \|\boldsymbol{y}\|_\infty + \|\boldsymbol{x}'\|_\infty + \|\boldsymbol{y}'\|_\infty$$

on the domain and the range of T. Prove $\|T\|$ is bounded by 1.

Exercise 6.3.3 *Define T on the set of two dimensional continuous vector functions on the interval $[a, b]$ by*

$$T\left(\begin{bmatrix} \boldsymbol{x} \\ \boldsymbol{y} \end{bmatrix}\right) = \begin{bmatrix} \int_a^t x(s)ds \\ \int_a^t y(s)ds \end{bmatrix}$$

for any such functions $\boldsymbol{x}$ and $\boldsymbol{y}$.

- *Prove T is a linear map between vector spaces whether or not we impose a norm structure.*
- *Use the norm*

$$\left\| \begin{bmatrix} \boldsymbol{x} \\ \boldsymbol{y} \end{bmatrix} \right\| = \|\boldsymbol{x}\|_\infty + \|\boldsymbol{y}\|_\infty$$

on the domain and the range of T. Prove $\|T\| \leq b - a$.

Exercise 6.3.4 *Define T on the set of two dimensional continuously differentiable $2 times 2$ matrix functions on the interval $[a, b]$ by*

$$T\left(\begin{bmatrix} \boldsymbol{x}_{11} & \boldsymbol{x}_{12} \\ \boldsymbol{x}_{21} & \boldsymbol{x}_{22} \end{bmatrix}\right) = \begin{bmatrix} \boldsymbol{x}_{11}' & \boldsymbol{x}_{12}' \\ \boldsymbol{x}_{21}' & \boldsymbol{x}_{22}' \end{bmatrix}$$

for any such functions $\boldsymbol{x_{ij}}$.

- *Prove* T *is a linear map between vector spaces whether or not we impose a norm structure.*
- *Use an appropriate norm involving* $\|\boldsymbol{u}\|_\infty$ *for a continuously differentiable* $\boldsymbol{u}$ *and show* $\|T\|$ *is not bounded.*

Exercise 6.3.5 *Define* T *on the set of two dimensional continuously differentiable* $2times2$ *matrix functions on the interval* $[a, b]$ *by*

$$T\left(\begin{bmatrix} \boldsymbol{x_{11}} & \boldsymbol{x_{12}} \\ \boldsymbol{x_{21}} & \boldsymbol{x_{22}} \end{bmatrix}\right) = \begin{bmatrix} \boldsymbol{x_{11}}' & \boldsymbol{x_{12}}' \\ \boldsymbol{x_{21}}' & \boldsymbol{x_{22}}' \end{bmatrix}$$

for any such functions $\boldsymbol{x_{ij}}$.

- *Prove* T *is a linear map between vector spaces whether or not we impose a norm structure.*
- *Use an appropriate norm involving* $\|\boldsymbol{u}\|_\infty$ *and* $\|\boldsymbol{u}'\|_\infty$ *for a continuously differentiable* $\boldsymbol{u}$ *and show* $\|T\|$ *is bounded.*

Exercise 6.3.6 *Define* T *on the set of two dimensional continuously* $2times2$ *matrix functions on the interval* $[a, b]$ *by*

$$T\left(\begin{bmatrix} \boldsymbol{x_{11}} & \boldsymbol{x_{12}} \\ \boldsymbol{x_{21}} & \boldsymbol{x_{22}} \end{bmatrix}\right) = \begin{bmatrix} \int_a^t x_{11}(s)ds & \int_a^t x_{12}(s)ds \\ \int_a^t x_{21}(s)ds & \int_a^t x_{22}(s)ds \end{bmatrix}$$

for any such functions $\boldsymbol{x_{ij}}$.

- *Prove* T *is a linear map between vector spaces whether or not we impose a norm structure.*
- *Use an appropriate norm involving* $\|\boldsymbol{u}\|_\infty$ *for a continuously differentiable* $\boldsymbol{u}$ *and show* $\|T\|$ *is bounded.*

Exercise 6.3.7 *Redo the exercises above using the* $\|\cdot\|_1$ *on the functions.*

Exercise 6.3.8 *Redo the exercises above using the* $\|\cdot\|_2$ *on the functions.*

Thus, any linear operator $\boldsymbol{T}$ such as this is associated with an $m \times n$ matrix once the bases $\boldsymbol{E}$ and $\boldsymbol{F}$ are chosen. What happens if we change to a new basis $\boldsymbol{G}$ for $\boldsymbol{X}$ and a new basis $\boldsymbol{H}$ for $\boldsymbol{Y}$? The change of an orthonormal basis determines an equivalence relation on the set of all matrices.

Theorem 6.3.2 A Linear Operator between Finite Dimensional Vector Spaces Determines an Equivalence Class of Matrices

Let $\boldsymbol{X}$ *and* $\boldsymbol{Y}$ *be two finite dimensional vector spaces over* $\Re$. *Assume* $\boldsymbol{X}$ *has dimension* n *and* $\boldsymbol{Y}$ *has dimension* m. *Let* $\boldsymbol{T}$ *be a linear transformation between the spaces. Then* $\boldsymbol{T}$ *is associated with an equivalence class in the set of* $m \times n$ *matrices under the equivalence relation* $\sim$ *where* $\boldsymbol{A} \sim \boldsymbol{B}$ *if and only if there are invertible matrices* $\boldsymbol{P}$ *and* $\boldsymbol{Q}$ *so that* $\boldsymbol{A} = \boldsymbol{P}^{-1}\boldsymbol{B}\boldsymbol{Q}$.

Proof 6.3.2

First, it is easy to see defines an equivalence relation on the set of all $m \times n$ *matrices. We leave that to you. Any such linear transformation* $\boldsymbol{T}$ *is associated with an* $m \times n$ *matrix once the bases* $\boldsymbol{E}$ *and* $\boldsymbol{F}$ *are chosen. What happens if we change to a new basis* $\boldsymbol{G}$ *for* $\boldsymbol{X}$ *and a new basis* $\boldsymbol{H}$ *for* $\boldsymbol{Y}$? *We know from earlier calculations that there are invertible matrices* $\boldsymbol{A_{EG}}$ *and* $\boldsymbol{B_{FH}}$ *so that*

$$\left[\boldsymbol{x}\right]_{\boldsymbol{E}} = \boldsymbol{A_{GE}} \left[\boldsymbol{x}\right]_{\boldsymbol{G}}, \quad \left[\boldsymbol{y}\right]_{\boldsymbol{F}} = \boldsymbol{B_{HF}} \left[\boldsymbol{y}\right]_{\boldsymbol{H}}$$

Using all this, we have

$$\left[\boldsymbol{Tx}\right]_{\boldsymbol{F}} = \boldsymbol{T_{EF}}\left[\boldsymbol{x}\right]_{\boldsymbol{E}} \Longrightarrow \boldsymbol{B_{HF}}\left[\boldsymbol{Tx}\right]_{\boldsymbol{H}} = \boldsymbol{T_{EF}}\,\boldsymbol{A_{GE}}\left[\boldsymbol{x}\right]_{\boldsymbol{G}}$$

Since, $\boldsymbol{B_{HF}}$ is invertible, this tells us

$$\left[\boldsymbol{Tx}\right]_{\boldsymbol{H}} = \boldsymbol{B_{HF}}^{-1}\,\boldsymbol{T_{EF}}\,\boldsymbol{A_{GE}}\left[\boldsymbol{x}\right]_{\boldsymbol{G}}$$

We also know $\left[\boldsymbol{Tx}\right]_{\boldsymbol{H}} = \boldsymbol{T_{GH}}\left[\boldsymbol{x}\right]_{\boldsymbol{G}}$ and thus we must have $\boldsymbol{T_{GH}} = \boldsymbol{B_{HF}}^{-1}\,\boldsymbol{T_{EF}}\,\boldsymbol{A_{GE}}$. This tells us $\boldsymbol{T_{GH}} \sim \boldsymbol{T_{EF}}$. ■

Comment 6.3.1 *Thus any linear transformation between two finite dimensional vector spaces is identified with an equivalence class of $m \times n$ matrices. When we pick basis $\boldsymbol{E}$ and $\boldsymbol{F}$ for $\boldsymbol{X}$ and $\boldsymbol{Y}$ respectively, we choose a particular representative from this equivalence class we denote by $\boldsymbol{T_{EF}}$. Note we can say more about the structure of $\boldsymbol{T_{EF}}$ if these bases are orthonormal.*

We have additional information here as both X and Y have a norm associated with them. Consider the ratio $\frac{\|T(\boldsymbol{x})\|_Y}{\|\boldsymbol{x}\|}$ for any $\boldsymbol{x} \in X$ which is not zero. We know

$$\begin{aligned}
\boldsymbol{x} &= x_1\boldsymbol{E_1} + \ldots + x_n\boldsymbol{E_n} \\
\boldsymbol{T}(\boldsymbol{x}) &= y_1\boldsymbol{F_1} + \ldots + y_m\boldsymbol{F_m} \\
&= x_1\boldsymbol{T}(\boldsymbol{E_1}) + \ldots + x_n\boldsymbol{T}(\boldsymbol{E_n}) \\
&= x_1\left(\sum_{j=1}^{m}\beta_{j1}\boldsymbol{F_j}\right) + \ldots + x_n\left(\sum_{j=1}^{m}\beta_{jn}\boldsymbol{F_j}\right) \\
&= \sum_{j=1}^{m}\left(\sum_{i=1}^{n}\beta_{ji}x_i\right)\boldsymbol{F_j}
\end{aligned}$$

Let $B_F = \max\{\|\boldsymbol{F_1}\|_Y, \ldots, \|\boldsymbol{F_m}\|_Y\}$. Then by the Linear Combination Theorem applied to $\boldsymbol{E}$ there is $\boldsymbol{c_E} > 0$ so that

$$\left\|\sum_{i=1}^{n} x_i\boldsymbol{E_i}\right\| \geq \boldsymbol{c_E}\sum_{i=1}^{n}|x_i| \Longrightarrow \|\boldsymbol{x}\|_X \geq \boldsymbol{c_E}\sum_{i=1}^{n}|x_i|$$

Now

$$\|T(\boldsymbol{x})\|_Y \leq \sum_{j=1}^{m}\sum_{i=1}^{n}|\beta_{ji}|\,|x_i|\,\|\boldsymbol{F_j}\|_Y \leq B_F\sum_{j=1}^{m}\sum_{i=1}^{n}|\beta_{ji}|\,|x_i|$$

Let $D_{EF}^i = \sum_{j=1}^{m}|\beta_{ji}|$. Note this is the sum of the absolute values of the entries in column i of the matrix $[T]_{EF}$. Then $\|T(\boldsymbol{x})\|_Y \leq B_F\sum_{i=1}^{n} D_{EF}^i\,|x_i|$. Now let $D_{EF} = \max\{D_{EF}^1, \ldots, D_{EF}^n\}$. Then $\|T(\boldsymbol{x})\|_Y \leq B_F\,D_{EF}\sum_{i=1}^{n}|x_i|$. Combining all these estimates, we have for $\boldsymbol{x} \neq \boldsymbol{0}$,

$$\frac{\|T(\boldsymbol{x})\|_Y}{\|\boldsymbol{x}\|_X} \leq \frac{B_F\,D_{EF}\sum_{i=1}^{n}|x_i|}{\boldsymbol{c_E}\sum_{i=1}^{n}|x_i|} = \frac{B_F\,D_{EF}\sum_{i=1}^{n}|x_i|}{\boldsymbol{c_E}\sum_{i=1}^{n}|x_i|} = \frac{B_F\,D_{EF}}{\boldsymbol{c_E}}$$

We see we have found a constant so that

$$\frac{\|T(\boldsymbol{x})\|_Y}{\|\boldsymbol{x}\|_X} \leq \frac{B_F\,D_{EF}}{\boldsymbol{c_E}}, \quad \forall \boldsymbol{x} \neq \boldsymbol{0}$$

where

- B_F depends on $\|\cdot\|_Y$ and $\boldsymbol{F}$, the basis for Y.
- $c_{\boldsymbol{E}}$ depends on $\|\cdot\|_X$ and $\boldsymbol{E}$, the basis for X.
- D_{EF} depends on $\boldsymbol{E}$ and $\boldsymbol{F}$.

Hence, $\sup_{\boldsymbol{x}\neq 0} \frac{\|T(\boldsymbol{x})\|_Y}{\|\boldsymbol{x}\|_X} < \infty$ for such linear operators.

Homework:

Exercise 6.3.9 *Let*

$$A = \begin{bmatrix} 1 & 3 \\ 4 & -2 \end{bmatrix}$$

be the matrix representation of the linear operator $T : \Re^2 \rightarrow \Re^2$ for the standard basis which we denote by $\boldsymbol{e} = \{\boldsymbol{e_1}, \boldsymbol{e_2}\}$ as usual. Hence, this is the matrix representation T_{ee}. So $\boldsymbol{F} = \boldsymbol{e}$ here. We will use this basis for both the range and the domain of T and we will use $\|\cdot\|_2$ on both. Let E be the set of two linearly independent vectors

$$\boldsymbol{E} = \left\{ \begin{bmatrix} -1 \\ 2 \end{bmatrix}, \begin{bmatrix} 2 \\ 4 \end{bmatrix} \right\}$$

- *Find B_F and D_{ee}.*
- *For x_1 and x_2 both positive or both negative, show*

$$\left(|x_1| \begin{bmatrix} -1 \\ 2 \end{bmatrix} + |x_2| \begin{bmatrix} 2 \\ 4 \end{bmatrix} \right)^2 \geq 5(|x_1| + |x_2|)^2$$

- *For x_1 and x_2 different signs, it is enough to look at the problem of finding c^2 so that*

$$\frac{5x_1^2 - 18x_1x_2 + 20x_2^2}{(|x_1| + |x_2|)^2} \geq c^2$$

and for $u = \frac{|x_1|}{|x_2|}$, this is equivalent to finding c^2 so that

$$\frac{5u^2 - 18u + 20}{1 + u^2} \geq c^2$$

Show using standard calculus that $c^2 = 2.53$.

- *Combining these arguments, we see we have estimated $c_{\boldsymbol{E}}$. and so we can estimate*

$$\frac{\|T(\boldsymbol{x})\|_2}{\|\boldsymbol{x}\|_2} \leq \frac{B_F \, D_{EF}}{c_{\boldsymbol{E}}}, \ \forall \boldsymbol{x} \neq \boldsymbol{0}$$

- *Now redo the above assuming the range has the basis*

$$\boldsymbol{F} = \left\{ \begin{bmatrix} 3 \\ -2 \end{bmatrix}, \begin{bmatrix} 1 \\ 5 \end{bmatrix} \right\}$$

Hence, we still use the representation of T but now it is interpreted as T_{eF} instead of T_{ee}. so you have to find D_{eF}.

Exercise 6.3.10 *Let*

$$A = \begin{bmatrix} 8 & 2 \\ 9 & -3 \end{bmatrix}$$

be the matrix representation of the linear operator $T : \Re^2 \rightarrow \Re^2$ *for the standard basis which we denote by* $\mathbf{e} = \{\mathbf{e_1}, \mathbf{e_2}\}$ *as usual. So* $\boldsymbol{F} = \mathbf{e}$ *here. Hence, this is the matrix representation* T_{ee}*. We will use this basis for both the range and the domain of* T *and we will use* $\|\cdot\|_2$ *on both. Let* E *be the set of two linearly independent vectors*

$$\boldsymbol{E} = \left\{ \begin{bmatrix} -1 \\ -2 \end{bmatrix}, \begin{bmatrix} 1 \\ 1 \end{bmatrix} \right\}$$

- *Find* B_F *and* D_{ee}.
- *Estimate* $\mathbf{c}_{\boldsymbol{E}}$ *as you did in the previous exercise.*
- *Combining these arguments, we see we have estimated* $\mathbf{c}_{\boldsymbol{E}}$. *and so we can estimate*

$$\frac{\|T(\boldsymbol{x})\|_2}{\|\boldsymbol{x}\|_2} \leq \frac{B_F \, D_{EF}}{\mathbf{c}_{\boldsymbol{E}}}, \quad \forall \boldsymbol{x} \neq \mathbf{0}$$

- *Now redo the above assuming the range has the basis*

$$\boldsymbol{F} = \left\{ \begin{bmatrix} 6 \\ 2 \end{bmatrix}, \begin{bmatrix} 3 \\ -4 \end{bmatrix} \right\}$$

  Hence, we still use the representation of T *but now it is interpreted as* T_{eF} *instead of* T_{ee}*. so you have to find* D_{eF}.

However, we need not consider linear mappings between finite dimensional normed linear spaces. We can explore the idea of a bounded linear operator with other mappings.

- $T : (C([0,1]), \|\cdot\|_\infty) \rightarrow (\Re, |\cdot|)$ by $T(\boldsymbol{x}) = \int_0^1 x(t)dt$. Note for $\boldsymbol{x} \neq \mathbf{0}$, we have

$$\frac{\|T(\boldsymbol{x})\|_Y}{\|\boldsymbol{x}\|_X} = \frac{\left|\int_0^1 x(t)\, dt\right|}{\|\boldsymbol{x}\|_\infty} \leq \frac{\int_0^1 |x(t)|\, dt}{\|\boldsymbol{x}\|_\infty} \leq \frac{\|\boldsymbol{x}\|_\infty| : \int_0^1 \mathbf{1}\, dt}{\|\boldsymbol{x}\|_\infty} = 1$$

  Thus,

$$\sup_{\|\boldsymbol{x}\|_\infty \neq 0} \frac{\left|\int_0^1 x(t)\, dt\right|}{\|\boldsymbol{x}\|_\infty} \leq 1$$

  and this ratio is bounded when we take the supremum over all possible ratios.

- $T : (C^1([0,1]), \|\cdot\|) \rightarrow (C([0,1]), \|\cdot\|_\infty)$ by $T(\boldsymbol{x}) = \boldsymbol{x}'$. Recall

$$C^1([0,1]) = \{\boldsymbol{x} : [0,1] \rightarrow \Re \mid \boldsymbol{x}' \in C([0,1])\}$$

  and here $\|\boldsymbol{x}\| = \|\boldsymbol{x}\|_\infty + \|\boldsymbol{x}'\|_\infty$. For $\boldsymbol{x} \neq \mathbf{0}$, we have

$$\frac{\|T(\boldsymbol{x})\|_Y}{\|\boldsymbol{x}\|_X} = \frac{\|\boldsymbol{x}'\|_\infty}{\|\boldsymbol{x}\|_\infty + \|\boldsymbol{x}'\|_\infty} \leq 1$$

So

$$\sup_{\|\boldsymbol{x}\|_\infty \neq 0} \frac{\|\boldsymbol{x}'\|_\infty}{\|\boldsymbol{x}\|_\infty + \|\boldsymbol{x}'\|_\infty} \leq 1$$

and this ratio is bounded when we take the supremum over all possible ratios.

- $T : (C^1([0,1]), \|\cdot\|) \rightarrow (C([0,1]), \|\cdot\|_\infty)$ by $T(\boldsymbol{x}) = \boldsymbol{x}'$. Recall

$$C^1([0,1]) = \{\boldsymbol{x} : [0,1] \rightarrow \Re \mid \boldsymbol{x}' \in C([0,1])\}$$

and now $\|\boldsymbol{x}\| = \|\boldsymbol{x}\|_\infty$ only, for $\boldsymbol{x} \neq \boldsymbol{0}$, we have

$$\frac{\|T(\boldsymbol{x})\|_Y}{\|\boldsymbol{x}\|_X} = \frac{\|\boldsymbol{x}'\|_\infty}{\|\boldsymbol{x}\|_\infty}$$

On $[0,1]$, $x_n(t) = t^n$ has sup norm 1 and for $n \geq 1$.

$$\frac{\|\boldsymbol{x_n}'\|_\infty}{\|\boldsymbol{x_n}\|_\infty} = \frac{\max_{0 \leq t \leq 1} |nt^{n-1}|}{\max_{0 \leq t \leq 1} |t^n|} = n$$

So

$$\sup_{\|\boldsymbol{x}\|_\infty \neq 0} \frac{\|\boldsymbol{x}'\|_\infty}{\|\boldsymbol{x}\|_\infty} = \infty$$

and this ratio becomes unbounded when we take the supremum over all possible ratios.

We are now ready to define what we mean by a bounded linear operator.

Definition 6.3.3 Bounded Linear Operators on Normed Linear Spaces

Let $T : dom(T) \subset (X, \|\cdot\|_X) \rightarrow (Y, \|\cdot\|_Y)$ be a linear operator (i.e. mapping) on the subspace $dom(T)$. We say T is a bounded linear operator if

$$B(T) = \sup_{\boldsymbol{x} \in dom(T),\, \boldsymbol{x} \neq \boldsymbol{0}} \frac{\|T(\boldsymbol{x})\|_Y}{\|\boldsymbol{x}\|_X} < \infty$$

Since T is linear

$$\frac{\|T(\boldsymbol{x})\|_Y}{\|\boldsymbol{x}\|_X} = T\left(\frac{\boldsymbol{x}}{\|\boldsymbol{x}\|_X}\right)$$

and so an equivalent way to define $B(T)$ is

$$B(T) = \sup_{\boldsymbol{x} \in dom(T),\, \|\boldsymbol{x}\|_X = 1} \|T(\boldsymbol{x})\|_Y$$

If the range is the real or complex numbers, T is a linear functional on X. We have

Definition 6.3.4 Bounded Real Linear Functionals on Normed Linear Spaces

Let $T : dom(T) \subset (X, \|\cdot\|_X) \rightarrow (\Re, \|\cdot\|)$ be a linear mapping on the subspace $dom(T)$. We say T is a bounded real linear functional if

$$B(T) = \sup_{\boldsymbol{x} \in dom(T),\, \boldsymbol{x} \neq \boldsymbol{0}} \frac{|T(\boldsymbol{x})|}{\|\boldsymbol{x}\|_X} = \sup_{\boldsymbol{x} \in dom(T),\, \|\boldsymbol{x}\|_X = 1} |T(\boldsymbol{x})| < \infty$$

If the scalar field is $\mathbb{C}$, the definitions are the same.

Homework

Exercise 6.3.11 *for any $N \geq 1$, define the operator $T_N : \ell^2 \rightarrow \ell^2$ by*

$$T(x_1, x_2, \ldots, x_N, x_{N+1}, \ldots) = (2x_1, 2x_2, \ldots 2x_N.3x_{N+1}, 3x_{N+2} \ldots)$$

for any $\boldsymbol{x} = (x_i)_{i \geq 1}$.

- *Prove T is linear.*
- *Prove T is a bounded linear operator with $B(T_N) = \sup_{\boldsymbol{x} \neq \boldsymbol{0}} \frac{\|T(\boldsymbol{x})\|_2}{\|\boldsymbol{x}\|_2} = 3$.*

Exercise 6.3.12 *Let L be the differentiable operator on the set of continuously twice differentiable functions on $[0,1]$, $C^2([0,1])$ given by $L(\boldsymbol{x}) = \boldsymbol{x}'' + 3\boldsymbol{x}' + 4\boldsymbol{x}$. Let the norm on $C^2([0,1])$ be $\|\boldsymbol{x}\| = \|\boldsymbol{x}\|_\infty + \|\boldsymbol{x}'\|_\infty + \|\boldsymbol{x}''\|_\infty$.*

- *Prove L is linear.*
- *Prove L is a bounded linear operator with $B(T_N) = \sup_{\boldsymbol{x} \neq \boldsymbol{0}} \frac{\|L(\boldsymbol{x})\|}{\|\boldsymbol{x}\|} = 4$.*

Exercise 6.3.13 *Let L be the differentiable operator on the set of continuously twice differentiable functions on $[0,1]$, $C^2([0,1])$ given by $L(\boldsymbol{x}) = \boldsymbol{x}'' + 3\boldsymbol{x}' + 4\boldsymbol{x}$. Let the norm on $C^2([0,1])$ be $\|\boldsymbol{x}\| = \|\boldsymbol{x}''\|_\infty$.*

- *Prove L is linear.*
- *Prove L is a not a bounded linear operator.*

Exercise 6.3.14 *Let the set of functions of two variables on $[0,1] \times [0,1]$, $u(t,x)$, with the first and second order partial derivatives of u continuous on $[0,1] \times [0,1]$ be X. The partial differential operator L is given by $L(\boldsymbol{u}) = u_t - 5u_{xx}$. Let the norm on X be*

$$\|\boldsymbol{u}\| = \|\boldsymbol{u}\|_\infty + \|\boldsymbol{u_t}\|_\infty + \|\boldsymbol{u_x}\|_\infty + \|\boldsymbol{u_{tt}}\|_\infty + \|\boldsymbol{u_{xt}}\|_\infty + \|\boldsymbol{u_{xx}}\|_\infty$$

where the supremum is taken over $[0,1] \times [0,1]$.

- *Prove L is linear.*
- *Prove L is a bounded linear operator with $B(T_N) = \sup_{\boldsymbol{x} \neq \boldsymbol{0}} \frac{\|L(\boldsymbol{u})\|}{\|\boldsymbol{u}\|} = 5$.*

Exercise 6.3.15 *Let the set of functions of two variables on $[0,1] \times [0,1]$, $u(t,x)$, with the first and second order partial derivatives of u continuous on $[0,1] \times [0,1]$ be X. The partial differential operator L is given by $L(\boldsymbol{u}) = u_t - 5u_{xx}$. Let the norm on X be $\|\boldsymbol{u}\| = \|\boldsymbol{u}\|_\infty$ where the supremum is taken over $[0,1] \times [0,1]$.*

- *Prove L is linear.*
- *Prove L is not a bounded linear operator*

The mappings T we are studying can be continuous at a given $\boldsymbol{x}_0$ in $dom(T) \subset X$.

Definition 6.3.5 Continuous Mappings between Normed Linear Spaces

Let $T : \Omega \subset (X, \|\cdot\|_X) \to (Y, \|\cdot\|_Y)$ be an operator. Assume T is locally defined at $\boldsymbol{x}_0 \in \Omega$; i.e. there is $r > 0$ so that $B(\boldsymbol{x}_0; r) \subset \Omega$. We say T is continuous at $\boldsymbol{x}_0 \in \Omega$ if given $\epsilon > 0$, there is a $\delta > 0$ with $\delta < r$ so that

$$\|\boldsymbol{x} - \boldsymbol{x}_0\|_X < \delta \implies \|T(\boldsymbol{x}) - T(\boldsymbol{x}_0)\|_Y < \epsilon$$

Linear operators have special properties.

Theorem 6.3.3 Properties of Linear Operators

Let $T : dom(T) \subset (X, \|\cdot\|_X) \to (Y, \|\cdot\|_Y)$ be a linear operator on the subspace $dom(T)$. Then

- *If T is continuous at $\boldsymbol{x}_0 \in dom(T)$, T is continuous on all $dom(T)$.*
- *If T is continuous at $\boldsymbol{x}_0$ in $dom(T)$, T is a bounded linear operator.*
- *If T is a bounded linear operator, T is continuous on $dom(T)$.*

Proof 6.3.3
If T is continuous at $\boldsymbol{x}_0 \in dom(T)$, T is continuous on all $dom(T)$:
Let T be continuous at $\boldsymbol{x}_0 \in dom(T)$. Then, there is an $r > 0$ so that $B(\boldsymbol{x}_0; r) \subset dom(T)$ and for $\epsilon > 0$, there is a positive $\delta < r$ so that $\|T(\boldsymbol{x}) - T(\boldsymbol{x}_0)\|_Y < \epsilon$ if $\|\boldsymbol{x} - \boldsymbol{x}_0\|_X < \delta$. But T is linear, so we can say

$$\|\boldsymbol{x} - \boldsymbol{x}_0\|_X < \delta \implies \|T(\boldsymbol{x} - \boldsymbol{x}_0)\|_Y < \epsilon$$

Let $\boldsymbol{y} = \boldsymbol{x} - \boldsymbol{x}_0$, then we can say

$$\|\boldsymbol{y}\|_X < \delta \implies \|T(\boldsymbol{y})\|_Y < \epsilon$$

Since $T(\boldsymbol{0}) = \boldsymbol{0}$, we have

$$\|\boldsymbol{y} - \boldsymbol{0}\|_X < \delta \implies \|T(\boldsymbol{y}) - T(\boldsymbol{0})\|_Y < \epsilon$$

and hence T is continuous at $\boldsymbol{0}$.

Now let $\boldsymbol{x}_1 \in dom(T)$. If $\|\boldsymbol{x} - \boldsymbol{x}_1\|_X < \delta$, letting $\boldsymbol{z} = \boldsymbol{x} - \boldsymbol{x}_1$, since T is continuous at $\boldsymbol{0}$.

$$\|\boldsymbol{x} - \boldsymbol{x}_1\|_X = \|\boldsymbol{z}\|_X < \delta \implies \|T(\boldsymbol{x}) - T(\boldsymbol{x}_1)\|_Y = \|T(\boldsymbol{x} - \boldsymbol{x}_1)\|_Y = \|T(\boldsymbol{z})\|_Y < \epsilon$$

and so T is continuous at $\boldsymbol{x}_1$ also.

If T is continuous at $\boldsymbol{x}_0$ in $dom(T)$, T is a bounded linear operator.:
Since T is continuous at $\boldsymbol{x}_0 \in dom(T)$, there is an $r > 0$ so that $B(\boldsymbol{x}_0; r) \subset dom(T)$ and for $\epsilon > 0$, there is a positive $\delta < r$ so that $\|T(\boldsymbol{x}) - T(\boldsymbol{x}_0)\|_Y < \epsilon$ if $\|\boldsymbol{x} - \boldsymbol{x}_0\|_X < \delta$. Let $\boldsymbol{y} \neq \boldsymbol{0} \in dom(T)$ and let $\boldsymbol{x} = \boldsymbol{x}_0 + \frac{\delta}{2\|\boldsymbol{y}\|_X}\boldsymbol{y}$. Then $\boldsymbol{x} - \boldsymbol{x}_0 = \frac{\delta}{2\|\boldsymbol{y}\|_X}\boldsymbol{y}$ and $\|\boldsymbol{x} - \boldsymbol{x}_0\|_X = \frac{\delta}{2} < \delta$. Thus, we can say

$\|T(\boldsymbol{x}) - T(\boldsymbol{x_0})\|_Y < \epsilon$. *But*

$$\begin{aligned} T(\boldsymbol{x}) - T(\boldsymbol{x_0}) &= T\left(\boldsymbol{x_0} + \frac{\delta}{2\|\boldsymbol{y}\|_X}\boldsymbol{y}\right) - T(\boldsymbol{x_0}) \\ &= T(\boldsymbol{x_0}) + \frac{\delta}{2\|\boldsymbol{y}\|_X} T(\boldsymbol{y}) - T(\boldsymbol{x_0}) = \frac{\delta}{2\|\boldsymbol{y}\|_X} T(\boldsymbol{y}) \end{aligned}$$

and therefore, we must have

$$\begin{aligned} \|T(\boldsymbol{x}) - T(\boldsymbol{x_0})\|_Y &= \frac{\delta}{2\|\boldsymbol{y}\|_X} \|T(\boldsymbol{y})\|_Y < \epsilon \Longrightarrow \frac{\|T(\boldsymbol{y})\|_Y}{\|\boldsymbol{y}\|_X} < \frac{2\epsilon}{\delta} \\ &\Longrightarrow \sup_{\boldsymbol{y}\neq 0, \boldsymbol{y}\in dom(T)} \frac{\|T(\boldsymbol{y})\|_Y}{\|\boldsymbol{y}\|_X} \leq \frac{2\epsilon}{\delta} < \infty \end{aligned}$$

We conclude T is a bounded linear operator.

If T is a bounded linear operator, T is continuous on $dom(T)$*:*
Since T is bounded, let

$$B(T) = \sup_{\boldsymbol{x}\neq 0, \boldsymbol{x}\in dom(T)} \frac{\|T(\boldsymbol{x})\|_Y}{\|\boldsymbol{x}\|_X}$$

If T is the zero map, then it is clearly continuous, so we can assume T is not always zero. Thus $B(T) > 0$. Pick any $\boldsymbol{x_0} \in dom(T)$. Then for any $\epsilon > 0$, if $\|\boldsymbol{x} - \boldsymbol{x_0}\|_X < \frac{\epsilon}{B(T)}$, from the definition of $B(T)$, we have

$$\|T(\boldsymbol{x}) - T(\boldsymbol{x_0})\|_Y = \|T(\boldsymbol{x} - \boldsymbol{x_0})\|_Y \leq B(T)\, \|\boldsymbol{x} - \boldsymbol{x_0}\| < \epsilon$$

as long as $\boldsymbol{x} \neq \boldsymbol{x_0}$. Of course, if they do match, we satisfy the inequality right away as we get $\|T(\boldsymbol{x}) - T(\boldsymbol{x_0})\|_Y = 0 < \epsilon$. This shows T is continuous at $\boldsymbol{x_0}$. Earlier we proved this also shows T is continuous on $dom(T)$. ■

We can explore how we use these ideas to see how much a solution to a problem changes when we alter data. Let's do this with some exercises.

Homework

Exercise 6.3.16 *Suppose A is an $n \times n$ invertible matrix and we are solving the problem $A\boldsymbol{x} = \mathbf{b}$. Let $\boldsymbol{x}(\boldsymbol{b})$ denote this solution which depends on the data $\mathbf{b}$. Then $\boldsymbol{x}(\boldsymbol{b}) = A^{-1}(\boldsymbol{b})$. Thus,*

$$\boldsymbol{\Delta x}(\boldsymbol{\Delta b}) = \boldsymbol{x}(\boldsymbol{b} + \boldsymbol{\Delta b}) - \boldsymbol{x}(\boldsymbol{b}) = A^{-1}(\boldsymbol{\Delta b})$$

This gives us a way to prove that the solution depends continuously on the data: i.e. given $\epsilon > 0$, there is a δ so that if $\|\mathbf{b}\| < \delta$, then $\|\boldsymbol{\Delta x}\| < \epsilon$. Your task is to prove this assertion using the fact that A^{-1} is a bounded linear operator.

Exercise 6.3.17 *Solve the exercise above for the case $A = \begin{bmatrix} 1 & -3 \\ -2 & 4 \end{bmatrix}$. Use the Frobenius norm for A^{-1}.*

Exercise 6.3.18 *We will prove the same sort of continuous dependence on the data in the context of an ODE. We solve $u'' + 5u' + 6u = f$. The differential operator is $L : C^2([0,1]) \to C([0,1])$ and* *we will $\|u\|_\infty$ as the norm on both the domain and the range. We already know L is an unbounded linear operator with this choice of norm.*

- *Use the initial conditions $u(0) = 0$ and $u'(0) = 0$ and find the homogeneous solution.*
- *Now find the particular solution using variation of parameters.*
- *Now find the general solution and use the initial conditions to find the solution $u(f)$ to the ODE.*
- *Show the inverse of L, $L^{-1} = T$ given by $u(f) = L^{-1}(f) = T(f)$ is defined by*

$$(T(f))(t) = \int_0^t f(s)(e^{-2(t-s)} - e^{-3(t-s)})ds$$

- *Show L^{-1} is a bounded linear operator on $C([0,1])$ even though L is not bounded.*
- *Find the operator norm of L^{-1} and use it to prove the solutions to this ODE depend continuously on the data. for this case of zero initial conditions.*
- *If the initial conditions are not zero, show A and B in the general solution satisfy*

$$\begin{bmatrix} 1 & 1 \\ -2 & -3 \end{bmatrix} \begin{bmatrix} A \\ B \end{bmatrix} = \begin{bmatrix} x_0 \\ x_1 \end{bmatrix}$$

- *We know how the solution to the A, B problem changes with a change in initial condition to $x_0 + \Delta x_0$ and $x_1 + \Delta x_1$. Using this prove the solution to*

$$u'' + 5u' + 6u = f, \quad u(0) = x_0, \quad u'(0) = x_1$$

varies continuously with the change to $f + \Delta f$, $x_0 + \Delta x_0$ and $x_1 + \Delta x_1$.

Exercise 6.3.19 *Prove the same sort of continuous dependence on the data as we did in the previous problem for the ODE $u'' - u' - 6u = f$.*

6.4 Linear Operators on $\Re^n$ to $\Re^m$

Let's specialize these results to the case of $(X, \|\cdot\|_X) = (\Re^n, \|\cdot\|)$ and $(Y, \|\cdot\|_Y) = (\Re^m, \|\cdot\|)$ where we use the same norm on both spaces and our choices for the norm will be $\|\cdot\|_\infty$, $\|\cdot\|_1$ or $\|\cdot\|_2$. Given a basis $\boldsymbol{E}$ for $\Re^n$ and a basis $\boldsymbol{F}$ for $\Re^m$, we know

$$[\boldsymbol{x}]_{\boldsymbol{E}} = \begin{bmatrix} x_1 \\ \vdots \\ x_n \end{bmatrix}_{\boldsymbol{E}}, \quad [\boldsymbol{T}(\boldsymbol{x})]_{\boldsymbol{F}} = \begin{bmatrix} (T(\boldsymbol{x}))_1 \\ \vdots \\ (T(\boldsymbol{x}))_m \end{bmatrix}_{\boldsymbol{F}}, \quad \boldsymbol{T_{EF}} = \begin{bmatrix} \beta_{11} & \beta_{12} & \beta_{13} & \cdots & \beta_{1n} \\ \beta_{21} & \beta_{22} & \beta_{23} & \cdots & \beta_{2n} \\ \vdots & \vdots & \vdots & \vdots & \vdots \\ \beta_{m1} & \beta_{m2} & \beta_{m3} & \cdots & \beta_{mn} \end{bmatrix}_{\boldsymbol{EF}}$$

with

$$[\boldsymbol{T}(\boldsymbol{x})]_{\boldsymbol{F}} = \boldsymbol{T_{EF}} \, [\boldsymbol{x}]_{\boldsymbol{E}}$$

For convenience of exposition, we will just use T to denote the matrix $\boldsymbol{T_{EF}}$.

Using $\|\cdot\|_1$ on both spaces: We have

$$\|T(\boldsymbol{x})\|_1 = \sum_{j=1}^{m} |(T(\boldsymbol{x}))_j| = \sum_{j=1}^{m} \left| \sum_{i=1}^{n} \beta_{ji} x_i \right| \leq \sum_{j=1}^{m} \sum_{i=1}^{n} |\beta_{ji}|\, |x_i|$$

Let $D^i_{EF} = \sum_{j=1}^{m} |\beta_{ji}|$ which is the sum of the absolute values of the jth row of the matrix $T = \boldsymbol{T_{EF}}$. Then, $\|T(\boldsymbol{x})\|_1 \leq \sum_{i=1}^{n} D^i_{EF}\, |x_i|$. Let $D_{EF} = \max_{1 \leq i \leq n} D^i_{EF}$. Then,

$$\|T(\boldsymbol{x})\|_1 \leq D_{EF} \quad \|\boldsymbol{x}\|_1 \Longrightarrow \sup_{\boldsymbol{x} \neq \boldsymbol{0}} \frac{\|T(\boldsymbol{x})\|_1}{\|\boldsymbol{x}\|_1} \leq D_{EF}$$

Let the index i_0 be an index where the maximum D_{EF} is achieved. Then $D_{Ef} = D^{i_0}_{EF}$. Let $\boldsymbol{x} = \boldsymbol{E_{i_0}}$. Note $(T(\boldsymbol{E_{i_0}}))_j = \beta_{j,i_0}$ as

$$\begin{bmatrix} \beta_{11} & \ldots & \boxed{\beta_{1i_0}} & \ldots & \beta_{1n} \\ \beta_{21} & \ldots & \beta_{2i_0} & \ldots & \beta_{2n} \\ \vdots & \ldots & \vdots & \ldots & \vdots \\ \beta m1 & \ldots & \beta m i_0 & \ldots & \beta mn \end{bmatrix} \begin{bmatrix} 0 \\ \vdots \\ 1,\ i_0^{th} \text{ slot} \\ \vdots \\ 0 \end{bmatrix} = \begin{bmatrix} \beta_{1i_0} \\ \vdots \\ \beta_{ji_0} \\ \vdots \\ \beta_{mi_0} \end{bmatrix}$$

and then we have $\|\boldsymbol{E_{i_0}}\|_1 = 1$ and

$$\|T(\boldsymbol{E_{i_0}})\|_1 = \sum_{j=1}^{m} |\beta_{ji_0}| = D^{i_0}_{EF} = D_{EF}$$

Thus, we have

$$D_{EF} = \frac{\|T(\boldsymbol{E_{i_0}})\|_1}{\|\boldsymbol{E_{i_0}}\|_1} \leq \sup_{\boldsymbol{x} \neq \boldsymbol{0}} \frac{\|T(\boldsymbol{x})\|_1}{\|\boldsymbol{x}\|_1} \leq D_E F$$

Thus we can characterize this operator bound as

$$B(T) = \sup_{\boldsymbol{x} \neq \boldsymbol{0}} \frac{\|T(\boldsymbol{x})\|_1}{\|\boldsymbol{x}\|_1} = \max_{1 \leq i \leq n} D^i_{EF} = \max_{1 \leq i \leq n} \sum_{j=1}^{m} |\beta_{ji}|$$

If we start using the notation

$$B_1(T) = \|T_{EF}\|_1 = \sup_{\boldsymbol{x} \neq \boldsymbol{0}} \frac{\|T(\boldsymbol{x})\|_1}{\|\boldsymbol{x}\|_1} = \max_{1 \leq i \leq n} \sum_{j=1}^{m} |\beta_{ji}|$$

we say the operator norm of the matrix T with respect to $\|\cdot\|_1$ on both the domain and range of T is the largest absolute column sum where by absolute column sum we mean the sum of the absolute values of T in that column.

Using $\|\cdot\|_\infty$ On both spaces: Using the same notation as before we have

$$\|T(\boldsymbol{x})\|_\infty = \max_{1 \leq j \leq m} |(T(\boldsymbol{x}))_j| = \max_{1 \leq j \leq m} \left| \sum_{i=1}^{n} \beta_{ji} x_i \right| \leq \max_{1 \leq j \leq m} \sum_{i=1}^{n} |\beta_{ji}|\, |x_i|$$

But $|x_i| \leq \|\boldsymbol{x}\|_\infty$, so for $\boldsymbol{x} \neq \boldsymbol{0}$,

$$\|T(\boldsymbol{x})\|_\infty \leq \left(\max_{1 \leq j \leq m} \sum_{i=1}^{n} |\beta_{ji}| \right) \|\boldsymbol{x}\|_\infty \Longrightarrow \frac{\|T(\boldsymbol{x})\|_\infty}{\|\boldsymbol{x}\|_\infty} \leq \max_{1 \leq j \leq m} \sum_{i=1}^{n} |\beta_{ji}|$$

If we let $B_\infty(T) = \sup_{\boldsymbol{x} \neq \boldsymbol{0}} \frac{\|T(\boldsymbol{x})\|_\infty}{\|\boldsymbol{x}\|_\infty}$, the above shows us $B_\infty(T) \leq \max_{1 \leq j \leq m} \sum_{i=1}^{n} |\beta_{ji}|$. Let $d^j_{EF} = \sum_{i=1}^{n} |\beta_{ji}|$ which is the sum of the absolute values of the entries in T for the jth row. There is some index j_0 where the maximum above occurs. Hence, $\max_{1 \leq j \leq m} d^j_{EF} = d^{j_0}_{EF}$ and

$$\sum_{i=1}^{n} |\beta_{j_0 i}| = \max_{1 \leq j \leq m} \sum_{i=1}^{n} |\beta_{ji}|$$

We can call this the maximum absolute row sum for T. Define $\hat{\boldsymbol{x}}$ by

$$\hat{\boldsymbol{x}} = \begin{bmatrix} sign(\beta_{j_0 1}) \\ \vdots \\ sign(\beta_{j_0,n}) \end{bmatrix}$$

where $sign(y) = 1$ when $y \geq 0$ and -1 otherwise. Then,

$$\begin{bmatrix} \beta_{11} & \cdots & \beta_{1,n} \\ \vdots & \vdots & \vdots \\ \beta_{m1} & \cdots & \beta{mn} \end{bmatrix} \begin{bmatrix} sign(\beta_{j_0 1}) \\ \vdots \\ sign(\beta_{j_0,n}) \end{bmatrix} = \begin{bmatrix} \sum_{i=1}^{n} \beta_{1i}\, sign(\beta_{j_0,i}) \\ \vdots \\ \sum_{i=1}^{n} \beta_{mi}\, sign(\beta_{j_0,i}) \end{bmatrix}$$

For any index j, we have

$$\left| \sum_{i=1}^{n} \beta_{ji}\, sign(\beta_{j_0,i}) \right| \leq \sum_{i=1}^{n} |\beta_{ji}| \leq \sum_{i=1}^{n} |\beta_{j_0,i}|$$

as j_0 is the maximizing index. Further

$$\left| \sum_{i=1}^{n} \beta_{j_0 i}\, sign(\beta_{j_0,i}) \right| = \sum_{i=1}^{n} |\beta_{j_0 i}|$$

We see $\|\hat{\boldsymbol{x}}\|_\infty = 1$ and

$$\|T(\hat{\boldsymbol{x}})\|_\infty = \max_{1 \leq j \leq m} |(T(\hat{\boldsymbol{x}}))_j| = \sum_{i=1}^{n} |\beta_{j_0 i}| = \max_{1 \leq j \leq m} \sum_{i=1}^{n} |\beta_{ji}| = d^{j_0}_{EF}$$

This shows

$$\frac{\|T(\hat{\boldsymbol{x}})\|_\infty}{\|\hat{\boldsymbol{x}}\|_\infty} = \sum_{i=1}^{n} |\beta_{j_0 i}|$$

and so

$$\sum_{i=1}^{n} |\beta_{j_0 i}| = \frac{\|T(\hat{\boldsymbol{x}})\|_\infty}{\|\hat{\boldsymbol{x}}\|_\infty} \leq B_\infty(T) = \sup_{\boldsymbol{x} \neq \boldsymbol{0}} \frac{\|T(\boldsymbol{x})\|_\infty}{\|\boldsymbol{x}\|_\infty} \leq \sum_{i=1}^{n} |\beta_{j_0 i}|$$

We conclude $B_\infty(T) = \sum_{i=1}^n |\beta_{j_0 i}|$ which is the maximum absolute row sum.

Using $\| \cdot \|_2$ on both spaces: We find

$$\| \left[\boldsymbol{T}(\boldsymbol{x}) \right] \|_2^2 \quad = \quad \sum_{j=1}^m ((\boldsymbol{T}(\boldsymbol{x}))_j)^2 = \sum_{i=1}^m \left(\sum_{j=1}^n \beta_{ij} x_j \right)^2$$

Apply the Cauchy - Schwartz Inequality here:

$$\left(\sum_{j=1}^n \beta_{ij} x_j \right)^2 \quad \leq \quad \left(\sum_{j=1}^n \beta_{ij}^2 \right) \left(\sum_{j=1}^n x_j^2 \right) = \left(\sum_{j=1}^n \beta_{ij}^2 \right) \|\boldsymbol{x}\|_2^2$$

Thus,

$$\|\boldsymbol{T}\boldsymbol{x}\|_2^2 \quad \leq \quad \left(\sum_{i=1}^m \sum_{j=1}^n \beta_{ij}^2 \right) \|\boldsymbol{x}\|_2^2$$

which implies

$$\|\boldsymbol{T}(\boldsymbol{x})\|_2 \quad \leq \quad \sqrt{\sum_{i=1}^m \sum_{j=1}^n \beta_{ij}^2} \; \|\boldsymbol{x}\|_2$$

or

$$\sup_{\boldsymbol{x} \neq \boldsymbol{0}} \frac{\|\boldsymbol{T}(\boldsymbol{x})\|_2}{\|\boldsymbol{x}\|_2} \quad \leq \quad \sqrt{\sum_{i=1}^m \sum_{j=1}^n \beta_{ij}^2}$$

It is traditional to let $B_F(T)$ be defined by $B_F(T) = \sqrt{\sum_{i=1}^m \sum_{j=1}^n \beta_{ij}^2}$. Then using the notation $B_2(T) = \sup_{\boldsymbol{x} \neq \boldsymbol{0}} \frac{\|\boldsymbol{T}(\boldsymbol{x})\|_2}{\|\boldsymbol{x}\|_2}$, we see we have found that $B_2(T) \leq B_F(T)$. The term $B_F(T)$ is called the **Frobenius Norm** of T. You can see it is an upper bound on the operator norm induced by $\| \cdot \|_2$.

Let's summarize here. We have found for a $m \times n$ matrix T

- $B_1(T)$ is called the **column norm of** T which is the largest absolute column sum.
- $B_\infty(T)$ is called the **row norm of** T which is the largest absolute row sum.
- $B_F(T)$ is called the **Frobenius norm of** T and is the $\| \cdot \|_2$ of the absolute values of the elements of T.
- $B_2(T)$ is the **two norm on** T. It is overestimated by $B_F(T)$.

In general, Definition 6.3.3 tells how to define the operator norm $B(T)$ for a linear operator between normed linear spaces. Recall for linear $T : dom(T) \subset (X, \| \cdot \|_X) \rightarrow (Y, \| \cdot \|_Y)$

$$B(T) \quad = \quad \sup_{\boldsymbol{x} \in dom(T), \|\boldsymbol{x}\|_X = 1} \|T(\boldsymbol{x})\|_Y$$

We say T is a bounded linear operator if $B(T) < \infty$. Let's call this $B_{op}(T)$ from now on. As we have seen, the particular form the operator norm takes depends on the choice of norm on the domain and the range. We have found some of these norms for finite dimensional linear operators. Note for a bounded linear operator, we always have

$$\|T(\boldsymbol{x})\|_Y \quad \leq \quad B_{op}(T)\ \|\boldsymbol{x}\|_X$$

We often let $B_{op} = \|\cdot\|_{op}$ also so that the fundamental inequality becomes

$$\|T(\boldsymbol{x})\|_Y \quad \leq \quad \|T\|_{op}\ \|\boldsymbol{x}\|_X$$

Homework

Exercise 6.4.1 *Show the solution to* $A\boldsymbol{x} = \boldsymbol{b}$ *depends continuously on the data for* $A = \begin{bmatrix} 1 & -3 \\ -2 & 4 \end{bmatrix}$ *using the* $\|\cdot\|_\infty$ *for the matrix norm.*

Exercise 6.4.2 *Show the solution to* $A\boldsymbol{x} = \boldsymbol{b}$ *depends continuously on the data for* $A = \begin{bmatrix} 4 & -3 \\ -2 & 4 \end{bmatrix}$ *using the* $\|\cdot\|_1$ *for the matrix norm.*

Exercise 6.4.3 *Show the solution to* $A\boldsymbol{x} = \boldsymbol{b}$ *depends continuously on the data for* $A = \begin{bmatrix} -6 & -3 \\ -2 & 4 \end{bmatrix}$ *using the* $\|\cdot\|_2$ *for the matrix norm.*

Exercise 6.4.4 *Show the solution to* $A\boldsymbol{x} = \boldsymbol{b}$ *depends continuously on the data for* $A = \begin{bmatrix} 1 & -2 & 3 \\ -2 & 4 & 5 \\ 6 & 2 & 1 \end{bmatrix}$ *using the* $\|\cdot\|_\infty$ *for the matrix norm.*

Exercise 6.4.5 *Show the solution to* $A\boldsymbol{x} = \boldsymbol{b}$ *depends continuously on the data for* $A = \begin{bmatrix} 1 & -2 & 3 \\ -2 & 4 & 5 \\ 6 & 2 & 1 \end{bmatrix}$ *using the* $\|\cdot\|_1$ *for the matrix norm.*

Exercise 6.4.6 *Show the solution to* $A\boldsymbol{x} = \boldsymbol{b}$ *depends continuously on the data for* $A = \begin{bmatrix} 1 & -2 & 3 \\ -2 & 4 & 5 \\ 6 & 2 & 1 \end{bmatrix}$ *using the* $\|\cdot\|_F$ *for the matrix norm.*

These norms are indeed norms in the sense of properties $N1$, $N2$, $N3$ and $N4$.

If we let M_{mn} denote the set of all $m \times n$ matrices, then it is a vector space over either the reals or the complex numbers. We can endow M_{mn} with many norms.

- $(M_{mn}, \widehat{B}_1)$ uses the norm $\widehat{B}_1(T) = B_1(T)$.
- $(M_{mn}, \widehat{B}_\infty)$ uses the norm $\widehat{B}_\infty(T) = B_\infty(T)$.
- $(M_{mn}, \widehat{B}_2)$ uses the norm $\widehat{B}_2(T) = B_2(T)$.
- $(M_{mn}, \widehat{B}_F)$ uses the norm $\widehat{B}_F(T) = B_F(T)$.

6.5 Eigenvalues and Eigenvectors for Operators

Recall the definitions from finite dimensional matrices. Let $[T]_{\boldsymbol{E},\boldsymbol{F}} \in M_{mn}$ be the matrix representation of a linear operator $T : (X, \|\cdot\|_X) \to (Y, \|\cdot\|_Y)$ where the dimension of x is n and the dimension of Y is m. Here, $\boldsymbol{E}$ is a basis for X and $\boldsymbol{F}$ is a basis for Y. For convenience, we will simply use T to denote this representation from now on. Consider $T(\boldsymbol{x}) = \boldsymbol{0}$. We let the kernel of T be $ker(T) = \{\boldsymbol{x} \in X | T(\boldsymbol{x}) = \boldsymbol{0}\}$. If $ker(T) = \boldsymbol{0}$, then T is a $1-1$ mapping onto its range and its range is all of Y. This is easy to prove. Note the range of T is a subspace of Y. So for $\boldsymbol{y_1}$ and $\boldsymbol{y_2}$ in the range of T, there are $\boldsymbol{x_1}$ and $\boldsymbol{x_2}$ in X with $T(\boldsymbol{x_1}) = \boldsymbol{y_1}$ and $T(\boldsymbol{x_2}) = \boldsymbol{y_2}$. Thus, $\alpha\boldsymbol{y_1} + \beta\boldsymbol{y_2}$ is the same as $\alpha T(\boldsymbol{x_1}) + \beta T(\boldsymbol{x_2})$ or $T(\alpha\boldsymbol{x_1} + \beta\boldsymbol{x_2})$. This tells us a linear combination of elements in the range is also in the range and so the range must be a subspace of Y.

Since $\boldsymbol{E} = \{\boldsymbol{E_1}, \ldots, \boldsymbol{E_n}\}$ is a basis of X, consider $T(\boldsymbol{E}) = \{T(\boldsymbol{E_1}), \ldots, T(\boldsymbol{E_n})\}$. We claim this is a basis for Y. If we assume

$$\alpha_1 T(\boldsymbol{E_1}) + \ldots + \alpha_n T(\boldsymbol{E_n}) = \boldsymbol{0} \Longrightarrow T(\alpha_1 \boldsymbol{E_1} + \ldots + \alpha_n \boldsymbol{E_n}) = \boldsymbol{0}$$

But since the kernel of T is zero, this means $\alpha_1 \boldsymbol{E_1} + \ldots + \alpha_n \boldsymbol{E_n} = \boldsymbol{0}$. And since $\boldsymbol{E}$ is a basis, this means $\alpha_1 = \ldots = \alpha_n = 0$. Hence $T(\boldsymbol{E})$ is a linearly independent set.

Since any element in the range of T is clearly in the span of $T(\boldsymbol{E})$, we see $T(\boldsymbol{E})$ is a basis for Y. Hence, since the kernel of T is zero, given any $\boldsymbol{y} \in Y$, there is a unique $\boldsymbol{x} \in X$ so that $T(\boldsymbol{x}) = \boldsymbol{y}$. This allows us to define the inverse operator $T^{-1} : Y \rightarrow X$ by $T^{-1}(\boldsymbol{y}) = \boldsymbol{x}$ where $T(\boldsymbol{x}) = \boldsymbol{y}$. Another way of looking at this is that the equation $T(\boldsymbol{x}) = \boldsymbol{y}$ has a unique solution $T^{-1}(\boldsymbol{y})$ in this case.

Now consider $T_\lambda : X \rightarrow Y$ where $T_\lambda = T - \lambda I$ for any complex number λ. The spaces X and Y are arbitrary normed linear spaces. If the kernel of T_λ is zero, then the equation $T_\lambda(\boldsymbol{x}) = \boldsymbol{y}$ will have a unique solution we will denote by $L_\lambda^{-1}(\boldsymbol{y}) = (T - \lambda I)^{-1}(\boldsymbol{y})$.
If the kernel of T_λ is not zero then it is straightforward to show $ker(T_\lambda)$ is a nontrivial subspace of X; i.e. it has at least dimension one. Hence, there is a nonzero $\boldsymbol{v}$ so that

$$T_\lambda(\boldsymbol{v}) = \boldsymbol{0} \Longrightarrow T(\boldsymbol{v}) = \lambda\boldsymbol{v}$$

In this case, we call $\boldsymbol{v}$ an **eigenvector** associated with the **eigenvalue** λ.

Definition 6.5.1 Spectrum of a Linear Operator

*Let $T : dom(T) \subset (X, \|\cdot\|_X) \rightarrow (Y, \|\cdot\|_Y)$ be a linear operator. The eigenvalues of T are the complex numbers λ so that $ker(T - \lambda I) \neq \boldsymbol{0}$ and any nonzero $\boldsymbol{v} \in ker(T - \lambda I)$ is called an eigenvector associated with the eigenvalue λ. The set of all eigenvalues of T is called the **spectrum** of T given by*

$$spec(T) = \{\lambda \in \mathbb{C} \mid ker(T_\lambda) \neq \boldsymbol{0}\}$$

Here is a nice example of eigenvalues and eigenvectors (here eigenfunctions) that arise in an ODE setting. Consider the simple homogeneous model

$$\begin{aligned} u'' + \Theta u &= 0, \; 0 \leq x \leq L, \\ \ell_1(u(0), u'(0)) &= \alpha_1 u(0) + \alpha_2 u'(0) = 0 \\ \ell_2(u(L), u'(L)) &= \beta_1 u(L) + \beta_2 u'(L) = 0 \end{aligned}$$

We define the differential operator $\mathscr{L}$ by $\mathscr{L}(u) = u''$ and then rewrite the ODE as

$$\begin{aligned} \mathscr{L}(u) &= -\Theta u \\ \ell_1(u(0), u'(0)) &= 0 \end{aligned}$$

$$\ell_2(u(L), u'(L)) = 0$$

which corresponds to the *eigenvalue problem* for $\mathscr{L}$.

We generally investigate an eigenvalue problem of this sort in three stages: we look at what happens when $\Theta > 0$, $\Theta = 0$ and $\Theta < 0$. In these cases, we then rewrite as $\Theta = \omega^2$, $\omega \neq 0$, $\omega = 0$ and $\Theta < -\omega^2$, $\omega \neq 0$. This gives three separate models to examine:

1. $\Theta = \omega^2, \omega \neq 0$:

$$\begin{aligned} u'' + \omega^2 u &= 0, \; 0 \leq x \leq L, \\ \ell_1(u(0), u'(0)) &= \alpha_1 \, u(0) + \alpha_2 \, u'(0) = 0 \\ \ell_2(u(L), u'(L)) &= \beta_1 \, u(L) + \beta_2 \, u'(L) = 0 \end{aligned}$$

2. $\Theta = 0, \omega = 0$:

$$\begin{aligned} u'' &= 0, \; 0 \leq x \leq L, \\ \ell_1(u(0), u'(0)) &= \alpha_1 \, u(0) + \alpha_2 \, u'(0) = 0 \\ \ell_2(u(L), u'(L)) &= \beta_1 \, u(L) + \beta_2 \, u'(L) = 0 \end{aligned}$$

3. $\Theta = -\omega^2, \omega \neq 0$:

$$\begin{aligned} u'' - \omega^2 u &= 0, \; 0 \leq x \leq L, \\ \ell_1(u(0), u'(0)) &= \alpha_1 \, u(0) + \alpha_2 \, u'(0) = 0 \\ \ell_2(u(L), u'(L)) &= \beta_1 \, u(L) + \beta_2 \, u'(L) = 0 \end{aligned}$$

Now, in any of these cases, if we can find nonzero solutions to the model, this will give us eigenvalue - eigenfunction pairs. From this discussion, we can see that the solvability of $\mathscr{L}(u) + \Theta u = f$ depends on whether or not $-\Theta$ is an eigenvalue of $\mathscr{L}$ for the specified boundary conditions. If so, then $\mathscr{L} + \Theta I$ is not invertible, its nullspace is at least one dimensional.

Homework

Exercise 6.5.1 *Find the eigenvalues and eigenvectors of the right shift operator $T : \ell^2 \to \ell^2$ defined by $T(x_1, x_2, \ldots) = (x_2, x_2,)$. What is the spectrum of T?*

Exercise 6.5.2 *Find the eigenvalues and eigenvectors of the right shift operator $T : \ell^1 \to \ell^1$ defined by $T(x_1, x_2, \ldots) = (x_2, x_2,)$. What is the spectrum of T?*

Exercise 6.5.3 *Write the differential equation $u' = 5u$ as a differential operator on the appropriate space and discuss how you can think of this as an eigenvalue problem.*

Exercise 6.5.4 *Write the differential equation $u' = -6u$ as a differential operator on the appropriate space and discuss how you can think of this as an eigenvalue problem.*

6.6 A Differential Operator Example

Let's look at an extended example from differential equations. The model we want to study is on the interval $[0, L]$ for convenience. Here is the model. Since the boundary conditions involve derivatives, we call these *derivative boundary conditions*, of course.

$$\begin{aligned} u'' + \Theta u &= f, \; 0 \leq x \leq L, \\ u'(0) &= 0 \\ u'(L) &= 0 \end{aligned}$$

There are three separate possibilities to consider. These are Θ is a positive or negative number or just zero. If $\Theta > 0$, we model it as $\Theta = \omega^2$ with $\omega \neq 0$ and if $\Theta < 0$, it is written as $-\omega^2$. Each case is quite involved in its analysis. We have discussed these in (Peterson (18) 2020) when we were studying Fourier series. In all of these cases, we are interested in solutions which are nonzero on the interval $[0, L]$ as the zero solution is not physically very interesting. For these boundary conditions, the only case that gives nonzero solutions is when $\Theta = \omega^2$ for $\omega \neq 0$. For convenience of exposition, we will include the analysis of all cases here.

In addition, to take some derivatives here, we will need Leibnitz's Rule. This is discussed in (Peterson (18) 2020) as well and we just state it for convenience here.

Theorem 6.6.1 Leibnitz's Rule

Let $f(s,t)$ be continuous on the rectangle $[a,b] \times [c,d]$ and let u and v be continuous functions on some intervals whose range is a subset of $[a,b]$. This means $u(t)$ is in $[a,b]$ for each t in its domain and $v(t)$ is in $[a,b]$ for each t in its domain. Then, define F on the interval $[c,d]$ by

$$F(t) = \int_{u(t)}^{v(t)} f(s,t)\, ds.$$

Note that because of our assumptions, the lower and upper limit of integration always give points in the interval $[a,b]$ where $f(s,t)$ is defined and so the Riemann integral makes sense to calculate. Then, we have

$$F'(t) = \int_{u(t)}^{v(t)} \frac{\partial f}{\partial t}(s,t)\, ds + f(v(t),t)\, v'(t) - f(u(t),t)\, u'(t).$$

Proof 6.6.1
See (Peterson (18) 2020). ■

6.6.1 The Separation Constant is Positive

The nonhomogeneous model to solve is then

$$\begin{aligned} u'' + \omega^2 u &= f \\ u'(0) &= 0, \\ u'(L) &= 0. \end{aligned}$$

The general solution to the homogeneous equation is given by

$$u_h(x) = A\cos(\omega x) + B\sin(\omega x).$$

Applying Variation of Parameters, we look for a particular solution of the form

$$u_p(x) \quad = \quad \phi(x)\,\cos(\omega x) \,+\, \psi(x)\,\sin(\omega x).$$

The theory of ordinary differential equations tells us we seek functions ϕ and ψ which must satisfy

$$\begin{bmatrix} \cos(\omega x) & \sin(\omega x) \\ -\omega\,\sin(\omega x) & \omega\cos(\omega x) \end{bmatrix} \begin{bmatrix} \phi'(x) \\ \psi'(x) \end{bmatrix} = \begin{bmatrix} 0 \\ f(x) \end{bmatrix}$$

Let W denote the 2×2 matrix above. Note that $\det(W) = \omega$ which is not zero by assumption. Hence, by Cramer's rule, we have

$$\begin{aligned} \phi'(x) &= \frac{\begin{vmatrix} 0 & \sin(\omega x) \\ f(x) & \omega\cos(\omega x) \end{vmatrix}}{\omega} = -\frac{f(x)\sin(\omega x)}{\omega} \\ \psi'(x) &= \frac{\begin{vmatrix} \cos(\omega x) & 0 \\ -\omega\,\sin(\omega x) & f \end{vmatrix}}{\omega} = \frac{f(x)\cos(\omega x)}{\omega} \end{aligned}$$

Hence,

$$\begin{aligned} \phi(x) &= -\frac{1}{\omega}\int_0^x f(s)\sin(\omega s)\,ds \\ \psi(x) &= \frac{1}{\omega}\int_0^x f(s)\cos(\omega s)\,ds \end{aligned}$$

The general solution $u(x) = u_h(x) + u_p(x)$ and so

$$\begin{aligned} u(x) &= A\,\cos(\omega x) \,+\, B\,\sin(\omega x) \\ &\quad + \left(-\frac{1}{\omega}\int_0^x f(s)\sin(\omega s)\,ds\right)\cos(\omega x) \,+\, \left(\frac{1}{\omega}\int_0^x f(s)\cos(\omega s)\,ds\right)\sin(\omega x) \\ &= A\,\cos(\omega x) \,+\, B\,\sin(\omega x) \,+\, \frac{1}{\omega}\int_0^x f(s)\Big(\cos(\omega s)\sin(\omega x) \,-\, \sin(\omega s)\cos(\omega x)\Big)ds \\ &= A\,\cos(\omega x) \,+\, B\,\sin(\omega x) \,+\, \frac{1}{\omega}\int_0^x f(s)\sin\Big(\omega(x-s)\Big)\,ds \end{aligned}$$

The last simplification arises from the use of the standard sin addition and subtraction of angles formulae. Next, apply the boundary conditions, $u'(0) = 0$ and $u'(L) = 0$. We find first, using Leibnitz's rule for derivatives of functions defined by integrals, that

$$\begin{aligned} u'(x) &= -\omega A\sin(\omega x) \,+\, \omega B\cos(\omega x) \\ &\quad + \frac{1}{\omega}\,f(x)\sin\Big(\omega(x-x)\Big) \,+\, \frac{1}{\omega}\int_0^x f(s)\omega\,\cos\Big(\omega(x-s)\Big)\,ds \\ &= -\omega A\sin(\omega x) \,+\, \omega\cos(\omega x) \,+\, \int_0^x f(s)\,\cos\Big(\omega(x-s)\Big)\,ds. \end{aligned}$$

Hence,

$$\begin{aligned} u'(0) &= 0 = \omega\,B \\ u'(L) &= 0 = -\omega A\sin(L\omega) \,+\, B\omega\cos(L\omega) \,+\, \int_0^L f(s)\,\cos\Big(\omega(L-s)\Big)\,ds. \end{aligned}$$

It immediately follows that $B = 0$ and to satisfy our boundary conditions, we must have

$$0 = -\omega A \sin(L\omega) + \int_0^L f(s) \cos\Big(\omega(L-s)\Big)\, ds.$$

We now see an interesting result. We can determine a unique value of A only if $\sin(L\omega) \neq 0$. Otherwise, since we assume $\omega \neq 0$, if $\omega L = n\pi$ for any integer $n \neq 0$, we find the value of A cannot be determined as we have the equation

$$0 = -\omega A \times 0 + \int_0^L f(s) \cos\Big(\frac{n\pi}{L}(L-s)\Big)\, ds.$$

In this case, we see that f must satisfy

$$\int_0^L f(s) \cos\Big(\frac{n\pi}{L}(L-s)\Big)\, ds = 0.$$

Otherwise, if $\omega L \neq n\pi$, we can solve for A to obtain

$$A = \frac{1}{\omega \sin(L\omega)} \int_0^L f(s) \cos\Big(\omega(L-s)\Big)\, ds.$$

This leads to the solution

$$u(x) = \frac{\cos(\omega x)}{\omega \sin(L\omega)} \int_0^L f(s) \cos\Big(\omega(L-s)\Big)\, ds + \frac{1}{\omega}\int_0^x f(s) \sin\Big(\omega(x-s)\Big)\, ds$$

6.6.1.1 The Kernel Function

We can manipulate this solution quite a bit. The following derivation is a bit tedious, so be patient.

$$\begin{aligned}
u(x) &= \frac{1}{\omega \sin(L\omega)} \int_0^L f(s) \cos(\omega x) \cos\Big(\omega(L-s)\Big) ds + \frac{1}{\omega}\int_0^x f(s) \sin\Big(\omega(x-s)\Big) ds \\
&= \frac{1}{\omega \sin(L\omega)} \int_0^x f(s)\bigg(\cos(\omega x) \cos\Big(\omega(L-s)\Big) + \sin(L\omega)\sin\Big(\omega(x-s)\Big)\bigg)\, ds \\
&\quad + \frac{1}{\omega \sin(L\omega)} \int_x^L f(s) \cos(\omega x) \cos\Big(\omega(L-s)\Big)\, ds
\end{aligned}$$

Now we can use trigonometric identities to simplify these expressions.

$$\begin{aligned}
&\cos(\omega x)\cos(\omega(L-s)) + \sin(L\omega)\sin(\omega(x-s)) \\
&= \cos(\omega x)\Big(\cos(\omega L)\cos(\omega s) + \sin(\omega L)\sin(\omega s)\Big) \\
&\quad + \sin(L\omega)\Big(\sin(\omega x)\cos(\omega s) - \sin(\omega s)\cos(\omega x)\Big) \\
&= \cos(\omega x)\cos(\omega L)\cos(\omega s) + \cos(\omega x)\sin(\omega L)\sin(\omega s) \\
&\quad + \sin(L\omega)\sin(\omega x)\cos(\omega s) - \sin(\omega L)\sin(\omega s)\cos(\omega x) \\
&= \cos(\omega x)\cos(\omega L)\cos(\omega s) + \sin(L\omega)\sin(\omega x)\cos(\omega s) \\
&= \cos(\omega s)\Big(\cos(\omega x)\cos(\omega L) + \sin(L\omega)\sin(\omega x)\Big) = \cos(\omega s)\cos\Big(\omega(L-x)\Big).
\end{aligned}$$

Using this rewrite of the first integral's term, we find

$$\begin{aligned} u(x) &= \frac{1}{\omega \sin(L\omega)} \int_0^x f(s) \cos(\omega s) \cos\Big(\omega(L-x)\Big) \, ds \\ &+ \frac{1}{\omega \sin(L\omega)} \int_x^L f(s) \cos(\omega x) \cos\Big(\omega(L-s)\Big) \, ds \end{aligned}$$

We can then define the *kernel* function k_ω by

$$k_\omega(x,s) = \frac{1}{\omega \sin(L\omega)} \begin{cases} \cos(\omega s) \cos(\omega(L-x)) & 0 \le s \le x \\ \cos(\omega x) \cos(\omega(L-s)) & x < s \le L \end{cases}$$

Note that k_ω is continuous on the square $[0, L] \times [0, L]$ and that it is also a symmetric functions as $k_\omega(x,s) = k_\omega(s,x)$. We can thus say that for any $\omega L \neq n\pi$, we find the solution to our nonhomogeneous boundary value problem can be written as

$$u(x) = \int_0^L f(s) \, k_\omega(x,s) ds$$

6.6.1.2 Finding the Kernel Function Another Way

We can determine the kernel function another way as well. Look at the functions

$$\begin{aligned} x_1(x) &= \cos(\omega x) \\ x_2(x) &= \cos(\omega(L-x)). \end{aligned}$$

Note that x_1 satisfies the homogeneous model with the first boundary condition

$$\begin{aligned} u'' + \omega^2 u &= 0, \ 0 \le x \le L \\ u'(0) &= 0 \end{aligned}$$

and x_2 satisfies the homogeneous model with the second boundary condition

$$\begin{aligned} u'' + \omega^2 u &= 0, \ 0 \le x \le L \\ u'(L) &= 0 \end{aligned}$$

Let $W(x_1, x_2)$ be the determinant

$$W(x_1,x_2) = \begin{vmatrix} x_1 & x_2 \\ x_1' & x_2' \end{vmatrix} = \begin{vmatrix} \cos(\omega x) & \cos(\omega(L-x)) \\ -\omega \sin(\omega x) & \omega \sin(\omega(L-x)) \end{vmatrix} = \omega \sin(L\omega).$$

Hence, we can rewrite the kernel function as

$$k_\omega(x,s) = \frac{1}{W(x_1,x_2)} \begin{cases} x_1(s) \, x_2(x) & 0 \le s \le x \\ x_1(x) \, x_2(s) & x < s \le L \end{cases}$$

This new way of finding the kernel function turns out to be a valid way to do things in our models.

Further, we can show the two functions x_1 and x_2 are linearly independent on $[0, L]$. If for all x in $[0, L]$, we had

$$c_1 x_1(x) + c_2 x_2(x) = 0,$$

then this would imply

$$c_1 x_1'(x) + c_2 x_2'(x) = 0,$$

then evaluating at 0 and L, we find the system

$$\begin{aligned} c_1 x_1'(0) + c_2 x_2'(0) &= 0, \\ c_1 x_1'(L) + c_2 x_2'(L) &= 0. \end{aligned}$$

The determinant of this system is $x_1'(0)x_2'(L) - x_2'(0)x_1'(L)$. This evaluates to $-\omega^2 \sin^2(\omega L)$ which is not zero since $\omega L \neq npi$. Hence, both c_1 and c_2 are zero and these two functions are independent.

6.6.1.3 Conclusions

We can say all of this more strongly. If $\omega L \neq n\pi$ for any integer $n \neq 0$, we have

$$\begin{pmatrix} u'' + \omega^2 u & = f, & 0 \leq x \leq L \\ u'(0) & = 0 & \\ u'(L) & = 0 & \end{pmatrix} \iff u(x) = \int_0^L f(s)\, k_\omega(x,s) ds.$$

We also know that for $f = 0$, the homogeneous problem has nonzero solutions

$$u_n(x) = \cos(\omega_n(x))$$

where $\omega_n L = n\pi$ for any integer $n \neq 0$. Hence, the nonzero functions u_n satisfy the boundary value problem

$$\begin{aligned} u'' + \omega_n^2 u &= 0, \; 0 \leq x \leq L \\ u'(0) &= 0 \\ u'(L) &= 0 \end{aligned}$$

Note we can rewrite this as

$$\begin{aligned} u_n'' &= -\omega_n^2 u_n, \; 0 \leq x \leq L \\ u_n'(0) &= 0 \\ u_n'(L) &= 0 \end{aligned}$$

6.6.1.4 The Abstract Version

The model above can be rewritten in a more abstract form by defining the operator $\mathscr{L}$ acting on a suitable domain of functions. Let $\mathscr{D}(\mathscr{L})$ denote the domain of $\mathscr{L}$ which for us will be

$$\mathscr{D}(L) = C^2([0,L]) \cap \{x \in C^2([0,L]) \text{ with } x'(0) = 0; x'(L) = 0\}$$

where $C^2([0,L])$ is the set of functions that are twice differentiable on $[0,L]$. We only assume the second derivative exists, but of course, if f is continuous and u is also continuous, since $u'' = f - \Theta u$, we actually know we want the second derivative to be continuous also. However, we can relax the *smoothness* of f and get situations where we lose the continuity of the second derivative at various points and even its existence itself. But that is a more advanced topic for later. For convenience, we will denote this collection of functions by $C^2[0,L] \cap \{BC\}$. The operator $\mathscr{L}$ is then defined on this domain to be

$$\mathscr{L}(u) = u''.$$

So we can rewrite our model again on $C^2([0,L]) \cap \{BC\}$ as the eigenvalue equation $\mathscr{L}(u_n) = -\omega_n^2 u$. As usual, since eigenvectors must be nonzero, we want our eigenfunctions to be nonzero functions as well. You can see our eigenfunctions here are indeed nonzero. So our work so far tells us the eigenvalues of $\mathscr{L}$ are $-\omega_n^2 = -\frac{n^2\pi^2}{L^2}$ for any nonzero integer n with associated eigenfunctions $u_n(x) = \cos(\omega_n\, x)$. Furthermore, there are infinitely many of them! There can never be more than a finite number of eigenvalues in a $Ax = xx$ eigenvalue problem, so this is a very new turn of events. Note we can also think of the solution $u(x) = \int_0^L f(s)\, k_\omega(x,s)ds$ as a way of defining the inverse of the operator $\mathscr{L} + \omega^2$, $(\mathscr{L} + \omega^2)^{-1}$ when ω^2 is not an eigenvalue. Finally, this inverse operator is very interesting. We can define the new operator $\mathscr{J}_\omega$ on the domain $C([0,L])$, the set of all functions continuous on the interval $[0,L]$, by

$$\mathscr{J}_\omega(u) = \int_0^L f(s)\, k_\omega(x,s)ds.$$

Hence $(\mathscr{L} + \omega^2)^{-1} = \mathscr{J}_\omega(u)$ for $\omega \neq \omega_n^2$. This will always transform the external function f into a continuous function u which solves the original model. However, it is *much* more general. As long as there is some notion of the integrability of f, we can do this transformation and that means our way of solving the model using a kernel function leads us to more general ways of looking at solutions. This actually brings us lots of new insights, but that discussion will come later. Finally note $(\mathscr{L} + \omega^2)^{-1}$ for ω an eigenvalue simply does not exist.

Homework

Exercise 6.6.1 *Find the kernel function and general solution to*

$$\begin{aligned} u'' + 9u &= f \\ u'(0) &= 0, \\ u'(3) &= 0. \end{aligned}$$

and find the inverse of the differential operator when it exists.

Exercise 6.6.2 *Find the kernel function and general solution to*

$$\begin{aligned} u'' + 4u &= f \\ u'(0) &= 0, \\ u'(6) &= 0. \end{aligned}$$

and find the inverse of the differential operator when it exists.

6.6.2 Case II: The Separation Constant is Zero

The nonhomogeneous model to solve is now

$$\begin{aligned} u'' &= f \\ u'(0) &= 0, \\ u'(L) &= 0. \end{aligned}$$

The general solution to the homogeneous equation is given by

$$u_h(x) = A\,\mathbf{1} + B\,\boldsymbol{x}.$$

where we temporarily put the two solutions of this second order model in boldface since one of the solutions is the constant function whose value is always 1. Applying Variation of Parameters, we look for a particular solution of the form

$$u_p(x) \;=\; \phi(x)\,\mathbf{1} \,+\, \psi(x)\,\boldsymbol{x}.$$

The theory of ordinary differential equations tells us we seek functions ϕ and ψ which must satisfy

$$\begin{bmatrix} 1 & x \\ 0 & 1 \end{bmatrix} \begin{bmatrix} \phi'(x) \\ \psi'(x) \end{bmatrix} \;=\; \begin{bmatrix} 0 \\ f(x) \end{bmatrix}$$

Let W denote the 2×2 matrix above. Note that $\det(W) = 1$ which is not zero. Hence, by Cramer's rule, we have

$$\begin{aligned} \phi'(x) &= \begin{vmatrix} 0 & x \\ f(x) & 1 \end{vmatrix} = -f(x)x \\ \psi'(x) &= \begin{vmatrix} 1 & 0 \\ 0 & f \end{vmatrix} = f(x) \end{aligned}$$

Hence,

$$\begin{aligned} \phi(x) &= -\int_0^x f(s)s\,ds \\ \psi(x) &= \int_0^x f(s)\,ds \end{aligned}$$

The general solution $u(x)$ is then

$$\begin{aligned} u(x) &= A + Bx \,-\, \int_0^x sf(s)\,ds \,+\, \left(\int_0^x f(s)\,ds\right) x \\ &= A + Bx \,+\, \int_0^x f(s)(x-s)\,ds \end{aligned}$$

Next, apply the boundary conditions, $u'(0) = 0$ and $u'(L) = 0$. We find

$$\begin{aligned} u'(x) &= B \,+\, f(x)(x-x) \,+\, \int_0^x f(s)\,ds \\ &= B \,+\, \int_0^x f(s)\,ds \end{aligned}$$

Hence,

$$\begin{aligned} u'(0) &= 0 = B \\ u'(L) &= 0 = B \,+\, \int_0^L f(s)\,ds. \end{aligned}$$

It immediately follows that $B = 0$ and to satisfy our boundary conditions, we must have

$$0 \;=\; \int_0^L f(s)\,ds.$$

It is clear that A is not determined. Hence, we cannot solve our problem here in general. The set of all functions v for which $\mathscr{L}(v) = 0$ is called the *nullspace* of $\mathscr{L}$. So the work above has shown us the operator $\mathscr{L}$ has the nonzero nullspace which consists of all the constant functions. And this is another way of saying the nonhomogeneous model is not solvable! Earlier, we found $u_n(x) = \cos(\omega_n x)$ were eigenfunctions of $\mathscr{L}$ for all $n \neq 0$. We have just added the new eigenfunction which corresponds to $u_0(x) = \cos(0\,x) = 1$ as $\mathscr{L}(u_0) = 0\,u_0$.

Homework

Exercise 6.6.3 *Find the kernel function and general solution to*

$$\begin{aligned} u'' &= f \\ u'(0) &= 0, \\ u'(3) &= 0. \end{aligned}$$

and find the inverse of the differential operator when it exists.

Exercise 6.6.4 *Find the kernel function and general solution to*

$$\begin{aligned} u'' &= f \\ u'(0) &= 0, \\ u'(6) &= 0. \end{aligned}$$

and find the inverse of the differential operator when it exists.

6.6.3 Case III: The Separation Constant is Negative

The nonhomogeneous model to solve is then

$$\begin{aligned} u'' - \omega^2 u &= f \\ u'(0) &= 0, \\ u'(L) &= 0. \end{aligned}$$

The general solution to the homogeneous equation is given by

$$u_h(x) = A\,\cosh(\omega x) + B\,\sinh(\omega x).$$

Applying Variation of Parameters, we look for a particular solution of the form

$$u_p(x) = \phi(x)\,\cosh(\omega x) + \psi(x)\,\sinh(\omega x).$$

The theory of ordinary differential equations tells us we seek functions ϕ and ψ which must satisfy

$$\begin{bmatrix} \cosh(\omega x) & \sinh(\omega x) \\ \omega\,\sinh(\omega x) & \omega\cosh(\omega x) \end{bmatrix} \begin{bmatrix} \phi'(x) \\ \psi'(x) \end{bmatrix} = \begin{bmatrix} 0 \\ f(x) \end{bmatrix}$$

Let W denote the 2×2 matrix above. Note that $\det(W) = \omega$ which is not zero by assumption. Hence, by Cramer's rule, we have

$$\phi'(x) = \frac{\begin{vmatrix} 0 & \sinh(\omega x) \\ f(x) & \omega\cosh(\omega x) \end{vmatrix}}{\omega} = -\frac{f(x)\sinh(\omega x)}{\omega}$$

$$\psi'(x) = \frac{\begin{vmatrix} \cosh(\omega x) & 0 \\ \omega \sinh(\omega x) & f \end{vmatrix}}{\omega} = \frac{f(x)\cosh(\omega x)}{\omega}$$

Hence,

$$\begin{aligned} \phi(x) &= -\frac{1}{\omega}\int_0^x f(s)\sinh(\omega s)\,ds \\ \psi(x) &= \frac{1}{\omega}\int_0^x f(s)\cosh(\omega s)\,ds \end{aligned}$$

The general solution $u(x) = u_h(x) + u_p(x)$ and so

$$\begin{aligned} u(x) &= A\cosh(\omega x) + B\sinh(\omega x) \\ &+\left(-\frac{1}{\omega}\int_0^x f(s)\sinh(\omega s)ds\right)\cosh(\omega x) + \left(\frac{1}{\omega}\int_0^x f(s)\cosh(\omega s)ds\right)\sinh(\omega x) \\ &= A\cosh(\omega x) + B\sinh(\omega x) + \frac{1}{\omega}\int_0^x f(s)\Big(\cosh(\omega s)\sinh(\omega x) - \sinh(\omega s)\cosh(\omega x)\Big)ds \\ &= A\cosh(\omega x) + B\sinh(\omega x) + \frac{1}{\omega}\int_0^x f(s)\sinh\Big(\omega(x-s)\Big)ds \end{aligned}$$

Note, we have used identities for the addition of angles for the hyperbolic function sinh to simplify the last equation. Next, apply the boundary conditions, $u'(0) = 0$ and $u'(L) = 0$. We find

$$\begin{aligned} u'(x) &= \omega A\sinh(\omega x) + \omega B\cosh(\omega x) \\ &\quad + \frac{1}{\omega} f(x)\sinh\Big(\omega(x-x)\Big) + \frac{1}{\omega}\int_0^x f(s)\omega\cosh\Big(\omega(x-s)\Big)\,ds \\ &= \omega A\sinh(\omega x) + \omega\cosh(\omega x) + \int_0^x f(s)\cosh\Big(\omega(x-s)\Big)\,ds. \end{aligned}$$

Hence,

$$\begin{aligned} u'(0) &= 0 = \omega B \\ u'(L) &= 0 = \omega A\sinh(L\omega) + B\omega\cosh(L\omega) + \int_0^L f(s)\cosh\Big(\omega(L-s)\Big)\,ds. \end{aligned}$$

It immediately follows that $B = 0$ and to satisfy our boundary conditions, we must have

$$0 = \omega A\sinh(L\omega) + \int_0^L f(s)\cosh\Big(\omega(L-s)\Big)\,ds.$$

Since ω is not zero, the term $\sinh(L\omega) \neq 0$ always; hence, this non homogeneous model is always solvable. This also tells us the operator $\mathscr{L}$ does not have eigenvalues with values ω^2. Hence, we can solve for A to obtain

$$A = -\frac{1}{\omega\sinh(L\omega)}\int_0^L f(s)\cosh\Big(\omega(L-s)\Big)\,ds.$$

This leads to the solution

$$u(x) = -\frac{\cosh(\omega x)}{\omega \sinh(L\omega)} \int_0^L f(s)\cosh\Big(\omega(L-s)\Big)\,ds + \frac{1}{\omega}\int_0^x f(s)\sinh\Big(\omega(x-s)\Big)\,ds$$

We can then manipulate this solution as usual.

$$\begin{aligned} u(x) &= -\frac{1}{\omega \sinh(L\omega)} \int_0^L f(s)\cosh(\omega x)\cosh\Big(\omega(L-s)\Big)ds + \frac{1}{\omega}\int_0^x f(s)\sinh\Big(\omega(x-s)\Big)\,ds \\ &= \frac{1}{\omega \sinh(L\omega)} \int_0^x f(s)\Big(-\cosh(\omega x)\cosh\Big(\omega(L-s)\Big) + \sinh(L\omega)\sinh\Big(\omega(x-s)\Big)\Big)\,ds \\ &- \frac{1}{\omega \sinh(L\omega)} \int_x^L f(s)\cosh(\omega x)\cosh\Big(\omega(L-s)\Big)\,ds \end{aligned}$$

Now we can use hyperbolic trigonometric identities to simplify these expressions.

$$\begin{aligned} &-\cosh(\omega x)\cosh(\omega(L-s)) \\ &+\sinh(L\omega)\sinh(\omega(x-s)) \\ &= -\cosh(\omega x)\Big(\cosh(\omega L)\cosh(\omega s) - \sinh(\omega L)\sinh(\omega s)\Big) \\ &+ \sinh(L\omega)\Big(\sinh(\omega x)\cosh(\omega s) - \sinh(\omega s)\cosh(\omega x)\Big) \\ &= -\cosh(\omega x)\cosh(\omega L)\cosh(\omega s) + \cosh(\omega x)\sinh(\omega L)\sinh(\omega s) \\ &+ \sinh(L\omega)\sinh(\omega x)\cosh(\omega s) - \sinh(\omega L)\sinh(\omega s)\cosh(\omega x) \\ &= -\cosh(\omega x)\cosh(\omega L)\cosh(\omega s) + \sinh(L\omega)\sinh(\omega x)\cosh(\omega s) \\ &= -\cosh(\omega s)\Big(\cosh(\omega x)\cosh(\omega L) - \sinh(L\omega)\sinh(\omega x)\Big) \\ &= -\cosh(\omega s)\cosh\Big(\omega(L-x)\Big). \end{aligned}$$

Using this rewrite of the first integral's term, we find

$$\begin{aligned} u(x) &= -\frac{1}{\omega \sinh(L\omega)} \int_0^x f(s)\cosh(\omega s)\cosh\Big(\omega(L-x)\Big)\,ds \\ &- \frac{1}{\omega \sinh(L\omega)} \int_x^L f(s)\cosh(\omega x)\cosh\Big(\omega(L-s)\Big)\,ds \end{aligned}$$

6.6.3.1 The Kernel Function

We define the *kernel* function k_ω by

$$k_\omega(x,s) = -\frac{1}{\omega \sinh(L\omega)} \begin{cases} \cosh(\omega s)\cosh(\omega(L-x)) & 0 \le s \le x \\ \cosh(\omega x)\cosh(\omega(L-s)) & x < s \le L \end{cases}$$

Note that k_ω is continuous and symmetric on the square $[0, L] \times [0, L]$. We can thus say that for any $\omega \neq 0$, the solution to our nonhomogeneous boundary value problem can be written as

$$u(x) = \int_0^L k_\omega(x,s)ds$$

6.6.3.2 Finding the Kernel Function Another Way

Look at the functions

$$\begin{aligned} x_1(x) &= \cosh(\omega x) \\ x_2(x) &= \cosh(\omega(L-x)). \end{aligned}$$

We see x_1 satisfies the homogeneous model with the first boundary condition

$$\begin{aligned} u'' - \omega^2 u &= 0, \ 0 \le x \le L \\ u'(0) &= 0 \end{aligned}$$

and x_2 satisfies the homogeneous model with the second boundary condition

$$\begin{aligned} u'' - \omega^2 u &= 0, \ 0 \le x \le L \\ u'(L) &= 0 \end{aligned}$$

Note, this is exactly the approach we did before in the case $\Theta = \omega^2$. Let $W(x_1, x_2)$ be the determinant

$$W(x_1, x_2) = \begin{vmatrix} x_1 & x_2 \\ x_1' & x_2' \end{vmatrix} = \begin{vmatrix} \cosh(\omega x) & \cosh(\omega(L-x)) \\ \omega \sinh(\omega x) & -\omega \sinh(\omega(L-x)) \end{vmatrix} = -\omega \sinh(L\omega).$$

Hence, the kernel function can be rewritten as

$$k_\omega(x,s) = \frac{1}{W(x_1,x_2)} \begin{cases} x_1(s)\, x_2(x) & 0 \le s \le x \\ x_1(x)\, x_2(s) & x < s \le L \end{cases}$$

We can easily show x_1 and x_2 are linearly independent on $[0, L]$.

6.6.3.3 Conclusions

If $\omega \neq 0$ for any integer $n \neq 0$, we have

$$\begin{pmatrix} u'' - \omega^2 u = f, & 0 \le x \le L \\ u'(0) = 0 & \\ u'(L) = 0 & \end{pmatrix} \iff u(x) = \int_0^L f(s)\, k_\omega(x,s) ds.$$

We also know for $f = 0$, the homogeneous problem has only the zero solution and hence $-\omega^2$ for nonzero ω is not an eigenvalue for $\mathscr{L}$.

Homework

Exercise 6.6.5 *Find the kernel function and general solution to*

$$\begin{aligned} u'' - 4u &= f \\ u'(0) &= 0, \\ u'(6) &= 0. \end{aligned}$$

Exercise 6.6.6

$$\begin{aligned} u'' - 9u &= f \\ u'(0) &= 0, \\ u'(16) &= 0. \end{aligned}$$

All of this work shows us that the spectrum of L is

$$sp(L) = \left\{0, -\frac{\pi^2}{L^2}, -\frac{4\pi^2}{L^2}, \ldots, -\frac{n^2\pi^2}{L^2}, \ldots\right\} = \left\{-\frac{n^2\pi^2}{L^2} \mid n \geq 0\right\}$$

and the corresponding eigenfunctions are

$$u_n(t) = \left\{1, \cos\left(\frac{\pi\, t}{L}\right), \cos\left(\frac{2\pi\, t}{L}\right), \cos\left(3\frac{\pi\, t}{L}\right), \ldots, \cos\left(n\frac{\pi\, t}{L}\right), \ldots\right\}$$

We also see all the eigenfunctions are linearly independent, are easily shown to be mutually orthogonal and can be normalized. Hence, we have found an orthonormal sequence in $\mathbb{L}_2([a,b])$. The computations we have done here are the same ones we did in (Peterson (18) 2020). However, there our interests were in handling the data function f using Fourier series. Our choice of expansion for the data function f depended on the eigenfunctions we found for the differential equation obtained using separation of variables on an appropriate linear partial differential equation. Our focus now is on the abstract structure we find by thinking of these as operator equations on function spaces. We have found this out here the hard way and we do not see all of the interesting structure here. In Chapter 13, we explore differential operators of this type more closely and find more efficient ways to find the spectrum and the sequence of eigenfunctions along with more general properties. Stay tuned!

6.6.4 Homework

Exercise 6.6.7 *Determine the eigenvalues and eigenfunctions for the ODE below with full details provided for your work.*

$$\begin{aligned} u'' + \Theta u &= f, \; 0 \leq x \leq 10, \\ u'(0) &= 0 \\ u'(10) &= 0 \end{aligned}$$

Exercise 6.6.8 *Determine the eigenvalues and eigenfunctions for the ODE below with full details provided for your work.*

$$\begin{aligned} u'' + \Theta u &= f, \; 0 \leq x \leq 25, \\ u'(0) &= 0 \\ u'(25) &= 0 \end{aligned}$$

Exercise 6.6.9 *Determine the eigenvalues and eigenfunctions for the ODE below with full details provided for your work.*

$$\begin{aligned} u'' + \Theta u &= f, \; 0 \leq x \leq 16, \\ u(0) &= 0 \end{aligned}$$

$$u(16) = 0$$

Exercise 6.6.10 *Determine the eigenvalues and eigenfunctions for the ODE below with full details provided for your work.*

$$\begin{aligned} u'' + \Theta u &= f, \; 0 \leq x \leq 16, \\ u'(0) &= 0 \\ u(16) &= 0 \end{aligned}$$

Exercise 6.6.11 *Determine the eigenvalues and eigenfunctions for the ODE below with full details provided for your work.*

$$\begin{aligned} u'' + \Theta u &= f, \; 0 \leq x \leq 16, \\ u(0) &= 0 \\ u'(16) &= 0 \end{aligned}$$

6.7 Spaces of Linear Operators

The set of bounded linear operators between normed linear spaces is a normed linear space itself.

Theorem 6.7.1 The Set of Bounded Linear Operators is a Normed Linear Space

Let $(X, \|\cdot\|_X)$ and $(Y, \|\cdot\|_Y)$ be normed linear spaces. Define

$$B(X,Y) = \{T : X \to Y \mid T \text{ is linear and } B(T) < \infty\}$$

Then $B(T)$ defines a norm on $B(X,Y)$ making $(B(X,Y), B(T))$ a normed linear space. We often denote $B(T)$ by $\|T\|_{op}$.

Proof 6.7.1

$(N1)$: This property is clear.

$(N2)$: If $B(T) = 0$, then $\sup_{\boldsymbol{x} \neq 0} \frac{\|T(\boldsymbol{x})\|_X}{\|\boldsymbol{x}\|_X} = 0$. This implies $\frac{\|T(\boldsymbol{x})\|_X}{\|\boldsymbol{x}\|_X} = 0$ for all nonzero $\boldsymbol{x}$. Hence, $\|T(\boldsymbol{x})\|_X = 0$ for all nonzero $\boldsymbol{x} \in X$. But this tells us $T(\boldsymbol{x}) = 0$ for all $\boldsymbol{x}$ and so T is the zero operator. The converse is easy to see.

$(N3)$: We have

$$\begin{aligned} B(\alpha T) &= \sup_{\boldsymbol{x} \neq 0} \frac{\|(\alpha T)(\boldsymbol{x})\|_X}{\|\boldsymbol{x}\|_X} = \sup_{\boldsymbol{x} \neq 0} \frac{\|\alpha(T(\boldsymbol{x}))\|_X}{\|\boldsymbol{x}\|_X} \\ &= |\alpha| \sup_{\boldsymbol{x} \neq 0} \frac{\|T(\boldsymbol{x})\|_X}{\|\boldsymbol{x}\|_X} = |\alpha| B(T) \end{aligned}$$

$(N4)$:

$$\begin{aligned} \|(T_1 + T_2)(\boldsymbol{x})\|_Y &\leq \|T_1(\boldsymbol{x})\|_Y + \|T_2(\boldsymbol{x})\|_Y \\ &\Longrightarrow \frac{\|(T_1 + T_2)(\boldsymbol{x})\|_Y}{\|\boldsymbol{x}\|_X} \leq \frac{\|T_1(\boldsymbol{x})\|_Y}{\|\boldsymbol{x}\|_X} + \frac{\|T_2(\boldsymbol{x})\|_Y}{\|\boldsymbol{x}\|_X}, \; \boldsymbol{x} \neq \boldsymbol{0} \end{aligned}$$

$$\Longrightarrow \quad \frac{\|(T_1 + T_2)(\boldsymbol{x})\|_Y}{\|\boldsymbol{x}\|_X} \leq \sup_{\boldsymbol{x} \neq \boldsymbol{0}} \frac{\|T_1(\boldsymbol{x})\|_Y}{\|\boldsymbol{x}\|_X} + \sup_{\boldsymbol{x} \neq \boldsymbol{0}} \frac{\|T_2(\boldsymbol{x})\|_Y}{\|\boldsymbol{x}\|_X}, \boldsymbol{x} \neq \boldsymbol{0}$$

Thus,

$$\begin{aligned} \frac{\|(T_1 + T_2)(\boldsymbol{x})\|_Y}{\|\boldsymbol{x}\|_X} &\leq B(T_1) + B(T_2), \boldsymbol{x} \neq \boldsymbol{0} \\ &\Longrightarrow B(T_1 + T_2) \leq B(T_1) + B(T_2) \end{aligned}$$

■

The completeness of the range space determines the completeness of $B(X, Y)$.

Theorem 6.7.2 $B(X, Y)$ **is complete if** Y **is complete**

Let $(X, \|\cdot\|_X)$ and $(Y, \|\cdot\|_Y)$ be normed linear spaces and assume Y is complete with respect to its norm. Then $B(X, Y)$ is complete with respect to the norm $\|\cdot\|_{op}$.

Proof 6.7.2
Let (T_n) be a Cauchy sequence in $B(X, Y)$. Fix $\boldsymbol{x_0} \in X$ which is nonzero. Then for any $\epsilon > 0$, there is N so that

$$n, m > N \quad \Longrightarrow \quad \|T_n - T_m\|_{op} < \frac{\epsilon}{\|\boldsymbol{x_0}\|_X}$$

Thus,

$$\begin{aligned} n, m > N \quad &\Longrightarrow \quad \sup_{\boldsymbol{x} \neq \boldsymbol{0}} \frac{\|T_n(\boldsymbol{x}) - T_m(\boldsymbol{x})\|_Y}{\|\boldsymbol{x}\|_X} < \frac{\epsilon}{\|\boldsymbol{x_0}\|_X} \\ &\Longrightarrow \quad \frac{\|T_n(\boldsymbol{x_0}) - T_m(\boldsymbol{x_0})\|_Y}{\|\boldsymbol{x_0}\|_X} < \frac{\epsilon}{\|\boldsymbol{x_0}\|_X} \\ &\Longrightarrow \quad \|T_n(\boldsymbol{x_0}) - T_m(\boldsymbol{x_0})\|_Y < \epsilon \end{aligned}$$

This says the sequence $(T_n(\boldsymbol{x_0}))$ is a Cauchy sequence in Y. Since Y is complete, there is an element $\boldsymbol{y_{x_0}}$ in Y so that $T_n(\boldsymbol{x_0}) \rightarrow \boldsymbol{y_{x_0}}$ in $\|\cdot\|_Y$. We can do this for any $\boldsymbol{x_0}$. This means we can define a mapping $T : X \rightarrow Y$ by $T(\boldsymbol{x_0}) = \boldsymbol{y_{x_0}}$ when $\boldsymbol{x_0}$ is not zero and $T(\boldsymbol{0}) = \boldsymbol{0}$.

(Is T linear?)
For any linear combination $\alpha\boldsymbol{x_0} + \beta\boldsymbol{x_1}$, we have $T_n(\boldsymbol{x_0}) \rightarrow T(\boldsymbol{x_0})$ and $T_n(\boldsymbol{x_1}) \rightarrow T(\boldsymbol{x_1})$. Thus,

$$T_n(\alpha\boldsymbol{x_0} + \beta\boldsymbol{x_1}) \quad = \quad \alpha T_n(\boldsymbol{x_0}) + \beta T_n(\boldsymbol{x_1}) \rightarrow \alpha T(\boldsymbol{x_0}) + \beta T(\boldsymbol{x_1})$$

But $T_n(\alpha\boldsymbol{x_0} + \beta\boldsymbol{x_1})$ also converges to $\boldsymbol{y_{\alpha x_0 + \beta x_1}}$ which is defined to be $T(\alpha\boldsymbol{x_0} + \beta\boldsymbol{x_1})$. Since these two limit values match, we see T is linear.

(Is T bounded?)
Given $\epsilon > 0$, there is N so that

$$m, n > N \quad \Longrightarrow \quad \sup_{\boldsymbol{x_0} \neq \boldsymbol{0}} \frac{\|T_n(\boldsymbol{x_0}) - T_m(\boldsymbol{x_0})\|_Y}{\|\boldsymbol{x_0}\|_X} < \frac{\epsilon}{2}$$

For $\boldsymbol{x_0}$ not zero, it then follows

$$m, n > N \implies \frac{\|T_n(\boldsymbol{x_0}) - T_m(\boldsymbol{x_0})\|_Y}{\|\boldsymbol{x_0}\|_X} < \frac{\epsilon}{2}$$

It is straightforward to prove the norm $\|\cdot\|_Y$ is continuous on its domain so for $n > N$,

$$\begin{aligned} \lim_{m\to\infty} \frac{\|T_n(\boldsymbol{x_0}) - T_m(\boldsymbol{x_0})\|_Y}{\|\boldsymbol{x_0}\|_X} &\leq \frac{\epsilon}{2} \implies \frac{\|T_n(\boldsymbol{x_0}) - \lim_{m\to\infty} T_m(\boldsymbol{x_0})\|_Y}{\|\boldsymbol{x_0}\|_X} \leq \frac{\epsilon}{2} \\ &\implies \frac{\|T_n(\boldsymbol{x_0}) - T(\boldsymbol{x_0})\|_Y}{\|\boldsymbol{x_0}\|_X} \leq \frac{\epsilon}{2} \end{aligned}$$

This holds for all nonzero $\boldsymbol{x_0}$. Thus, if $n > N$

$$\|T_n - T\|_{op} = \sup_{\boldsymbol{x_0}\neq \boldsymbol{0}} \frac{\|T_n(\boldsymbol{x_0}) - T(\boldsymbol{x_0})\|_Y}{\|\boldsymbol{x_0}\|_X} \leq \frac{\epsilon}{2} \tag{6.1}$$

Define the mapping $S_n = T_n - T$ for $n > N$. Then S_n is linear because both T and T_n are linear. Further, S_n is bounded by Equation 6.1. Now pick any $\hat{n} > N$. Then $T(\boldsymbol{x}) = T_{\hat{n}} - S_{\hat{n}}$. Hence,

$$\|T\|_{op} = \|T_{\hat{n}} - S_{\hat{n}}\|_{op} \leq \|T_{\hat{n}}\|_{op} + \|S_{\hat{n}}\|_{op} < \infty$$

We conclude T is bounded and since it is linear $T \in B(X, Y)$.

(Does $T_n \to T$ in $\|\cdot\|_{op}$?)
Equation 6.1 says given $\epsilon > 0$ there is N so that $\|T_n - T\|_{op} < \epsilon$ when $n > N$. Thus $T_n \to T$ in $\|\cdot\|_{op}$. ■

There is a big consequence to this theorem we use a lot.

Lemma 6.7.3 $B(X, \Re)$ and $B(X, \mathbb{C})$ are Complete

The set of bounded linear operators from the normed linear space $(X, \|\cdot\|_X)$ to $(\Re, |\cdot|)$, $B(X, \Re)$, is complete. The same is true for $B(X, \mathbb{C})$.

Proof 6.7.3
This is true because the range in both cases is complete. ■

Homework

Exercise 6.7.1 *Let (a_n), (b_n), (c_n) and (d_n) be Cauchy Sequences in $\Re$. Define the sequence of 2×2 matrices (A_p) by $A_p = \begin{bmatrix} a_{np} & b_{np} \\ c_{np} & d_{np} \end{bmatrix}$. Prove (A_p) is a Cauchy sequence in $B(\Re^2, \Re^2)$ using the matrix norm $\|\cdot\|_\infty$.*

Exercise 6.7.2 *Let (a_n), (b_n), (c_n) and (d_n) be Cauchy Sequences in $\Re$. Define the sequence of 2×2 matrices (A_p) by $A_p = \begin{bmatrix} a_{np} & b_{np} \\ c_{np} & d_{np} \end{bmatrix}$. Prove (A_p) is a Cauchy sequence in $B(\Re^2, \Re^2)$ using the matrix norm $\|\cdot\|_1$.*

Exercise 6.7.3 *Let (a_n), (b_n), (c_n) and (d_n) be Cauchy Sequences in $\Re$. Define the sequence of 2×2 matrices (A_p) by $A_p = \begin{bmatrix} a_{np} & b_{np} \\ c_{np} & d_{np} \end{bmatrix}$. Prove (A_p) is a Cauchy sequence in $B(\Re^2, \Re^2)$ using the matrix norm $\|\cdot\|_F$.*

Exercise 6.7.4 *Let (a_{ij}) be Cauchy sequences in $\Re$ for $1 \leq i, j \leq n$. Define the sequence of $n \times n$ matrices $A_p = (\,(a_{ijp})\,)$. Prove (A_p) is a Cauchy sequence in $B(\Re^n, \Re^n)$ using the matrix norm $\|\cdot\|_\infty$.*

Exercise 6.7.5 *Let (a_{ij}) be Cauchy sequences in $\Re$ for $1 \leq i, j \leq n$. Define the sequence of $n \times n$ matrices $A_p = (\,(a_{ijp})\,)$. Prove (A_p) is a Cauchy sequence in $B(\Re^n, \Re^n)$ using the matrix norm $\|\cdot\|_1$.*

Exercise 6.7.6 *Let (a_{ij}) be Cauchy sequences in $\Re$ for $1 \leq i, j \leq n$. Define the sequence of $n \times n$ matrices $A_p = (\,(a_{ijp})\,)$. Prove (A_p) is a Cauchy sequence in $B(\Re^n, \Re^n)$ using the matrix norm $\|\cdot\|_F$.*

Exercise 6.7.7 *Let $X = C([0,1])$ with norm $\|\cdot\|_\infty$ and define $T : X \to \Re$ by $T(\boldsymbol{x})(t) = \int_0^t x(s)ds$. If $(\boldsymbol{x_n})$ is a Cauchy sequence in $(X, \|\cdot\|_\infty)$, prove $(T(\boldsymbol{x_n}))$ is a Cauchy sequence in $\Re$.*

Exercise 6.7.8 *Let $X = C([0,1])$ with norm $\|\cdot\|_1$ and define $T : X \to \Re$ by $T(\boldsymbol{x})(t) = \int_0^t x(s)ds$. If $(\boldsymbol{x_n})$ is a Cauchy sequence in $(X, \|\cdot\|_1)$, prove $(T(\boldsymbol{x_n}))$ is a Cauchy sequence in $\Re$.*

Exercise 6.7.9 *Let $X = C([0,1])$ with norm $\|\cdot\|_2$ and define $T : X \to \Re$ by $T(\boldsymbol{x})(t) = \int_0^t x(s)ds$. If $(\boldsymbol{x_n})$ is a Cauchy sequence in $(X, \|\cdot\|_2)$, prove $(T(\boldsymbol{x_n}))$ is a Cauchy sequence in $\Re$.*

Part IV

Inner Product Spaces

Chapter 7

Inner Product Spaces

A vector space can also have a special mapping called an inner product.

7.1 Inner Products

Now there is an important idea that we use a lot in applied work. If we have an object $\boldsymbol{u}$ in a Vector Space $\mathscr{V}$, we often want to find to *approximate* $\boldsymbol{u}$ using an element from a given subspace $\mathscr{W}$ of the vector space. To do this, we need to add another property to the vector space. This is the notion of an **inner product**. A vector space with an inner product is called an **Inner Product Space**. We already know what an inner product is in a simple vector space like $\Re^2$. Many vector spaces can have an inner product structure added easily. For example, in $C([a, b])$, since each object is continuous, each object is Riemann Integrable. Hence, given two functions f and g from $C([a, b])$, the real number given by $\int_a^b f(s)g(s)ds$ is well-defined. It satisfies all the usual properties that the inner product for finite dimensional vectors in $\Re^n$ does also. These properties are so common we will codify them into a definition for what an inner product for a vector space $\mathscr{V}$ should behave like.

Definition 7.1.1 Real Inner Product

Let $\mathscr{V}$ be a vector space with the reals as the scalar field. Then a mapping ω which assigns a pair of objects to a real number is called an inner product on $\mathscr{V}$ if

IP1: $\omega(\boldsymbol{u}, \boldsymbol{v}) = \omega(\boldsymbol{v}, \boldsymbol{u})$*; that is, the order is not important for any two objects.*

IP2: $\omega(c \odot \boldsymbol{u}, \boldsymbol{v}) = c\omega(\boldsymbol{u}, \boldsymbol{v})$*; that is, scalars in the* first slot *can be pulled out.*

IP3: $\omega(\boldsymbol{u} \oplus \boldsymbol{w}, \boldsymbol{v}) = \omega(\boldsymbol{u}, \boldsymbol{v}) + \omega(\boldsymbol{w}, \boldsymbol{v})$*, for any three objects.*

IP4: $\omega(\boldsymbol{u}, \boldsymbol{u}) \geq 0$ *and* $\omega(\boldsymbol{u}, \boldsymbol{u}) = 0$ *if and only if* $u = 0$.

These properties imply that $\omega(\boldsymbol{u}, c \odot \boldsymbol{v}) = c\omega(\boldsymbol{u}, \boldsymbol{v})$ as well. A vector space $\mathscr{V}$ with an inner product is called an **Inner Product Space**.

Comment 7.1.1 *The inner product is usually denoted with the symbol $< \, , \, >$ instead of $\omega(\, , \,)$. We will use this notation from now on.*

Comment 7.1.2 *When we have an inner product, we can* measure *the* size *or* magnitude *of an object, as follows. We define the analogue of the Euclidean norm of an object $\boldsymbol{u}$ using the usual $|| \; ||$ symbol* *as*

$$\|\boldsymbol{u}\| = \sqrt{<\boldsymbol{u},\boldsymbol{u}>}.$$

This is called the norm induced by the inner product *of the object. It is possible to prove the Cauchy - Schwartz Inequality in this more general setting also.*

Theorem 7.1.1 Cauchy - Schwartz Inequality for Real Inner Product Spaces

If $\mathscr{V}$ is an inner product space with real inner product $< \cdot, \cdot >$ and induced norm $\|\cdot\|$, then

$$|<\boldsymbol{u},\boldsymbol{v}>| \leq \|\boldsymbol{u}\|\,\|\boldsymbol{v}\|$$

with equality occurring if and only if $\boldsymbol{u}$ and $\boldsymbol{v}$ are linearly dependent.

Proof 7.1.1
If $\boldsymbol{u} = t\boldsymbol{v}$, then

$$\begin{aligned} |<\boldsymbol{u},\boldsymbol{v}>| &= |<t\boldsymbol{v},\boldsymbol{v}>| = |t<\boldsymbol{v},\boldsymbol{v}>| = |t|\|\boldsymbol{v}\|^2| \\ &= (|t|\|\boldsymbol{v}\|)\,(\|\boldsymbol{v}\|) = \|\boldsymbol{u}\|\,\|\boldsymbol{v}\| \end{aligned}$$

Hence, the result holds if $\boldsymbol{u}$ and $\boldsymbol{v}$ are linearly dependent. In general, if $\boldsymbol{u}$ and $\boldsymbol{v}$ are linearly independent, look at the subspace spanned by $\boldsymbol{v}$. Any element in this subspace can be written as $a\boldsymbol{v}/\|\boldsymbol{v}\|$. Consider the function

$$f(a) = <\boldsymbol{u} - a\boldsymbol{v}/\|\boldsymbol{v}\|, \boldsymbol{u} - a\boldsymbol{v}/\|\boldsymbol{v}>= \|\boldsymbol{u}\|^2 - 2a<\boldsymbol{u},\boldsymbol{v}/\|\boldsymbol{v}\|> + a^2$$

Then

$$f'(a) = -2<\boldsymbol{u},\boldsymbol{v}/\|\boldsymbol{v}\|> + 2a$$

which implies the critical point is $a =<\boldsymbol{u},\boldsymbol{v}/\|\boldsymbol{v}\|>$. Since $f''(a) = 2 > 0$, the critical point is a global minimum. Thus, the vector whose minimum norm distance to the subspace generated by $\boldsymbol{v}$ is $\boldsymbol{P_{uv}} =<\boldsymbol{u},\boldsymbol{v}/\|\boldsymbol{v}\|>\ \boldsymbol{v}/\|\boldsymbol{v}\|$ of length $|<\boldsymbol{u},\boldsymbol{v}\|\boldsymbol{v}\|>|$. As usual the vector $\boldsymbol{P_{uv}}$ is called the projection of $\boldsymbol{u}$ onto $\boldsymbol{v}$. We can then use GSO to find the orthonormal basis for the subspace spanned by $\boldsymbol{u}$ and $\boldsymbol{v}$.

$$\begin{aligned} \boldsymbol{Q_1} &= \boldsymbol{v}/\|\boldsymbol{v}\| \\ \boldsymbol{W} &= \boldsymbol{u}-<\boldsymbol{u},\boldsymbol{Q_1}>\boldsymbol{Q_1}, \quad \boldsymbol{Q_2} = \boldsymbol{W}/\|\boldsymbol{W}\| \end{aligned}$$

Thus we have the decomposition $\boldsymbol{u} =<\boldsymbol{u},\boldsymbol{Q_1}>\boldsymbol{Q_1}+<\boldsymbol{u},\boldsymbol{Q_2}>\boldsymbol{Q_2}$. Hence, if we assumed $\boldsymbol{u}$ and $\boldsymbol{v}$ were linearly independent with

$$|<\boldsymbol{u},\boldsymbol{v}>= \|\boldsymbol{u}\|\,\|\boldsymbol{v}\| \implies \|\boldsymbol{P_{uv}}\| = \left|\left\langle \boldsymbol{u}, \frac{\boldsymbol{v}}{\|\boldsymbol{v}\|}\right\rangle\right| = \|\boldsymbol{u}\|$$

telling us $\boldsymbol{u}$ is a multiple of $\boldsymbol{v}$ which is not possible as they are assumed independent. Hence, we conclude if $<\boldsymbol{u},\boldsymbol{v}>= \|\boldsymbol{u}\|\,\|\boldsymbol{v}\|$, $\boldsymbol{u}$ and $\boldsymbol{v}$ must be dependent.
However, if $\boldsymbol{u}$ and $\boldsymbol{v}$ are linearly independent, we have the decomposition $\boldsymbol{u} =<\boldsymbol{u},\boldsymbol{Q_1}>\boldsymbol{Q_1}+<\boldsymbol{u},\boldsymbol{Q_2}>\boldsymbol{Q_2}$. which implies $\|\boldsymbol{u}\|^2 = |<\boldsymbol{u},\boldsymbol{Q_1}>|^2 + |<\boldsymbol{u},\boldsymbol{Q_2}>|^2$. This immediately tells us

$$|<\boldsymbol{u},\boldsymbol{v}/\|\boldsymbol{v}\|>|^2 \leq \|\boldsymbol{u}\|^2 \implies |<\boldsymbol{u},\boldsymbol{v}/\|\boldsymbol{v}\|>| \leq \|\boldsymbol{u}\| \implies |<\boldsymbol{u},\boldsymbol{v}>| \leq \|\boldsymbol{u}\|\,/\|\boldsymbol{v}\|$$

which is the result we wanted to show. ■

If the vector space is over the complex field, the inner product needs to be defined a bit differently.

Definition 7.1.2 Complex Inner Product

Let $\mathscr{V}$ be a vector space with the complex numbers as the scalar field. Then a mapping ω which assigns a pair of objects to a complex number is called an inner product on $\mathscr{V}$ if

IPC1: $\omega(\boldsymbol{u}, \boldsymbol{v}) = \overline{\omega(\boldsymbol{v}, \boldsymbol{u})}$; *that is, the order is important. Switching the order conjugates the result.*

IPC2: $\omega(c \odot \boldsymbol{u}, \boldsymbol{v}) = c\omega(\boldsymbol{u}, \boldsymbol{v})$; *that is, scalars in the* first slot *can be pulled out. Using $IPC1$, this implies*

$$\omega(\boldsymbol{u}, c \odot \boldsymbol{v}) = \overline{\omega(c \odot \boldsymbol{v}, \boldsymbol{u})} = \overline{c\omega(\boldsymbol{v}, \boldsymbol{u})} = \overline{c}\omega(\boldsymbol{u}, \boldsymbol{v})$$

IPC3: $\omega(\boldsymbol{u} \oplus \boldsymbol{w}, \boldsymbol{v}) = \omega(\boldsymbol{u}, \boldsymbol{v}) + \omega(\boldsymbol{w}, \boldsymbol{v})$, *for any three objects.*

IPC4: $\omega(\boldsymbol{u}, \boldsymbol{u}) \geq 0$ *and* $\omega(\boldsymbol{u}, \boldsymbol{u}) = 0$ *if and only if* $u = 0$.

A vector space $\mathscr{V}$ with a complex inner product is called an **Complex Inner Product Space**.

Comment 7.1.3 *When we have a complex inner product, we can still* measure *the* size *or* magnitude *of an object*

$$||\boldsymbol{u}|| = \sqrt{< \boldsymbol{u}, \boldsymbol{u} >}.$$

This is called the norm induced by the complex inner product *of the object. It is possible to prove the Cauchy - Schwartz Inequality for complex vector spaces with induced norm.*

Theorem 7.1.2 Cauchy - Schwartz Inequality for Complex Vector Spaces

If $\mathscr{V}$ is an inner product space with complex inner product $< , >$ and induced norm $|| \ ||$, then

$$| < \boldsymbol{u}, \boldsymbol{v} > | \leq ||\boldsymbol{u}|| \, ||\boldsymbol{v}||$$

where $|\cdot|$ here is the complex magnitude. Equality occurs if and only if $\mathbf{u}$ and $\mathbf{v}$ are linearly dependent.

Proof 7.1.2
If $\boldsymbol{u} = t\boldsymbol{v}$ for a complex scalar t, then

$$\begin{aligned} | < \boldsymbol{u}, \boldsymbol{v} > | &= | < t\boldsymbol{v}, \boldsymbol{v} > | = |t| < \boldsymbol{v}, \boldsymbol{v} >= |t| \|\boldsymbol{v}\|^2 | \\ &= (|t| \|\boldsymbol{v}\|) \, (\|\boldsymbol{v}\|) = \|\boldsymbol{u}\| \, \|\boldsymbol{v}\| \end{aligned}$$

Hence, the result holds if $\mathbf{u}$ and $\mathbf{v}$ are linearly dependent. In general, if $\mathbf{u}$ and $\mathbf{v}$ are linearly independent, look at the subspace spanned by $\boldsymbol{v}$. Any element in this subspace can be written as $a\boldsymbol{v}\|\boldsymbol{v}\|$. Consider the complex function

$$f(a) = < \boldsymbol{u} - a\boldsymbol{v}/\|\boldsymbol{v}\|, \boldsymbol{u} - a\boldsymbol{v}/\|\boldsymbol{v} >= \|\boldsymbol{u}\|^2 - \overline{a} < \boldsymbol{u}, \boldsymbol{v}/\|\boldsymbol{v}\| > -a\overline{< \boldsymbol{u}, \boldsymbol{v}/\|\boldsymbol{v}\| >} + a^2$$

Let $a = \alpha + i\beta$ *and* $< \boldsymbol{u}, \boldsymbol{v}/\|\boldsymbol{v}\| >= C + \boldsymbol{i}D$. *Then we can rewrite* $f(a)$ *as*

$$g(\alpha,\beta) = \|\boldsymbol{u}\|^2 + \alpha^2 + \beta^2 - \{\alpha(\overline{< \boldsymbol{u}, \boldsymbol{w} >} + < \boldsymbol{u}, \boldsymbol{w} >) - \boldsymbol{i}\beta(\overline{< \boldsymbol{u}, \boldsymbol{w} >} - < \boldsymbol{u}, \boldsymbol{w} >)\}$$

where $\boldsymbol{w} = \boldsymbol{v}/\|\boldsymbol{v}\|$. *Then*

$$\begin{aligned} g(\alpha,\beta) &= \|\boldsymbol{u}\|^2 + \alpha^2 + \beta^2 - \{\alpha(C - \boldsymbol{i}D + C + \boldsymbol{i}D) - \boldsymbol{i}\beta(C - \boldsymbol{i}D - C - \boldsymbol{i}D)\} \\ &= \|\boldsymbol{u}\|^2 + \alpha^2 + \beta^2 - 2\alpha C - 2\beta D \end{aligned}$$

To fit the minimum distance, note

$$g_\alpha = 2\alpha - 2C, \quad g_\beta = 2\beta - 2D$$

which implies the critical point is

$$a = \alpha + i\beta = C + \boldsymbol{i}D =< \boldsymbol{u}, \boldsymbol{v}/\|\boldsymbol{v}\| >$$

Since the Hessian of g *is* 4, *this is an absolute minimum. Thus, the vector whose minimum norm distance to the subspace generated by* $\boldsymbol{v}$ *is* $\boldsymbol{P_{uv}} =< \boldsymbol{u}, \boldsymbol{v}\|\boldsymbol{v}\| > \ \boldsymbol{v}\|\boldsymbol{v}\|$ *of length* $| < \boldsymbol{u}, \boldsymbol{v}\|\boldsymbol{v}\| > |$. *As usual the vector* $\boldsymbol{P_{uv}}$ *is called the projection of* $\boldsymbol{u}$ *onto* $\boldsymbol{v}$. *The rest of the proof is the same as the one we used for a real vector space and is not repeated.* ∎

An even easier proof that works in either the real or the complex case is the following. We just think you should see different ways of doing things.

Theorem 7.1.3 Cauchy - Schwartz: the Easy Proof

> *If* $\mathscr{V}$ *is an inner product space with real inner product* $< \cdot, \cdot >$ *and induced norm* $\|\cdot\|$, *then*
>
> $$| < \boldsymbol{u}, \boldsymbol{v} > | \leq \|\boldsymbol{u}\| \, \|\boldsymbol{v}\|$$
>
> *with equality occurring if and only if* $\boldsymbol{u}$ *and* $\boldsymbol{v}$ *are linearly dependent.*

Proof 7.1.3
If $\boldsymbol{y} = \boldsymbol{0}$, *this is true. If* $\boldsymbol{y} \neq \boldsymbol{0}$, *consider* $\boldsymbol{x} - \alpha\boldsymbol{y}$ *for all* α *in the field of scalars. Now*

$$\begin{aligned} 0 &\leq \|\boldsymbol{x} - \alpha\boldsymbol{y}\|^2 =< \boldsymbol{x} - \alpha\boldsymbol{y}, \boldsymbol{x} - \alpha\boldsymbol{y} > \\ &= \|\boldsymbol{x}\|^2 - \overline{\alpha} < \boldsymbol{x}, \boldsymbol{y} > - \alpha < \boldsymbol{y}, \boldsymbol{x} > + \alpha\,\overline{\alpha}\|\boldsymbol{y}\|^2 \\ &= \|\boldsymbol{x}\|^2 - \overline{\alpha} < \boldsymbol{x}, \boldsymbol{y} > - \alpha\left(< \boldsymbol{y}, \boldsymbol{x} > - \overline{\alpha}\|\boldsymbol{y}\|^2\right) \end{aligned}$$

Choose $\overline{\alpha}$ *so that* $< \boldsymbol{y}, \boldsymbol{x} > - \overline{\alpha}\|\boldsymbol{y}\|^2 = 0$; *i.e. let* $\overline{\alpha_0} = \frac{<\boldsymbol{y},\boldsymbol{x}>}{\|\boldsymbol{y}\|^2}$. *Then,*

$$\begin{aligned} 0 &\leq \|\boldsymbol{x} - \alpha_0\boldsymbol{y}\|^2 = \|\boldsymbol{x}\|^2 - \overline{\alpha_0} < \boldsymbol{x}, \boldsymbol{y} >= \|\boldsymbol{x}\|^2 - \frac{< \boldsymbol{y}, \boldsymbol{x} >< \boldsymbol{x}, \boldsymbol{y} >}{\|\boldsymbol{y}\|^2} \\ &= \|\boldsymbol{x}\|^2 - \frac{\overline{< \boldsymbol{x}, \boldsymbol{y} >} \ < \boldsymbol{x}, \boldsymbol{y} >}{\|\boldsymbol{y}\|^2} = \|\boldsymbol{x}\|^2 - \frac{| < \boldsymbol{x}, \boldsymbol{y} > |}{\|\boldsymbol{y}\|^2} \end{aligned}$$

which implies $| < \boldsymbol{x}, \boldsymbol{y} > | \leq \|\boldsymbol{x}\| \, \|\boldsymbol{y}\|$ *for all choices of* $\boldsymbol{x}$ *and* $\boldsymbol{y}$.

Finally, if $\boldsymbol{x} = \xi\boldsymbol{y}$, *then we have* $| < \boldsymbol{x}, \boldsymbol{y} > | = |\overline{\xi}| \, | < \boldsymbol{x}, \boldsymbol{x} > | = |\xi|\|\boldsymbol{x}\|^2$. *Also,* $\|\boldsymbol{x}\| \, \|\boldsymbol{y}\| =$

$|\xi| \|\boldsymbol{x}\|^2$. *These expressions are the same and so if the two vectors are equal, we have* $| < \boldsymbol{x}, \boldsymbol{y} > | = \|\boldsymbol{x}\| \, \|\boldsymbol{y}\|$. *On the other hand, if* $| < \boldsymbol{x}, \boldsymbol{y} > | = \|\boldsymbol{x}\| \, \|\boldsymbol{y}\|$, *then the projection of* $\boldsymbol{x}$ *along* $\boldsymbol{y}$ *is* $\frac{<\boldsymbol{x},\boldsymbol{y}>}{\|\boldsymbol{y}\|} = \boldsymbol{x}\|$. *The component of* $\boldsymbol{x}$ *perpendicular to* $\boldsymbol{y}$ *is then zero and* $\boldsymbol{x}$ *is a multiple of* $\boldsymbol{y}$. ■

Homework

Exercise 7.1.1 *Let* $\mathbf{V_1}$ *and* $\mathbf{V_2}$ *be two linearly independent vectors in* $\Re^2$. *Define* $\omega(\mathbf{V_1}, \mathbf{V_1}) = \alpha$, $\omega(\mathbf{V_2}, \mathbf{V_2}) = \beta$ *and* $\omega(\mathbf{V_1}, \mathbf{V_2}) = \gamma$. *Extend* ω *to* $\Re^2 \times \Re^2$ *linearly.*

- *Prove since* $\mathbf{V_1}$ *and* $\mathbf{V_2}$ *are linearly independent* $\det\left(\begin{bmatrix} \alpha & \gamma \\ \gamma & \beta \end{bmatrix}\right)$ *must be nonzero.*
- *Prove* ω *defines a real inner product as long as* $\begin{bmatrix} \alpha & \gamma \\ \gamma & \beta \end{bmatrix}$ *is a positive definite matrix which implies its eigenvalues are positive.*
- *Prove* ω *defines a real inner product as long as* $\det\left(\begin{bmatrix} \alpha & \gamma \\ \gamma & \beta \end{bmatrix}\right)$ *is positive.*

Exercise 7.1.2 *For* $\boldsymbol{x} = \begin{bmatrix} x_1 \\ x_2 \end{bmatrix}$ *in* $\Re^2$ *and* $D = \begin{bmatrix} 2 & 0 \\ 0 & 4 \end{bmatrix}$, *define* $\omega(\boldsymbol{x}, \boldsymbol{y}) = \boldsymbol{x}^T D \boldsymbol{y}$. *Prove* ω *is an inner product, find the induced norm and the induced metric.*

Exercise 7.1.3 *For* $\boldsymbol{x} = \begin{bmatrix} x_1 \\ x_2 \end{bmatrix}$ *in* $\Re^2$ *and* $D = \begin{bmatrix} 3 & 0 \\ 0 & 8 \end{bmatrix}$, *define* $\omega(\boldsymbol{x}, \boldsymbol{y}) = \boldsymbol{x}^T D \boldsymbol{y}$. *Prove* ω *is an inner product, find the induced norm and the induced metric.*

Exercise 7.1.4 *For* $\boldsymbol{x} = \begin{bmatrix} x_1 \\ x_2 \end{bmatrix}$ *in* $\Re^2$ *and* $D = \begin{bmatrix} \lambda_1^2 & 0 \\ 0 & \lambda_2^2 \end{bmatrix}$ *where* λ_1 *and* λ_2 *are both positive. Define* $\omega(\boldsymbol{x}, \boldsymbol{y}) = \boldsymbol{x}^T D \boldsymbol{y}$. *Prove* ω *is an inner product, find the induced norm and the induced metric.*

Exercise 7.1.5 *For* $\boldsymbol{x} = \begin{bmatrix} x_1 \\ x_2 \\ \vdots \\ x_n \end{bmatrix}$ *in* $\Re^n$ *and* D *an* $n \times n$ *diagonal matrix whose diagonal entries are* λ_i^2 *with each* λ_i *positive. Define* $\omega(\boldsymbol{x}, \boldsymbol{y}) = \boldsymbol{x}^T D \boldsymbol{y}$. *Prove* ω *is an inner product, find the induced norm and the induced metric.*

Exercise 7.1.6 *Let* $\boldsymbol{x}$ *and* $\boldsymbol{y}$ *be in* $\Re^n$ *and let* A *be an* $n \times n$ *positive definite symmetric matrix. Define* $\omega(\boldsymbol{x}, \boldsymbol{y}) = \boldsymbol{x}^T A \boldsymbol{y}$. *Prove* ω *is an inner product, find the induced norm and the induced metric.*

Exercise 7.1.7 *Let* $X = C([0,1])$ *and let* $\omega(\boldsymbol{x}, \boldsymbol{y}) = \int_0^1 e^{-2s} x(s) y(s) ds$. *Prove* ω *is an inner product on* X, *find the induced norm and the induced metric.*

Exercise 7.1.8 *Let* $X = C([0,1]) \times C([0,1])$ *and define* ω *on pairs of vector functions by*

$$\omega\left(\begin{bmatrix} \boldsymbol{x_1} \\ \boldsymbol{x_2} \end{bmatrix}, \begin{bmatrix} \boldsymbol{y_1} \\ \boldsymbol{y_2} \end{bmatrix}\right) = \int_0^1 \begin{bmatrix} x_1(s) \\ x_2(s) \end{bmatrix}^T \begin{bmatrix} 4 & 0 \\ 0 & 5 \end{bmatrix} \begin{bmatrix} y_1(s) \\ y_2(s) \end{bmatrix} ds$$

Prove ω *is an inner product, find the induced norm and the induced metric.*

Now let's look more closely at the induced norms.

Theorem 7.1.4 Norm Induced by a Complex Inner Product

If $(\mathscr{V}, < \cdot, \cdot >)$ *is a complex inner product space, then* $\| \cdot \|$ *defined by* $\|\boldsymbol{x}\| = \sqrt{< \boldsymbol{x}, \boldsymbol{x} >}$ *is an inner product on* $\mathscr{V}$ *and* $(\mathscr{V}, \| \cdot \|)$ *is a normed linear space.*

Proof 7.1.4
We need to show the properties of a norm are satisfied.

N1: $\|\boldsymbol{x}\| = \sqrt{<\boldsymbol{x},\boldsymbol{x}>} \geq 0$ *as* $<\cdot,\cdot>$ *is an inner product.*

N2: *If* $\|\boldsymbol{x}\| = \sqrt{<\boldsymbol{x},\boldsymbol{x}>} = 0$*, then* $\boldsymbol{x} = 0$ *by* $IP4$*. Conversely, if* $\boldsymbol{x} = 0$, $<\boldsymbol{x},\boldsymbol{x}>= 0$ *by* $IP4$ *also.*

N3: $\|\alpha\boldsymbol{x}\| = \sqrt{<\alpha\boldsymbol{x},\alpha\boldsymbol{x}>}$*. In a real vector space, this gives* $\|\alpha\boldsymbol{x}\| = \sqrt{\alpha^2 <\boldsymbol{x},\boldsymbol{x}>}$*. Thus,* $\|\boldsymbol{\alpha x}\| = |\alpha|\|\boldsymbol{x}\|$.

In a complex vector space, we have $\|\alpha\boldsymbol{x}\| = \sqrt{\alpha\overline{\alpha} <\boldsymbol{x},\boldsymbol{x}>}$*. Thus,* $\|\boldsymbol{\alpha x}\| = |\alpha|\|\boldsymbol{x}\|$ *where* $|\alpha|$ *is the complex magnitude of* α*.*

N4: *For real vector spaces,*

$$\begin{aligned}\|\boldsymbol{x}+\boldsymbol{y}\|^2 &= <\boldsymbol{x}+\boldsymbol{y},\boldsymbol{x}+\boldsymbol{y}>=<\boldsymbol{x},\boldsymbol{x}> +2<\boldsymbol{x},\boldsymbol{y}> + <\boldsymbol{y},\boldsymbol{y}> \\ &\leq <\boldsymbol{x},\boldsymbol{x}> +2<\boldsymbol{x},\boldsymbol{y}> |+ <\boldsymbol{y},\boldsymbol{y}> \leq \|\boldsymbol{x}\|^2 + 2\|\boldsymbol{x}\|\,\|\boldsymbol{y}\| + \|\boldsymbol{y}\|^2 \\ &= (\|\boldsymbol{x}\| + \|\boldsymbol{y}\|)^2\end{aligned}$$

which implies the result. For complex spaces we have for $<\boldsymbol{x},\boldsymbol{y}>= \alpha + \boldsymbol{i}\beta$*,*

$$\begin{aligned}\|\boldsymbol{x}+\boldsymbol{y}\|^2 &= <\boldsymbol{x}+\boldsymbol{y},\boldsymbol{x}+\boldsymbol{y}>=<\boldsymbol{x},\boldsymbol{x}> + <\boldsymbol{x},\boldsymbol{y}> +\overline{<\boldsymbol{x},\boldsymbol{y}>}+ <\boldsymbol{y},\boldsymbol{y}> \\ &= <\boldsymbol{x},\boldsymbol{x}> +2\alpha+ <\boldsymbol{y},\boldsymbol{y}> +\|\boldsymbol{y}\|^2\end{aligned}$$

where α *is the real part of* $<\boldsymbol{x},\boldsymbol{y}>$*. Now if* $\boldsymbol{x} = A + \boldsymbol{i}B$ *and* $\boldsymbol{y} = C + \boldsymbol{i}D$*, the real part of* $<\boldsymbol{x},\boldsymbol{y}>$ *is* $<A,C> - <B,D>$*. The real and imaginary parts of a vector in a complex vector space belong to a real vector space in which the Cauchy - Schwartz Inequality hold. Thus,*

$$2|\alpha| \leq |<A,C> - <B,D>| \leq \|A\|\,\|C\| + \|B\|\,\|D\| \leq 2\|\boldsymbol{x}\|\,\|\boldsymbol{y}\|$$

We conclude

$$\|\boldsymbol{x}+\boldsymbol{y}\|^2 \leq \|\boldsymbol{x}\|^2 + 2\|\boldsymbol{x}\|\,\|\boldsymbol{y}\| + \|\boldsymbol{y}\|^2 = (\|\boldsymbol{x}\| + \|\boldsymbol{y}\|)^2$$

and the result follows.

Thus, the induced norm is a norm for the $\mathscr{V}$*.* ■

The norm induced by an inner product has special properties which allows us to show norms cannot come from inner products.

Theorem 7.1.5 The Parallelogram Equality

Let $(X,<\cdot,\cdot>)$ *be an inner product space. Then the norm induced by the inner product satisfies the parallelogram equality:*

$$\|(\boldsymbol{x}+\boldsymbol{y})\|^2 + \|(\boldsymbol{x}-\boldsymbol{y})\|^2 = 2(\|\boldsymbol{x}\|^2 + \|\boldsymbol{y}\|^2)$$

for all $\boldsymbol{x}$ *and* $\boldsymbol{y}$ *in* X*.*

Proof 7.1.5

$$\begin{aligned}
\|(\boldsymbol{x}+\boldsymbol{y})\|^2 + \|(\boldsymbol{x}-\boldsymbol{y})\|^2 &= <\boldsymbol{x}+\boldsymbol{y}, \boldsymbol{x}+\boldsymbol{y}> + <\boldsymbol{x}-\boldsymbol{y}, \boldsymbol{x}-\boldsymbol{y}> \\
&= <\boldsymbol{x}, \boldsymbol{x}+\boldsymbol{y}> + <\boldsymbol{y}, \boldsymbol{x}+\boldsymbol{y}> + <\boldsymbol{x}, \boldsymbol{x}-\boldsymbol{y}> \\
&- <\boldsymbol{y}, \boldsymbol{x}-\boldsymbol{y}> \\
&= \|\boldsymbol{x}\|^2 + <\boldsymbol{x}, \boldsymbol{y}> + <\boldsymbol{y}, \boldsymbol{x}> + \|\boldsymbol{y}\|^2 \\
&+ \|\boldsymbol{x}\|^2 - <\boldsymbol{x}, \boldsymbol{y}> - <\boldsymbol{y}, \boldsymbol{x}> + \|\boldsymbol{y}\|^2 \\
&= 2(\|\boldsymbol{x}\|^2 + \|\boldsymbol{y}\|^2)
\end{aligned}$$

■

The Parallelogram Equality allows us to show some spaces are not inner product spaces.

Lemma 7.1.6 $C([a,b])$ is not an Inner Product Space with the Infinity Norm

> *The continuous functions on $[a,b]$ with the supremum norm is not an inner product space.*

Proof 7.1.6
We will show this on the interval $[0,1]$ for convenience. Let $x(t) = t$ and $y(t) = 1-t$. Then $\|\boldsymbol{x}\|_\infty = \|\boldsymbol{y}\|_\infty = 1$, $\|(\boldsymbol{x}+\boldsymbol{y})\|_\infty = 1$ and $\|(\boldsymbol{x}-\boldsymbol{y})\|_\infty = 1$. If $\|\cdot\|_\infty$ was induced by an inner product, then the Parallelogram Equality should hold. We see

$$\begin{aligned}
\|(\boldsymbol{x}+\boldsymbol{y})\|_\infty + \|(\boldsymbol{x}-\boldsymbol{y})\|_\infty &= 2 \\
2(\|\boldsymbol{x}\|_\infty^2 + \|\boldsymbol{y}\|_\infty^2) &= 4
\end{aligned}$$

These are not equal so this norm is not induced by an inner product. ■

Lemma 7.1.7 ℓ^p is not an Inner Product Space with the p norm for $p \neq 2$

> *The sequence space ℓ^p is not an inner product space unless $p = 2$.*

Proof 7.1.7
If p is finite, let $\boldsymbol{x} = \{1, 1, 0, 0, 0, \ldots\}$ and $\boldsymbol{y} = \{1-1, 0, 0, 0, \ldots\}$. Then

$$\begin{aligned}
\|\boldsymbol{x}\|_p &= (1^p + 1^p)^{\frac{1}{p}} = 2^{\frac{1}{p}} = \|\boldsymbol{y}\|_p \\
\|\boldsymbol{x}+\boldsymbol{y}\|_p &= (2^p + 0^p)^{\frac{1}{p}} = 2 = \|\boldsymbol{x}-\boldsymbol{y}\|_p
\end{aligned}$$

It follows

$$\begin{aligned}
\|\boldsymbol{x}+\boldsymbol{y}\|_p^2 + \|\boldsymbol{x}-\boldsymbol{y}\|_p^2 &= 8 \\
2(\|\boldsymbol{x}\|_p^2 + \|\boldsymbol{y}\|_p^2) = 2(2^{\frac{1}{p}} + 2^{\frac{2}{p}}) = (4)\, 2^{\frac{2}{p}}
\end{aligned}$$

These are equal when $2 = 2^{\frac{2}{p}}$ or $p = 2$.

When $p = \infty$, the same $\boldsymbol{x}$ and $\boldsymbol{y}$ give a case where the Parallelogram Equality fails. ■

Homework

Exercise 7.1.9 *Prove ℓ^∞ fails the parallelogram equality also.*

Exercise 7.1.10 *Define $\phi : \ell^1 \times \ell^\infty \to \Re$ by $\phi(\boldsymbol{x}, \boldsymbol{y}) = \sum_{i=1}^\infty x_i y_i$ for any $\boldsymbol{x} \in \ell^1$ and $\boldsymbol{y} \in \ell^\infty$.*

- *Prove ϕ is well-defined by using Hölder's Inequality.*
- *Prove ϕ is linear in each slot; i.e. ϕ is a bilinear map.*
- *Fix $\boldsymbol{x} \in \ell^1$ and define $T : \ell^\infty \to \Re$ by $T(\boldsymbol{y}) = \sum_{i=1}^\infty x_i y_i$ for any $\boldsymbol{y} \in \ell^\infty$. Prove T is a bounded linear functional with $\|T\| = \|\boldsymbol{x}\|_1$.*
- *Fix $\boldsymbol{y} \in \ell^\infty$ and define $S : \ell^1 \to \Re$ by $S(\boldsymbol{x}) = \sum_{i=1}^\infty x_i y_i$ for any $\boldsymbol{x} \in \ell^1$. Prove S is a bounded linear functional with $\|S\| = \|\boldsymbol{x}\|_\infty$. Note, for $\boldsymbol{x_n} = (0, \ldots, \|\boldsymbol{x}\|_1 y_n, 0, \ldots)$, $|S(\boldsymbol{x_n})| = \|\boldsymbol{x}\|_1 |y_n|$. Hence $\|S\| \geq |y_n|$ for all n.*

Exercise 7.1.11 *Define $\phi : \ell^1 \times \ell^\infty \to \Re$ by $\phi(\boldsymbol{x}, \boldsymbol{y}) = \sum_{i=1}^\infty x_i y_i$ for any $\boldsymbol{x} \in \ell^1$ and $\boldsymbol{y} \in \ell^\infty$.*

- *Prove ϕ is well-defined by using Hölder's Inequality.*
- *Prove ϕ is linear in each slot; i.e. ϕ is a bilinear map.*
- *Fix $\boldsymbol{x} \in \ell^1$ and define $T : \ell^\infty \to \Re$ by $T(\boldsymbol{y}) = \sum_{i=1}^\infty x_i y_i$ for any $\boldsymbol{y} \in \ell^\infty$. Prove T is a bounded linear functional with $\|T\| = \|\boldsymbol{x}\|_1$.*
- *Fix $\boldsymbol{y} \in \ell^\infty$ and define $S : \ell^1 \to \Re$ by $S(\boldsymbol{x}) = \sum_{i=1}^\infty x_i y_i$ for any $\boldsymbol{x} \in \ell^1$. Prove S is a bounded linear functional with $\|S\| = \|\boldsymbol{y}\|_\infty$. Note, for $\boldsymbol{x_n} = (0, \ldots, \|\boldsymbol{x}\|_1 y_n, 0, \ldots)$, $|S(\boldsymbol{x_n})| = \|\boldsymbol{x}\|_1 |y_n|$. Hence $\|S\| \geq |y_n|$ for all n.*

Exercise 7.1.12 *Define $\phi : \ell^2 \times \ell^2 \to \Re$ by $\phi(\boldsymbol{x}, \boldsymbol{y}) = \sum_{i=1}^\infty x_i y_i$ for any $\boldsymbol{x} \in \ell^2$ and $\boldsymbol{y} \in \ell^2$.*

- *Prove ϕ is well-defined by using Hölder's Inequality.*
- *Prove ϕ is linear in each slot; i.e. ϕ is a bilinear map.*
- *Fix $\boldsymbol{x} \in \ell^2$ and define $T : \ell^2 \to \Re$ by $T(\boldsymbol{y}) = \sum_{i=1}^\infty x_i y_i$ for any $\boldsymbol{y} \in \ell^2$. Prove T is a bounded linear functional with $\|T\| = \|\boldsymbol{x}\|_2$.*
- *Fix $\boldsymbol{y} \in \ell^2$ and define $S : \ell^2 \to \Re$ by $S(\boldsymbol{x}) = \sum_{i=1}^\infty x_i y_i$ for any $\boldsymbol{x} \in \ell^2$. Prove S is a bounded linear functional with $\|S\| = \|\boldsymbol{y}\|_2$.*
- *Prove ϕ is an inner product on ℓ^2.*

Exercise 7.1.13 *Define $\phi : \ell^p \times \ell^q \to \Re$ by $\phi(\boldsymbol{x}, \boldsymbol{y}) = \sum_{i=1}^\infty x_i y_i$ for any $\boldsymbol{x} \in \ell^p$ and $\boldsymbol{y} \in \ell^q$ for p and q finite conjugate indices.*

- *Prove ϕ is well-defined by using Hölder's Inequality.*
- *Prove ϕ is linear in each slot; i.e. ϕ is a bilinear map.*
- *Fix $\boldsymbol{x} \in \ell^p$ and define $T : \ell^q \to \Re$ by $T(\boldsymbol{y}) = \sum_{i=1}^\infty x_i y_i$ for any $\boldsymbol{y} \in \ell^q$. Prove T is a bounded linear functional with $\|T\| = \|\boldsymbol{x}\|_p$.*
- *Fix $\boldsymbol{y} \in \ell^q$ and define $S : \ell^p \to \Re$ by $S(\boldsymbol{x}) = \sum_{i=1}^\infty x_i y_i$ for any $\boldsymbol{x} \in \ell^p$. Prove S is a bounded linear functional with $\|S\| = \|\boldsymbol{y}\|_q$.*

Exercise 7.1.14 *Let $X = C([0,1])$ with the supremum norm. Define $\phi : X \times X \to \Re$ by $\phi(\boldsymbol{x}, \boldsymbol{y}) = \int_0^1 x(s) y(s) ds$ for any $\boldsymbol{x} \in X$ and $\boldsymbol{y} \in X$.*

- *Prove ϕ is well-defined by using Hölder's Inequality.*

- *Prove ϕ is linear in each slot; i.e. ϕ is a bilinear map.*
- *Fix $\boldsymbol{x} \in X$ and define $T : X \rightarrow \Re$ by $T(\boldsymbol{y}) = \int_0^1 x(s)y(s)ds$ for any $\boldsymbol{y} \in X$. Prove T is a bounded linear functional with $\|T\| = \|\boldsymbol{x}\|_\infty$.*
- *Fix $\boldsymbol{y} \in X$ and define $S : X \rightarrow \Re$ by $S(\boldsymbol{x}) = \int_0^1 x(s)y(s)ds$ for any $\boldsymbol{x} \in X$. Prove S is a bounded linear functional with $\|S\| = \|\boldsymbol{y}\|_\infty$.*

Exercise 7.1.15 *Let $X = C([0,1])$ with the $\|\cdot\|_2$ norm. Define $\phi : X \times X \rightarrow \Re$ by $\phi(\boldsymbol{x}, \boldsymbol{y}) = \int_0^1 x(s)y(s)ds$ for any $\boldsymbol{x} \in X$ and $\boldsymbol{y} \in X$.*

- *Prove ϕ is well-defined by using Hölder's Inequality.*
- *Prove ϕ is linear in each slot; i.e. ϕ is a bilinear map.*
- *Fix $\boldsymbol{x} \in X$ and define $T : X \rightarrow \Re$ by $T(\boldsymbol{y}) = \int_0^1 x(s)y(s)ds$ for any $\boldsymbol{y} \in X$. Prove T is a bounded linear functional with $\|T\| = \|\boldsymbol{x}\|_2$.*
- *Fix $\boldsymbol{y} \in X$ and define $S : X \rightarrow \Re$ by $S(\boldsymbol{x}) = \int_0^1 x(s)y(s)ds$ for any $\boldsymbol{x} \in X$. Prove S is a bounded linear functional with $\|S\| = \|\boldsymbol{y}\|_2$.*

7.2 Hermitian Matrices

Now let's specialize to **Hermitian** matrices. We have gone through these arguments for symmetric matrices in (Peterson (20) 2020). The setting there was real vector spaces and now we will work in complex vector spaces.

Definition 7.2.1 Hermitian Matrices

An $n \times n$ complex-valued matrix $\boldsymbol{A}$ is called Hermitian if $A_{ij} = \overline{A_{ij}}$. This forces the diagonal elements to be real numbers and means the complex conjugate $\overline{\boldsymbol{A}}$ is the same as the transpose $\boldsymbol{A}^T$.

Example 7.2.1 *If*

$$\begin{aligned} \boldsymbol{A} &= \begin{bmatrix} 1 & 2+3\boldsymbol{i} \\ 2-3\boldsymbol{i} & 3 \end{bmatrix} \Longrightarrow \overline{\boldsymbol{A}} = \begin{bmatrix} 1 & 2-3\boldsymbol{i} \\ 2+3\boldsymbol{i} & 3 \end{bmatrix} \\ \boldsymbol{A}^T &= \begin{bmatrix} 1 & 2-3\boldsymbol{i} \\ 2+3\boldsymbol{i} & 3 \end{bmatrix} \Longrightarrow \overline{\boldsymbol{A}} = \boldsymbol{A}^T \end{aligned}$$

Hence $\boldsymbol{A}$ is Hermitian. Note

$$\begin{aligned} \boldsymbol{B} &= \begin{bmatrix} 1 & 4+3\boldsymbol{i} \\ 2-3\boldsymbol{i} & 3 \end{bmatrix} \Longrightarrow \overline{\boldsymbol{B}} = \begin{bmatrix} 1 & 4-3\boldsymbol{i} \\ 2+3\boldsymbol{i} & 3 \end{bmatrix} \\ \boldsymbol{B}^T &= \begin{bmatrix} 1 & 2-3\boldsymbol{i} \\ 4+3\boldsymbol{i} & 3 \end{bmatrix} \Longrightarrow \overline{\boldsymbol{B}} \neq \boldsymbol{B}^T \end{aligned}$$

and so $\boldsymbol{B}$ is not Hermitian.

Now for the calculations we are going to do now, we will be using the complex inner product. An easy calculation shows

$$< \boldsymbol{A}(\boldsymbol{x}), \boldsymbol{y} > \; = \; \sum_{i=1}^{n} (\boldsymbol{A}\boldsymbol{x})_i \, \overline{y_i} = \sum_{i=1}^{n} \sum_{j=1}^{n} A_{ij} x_j \overline{y_j} = \sum_{j=1}^{n} \overline{\boldsymbol{A}^T}_{ij} (\overline{\boldsymbol{y}})_j x_j$$

$$= \sum_{j=1}^{n} \left(\sum_{i=1}^{n} \overline{A_{ji} y_i} \right) x_j = \sum_{j=1}^{n} x_j \left(\overline{\boldsymbol{A}^T(\boldsymbol{y})} \right)_j =< \boldsymbol{x}, \boldsymbol{A}(\boldsymbol{y}) >$$

Since $< \boldsymbol{A}(\boldsymbol{x}), \boldsymbol{x} >=< \boldsymbol{x}, \boldsymbol{A}(\boldsymbol{x}) >= \overline{< \boldsymbol{A}(\boldsymbol{x}), \boldsymbol{x} >}$, we know $< \boldsymbol{A}(\boldsymbol{x}), \boldsymbol{x} >$ is a real number. We can also show

$$\|\boldsymbol{A}\| = \max_{\|\boldsymbol{x}\|=1} | < \boldsymbol{A}(\boldsymbol{x}), \boldsymbol{x} > |$$

We can use a maximum here as the function $| < \boldsymbol{A}(\boldsymbol{x}), \boldsymbol{x} > |$ is a continuous function on a compact domain and so the supremum is achieved. Let the maximum on the right hand side be denoted by $\boldsymbol{J}$ for convenience. For any vector $\boldsymbol{y}$

$$| < \boldsymbol{A}(\boldsymbol{y}), \boldsymbol{y} > | \leq \|\boldsymbol{A}(\boldsymbol{y})\| \, \|\boldsymbol{y}\|$$

Next, from the way $\|\boldsymbol{A}\|$ is defined, for any particular nonzero vector $\boldsymbol{y}$, we have $\|\boldsymbol{A}(\boldsymbol{y})\| \leq \|\boldsymbol{A}\| \|\boldsymbol{y}\|$. Thus, combining these ideas, we have for any vector $\boldsymbol{x}$ with norm 1,

$$| < \boldsymbol{A}(\boldsymbol{x}), \boldsymbol{x} > | \leq \|\boldsymbol{A}(\boldsymbol{x})\| \, \|\boldsymbol{x}\| \leq \|\boldsymbol{A}\| \, \|\boldsymbol{x}\| \, \|\boldsymbol{x}\| = \|\boldsymbol{A}\|.$$

Since this is true for all such vectors, we must have the maximum $\boldsymbol{J}$ satisfies $\boldsymbol{J} \leq \|\boldsymbol{A}\|$. Next, we will show the reverse inequality and the two pieces together then show $\boldsymbol{J} = \|\boldsymbol{A}\|$. Now for any nonzero $\boldsymbol{y}$ we have

$$\left\langle \boldsymbol{A}\left(\frac{\boldsymbol{y}}{\|\boldsymbol{y}\|} \right), \frac{\boldsymbol{y}}{\|\boldsymbol{y}\|} \right\rangle \leq \boldsymbol{J}$$

which implies $< \boldsymbol{A}(\boldsymbol{y}), \boldsymbol{y} >\leq \boldsymbol{J} \, \|\boldsymbol{y}\|^2$. Now do the following calculations. We have for any $\boldsymbol{x}$ and $\boldsymbol{y}$ that

$$\begin{aligned} < \boldsymbol{A}(\boldsymbol{x}+\boldsymbol{y}), \boldsymbol{x}+\boldsymbol{y} > &= < \boldsymbol{A}(\boldsymbol{x}), \boldsymbol{x} > + < \boldsymbol{A}(\boldsymbol{x}), \boldsymbol{y} > + < \boldsymbol{A}(\boldsymbol{y}), \boldsymbol{x} > + < \boldsymbol{A}(\boldsymbol{y}), \boldsymbol{y} > \\ &\leq \boldsymbol{J}\|(\boldsymbol{x}+\boldsymbol{y})\|^2 \end{aligned}$$

We also have

$$\begin{aligned} < \boldsymbol{A}(\boldsymbol{x}-\boldsymbol{y}), \boldsymbol{x}-\boldsymbol{y} > &= < \boldsymbol{A}(\boldsymbol{x}), \boldsymbol{x} > - < \boldsymbol{A}(\boldsymbol{x}), \boldsymbol{y} > - < \boldsymbol{A}(\boldsymbol{y}), \boldsymbol{x} > + < \boldsymbol{A}(\boldsymbol{y}), \boldsymbol{y} > \\ &\leq \boldsymbol{J}\|(\boldsymbol{x}-\boldsymbol{y})\|^2 \end{aligned}$$

Now subtract the second inequality from the first to get

$$\begin{aligned} 2 < \boldsymbol{A}(\boldsymbol{x}), \boldsymbol{y} > +2 < \boldsymbol{A}(\boldsymbol{y}), \boldsymbol{x} > &\leq |< \boldsymbol{A}(\boldsymbol{x}+\boldsymbol{y}), \boldsymbol{x}+\boldsymbol{y} >| + |< \boldsymbol{A}(\boldsymbol{x}-\boldsymbol{y}), \boldsymbol{x}-\boldsymbol{y} >| \\ &\leq \boldsymbol{J}\Big(\|\boldsymbol{x}+\boldsymbol{y}\|^2 + \|\boldsymbol{x}-\boldsymbol{y}\|^2 \Big) = \boldsymbol{J}(2\|\boldsymbol{x}\|^2 + 2\|\boldsymbol{y}\|^2) = 4\boldsymbol{J} \end{aligned}$$

if $\boldsymbol{x}$ and $\boldsymbol{y}$ are unit norm. But

$$< \boldsymbol{A}(\boldsymbol{y}), \boldsymbol{x} > = < \boldsymbol{y}, \boldsymbol{A}(\boldsymbol{x}) >= \overline{< \boldsymbol{A}(\boldsymbol{x}), \boldsymbol{y} >}$$

and so for $\boldsymbol{x}$ and $\boldsymbol{y}$ of unit norm

$$Re(< \boldsymbol{A}(\boldsymbol{x}), \boldsymbol{y} >) \leq \boldsymbol{J}$$

In particular, if $A(x) = 0$, we would have

$$\|A(x)\|^2 \quad = \quad < A(x), A(x) >=< 0, 0 >= 0,$$

and so by definition of J, $|A(x)\| \leq J$. Next, if $A(x) \neq 0$,

$$Re\left\langle A(x), \frac{A(x)}{\|A(x)\|}\right\rangle \quad = \quad \left\langle A(x), \frac{A(x)}{\|A(x)\|}\right\rangle \leq J$$

or

$$\frac{\|A(x)\|^2}{\|A(x)\|} \quad = \quad \|A(x)\| \leq J$$

This implies the maximum over all $\|x\| = 1$, also satisfies this inequality and so

$$\|A\| \quad = \quad \sup_{\|x\|=1} \|A(x)\| \leq J.$$

With the reverse inequality established, we have proven the result we want. Let's summarize our technical discussion. We have proven the following.

Theorem 7.2.1 Two Equivalent Ways to Calculate the Norm of a Hermitian Matrix

If A is a Hermitian $n \times n$ matrix, then

$$\|A\| \quad = \quad \max_{\|x\|=1} \|A(x)\| = \max_{\|x\|=1} |< A(x), x >| = J.$$

Proof 7.2.1
See the discussions above. ■

Homework

Exercise 7.2.1 *Let*

$$A \quad = \quad \begin{bmatrix} 2 & 5+3i \\ 5-3i & 4 \end{bmatrix}$$

Use MATLAB to find the two eigenvalues and eigenvectors of A.

- *Are the eigenvalues real?*
- *Are the eigenvectors orthogonal?*
- *Use the eigenvectors to find an orthonormal basis for $\mathbb{C}^2$; i.e. the basis vectors are orthogonal and length one.*
- *Let P be the matrix whose columns are the vectors from the orthonormal basis. Calculate $\overline{P}^T AP$. It should be interesting!*
- *Find $\|A\|_\infty$.*
- *Find $\|A\|_1$.*
- *Find $\|A\|_F$.*

- *In a bit, we will find the largest eigenvalue magnitude is the value of* $\|A\|_2$. *So what is* $\|A\|_2$?

Exercise 7.2.2 *Let*

$$A = \begin{bmatrix} -3 & -2+3i \\ -2-3i & 5 \end{bmatrix}$$

Use MATLAB to find the two eigenvalues and eigenvectors of A.

- *Are the eigenvalues real?*
- *Are the eigenvectors orthogonal?*
- *Use the eigenvectors to find an orthonormal basis for* $\mathbb{C}^2$*; i.e. the basis vectors are orthogonal and length one.*
- *Let* P *be the matrix whose columns are the vectors from the orthonormal basis. Calculate* $\overline{P}^T AP$. *It should be interesting!*
- *Find* $\|A\|_\infty$.
- *Find* $\|A\|_1$.
- *Find* $\|A\|_F$.
- *In a bit, we will find the largest eigenvalue magnitude is the value of* $\|A\|_2$. *So what is* $\|A\|_2$?

Exercise 7.2.3 *Let*

$$A = \begin{bmatrix} 3 & 2+7i & 6-8i \\ 2-7i & 4 & -7+2i \\ 6+8i & -7-2i & -3 \end{bmatrix}$$

Use MATLAB to find the three eigenvalues and eigenvectors of A.

- *Are the eigenvalues real?*
- *Are the eigenvectors orthogonal?*
- *Use the eigenvectors to find an orthonormal basis for* $\mathbb{C}^2$*; i.e. the basis vectors are orthogonal and length one.*
- *Let* P *be the matrix whose columns are the vectors from the orthonormal basis. Calculate* $\overline{P}^T AP$. *It should be interesting!*
- *Find* $\|A\|_\infty$.
- *Find* $\|A\|_1$.
- *Find* $\|A\|_F$.
- *In a bit, we will find the largest eigenvalue magnitude is the value of* $\|A\|_2$. *So what is* $\|A\|_2$?

Use MATLAB to find the two eigenvalues and eigenvectors of A.

- *Are the eigenvalues real?*
- *Are the eigenvectors orthogonal?*

Exercise 7.2.4 *Let*

$$A \;=\; \begin{bmatrix} 1 & 12+3i & 6-2i \\ 12-3i & 3 & 3+5i \\ 6+2i & 3-5i & 8 \end{bmatrix}$$

Use MATLAB to find the three eigenvalues and eigenvectors of A.

- *Are the eigenvalues real?*

- *Are the eigenvectors orthogonal?*

- *Use the eigenvectors to find an orthonormal basis for* $\mathbb{C}^2$; *i.e. the basis vectors are orthogonal and length one.*

- *Let* P *be the matrix whose columns are the vectors from the orthonormal basis. Calculate* $\overline{P}^T A P$. *It should be interesting!*

- *Find* $\|A\|_\infty$.

- *Find* $\|A\|_1$.

- *Find* $\|A\|_F$.

- *In a bit, we will find the largest eigenvalue magnitude is the value of* $\|A\|_2$. *So what is* $\|A\|_2$?

Use MATLAB to find the two eigenvalues and eigenvectors of A.

- *Are the eigenvalues real?*

- *Are the eigenvectors orthogonal?*

7.2.1   Constructing Eigenvalues

For the Hermitian matrix, we can construct its eigenvalues, just as we did in the symmetric matrix case, using a procedure which also gives us useful estimates. Now let's find eigenvalues:

7.2.1.1   Eigenvalue One

Theorem 7.2.2 The First Eigenvalue

Either $\|A\|$ *or* $\|A\|$ *is an eigenvalue for* A. *Letting the eigenvalue be* λ_1, *then* $|\lambda_1| = \|A\|$ *with an associated eigenvector of norm* 1, E_1.

Proof 7.2.2
We know that the maximum value of $|<A(x),x>|$ *over all* $\|x\| = 1$ *occurs at some unit vector. Call this unit vector* E_1. *For convenience, let* $\alpha = \|A\|$. *Then we know* $|<A(E_1),E_1>| = \alpha$.
Case I: *We assume* $<A(E_1),E_1> = -\alpha$. *Then,*

$$<A(E_1)-(-\alpha)E_1, A(E_1)-(-\alpha)E_1> \;=\; <A(E_1)+\alpha E_1, A(E_1)+\alpha E_1>$$
$$=\; <A(E_1),A(E_1)> +\alpha<A(E_1),E_1>$$
$$+\alpha\overline{<A(E_1),E_1>}+\alpha^2<E_1,E_1>$$

as $<A(E_1),E_1>$ *is real. So we have*

$$<A(E_1)-(-\alpha)E_1, A(E_1)-(-\alpha)E_1> \;=\; \|A(E_1)\|^2 + 2\alpha<A(E_1),E_1> +\alpha^2$$

$$\begin{aligned} &= \; < \boldsymbol{A}(\boldsymbol{E_1}), \boldsymbol{A}(\boldsymbol{E_1}) > -2\alpha^2 + \alpha^2 \\ &= \; \|\boldsymbol{A}(\boldsymbol{E_1})\|^2 - \alpha^2 \end{aligned}$$

Then, overestimating, we have

$$< \boldsymbol{A}(\boldsymbol{E_1}) - (-\alpha)\boldsymbol{E_1}, \boldsymbol{A}(\boldsymbol{E_1}) - (-\alpha)\boldsymbol{E_1} > \;\; \leq \;\; \|\boldsymbol{A}\|^2 \, \|\boldsymbol{E_1}\|^2 - \alpha^2.$$

But $\alpha = \|\boldsymbol{A}\|$, *so we have*

$$< \boldsymbol{A}(\boldsymbol{E_1}) - (-\alpha)\boldsymbol{E_1}, \boldsymbol{A}(\boldsymbol{E_1}) - (-\alpha)\boldsymbol{E_1} > \;\; \leq \;\; \alpha^2 - \alpha^2 \; = \; 0.$$

We conclude $\|\boldsymbol{A}(\boldsymbol{E_1}) - (-\alpha)\boldsymbol{E_1}\| = 0$ *which tells us* $\boldsymbol{A}(\boldsymbol{E_1}) = -\alpha\boldsymbol{E_1}$. *Hence,* $-\alpha$ *is an eigenvalue of* $\boldsymbol{A}$ *and we can pick* $\boldsymbol{E_1}$ *as the associated eigenvector of norm* 1.

Case II*: We assume* $< \boldsymbol{A}(\boldsymbol{E_1}), \boldsymbol{E_1} >= -\alpha$.*. The argument is then quite similar. We conclude* $\|\boldsymbol{A}(\boldsymbol{E_1}) - \alpha\boldsymbol{E_1}\| = 0$ *which tells us* $\boldsymbol{A}(\boldsymbol{E_1}) = \alpha\boldsymbol{E_1}$. *Hence,* α *is an eigenvalue of* $\boldsymbol{A}$ *and we can pick* $\boldsymbol{E_1}$ *as the associated eigenvector of norm* 1. ■

Homework

We will look at the previous exercises again in a new context.

Exercise 7.2.5 *Let*

$$\boldsymbol{A} \;\; = \;\; \begin{bmatrix} 2 & 5+3\boldsymbol{i} \\ 5-3\boldsymbol{i} & 4 \end{bmatrix}$$

We know the largest absolute eigenvalue is $\|\boldsymbol{A}\|$ *by our theorem for the first eigenvalue. Write down the optimization problem you need to solve to find this number. Let* $\boldsymbol{u} = \boldsymbol{x} + \boldsymbol{i}\boldsymbol{y}$ *be the complex vector*

$$\boldsymbol{x} + \boldsymbol{i}\boldsymbol{y} \;\; = \;\; \begin{bmatrix} x_1 \\ y_1 \end{bmatrix} + \boldsymbol{i} \begin{bmatrix} x_2 \\ y_2 \end{bmatrix}$$

Then to find the first eigenvalue and eigenvector, we need to find

$$\begin{aligned} |\lambda_1| \;\; &= \;\; \max_{\|\boldsymbol{u}\|=1} \, | < \boldsymbol{A}(\boldsymbol{u}), \boldsymbol{u} > | = \max_{\|\boldsymbol{u}\|=1} |\boldsymbol{u}^T \boldsymbol{A} \overline{\boldsymbol{u}}| \\ &= \;\; \max_{\|\boldsymbol{u}\|=1} \left| \begin{bmatrix} x_1 + x_2\boldsymbol{i} & y_1 + y_2\boldsymbol{i} \end{bmatrix} \begin{bmatrix} 2 & 5+3\boldsymbol{i} \\ 5-3\boldsymbol{i} & 2 \end{bmatrix} \begin{bmatrix} x_1 - x_2\boldsymbol{i} \\ y_1 - y_2\boldsymbol{i} \end{bmatrix} \right| \end{aligned}$$

Simplify this until you can write the optimization problem in terms of x_1, x_2, y_3 *and* y_4. *This is a fair bit of algebra! It is easy to see solving this optimization problem is not easy, so it is remarkable our other techniques for finding eigenvalues and eigenvectors such as finding roots of polynomials furnish quick solutions via MATLAB!*

Exercise 7.2.6 *Let*

$$\boldsymbol{A} \;\; = \;\; \begin{bmatrix} -3 & -2+3\boldsymbol{i} \\ -2-3\boldsymbol{i} & 5 \end{bmatrix}$$

Let $\boldsymbol{u} = \boldsymbol{x} + \boldsymbol{i}\boldsymbol{y}$ *be the complex vector*

$$\boldsymbol{x} + \boldsymbol{i}\boldsymbol{y} = \begin{bmatrix} x_1 \\ y_1 \end{bmatrix} + \boldsymbol{i} \begin{bmatrix} x_2 \\ y_2 \end{bmatrix}$$

Then to find the first eigenvalue and eigenvector, we need to find

$$\begin{aligned} |\lambda_1| &= \max_{\|\boldsymbol{u}\|=1} |<\boldsymbol{A}(\boldsymbol{u}), \boldsymbol{u}>| = \max_{\|\boldsymbol{u}\|=1} |\boldsymbol{u}^T \boldsymbol{A}\overline{\boldsymbol{u}}| \\ &= \max_{\|\boldsymbol{x}\|=1} \left| \begin{bmatrix} x_1 + x_2\boldsymbol{i} & y_1 + y_2\boldsymbol{i} \end{bmatrix} \begin{bmatrix} -3 & -2+3\boldsymbol{i} \\ -2-3\boldsymbol{i} & 5 \end{bmatrix} \begin{bmatrix} x_1 - x_2\boldsymbol{i} \\ y_1 - y_2\boldsymbol{i} \end{bmatrix} \right| \end{aligned}$$

Simplify this until you can write the optimization problem in terms of x_1, x_2, y_3 *and* y_4. *Again, this is a fair bit of algebra!*

Exercise 7.2.7 *Let*

$$\boldsymbol{A} = \begin{bmatrix} 3 & 2+7\boldsymbol{i} & 6-8\boldsymbol{i} \\ 2-7\boldsymbol{i} & 4 & -7+2\boldsymbol{i} \\ 6+8\boldsymbol{i} & -7-2\boldsymbol{i} & -3 \end{bmatrix}$$

Let $\boldsymbol{u} = \boldsymbol{x} + \boldsymbol{i}\boldsymbol{y}$ *be the complex vector*

$$\boldsymbol{u} = \boldsymbol{x} + \boldsymbol{i}\boldsymbol{y} = \begin{bmatrix} x_1 \\ y_1 \\ z_1 \end{bmatrix} + \boldsymbol{i} \begin{bmatrix} x_2 \\ y_2 \\ z_2 \end{bmatrix}$$

Then to find the first eigenvalue and eigenvector, we need to find

$$\begin{aligned} |\lambda_1| &= \max_{\|\boldsymbol{u}\|=1} |<\boldsymbol{A}(\boldsymbol{u}), \boldsymbol{u}>| = \max_{\|\boldsymbol{u}\|=1} |\boldsymbol{u}^T \boldsymbol{A}\overline{\boldsymbol{u}}| \\ &= \max_{\|\boldsymbol{u}\|=1} \left| \begin{bmatrix} x_1 + x_2\boldsymbol{i} & y_1 + y_2\boldsymbol{i} & z_1 + z_2\boldsymbol{i} \end{bmatrix} \begin{bmatrix} 3 & 2+7\boldsymbol{i} & 6-8\boldsymbol{i} \\ 2-7\boldsymbol{i} & 4 & -7+2\boldsymbol{i} \\ 6+8\boldsymbol{i} & -7-2\boldsymbol{i} & -3 \end{bmatrix} \begin{bmatrix} x_1 - x_2\boldsymbol{i} \\ y_1 - y_2\boldsymbol{i} \\ z_1 - z_2\boldsymbol{i} \end{bmatrix} \right| \end{aligned}$$

Simplify this until you can write the optimization problem in terms of x_i, y_i *and* z_i. *Be warned, this is a horrible amount of algebra!*

Exercise 7.2.8 *Let*

$$\boldsymbol{A} = \begin{bmatrix} 1 & 12+3\boldsymbol{i} & 6-2\boldsymbol{i} \\ 12-3\boldsymbol{i} & 3 & 3+5\boldsymbol{i} \\ 6+2\boldsymbol{i} & 3-5\boldsymbol{i} & 8 \end{bmatrix}$$

Let $\boldsymbol{u} = \boldsymbol{x} + \boldsymbol{i}\boldsymbol{y}$ *be the complex vector*

$$\boldsymbol{u} = \boldsymbol{x} + \boldsymbol{i}\boldsymbol{y} = \begin{bmatrix} x_1 \\ y_1 \\ z_1 \end{bmatrix} + \boldsymbol{i} \begin{bmatrix} x_2 \\ y_2 \\ z_2 \end{bmatrix}$$

Then to find the first eigenvalue and eigenvector, we need to find

$$|\lambda_1| = \max_{\|\boldsymbol{u}\|=1} |<\boldsymbol{A}(\boldsymbol{u}), \boldsymbol{u}>| = \max_{\|\boldsymbol{u}\|=1} |\boldsymbol{u}^T \boldsymbol{A}\overline{\boldsymbol{u}}|$$

$$= \max_{||\boldsymbol{u}||=1} \left| \begin{bmatrix} x_1 + x_2\boldsymbol{i} & y_1 + y_2\boldsymbol{i} & z_1 + z_2\boldsymbol{i} \end{bmatrix} \begin{bmatrix} 1 & 12+3\boldsymbol{i} & 6-2\boldsymbol{i} \\ 12-3\boldsymbol{i} & 3 & 3+5\boldsymbol{i} \\ 6+2\boldsymbol{i} & 3-5\boldsymbol{i} & 8 \end{bmatrix} \begin{bmatrix} x_1 - x_2\boldsymbol{i} \\ y_1 - y_2\boldsymbol{i} \\ z_1 - z_2\boldsymbol{i} \end{bmatrix} \right|$$

Simplify this until you can write the optimization problem in terms of x_i, y_i and z_i. Again, be warned, this is a horrible amount of algebra!

7.2.1.2 Eigenvalue Two

We can now find the next eigenvalue using a similar argument.

Theorem 7.2.3 The Second Eigenvalue

There is a second eigenvalue λ_2 which satisfies $|\lambda_2| \leq |\lambda_1|$. Let the new Hermitian matrix $\boldsymbol{A_2}$ be defined by

$$\boldsymbol{A_2} = \boldsymbol{A} - \lambda_1 \, \boldsymbol{E_1} \, \overline{\boldsymbol{E_1}}^T$$

Then, $|\lambda_2| = \|\boldsymbol{A}_2\|$ and the associated eigenvector $\boldsymbol{E_2}$ is orthogonal to $\boldsymbol{E_1}$, i.e., $< \boldsymbol{E_1}, \boldsymbol{E_1} >= 0$.

Proof 7.2.3
The reasoning here is virtually identical to what we did for Theorem 7.2.2. First, note, if $\boldsymbol{x}$ is a multiple of $\boldsymbol{E_1}$, we have $\boldsymbol{x} = \mu\boldsymbol{E_1}$ for some μ giving

$$\begin{aligned} \boldsymbol{A}_2(\boldsymbol{x}) &= \boldsymbol{A}(\mu\boldsymbol{E}_1) - \lambda_1\mu\boldsymbol{E_1}\,\overline{\boldsymbol{E_1}}^T\boldsymbol{E_1} \\ &= \mu\lambda_1\boldsymbol{E_1} - \mu\lambda_1\boldsymbol{E_1} = 0. \end{aligned}$$

since $\boldsymbol{E_1}$ is an eigenvector with eigenvalue λ_1. Hence, it is easy to see that

$$\max_{\|\boldsymbol{x}\|=1} |< \boldsymbol{A_2}(\boldsymbol{x}), \boldsymbol{x} >| = \max_{\|\boldsymbol{x}\|=1, \boldsymbol{x}\in \boldsymbol{W_1}} |< \boldsymbol{A_2}(\boldsymbol{x}), \boldsymbol{x} >|$$

where $\boldsymbol{W_1}$ is the set of vectors in that are orthogonal to eigenvector $\boldsymbol{E_1}$. Since this is a smaller set of vectors, the maximum we obtain will, of course, be possibly smaller. Hence, we know $\max_{\|\boldsymbol{x}\|=1} |< \boldsymbol{A_2}(\boldsymbol{x}), \boldsymbol{x} >| \geq \max_{\|\boldsymbol{x}\|=1,\ \boldsymbol{x}\in\boldsymbol{W_1}} |< \boldsymbol{A_2}(\boldsymbol{x}), \boldsymbol{x} >|$.
The rest of the argument, applied to the new operator $\boldsymbol{A}_2$ is identical. Thus, we find

$$|\lambda_2| = \|\boldsymbol{A_2}\| = \max_{\|\boldsymbol{x}\|=1, \boldsymbol{x}\in\boldsymbol{W_1}} |< \boldsymbol{A_2}(\boldsymbol{x}), \boldsymbol{x} >|.$$

Further, $|\lambda_2| \leq |\lambda_1|$ and eigenvector $\boldsymbol{E_2}$ is orthogonal to $\boldsymbol{E_1}$ because it comes from $\boldsymbol{W_1}$. ■

Homework

We will look at the previous exercises again but now we will set up the optimization problems for the second eigenvalue.

Exercise 7.2.9 *Let*

$$\boldsymbol{A} = \begin{bmatrix} 2 & 5+3\boldsymbol{i} \\ 5-3\boldsymbol{i} & 4 \end{bmatrix}$$

Use MATLAB to find the largest absolute value eigenvalue and an associated unit length eigenvector. Let the eigenvalue be λ_1 and the unit eigenvector be $\boldsymbol{E_1}$. Calculate

$$\boldsymbol{A_2} = \boldsymbol{A} - \lambda_1\, \boldsymbol{E_1}\, \overline{\boldsymbol{E_1}}^T,$$

Write down the optimization problem you need to solve to find the second eigenvalue. Let $\boldsymbol{u} = \boldsymbol{x}+\boldsymbol{i}\boldsymbol{y}$ be the complex vector

$$\boldsymbol{x}+\boldsymbol{i}\boldsymbol{y} = \begin{bmatrix} x_1 \\ y_1 \end{bmatrix} + \boldsymbol{i} \begin{bmatrix} x_2 \\ y_2 \end{bmatrix}$$

Let W be the set of vectors orthogonal to $\boldsymbol{E_1}$. Then to find the second eigenvalue and eigenvector, we need to find

$$|\lambda_2| = \max_{\|\boldsymbol{u}\|=1 \in W} | < \boldsymbol{A_2}(\boldsymbol{u}), \boldsymbol{u} > | = \max_{\|\boldsymbol{u}\|=1 \in W} |\boldsymbol{u}^T \boldsymbol{A_2} \overline{\boldsymbol{u}}|$$

Simplify this until you can write the optimization problem in terms of x_1, x_2, y_3 and y_4. You don't need to find how to describe W as we are just looking for the form of the optimization. It is easy to see finding W and solving this optimization problem is not easy , so it is again remarkable our other techniques for finding eigenvalues and eigenvectors such as finding roots of polynomials furnish quick solutions via MATLAB!

Exercise 7.2.10 *Let*

$$\boldsymbol{A} = \begin{bmatrix} -3 & -2+3\boldsymbol{i} \\ -2-3\boldsymbol{i} & 5 \end{bmatrix}$$

Use MATLAB to find the largest absolute value eigenvalue and an associated unit length eigenvector. Let the eigenvalue be λ_1 and the unit eigenvector be $\boldsymbol{E_1}$. Calculate

$$\boldsymbol{A_2} = \boldsymbol{A} - \lambda_1\, \boldsymbol{E_1}\, \overline{\boldsymbol{E_1}}^T,$$

Write down the optimization problem you need to solve to find the second eigenvalue. Let $\boldsymbol{u} = \boldsymbol{x}+\boldsymbol{i}\boldsymbol{y}$ be the complex vector

$$\boldsymbol{x}+\boldsymbol{i}\boldsymbol{y} = \begin{bmatrix} x_1 \\ y_1 \end{bmatrix} + \boldsymbol{i} \begin{bmatrix} x_2 \\ y_2 \end{bmatrix}$$

Let W be the set of vectors orthogonal to $\boldsymbol{E_1}$. Then to find the second eigenvalue and eigenvector, we need to find

$$|\lambda_2| = \max_{\|\boldsymbol{u}\|=1 \in W} | < \boldsymbol{A_2}(\boldsymbol{u}), \boldsymbol{u} > | = \max_{\|\boldsymbol{u}\|=1 \in W} |\boldsymbol{u}^T \boldsymbol{A_2} \overline{\boldsymbol{u}}|$$

Simplify this until you can write the optimization problem in terms of x_1, x_2, y_3 and y_4. You don't need to find how to describe W as we are just looking for the form of the optimization.

Exercise 7.2.11 *Let*

$$\boldsymbol{A} = \begin{bmatrix} 3 & 2+7\boldsymbol{i} & 6-8\boldsymbol{i} \\ 2-7\boldsymbol{i} & 4 & -7+2\boldsymbol{i} \\ 6+8\boldsymbol{i} & -7-2\boldsymbol{i} & -3 \end{bmatrix}$$

Use MATLAB to find the largest absolute value eigenvalue and an associated unit length eigenvector. Let the eigenvalue be λ_1 and the unit eigenvector be $\boldsymbol{E_1}$. Calculate

$$\boldsymbol{A_2} \quad = \quad \boldsymbol{A} - \lambda_1 \, \boldsymbol{E_1} \, \overline{\boldsymbol{E_1}}^T,$$

Let W be the set of vectors orthogonal to $\boldsymbol{E_1}$. Write down the optimization problem you need to solve to find the second eigenvalue. Let $\boldsymbol{u} = \boldsymbol{x} + \boldsymbol{iy}$ be the complex vector

$$\boldsymbol{u} = \boldsymbol{x} + \boldsymbol{iy} \quad = \quad \begin{bmatrix} x_1 \\ y_1 \\ z_1 \end{bmatrix} + \boldsymbol{i} \begin{bmatrix} x_2 \\ y_2 \\ z_2 \end{bmatrix}$$

Then to find the second eigenvalue and eigenvector, we need to find

$$|\lambda_2| \quad = \quad \max_{\|\boldsymbol{u}\|=1 \in W} | < \boldsymbol{A_2}(\boldsymbol{u}), \boldsymbol{u} > | = \max_{\|\boldsymbol{u}\|=1 \in W} |\boldsymbol{u}^T \boldsymbol{A_2} \overline{\boldsymbol{u}}|$$

Simplify this until you can write the optimization problem in terms of x_i, y_i and z_i. You don't need to find how to describe W as we are just looking for the form of the optimization.

Exercise 7.2.12 *Let*

$$\boldsymbol{A} \quad = \quad \begin{bmatrix} 1 & 12 + 3\boldsymbol{i} & 6 - 2\boldsymbol{i} \\ 12 - 3\boldsymbol{i} & 3 & 3 + 5\boldsymbol{i} \\ 6 + 2\boldsymbol{i} & 3 - 5\boldsymbol{i} & 8 \end{bmatrix}$$

Use MATLAB to find the largest absolute value eigenvalue and an associated unit length eigenvector. Let the eigenvalue be λ_1 and the unit eigenvector be $\boldsymbol{E_1}$. Calculate

$$\boldsymbol{A_2} \quad = \quad \boldsymbol{A} - \lambda_1 \, \boldsymbol{E_1} \, \overline{\boldsymbol{E_1}}^T,$$

Let W be the set of vectors orthogonal to $\boldsymbol{E_1}$. Write down the optimization problem you need to solve to find the second eigenvalue. Let $\boldsymbol{u} = \boldsymbol{x} + \boldsymbol{iy}$ be the complex vector

$$\boldsymbol{u} = \boldsymbol{x} + \boldsymbol{iy} \quad = \quad \begin{bmatrix} x_1 \\ y_1 \\ z_1 \end{bmatrix} + \boldsymbol{i} \begin{bmatrix} x_2 \\ y_2 \\ z_2 \end{bmatrix}$$

Then to find the second eigenvalue and eigenvector, we need to find

$$|\lambda_2| \quad = \quad \max_{\|\boldsymbol{u}\|=1 \in W} | < \boldsymbol{A_2}(\boldsymbol{u}), \boldsymbol{u} > | = \max_{\|\boldsymbol{u}\|=1 \in W} |\boldsymbol{u}^T \boldsymbol{A_2} \overline{\boldsymbol{u}}|$$

Simplify this until you can write the optimization problem in terms of x_i, y_i and z_i. You don't need to find how to describe W as we are just looking for the form of the optimization.

7.2.1.3 Eigenvalue Three

We can now find the next eigenvalue.

Theorem 7.2.4 The Third Eigenvalue

There is a third eigenvalue λ_3 which satisfies $|\lambda_3| \leq |\lambda_2|$. Let the new Hermitian matrix $\boldsymbol{A_3}$ be defined by

$$\boldsymbol{A_3} \quad = \quad \boldsymbol{A} - \lambda_1 \, \boldsymbol{E_1} \, \overline{\boldsymbol{E_1}}^T - \lambda_2 \, \boldsymbol{E_2} \, \overline{\boldsymbol{E_2}}^T$$

Then, $|\lambda_3| = \|\boldsymbol{A}_3\|$ and the associated eigenvector $\boldsymbol{E_3}$ is orthogonal to $\boldsymbol{E_1}$ and $\boldsymbol{E_1}$.

Proof 7.2.4
The reasoning here is virtually identical to what we did for the first two eigenvalues. First, note, if $\boldsymbol{x}$ is in the plane determined by $\boldsymbol{E_1}$ and $\boldsymbol{E_2}$, we have $\boldsymbol{x} = \mu_1 \boldsymbol{E_1} + \mu_2 \boldsymbol{E_2}$ for some constants μ_1 and μ_2 giving

$$\begin{aligned} \boldsymbol{A}_3(\boldsymbol{x}) &= \boldsymbol{A}(\mu_1 \boldsymbol{E_1} + \mu_2 \boldsymbol{E_2}) - \lambda_1 \mu_1 \boldsymbol{E_1} \, \overline{\boldsymbol{E_1}}^T \boldsymbol{E_1} - \lambda_2 \mu_2 \boldsymbol{E_2} \, \overline{\boldsymbol{E_2}}^T \boldsymbol{E_2} \\ &= \mu_1 \lambda_1 \boldsymbol{E_1} + \mu_2 \lambda_2 \boldsymbol{E_2} - \mu_1 \lambda_1 \boldsymbol{E_1} - \mu_2 \lambda_2 \boldsymbol{E_2} = 0. \end{aligned}$$

since $\boldsymbol{E_1}$ and $\boldsymbol{E_2}$ are eigenvectors. Hence, it is easy to see that

$$\max_{\|\boldsymbol{x}\|=1} | < \boldsymbol{A_3}(\boldsymbol{x}), \boldsymbol{x} > | \quad = \quad \max_{\|\boldsymbol{x}\|=1, \boldsymbol{x} \in \boldsymbol{W_2}} | < \boldsymbol{A_3}(\boldsymbol{x}), \boldsymbol{x} > |$$

where $\boldsymbol{W_2}$ is the set of vectors in that are orthogonal to the plane determined by the first two eigenvectors. Since this is a smaller set of vectors, the maximum we obtain will, of course, be possibly smaller. Hence, we know $\max_{\|\boldsymbol{x}\|=1} | < \boldsymbol{A_2}(\boldsymbol{x}), \boldsymbol{x} > | \geq \max_{\|\boldsymbol{x}\|=1, \, \boldsymbol{x} \in \boldsymbol{W_2}} | < \boldsymbol{A_3}(\boldsymbol{x}), \boldsymbol{x} > |$. The rest of the argument, applied to the new operator $\boldsymbol{A}_3$ is identical. Thus, we find

$$|\lambda_2| \quad = \quad \|\boldsymbol{A_3}\| \quad = \quad \max_{\|\boldsymbol{x}\|=1, \boldsymbol{x} \in \boldsymbol{W_2}} | < \boldsymbol{A_3}(\boldsymbol{x}), \boldsymbol{x} > |.$$

Further, $|\lambda_3| \leq |\lambda_2|$ and eigenvector $\boldsymbol{E_3}$ is orthogonal to both $\boldsymbol{E_1}$ and $\boldsymbol{E_2}$ because it comes from $\boldsymbol{W_2}$. ■

Homework
We will look at the previous exercises again but now we will set up the optimization problems for the third eigenvalue.

Exercise 7.2.13 *Let*

$$\boldsymbol{A} \quad = \quad \begin{bmatrix} 2 & 5+3\boldsymbol{i} \\ 5-3\boldsymbol{i} & 4 \end{bmatrix}$$

Use MATLAB to find the largest absolute value eigenvalue and an associated unit length eigenvector and the next largest and its eigenvector. Let the eigenvalues be λ_1 and λ_2 and the unit eigenvectors be $\boldsymbol{E_1}$ and $\boldsymbol{E_2}$. Calculate

$$\boldsymbol{A_3} \quad = \quad \boldsymbol{A} - \lambda_1 \, \boldsymbol{E_1} \, \overline{\boldsymbol{E_1}}^T - \lambda_2 \, \boldsymbol{E_2} \, \overline{\boldsymbol{E_2}}^T,$$

Write down the optimization problem you need to solve to find the third eigenvalue. Let $\boldsymbol{u} = \boldsymbol{x} + \boldsymbol{i}\boldsymbol{y}$ be the complex vector

$$\boldsymbol{x} + \boldsymbol{i}\boldsymbol{y} \quad = \quad \begin{bmatrix} x_1 \\ y_1 \end{bmatrix} + \boldsymbol{i} \begin{bmatrix} x_2 \\ y_2 \end{bmatrix}$$

Let W be the set of vectors orthogonal to the span of $\boldsymbol{E_1}$ and $\boldsymbol{E_2}$. Note this is one dimensional now. Then to find the third eigenvalue and eigenvector, we need to find

$$|\lambda_3| = \max_{\|\boldsymbol{u}\|=1 \in W} | < \boldsymbol{A_3}(\boldsymbol{u}), \boldsymbol{u} > | = \max_{\|\boldsymbol{u}\|=1 \in W} |\boldsymbol{u}^T \boldsymbol{A_3} \overline{\boldsymbol{u}}|$$

Simplify this until you can write the optimization problem in terms of x_1, x_2, y_3 and y_4. You don't need to find how to describe W as we are just looking for the form of the optimization. It is easy to see finding W and solving this optimization problem is not easy , so it is again remarkable our other techniques for finding eigenvalues and eigenvectors such as finding roots of polynomials furnish quick solutions via MATLAB!

Exercise 7.2.14 *Let*

$$\boldsymbol{A} = \begin{bmatrix} -3 & -2+3\boldsymbol{i} \\ -2-3\boldsymbol{i} & 5 \end{bmatrix}$$

Use MATLAB to find the largest absolute value eigenvalue and an associated unit length eigenvector and the next largest and its eigenvector. Let the eigenvalues be λ_1 and λ_2 and the unit eigenvectors be $\boldsymbol{E_1}$ and $\boldsymbol{E_2}$. Calculate

$$\boldsymbol{A_3} = \boldsymbol{A} - \lambda_1 \boldsymbol{E_1} \overline{\boldsymbol{E_1}}^T - \lambda_2 \boldsymbol{E_2} \overline{\boldsymbol{E_2}}^T,$$

Write down the optimization problem you need to solve to find the third eigenvalue. Let $\boldsymbol{u} = \boldsymbol{x} + \boldsymbol{i}\boldsymbol{y}$ be the complex vector

$$\boldsymbol{x} + \boldsymbol{i}\boldsymbol{y} = \begin{bmatrix} x_1 \\ y_1 \end{bmatrix} + \boldsymbol{i} \begin{bmatrix} x_2 \\ y_2 \end{bmatrix}$$

Let W be the set of vectors orthogonal to the span of $\boldsymbol{E_1}$ and $\boldsymbol{E_2}$. Note this is one dimensional now. Then to find the third eigenvalue and eigenvector, we need to find

$$|\lambda_3| = \max_{\|\boldsymbol{u}\|=1 \in W} | < \boldsymbol{A_3}(\boldsymbol{u}), \boldsymbol{u} > | = \max_{\|\boldsymbol{u}\|=1 \in W} |\boldsymbol{u}^T \boldsymbol{A_3} \overline{\boldsymbol{u}}|$$

Simplify this until you can write the optimization problem in terms of x_1, x_2, y_3 and y_4. You don't need to find how to describe W as we are just looking for the form of the optimization.

Exercise 7.2.15 *Let*

$$\boldsymbol{A} = \begin{bmatrix} 3 & 2+7\boldsymbol{i} & 6-8\boldsymbol{i} \\ 2-7\boldsymbol{i} & 4 & -7+2\boldsymbol{i} \\ 6+8\boldsymbol{i} & -7-2\boldsymbol{i} & -3 \end{bmatrix}$$

Use MATLAB to find the largest absolute value eigenvalue and an associated unit length eigenvector and the next largest and its eigenvector. Let the eigenvalues be λ_1 and λ_2 and the unit eigenvectors be $\boldsymbol{E_1}$ and $\boldsymbol{E_2}$. Calculate

$$\boldsymbol{A_3} = \boldsymbol{A} - \lambda_1 \boldsymbol{E_1} \overline{\boldsymbol{E_1}}^T - \lambda_2 \boldsymbol{E_2} \overline{\boldsymbol{E_2}}^T,$$

Write down the optimization problem you need to solve to find the third eigenvalue. Let $\boldsymbol{u} = \boldsymbol{x} + \boldsymbol{i}\boldsymbol{y}$ *be the complex vector*

$$\boldsymbol{u} = \boldsymbol{x} + \boldsymbol{i}\boldsymbol{y} \quad = \quad \begin{bmatrix} x_1 \\ y_1 \\ z_1 \end{bmatrix} + \boldsymbol{i} \begin{bmatrix} x_2 \\ y_2 \\ z_2 \end{bmatrix}$$

Then to find the second eigenvalue and eigenvector, we need to find

$$|\lambda_2| \quad = \quad \max_{\|\boldsymbol{u}\|=1 \in W} |< \boldsymbol{A_3}(\boldsymbol{u}), \boldsymbol{u} > | = \max_{\|\boldsymbol{u}\|=1 \in W} |\boldsymbol{u}^T \boldsymbol{A_3} \overline{\boldsymbol{u}}|$$

Simplify this until you can write the optimization problem in terms of x_i, y_i *and* z_i. *You don't need to find how to describe* W *as we are just looking for the form of the optimization.*

Exercise 7.2.16 *Let*

$$\boldsymbol{A} \quad = \quad \begin{bmatrix} 1 & 12 + 3\boldsymbol{i} & 6 - 2\boldsymbol{i} \\ 12 - 3\boldsymbol{i} & 3 & 3 + 5\boldsymbol{i} \\ 6 + 2\boldsymbol{i} & 3 - 5\boldsymbol{i} & 8 \end{bmatrix}$$

Use MATLAB to find the largest absolute value eigenvalue and an associated unit length eigenvector and the next largest and its eigenvector. Let the eigenvalues be λ_1 *and* λ_2 *and the unit eigenvectors be* $\boldsymbol{E_1}$ *and* $\boldsymbol{E_2}$. *Calculate*

$$\boldsymbol{A_3} \quad = \quad \boldsymbol{A} - \lambda_1 \, \boldsymbol{E_1} \, \overline{\boldsymbol{E_1}}^T - \lambda_2 \, \boldsymbol{E_2} \, \overline{\boldsymbol{E_2}}^T,$$

Write down the optimization problem you need to solve to find the third eigenvalue. Let $\boldsymbol{u} = \boldsymbol{x} + \boldsymbol{i}\boldsymbol{y}$ *be the complex vector*

$$\boldsymbol{u} = \boldsymbol{x} + \boldsymbol{i}\boldsymbol{y} \quad = \quad \begin{bmatrix} x_1 \\ y_1 \\ z_1 \end{bmatrix} + \boldsymbol{i} \begin{bmatrix} x_2 \\ y_2 \\ z_2 \end{bmatrix}$$

Then to find the second eigenvalue and eigenvector, we need to find

$$|\lambda_2| \quad = \quad \max_{\|\boldsymbol{u}\|=1 \in W} |< \boldsymbol{A_3}(\boldsymbol{u}), \boldsymbol{u} > | = \max_{\|\boldsymbol{u}\|=1 \in W} |\boldsymbol{u}^T \boldsymbol{A_3} \overline{\boldsymbol{u}}|$$

Simplify this until you can write the optimization problem in terms of x_i, y_i *and* z_i. *You don't need to find how to describe* W *as we are just looking for the form of the optimization.*

Since $\boldsymbol{A}$ is a $n \times n$ matrix, this process terminates after n steps. We can now state the full result.

Theorem 7.2.5 All Eigenvalues

There is a sequence of eigenvalues λ_i which satisfies $|\lambda_1| \geq |\lambda_2| \geq \ldots \geq |\lambda_n|$ with associated eigenvectors $\boldsymbol{E_i}$ which form an orthonormal basis for $\mathbb{C}^n$. Let the new Hermitian matrix $\boldsymbol{A_i}$ be defined by

$$\boldsymbol{A_i} = \boldsymbol{A} - \sum_{j=1}^{i} \lambda_j \, \boldsymbol{E_j} \overline{\boldsymbol{E_j}}^T$$

Then, $|\lambda_i| = \|\boldsymbol{A_i}\|$ and the associated eigenvector $\boldsymbol{E_i}$ is orthogonal to $\boldsymbol{E_j}$ for all $j \leq i$. Moreover, we can say

$$\boldsymbol{A} = \sum_{j=1}^{n} \lambda_j \, \boldsymbol{E_j} \overline{\boldsymbol{E_j}}^T,$$

7.2.2 What Does This Mean?

Let's put all of this together. Our Hermitian $n \times n$ matrix always has n real eigenvalues and their respective eigenvectors $\{\boldsymbol{E_1}, \ldots \boldsymbol{E_n}\}$ are mutually orthogonal and so define an orthonormal basis for $\mathbb{C}^n$. We can use these eigenvectors to write $\boldsymbol{A}$ is a canonical form just like we did for the 2×2 case. We define the two $n \times n$ complex matrices

$$\boldsymbol{P} = \begin{bmatrix} \boldsymbol{E_1} & \boldsymbol{E_2} & \ldots & \boldsymbol{E_n} \end{bmatrix}$$

which has conjugate transpose

$$\overline{\boldsymbol{P}}^T = \begin{bmatrix} \overline{\boldsymbol{E_1}}^T \\ \overline{\boldsymbol{E_2}}^T \\ \ldots \\ \overline{\boldsymbol{E_n}}^T \end{bmatrix}$$

We know the eigenvectors are mutually orthogonal, so we must have $\overline{\boldsymbol{P}}^T \boldsymbol{P} = \boldsymbol{I}$, $\boldsymbol{P}\overline{\boldsymbol{P}}^T = \boldsymbol{I}$ and

$$\begin{aligned}
\overline{\boldsymbol{P}}^T \boldsymbol{A} \, \boldsymbol{P} &= \begin{bmatrix} \lambda_1 < \boldsymbol{E_1}, \boldsymbol{E_1} > & 0 & 0 & \ldots & 0 \\ 0 & \lambda_2 < \boldsymbol{E_2}, \boldsymbol{E_2} > & 0 & \ldots & 0 \\ \vdots & \vdots & \ddots & \ldots & 0 \\ \vdots & \vdots & 0 & \ddots & 0 \\ 0 & 0 & 0 & 0 & \lambda_n < \boldsymbol{E_n}, \boldsymbol{E_n} > \end{bmatrix} \\
&= \begin{bmatrix} \lambda_1 & 0 & 0 & \ldots & 0 \\ 0 & \lambda_2 & 0 & \ldots & 0 \\ \vdots & \vdots & \ddots & \ldots & 0 \\ \vdots & \vdots & 0 & \ddots & 0 \\ 0 & 0 & 0 & 0 & \lambda_n \end{bmatrix}
\end{aligned}$$

which can be rewritten as

$$A = P \begin{bmatrix} \lambda_1 & 0 & 0 & \dots & 0 \\ 0 & \lambda_2 & 0 & \dots & 0 \\ \vdots & \vdots & \ddots & \dots & 0 \\ \vdots & \vdots & 0 & \ddots & 0 \\ \vdots & \vdots & 0 & 0 & \lambda_n \end{bmatrix} \overline{P}^T$$

And we have a nice representation for an arbitrary $n \times n$ Hermitian matrix A. Now at this point, we know all the eigenvalues are real and we can write them in descending order, but eigenvalues can be repeated and can even be zero.

Also, we can represent the norm of the Hermitian matrix A in terms of its eigenvalues.

Theorem 7.2.6 The Eigenvalue Representation of the norm of an Hermitian matrix

Let A be an $n \times n$ Hermitian matrix with eigenvalues $|\lambda_1| \geq |\lambda_2| \geq \dots \geq |\lambda_n|$. Then

$$\|A\| = |\lambda_1| = \max_{1 \leq i \leq n} |\lambda_i|$$

This also implies

$$\|Ax\| \leq \|A\| \, \|x\|$$

for all x.

Proof 7.2.5
We know $\|A\| \geq | < AE_1, E_1 > |$ where E_1 is the unit norm eigenvector corresponding to eigenvalue λ_1. Thus,

$$\|A\| \geq | < \lambda_1 E_1, E_1 > | = |\lambda_1|$$

Any x has representation $\sum_{i=1}^{n} c_i E_i$ in terms of the orthonormal basis of eigenvectors of A. If $\|x\| = 1$, this implies $\sum_{i=1}^{n} c_i^2 = 1$. Then

$$\begin{aligned} | < Ax, x > | &= \left| \sum_{i=1}^{n} \sum_{j=1}^{n} c_i c_j \lambda_i < E_i, E_j > \right| \\ &\leq \sum_{i=1}^{n} c_i^2 |\lambda_i| \leq (1) \, |\lambda_1| \end{aligned}$$

Thus, $\|A\| = \max_{\|x\|=1} | < Ax, x > | \leq |\lambda_1|$. Combining, we see $\|A\| = |\lambda_1|$. ■

7.2.2.1 A Worked Out Example

Let A be defined to be

$$A = \begin{bmatrix} 1 & 3 + 2i \\ 3 - 2i & 2 \end{bmatrix}$$

Let $\boldsymbol{x} + \boldsymbol{i}\boldsymbol{y}$ be the complex vector

$$\boldsymbol{x} + \boldsymbol{i}\boldsymbol{y} = \begin{bmatrix} x_1 \\ y_1 \end{bmatrix} + \boldsymbol{i} \begin{bmatrix} x_2 \\ y_2 \end{bmatrix}$$

Then to find the first eigenvalue and eigenvector, we need to find

$$\begin{aligned} \mu &= \max_{\|\boldsymbol{x}\|=1} | < \boldsymbol{A}(\boldsymbol{x}), \boldsymbol{x} > | = \max_{\|\boldsymbol{x}\|=1} |\boldsymbol{x}^T \boldsymbol{A}\overline{\boldsymbol{x}}| \\ &= \max_{\|\boldsymbol{x}\|=1} \left| \begin{bmatrix} x_1 + x_2\boldsymbol{i} & y_1 + y_2\boldsymbol{i} \end{bmatrix} \begin{bmatrix} 1 & 3+2\boldsymbol{i} \\ 3-2\boldsymbol{i} & 2 \end{bmatrix} \begin{bmatrix} x_1 - x_2\boldsymbol{i} \\ y_1 - y_2\boldsymbol{i} \end{bmatrix} \right| \end{aligned}$$

After a fair bit of calculation, this becomes

$$\begin{aligned} \mu &= \max_{\|\boldsymbol{x}\|=1} |x_1^2 + 6x_2x_2 + 2x_1^2 + 4x_1y_2 - 4x_2y_1 + y_1^2 + 6y_1y_2 + 2y_2^2| \\ &= \max_{\|\boldsymbol{x}\|=1} \left| \begin{bmatrix} x_1 \\ x_2 \\ y_1 \\ y_2 \end{bmatrix}^T \begin{bmatrix} 1 & 3 & 0 & 2 \\ 3 & 2 & -2 & 0 \\ 0 & -2 & 1 & 3 \\ 2 & 0 & 3 & 2 \end{bmatrix} \begin{bmatrix} x_1 \\ x_2 \\ y_1 \\ y_2 \end{bmatrix} \right| \end{aligned}$$

Now this was the hard way to set up this optimization. In general, the Hermitian matrix $\boldsymbol{A}$ has the form $\boldsymbol{A} = \Phi + \Psi\boldsymbol{i}$ where Φ and Ψ are real matrices, Φ is symmetric and Ψ has the form

$$\Psi = \begin{bmatrix} 0 & \boldsymbol{D} \\ -\boldsymbol{D}^T & 0 \end{bmatrix}$$

where $\boldsymbol{D}$ is an upper triangular matrix which is zero on the main diagonal . For example, if

$$\boldsymbol{A} = \begin{bmatrix} 1 & 3+2\boldsymbol{i} \\ 3-2\boldsymbol{i} & 2 \end{bmatrix}$$

then

$$\boldsymbol{\Phi} = \begin{bmatrix} 1 & 3 \\ 3 & 1 \end{bmatrix}, \quad \boldsymbol{\Psi} = \begin{bmatrix} 0 & 2 \\ -2 & 0 \end{bmatrix}$$

so that $\boldsymbol{D} = 2$. If

$$\boldsymbol{A} = \begin{bmatrix} 1 & 3+2\boldsymbol{i} & -3+8\boldsymbol{i} \\ 3-2\boldsymbol{i} & 2 & 5+6\boldsymbol{i} \\ -3-8\boldsymbol{i} & 5-6\boldsymbol{i} & -4 \end{bmatrix}$$

then

$$\boldsymbol{\Phi} = \begin{bmatrix} 1 & 3 & -3 \\ 3 & 2 & 5 \\ -3 & 5 & -4 \end{bmatrix}, \quad \boldsymbol{\Psi} = \begin{bmatrix} 0 & 2 & 8 \\ -2 & 0 & 6 \\ -8 & -6 & 0 \end{bmatrix}$$

so that

$$\boldsymbol{D} = \begin{bmatrix} 0 & 2 & 8 \\ 0 & 0 & 6 \\ 0 & 0 & 0 \end{bmatrix}$$

Then, the calculation we need to perform is

$$\begin{aligned}
\left[\boldsymbol{x} + \boldsymbol{y}\boldsymbol{i}\right]^T (\boldsymbol{\Phi} + \boldsymbol{\Psi}\boldsymbol{i}) \left[\boldsymbol{x} - \boldsymbol{y}\boldsymbol{i}\right] &= \left[\boldsymbol{x} + \boldsymbol{y}\boldsymbol{i}\right]^T \left\{ \Big(\boldsymbol{\Phi}(\boldsymbol{x}) + \boldsymbol{\Psi}(\boldsymbol{y}) \Big) + \boldsymbol{i} \Big(\boldsymbol{\Psi}(\boldsymbol{x}) - \boldsymbol{\Phi}(\boldsymbol{y}) \Big) \right\} \\
&= \Big(\boldsymbol{x}^T \boldsymbol{\Phi} \boldsymbol{x} + \boldsymbol{x}^T \boldsymbol{\Psi} \boldsymbol{y} - \boldsymbol{y}^T \boldsymbol{\Psi} \boldsymbol{x} + \boldsymbol{y}^T \boldsymbol{\Phi} \boldsymbol{y} \Big) \\
&\quad + \boldsymbol{i} \Big(\boldsymbol{y}^T \boldsymbol{\Phi} \boldsymbol{x} + \boldsymbol{y}^T \boldsymbol{\Psi} \boldsymbol{y} + \boldsymbol{x}^T \boldsymbol{\Psi} \boldsymbol{x} - \boldsymbol{x}^T \boldsymbol{\Phi} \boldsymbol{y} \Big)
\end{aligned}$$

It is straightforward to see $\boldsymbol{z}^T \boldsymbol{\Psi} \boldsymbol{z} = 0$ for all $\boldsymbol{z}$. Thus, we have

$$\left[\boldsymbol{x} + \boldsymbol{y}\boldsymbol{i}\right]^T (\boldsymbol{\Phi} + \boldsymbol{\psi}\boldsymbol{i}) \left[\boldsymbol{x} - \boldsymbol{y}\boldsymbol{i}\right] = \Big(\boldsymbol{x}^T \boldsymbol{\Phi} \boldsymbol{x} + \boldsymbol{x}^T \boldsymbol{\Psi} \boldsymbol{y} - \boldsymbol{y}^T \boldsymbol{\Psi} \boldsymbol{x} + \boldsymbol{y}^T \boldsymbol{\Phi} \boldsymbol{y} \Big) + \boldsymbol{i} \Big(\boldsymbol{y}^T \boldsymbol{\Phi} \boldsymbol{x} - \boldsymbol{x}^T \boldsymbol{\Phi} \boldsymbol{y} \Big)$$

Since $\boldsymbol{\Phi}$ is symmetric, $\boldsymbol{x}^T \boldsymbol{\Phi} \boldsymbol{y} = \boldsymbol{y}^T \boldsymbol{\Phi} \boldsymbol{x}$. Therefore

$$\left[\boldsymbol{x} + \boldsymbol{y}\boldsymbol{i}\right]^T (\boldsymbol{\Phi} + \boldsymbol{\psi}\boldsymbol{i}) \left[\boldsymbol{x} - \boldsymbol{y}\boldsymbol{i}\right] = \Big(\boldsymbol{x}^T \boldsymbol{\Phi} \boldsymbol{x} + \boldsymbol{x}^T \boldsymbol{\Psi} \boldsymbol{y} - \boldsymbol{y}^T \boldsymbol{\Psi} \boldsymbol{x} + \boldsymbol{y}^T \boldsymbol{\Phi} \boldsymbol{y} \Big)$$

which can be written in matrix form as

$$\left[\boldsymbol{x} + \boldsymbol{y}\boldsymbol{i}\right]^T (\boldsymbol{\Phi} + \boldsymbol{\psi}\boldsymbol{i}) \left[\boldsymbol{x} + \boldsymbol{y}\boldsymbol{i}\right] = \begin{bmatrix} \boldsymbol{x} \\ \boldsymbol{y} \end{bmatrix}^T \begin{bmatrix} \boldsymbol{\Phi} & \boldsymbol{D} \\ -\boldsymbol{D}^T & \boldsymbol{\Phi} \end{bmatrix} \begin{bmatrix} \boldsymbol{x} \\ \boldsymbol{y} \end{bmatrix}$$

which clearly shows this is a $2n$ dimensional optimization of a quadratic form based on the matrix

$$\boldsymbol{B} = \begin{bmatrix} \boldsymbol{\Phi} & \boldsymbol{D} \\ -\boldsymbol{D}^T & \boldsymbol{\Phi} \end{bmatrix}$$

Hence, although all of our computations here are guaranteed to have absolute maximums as we are maximizing a continuous quadratic form over the compact set $\|\boldsymbol{x}\|^2 + \|\boldsymbol{y}\|^2 = 1$, this is clearly a nontrivial thing to do. The value of our work here is that it shows us how to recursively prove the eigenvalues of a Hermitian matrix have associated eigenvectors that are mutually orthogonal and hence give us an orthonormal basis.

The optimization problem to find the largest eigenvalue for

$$\boldsymbol{A} = \begin{bmatrix} 1 & 3 + 2\boldsymbol{i} & -3 + 8\boldsymbol{i} \\ 3 - 2\boldsymbol{i} & 2 & 5 + 6\boldsymbol{i} \\ -3 - 8\boldsymbol{i} & 5 - 6\boldsymbol{i} & -4 \end{bmatrix}$$

would then use the matrix

$$\boldsymbol{B} = \begin{bmatrix} \begin{bmatrix} 1 & 3 & -3 \\ 3 & 2 & 5 \\ -3 & 5 & -4 \end{bmatrix} & \begin{bmatrix} 0 & 2 & 8 \\ 0 & 0 & 6 \\ 0 & 0 & 0 \end{bmatrix} \\ \begin{bmatrix} 0 & 0 & 0 \\ -2 & 0 & 0 \\ -8 & -6 & 0 \end{bmatrix} & \begin{bmatrix} 1 & 3 & -3 \\ 3 & 2 & 5 \\ -3 & 5 & -4 \end{bmatrix} \end{bmatrix}$$

This also shows us how to apply these ideas to the extension of Hermitian matrices to the right kind of operator on an inner product space: the self-adjoint operator.

Homework

Use the techniques of the worked out example to find the optimization problem for the largest absolute eigenvalue and its unit eigenvector. Also, we will find the usual decomposition of $\boldsymbol{A}$.

Exercise 7.2.17 *Let*

$$\boldsymbol{A} = \begin{bmatrix} 2 & 5+3i \\ 5-3i & 4 \end{bmatrix}$$

- *Use the techniques of the worked out example to find the optimization problem for the largest absolute eigenvalue and its unit eigenvector.*
- *Let $\boldsymbol{P}$ be the matrix whose columns are the eigenvectors of $\boldsymbol{A}$. Calculate these using MATLAB making sure the eigenvectors are unit vectors. Verify the decomposition $\boldsymbol{A} = \boldsymbol{P}\boldsymbol{D}\overline{\boldsymbol{P}}^T$ where D is the diagonal matrix of the real eigenvalues of $\boldsymbol{A}$. Organize this so the entries in $\boldsymbol{D}$ are ranked from largest absolute value to smallest absolute value eigenvalue.*

Exercise 7.2.18 *Let*

$$\boldsymbol{A} = \begin{bmatrix} -3 & -2+3i \\ -2-3i & 5 \end{bmatrix}$$

- *Use the techniques of the worked out example to find the optimization problem for the largest absolute eigenvalue and its unit eigenvector.*
- *Let $\boldsymbol{P}$ be the matrix whose columns are the eigenvectors of $\boldsymbol{A}$. Calculate these using MATLAB making sure the eigenvectors are unit vectors. Verify the decomposition $\boldsymbol{A} = \boldsymbol{P}\boldsymbol{D}\overline{\boldsymbol{P}}^T$ where D is the diagonal matrix of the real eigenvalues of $\boldsymbol{A}$. Organize this so the entries in $\boldsymbol{D}$ are ranked from largest absolute value to smallest absolute value eigenvalue.*

Exercise 7.2.19 *Let*

$$\boldsymbol{A} = \begin{bmatrix} 3 & 2+7i & 6-8i \\ 2-7i & 4 & -7+2i \\ 6+8i & -7-2i & -3 \end{bmatrix}$$

- *Use the techniques of the worked out example to find the optimization problem for the largest absolute eigenvalue and its unit eigenvector.*
- *Let $\boldsymbol{P}$ be the matrix whose columns are the eigenvectors of $\boldsymbol{A}$. Calculate these using MATLAB making sure the eigenvectors are unit vectors. Verify the decomposition $\boldsymbol{A} = \boldsymbol{P}\boldsymbol{D}\overline{\boldsymbol{P}}^T$ where D is the diagonal matrix of the real eigenvalues of $\boldsymbol{A}$. Organize this so the entries in $\boldsymbol{D}$ are ranked from largest absolute value to smallest absolute value eigenvalue.*

Exercise 7.2.20 *Let*

$$\boldsymbol{A} = \begin{bmatrix} 1 & 12+3i & 6-2i \\ 12-3i & 3 & 3+5i \\ 6+2i & 3-5i & 8 \end{bmatrix}$$

- *Use the techniques of the worked out example to find the optimization problem for the largest absolute eigenvalue and its unit eigenvector.*
- *Let $\boldsymbol{P}$ be the matrix whose columns are the eigenvectors of $\boldsymbol{A}$. Calculate these using MATLAB making sure the eigenvectors are unit vectors. Verify the decomposition $\boldsymbol{A} = \boldsymbol{P}\boldsymbol{D}\overline{\boldsymbol{P}}^T$ where*

D is the diagonal matrix of the real eigenvalues of $\boldsymbol{A}$. Organize this so the entries in $\boldsymbol{D}$ are ranked from largest absolute value to smallest absolute value eigenvalue.

The inner product space can be complete with respect to the induced norm. Such a space is called a **Hilbert Space**.

Definition 7.2.2 Hilbert Space

Let $\mathscr{V}$ be a vector space over the real or the complex numbers as the scalar field with inner product $<\ ,\ >$. Let $||\ ||$ denote the norm induced by the inner product defined by $||x|| = \sqrt{<x, x>}$. If the inner product space $\mathscr{V}$ is complete, we say it is a Hilbert Space.

7.3 Examples of Hilbert Spaces

We need good examples of Hilbert Spaces so we can have some intuition. Here are some standard ones.

$\Re^n$: The metric here is the usual Euclidean norm and the inner product the usual one we use for vectors. We have shown this inner product space is complete with respect to this induced norm. Hence, it is a Hilbert Space which is finite dimensional.

$\mathbb{C}^n$: The metric here is the usual Euclidean norm for complex vectors and the inner product the usual one we use for complex vectors. We have shown this inner product space is complete with respect to this induced norm. Hence, it is a complex Hilbert Space which is finite dimensional.

ℓ^2: The inner product here for $\boldsymbol{x} = (x_i)$ and $\boldsymbol{y} = (y_i)$ with $\sum_i |x_i^2 < \infty$ and $\sum_i |y_i^2 < \infty$ is

$$<\boldsymbol{x}, \boldsymbol{y}> \quad = \quad \sum_i x_i y_i$$

which we know converges because of the Hölder's Inequality. The induced norm is then the usual $\|\cdot\|_2$ which induces the usual d_2 metric. We have shown ℓ^2 is complete with respect to this metric, so ℓ^2 is an infinite dimensional Hilbert Space.

We often use a set of functions which is complete with respect to its metric. Cauchy sequences of continuous functions with respect to the d_2 metric do not have to converge to a continuous function and so they do not form a complete space. The inner product for continuous functions on $[a, b]$ is the usual $\int_a^b f(t)g(t)dt$ which induces the $\|\cdot\|_2$ norm which in turn induces the d_2 metric. Let's look closely at the completion process of this space of functions.

7.4 Completing the Integrable Functions

Let's first look at continuous functions which are a subset of all Riemann Integrable functions. Let's go through the steps of constructing $(\widetilde{C}([a, b]), \widetilde{d}_2)$.

Step 1: Define

$$S \quad = \quad \{(x_n) \mid x_n \in C([a, b]), (x_n) \text{ is a Cauchy sequence in } (C([a, b]), d_2)\}$$

Step 2: Define an equivalence relation on S by

$$(x_n) \sim (y_n) \quad \Longleftrightarrow \quad \lim_{n \to \infty} d_2(x_n, y_n) = 0$$

Step 3: Let $\widetilde{C}([a,b]) = S/\sim$, the set of all equivalence classes in S under $\sim$. Let the elements of $\widetilde{C}([a,b])$ be denoted by $[(x_n)]$.

Step 4: Extend the metric d_2 on $\widetilde{C}([a,b])$ to the metric $\tilde{d}_2$ on $\widetilde{C}([a,b])$ by defining

$$\tilde{d}_2(\tilde{x},\tilde{y}) = \lim_{n\to\infty} d_2(x_n,y_n)$$

for any equivalence classes $\tilde{x}$ and $\tilde{y}$ in $\widetilde{X}$ and any choice of representatives $(x_n) \in \tilde{x}$ and $(y_n) \in \tilde{y}$. The proof that $\tilde{d}_2$ is a metric is just like the general case.

Step 5: Define the mapping $T : C([a,b]) \to \widetilde{C}([a,b])$ by $T(x) = [(\bar{x})]$ where $\bar{x} = (x,x,x,\ldots)$ is the constant sequence. Note each entry x is actually a continuous function on $[a,b]$. Then T is 1-1, onto and **isometric**.

Step 6: $T(C([a,b]))$ is dense in $(\widetilde{C}([a,b]),\tilde{d}_2)$.

Step 7: $(\widetilde{C}([a,b]),\tilde{d}_2)$ is complete.

So going back to a nice example in Chapter 2.12, the sequence in $C([0,1])$ with metric d_2 given by (x_n) given by

$$x_n(t) = \begin{cases} 0, & 0 \le t \le \frac{1}{2} \\ n(t-\frac{1}{2}), \frac{1}{2} < t < \frac{1}{2}+\frac{1}{n} & \\ 1, & \frac{1}{2}+\frac{1}{n} \le t \le 1 \end{cases}$$

is a Cauchy sequence which does not converge. It does converge pointwise to the discontinuous function

$$x(t) = \begin{cases} 0, 0 \le t \le \frac{1}{2} \\ 1, \frac{1}{2} < t \le 1 \end{cases}$$

This Cauchy sequence determines an equivalence in $(\widetilde{C}([0,1]),\tilde{d}_2)$ called $[(\bar{x_n})]$. We also know $\tilde{d}_2([(\bar{x_n})],[(\bar{x})]) \to 0$. Thus, from a certain point of view, the pointwise limit function x is a member of this equivalence class. Of course, if we look at (y_n) given by

$$y_n(t) = \begin{cases} 0, & 0 \le t < \frac{1}{2} \\ n(t-\frac{1}{2}), \frac{1}{2} \le t < \frac{1}{2}+\frac{1}{n} & \\ 1, & \frac{1}{2}+\frac{1}{n} \le t \le 1 \end{cases}$$

this is a Cauchy sequence which also does not converge. It does converge pointwise to the discontinuous function

$$y(t) = \begin{cases} 0, 0 \le t < \frac{1}{2} \\ 1, \frac{1}{2} \le t \le 1 \end{cases}$$

Clearly (y_n) is in $[(x_n)]$ as $\lim_n d_2(x_n,y_n) = 0$.

Let's look at Riemann Integrable functions using the d_2 metric. As we have discussed, d_2 is not a metric really as there are Riemann Integrable functions with $d_2(x,y) = 0$ even though $x \neq y$. The functions x and y above are examples of that. However, we can think of d_2 as a metric here if we switch to equivalence classes. Let $RI([a,b])$ be the set of all Riemann Integrable functions on $[a,b]$. Define the equivalence relation $\sim$ by $x \sim y$ if $d_2(x,y) = 0$. Let $\mathbb{RI}([a,b]) = RI([a,b])/\sim$ and denote an equivalence class by $[x]$ as usual for any Riemann Integrable function x on $[a,b]$. Extend

d_2 to equivalence classes by

$$\mathbb{D}_2([x],[y]) = d_2(x,y)$$

where x and y are any two representatives of these equivalence classes. This is always a non-negative number and if x' and y' are two other representatives, then by Minkowski's Inequality for integrable functions

$$\begin{aligned}\sqrt{\int_a^b |x'(t)-y'(t)|^2dt} &\leq \sqrt{\int_a^b |x'(t)-x(t)|^2dt} + \sqrt{\int_a^b |x(t)-y(t)|^2dt} \\ &+ \sqrt{\int_a^b |y(t)-y'(t)|^2dt} = \sqrt{\int_a^b |x(t)-y(t)|^2dt}\end{aligned}$$

A similar argument shows $\sqrt{\int_a^b |x(t)-y(t)|^2dt} \leq \sqrt{\int_a^b |x'(t)-y'(t)|^2dt}$. It follows immediately that $d_2(x,y) = d_2(x',y')$. Hence, this value is independent of the choice of representatives and so $\mathbb{D}_2$ is well-defined. This shows property $M1$ for a metric. For property $M2$, note

$$\mathbb{D}_2([x],[y]) = d_2(x,y) = 0 \Longrightarrow |x-y| = 0 \text{ a.e.}$$

which tells us $x-y \in [0]$. The other properties $M3$ and $M4$ are easily established. Thus, $(\mathbb{RI}([a,b]), \mathbb{D}_2)$ is a metric space.

Let's go through the steps of constructing $(\mathbb{L}_2([a,b]), \widetilde{\mathbb{D}}_2) \equiv (\widetilde{\mathbb{RI}}([a,b]), \widetilde{\mathbb{D}}_2)$.

Step 1: Define

$$\mathbb{S} = \{([x_n]) \mid [x_n] \in \mathbb{RI}([a,b]), ([x_n]) \text{ is a Cauchy sequence in } (\mathbb{RI}([a,b]), \mathbb{D}_2)\}$$

Step 2: Define an equivalence relation on $\mathbb{S}$ by

$$([x_n]) \sim ([y_n]) \iff \mathbb{D}_2([x_n],[y_n]) = 0$$

Step 3: Let $\mathbb{L}_2([a,b]) = \mathbb{S}/\sim$, the set of all equivalence classes in $\mathbb{S}$ under $\sim$. Let the elements of $\mathbb{L}_2([a,b])$ be denoted by $[\,[(x_n)]\,]$. This is a messy notation, so let's use $\mathscr{X} = [\,[(x_n)]\,]$.

Step 4: Extend the metric $\mathbb{D}_2$ on $\mathbb{RI}([a,b])$ to the metric $\widetilde{\mathbb{D}}_2$ on $\mathbb{L}_2([a,b])$ by defining

$$\widetilde{\mathbb{D}}_2(\mathscr{X},\mathscr{Y}) = \lim_{n\to\infty} \mathbb{D}_2([(x_n)],[(y_n)]) = \lim_{n\to\infty} d_2(x_n,y_n)$$

for any equivalence classes $\mathscr{X}$ and $\mathscr{Y}$ in $\mathbb{L}_2([a,b])$ and any choice of representatives $([x_n]) \in \mathscr{X}$ and $([y_n]) \in \mathscr{Y}$. The proof $\widetilde{\mathbb{D}}_2$ is a metric is just like the general case.

Step 5: Define the mapping $T : \mathbb{RI}([a,b]) \to \mathbb{L}_2([a,b])$ by $T([x]) = [\,[\bar{x}]\,] = \bar{\mathscr{X}}$ where $[\bar{x}] = ([x],[x],[x],\ldots)$ is the constant sequence. Note each entry $[x]$ is actually an equivalence class in $\mathbb{RI}([a,b])$. The T is 1-1, onto and **isometric**.

Step 6: $T(\mathbb{RI}([a,b]))$ is dense in $(\mathbb{L}_2([a,b]), \widetilde{\mathbb{D}}_2)$.

Step 7: $(\mathbb{L}_2([a,b]), \widetilde{\mathbb{D}}_2)$ is complete.

So the discontinuous function x given by

$$x(t) = \begin{cases} 0, 0 \leq t \leq \frac{1}{2} \\ 1, \frac{1}{2} < t \leq 1 \end{cases}$$

generates the equivalence class $[x]$ in $\mathbb{RI}([0,1])$. This in turn is embedded in $\mathbb{L}_2([0,1])$ as the constant sequence $T([x]) = [\,[\bar{x}]\,] = \mathscr{X}$. The sequence of equivalence classes $[x_n]$ where

$$x_n(t) = \begin{cases} 0, & 0 \leq t \leq \frac{1}{2} \\ n(t-\frac{1}{2}), \frac{1}{2} < t < \frac{1}{2} + \frac{1}{n} & \\ 1, & \frac{1}{2} + \frac{1}{n} \leq t \leq 1 \end{cases}$$

is a Cauchy sequence of equivalence classes in $\mathbb{RI}([0,1])$ which are embedded as the sequence $(T([x_n]))$ into $\mathbb{L}_1([0,1])$ as constant sequences. This Cauchy sequence of equivalence classes in $\mathbb{L}_2([0,1])$ converges to $T([x])$. Note

$$\begin{aligned} \widetilde{\mathbb{D}}_2(T([x_n]), T([x])) &= \lim_n \mathbb{D}_2([x_n],[x]) = \lim_n d_2(x_n, x) \\ &= \lim_n \sqrt{\int_0^1 |x_n(t) - x(t)|^2 dt} = 0 \end{aligned}$$

Thus, the function which the Cauchy sequence (x_n) in $(RI([0,1]), d_2)$ converges to is the equivalence class $[T([x])]$ in $\mathbb{L}_2([0,1])$. Let y be another function with $[T(y)]$ in $[T(x)]$. Note $\mathscr{Y} = [\,[y_n]\,]$ is in $[T([x])]$ if $\mathbb{D}_2([\,[y_n]\,],[\,[\bar{y}]\,]) = 0$. This says $\lim_n d_2(y_n, y) = 0$ and $d_1(x,y) = 0$. So any Riemann Integrable function y which is similar to x under d_2 is similar to the Cauchy sequence $(T([x_n]))$. Hence the limiting equivalence class here is the set of all Riemann Integrable functions y with $d_2(x,y) = 0$.

We extend the usual inner product on Riemann Integrable functions to an inner product on $\mathbb{L}_2([a,b])$ is a similar way. The inner product space $(C([a,b]), <\cdot,\cdot>)$ where $<f,g> = \int_a^b f(t)g(t)dt$ becomes a normed linear space using $\|\cdot\|_2 = \sqrt{<\cdot,\cdot>}$ and that in turn becomes a metric space using $d_2(\cdot,\cdot) = \|\cdot - \cdot\|_2$. The same is true for the space of Riemann Integrable functions on $[a,b]$ except we have to use equivalence classes in one more step. This metric space $(RI([a,b]), d_2)$ is completed, following the outline above using

- $RI([a,b])$ be the set of all Riemann Integrable functions on $[a,b]$.
- $\mathbb{RI}([a,b]) = RI([a,b])/\sim$ is the set of equivalence classes under the equivalence relation $\sim$ where under $f \sim g$ means $d_2(f,g) = 0$. Extend d_2 to equivalence classes by $\mathbb{D}_2([x],[y]) = d_2(x,y)$ for any x a representative of $[x]$ and any y a representative of $[y]$.
- Let $\mathbb{S}$ be the set of all Cauchy sequences in $\mathbb{RI}([a,b])$ under $\sim$ where $([x_n]) \sim ([y_n])$ means $\mathbb{D}_2([x_n],[y_n]) = 0$. Define $\mathbb{L}_2([a,b]) = \mathbb{S}/\sim$. Extend the metric $\mathbb{D}_2$ on $\mathbb{RI}([a,b])$ to the metric $\widetilde{\mathbb{D}}_2$ on $\mathbb{L}_2([a,b])$ by

$$\widetilde{\mathbb{D}}_2(\mathscr{X}, \mathscr{Y}) = \lim_{n\to\infty} \mathbb{D}_2([(x_n)],[(y_n)]) = \lim_{n\to\infty} d_2(x_n, y_n)$$

  for any equivalence classes $\mathscr{X}$ and $\mathscr{Y}$ in $\mathbb{L}_2([a,b])$ and any choice of representatives $([x_n]) \in \mathscr{X}$ and $([y_n]) \in \mathscr{Y}$.
- Then $(\mathbb{L}_2([a,b]), \widetilde{\mathbb{D}}_2)$ is complete.

We still need to extend the inner product and the norm from $RI([a,b])$. Let's go through the chain again:

- $RI([a,b])$ be the set of all Riemann Integrable functions on $[a,b]$.
- $\mathbb{RI}([a,b]) = RI([a,b])/\sim$ is the set of equivalence classes under the equivalence relation $\sim$ where under $f \sim g$ means $d_2(f,g) = 0$. Extend d_2 to equivalence classes by $\mathbb{D}_2([x],[y]) =$

$d_2(x, y)$ for any x a representative of $[x]$ and any y a representative of $[y]$. Thus, we extend $\| \cdot \|_2$ to $\widetilde{\|} [x] \widetilde{\|}_2 = \|x\|_2$ for any representative x of $[x]$. We also extend the inner product similarly: $\widetilde{<} [x], [y] \widetilde{>} =< x, y >$ for any representative x of $[x]$ and y of $[y]$.

- Let $\mathbb{S}$ be the set of all Cauchy sequences in $\mathbb{RI}([a, b])$ under $\sim$ where $([x_n]) \sim ([y_n])$ means $\mathbb{D}_2([x_n], [y_n]) = 0$. Define $\mathbb{L}_2([a, b]) = \mathbb{S}/\sim$. Extend the metric $\mathbb{D}_2$ on $\mathbb{RI}([a, b])$ to the metric $\widetilde{\mathbb{D}}_2$ on $\mathbb{L}_2([a, b])$ by

$$\widetilde{\mathbb{D}}_2(\mathscr{X}, \mathscr{Y}) = \lim_{n \to \infty} \mathbb{D}_2([(x_n)], [(y_n)]) = \lim_{n \to \infty} d_2(x_n, y_n)$$

for any equivalence classes $\mathscr{X}$ and $\mathscr{Y}$ in $\mathbb{L}_2([a, b])$ and any choice of representatives $([x_n]) \in \mathscr{X}$ and $([y_n]) \in \mathscr{Y}$. Thus, we extend $\widetilde{\|} \cdot \widetilde{\|}_2$ to $\widetilde{\widetilde{\|}} \mathscr{X} \widetilde{\widetilde{\|}}_2 = \widetilde{\|}([x_n])\widetilde{\|}_2$ for any representative $([x_n])$ of $\mathscr{X}$. We also extend the inner product similarly: $\widetilde{\widetilde{<}} \mathscr{X}, \mathscr{Y} \widetilde{\widetilde{>}} = \widetilde{<}([x_n]), ([y_n])\widetilde{>}$ for any representative $([x_n])$ of $\mathscr{X}$ and $([y_n])$ of $\mathscr{Y}$.

- Then $(\mathbb{L}_2([a, b]), \widetilde{\widetilde{<}} \cdot, \cdot \widetilde{\widetilde{>}})$ is a complete inner product space.

Clearly, the extended inner product notation is pretty bad. We typically simply say $\mathbb{L}_2([a, b])$ is a complete inner product space using the standard L_2 inner product $< f, g >= \int_a^b f(t)g(t)dt$ with the understanding that objects in $\mathbb{L}_2([a, b])$ are functions x whose pointwise representation on $[a, b]$ is not a unique description of x. Instead, there are many pointwise representations of x via the two steps of equivalence relations we have used. That is x is an equivalence class of Cauchy sequences of equivalence classes in $RI([a, b])$ under $f \sim g$ means $d_2(f, g) = \|f - g\|_2 = \sqrt{< f - g, f - g >} = 0$. We don't really stress about this normally any more than we stress about the true meaning of the irrational number $\sqrt{3}$. The number $\sqrt{3}$ is also properly interpreted as a Cauchy sequence of rational numbers but we normally do not need to examine those details closely.

However, an important takeaway here is that the set of Riemann Integrable functions on $[a, b]$ is dense in $\mathbb{L}_2([a, b])$ using this extended metric. We typically start this complicated construction process with $(C([a, b]), d_2)$ instead of $(RI([a, b]), d_2)$ which simply means we do not need the first equivalence relation for $f \sim g$ means $d_2(f, g) = 0$ as if the functions are continuous, the only way $d_2(f, g) = 0$ is if $f = g$ and so this equivalence relation is not needed. Call the resulting Hilbert Space $\mathbb{L}_2([a, b])$ again and we see continuous functions on $[a, b]$ are dense in $\mathbb{L}_2([a, b])$.

If we define a new notion of integration called Lebesgue Integration, we can interpret the $\int$ as Lebesgue Integration instead of Riemann Integration and use the same completion process as we did for $(RI([a, b]), d_2)$ to construct $\mathbb{L}_2([a, b])$. In (Peterson (16) 2019), we see we can define $\mathbb{L}_2([a, b])$ more directly by letting it consist of all the equivalence classes of Square Lebesgue Integrable functions where $\sim$ is the usual $f \sim g$ means $\int_a^b |f(t) - g(t)|^2 dt = 0$ except now the integration is Lebesgue Integration. In either method, we end up with a complete inner product space. So for us from now on, we will assume we have a function space $\mathbb{L}_2([a, b])$ which is complete with respect to the usual inner product as long as we interpret $< f, g >$ as using representatives of equivalence classes.

Let's make some additional comments about the Hilbert Space $\mathbb{L}_2([a, b])$. We have discussed two ways to construct this space: one starts from the Riemann Integrable functions and the other from the continuous functions. Let's temporarily call the one that comes from the Riemann Integrable functions X_{RI} and the one from the continuous functions X_C. Here is what we know:

- If we take the X_C route, the continuous functions, embedded into X_C, are dense. So for any $\epsilon > 0$ and for any element we choose in X_C, we can find a constant Cauchy sequence formed by a continuous function that is ϵ away in the two norm. The space X_C contains all Riemann Integrable functions too. The reasoning is this: If f is a Riemann Integrable function on $[a,b]$, then any sequence of partitions whose norm goes to zero gives us a sequence of Riemann Sums that converge to the value of the Riemann Integrable. Each Riemann sum defines a function which is constant on subintervals of the partition used to define the Riemann sum and which has jump discontinuities at each of the points t_i in the partition. It is not too hard to imagine how to build a sequence of continuous functions which converge to this piecewise continuous step function. This sequence will be a Cauchy sequence in our two norm and so the Riemann Integrable function will be in the equivalent class associated with this Cauchy sequence. Thus, X_C contains elements we can identify with all Riemann Integrable functions and we know this is a lot of possible functions.

- If we take the X_{RI} route, the Riemann Integrable functions are dense in X_{RI} in the usual sense described above. Since each Riemann Integrable function is interpreted as an equivalence class of functions under our definition of $\sim$, for each $\epsilon > 0$ we can find a continuous function constant sequence which is within $\epsilon > 0$ of the given Riemann Integrable function in the two norm.

- So either way X_C or X_{RI} has the continuous functions as a dense subset under the appropriate interpretation.

We can discuss this further and add some more details. First, let's look at an arbitrary Riemann Integrable function f. The function f is defined pointwise on $[a,b]$ in some way and we often do not know the exact description. We know if a bounded function f on $[a,b]$ is Riemann Integrable, then $\int_a^b f(t)dt$ can be approximated by any sequence of partitions $(\boldsymbol{\pi_n})$ and evaluation sets $\boldsymbol{\sigma_n}$ from $\boldsymbol{\pi_n}$ with $S(f, \boldsymbol{\pi_n}, \boldsymbol{\sigma_n}) \rightarrow \int_a^b f(t)dt$. Let's focus on a given Riemann Sum $S(f, \boldsymbol{\pi_n}, \boldsymbol{\sigma_n})$. This sum is determined by a special type of function called **step** function.

Definition 7.4.1 Step Functions

Let $\boldsymbol{\pi}$ be a partition of $[a,b]$; $\boldsymbol{\pi} = \{t_0 = a, t_1, \ldots, t_{n-1}, t_n = b\}$ for some n with $t_0 < t_1 < \ldots < t_{n-1} < t_n$. A step function $\boldsymbol{s_\pi}$ determined by $\boldsymbol{\pi}$ is defined by

$$(\boldsymbol{s_\pi})(t) = \begin{cases} c_1, & t_0 \le t \le t_1 \\ c_2, & t_1 < t \le t_2 \\ \vdots & \vdots \\ c_{n-1}, & t_{n-1} < t \le t_n \end{cases}$$

It is then clear $\int_a^b (\boldsymbol{s_\pi})(t)dt = \sum_{i \in \boldsymbol{\pi}} c_i \Delta_i$. A Riemann sum thus has associated with it a step function $S(f, \boldsymbol{\pi_n}, \boldsymbol{\sigma_n})$ which we will denote by

$$g(f, \boldsymbol{\pi}, \boldsymbol{\sigma_n}) = \begin{cases} f(s_1), & t_0 \le t \le t_1 \\ f(s_2), & t_1 < t \le t_2 \\ \vdots & \vdots \\ f(s_{n-1}), & t_{n-1} < t \le t_n \end{cases}$$

and $\int_a^b g(f, \boldsymbol{\pi}, \boldsymbol{\sigma_n})(t)dt = S(f, \boldsymbol{\pi_n}, \boldsymbol{\sigma_n})$. For the function which is

$$x(t) = \begin{cases} 0, & 0 \le t \le \frac{1}{2} \\ 1, & \frac{1}{2} < t \le 1 \end{cases}$$

the sequence $(\boldsymbol{x_n})$ in $C([0,1])$ defined by

$$x_n(t) = \begin{cases} 0, & 0 \leq t \leq \frac{1}{2} \\ n(t - \frac{1}{2}), \frac{1}{2} < t < \frac{1}{2} + \frac{1}{n} & \\ 1, & \frac{1}{2} + \frac{1}{n} \leq t \leq 1 \end{cases}$$

is a Cauchy sequence in $(C([a,b]), \|\cdot\|_2)$ which converges to $\boldsymbol{x}$. Hence $\boldsymbol{x}$ is a representative of the equivalence class $[(\boldsymbol{x_n})]$. We can easily construct such an equivalence class of functions for a function that looks like this

$$x(t) = \begin{cases} 0, & a \leq t \leq b \\ \gamma, & b < t \leq c \\ 0, & c < t \leq b \end{cases}$$

by *glueing* two such Cauchy sequences together. Since adding two Cauchy sequences gives another Cauchy sequence, we see there is an equivalence class $[(sx_n)]$ associated with this step function.

We can therefore construct an equivalence class $[(sx_n)]$ associated with $g(f, \boldsymbol{\pi}, \boldsymbol{\sigma_n})$ and from the Riemann Integral approximation theorem, we can see given $\epsilon > 0$, there is an N so that $n > N$ implies

$$\left| \int_a^b (f(t) - g(f, \boldsymbol{\pi}, \boldsymbol{\sigma_n})(t))dt \right| < \epsilon$$

This tells us the step functions are dense in $\mathbb{L}_2([a,b])$.

What does this mean? It means we can understand what a Riemann Integrable function is a bit better. We know a Riemann Integrable function is continuous a.e. but getting a pointwise description of an arbitrary Riemann Integrable function is hard. What we do know is that there is a sequence of equivalence classes $([g(f, \boldsymbol{\pi}, \boldsymbol{\sigma_n})])$ which converges in the norm of $\mathbb{L}_2([a,b])$ to the equivalence class $([f])$. We also know we can construct Cauchy sequences of continuous functions which converge in $\|\cdot\|_2$ to each $g(f, \boldsymbol{\pi}, \boldsymbol{\sigma_n})$ and these sequences are representatives of $([g(f, \boldsymbol{\pi}, \boldsymbol{\sigma_n})])$.

The Weierstrass Approximation Theorem tells us polynomials are dense in $(C([a,b]), \|\cdot\|_2)$, so each of the continuous functions we used in our original construction which were piecewise differentiable can be approximated by functions having as many derivatives as we like. So the Riemann Integrable functions are dense in $\mathbb{L}_2([a,b])$ and the step functions, continuous functions and polynomials are as well. So an arbitrary function f in $\mathbb{L}_2([a,b])$ has a pointwise representation taken from an equivalence class and we can approximate the representatives of this equivalence class as close as we like by step functions, continuous functions, polynomials or Riemann Integrable functions.

So from now on, when we use the Hilbert Space $\mathbb{L}_2([a,b])$ we will not worry about how we decided to start the metric space completion process. We will know the continuous functions and the Riemann Integrable functions and others are dense in the Hilbert Space.

A last comment should help. An element in $\mathbb{L}_2([a,b])$ is somewhat difficult to nail down. If you choose a representation $x : [a,b] \rightarrow \Re$ of an element in $\mathbb{L}_2([a,b])$, what we know is that $\|f - x\|_2 = 0$ for any other choice of representation. Now if we apply an algorithm to solve a problem and it turns out we are generating approximations x_n at each step n which converge in $\|\cdot\|_2$ to a function x, note it is helpful if we could also say this sequence (x_n) converged pointwise to x at least a.e. If we have that, we would have a better idea of how to describe the equivalence class of solutions in such a way as it is useful to us. A lot of work in applied analysis is devoted to trying to understand this pointwise limit better.

Homework

Exercise 7.4.1 *For*

$$x(t) = \begin{cases} 0, & 1 \leq t \leq 2 \\ 10, & 2 < t \leq 3 \\ 0, & 3 < t \leq 4 \end{cases}$$

find an equivalence class of functions which converge in $\| \cdot \|_2$ *to* $\boldsymbol{x}$.

Exercise 7.4.2 *For*

$$x(t) = \begin{cases} 5, & 1 \leq t \leq 2 \\ 10, & 2 < t \leq 3 \\ 3, & 3 < t \leq 4 \end{cases}$$

find an equivalence class of functions which converge in $\| \cdot \|_2$ *to* $\boldsymbol{x}$.

Exercise 7.4.3 *For*

$$x(t) = \begin{cases} 5, & 1 \leq t \leq 2 \\ 3, & 2 < t \leq 3 \\ 6, & 3 < t \leq 4 \\ 1, & 4 < t \leq 5 \end{cases}$$

find an equivalence class of functions which converge in $\| \cdot \|_2$ *to* $\boldsymbol{x}$.

Exercise 7.4.4 *For*

$$x(t) = \begin{cases} 0, & 1 \leq t \leq 2 \\ 10, & 2 < t \leq 3 \\ 0, & 3 < t \leq 4 \end{cases}$$

Use standard MATLAB code as described in (Peterson (18) 2020) to plot the Bernstein polynomial which approximates $\boldsymbol{x}$ *for* $\epsilon = 0.1, 0.01, 0.001$ *and* 0.0001.

Exercise 7.4.5 *For*

$$x(t) = \begin{cases} 5, & 1 \leq t \leq 2 \\ 10, & 2 < t \leq 3 \\ 3, & 3 < t \leq 4 \end{cases}$$

Use standard MATLAB code as described in (Peterson (18) 2020) to plot the Bernstein polynomial which approximates $\boldsymbol{x}$ *for* $\epsilon = 0.1, 0.01, 0.001$ *and* 0.0001.

In the earlier courses, we always handle improper Riemann Integrals in a sort of ad hoc manner. Consider this simple example. The function $f(t) = \frac{1}{\sqrt{t}}$ is not bounded on $(0, 1)$ and so it cannot be Riemann Integrable. Of course, its improper integral does exist. However, it is not in $\mathbb{L}_2$ because the improper integral of it's square does not exist. Here is a good way to think about it in the context of $\mathbb{L}_2([a, b])$. If we want the function t^{-p} to be square integrable in this sense, we can see we want $0 < p < 1/2$. Here is an example. Consider the functions

$$x_n(t) = \begin{cases} 0, & 0 \leq 0 < \frac{1}{n} \\ \frac{1}{t^{1/4}}, & \frac{1}{n} \leq t \leq 1 \end{cases}$$

It is easy to see each $\boldsymbol{x_n}$ is Riemann Integrable so it is a representative of an equivalence class which is an element of $\mathbb{L}_2([a,b])$. Consider for $n > m$,

$$\int_0^1 |x_n(t) - x_m(t)|^2 dt \quad = \quad \int_{\frac{1}{n}}^{\frac{1}{m}} \left|\frac{1}{t^{1/4}}\right|^2 dt = \int_{\frac{1}{n}}^{\frac{1}{m}} \frac{1}{\sqrt{t}}\, dt = 2\sqrt{1/m} - 2\sqrt{1/n} \leq 2\sqrt{1/m}$$

Clearly, given $\epsilon > 0$, we can choose M so that $2\sqrt{1/M} < \epsilon$ and this is a Cauchy sequence in $\|\cdot\|_2$ which is a representative of the function $\frac{1}{\sqrt{t}}$ on $(0,1)$. From the way we defined the inner product on the equivalence classes, we know $\langle [\boldsymbol{x_n}], [\boldsymbol{f}]\rangle = 0$ and so it seems reasonable to define

$$\int_a^b f(t)dt \quad = \quad \lim_{n\to\infty} \int_a^b x_n(t)dt$$

for any representative of $[\boldsymbol{f}]$ which is a Cauchy sequence of Riemann Integrable functions. We have shown one particular such sequence. So this suggests $\mathbb{L}_2([a,b])$ also contains the functions which are improperly Riemann Integrable in various ways. Hence, $\mathbb{L}_2([a,b])$ is a great way to make more quantitative a number of ad hoc constructions.

Homework

Exercise 7.4.6 *Find a Cauchy sequence of Riemann Integrable functions which gives a representative for* $f(t) = t^{-\frac{1}{3}}$ *on* $[0,1]$.

Exercise 7.4.7 *Find a Cauchy sequence of Riemann Integrable functions which gives a representative for* $f(t) = t^{-\frac{3}{7}}$ *on* $[0,1]$.

7.5 Properties of Inner Product Spaces

There are interesting properties that an inner product space has.

Theorem 7.5.1 The Minimizing Vector Theorem

Let $(X, <\cdot,\cdot>)$ be a nonempty inner product space. Let $M \neq 0$ be a convex subset of X which means $t\boldsymbol{x} + (1-t)\boldsymbol{y} \in X$ for all $t \in [0,1]$ and $\boldsymbol{x}$ and $\boldsymbol{y}$ in M. This says the line segment $[\boldsymbol{x}, \boldsymbol{y}]$ lies in M if $\boldsymbol{x}$ and $\boldsymbol{y}$ in M. Assume M is complete in the metric induced by $<\cdot,\cdot>$. Then for all $\boldsymbol{x} \in X$, there is a unique $\boldsymbol{y_x} \in M$ so that $\inf_{\boldsymbol{y}\in M} \|\boldsymbol{x} - \boldsymbol{y}\| = \|\boldsymbol{x} - \boldsymbol{y_x}\|$

Proof 7.5.1

Let $\delta = \inf_{\boldsymbol{y}\in M} \|\boldsymbol{x} - \boldsymbol{y}\|$. Since M is not empty and all the norms are bounded below by zero, there is a finite non-negative infimum. Let $(\boldsymbol{y_n}) \subset M$ be a minimizing sequence so that $\|\boldsymbol{y_n} - \boldsymbol{x}\| \to \delta$ with $\delta \leq \|\boldsymbol{x} - \boldsymbol{y_n}\| < \delta + \frac{1}{n}$.

($(\boldsymbol{y_n})$ is a Cauchy sequence in M):
Since $\|\cdot\|$ is induced by the inner product, it satisfies the parallelogram law. Therefore

$$\|(\boldsymbol{y_n} - \boldsymbol{x}) - (\boldsymbol{y_m} - \boldsymbol{x})\|^2 + \|(\boldsymbol{y_n} - \boldsymbol{x}) + (\boldsymbol{y_m} - \boldsymbol{x})\|^2 \quad = \quad 2\left(\|\boldsymbol{y_n} - \boldsymbol{x}\|^2 + \|\boldsymbol{y_m} - \boldsymbol{x}\|^2\right)$$

Now,

$$\|(\boldsymbol{y_n} - \boldsymbol{x}) - (\boldsymbol{y_m} - \boldsymbol{x})\|^2 \quad = \quad \|\boldsymbol{y_n} - \boldsymbol{y_m}\|^2$$

$$\|(\boldsymbol{y_n} - \boldsymbol{x}) + (\boldsymbol{y_m} - \boldsymbol{x})\|^2 = \|\boldsymbol{y_n} + \boldsymbol{y_m} - 2\boldsymbol{x}\|^2 = 4\left\|\frac{\boldsymbol{y_n} + \boldsymbol{y_m}}{2} - \boldsymbol{x}\right\|^2$$

But M is convex and so $\frac{\boldsymbol{y_n}+\boldsymbol{y_m}}{2} \in M$. Hence, $\|\frac{\boldsymbol{y_n}+\boldsymbol{y_m}}{2} - \boldsymbol{x}\|^2 \geq \delta^2$. Combining, we have

$$\|\boldsymbol{y_n} + \boldsymbol{y_m} - 2\boldsymbol{x}\|^2 \geq 4\delta^2$$

This gives

$$\begin{aligned} 2\left(\|(\boldsymbol{y_n} - \boldsymbol{x})\|^2 + \|(\boldsymbol{y_m} - \boldsymbol{x})\|^2\right) &= \|(\boldsymbol{y_n} - \boldsymbol{x}) - (\boldsymbol{y_m} - \boldsymbol{x})\|^2 + \|(\boldsymbol{y_n} - \boldsymbol{x}) + (\boldsymbol{y_m} - \boldsymbol{x})\|^2 \\ &\geq \|\boldsymbol{y_n} - \boldsymbol{y_m}\|^2 + 4\delta^2 \end{aligned}$$

So if $m > n$,

$$\begin{aligned} \|\boldsymbol{y_n} - \boldsymbol{y_m}\|^2 &\leq 2\left(\|(\boldsymbol{y_n} - \boldsymbol{x})\|^2 - \delta^2\right) + 2\left(\|(\boldsymbol{y_m} - \boldsymbol{x})\|^2 - \delta^2\right) \\ &< 2\frac{\delta}{n} + \frac{1}{n^2} + 2\frac{\delta}{m} + \frac{1}{m}^2 < 4\frac{\delta}{m} + \frac{2}{m}^2 \end{aligned}$$

For $\epsilon > 0$, choose N so that $4\frac{\delta}{N} + \frac{2}{N}^2 < \epsilon^2$. Then, if $n > m > N$, we have $\|\boldsymbol{y_n} - \boldsymbol{y_m}\| < \epsilon$ which shows $(\boldsymbol{y_n})$ is a Cauchy sequence in M.

(There is a unique $\boldsymbol{y_x}$ in M with $\|\boldsymbol{x} - \boldsymbol{y_x}\| = \delta$):
Since $(\boldsymbol{y_n})$ in M is a Cauchy sequence and M is complete, there is a $\boldsymbol{y_x}$ so that $\boldsymbol{y_n} \to \boldsymbol{y_x}$ in norm. So for any $\boldsymbol{y} \in M$, we also have $\|\boldsymbol{x} - \boldsymbol{y_x}\| \geq \inf_{\boldsymbol{y} \in M} \|\boldsymbol{x} - \boldsymbol{y}\| = \delta$. However, we also know for all n

$$\|\boldsymbol{x} - \boldsymbol{y_x}\| \leq \|\boldsymbol{x} - \boldsymbol{y_n}\| + \boldsymbol{y_n} - \boldsymbol{y_x}\| \leq \delta + \frac{1}{n} + \|\boldsymbol{y_n} - \boldsymbol{y_x}\|$$

Letting $n \to \infty$, we see $\|\boldsymbol{x} - \boldsymbol{y_x}\| \leq \delta$. Combining inequalities, we see $\boldsymbol{y_x}$ is a minimizer.

To see this minimizer is unique, if there was $\boldsymbol{z} \in M$ so that $\|\boldsymbol{x} - \boldsymbol{z}\| = \delta$, then by the parallelogram law

$$\|(\boldsymbol{x} - \boldsymbol{y_x}) - (\boldsymbol{x} - \boldsymbol{z})\|^2 + \|(\boldsymbol{x} - \boldsymbol{y_x}) + (\boldsymbol{x} - \boldsymbol{z})\|^2 = 2\left(\|\boldsymbol{x} - \boldsymbol{z}\|^2 + \|\boldsymbol{x} - \boldsymbol{y_x}\|^2\right) = 4\delta^2$$

Now

$$\|(\boldsymbol{x} - \boldsymbol{y_x}) + (\boldsymbol{x} - \boldsymbol{z})\|^2 = \|2\boldsymbol{x} - (\boldsymbol{y_x} + \boldsymbol{z})\|^2 = 4\left\|\boldsymbol{x} - \frac{\boldsymbol{y_x} + \boldsymbol{z}}{2}\right\| \geq 4\delta^2$$

because $\frac{\boldsymbol{y_x}+\boldsymbol{z}}{2} \in M$. This implies

$$\|\boldsymbol{z} - \boldsymbol{y_x}\|^2 \leq 4\delta^2 - 4\delta^2 = 0$$

and so $\boldsymbol{y_x} = \boldsymbol{z}$ and the minimizer is unique. ■

We can now look at minimization ideas over other sorts of sets other than convex complex subsets of the inner product space.

Theorem 7.5.2 Minimization Over a Complete Subspace of an Inner Product Space

Let $(X, < \cdot, \cdot >)$ be an inner product space and let Y be a complete subspace of X. Let the induced norm be $\| \cdot \|$. Pick $\boldsymbol{x} \in X$. Then there is a unique $\boldsymbol{y_x} \in Y$ so that $\inf_{\boldsymbol{y} \in Y} \|\boldsymbol{x} - \boldsymbol{y}\| = \|\boldsymbol{x} - \boldsymbol{y_x}\|$

Proof 7.5.2

Since Y is a subspace, it is convex. Then by the Minimizing Vector Theorem, $\boldsymbol{y_x} \in Y$ exists and is unique. ■

If the minimization is over a finite dimensional subspace, it is easy to prove the following:

Theorem 7.5.3 Minimization Over a Finite Dimensional Subspace of an Inner Product Space

Let $(X, < \cdot, \cdot >)$ be an inner product space and let Y be a finite dimensional subspace of X. Let the induced norm be $\| \cdot \|$. Pick $\boldsymbol{x} \in X$. Then there is a unique $\boldsymbol{y_x} \in Y$ so that $\inf_{\boldsymbol{y} \in Y} \|\boldsymbol{x} - \boldsymbol{y}\| = \|\boldsymbol{x} - \boldsymbol{y_x}\|$.

Proof 7.5.3

Y is convex as it is a subspace and Y is complete as it is finite dimensional. Hence there is a unique $\boldsymbol{y_x} \in Y$ so that $\inf_{\boldsymbol{y} \in Y} \|\boldsymbol{x} - \boldsymbol{y}\| = \|\boldsymbol{x} - \boldsymbol{y_x}\|$. ■

Homework

Exercise 7.5.1 *Let Y be the span of the functions 1 and t on $[0, 1]$ in $C([0, 1])$. Prove there is a unique a_0 and a_1 so that*

$$\inf_{\boldsymbol{y} \in Y} \|\sin(t) - a - bt\|_2 = \|\sin(t) - a_0 - a_1 t\|_2$$

- *Can you interpret the space $RI([0, 1])$ as an inner product space with the d_2 metric?*
- *If you can think of $RI([0, 1])$ as an inner product space with the d_2 metric, can you find a unique a_0 and a_1 as before?*
- *Can you interpret the space $\mathbb{L}_2([a, b])$ as an inner product space with the right metric?*
- *If you can think of $\mathbb{L}_2([a, b])$ as an inner product space with the right metric, can you find a unique a_0 and a_1 as before?*
- *Note that by extending the class of functions you are looking at you are forced to consider all of this from the point of view of equivalence classes of functions.*

Exercise 7.5.2 *Let M be closed unit ball about zero in $C([0, 1])$ with the d_2 metric. Prove there is a unique z in M so that*

$$\inf_{\boldsymbol{y} \in M} \|\sin(t) - y\|_2 = \|\sin(t) - z\|_2$$

Exercise 7.5.3 *Let Y be the span of the functions 1 and t on $[0, 1]$ in $C([0, 1])$.*

- *If you use with the infinity metric what is the optimization problem we have to solve to find the best a and b that minimizes $\|sin(t) - a - bt\|_\infty$?*
- *If you use with metric d_1 what is the optimization problem we have to solve to find the best a and b that minimizes $\|sin(t) - a - bt\|_1$?*

- *If you use with metric d_2 what is the optimization problem we have to solve to find the best a and b that minimizes $\|sin(t) - a - bt\|_2$? Note this is the case where we can prove there is a unique a_0 and b_0 that solves the minimization problem using the minimizing vector theorem.*

Exercise 7.5.4 *This is a nice set of problems about minimization using different norms.*

- *Let M be a closed unit ball about zero in $C([0,1])$ with the d_∞ metric. If you use with the infinity metric what is the optimization problem we have to solve to find the best z that minimizes $\inf_{\boldsymbol{y} \in M} \| \sin(t) - y\|_2 = \| \sin(t) - z\|_\infty$*

- *Let M be a closed unit ball in $C([0,1])$ with the d_1 metric. If you use with metric d_1 what is the optimization problem we have to solve to find the best z that minimizes $\inf_{\boldsymbol{y} \in M} \| \sin(t) - y\|_2 = \| \sin(t) - z\|_\infty$*

- *Let M be a closed unit ball in $C([0,1])$ with the d_2 metric. If you use with metric d_2 what is the optimization problem we have to solve to find the best z that minimizes $\inf_{\boldsymbol{y} \in M} \| \sin(t) - y\|_2 = \| \sin(t) - z\|_\infty$ Note this is the case where we can prove there is a unique z that solves the minimization problem using the minimizing vector theorem.*

Exercise 7.5.5 *Do all of the problems above for $\cos(t)$ instead of $\sin(t)$.*

Exercise 7.5.6 *Do all of the problems above for $\cos(2t)$ and $\sin(3t)$ and the span of the functions 1, t and t^2.*

Now let's talk about orthogonality.

Definition 7.5.1 Orthogonal Vectors in an Inner Product Space

Let $(X, < \cdot, \cdot >)$ be an inner product space. If $< \boldsymbol{x}, \boldsymbol{y} >= 0$ for a pair $\boldsymbol{x}$ and $\boldsymbol{y}$ in X, we say $\boldsymbol{x}$ is orthogonal to $\boldsymbol{y}$ and vice - versa.

Comment 7.5.1 *Nonzero orthogonal vectors must be linearly independent as $\alpha \boldsymbol{x} + \beta \boldsymbol{y} = \boldsymbol{0}$ implies $\boldsymbol{0} = \alpha < \boldsymbol{x}, \boldsymbol{y} > + \beta < \boldsymbol{y}, \boldsymbol{y} = \beta \|\boldsymbol{y}\|^2$. This says $\beta = 0$. A similar argument shows $\alpha = 0$ and so $\boldsymbol{x}$ and $\boldsymbol{y}$ are linearly independent.*

Comment 7.5.2 *Note if $\boldsymbol{x}$ and $\boldsymbol{y}$ are not zero, we can define the angle between $\boldsymbol{x}$ and $\boldsymbol{y}$ using the Cauchy - Schwartz Inequality by*

$$\cos(\theta) = \frac{< \boldsymbol{x}, \boldsymbol{y} >}{\|\boldsymbol{x}\| \, \|\boldsymbol{y}\|}$$

and orthogonal vectors, as usual are 90° or 270° apart.

We can extend this idea to vectors orthogonal to sets and sets orthogonal to sets.

Definition 7.5.2 Vectors Orthogonal to a Set in an Inner Product Space

Let $(X, < \cdot, \cdot >)$ be an inner product space and $D \subset X$. We say $\boldsymbol{x}$ is orthogonal to D if $< \boldsymbol{x}, \boldsymbol{y} >= 0$ for all $\boldsymbol{y}$ in D,

If D and E are two sets, we say D is orthogonal to E if $< \boldsymbol{x}, \boldsymbol{y} >= 0$ for all $\boldsymbol{x} \in D$ and $\boldsymbol{y} \in E$.

Now we can look at the orthogonality of the minimizing vector relative to the subspace Y.

Theorem 7.5.4 The Minimizing Vector Theorem and Orthogonality

Let $(X, < \cdot, \cdot >)$ be a nonempty inner product space. Let Y be a subspace of X which is complete in the metric induced by $< \cdot, \cdot >$. Then for all $\boldsymbol{x} \in X$, there is a unique $\boldsymbol{y_x} \in Y$ so that $\inf_{\boldsymbol{y} \in Y} \|\boldsymbol{x} - \boldsymbol{y}\| = \|\boldsymbol{x} - \boldsymbol{y_x}\|$ and $< \boldsymbol{x} - \boldsymbol{y}, \boldsymbol{u} >= 0$ for all $\boldsymbol{u} \in Y$; i.e. $\boldsymbol{x} - \boldsymbol{y}$ is orthogonal to Y.

Proof 7.5.4
Since Y is a subspace which is complete, there is a unique $\boldsymbol{y_x} \in Y$ so that $\inf_{\boldsymbol{y} \in Y} \|\boldsymbol{x} - \boldsymbol{y}\| = \|\boldsymbol{x} - \boldsymbol{y_x}\|$. Now assume there is at least one element $\boldsymbol{u_1} \in Y$ with $< \boldsymbol{x} - \boldsymbol{y_x}, \boldsymbol{u_1} > \neq 0$. Clearly, $\boldsymbol{u_1} \neq \boldsymbol{0}$. Consider for all α, letting $\boldsymbol{z} = \boldsymbol{x} - \boldsymbol{y_x}$,

$$0 \leq \|\boldsymbol{z} - \alpha \boldsymbol{u_1}\|^2 = < \boldsymbol{z} - \alpha \boldsymbol{u_1}, \boldsymbol{z} - \alpha \boldsymbol{u_1} >= \|\boldsymbol{z}\|^2 - \alpha < \boldsymbol{u_1}, \boldsymbol{z} > - \overline{\alpha} < \boldsymbol{z}, \boldsymbol{u_1} > + \alpha\, \overline{\alpha} \|\boldsymbol{u_1}\|^2$$

Define the scalar $\beta =< \boldsymbol{u_1}, \boldsymbol{z} >$ which implies $\overline{\beta} =< \boldsymbol{z}, \boldsymbol{u_1} >$. Then, for all α

$$\begin{aligned} 0 \leq \|\boldsymbol{z} - \alpha \boldsymbol{u_1}\|^2 &= \|\boldsymbol{z}\|^2 - \alpha\beta - \overline{\alpha}\overline{\beta} + \alpha\, \overline{\alpha} \|\boldsymbol{u_1}\|^2 \\ &= \|\boldsymbol{z}\|^2 - \overline{\alpha}\overline{\beta} - \alpha(\beta - \overline{\alpha}\|\boldsymbol{u_1}\|^2) \end{aligned}$$

Now choose $\overline{\alpha_0} = \frac{\beta}{\|\boldsymbol{u_1}\|^2}$. Then

$$0 \leq \|\boldsymbol{z} - \alpha_0 \boldsymbol{u_1}\|^2 = \|\boldsymbol{z}\|^2 - \overline{\alpha_0}\overline{\beta} - \alpha_0(\beta - \overline{\alpha_0}\|\boldsymbol{u_1}\|^2) = \|\boldsymbol{z}\|^2 - \overline{\alpha_0}\overline{\beta} = \|\boldsymbol{z}\|^2 - \frac{\beta\, \overline{\beta}}{\|\boldsymbol{u_1}\|^2}$$

Now let $\delta = \inf_{\boldsymbol{y} \in Y} \|\boldsymbol{x} - \boldsymbol{y}\| = \|\boldsymbol{x} - \boldsymbol{y_x}\| = \|\boldsymbol{z}\|$.

Now here is the contradiction. The element $\boldsymbol{y_x} - \alpha_0 \boldsymbol{u_1}$ is in Y because Y is a subspace and so we must have $\|\boldsymbol{x} - (\boldsymbol{y_x} - \alpha_0 \boldsymbol{u_1})\| \geq \delta$. However,

$$\|\boldsymbol{z} - \alpha_0 \boldsymbol{u_1}\|^2 = \|\boldsymbol{z}\|^2 - \frac{\beta\, \overline{\beta}}{\|\boldsymbol{u_1}\|^2} = \delta^2 - \frac{\beta\, \overline{\beta}}{\|\boldsymbol{u_1}\|^2} < \delta^2$$

which is not possible. So such an element $\boldsymbol{u_1}$ does not exist and $< \boldsymbol{x} - \boldsymbol{y}, \boldsymbol{u} >= 0$ for all $\boldsymbol{u} \in Y$; i.e. $\boldsymbol{x} - \boldsymbol{y}$ is orthogonal to Y. ■

Orthogonality is a very useful concept in an inner product space. We use the following notation:

- If $< \boldsymbol{x}, \boldsymbol{y} >= 0$, we write $\boldsymbol{x} \perp \boldsymbol{y}$.
- If $\boldsymbol{x}$ is orthogonal to D, we write $\boldsymbol{x} \perp D$.
- If D is orthogonal to E, we write $D \perp E$.

We can now define orthogonal complements.

Definition 7.5.3 The Orthogonal Complement of a Set in an Inner Product Space

Let $(X, < \cdot, \cdot >)$ be an inner product space. Let D be a subset of X. The **orthogonal complement** *of D is denoted $D^\perp$ and is defined by $D^\perp = \{\boldsymbol{x} \in X \mid \boldsymbol{x} \perp D\}$.*

We can then restate Theorem 7.5.4 as follows:

Theorem 7.5.5 The Minimizing Vector Theorem and Orthogonal Decompositions

Let $(X, < \cdot, \cdot >)$ be a nonempty inner product space. Let Y be a subspace of X which is complete in the metric induced by $< \cdot, \cdot >$. Then for each $\boldsymbol{x} \in X$, there is a unique representation $\boldsymbol{x} = \boldsymbol{y} + \boldsymbol{z}$ where $\boldsymbol{y} \in Y$ and $\boldsymbol{z} \in Y^\perp$ with $\boldsymbol{y}$ satisfying $\inf_{\boldsymbol{u} \in Y} \|\boldsymbol{x} - \boldsymbol{u}\| = \|\boldsymbol{x} - \boldsymbol{y}\|$

Proof 7.5.5

Given $\boldsymbol{x} \in X$, we already know such a unique $\boldsymbol{y}$ exists and $\boldsymbol{x} - \boldsymbol{y} \in Y^\perp$. Let $\boldsymbol{z} = \boldsymbol{x} - \boldsymbol{y}$, we have $\boldsymbol{x} = \boldsymbol{y} + \boldsymbol{z}$ which is the representation we seek.

Is $\boldsymbol{z}$ unique? Let $\boldsymbol{x} = \boldsymbol{y} + \boldsymbol{z}_1$ where $\boldsymbol{z}_1 \in Y^\perp$. Then $\boldsymbol{x} - \boldsymbol{y} = \boldsymbol{z}_1 = \boldsymbol{z}$ showing the choice $\boldsymbol{z}$ is unique. ■

From (Peterson (18) 2020), we already have a computational method for finding the best approximation of an element in an inner product space to a finite dimensional subspace of the inner product space. This is a complete subspace so our theorems do apply, but we can compute this element efficiently. Let's go through this argument again.

Theorem 7.5.6 The Finite Dimensional Approximation Theorem

Let p be any object in the inner product space $\mathscr{V}$ with inner product $<, >$ and induced norm $\| \cdot \|$. Let $\mathscr{W}$ be a finite dimensional subspace with an orthonormal basis $\{\boldsymbol{w}_1, \ldots \boldsymbol{w}_N\}$ where N is the dimension of the subspace. Then there is an unique object p^* in $\mathscr{W}$ which satisfies

$$||p - p^*|| = \min_{u \in \mathscr{W}} ||u - p||$$

with

$$p^* = \sum_{i=1}^{N} < \boldsymbol{p}, \boldsymbol{w}_i > \boldsymbol{w}_i.$$

Further, $p - p^*$ is orthogonal to the subspace $\mathscr{W}$.

Proof 7.5.6

Any object in the subspace has the representation $\sum_{i=1}^{N} a_i \boldsymbol{w}_i$ for some scalars a_i. Consider the function of N variables

$$\begin{aligned} E(a_1, \ldots, a_N) &= \left\langle \boldsymbol{p} - \sum_{i=1}^{N} a_i \boldsymbol{w}_i, \boldsymbol{p} - \sum_{j=1}^{N} a_j \boldsymbol{w}_j \right\rangle \\ &= < \boldsymbol{p}, \boldsymbol{p} > -2 \sum_{i=1}^{N} a_i < \boldsymbol{p}, \boldsymbol{w}_i > \\ &\quad + \sum_{i=1}^{N} \sum_{j=1}^{N} a_i a_j < \boldsymbol{w}_i, \boldsymbol{w}_j > . \end{aligned}$$

Simplifying using the orthonormality of the basis, we find

$$E(a_1, \ldots, a_N) = < \boldsymbol{p}, \boldsymbol{p} > -2 \sum_{i=1}^{N} a_i < \boldsymbol{p}, \boldsymbol{w}_i > + \sum_{i=1}^{N} a_i^2.$$

This is a quadratic expression and setting the gradient of E to zero, we find the critical points $a_j = <\boldsymbol{p}, \boldsymbol{w}_j>$. *It is easy to see the Hessian here is positive definite so it is clear this gives a global minimum for the function E. Hence, the optimal* p^* *has the form*

$$p^* = \sum_{i=1}^{N} <\boldsymbol{p}, \boldsymbol{w}_i> \boldsymbol{w}_i.$$

Finally, we see

$$\begin{aligned} <p - p^*, \boldsymbol{w}_j> &= <p, \boldsymbol{w}_j> - \sum_{k=1}^{N} <p, \boldsymbol{w}_k><\boldsymbol{w}_k, \boldsymbol{w}_j> \\ &= <p, \boldsymbol{w}_j> - <p, \boldsymbol{w}_j> = 0, \end{aligned}$$

and hence, $p - p^*$ *is orthogonal of* $\mathscr{W}$. ■

Homework

Exercise 7.5.7 *Let* Y *be the span of the functions* 1 *and* t *on* $[0, 1]$ *in* $C([0, 1])$.

- *Use GSO to find an orthonormal basis for* Y. *Use MATLAB for this.*
- *Find the unique* a_0 *and* a_1 *so that*

$$\inf_{\boldsymbol{y} \in Y} \| \sin(t) - a - bt \|_2 = \| \sin(t) - a_0 - a_1 t \|_2$$

- *How does this solution compare to the Taylor Series solution?*
- *What is the orthogonal complement to* Y*?*
- *Find the decomposition* $\sin(t) = \boldsymbol{u} + \boldsymbol{v}$ *with* $\boldsymbol{u} \in Y$ *and* $\boldsymbol{v} \in Y^{\perp}$.

Exercise 7.5.8 *Let* Y *be the span of the functions* 1, t *and* t^2 *on* $[0, 1]$ *in* $C([0, 1])$.

- *Use GSO to find an orthonormal basis for* Y. *Use MATLAB for this.*
- *Find the unique* a_0, a_1 *and* a_2 *so that*

$$\inf_{\boldsymbol{y} \in Y} \| \sin(t) - a - bt - ct62 \|_2 = \| \sin(t) - a_0 - a - 1t - a_2 t^2 \|_2$$

- *How does this solution compare to the Taylor Series solution?*
- *What is the orthogonal complement to* Y*?*
- *Find the decomposition* $\sin(t) = \boldsymbol{u} + \boldsymbol{v}$ *with* $\boldsymbol{u} \in Y$ *and* $\boldsymbol{v} \in Y^{\perp}$.

Exercise 7.5.9 *Let* Y *be the span of the functions* 1 *and* t *on* $[-1, 1]$ *in* $C([-1, 1])$.

- *Use GSO to find an orthonormal basis for* Y. *Use MATLAB for this.*
- *Find the unique* a_0 *and* a_1 *so that*

$$\inf_{\boldsymbol{y} \in Y} \| \sin(t) - a - bt \|_2 = \| \sin(t) - a_0 - a_1 t \|_2$$

- *How does this solution compare to the Taylor Series solution?*

- *What is the orthogonal complement to Y?*
- *Find the decomposition* $\sin(t) = \boldsymbol{u} + \boldsymbol{v}$ *with* $\boldsymbol{u} \in Y$ *and* $\boldsymbol{v} \in Y^{\perp}$.

Exercise 7.5.10 *Let Y be the span of the functions 1, t and t^2 on $[-2, 1]$ in $C([-2, 1])$.*

- *Use GSO to find an orthonormal basis for Y. Use MATLAB for this.*
- *Find the unique a_0, a_1 and a_2 so that*

$$\inf_{\boldsymbol{y} \in Y} \| \sin(t) - a - bt - ct^2 \|_2 = \| \sin(t) - a_0 - a_1 t - a_2 t^2 \|_2$$

- *How does this solution compare to the Taylor Series solution?*
- *What is the orthogonal complement to Y?*
- *Find the decomposition* $\sin(t) = \boldsymbol{u} + \boldsymbol{v}$ *with* $\boldsymbol{u} \in Y$ *and* $\boldsymbol{v} \in Y^{\perp}$.

Exercise 7.5.11 *Let Y be the span of the functions $\sin(\pi t)$, $\sin(2\pi t)$ and $\sin(3\pi t)$ on $[0, 1]$ in $C([0, 1])$.*

- *Use GSO to find an orthonormal basis for Y. Use MATLAB for this.*
- *Do you really need GSO here? Look back at the treatment of Fourier Series in (Peterson (18) 2020).*
- *Find the unique constants a_i so that*

$$\inf_{\boldsymbol{y} \in Y} \|2t^2 - y\|_2 = \|t^2 - a_1 \sin(\pi t) - a_2 \sin(2\pi t) - a_3 \sin(3\pi t)\|_2$$

- *What is the orthogonal complement to Y?*
- *Find the decomposition* $2t^2 = \boldsymbol{u} + \boldsymbol{v}$ *with* $\boldsymbol{u} \in Y$ *and* $\boldsymbol{v} \in Y^{\perp}$.

Exercise 7.5.12 *Let Y be the span of the functions 1, $\cos(\pi t)$, $\cos(2\pi t)$ and $\cos(3\pi t)$ on $[0, 1]$ in $C([0, 1])$.*

- *Use GSO to find an orthonormal basis for Y. Use MATLAB for this.*
- *Do you really need GSO here? Look back at the treatment of Fourier Series in (Peterson (18) 2020).*
- *Find the unique constants b_i so that*

$$\inf_{\boldsymbol{y} \in Y} \|2t^2 - y\|_2 = \|t^2 - b_0 - b_1 \cos(\pi t) - b_2 \cos(2\pi t) - b_3 \cos(3\pi t)\|_2$$

- *What is the orthogonal complement to Y?*
- *Find the decomposition* $2t^2 = \boldsymbol{u} + \boldsymbol{v}$ *with* $\boldsymbol{u} \in Y$ *and* $\boldsymbol{v} \in Y^{\perp}$.

We are now ready to talk about direct sum decompositions of inner product spaces.

Definition 7.5.4 The Direct Sum Decomposition of an Inner Product Space

Let $(X, < \cdot, \cdot >)$ be an inner product space and Y and Z subspaces of X. We say X is the direct sum of Y and Z and write $X = Y \oplus Z$ if each $\boldsymbol{x} \in X$ can be written as $\boldsymbol{x} = \boldsymbol{y} + \boldsymbol{z}$ for a unique $\boldsymbol{y} \in Y$ and $\boldsymbol{z} \in Z$.

We can therefore restate Theorem 7.5.4 and Theorem 7.5.5 again as

Theorem 7.5.7 The First Projection Theorem and Direct Sum Decompositions

Let $(X, < \cdot, \cdot >)$ be a nonempty inner product space. Let Y be a subspace of X which is complete in the metric induced by $< \cdot, \cdot >$. Then $X = Y \oplus Y^\perp$.

Proof 7.5.7
We know given $\boldsymbol{x} \in X$, there is a unique $\boldsymbol{y} \in Y$ and a unique $\boldsymbol{z} \in Y^\perp$ so that $\boldsymbol{x} = \boldsymbol{y} + \boldsymbol{z}$ with $\inf_{\boldsymbol{u} \in Y} \|\boldsymbol{x} - \boldsymbol{u}\| = \|\boldsymbol{x} - \boldsymbol{y}\|$. This says $X = Y \oplus Y^\perp$. ■

Homework

Exercise 7.5.13 *Let Y be the span of the functions 1 and t on $[0, 1]$ in $C([0, 1])$.*

- *Use GSO to find an orthonormal basis for Y. Use MATLAB for this.*
- *Find the unique a_2 and b_2 so that*

$$\inf_{\boldsymbol{y} \in Y} \| \sin(t) - a - bt\|_2 \quad = \quad \| \sin(t) - a_2 - b_2 t\|_2$$

- *How does this solution compare to the Taylor Series solution?*
- *What is the orthogonal complement to Y?*
- *Find the decomposition $\sin(t) = \boldsymbol{u} + \boldsymbol{v}$ with $\boldsymbol{u} \in Y$ and $\boldsymbol{v} \in Y^\perp$.*

Exercise 7.5.14 *Let Y be the span of the functions 1, t and t^2 on $[0, 1]$ in $C([0, 1])$. Find $Y^\perp$ so that $C([0, 1]) = Y \oplus Y^\perp$.*

Exercise 7.5.15 *Let Y be the span of the functions 1, t and t^2 on $[-1, 1]$ in $C([-1, 1])$. Find $Y^\perp$ so that $C([0, 1]) = Y \oplus Y^\perp$.*

Exercise 7.5.16 *Let Y be the span of the functions 1, t and t^2 on $[-2, 1]$ in $C([-2, 1])$. Find $Y^\perp$ so that $C([0, 1]) = Y \oplus Y^\perp$.*

Exercise 7.5.17 *Let Y be the span of the functions $\sin(\pi t)$, $\sin(2\pi t)$ and $\sin(3\pi t)$ on $[0, 1]$ in $C([0, 1])$. Find $Y^\perp$ so that $C([0, 1]) = Y \oplus Y^\perp$.*

Exercise 7.5.18 *Let Y be the span of the functions 1, $\cos(\pi t)$, $\cos(2\pi t)$ and $\cos(3\pi t)$ on $[0, 1]$ in $C([0, 1])$. Find $Y^\perp$ so that $C([0, 1]) = Y \oplus Y^\perp$.*

Chapter 8

Hilbert Spaces

As you know, a Hilbert Space is an inner product space which is complete with respect to the metric induced by the inner product. If we have that additional information, we can establish more things about the inner product space.

8.1 Completeness and Projections

Let's look at some completeness results.

Theorem 8.1.1 The Completeness of Subsets of a Complete Normed Linear Space

Let $(X, \|\cdot\|)$ be a complete normed linear space and Y a closed subset of X. Then Y is complete if and only if Y is closed in X.

Proof 8.1.1
First, if Y is complete and $(\boldsymbol{y_n}) \subset Y$ converges to an element $\boldsymbol{y_0} \in X$, then $(\boldsymbol{y_n})$ is a Cauchy sequence. Since Y is complete, this sequence must converge to an element $\boldsymbol{y_1} \in Y$. Since limits are unique, we have $\boldsymbol{y_1} = \boldsymbol{y_0}$ which tells us Y is closed as it contains its limit points.

On the other hand, if Y is closed, if $(\boldsymbol{y_n})$ is a Cauchy sequence in $Y \subset H$ it must converge to an element $\boldsymbol{y_0} \in X$. But since Y is closed, $\boldsymbol{y_0} \in Y$ as well. This says Y is complete.
■

Comment 8.1.1 *Of course, this result holds if Y is a closed subspace of X as well.*

If we add an inner product, we can say a bit more.

Theorem 8.1.2 The Completeness of Subspaces in a Hilbert Space

Let $(H, <\cdot, \cdot>)$ be a Hilbert Space and Y a subspace of H. Then

1. *Y is complete if and only if Y is closed in H.*
2. *If Y is finite dimensional, then Y is complete.*

Proof 8.1.2
The first part has already been proved in the previous theorem. Then the statement follows easily as if Y is finite dimensional as a subspace, it must be closed and complete. ■

Now we can state a major result.

Theorem 8.1.3 Fundamental Decomposition Result for a Hilbert Space

Let $(H, < \cdot, \cdot >)$ be a Hilbert Space and Y be a closed subspace of H. Then $H = Y \oplus Y^\perp$. Thus, given $\boldsymbol{x} \in H$, there are unique elements $\boldsymbol{y} \in Y$ and $\boldsymbol{z} \in Y^\perp$ with $\boldsymbol{x} = \boldsymbol{y} + \boldsymbol{z}$ and $\inf_{\boldsymbol{u} \in Y} \|\boldsymbol{x} - \boldsymbol{u}\| = \|\boldsymbol{x} - \boldsymbol{y}\|$.

Proof 8.1.3
Since a closed subspace of a Hilbert Space is complete, the result follows from the previous theorem.
■

Homework

Exercise 8.1.1 *Let P_n be the set of all polynomials of degree p where $0 \leq p \leq n$ on $[0, 1]$,*

- *Prove the inner product space $C([0, 1])$ with norm $\|\cdot\|_2$ has the direct sum decomposition $C([0, 1]) = P_n \oplus P_n^\perp$.*
- *Prove P_n is complete and closed with respect to the norm $\|\cdot\|_2$.*

Exercise 8.1.2 *Let P_n be the set of all polynomials of degree p where $0 \leq p \leq n$ on $[0, 1]$,*

- *Prove the inner product space $RI([0, 1])$ with norm $\|\cdot\|_2$ has the direct sum decomposition $RI([0, 1]) = P_n \oplus P_n^\perp$.*
- *Prove P_n is complete and closed with respect to the norm $\|\cdot\|_2$.*

Exercise 8.1.3 *We know ℓ^2 is a Hilbert Space. Let Y_n be the finite dimensional subspace given by the span of $\mathbf{e_i}$ for $1 \leq i \leq n$ where $\mathbf{e_i}$ is the sequence of all zeros except for a 1 in slot i. Clearly Y_n is equivalent to $\Re^n$.*

- *Prove Y_n a closed subspace of ℓ^2.*
- *Prove Y_n is a complete subspace of ℓ^2.*
- *Find $Y_n^\perp$ and given any $\boldsymbol{x} \in \ell^2$. Use the Schauder Basis representation of $\boldsymbol{x} \in \ell^2$ to do this. Find its direct sum decomposition.*

Exercise 8.1.4 *Let $(H, < \cdot, \cdot >)$ be a Hilbert Space and P a closed subspace of H. We know $H = P \oplus P^\perp$. Prove $P^\perp$ is also closed.*

We now have the machinery to define projection maps.

8.2 Projections and Consequences

Definition 8.2.1 Hilbert Space Projections

Let $(H, < \cdot, \cdot >)$ be a Hilbert Space. If Y be a closed subspace of H, then $H = Y \oplus Y^\perp$ and any $\boldsymbol{x} \in H$ has representation $\boldsymbol{x} = \boldsymbol{y} + \boldsymbol{z}$ for a unique $\boldsymbol{y} \in Y$ and a unique $\boldsymbol{z} \in Y^\perp$. The unique element $\boldsymbol{y}$ defines a mapping $P : H \to Y$ by $P(\boldsymbol{x}) = \boldsymbol{y}$ and the unique element $\boldsymbol{z}$ defines a mapping $Q : H \to Y^\perp$ by $Q(\boldsymbol{x}) = \boldsymbol{z}$. P is called the **projection** *of H onto Y and Q is the* **projection** *of H onto $Y^\perp$.*

It is straightforward to prove the following properties for projections.

Theorem 8.2.1 Hilbert Space Projection Properties

Let $(H, < \cdot, \cdot >)$ be a Hilbert Space and Y be a closed subspace of H. Then the projections P and Q satisfy

- *P and Q are linear.*
- *P and Q are bounded with $\|P\|_{op} = 1$ and $\|Q\|_{op} = 1$.*
- *$P + Q = I$*
- *$P^2 = P$ and $Q^2 = Q$; such mappings are called* **idempotent**.
- *$P|_Y = I$ and $Q|_{Y^\perp} = I$.*
- *$Y^\perp = ker(P)$ and $Y = ker(Q)$.*

Proof 8.2.1
Most of these properties are easy to establish. Let's look at the operator norms.

$$\begin{aligned} \|\boldsymbol{x}\|^2 &= \|\boldsymbol{y} + \boldsymbol{z}\|^2 =< \boldsymbol{y} + \boldsymbol{z}, \boldsymbol{y} + \boldsymbol{z} > \\ &= \|\boldsymbol{y}\|^2 =< \boldsymbol{y}, \boldsymbol{z} > + < \boldsymbol{z}, \boldsymbol{y} > + \|\boldsymbol{z}\|^2 \\ &= \|\boldsymbol{y}\|^2 + \|\boldsymbol{z}\|^2 \end{aligned}$$

as $\boldsymbol{z} \in Y^\perp$. Hence

$$\|P(\boldsymbol{x})\|^2 = \|\boldsymbol{y}\|^2 \leq \|\boldsymbol{y}\|^2 + \|\boldsymbol{z}\|^2 = \|\boldsymbol{x}\|^2$$

which implies for $\boldsymbol{x} \neq \boldsymbol{0}$

$$\frac{\|P(\boldsymbol{x})\|}{\|\boldsymbol{x}\|} \leq 1$$

We see P is a bounded linear operator with $\|P\|_{op} \leq 1$. Now if $\boldsymbol{x_0} \neq \boldsymbol{0}$ is in Y, $\boldsymbol{z} = \boldsymbol{0}$ and $\|P(\boldsymbol{x_0})\| = \|\boldsymbol{x_0}\|$. This tells us $\|P\|_{op} \geq 1$. Combining, we have $\|P\|_{op} = 1$. The argument to show Q is bounded with $\|Q\|_{op} = 1$ is similar.

The rest are simple calculations.

$$\begin{aligned} I(\boldsymbol{x}) = \boldsymbol{x} &= \boldsymbol{y} + \boldsymbol{z} = P(\boldsymbol{x}) + Q(\boldsymbol{x}) \Longrightarrow I = P + Q \\ P(P(\boldsymbol{x})) &= P(\boldsymbol{y}) = \boldsymbol{y} = P(\boldsymbol{x}) \Longrightarrow P^2 = P \\ Q(Q(\boldsymbol{x})) &= Q(\boldsymbol{z}) = \boldsymbol{z} = Q(\boldsymbol{x}) \Longrightarrow Q^2 = Q \end{aligned}$$

The $ker(P)$ is clearly $Y^\perp$ and $ker(Q)$ is clearly Y. ■

Homework

Exercise 8.2.1 *Let Y be the span of the functions 1 and t on $[0, 1]$ in $C([0, 1])$.*

- *Use GSO to find an orthonormal basis for Y. Use MATLAB for this.*

- *Find the unique* a_0 *and* a_1 *so that*

$$\inf_{\boldsymbol{y}\in Y} \|\sin(t) - a - bt\|_2 \quad = \quad \|\sin(t) - a_0 - a_1 t\|_2$$

- *How does this solution compare to the Taylor Series solution?*
- *What is the orthogonal complement to* Y*?*
- *Find the decomposition* $\sin(t) = \boldsymbol{u} + \boldsymbol{v}$ *with* $\boldsymbol{u} \in Y$ *and* $\boldsymbol{v} \in Y^\perp$.

Exercise 8.2.2 *Let* Y *be the span of the functions* 1, t *and* t^2 *on* $[0, 1]$ *in* $C([0, 1])$.

- *Use GSO to find an orthonormal basis for* Y. *Use MATLAB for this.*
- *Find the unique* a_2 *and* b_2 *so that*

$$\inf_{\boldsymbol{y}\in Y} \|\sin(t) - a - bt\|_2 \quad = \quad \|\sin(t) - a_2 - b_2 t\|_2$$

- *How does this solution compare to the Taylor Series solution?*
- *What is the orthogonal complement to* Y*?*
- *Find the decomposition* $\sin(t) = \boldsymbol{u} + \boldsymbol{v}$ *with* $\boldsymbol{u} \in Y$ *and* $\boldsymbol{v} \in Y^\perp$.

Exercise 8.2.3 *Let* Y *be the span of the functions* 1 *and* t *on* $[-1, 1]$ *in* $C([-1, 1])$.

- *Use GSO to find an orthonormal basis for* Y. *Use MATLAB for this.*
- *Find the unique* a_2 *and* b_2 *so that*

$$\inf_{\boldsymbol{y}\in Y} \|\sin(t) - a - bt\|_2 \quad = \quad \|\sin(t) - a_2 - b_2 t\|_2$$

- *How does this solution compare to the Taylor Series solution?*
- *What is the orthogonal complement to* Y*?*
- *Find the decomposition* $\sin(t) = \boldsymbol{u} + \boldsymbol{v}$ *with* $\boldsymbol{u} \in Y$ *and* $\boldsymbol{v} \in Y^\perp$.

Exercise 8.2.4 *Let* Y *be the span of the functions* 1, t *and* t^2 *on* $[-2, 1]$ *in* $C([-2, 1])$.

- *Use GSO to find an orthonormal basis for* Y. *Use MATLAB for this.*
- *Find the unique* a_2 *and* b_2 *so that*

$$\inf_{\boldsymbol{y}\in Y} \|\sin(t) - a - bt\|_2 \quad = \quad \|\sin(t) - a_2 - b_2 t\|_2$$

- *How does this solution compare to the Taylor Series solution?*
- *What is the orthogonal complement to* Y*?*
- *Find the decomposition* $\sin(t) = \boldsymbol{u} + \boldsymbol{v}$ *with* $\boldsymbol{u} \in Y$ *and* $\boldsymbol{v} \in Y^\perp$.

Exercise 8.2.5 *Let* Y *be the span of the functions* $\sin(\pi t)$, $\sin(2\pi t)$ *and* $\sin(3\pi t)$ *on* $[0, 1]$ *in* $C([0, 1])$.

- *Use GSO to find an orthonormal basis for* Y. *Use MATLAB for this.*
- *Do you really need GSO here? Look back at the treatment of Fourier Series in (Peterson (18) 2020).*

- *Find the unique constants a_i so that*

$$\inf_{\boldsymbol{y} \in Y} \|2t^2 - y\|_2 = \|t^2 - a_1 \sin(\pi t) - a_2 \sin(2\pi t) - a_3 \sin(3\pi t)\|_2$$

- *What is the orthogonal complement to Y?*
- *Find the decomposition $2t^2 = \boldsymbol{u} + \boldsymbol{v}$ with $\boldsymbol{u} \in Y$ and $\boldsymbol{v} \in Y^\perp$.*

Exercise 8.2.6 *Let Y be the span of the functions 1, $\cos(\pi t)$, $\cos(2\pi t)$ and $\cos(3\pi t)$ on $[0, 1]$ in $C([0, 1])$.*

- *Use GSO to find an orthonormal basis for Y. Use MATLAB for this.*
- *Do you really need GSO here? Look back at the treatment of Fourier Series in (Peterson (18) 2020).*
- *Find the unique constants b_i so that*

$$\inf_{\boldsymbol{y} \in Y} \|2t^2 - y\|_2 = \|t^2 - b_0 - b_1 \cos(\pi t) - b_2 \cos(2\pi t) - b_3 \cos(3\pi t)\|_2$$

- *What is the orthogonal complement to Y?*
- *Find the decomposition $2t^2 = \boldsymbol{u} + \boldsymbol{v}$ with $\boldsymbol{u} \in Y$ and $\boldsymbol{v} \in Y^\perp$.*

It is very helpful to have a way to determine if the norm closure of span of a sequence of linearly independent elements in the Hilbert Space H is the same as H. We can prove this test.

Theorem 8.2.2 Testing When the Norm Closure of the Span of a Set in a Hilbert Space is the Same as the Hilbert Space

Let $(H, < \cdot, \cdot >)$ be a Hilbert Space. For the subset M of H which is non empty,

$$\overline{span(M)} = H \iff M^\perp = \boldsymbol{0}$$

Proof 8.2.2

($\Longrightarrow$):

Let's assume $\overline{span(M)} = H$. This means $span(H)$ is dense in H as every element in H is the limit in norm of a sequence of elements from M. Let $\boldsymbol{x} \in M^\perp$. Since $M^\perp$ is in H, $\boldsymbol{x} \in \overline{span(M)}$. The first possibility is that if $\boldsymbol{x} \in sp(M)$, then there is a finite linear combination of elements from M which represents $\boldsymbol{x}$. Then $\boldsymbol{x} = \alpha_1 \boldsymbol{m_1} + \ldots \alpha_n \boldsymbol{m_n}$ for some positive integer n, elements $\boldsymbol{m_i}$ from M and scalars α_i. Since $\boldsymbol{x} \in M^\perp$, $< \boldsymbol{x}, \boldsymbol{m_i} >= 0$ for all indices i. Then

$$\begin{aligned} < \boldsymbol{x}, \boldsymbol{x} > &= < \alpha_1 \boldsymbol{m_1} + \ldots \alpha_n \boldsymbol{m_n}, \boldsymbol{x} > \\ &= \alpha_1 < \boldsymbol{x}, \boldsymbol{m_1} > \ldots \alpha_n < \boldsymbol{x}, \boldsymbol{m_n} >= 0 \end{aligned}$$

Since $\|\boldsymbol{x}\|^2 = 0$, we have $\boldsymbol{x} = \boldsymbol{0}$.

The second possibility is $\boldsymbol{x} \in \overline{span(M)}$ but not in $span(M)$. Then there is a sequence $\boldsymbol{x} \in span(M)$ so that $\boldsymbol{x_n} \to \boldsymbol{x}$ in norm. Each $\boldsymbol{x_n}$ has a representation $\sum_{i=1}^{p_i} \alpha_{ni} \boldsymbol{m_{ni}}$ with the $\boldsymbol{m_{ni}}$ all in M. So

$$< \boldsymbol{x_n}, \boldsymbol{x} > = \sum_{i=1}^{p_i} \alpha_{ni} < \boldsymbol{m_{ni}}, \boldsymbol{x} >= 0$$

as $\boldsymbol{x} \in M^\perp$. Now by the Cauchy - Schwartz Inequality, we have

$$| < \boldsymbol{x}, \boldsymbol{x} > - < \boldsymbol{x_n}, \boldsymbol{x} > | \quad = \quad | < \boldsymbol{x} - \boldsymbol{x_n}, \boldsymbol{x} > | \leq \|\boldsymbol{x_n} - \boldsymbol{x}\| \, \|\boldsymbol{x}\|$$

and since $\boldsymbol{x_n} \to \boldsymbol{x}$ in norm, this says $< \boldsymbol{x_n}, \boldsymbol{x} > \to < \boldsymbol{x}, \boldsymbol{x} >$. But $< \boldsymbol{x_n}, \boldsymbol{x} >= 0$ always, so $\|\boldsymbol{x}\| = 0$ implying $\boldsymbol{x} = \boldsymbol{0}$.

($\Longleftarrow$):

We assume $M^\perp = \boldsymbol{0}$. Let $\boldsymbol{x} \in \left(\overline{span(M)}\right)^\perp$. In particular, this means $\boldsymbol{x} \in M^\perp$ and so $\boldsymbol{x} = \boldsymbol{0}$. We conclude $\left(\overline{span(M)}\right)^\perp = \boldsymbol{0}$. But then

$$H \quad = \quad \overline{span(M)} \oplus \left(\overline{span(M)}\right)^\perp = \overline{span(M)} \oplus \{\boldsymbol{0}\} = \overline{span(M)}$$

■

Homework

Exercise 8.2.7 *Let $S = \{1, t, \ldots, t^n, \ldots\}$. This is a subset of $C([0,1])$ with the usual inner product $< \cdot, \cdot >$.*

- *For any continuous $\boldsymbol{y}$ on $[0,1]$, prove there is a sequence of polynomials $(\boldsymbol{P_n})$ which converges uniformly to $\boldsymbol{y}$. This uses the Bernoulli approximation theorem from (Peterson (18) 2020).*
- *If $< y, t^i >= 0$ for all i, prove $< \boldsymbol{y}, \boldsymbol{Q} >= 0$ for all polynomials.*
- *Prove $\boldsymbol{P_n}$ converges in $\| \cdot \|_2$.*
- *Now prove $\boldsymbol{y} = \boldsymbol{0}$. This shows the only continuous function perpendicular to all powers of t is the zero function.*

Exercise 8.2.8 *Let $S = \{1, t, \ldots, t^n, \ldots\}$. This is a subset of $C([0,1])$ with the usual inner product $< \cdot, \cdot >$. However, this is not a complete inner product space or Hilbert Space, We know $\mathbb{L}_2([0,1])$ with the induced inner product we obtain by completing $C([0,1])$ or $RI([0,1])$ with the inner product $< \cdot, \cdot >$ is complete. Further both the continuous functions and the Riemann Integrable functions are dense in $\mathbb{L}_2([0,1])$. We are going to redo the previous exercise. We use the notation $< \cdot, \cdot >$ even though we are actually looking at equivalence classes of functions.*

- *Assume $< y, t^i >= 0$ for all i for $\boldsymbol{y} \in \mathbb{L}_2([0,1])$.*
- *For any continuous $\boldsymbol{y}$ on $[0,1]$, like before we can prove there is a sequence of polynomials $(\boldsymbol{P_n})$ which converges uniformly to $\boldsymbol{y}$. This uses the Bernoulli approximation theorem. We also know given any $\epsilon > 0$, there is a continuous function $\boldsymbol{z_\epsilon}$, so that $\|\boldsymbol{y} - \boldsymbol{z_\epsilon}\|_2 < \epsilon$.*
- *Use these facts to prove that $< \boldsymbol{y}, \boldsymbol{y} >= \boldsymbol{0}$ implying $\boldsymbol{y} = \boldsymbol{0}$. This shows the only $\mathbb{L}_2([0,1])$ function perpendicular to all powers of t is the zero function.*

Note how many identifications we are making here between the kinds of functions we are used to in $C([0,1])$ and the notions of equivalence classes of functions. Go back and reread the sections on completing the reals, completing an arbitrary metric space and the completion of $C([a,b])$ and $RI([a,b])$. Also note we do something very similar when we use Lebesgue measure of $\Re$ to build Lebesgue integration and its Hilbert Space of square Lebesgue integrable functions as detailed in (Peterson (16) 2019).

Exercise 8.2.9 *Let A be an $n \times n$ symmetric matrix. So we know it has n mutually orthogonal unit eigenvectors with associated real eigenvalues. These determine the basis $\boldsymbol{E} = \{\boldsymbol{E_1}, \ldots, \boldsymbol{E_n}\}$ and we have $A(\boldsymbol{E_i}) = \lambda_i \boldsymbol{E_i}$ for eigenvalues λ_i. For $\boldsymbol{x} = \sum_{i=1}^{n} x_i \boldsymbol{E_i}$*

- *Define the matrix P_i by $P_i(\boldsymbol{x}) = x_i \boldsymbol{E_i}$. Prove P is a projection operator.*
- *Prove $A(\boldsymbol{x}) = \left(\sum_{i=1}^{n} \lambda_i P_i\right)(\boldsymbol{x})$ and hence $A = \sum_{i=1}^{n} \lambda_i P_i$ is a decomposition of the operator A.*

Exercise 8.2.10 *Let A be an $n \times n$ Hermitian matrix. So we know it has n mutually orthogonal unit complex eigenvectors with associated complex eigenvalues. These determine the basis $\boldsymbol{E} = \{\boldsymbol{E_1}, \ldots, \boldsymbol{E_n}\}$ and we have $A(\boldsymbol{E_i}) = \lambda_i \boldsymbol{E_i}$ for eigenvalues λ_i. For $\boldsymbol{x} = \sum_{i=1}^{n} x_i \boldsymbol{E_i}$*

- *Define the matrix P_i by $P_i(\boldsymbol{x}) = x_i \boldsymbol{E_i}$. Prove P is a projection operator.*
- *Prove $A(\boldsymbol{x}) = \left(\sum_{i=1}^{n} \lambda_i P_i\right)(\boldsymbol{x})$ and hence $A = \sum_{i=1}^{n} \lambda_i P_i$ is a decomposition of the operator A.*

Note there is little difference in the argument here even though we have moved into the complex field.

Sets M satisfying $\overline{span(M)} = H$ are important and we need some notation for this kind of set.

8.3 Orthonormal Sequences

Definition 8.3.1 Orthogonal and Orthonormal Sets in an Inner Product Space

Let $(X, <\cdot,\cdot>)$ be an inner product space.

- *Let E be a subset of X satisfying $\boldsymbol{x} \perp \boldsymbol{y}$ for all $\boldsymbol{x} \neq \boldsymbol{y} \in M$. We say E is an* **orthogonal set** *in X.*
- *If E is an orthogonal set satisfying $\|\boldsymbol{x}\| = 1$ for all $\boldsymbol{x} \in M$, we say E is an* **orthonormal set** *in X.*
- *If E is a countable orthonormal set, we say E is an* **orthonormal sequence** *in X and write*

$$E = \{\boldsymbol{E_n} \,|\, n \geq 1,\ \|\boldsymbol{E_n}\| = 1, <\boldsymbol{E_n}, \boldsymbol{E_n}>= \delta_{nm}\}$$

We can then restate Theorem 8.2.2 as

Theorem 8.3.1 Testing When the Norm Closure of the Span of an Orthonormal Sequence in a Hilbert Space is the Same as the Hilbert Space

Let $(H, <\cdot,\cdot>)$ be a Hilbert Space. For the orthonormal sequence $(\boldsymbol{E_n})$ of H

$$\overline{span(\boldsymbol{E_n})} = H \iff (<\boldsymbol{x}, \boldsymbol{E_n}>= \boldsymbol{0}\ \forall n \Longrightarrow \boldsymbol{x} = \boldsymbol{0})$$

Proof 8.3.1
This is just a restatement using the orthonormal sequence for M. ∎

Orthonormal sequences like this are important and we say they are complete orthonormal sequences.

Definition 8.3.2 Complete Orthonormal Sequences in Hilbert Space

Let H be a Hilbert Space. We say the orthonormal sequence $(\boldsymbol{E_n})$ is complete if $span(\boldsymbol{E_n}) = H$ or equivalently if $< \boldsymbol{x}, \boldsymbol{E_n} >= \boldsymbol{0} \; \forall n \Longrightarrow \boldsymbol{x} = \boldsymbol{0}$.

Homework

Exercise 8.3.1 *Let H be a Hilbert Space and T a bounded linear operator from H to H. Let $\boldsymbol{E} = (\boldsymbol{e_n})$ an orthonormal sequence satisfying $E^{\perp} = \boldsymbol{0}$.*

- *Prove given $\boldsymbol{x} \in H$, there is a sequence $(\boldsymbol{y_n})$ with each $\boldsymbol{y_n} \in span(\boldsymbol{E})$ so that $\boldsymbol{y_n} \to \boldsymbol{x}$ in the norm of H.*
- *Assume also T is a compact operator which means if $\Omega \subset H$ is bounded, then $\overline{T(\Omega)}$ is sequentially compact. Prove $\overline{(T(\boldsymbol{E_i}))}$ is sequentially compact in H.*

Exercise 8.3.2 *This is a redo of the exercise we did earlier. Let $S = \{1, t, \ldots, t^n, \ldots\}$ be a subset of $(C([0,1], < \cdot, \cdot >))$ where the inner product is the usual one which induces $\| \cdot \|_2$.*

- *Prove using GSO there is an orthonormal sequence $G = (\boldsymbol{g_n})$ whose span is the same as the span of S. Thus, $(\boldsymbol{g_n})$ is an orthonormal sequence in the Hilbert Space $\mathbb{L}_2([0,1])$.*
- *In the exercise before, we showed the only $\mathbb{L}_2([0,1])$ function perpendicular to all powers of t is the zero function. Prove this means G is a complete orthonormal sequence.*

Exercise 8.3.3 *Let $S = \left\{\cos\left(\frac{n\pi x}{L}\right)\right\}_{0 \leq n < \infty}$.*

- *Revisit the arguments in (Peterson (18) 2020) in the discussion of Fourier series and prove S is an orthogonal sequence in $(C([0,1], < \cdot, \cdot >))$ where the inner product is the usual one which induces $\| \cdot \|_2$.*
- *Calculate the norm of each function in S and use that to find an orthonormal sequence in $\hat{S}$ in $(C([0,1],, < \cdot, \cdot >))$.*
- *Explain why GSO is not needed here.*
- *Explain why if we knew $< \boldsymbol{y}, \cos\left(\frac{n\pi x}{L}\right) >= 0$ for all $n \geq 0$ implies the only $\mathbb{L}_2([0,1])$ function perpendicular to all the functions in $\hat{S}$ is the zero function tells us $\hat{S}$ is a complete orthonormal sequence. This is hard to do and we will do this step later.*

Exercise 8.3.4 *Let $S = \left\{\sin\left(\frac{n\pi x}{L}\right)\right\}_{1 \leq n < \infty}$.*

- *Revisit the arguments in (Peterson (18) 2020) in the discussion of Fourier series and prove S is an orthogonal sequence in $(C([0,1], < \cdot, \cdot >))$ where the inner product is the usual one which induces $\| \cdot \|_2$.*
- *Calculate the norm of each function in S and use that to find an orthonormal sequence in $\hat{S}$ in $(C([0,1], < \cdot, \cdot >))$.*
- *Explain why GSO is not needed here.*
- *Explain why if we knew $< \boldsymbol{y}, \sin\left(\frac{n\pi x}{L}\right) >= 0$ for all $n \geq 0$ implies the only $\mathbb{L}_2([0,1])$ function perpendicular to all the functions in $\hat{S}$ is the zero function tells us $\hat{S}$ is a complete orthonormal sequence. Again, this is hard to do and we will do this step later.*

Exercise 8.3.5 *Let X be an inner product space.*

1. *Prove that* $\|x - y\| + \|y - z\| = \|x - z\|$ *iff* y *is a convex combination of* x *and* y.

2. *Prove that Apollonius' identity*

$$\|z - x\|^2 + \|z - y\|^2 = \frac{1}{2}\|x - y\|^2 + 2\|z - \frac{x+y}{2}\|^2$$

3. *Prove that the norm in an inner product space is strictly convex.*

Exercise 8.3.6 *Let* X *be the inner product space* $C([-1,1])$ *with the standard integral inner product. Let* $S = \{1, t, t^2\}$.

1. *Prove* S *is a linearly independent set in* X.

2. *Use GSO to compute the orthonormal basis* E *from* S.

3. *Plot the resulting o.n. basis in a common plot on* $[-1, 1]$.

4. *Let* $x(t) = \exp^{-4(t-0.5)^2}$. *Let* V *be the span of* S. *Solve* $\inf_{u \in V} \|x - u\|$ *for* y *in* V.

5. *Plot* x *and* y *on the same plot on* $[-1, 1]$.

Exercise 8.3.7 *Define the Legendre polynomials* P_n *on* $[-1, 1]$ *recursively by* $P_0(t) = 1$ *and for* $n > 0$ *by*

$$\begin{aligned} p_n(t) &= (t^2 - 1)^n \\ P_n(t) &= \frac{1}{2^n n!} \frac{d^n}{dt^n} p_n(t) \end{aligned}$$

We now show that $\{P_n\}$ *can be made into an orthonormal sequence.*

1. *Show* $\{P_n\}$ *is an orthogonal sequence.*

   (a) *Show for* $m < n$ *by recursion:*

$$\int_{-1}^{1} P_n(t) t^m dt = 0$$

   Hint: $\frac{d^k}{dt^k} p_n(t) = 0$ *at* $t = \pm 1$ *for* $0 \leq k < n$, *so we can use integration by parts repeatedly.*

   (b) *Show for* $n \neq m$,

$$\int_{-1}^{1} P_n(t) P_m dt = 0$$

2. *Show* $\|P_n\|_2^2$ *is* $\frac{2}{2n+1}$ *for all* n. *Hint: using integration by parts repeatedly, you can evaluate*

$$\begin{aligned} \int_{-1}^{1} (1 - t^2)^n dt &= \int_{-1}^{1} (1 - t)^n (1 + t)^n dt \\ &= \frac{n}{n+1} \int_{-1}^{1} (1 - t)^{n-1} (1 + t)^{n+1} dt \\ &\vdots \\ &= \frac{n!}{(n+1)(n+2)\cdots(2n)} \int_{-1}^{1} (1 + t)^{2n} dt \end{aligned}$$

and

$$\int_{-1}^{1} \left(\frac{d^n}{dt^n} p_n(t) \right)^2 dt \;=\; (2n)! \int_{-1}^{1} (1-t)^n \, (1+t)^n dt$$

3. *Find an orthonormal sequence $\{Q_n\}$ based on $\{P_n\}$.*
4. *Plot the first ten functions in this resulting o.n. sequence in a common plot on $[-1,1]$.*
5. *Compare the terms obtained in the Graham - Schmidt orthogonalization process in the previous with the first three terms of $\{Q_n\}$. What do you notice?*

We can now state the best approximation to a given $\boldsymbol{x}$ to a finite dimensional subspace.

Theorem 8.3.2 Finite Dimensional Approximation in an Inner Product Space

Let $(X, <\cdot,\cdot>)$ be an inner product space and let $(\boldsymbol{E_n})$ be a orthonormal sequence. Let $Y_n = span(\boldsymbol{E_1}, \ldots, \boldsymbol{E_n})$. Then given $\boldsymbol{x} \in X$, there is a unique $\boldsymbol{y_x} \in Y_n$ so that

- $(\boldsymbol{x} - \boldsymbol{y_x}) \perp Y_n$
- $\inf_{\boldsymbol{u} \in Y_n} \|\boldsymbol{x} - \boldsymbol{u}\| = \|\boldsymbol{x} - \boldsymbol{y_x}\|$
- $\boldsymbol{y_x} = \sum_{i=1}^n <\boldsymbol{x}, \boldsymbol{E_i}> \boldsymbol{E_i}$

Proof 8.3.2
Since $Y_n = span(\boldsymbol{E_1}, \ldots, \boldsymbol{E_n})$ is a finite dimensional subspace of X, we know Y_n is closed as a subspace and complete with respect to the norm. Hence, there is a unique $\boldsymbol{y_x} \in Y_n$ and $\boldsymbol{z_x} \in Y_n^\perp$ so that $\boldsymbol{x} = \boldsymbol{y_x} + \boldsymbol{z_x}$ and $\inf_{\boldsymbol{u} \in Y_n} \|\boldsymbol{x} - \boldsymbol{u}\| = \|\boldsymbol{x} - \boldsymbol{y_x}\|$. Let $\boldsymbol{y_0} = \sum_{i=1}^n <\boldsymbol{x}, \boldsymbol{E_i}> \boldsymbol{E_i} \in Y_n$ and let $\boldsymbol{z_0} = \boldsymbol{x} - \boldsymbol{y_0}$. Any $\boldsymbol{u} \in Y_n$ has representation $\boldsymbol{u} = \sum_{j=1}^n \beta_j \boldsymbol{E_j}$. Thus, letting $\alpha_i =< \boldsymbol{x}, \boldsymbol{E_i} >$, we have

$$\begin{aligned}
< \boldsymbol{z_0}, \boldsymbol{u} > &= < \boldsymbol{x} - \boldsymbol{y_0}, \boldsymbol{u} >= \left\langle \boldsymbol{x} - \sum_{i=1}^n \alpha_i \boldsymbol{E_i}, \sum_{j=1}^n \beta_j \boldsymbol{E_j} \right\rangle \\
&= \sum_{j=1}^n \overline{\beta_j} < \boldsymbol{x}, \boldsymbol{E_j} > - \sum_{i=1}^n \alpha_i \sum_{j=1}^n \overline{\beta_j} < \boldsymbol{E_i}, \boldsymbol{E_j} > \\
&= \sum_{j=1}^n \overline{\beta_j} < \boldsymbol{x}, \boldsymbol{E_j} > - \sum_{i=1}^n \alpha_i \sum_{j=1}^n \overline{\beta_j} \delta_{ij} \\
&= \sum_{j=1}^n \overline{\beta_j} \alpha_i - \sum_{=1}^n \alpha_j \, \overline{\beta_j} = 0
\end{aligned}$$

Thus, $\boldsymbol{z_0} \perp Y_n$ and $y_0 \in Y_n$. This gives two representations of $\boldsymbol{x}$: $\boldsymbol{x} = \boldsymbol{y_0} + \boldsymbol{z_0} = \boldsymbol{y_x} + \boldsymbol{z_x}$. But the representation is unique: thus $\boldsymbol{y_x} = \boldsymbol{y_0} = \sum_{i=1}^n < \boldsymbol{x}, \boldsymbol{E_i} > \boldsymbol{E_i}$ and $\boldsymbol{z_0} = \boldsymbol{z_x}$. We conclude $\boldsymbol{x} = \sum_{i=1}^n < \boldsymbol{x}, \boldsymbol{E_i} > \boldsymbol{E_i} + \boldsymbol{z_x}$ where $\boldsymbol{z_x} \in Y_n^\perp$ with

$$\inf_{\boldsymbol{u} \in Y_n} \|\boldsymbol{x} - \boldsymbol{u}\| = \left\| \boldsymbol{x} - \sum_{i=1}^n < \boldsymbol{x}, \boldsymbol{E_i} > \boldsymbol{E_i} \right\|$$

■

Homework

Exercise 8.3.8 *Let Y be the span of the functions* 1 *and* t *on* $[0,1]$ *in* $C([0,1])$.

- *Use GSO to find an orthonormal basis for Y. Use MATLAB for this.*
- *Find the projections to Y and $Y^{\perp}$.*

Exercise 8.3.9 *Let Y be the span of the functions* 1, t *and* t^2 *on* $[0,1]$ *in* $C([0,1])$.

- *Use GSO to find an orthonormal basis for Y. Use MATLAB for this.*
- *Find the projections to Y and $Y^{\perp}$.*

Exercise 8.3.10 *Let Y be the span of the functions* 1 *and* t *on* $[-1,1]$ *in* $C([-1,1])$.

- *Use GSO to find an orthonormal basis for Y. Use MATLAB for this.*

Exercise 8.3.11 *Let Y be the span of the functions* 1, t *and* t^2 *on* $[-2,1]$ *in* $C([-2,1])$.

- *Use GSO to find an orthonormal basis for Y. Use MATLAB for this.*
- *Find the projections to Y and $Y^{\perp}$.*

Exercise 8.3.12 *Let Y be the span of the functions* $\sin(\pi t)$, $\sin(2\pi t)$ *and* $\sin(3\pi t)$ *on* $[0,1]$ *in* $C([0,1])$.

- *Use GSO to find an orthonormal basis for Y. Use MATLAB for this.*
- *Find the projections to Y and $Y^{\perp}$.*
- *Do you really need GSO here? Look back at the treatment of Fourier Series in (Peterson (18) 2020).*

Exercise 8.3.13 *Let Y be the span of the functions* 1, $\cos(\pi t)$, $\cos(2\pi t)$ *and* $\cos(3\pi t)$ *on* $[0,1]$ *in* $C([0,1])$.

- *Use GSO to find an orthonormal basis for Y. Use MATLAB for this.*
- *Find the projections to Y and $Y^{\perp}$.*
- *Do you really need GSO here? Look back at the treatment of Fourier Series in (Peterson (18) 2020).*

8.4 Fourier Coefficients and Completeness

For a given orthonormal sequence $(\boldsymbol{E_n})$ on the interval $[a,b]$, the inner products $<\boldsymbol{x},\boldsymbol{E_n}>$ for each $\boldsymbol{X}$ give a sequence $(<\boldsymbol{x},\boldsymbol{E_n}>)$. The inner products are called the **Fourier Coefficients** of $\boldsymbol{x}$ with respect to the orthonormal sequence $(\boldsymbol{E_n})$. For convenience, denote these coefficients by $\mathscr{F}_n((\boldsymbol{E_n}),\boldsymbol{x},[a,b])$. We can prove these coefficients are in ℓ^2 and their sum has $\|\boldsymbol{x}\|^2$ as an upper bound.

Theorem 8.4.1 Bessel's Inequality

Let $(\boldsymbol{E_n})_{n=1}^{\infty}$ be an orthonormal sequence in the inner product space $(X,<\cdot,\cdot>)$. Then given $\boldsymbol{x}\in X$, $(<\boldsymbol{x},\boldsymbol{E_n}>)_{n=1}^{\infty}\in\ell^2$ and $\sum_{n=1}^{\infty}|<\boldsymbol{x},\boldsymbol{E_n}>|^2\leq\|\boldsymbol{x}\|^2$.

Proof 8.4.1
From the previous theorem, for $Y_n = span(\boldsymbol{E_1}, \ldots, \boldsymbol{E_n})$, we have the decomposition

$$\boldsymbol{x} = \sum_{i=1}^{n} < \boldsymbol{x}, \boldsymbol{E_i} > \boldsymbol{E_i} + \boldsymbol{z_n}$$

where $\boldsymbol{z_n} \in Y_n^\perp$. Further,

$$\begin{aligned} \|\boldsymbol{x}\|^2 &= \left\langle \sum_{i=1}^{n} < \boldsymbol{x}, \boldsymbol{E_i} > \boldsymbol{E_i} + \boldsymbol{z_n}, \sum_{j=1}^{n} < \boldsymbol{x}, \boldsymbol{E_j} > \boldsymbol{E_j} + \boldsymbol{z_n} \right\rangle \\ &= \sum_{i=1}^{n} < \boldsymbol{x}, \boldsymbol{E_i} > \sum_{j=1}^{n} \overline{< \boldsymbol{x}, \boldsymbol{E_j} >} < \boldsymbol{E_i}, \boldsymbol{E_j} > + \sum_{i=1}^{n} < \boldsymbol{x}, \boldsymbol{E_i} >< \boldsymbol{E_i}, \boldsymbol{z_n} > \\ &+ \sum_{j=1}^{n} \overline{< \boldsymbol{x}, \boldsymbol{E_j} >} < \boldsymbol{z_n}, \boldsymbol{E_j} > + \|\boldsymbol{z_n}\|^2 \end{aligned}$$

But $\boldsymbol{z_n} \perp Y_n$ and $(\boldsymbol{E_n})$ is orthonormal. So

$$\|\boldsymbol{x}\|^2 = \sum_{i=1}^{n} | < \boldsymbol{x}, \boldsymbol{E_i} > |^2 + \|\boldsymbol{z_n}\|^2 \geq \sum_{i=1}^{n} | < \boldsymbol{x}, \boldsymbol{E_i} > |^2$$

This tells us the sequence of increasing partial sums $\sum_{i=1}^{n} | < \boldsymbol{x}, \boldsymbol{E_i} > |^2$ is bounded above and thus converges. Thus the sequence of Fourier coefficients of $\boldsymbol{x}$ is in ℓ^2 and its sum is bounded above by $\|\boldsymbol{x}\|^2$. ■

Comment 8.4.1 *The result above tells us the Fourier coefficients of $\boldsymbol{x}$ always give a convergent ℓ^2 series. Hence, $\sum_{n=1}^{\infty} | < \boldsymbol{x}, \boldsymbol{E_n} > |^2$ always exists. However, we* **do not know** *$\boldsymbol{x}$ is the same as the expansion $\sum_{n=1}^{\infty} | < \boldsymbol{x}, \boldsymbol{E_n} > \boldsymbol{E_n}$ in any sense.*

Homework

Exercise 8.4.1 *We know the sequence $(\boldsymbol{x_n})_{n\geq 1}$ where $x_n(t) = \frac{2}{L} \sin\left(\frac{n\pi x}{L}\right)$ for $L > 0$ is orthonormal sequence in the inner product space $(C([0, L]), < \cdot, \cdot >)$ with the inner product $< f, g >= \int_0^L f(t)g(t)dt$ for any $\boldsymbol{f}$ and $\boldsymbol{g}$ in $C([0, L])$. Let h be defined by*

$$h(t) = \begin{cases} H, & 0 \leq t \leq \frac{L}{2} \\ 0, & \frac{L}{2} < t \leq L \end{cases}$$

Find the Fourier coefficients of **h** *for this orthonormal sequence.*

Exercise 8.4.2 *We know the sequence $(\boldsymbol{y_n})_{n\geq 0}$ where $y_n(t) = \frac{2}{L} \cos\left(\frac{n\pi x}{L}\right)$ for $n \geq 1$ and $y_0(t) = \frac{1}{L} \cos\left(\frac{n\pi x}{L}\right)$ for $L > 0$ is orthonormal sequence in the inner product space $(C([0, L]), < \cdot, \cdot >)$ with the inner product $< f, g >= \int_0^L f(t)g(t)dt$ for any $\boldsymbol{f}$ and $\boldsymbol{g}$ in $C([0, L])$. Let h be defined by*

$$h(t) = \begin{cases} H, & 0 \leq t \leq \frac{L}{2} \\ 0, & \frac{L}{2} < t \leq L \end{cases}$$

Find the Fourier coefficients of **h** *for this orthonormal sequence.*

Exercise 8.4.3 *Let $E = \{\boldsymbol{E_1}, \ldots, \boldsymbol{E_n}\}$ be any orthonormal basis for $\Re^n$. Let $\boldsymbol{f}$ be a continuous vector function $f : [0, L] \to \Re^n$.*

- *Prove* $f(t) = \sum_{i=1}^{n} < f(t), \boldsymbol{E_i} > \boldsymbol{E_i}$ *for all* t *in* $[0, L]$.
- *Prove* $\|\boldsymbol{f}\|_2 = \sum_{i=1}^{n} (< f(t), \boldsymbol{E_i} >)^2$.

Exercise 8.4.4 *Let's use the same notation as in the first two exercises giving us two orthonormal sequences* $(\boldsymbol{x_n})_{n\geq 1}$ *and* $(\boldsymbol{y_n})_{n\geq 0}$ *in the inner product space* $(C([0,L]), < \cdot, \cdot >)$ *with the inner product* $< f, g >= \int_0^L f(t)g(t)dt$ *for any* $\boldsymbol{f}$ *and* $\boldsymbol{g}$ *in* $C([0,L])$. *Assume* $\boldsymbol{h}$ *has a continuous derivative on* $[0, L]$.

- *Prove for* $n \geq 1$

$$\mathscr{F}_n((\boldsymbol{x_n}), \boldsymbol{h'}, [0,L]) = \sqrt{\frac{2}{L}} \frac{L}{n\pi} (-(-1)^n h(L) + h(0)) + \frac{L}{n\pi} \mathscr{F}_n((\boldsymbol{y_n}), \boldsymbol{h}, [0,L])$$

- *Prove for* $n \geq 1$

$$\mathscr{F}_n((\boldsymbol{y_n}), \boldsymbol{h'}, [0,L]) = \frac{L}{n\pi} \mathscr{F}_n(\boldsymbol{x_n}, \boldsymbol{h}, [0,L])$$

- *Prove for* $n = 0$

$$\mathscr{F}_0((\boldsymbol{y_n}), \boldsymbol{h'}, [0,L]) = \sqrt{\frac{1}{L}} (h(L) - h(0))$$

Exercise 8.4.5 *Let's use the same notation as before giving us two orthonormal sequences* $(\boldsymbol{x_n})_{n\geq 1}$ *and* $(\boldsymbol{y_n})_{n\geq 0}$ *in the inner product space* $(C([0,L]), < \cdot, \cdot >)$ *with the inner product* $< f, g >= \int_0^L f(t)g(t)dt$ *for any* $\boldsymbol{f}$ *and* $\boldsymbol{g}$ *in* $C([0,L])$. *Assume* $\boldsymbol{h}$ *has a continuous derivative on* $[0, L]$ *and now assume* $h(0) = h(L)$.

- *Prove for* $n \geq 1$

$$\mathscr{F}_n((\boldsymbol{x_n}), \boldsymbol{h'}, [0,L]) = \sqrt{\frac{2}{L}} \frac{L}{n\pi} ((1 - (-1)^n) h(0)) + \frac{L}{n\pi} \mathscr{F}_n((\boldsymbol{y_n}), \boldsymbol{h}, [0,L])$$

  and for $N \geq 1$,

$$\sum_{n=1}^{N} \left(\sqrt{\frac{2}{L}} \frac{L}{n\pi} (1 \quad (\ 1)^n) h(0) + \frac{L}{n\pi} \mathscr{F}_n((\boldsymbol{y_n}), \boldsymbol{h}, [0,L]) \right)^2 \leq \|\boldsymbol{h'}\|_2^2$$

- *Prove for* $n = 0$

$$\mathscr{F}_0((\boldsymbol{y_n}), \boldsymbol{h'}, [0,L]) = 0$$

- *Prove for* $n \geq 1$

$$\mathscr{F}_n((\boldsymbol{y_n}), \boldsymbol{h'}, [0,L]) = \frac{L}{n\pi} \mathscr{F}(\boldsymbol{x_n}, \boldsymbol{h'}, [0,L])$$

  and for $N \geq 0$,

$$\sum_{n=1}^{N} \left(\frac{L}{n\pi} \mathscr{F}_n((\boldsymbol{x_n}), \boldsymbol{h}, [0,L]) \right)^2 \leq \|\boldsymbol{h'}\|_2^2$$

Exercise 8.4.6 *Now let's double the interval length to* $[0, 2L]$. *Use the sequence* $(\boldsymbol{x_n})_{n\geq 1}$ *where* $x_n(t) = \frac{1}{L}\sin\left(\frac{n\pi x}{L}\right)$ *and the sequence* $(\boldsymbol{y_n})_{n\geq 0}$ *where* $y_n(t) = \frac{1}{L}\cos\left(\frac{n\pi x}{L}\right)$ *for* $n \geq 1$ *and* $y_0(t) = \frac{1}{2L}\cos\left(\frac{n\pi x}{L}\right)$ *This gives two orthonormal sequences in the inner product space* $(C([0, 2L]), <\cdot,\cdot>)$ *(note the doubled interval length) with the inner product* $< f, g >= \int_0^{2L} f(t)g(t)dt$ *for any* $\boldsymbol{f}$ *and* $\boldsymbol{g}$ *in* $C([0, 2L])$. *Assume* $\boldsymbol{h}$ *has a continuous derivative on* $[0, 2L]$ *and now assume* $h(0) = h(2L)$.

- *Prove for* $n \geq 1$

$$\mathscr{F}_n((\boldsymbol{x_n}), \boldsymbol{h'}, [0, 2L]) = \frac{L}{n\pi}\mathscr{F}_n((\boldsymbol{y_n}), \boldsymbol{h}, [0, 2L])$$

and for $N \geq 1$,

$$\sum_{n=1}^{N}\left(\frac{L}{n\pi}\mathscr{F}_n((\boldsymbol{y_n}), \boldsymbol{h}, [0, 2L])\right)^2 \leq \|\boldsymbol{h'}\|_2^2$$

- *Prove for* $n = 0$

$$\mathscr{F}_0((\boldsymbol{y_n}), \boldsymbol{h'}, [0, 2L]) = 0$$

- *Prove for* $n \geq 1$

$$\mathscr{F}_n((\boldsymbol{y_n}), \boldsymbol{h'}, [0, L]) = \frac{L}{n\pi}\mathscr{F}(\boldsymbol{x_n}, \boldsymbol{h'}, [0, 2L])$$

and for $N \geq 0$,

$$\sum_{n=1}^{N}\left(\frac{L}{n\pi}\mathscr{F}_n((\boldsymbol{x_n}), \boldsymbol{h}, [0, 2L])\right)^2 \leq \|\boldsymbol{h'}\|_2^2$$

Let's examine the behavior of orthonormal sequences in a Hilbert Space next.

Theorem 8.4.2 The Hilbert Space Convergence Theorem

Let $(H, <\cdot,\cdot>)$ *be a Hilbert Space and* $(\boldsymbol{E_n})_{n=1}^{\infty}$ *an orthonormal sequence.*

- *The series* $\sum_{n=1}^{\infty}\alpha_n\boldsymbol{E_n}$ *converges if and only if* $(\alpha_n)_{n=1}^{\infty}$ *is in* ℓ^2.
- *If* $\sum_{n=1}^{\infty}\alpha_n\boldsymbol{E_n}$ *converges to* $\boldsymbol{y} \in H$, *then* $\alpha_n =< \boldsymbol{y}, \boldsymbol{E_n} >$.
- *If* $\boldsymbol{x} \in H$, *then* $\sum_{n=1}^{\infty} < \boldsymbol{x}, \boldsymbol{E_n} > \boldsymbol{E_n}$ *converges to some* $\boldsymbol{y} \in H$.

Proof 8.4.2
First, recall a series of non-negative terms, $\sum_{n=1}^{\infty} a_n$ *converges if and only if it is a Cauchy sequence. Let* $\boldsymbol{S_n} = \sum_{i=1}^{n}\alpha_i\boldsymbol{E_i}$. *Then for* $n > m$, $\boldsymbol{S_n} - \boldsymbol{S_m} = \sum_{i=m+1}^{n}\alpha_i\boldsymbol{E_i}$. *Then, using the orthonormality of* $(\boldsymbol{E_n})$, *we have*

$$\|\boldsymbol{S_n} - \boldsymbol{S_m}\|^2 = \left\langle \sum_{i=m+1}^{n}\alpha_i\boldsymbol{E_i}, \sum_{j=m+1}^{n}\alpha_j\boldsymbol{E_j}\right\rangle = \sum_{i=m+1}^{n}\alpha_i|^2$$

If we know $\boldsymbol{S_n}$ *converges to* $\boldsymbol{S}$ *for some element* $\boldsymbol{S}$ *in* H, *then* $(\boldsymbol{S_n})$ *is a Cauchy sequence in the complete inner product space* H. *Hence, given* $\epsilon > 0$, *there is* N *so that* $\|\boldsymbol{S_n} - \boldsymbol{S_m}\|^2 < \epsilon$ *if*

$n > m > N$. This implies $\sum_{i=m+1}^{n} \alpha_i|^2 < \epsilon$ for $n > m > N$. Thus we see (α_n) is a Cauchy sequence and therefore converges. This shows (α_n) is in ℓ^2.

Conversely, if (α_n) is in ℓ^2, then we start with the partial sums $\sigma_n = \sum_{i=1}^{n} |\alpha_i|^2$ being a Cauchy sequence. Thus, given $\epsilon > 0$, there is N so that $n > m > N$ implies (α_n)

$$\sigma_n - \sigma_m = \sum_{i=m+1}^{n} |\alpha_i|^2 \quad < \quad \epsilon$$

But then $\|\boldsymbol{S_n} - \boldsymbol{S_m}\|^2 < \epsilon$ when $n > m > N$ which tells us $(\boldsymbol{S_n})$ is a Cauchy sequence in H and so must converge.

Next, if $\sum_{n=1}^{\infty} \alpha_n \boldsymbol{E_n}$ converges to $\boldsymbol{y} \in H$, let $\boldsymbol{S_n} = \sum_{i=1}^{n} \alpha_i \boldsymbol{E_i}$. Then

$$< \boldsymbol{S_m}, \boldsymbol{E_j} > \quad = \quad \left\langle \sum_{i=1}^{n} \alpha_i \boldsymbol{E_i}, \boldsymbol{E_j} > \right\rangle = \alpha_j$$

and so $\lim_{n\to\infty} < \boldsymbol{S_n}, \boldsymbol{E_j} >= \alpha_j$. This is the same as saying $\lim_{n\to\infty} < \boldsymbol{S_n}, \boldsymbol{E_j} >= \alpha_j$. Now as a side calculation,

$$| < \boldsymbol{y} - \boldsymbol{S_n}, \boldsymbol{E_j} > | \quad \leq \quad \|\boldsymbol{y} - \boldsymbol{E_n}\| \, \|\boldsymbol{E_j}\| = \|\boldsymbol{y} - \boldsymbol{E_n}\|$$

and so $< \boldsymbol{S_n}, \boldsymbol{E_j} > \to < \boldsymbol{y}, \boldsymbol{E_j} >$. Thus, we have $< \boldsymbol{y}, \boldsymbol{E_j} >= \alpha_j$ as desired.

Finally, if $\boldsymbol{x} \in H$, Bessel's Inequality tells us $\sum_{n=1}^{\infty} | < \boldsymbol{x}, \boldsymbol{E_n} > |^2 \leq \|\boldsymbol{x}\|^2$. Thus $(< \boldsymbol{x}, \boldsymbol{E_n} >)$ is in ℓ^2 which implies the partial sums $\sum_{i=1}^{n} < \boldsymbol{x}, \boldsymbol{E_i} > \boldsymbol{E_i}$ converge to some $\boldsymbol{y} \in H$. ■

We need to know more about the behavior of the orthonormal sequence in H in order to have $\boldsymbol{x} = \sum_{n=1}^{\infty} < \boldsymbol{x}, \boldsymbol{E_n} > \boldsymbol{E_n}$.

Theorem 8.4.3 When is an Orthonormal Sequence a Schauder Basis?

Let $(H, < \cdot, \cdot >)$ be a Hilbert Space and $(\boldsymbol{E_n})_{n=1}^{\infty}$ an orthonormal sequence satisfying $\boldsymbol{x} \perp \boldsymbol{E_n} >$ for all n implies $\boldsymbol{x} = 0$; that is the orthonormal sequence is complete. Then $\boldsymbol{x} = \sum_{n=1}^{\infty} < \boldsymbol{x}, \boldsymbol{E_n} > \boldsymbol{E_n}$. Further $(\boldsymbol{E_n})_{n=1}^{\infty}$ is a Schauder Basis for H and $H = (\boldsymbol{E_n})_{n=1}^{\infty}$.

Proof 8.4.3

Let $Y_n = span(\boldsymbol{E_1}, \ldots, \boldsymbol{E_n})$. Then given $\boldsymbol{x} \in H$, $\boldsymbol{x} = \sum_{i=1}^{n} < \boldsymbol{x}, \boldsymbol{E_i} > \boldsymbol{E_i} + \boldsymbol{z_n}$ for $\boldsymbol{z_n} \in Y_n^{\perp}$. By the convergence theorem for Hilbert Space, there is $\boldsymbol{y} \in H$ so that $\boldsymbol{y} = \lim_{n\to\infty} \sum_{i=1}^{n} < \boldsymbol{x}, \boldsymbol{E_i} > \boldsymbol{E_i}$. Since $\boldsymbol{z_n} = \boldsymbol{x} - \sum_{i=1}^{n} < \boldsymbol{x}, \boldsymbol{E_i} > \boldsymbol{E_i}$, $\lim_{n\to\infty} \boldsymbol{z_n}$ exists and equals $\boldsymbol{x} - \boldsymbol{y}$.

We also know

$$< \boldsymbol{z_n}, \boldsymbol{E_j} > \quad = \quad \left\langle \boldsymbol{x} - \sum_{i=1}^{n} < \boldsymbol{x}, \boldsymbol{E_i} > \boldsymbol{E_i}, \boldsymbol{E_j} \right\rangle =< \boldsymbol{x}, \boldsymbol{E_j} > - < \boldsymbol{x}, \boldsymbol{E_j} >= 0$$

$$< \boldsymbol{z} - \boldsymbol{z_n}, \boldsymbol{E_j} > \quad \leq \quad \|\boldsymbol{z} - \boldsymbol{z_n}\| \, \|\boldsymbol{E_j}\| = \|\boldsymbol{z} - \boldsymbol{z_n}\|$$

Since $z_n \to Z$*, this implies* $0 = \lim_{n\to\infty} < z_n, E_j > = < z, E_j >$*. Thus, by assumption* $z = x - y = 0$*. Of course, this also says* $H = \overline{(E_n)_{n=1}^{\infty}}$*. This then says*

$$x = \lim_{n\to\infty} \left(\sum_{i=1}^{n} < x, E_i > E_i + z_n \right) = y + z = y$$

Hence, $x = \sum_{n=1}^{\infty} < x, E_n > E_n$ *showing* x *has a unique expansion in terms of the orthonormal basis. We have shown* $(E_n)_{n=1}^{\infty}$ *is a Schauder Basis for* H. ■

Homework

Exercise 8.4.7 *Use Bessel's Inequality to show that if* (E_n) *is an orthonormal sequence in the inner product space* $C([a,b])$ *with the usual inner product* $< \cdot, \cdot >$*, then for any* $h \in \mathbb{L}_2([a,b])$*,* $\mathscr{F}_n((E_n), h, [a,b]) \to 0$*.*

Exercise 8.4.8 *From the last sections exercises, we know the sequence* $(x_n)_{n\geq 1}$ *where* $x_n(t) = \frac{2}{L}\sin\left(\frac{n\pi x}{L}\right)$ *and the sequence* $(y_n)_{n\geq 0}$ *where* $y_n(t) = \frac{2}{L}\cos\left(\frac{n\pi x}{L}\right)$ *for* $n \geq 1$ *and* $y_0(t) = \frac{1}{L}\cos\left(\frac{n\pi x}{L}\right)$ *for* $L > 0$ *are orthonormal sequences in the inner product space* $(C([0,L]), < \cdot, \cdot >)$ *with the inner product* $< f, g > = \int_0^L f(t)g(t)dt$ *for any* f *and* g *in* $C([0,L])$*. Note the following:*

- *The function* x_n *satisfies the ODE* $u'' + (\,)^2 u = 0$ *with* $u(0) = u(L) = 0$ *for all* $n \geq 1$*.*
- *The function* y_n *satisfies the ODE* $u'' + (\,)^2 u = 0$ *with* $u'(0) = u'(L) = 0$ *for all* $n \geq 1$*.*

Prove x_n *is orthogonal to* y_n *on* $[0,L]$ *two ways:*

- *Direct integration of* $\int_0^L \sin\left(\frac{n\pi t}{L}\right)\cos\left(\frac{n\pi t}{L}\right) dt$*.*
- *Integration by parts of* $\int_0^L x_n(t)x_n'(t)dt$*.*

Hence these two orthonormal sequences are special as they are related by an ODE with different boundary conditions.

Exercise 8.4.9 *Again, we know the sequence* $(x_n)_{n\geq 1}$ *where* $x_n(t) = \frac{2}{L}\sin\left(\frac{n\pi x}{L}\right)$ *and the sequence* $(y_n)_{n\geq 0}$ *where* $y_n(t) = \frac{2}{L}\cos\left(\frac{n\pi x}{L}\right)$ *for* $n \geq 1$ *and* $y_0(t) = \frac{1}{L}\cos\left(\frac{n\pi x}{L}\right)$ *for* $L > 0$ *are orthonormal sequences in the inner product space* $(C([0,L]), < \cdot, \cdot >)$ *with the inner product* $< f, g > = \int_0^L f(t)g(t)dt$ *for any* f *and* g *in* $C([0,L])$*. Assume* u *is a twice continuously differentiable function that satisfies* $< u, x_n > = 0$ *for all* $n \geq 1$*.*

- *Use integration by parts twice to show*

$$\int_0^L \left(u''(t) + \left(\frac{n\pi}{L}\right)^2 u(t) \right) x_n(t)dt = \frac{L}{n\pi}\left(u'(L)(-1)^n - u'(0)\right)$$

- *Then show* $u'(0) = u'(L) = 0$*.*

Exercise 8.4.10 *In* $\mathbb{L}_2([a,b])$ *with the induced inner product we have discussed, prove that if the orthonormal sequence* (E_n) *is complete, then the only function in* $\mathbb{L}_2([a,b])$ *with all zero Fourier coefficients in the zero function.*

Exercise 8.4.11 *In* $\mathbb{L}_2([a,b])$ *with the induced inner product we have discussed, prove that if the orthonormal sequence* (E_n) *is complete, then it is a Schauder basis.*

Exercise 8.4.12 *On the interval $[0, 1]$, define the Rademacher $R_m(t)$ functions for $m \geq 0$ as follows:*

$$R_m(t) = sgn(\sin(2^m \pi t))$$

where the signum function sgn is defined by

$$sgn(t) = \begin{cases} 1, & x > 0 \\ 0, & x = 0 \\ -1, & x < 0 \end{cases}$$

1. *Show the Rademacher functions form an orthogonal system in $C([0, 1])$ with the usual integral inner product. Hint, to show $\int_0^1 R_m(t)R_n(t)dt$ is zero if $n \neq m$, note that $\int_a^b R_m(t)dt = 0$ when $2^m(b - a)$ is even. Thus, if $m > n \geq 0$,*

$$\begin{aligned} \int_0^1 R_m(t)R_n(t)dt &= \sum_{k=1}^{2^n} \int_{\frac{k-1}{2^n}}^{\frac{k}{2^n}} R_m(t)R_n(t) \\ &= \sum_{k=1}^{2^n} sgn\left(R_n(\frac{2k-1}{2})\right) \int_{\frac{k-1}{2^n}}^{\frac{k}{2^n}} R_m(t)dt \end{aligned}$$

   Hint: if m is larger than n and satisfies the condition, R_n is constant on each subintegral whereas R_m cycles through its -1 and $+1$ states an even number of times. Thus, integral is always an even number of $+1$'s minus an even number of -1's. It helps to examine where the argument of the $\sin$ function resides in these cases.

2. *Show the Rademacher functions are not complete. Hint: let f be defined on $[0, 1]$ by*

$$f(t) = \begin{cases} 0 & 0 \leq x < \frac{1}{4} \\ 1 & \frac{1}{4} \leq x \leq \frac{3}{4} \\ 0 & \frac{3}{4} < x \leq 1 \end{cases}$$

   Then show $< f, R_0 > = 0.5$ with all other inner products 0.

3. *Plot the first ten functions in this resulting o.n. sequence in a common plot on $[0, 1]$.*

Chapter 9

Dual Spaces

We now want to look at the sets of mappings which are linear and continuous on a normed linear space whose range is the underlying field. These mappings are called continuous linear functionals on the normed linear space and are very important.

9.1 Linear Functionals

If $(X, \|\cdot\|)$ is a normed linear space, recall for the field F (which for us is $\Re$ or $\mathbb{C}$)

$B(X,F) = \{f : X \to F \mid f$ is continuous and linear $\} = \{f : X \to F \mid f$ is bounded and linear $\}$

is a normed linear space itself with norm $\|\cdot\|_{op}$ where

$$\|f\|_{op} \quad = \quad \sup_{\boldsymbol{x} \neq \boldsymbol{0} \in X} \frac{|f(\boldsymbol{x})|}{\|\boldsymbol{x}\|} = \sup_{\|\boldsymbol{x}\|=1} |f(\boldsymbol{x})|$$

Definition 9.1.1 The Dual of a Normed Linear Space

The dual of the normed linear space $(X, \|\cdot\|)$ is $X' = B(X,F)$ where F is the underlying field.

We can then define the **dual** of the dual space X'. The dual of the dual is the set

$$B(B(X,F),F) \quad = \quad \{g : X' \to F | g \text{ is continuous and linear}\} = \{g : X' \to F | g \text{ is bounded and linear}\}$$

is a normed linear space itself with norm $\|\cdot\|_{op}$ where

$$\|g\|_{op} \quad = \quad \sup_{\boldsymbol{f} \neq \boldsymbol{0} \in X'} \frac{|g(f)|}{\|f\|_{op}} = \sup_{\|f\|_{op}=1} |g(f)|$$

Definition 9.1.2 The Double Dual of a Normed Linear Space

The double dual of the normed linear space $(X, \|\cdot\|)$ is $X'' = B(B(X,F),F)$ where F is the underlying field.

Since our field F is either $\Re$ or $\mathbb{C}$, we know F is complete and so we also know X' and X'' are complete with respect to their norms $\|\cdot\|_{op}$.
Define J at $\boldsymbol{x} \in X$ by $(J(\boldsymbol{x}))(f) = f(\boldsymbol{x})$ for all $f \in X'$. We will show $J(x)$ is well-defined and a bounded linear functional on X'.

- Given $\boldsymbol{x}$ and $\boldsymbol{y}$, for all $f \in X'$, we have for any α and β in the field

  $$\alpha(J(\boldsymbol{x}))(f) + \beta(J(\boldsymbol{y}))(f) = \alpha f(\boldsymbol{x}) + \beta f(\boldsymbol{y}) = f(\alpha\boldsymbol{x} + \beta\boldsymbol{y})$$

  But by definition

  $$(J(\alpha\boldsymbol{x} + \beta\boldsymbol{y}))(f) = f(\alpha\boldsymbol{x} + \beta\boldsymbol{y})$$

  which is the same. Hence, J is a linear map.

- $$\sup_{\|f\|_{op}=1} |J(\boldsymbol{x})| = \sup_{\|f\|_{op}=1} |f(\boldsymbol{x})| \leq \sup_{\|f\|_{op}=1} \|f\|_{op}\|\boldsymbol{x}\| \leq \|\boldsymbol{x}\|$$

  Hence, $J(\boldsymbol{x})$ is a bounded linear operator. We see therefore $J(\boldsymbol{x}) \in X''$ for all $\boldsymbol{x} \in X$.

Hence the map J is well-defined.

Definition 9.1.3 The Canonical Embedding of a Normed Linear Space into its Double Dual

Let $(X, \|\cdot\|)$ be a normed linear space. The mapping $J : X \to X''$ defined by $J(\boldsymbol{x})$ is the element of X'' defined element wise by $(J(\boldsymbol{x}))(f) = f(\boldsymbol{x})$ is the canonical embedding map of X into X''.

We will be able to say more later. It turns out $\|J(\boldsymbol{x})\|_{op} = \|\boldsymbol{x}\|$ so that J preserves norm which tells us J is one to one. So $J(X) \subset X''$ is a $1-1$ embedding. However, it need not be onto as we will find out. However, we need more machinery to understand the canonical map fully. We will now work out how this works in $\Re^n$ in the exercises below.
Homework

Exercise 9.1.1 *Let $\boldsymbol{E} = \{\boldsymbol{E_1}, \ldots, \boldsymbol{E_n}\}$ be an orthonormal basis for $\Re^n$.*

- *As usual, we define the inner product here by $< \boldsymbol{E_i}, \boldsymbol{E_j} >= \delta_i^j$. Then, you should be able to easily prove that using this and extending to all of $\Re^n$ is both slots using linearity, we have*

  $$< \boldsymbol{x}, \boldsymbol{y} > = \sum_{i=1}^{n} x_i y_i$$

  where $\boldsymbol{x} = \sum_{i=1}^n x_i \boldsymbol{E_i}$ and $\boldsymbol{y} = \sum_{i=1}^n y_i \boldsymbol{E_i}$ are any two vectors in $\Re^n$.

- *Let $\boldsymbol{f} \in (\Re^n)'$. Define the vector $\boldsymbol{y_f}$ by $(y_f)_i = \boldsymbol{f}(\boldsymbol{E_i})$. Then prove $f(\boldsymbol{x}) =< \boldsymbol{x}, \boldsymbol{y_f} >$ for any $\boldsymbol{x} \in \Re^n$.*

- *For $\boldsymbol{f} \in (\Re^n)'$, prove $\|\boldsymbol{f}\| = \|\boldsymbol{y_f}\|_2$ where $\|\boldsymbol{f}\|$ is the usual operator norm for $\boldsymbol{f}$.*

- *Define the map $T : (\Re^n)' \to \Re^n$ by $T(\boldsymbol{f}) = \boldsymbol{y_f}$. Prove T is $1-1$ and onto and preserves norm; i.e. $\|T(\boldsymbol{f})\| = \|\boldsymbol{f}\|$.*

- *The above map is why we can identify the dual space of the $\Re^n$ thought of as column vectors as $\Re^n$ thought of as row vectors. Make sure you work out this identification.*

- *The canonical embedding in this case is* $(J(\boldsymbol{x}))(\boldsymbol{f}) = \boldsymbol{f}(\boldsymbol{x}) =< \boldsymbol{x}, \boldsymbol{y_f} >= T(\boldsymbol{x})$. *Prove* J *is* $1-1$ *and onto and preserves norm; i.e.* $\|J(\boldsymbol{x})\| = \|\boldsymbol{x}\|_2$. *This is a nice example of manipulating identifications.*

  - *We know* $\|J(\boldsymbol{x})\| \leq \|\boldsymbol{x}\|_2$. *Define* g *on* $\Re^n$ *by* $g(\boldsymbol{y}) =< \boldsymbol{x}, \boldsymbol{y} >$. *This is a bounded linear functional. Show* $|(J(\boldsymbol{x}))(\boldsymbol{g})| = \|\boldsymbol{x}\|_2^2$ *and so* $\|J(\boldsymbol{x})\| \geq \|\boldsymbol{x}\|_2$. *Note this is because* $\boldsymbol{g}\| = \|bsx\|_2$. *Also note this proof needs that we can find an element of the dual space whose norm matches* $\boldsymbol{x}$.
  - *Prove* J *is norm preserving and therefore* $1-1$ *and onto. Hence we can identify* $(\Re^n)''$ *with* $\Re^n$. *So the dual of columns is rows and the dual of the rows flips us back to columns. Note we can write the inner product as* $< \boldsymbol{x}, \boldsymbol{y} >= \boldsymbol{x}^T \boldsymbol{y}$ *which uses this row and column duality as well as matrix-vector multiplication. We have shown we can identify* $(\Re^n)''$ *with* $\Re^n$ *itself via the* $1-1$, *onto norm preserving map* J.

Exercise 9.1.2 *Let* $\boldsymbol{E} = \{\boldsymbol{E_1}, \ldots, \boldsymbol{E_n}\}$ *be an orthonormal basis for* $\mathbb{C}^n$. *Making obvious changes in the arguments, redo the first exercise for this case.*

Exercise 9.1.3 *Prove if* x *is a Riemann Integrable function on* $[a, b]$, *then* $f(\boldsymbol{x}) = \int_a^b x(t)dt$ *defines an element of* $(RI([a, b]), < \cdot, \cdot >)'$ *with* $\|\boldsymbol{f}\| = \sqrt{b-a}$.

Exercise 9.1.4 *This is essentially the same as the last exercise, but it is a different dual space. Prove if* x *is a continuous function on* $[a, b]$, *then* $f(\boldsymbol{x}) = \int_a^b x(t)dt$ *defines an element of* $(C([a, b]), < \cdot, \cdot >)'$ *with* $\|\boldsymbol{f}\| = \sqrt{b-a}$.

Exercise 9.1.5 *Prove if a mapping* $T : X \to Y$ *between two norm linear spaces with respective norms* $\|\cdot\|_X$ *and* $\|\cdot\|_Y$, *then* T *is* $1-1$.

Exercise 9.1.6 *Let* T *be a linear map from* $\Re^n$ *to* $\Re^n$ *which is onto. Prove* T *is also* $1-1$.

9.2 Weak Convergence

We now want to explore a new type of convergence called **weak convergence**.

Definition 9.2.1 Weak Convergence

Let $(X, \|\cdot\|)$ *be a normed linear space. and* $(X', \|\cdot\|_{op})$ *its dual. We say the sequence* $(\boldsymbol{x_n})$ *in* X *converges weakly to* $\boldsymbol{x}$ *in* X *if* $f(\boldsymbol{x_n}) \to f(\boldsymbol{x})$ *for all* $f \in X'$. *We usually write this as* $\boldsymbol{x_n} \xrightarrow{w} \boldsymbol{x}$.

Note weak convergence is in contrast to norm convergence which means $\|\boldsymbol{x_n} - \boldsymbol{x}\| \to 0$. We also want to look at convergence of sequences in X'.

Definition 9.2.2 Weak * Convergence

Let $(X, \|\cdot\|)$ *be a normed linear space. and* $(X', \|\cdot\|_{op})$ *its dual space and* $(X'', \|\cdot\|_{op})$ *its double dual. We say the sequence* $(\boldsymbol{f_n})$ *in* X' *converges weak * to* $\boldsymbol{f}$ *in* X' *if* $f_n(\boldsymbol{x}) \to f(\boldsymbol{x})$ *for all* $\boldsymbol{x} \in X$. *Note this is our usual definition of pointwise convergence of the sequence of functionals* (f_n) *on* X.

We will discuss weak * convergence later. For now, let's concentrate on weak convergence. It turns out norm convergence will imply weak convergence, but not the converse.

Theorem 9.2.1 Norm Convergence Implies Weak Convergence

Let $(X, \|\cdot\|)$ be a normed linear space. Then if $(\boldsymbol{x_n})$ converges in norm to $\boldsymbol{x} \in X$, it also converges weakly to $\boldsymbol{x}$.

Proof 9.2.1
Let f in X' be chosen. Pick any $\epsilon > 0$. Since $\boldsymbol{x_n} \rightarrow \boldsymbol{x}$ in norm, there is N so that $n > N$ implies that $\|\boldsymbol{x_n} - \boldsymbol{x}\| < \frac{\epsilon}{\|f\|_{op}+1}$. Now $|f(\boldsymbol{u})| \leq \|f\|_{op} \|\boldsymbol{u}\|$ for all $\boldsymbol{u} \in X$. Thus, for $n > N$,

$$|f(\boldsymbol{x_n}) - f(\boldsymbol{x})| = |f(\boldsymbol{x_n} - \boldsymbol{x})| \leq \|f\|_{op} \|\boldsymbol{x_n} - \boldsymbol{x}\| < \|f\|_{op} \frac{\epsilon}{\|f\|_{op} + 1} < \epsilon$$

Hence, $f(\boldsymbol{x_n}) \rightarrow f(\boldsymbol{x})$. Since f in X' was arbitrary, we have shown $\boldsymbol{x_n} \stackrel{w}{\rightarrow} \boldsymbol{x}$. ■

Homework

Exercise 9.2.1 *Let's work out what this means in $\Re^n$.*

- *Let $(\boldsymbol{x_n})$ be a sequence in $\Re^n$ which converges weakly to $\boldsymbol{x}$ in $\Re^n$. Prove $\boldsymbol{x_n} \stackrel{weak}{\rightarrow} \boldsymbol{x}$ means $< \boldsymbol{x_n}, \boldsymbol{y} > \rightarrow < \boldsymbol{x}, \boldsymbol{y} >$ for all $\boldsymbol{y} \in \Re^n$.*

- *Let $(\boldsymbol{f_n})$ be a sequence in $(\Re^n)'$ which converges weak $*$ to $\boldsymbol{f}$ in $(\Re^n)'$. Prove $\boldsymbol{f_n} \stackrel{weak*}{\rightarrow} \boldsymbol{f}$ means there is a sequence $(\boldsymbol{x_n})$ in $\Re^n$ so that $\boldsymbol{x_n} \stackrel{weak}{\rightarrow} \boldsymbol{x}$.*

Exercise 9.2.2 *Prove the same results as the previous exercise in $\mathbb{C}^n$.*

Exercise 9.2.3 *Let $(\boldsymbol{x_n})$ be a sequence in $(C([a, b]), \|\cdot\|_\infty)$.*

- *Define f_t by $f_t(\boldsymbol{x}) = x(t)$. This is the pointwise evaluation operator. Prove f_t is a bounded linear functional on this space with operator norm 1.*

- *If we change the norm to $\|\cdot\|_2$ prove f_t is linear but not bounded.*

- *Prove that if $\boldsymbol{x_n} \stackrel{weak}{\rightarrow} \boldsymbol{x}$ this means $x_n \stackrel{ptws}{\rightarrow} \boldsymbol{x}$.*

- *Prove that $\int_a^b x_n(t)dt \rightarrow \int_a^b x(t)dt$.*

Exercise 9.2.4 *Let $(\boldsymbol{x_n})$ be the sequence $x_n(t) = \left(\frac{2nt}{e^{nt^2}}\right)$ on $[0, 1]$. Look back at how we analyzed this sequence in (Peterson (18) 2020) to refresh your mind about it. We know $\int_0^1 x_n(t)dt$ does not converge to $\int_0^1 x(t)dt$ even though $x_n \stackrel{ptws}{\rightarrow} \boldsymbol{x}$ with $\boldsymbol{x} = \boldsymbol{0}$.*

- *Prove this implies $(\boldsymbol{x_n})$ does not converge weakly in $(C([0, 1]), \|\cdot\|_\infty)$ even though it does converge pointwise.*

- *Prove this is still true even if we use the norm $\|\cdot\|_2$.*

Exercise 9.2.5 *Let $(\boldsymbol{x_n})$ be the sequence $x_n(t) = t^n$. This converges pointwise to*

$$x(t) = \begin{cases} 0, & 0 \leq t < 1 \\ 1, & t = 1 \end{cases}$$

and $\int_0^1 x_n(t)dt \rightarrow \int_0^1 x(t)dt$. Explain why we still don't know that $\boldsymbol{x_n} \stackrel{weak}{\rightarrow} \boldsymbol{x}$ in $(C([a, b]), \|\cdot\|_\infty)$. Note this is because we don't know how to characterize the dual of $(C([a, b]), \|\cdot\|_\infty)$. This is hard to do. Look ahead to proof of this characterization in Section 11.2.

We can now characterize the dual space of familiar normed linear spaces.

9.2.1 The Dual of ℓ^1 Sequences

Let's characterize some dual spaces. First, we need what we mean by two normed linear spaces being the **same**.

Definition 9.2.3 Linear Bijective Isometries between Normed Linear Spaces

Let $(X, \|\cdot\|_X)$ and $(Y, \|\cdot\|_Y)$ be two normed linear spaces. We say $T : X \to Y$ is a linear bijective isometry if T is $1-1$, onto and preserves norm: i.e. $\|T(\boldsymbol{x})\|_Y = \|\boldsymbol{x}\|_X$ for all $\boldsymbol{x}$ in X. If such a map T exists, we write $X \equiv Y$.

Let's start by looking at the dual space of ℓ^1.

Theorem 9.2.2 The Dual Space of ℓ^1

Let $f \in (\ell^1)'$. Then there is a unique $\boldsymbol{z_f} \in \ell^\infty$ so that for any $\boldsymbol{x} \in \ell^1$

$$f(\boldsymbol{x}) = \sum_{n=1}^{\infty} x_n (z_f)_i$$

where $\boldsymbol{x} = (x_n)$ and $\boldsymbol{z_f} = ((z_f)_n)$. Further $\|\boldsymbol{z_f}\|_\infty = \|f\|_{op}$.

Proof 9.2.2
Let $(\boldsymbol{E_n})_{n=1}^\infty$ be the standard Schauder basis for ℓ^1. Let $f \in (\ell^1)'$. We have $\boldsymbol{x} = (x_n)$ has representation $\boldsymbol{x} = \sum_{n=1}^\infty x_n \boldsymbol{E_n}$. Note $\|\boldsymbol{x}\|_1 = \sum_{n=1}^\infty |x_n| < \infty$ and $\|\boldsymbol{E_n}\|_1 = 1$ for all n.

Let $\boldsymbol{S_n} = \sum_{i=1}^n x_i \boldsymbol{E_i}$. Then by the linearity of f, $f(\boldsymbol{S_n}) = \sum_{i=1}^n x_i f(\boldsymbol{E_i})$. We know $\lim_{n\to\infty} \boldsymbol{S_n} = \boldsymbol{x}$ because $(\boldsymbol{E_n})$ is a Schauder basis and we also know f is continuous. So

$$\lim_{n\to\infty} f(\boldsymbol{S_n}) = f(\lim_{n\to\infty} \boldsymbol{S_n}) = f(\boldsymbol{x})$$

Hence, $f(\boldsymbol{x}) = \lim_{n\to\infty} \sum_{i=1}^n x_i f(\boldsymbol{E_i})$ and we therefore have shown the series $\sum_{i=1}^\infty x_i f(\boldsymbol{E_i})$ converges.

Let the sequence $\boldsymbol{z_f}$ be defined by $(z_f)_n = f(\boldsymbol{E_n})$. Next, we must show the sequence $\boldsymbol{z_f} \in \ell^\infty$ and its norm is $\|\boldsymbol{z_f}\|_\infty = \|f\|_{op}$.

($\boldsymbol{z_f} \in \ell^\infty$):
To see this, note

$$|(z_f)_n| = |f(\boldsymbol{E_n})| \le \|f\|_{op} \, \|\boldsymbol{E_n}\|_1 = \|f\|_{op}$$

This implies all the entries of $\boldsymbol{z_f}$ are bounded and so $\boldsymbol{z_f} \in \ell^\infty$.

($\|\boldsymbol{z_f}\|_\infty = \|f\|_{op}$):
We already have $\boldsymbol{z_f}\|_\infty \le \|f\|_{op}$. To see the reverse inequality,

$$\begin{aligned} |f(\boldsymbol{x})| &= \left| f\left(\sum_{n=1}^\infty x_n \boldsymbol{E_n}\right)\right| = |f(\lim_{n\to\infty} \boldsymbol{S_n})| = |\lim_{n\to\infty} f(\boldsymbol{S_n})| \text{ by continuity of } f \\ &= \lim_{n\to\infty} |f(\boldsymbol{S_n})| \text{ by continuity of } |\cdot| \end{aligned}$$

$$\leq \lim_{n\to\infty} \sum_{i=1}^{n} |x_i f(\boldsymbol{E_i})| = \lim_{n\to\infty} \sum_{i=1}^{n} |x_i (z_f)_i| \leq \|\boldsymbol{z_f}\|_\infty \sum_{i=1}^{n} |x_i| \leq \|\boldsymbol{z_f}\|_\infty \, \|\boldsymbol{x}\|_1$$

Therefore

$$\|f\|_{op} = \sup_{\boldsymbol{x}\neq \boldsymbol{0}} \frac{|f(\boldsymbol{x})|}{\|\boldsymbol{x}\|_1} \leq \|\boldsymbol{z_f}\|_\infty$$

Combining, $\boldsymbol{z_f} \in \ell^\infty$.

(Is $\boldsymbol{z_f}$ *unique?):*
If we could find $\boldsymbol{z} \in \ell^\infty$ *with* $f(\boldsymbol{x}) = \sum_{n=1}^\infty x_n z_n$ *then* $\sum_{n=1}^\infty x_n z_n = \sum_{n=1}^\infty x_n (z_f)_n$*. In particular, we would have* $f(1\,\boldsymbol{E_i}) = 1\, z_i$ *and so* $\boldsymbol{z_f} = \boldsymbol{z}$. ■

Theorem 9.2.3 $(\ell^1)' \equiv \ell^\infty$

For $f \in (\ell^1)'$*, there is a unique* $\boldsymbol{z_f} = ((z_f)_n) \in \ell^\infty$ *so* $f(\boldsymbol{x}) = \sum_{n=1}^\infty x_n (z_f)_i$ *where* $\boldsymbol{x} = (x_n) \in \ell^1$ *and* $\|\boldsymbol{z_f}\|_\infty = \|f\|_{op}$*. Define* $T : (\ell^1)' \to \ell^\infty$ *by* $T(f) = \boldsymbol{z_f}$*. Then is a linear bijective isometry and* $(\ell^1)' \equiv \ell^\infty$.

Proof 9.2.3
We need to show T *is a linear bijective norm isometry. We already know* $\|T(f)\|_\infty = \|\boldsymbol{z_f}\|_\infty = \|f\|_{op}$ *so* T *preserves norms. We also know* $\boldsymbol{z_f}$ *is unique so* T *is* $1-1$*. Then we have* $T(\alpha f + \beta g) = \boldsymbol{z_{\alpha f + \beta g}}$ *and*

$$(\alpha f + \beta g)(\boldsymbol{x}) = \sum_{n=1}^\infty x_n \, (z_{\alpha f + \beta g})_n$$

But

$$(\alpha f)(\boldsymbol{x}) = \sum_{n=1}^\infty x_n (z_{\alpha f})_n, \quad (\beta g)(\boldsymbol{x}) = \sum_{n=1}^\infty x_n (z_{\beta g})_n$$
$$f(\boldsymbol{x}) = \sum_{n=1}^\infty x_n (z_f)_n, \quad g(\boldsymbol{x}) = \sum_{n=1}^\infty x_n (z_g)_n$$

Also

$$\sum_{n=1}^\infty x_n \, (z_{\alpha f})_n = (\alpha f)(\boldsymbol{x}) = \alpha \sum_{n=1}^\infty x_n (z_f)_n = \sum_{n=1}^\infty x_n \, \alpha (z_f)_n$$

So we have $(z_{\alpha f})_n = \alpha (z_f)_n$ *for all* n*. A similar argument shows* $(z_{\beta g})_n = \beta (z_g)_n$ *for all* n*. Thus,*

$$\begin{aligned}
\sum_{n=1}^\infty x_n (z_{\alpha f + \beta g})_n &= (\alpha f + \beta g)(\boldsymbol{x}) = \alpha f(\boldsymbol{x}) + \beta g(\boldsymbol{x}) \\
&= \sum_{n=1}^\infty x_n \, \alpha (z_f)_n + \sum_{n=1}^\infty x_n \, \beta (z_g)_n \\
&= \sum_{n=1}^\infty x_n (\, \alpha (z_f)_n + \beta (z_g)_n \,)
\end{aligned}$$

By uniqueness, we must have $(z_{\alpha f+\beta g})_n = \alpha(z_f)_n + \beta(z_g)_n$. *which tells us* $T(\alpha f + \beta g) = \alpha T(f) + \beta T(g)$. *We conclude* T *is linear.*

To see T *is onto, note if* $\boldsymbol{w} \in \ell^\infty$, *we can define* f *at any* $\boldsymbol{x} \in \ell^1$ *by* $f(\boldsymbol{x}) = \sum_{n=1}^{\infty} x_n w_n$ *which converges by Hölder's Inequality as* $|f(\boldsymbol{x})| \leq \|\boldsymbol{w}\|_\infty \|\boldsymbol{x}\|_1$. *This then tells us* $\|f\|_{op} \leq \|\boldsymbol{w}\|_\infty$ *and so* f *is bounded. It is clearly linear also. Thus* $f \in (\ell^1)'$. *Also, we have* $f(\boldsymbol{E_n}) = w_n$ *and so* $(z_f)_n = f(\boldsymbol{E_n}) = w_n$. *This shows* $T(f) = \boldsymbol{z_f} = \boldsymbol{w}$ *and therefore* T *is onto. We have shown* $(\ell^1)' \equiv \ell^\infty$. ■

To make sure you understand these arguments, in the exercises we will redo them in a vector and matrix setting. Yucky to be sure!

Homework

Exercise 9.2.6 *Let* M *be the space of all* 2×1 *vectors whose components are in* ℓ^1. *Let's work out a lot of stuff.*

- *Endow* M *with the norm* $\|A\| = \sum_{i=1}^{2} \|A_i\|_1$. *Prove this is a norm.*
- *Define* $(\boldsymbol{F_n})$ *as follows:*

$$\boldsymbol{F_n}^1 = \begin{bmatrix} \boldsymbol{E_n} \\ \boldsymbol{0} \end{bmatrix} \quad \boldsymbol{F_n}^2 = \begin{bmatrix} \boldsymbol{0} \\ \boldsymbol{E_n} \end{bmatrix}$$

  where $(\boldsymbol{E_n})$ *is the usual Schauder basis for* ℓ^1. *Prove this is a Schauder basis for* M.
- *Now go through all the steps of the proof to characterize the dual space of* M.
- *Prove this dual space is equivalent to the space* N *of two dimensional vectors whose components are in* ℓ^∞ *using the obvious norm.*

Exercise 9.2.7 *Let* M *be the space of all* 2×2 *vectors whose components are in* ℓ^1. *Let's work out a lot of stuff again.*

- *Endow* M *with the norm* $\|A\| = \sum_{i=1}^{2} \sum_{j=1}^{2} \|A_{ij}\|_1$. *Prove this is a norm.*
- *Define* $(\boldsymbol{F_n})$ *by*

$$\boldsymbol{F_n}^{11} = \begin{bmatrix} \boldsymbol{E_n} & \boldsymbol{0} \\ \boldsymbol{0} & \boldsymbol{0} \end{bmatrix} \quad \boldsymbol{F_n}^{12} = \begin{bmatrix} \boldsymbol{0} & \boldsymbol{E_n} \\ \boldsymbol{0} & \boldsymbol{0} \end{bmatrix}$$

$$\boldsymbol{F_n}^{21} = \begin{bmatrix} \boldsymbol{0} & \boldsymbol{0} \\ \boldsymbol{E_n} & \boldsymbol{0} \end{bmatrix} \quad \boldsymbol{F_n}^{22} = \begin{bmatrix} \boldsymbol{0} & \boldsymbol{0} \\ \boldsymbol{0} & \boldsymbol{E_n} \end{bmatrix}$$

  where $(\boldsymbol{E_n})$ *is the usual Schauder basis for* ℓ^1. *Prove this is a Schauder basis for* M.
- *Now go through all the steps of the proof to characterize the dual space of* M.
- *Prove this dual space is equivalent to the space* N *of* 2×2 *matrices whose components are in* ℓ^∞ *using the obvious norm.*

Exercise 9.2.8 *Let* M *be the space of all* $n \times n$ *vectors whose components are in* ℓ^1.

- *Endow* M *with the norm* $\|A\| = \sum_{i=1}^{n} \sum_{j=1}^{n} \|A_{ij}\|_1$. *Prove this is a norm.*
- *Define* $(\boldsymbol{F_n})$ *as usual. Prove this is a Schauder basis for* M.

- *Now go through all the steps of the proof to characterize the dual space of M.*
- *Prove this dual space is equivalent to the space N of $n \times n$ matrices whose components are in ℓ^∞ using the obvious norm.*

9.2.2 The Dual of ℓ^p Sequences for Finite p

Theorem 9.2.4 Characterizing a Linear Functional on ℓ^p, $1 < p < \infty$

Let $1 < p < \infty$. Let $f \in (\ell^p)'$. Then there is a unique $\boldsymbol{z_f} \in \ell^q$ where p and q are conjugate indices satisfying $f(\boldsymbol{x}) = \sum_{n=1}^\infty x_n (z_f)_n$ for $\boldsymbol{x} = (x_n) \in \ell^p$. Further, $\|\boldsymbol{z_f}\|_q = \|f\|_{op}$.

Proof 9.2.4

Let $(\boldsymbol{E_n})$ be the standard Schauder basis for ℓ^p. Then if $\boldsymbol{x} = (x_n) \in \ell^p$, $\boldsymbol{x}$ has unique representation $\boldsymbol{x} = \sum_{n=1}^\infty x_n \boldsymbol{E_n}$ and $\sum_{n=1}^\infty |x_n|^p < \infty$. Using the continuity and linearity of f as we did in the proof of Theorem 9.2.2, we have $f(\boldsymbol{x}) = \sum_{n=1}^\infty x_n \, f(\boldsymbol{E_n})$. Define the sequence $\boldsymbol{z_f}$ by $(z_f)_n = f(\boldsymbol{E_n})$. Then

($\boldsymbol{z_f} \in \ell^q$):
for each n define the sequence $\boldsymbol{y_n}$ by

$$y_{ni} = \begin{cases} \frac{|(z_f)_i|^q}{(z_f)_i}, & 1 \le i \le n \text{ and } (z_f)_i \neq 0 \\ 0, & i > n \text{ or } (z_f)_i = 0 \end{cases}$$

For example, at $n = 4$, letting $(z_f)_i = \gamma_i$ for convenience, if γ_1 through γ_4 are nonzero,

$$\boldsymbol{y_4} = \left(\frac{|\gamma_1|^q}{\gamma_1}, \frac{|\gamma_2|^q}{\gamma_2}, \frac{|\gamma_3|^q}{\gamma_3}, \frac{|\gamma_4|^q}{\gamma_4}, 0, 0, \ldots \right)$$

and if $\gamma_2 = 0$, we have

$$\boldsymbol{y_4} = \left(\frac{|\gamma_1|^q}{\gamma_1}, 0, \frac{|\gamma_3|^q}{\gamma_3}, \frac{|\gamma_4|^q}{\gamma_4}, 0, 0, \ldots \right)$$

Then,

$$f(\boldsymbol{y_n}) = \sum_{i=1}^n y_{ni} f(\boldsymbol{E_i}) = \sum_{\substack{i=1 \\ \gamma_i \neq 0}}^n \frac{|\gamma_i|^q}{\gamma_i} f(\boldsymbol{E_i})$$

But $\gamma_i = f(\boldsymbol{E_i})$ so we have

$$f(\boldsymbol{y_n}) = \sum_{i=1}^n y_{ni} f(\boldsymbol{E_i}) = \sum_{\substack{i=1 \\ \gamma_i \neq 0}}^n |\gamma_i^q| = \sum_{i=1}^n |\gamma_i|^q$$

as adding in the zero terms does not affect the sum. It is clear then that $f(\boldsymbol{y_n}) \ge 0$ and

$$|f(\boldsymbol{y_n})| = \sum_{i=1}^n |\gamma_i|^q \le \|f\|_{op} \, \|\boldsymbol{y_n}\|_p$$

Now

$$\|\boldsymbol{y_n}\|_p = \left(\sum_{\substack{i=1\\ \gamma_i\neq 0}}^{n} \left|\frac{|\gamma_i|^q}{\gamma_i}\right|^p\right)^{\frac{1}{p}} = \left(\sum_{\substack{i=1\\ \gamma_i\neq 0}}^{n} (|\gamma_i|^{q-1})^p\right)^{\frac{1}{p}}$$
$$= \left(\sum_{\substack{i=1\\ \gamma_i\neq 0}}^{n} |\gamma_i|^q\right)^{\frac{1}{p}} = \left(\sum_{i=1}^{n} |\gamma_i|^q\right)^{\frac{1}{p}}$$

Hence, we can now say

$$|f(\boldsymbol{y_n})| = \sum_{i=1}^{n} |\gamma_i|^q \leq \|f\|_{op}\, \|\boldsymbol{y_n}\|_p = \|f\|_{op} \left(\sum_{i=1}^{n} |\gamma_i|^q\right)^{\frac{1}{p}}$$
$$\Longrightarrow \left(\sum_{i=1}^{n} |\gamma_i|^q\right)^{1-\frac{1}{p}=\frac{1}{q}} \leq \|f\|_{op}$$

But $\gamma_i = (z_f)_i$, *so we conclude* $\|\boldsymbol{z_f}\|_q \leq \|f\|_{op}$. *This tells us* $\boldsymbol{z_f} \in \ell^q$.
($\|\boldsymbol{z_f}\|_q = \|f\|_{op}$*):*
We already have $\|\boldsymbol{z_f}\|_q \leq \|f\|_{op}$. *We need to show the reverse inequality. Now*

$$|f(\boldsymbol{x})| = \left|\sum_{i=1}^{\infty} x_i f(\boldsymbol{E_i})\right| \leq \|\boldsymbol{x}\|_p\, \|\boldsymbol{z_f}\|_q$$

by Hölder's Inequality. Hence, taking the supremum over all nonzero $\boldsymbol{x}$, *we find*

$$\|f\|_{op} = sup_{\boldsymbol{x}\neq \boldsymbol{0}} \frac{|f(\boldsymbol{x})|}{\|\boldsymbol{x}\|} \leq \|\boldsymbol{z_f}\|_q$$

This is the reverse inequality we need.

($\boldsymbol{z_f}$ *is unique):*
If there was another $\boldsymbol{z} \in \ell^q$ *with*

$$f(\boldsymbol{x}) = \sum_{i=1} x_i\, z_i = \sum_{i=1}^{\infty} x_i\, (z_f)_i$$

In particular, we would have $f(1\,\boldsymbol{E_i}) = 1\, z_i$ *and so* $\boldsymbol{z_f} = \boldsymbol{z}$. ■

Theorem 9.2.5 $(\ell^p)' \equiv \ell^q$**,** $1 < p < \infty$**,** p **and** q **Conjugate Indices**

Let $1 < p < \infty$. *Let* $f \in (\ell^p)'$. *There is a unique* $\boldsymbol{z_f} \in \ell^q$ *where p and q are conjugate indices satisfying* $f(\boldsymbol{x}) = \sum_{n=1}^{\infty} x_n (z_f)_n$ *for* $\boldsymbol{x} = (x_n) \in \ell^p$ *with* $\|\boldsymbol{z_f}\|_q = \|f\|_{op}$. *The mapping* T *defined by* $T(f) = \boldsymbol{z_f}$ *is a linear bijective norm isometry from* $(\ell^p)' \to \ell^q$. *Hence* $(\ell^p)' \equiv \ell^q$.

Proof 9.2.5
We need to show T *is a linear bijective norm isometry. We already know* $\|T(f)\|_q = \|\boldsymbol{z_f}\|_q = \|f\|_{op}$

so T preserves norms. We also know $\boldsymbol{z_f}$ is unique so T is $1-1$. The argument to show T is linear is the same as the one we used in the proof of Theorem 9.2.3 and so we won't repeat it here.
To see T is onto, note if $\boldsymbol{w} \in \ell^q$, we can define f at any $\boldsymbol{x} \in \ell^p$ by $f(\boldsymbol{x}) = \sum_{n=1}^{\infty} x_n w_n$ which converges by Hölder's Inequality as $|f(\boldsymbol{x})| \leq \|\boldsymbol{w}\|_q \|\boldsymbol{x}\|_p$. This then tells us $\|f\|_{op} \leq \|\boldsymbol{w}\|_q$ and so f is bounded. It is clearly linear also. Thus $f \in (\ell^p)'$. Also, we have $f(\boldsymbol{E_n}) = w_n$ and so $(z_f)_n = f(\boldsymbol{E_n}) = w_n$. This shows $T(f) = \boldsymbol{z_f} = \boldsymbol{w}$ and therefore T is onto. We have shown $(\ell^p)' \equiv \ell^q$. ■

Comment 9.2.1 *We can characterize weak convergence in ℓ^1 now. Every $f \in (\ell^1)'$ is identified with a unique $\boldsymbol{z_f} \in \ell^\infty$. So given $(\boldsymbol{x_n}) \in \ell^1$ which converges weakly to $\boldsymbol{x} \in \ell^1$, since each $\boldsymbol{E_k} \in \ell^\infty$ corresponds to a linear functional on ℓ^1, weak convergence implies*

$$\sum_{j=1}^{\infty} x_{nj}(E_{kj}) \rightarrow \sum_{j=1}^{\infty} x_j(E_{kj}) \Longrightarrow \lim_{n\rightarrow\infty} x_{nk} \rightarrow x_k$$

Comment 9.2.2 *We can characterize weak convergence in ℓ^p for $1 < p < \infty$ too. Every $f \in (\ell^p)'$ is identified with a unique $\boldsymbol{z_f} \in \ell^q$ where p and q are conjugate indices. So given $(\boldsymbol{x_n}) \in \ell^p$ which converges weakly to $\boldsymbol{x} \in \ell^p$, since each $\boldsymbol{E_k} \in \ell^q$ corresponds to a linear functional on ℓ^p, weak convergence implies*

$$\sum_{j=1}^{\infty} x_{nj}(E_{kj}) \rightarrow \sum_{j=1}^{\infty} x_j E_{kj} \Longrightarrow \lim_{n\rightarrow\infty} x_{nk} \rightarrow x_k$$

as well.

Homework

Exercise 9.2.9 *Let M be the space of all 2×1 vectors whose components are in ℓ^p. Let's work out a lot of stuff.*

- *Endow M with the norm $\|A\| = \sum_{i=1}^{2} \|A_i\|_p$. Prove this is a norm.*
- *Define $(\boldsymbol{F_n})$ by*

  $$\boldsymbol{F_n}^1 = \begin{bmatrix} \boldsymbol{E_n} \\ \boldsymbol{0} \end{bmatrix} \quad \boldsymbol{F_n}^2 = \begin{bmatrix} \boldsymbol{0} \\ \boldsymbol{E_n} \end{bmatrix}$$

  where $(\boldsymbol{E_n})$ is the usual Schauder basis for ℓ^p. Prove this is a Schauder basis for M.
- *Now go through all the steps of the proof to characterize the dual space of M.*
- *Prove this dual space is equivalent to the space N of two dimensional vectors whose components are in ℓ^q using the obvious norm.*

Exercise 9.2.10 *Let M be the space of all 2×2 vectors whose components are in ℓ^p.*

- *Endow M with the norm $\|A\| = \sum_{i=1}^{2} \sum_{j=1}^{2} \|A_{ij}\|_p$. Prove this is a norm.*
- *Define $(\boldsymbol{F_n})$ by*

  $$\boldsymbol{F_n}^{11} = \begin{bmatrix} \boldsymbol{E_n} & \boldsymbol{0} \\ \boldsymbol{0} & \boldsymbol{0} \end{bmatrix} \quad \boldsymbol{F_n}^{12} = \begin{bmatrix} \boldsymbol{0} & \boldsymbol{E_n} \\ \boldsymbol{0} & \boldsymbol{0} \end{bmatrix}$$

  $$\boldsymbol{F_n}^{21} = \begin{bmatrix} \boldsymbol{0} & \boldsymbol{0} \\ \boldsymbol{E_n} & \boldsymbol{0} \end{bmatrix} \quad \boldsymbol{F_n}^{22} = \begin{bmatrix} \boldsymbol{0} & \boldsymbol{0} \\ \boldsymbol{0} & \boldsymbol{E_n} \end{bmatrix}$$

where $(\boldsymbol{E_n})$ *is the usual Schauder basis for* ℓ^p. *Prove this is a Schauder basis for* M.

- *Now go through all the steps of the proof to characterize the dual space of* M.
- *Prove this dual space is equivalent to the space* N *of* 2×2 *matrices whose components are in* ℓ^q *using the obvious norm.*

Exercise 9.2.11 *Let* M *be the space of all* $n \times n$ *vectors whose components are in* ℓ^p.

- *Endow* M *with the norm* $\|A\| = \sum_{i=1}^{n} \sum_{j=1}^{n} \|A_{ij}\|_p$. *Prove this is a norm.*
- *Define* $(\boldsymbol{F_n})$ *as usual. Prove this is a Schauder basis for* M.
- *Now go through all the steps of the proof to characterize the dual space of* M.
- *Prove this dual space is equivalent to the space* N *of* $n \times n$ *matrices whose components are in* ℓ^q *using the obvious norm.*

Chapter 10

Hahn - Banach Results

We now prove the Hahn - Banach Theorem and several of its corollaries. The Hahn - Banach Theorem is a fundamental result in analysis. A typical but important application of the Hahn - Banach Theorem is its use in the proof that linear functionals on certain linear spaces can be represented in the useful "inner product" form, or as a pairing between the space and its dual space. There are many special cases of the Hahn - Banach Theorem, some of which we will prove as corollaries.

10.1 Linear Extensions

A nice application of Zorn's Lemma occurs in the study of extending a function defined on one set to a larger set in such a way that properties the function has are satisfied on both the original and the final set. Considered abstractly, we can state the extension problem more precisely as follows: Let X be a vector space of a field F and $Z \subset X$ a subspace. Further let $f : Z \to F$ be a linear function which satisfies some list of properties. Can we find a new function $\widetilde{f} : X \to F$ so that $\widetilde{f}|_Z = f$? Will the properties F has on Z still hold for $\widetilde{f}$ off Z? Recall the notation $\widetilde{f}|_Z = f$ means the restriction of $\widetilde{f}$ to the domain Z.

We can make one further remark about linear mappings on vector spaces. If $T : D \subset X \to Y$ is a linear map from a from set of D in the vector space X to another vector space Y, we could define a new function $\widetilde{T} : span(D) \subset X \to Y$ as follows: if α and β are in the field F and $\boldsymbol{x}$ and $\boldsymbol{y}$ are in D, we set $\widetilde{T}(\alpha \boldsymbol{x} + \beta \boldsymbol{y}) = \alpha T(\boldsymbol{x}) + \beta T(\boldsymbol{y})$ which is a well-defined element of Y because Y is a vector space. Note $\widetilde{T}|_D = T$ and $\widetilde{T}$ is linear on the subspace $span(D)$.

Hence, we may always assume, without any loss of generality, that a linear map such as T has a domain which is a subspace because if it did not, we would simply replace T by the extensions $\widetilde{T}$.

Definition 10.1.1 Subadditive Functionals

Let X be a vector space over the field F. Let $p : X \to \Re$. We say p is a **subadditive functional** *if $p(\boldsymbol{x} + \boldsymbol{y}) \leq p(\boldsymbol{x}) + p(\boldsymbol{y})$ for all $\boldsymbol{x}$ and $\boldsymbol{y}$ in X.*

We say p is **positive homogeneous** *on X if when $F = \Re$, $p(\alpha x) = \alpha\, p(\boldsymbol{x})$ for all real numbers $\alpha \geq 0$ and $\boldsymbol{x}$ in X. These are also called* **sublinear** *functionals.*

A good example of such a functional is a *seminorm* which is a mapping ρ that fails condition N2 for a norm but satisfies all the other conditions: i.e. $\rho(\boldsymbol{x}) = 0$ does not necessarily imply $x = 0$. Note a norm $\|\cdot\|$ is an additive map (and so subadditive), is positively homogeneous and is non-negative.

Homework

Exercise 10.1.1 *If $(X, \|\cdot\|)$ is a normed linear space, prove $\|\cdot\|$ is a sublinear functional which is positive homogeneous.*

10.2 The Hahn - Banach Theorem

Theorem 10.2.1 Hahn - Banach Theorem

Let V be a linear space over $\Re$, and let $p : V \to \Re$ satisfy

1. $p(u+v) \leq p(u) + p(v) \qquad \forall\, u, v \in V$
2. $p(\alpha v) = \alpha p(v) \qquad \forall\, v \in V,\ \alpha \geq 0.$

That is p is a **subadditive, positive homogeneous** *functional on X.*

Suppose V_0 is a subspace of V and λ_0 is a linear functional on V_0 such that $\lambda_0(v) \leq p(v)\ \forall\, v \in V_0$. Then there exists a linear functional, λ, on V such that $\lambda|_{V_0} = \lambda_0$ and $\lambda(v) \leq p(v)\ \forall\, v \in V$.

Proof 10.2.1
If $V_0 = V$, the result is trivial. So, we can assume that V_0 is a proper subspace of V. The proof follows by enlarging the domain of λ_0 successively until we extend it to all of V. It should be pointed out, however, that this is not *an inductive proof! Such a proof would assume that V is spanned by a countable set, which is not necessarily true. So, the proof requires implementing a variant of the axiom of choice which for us is Zorn's Lemma.*
Let v_1 be in ${V_0}^c$, and consider the set $V_1 = \{\alpha v_1 + v \ :\ \alpha \in \Re,\ v \in V_0\}$. This is a linear subspace of V that contains V_0. We want to extend λ_0 to a linear functional, λ_1, on V_1 so that λ_1 agrees with λ_0 on V_0 and satisfies $\lambda_1(\alpha v_1 + v) \leq p(\alpha v_1 + v)\ \forall\, v \in V_0$ and $\alpha \in \Re$. To define λ_1, we first need a preliminary result.

We claim that there is a $\beta \in \Re$ such that for all $\alpha \in \Re$ and $v \in V_0$ we have

$$\alpha\beta \leq p(\alpha v_1 + v) - \lambda_0(v). \tag{10.1}$$

For $\alpha = 0$, any $\beta \in \Re$ works. If $\alpha > 0$, then Equation 10.1 holds if and only if

$$\beta \leq \frac{1}{\alpha} p(\alpha v_1 + v) - \frac{1}{\alpha}\lambda_0(v) \quad = \quad p\Big(v_1 + \frac{1}{\alpha} v\Big) - \lambda_0\Big(\frac{1}{\alpha} v\Big) \qquad \forall\, v \in V_0.$$

Note that, for $\alpha \neq 0$, $\{(1/\alpha)v \ :\ v \in V_0\} = V_0$. Thus, the previous inequality implies that Equation 10.1 holds if and only if $\beta \leq p(v_1 + v) - \lambda_0(v)\ \forall\, v \in V_0$. That is, Equation 10.1 holds if and only if $\beta \leq \inf\{p(v_1 + v) - \lambda_0 \ :\ v \in V_0\}$.

If $\alpha < 0$, then Equation 10.1 holds if and only if

$$\beta \geq \frac{1}{\alpha} p(\alpha v_1 + v) - \frac{1}{\alpha}\lambda_0(v) \quad = \quad -p\Big(-v_1 - \frac{1}{\alpha} v\Big) + \lambda_0\Big(-\frac{1}{\alpha} v\Big) \qquad \forall\, v \in V_0.$$

Again, since $\{(1/\alpha)v \ :\ v \in V_0\} = V_0$, this inequality implies that Equation 10.1 holds if and only if $\beta \geq -p(-v_1 - v) + \lambda_0(v)$ for all $v \in V_0$. That is, Equation 10.1 holds if and only if

$\beta \geq \sup\{-p(-v_1 + v) + \lambda_0(v) \,:\, v \in V_0\}$.

So, in order for Equation 10.1 to hold for arbitrary $\alpha \in \Re$, we must choose a $\beta \in \Re$ such that

$$\sup\{-p(-v_1 + v) + \lambda_0(v) \,:\, v \in V_0\} \quad \leq \quad \beta \leq \inf\{p(v_1 + v) - \lambda_0(v) \,:\, v \in V_0\} \quad (10.2)$$

This will be possible if

$$\sup\{-p(-v_1 + v) + \lambda_0(v) \,:\, v \in V_0\} \quad \leq \quad \inf\{p(v_1 + v) - \lambda_0(v) \,:\, v \in V_0\}.$$

To see that this is true, let v and w be in V_0. Then

$$\begin{aligned} \lambda_0(v) + \lambda_0(w) &= \lambda_0(v + w) \\ &\leq p(v + v_1 - v_1 + w) \\ &\leq p(v + v_1) + p(-v_1 + w). \end{aligned}$$

Thus, $-p(-v_1 + w) + \lambda_0(w) \leq p(v + v_1) - \lambda_0(v)$. But v and w are arbitrary, so it follows that

$$\sup\{-p(-v_1 + v) + \lambda_0(v) \,:\, v \in V_0\} \quad \leq \quad \inf\{p(v_1 + v) - \lambda_0(v) \,:\, v \in V_0\}.$$

Hence, we can choose a $\beta \in \Re$ so that Equation 10.2 holds. Now, set $\lambda_1(v_1) = \beta$, and define $\lambda_1 : V_1 \rightarrow \Re$ by $\lambda_1(\alpha v_1 + v) = \alpha\beta + \lambda_0(v)$. For any two real scalars c and c' and elements $\alpha v_1 + v$ and $\alpha' v_1 + v'$ in V_1, we have

$$\begin{aligned} \lambda_1\big(c(\alpha v_1 + v) + c'(\alpha' v_1 + v')\big) &= \lambda_1\big((c\alpha + c'\alpha')v_1 + cv + c'v'\big) \\ &= \big(c\alpha + c'\alpha'\big)\beta + \lambda_0(cv + c'v') \\ &= c\alpha\lambda_1(v_1) + c'\alpha'\lambda_1(v_1) + \lambda_0(cv) + \lambda_0(c'v') \\ &= c\big(\alpha\lambda_1(v_1) + \lambda_0(v)\big) + c'\big(\alpha'\lambda_1(v_1) + \lambda_0(v')\big) \\ &= c\lambda_1(\alpha v_1 + v) + c'\lambda_1(\alpha' v_1 + v') \end{aligned}$$

Hence, λ_1 is a linear functional on V_1. Moreover, if $v \in V_0$, then $0v_1 + v \in V_1$ and $\lambda_1(v) = \lambda_1(0v_1 + v) = 0 \cdot \beta + \lambda_0(v) = \lambda_0(v)$. So, $\lambda_1|_{V_0} = \lambda_0$. Finally, let $\alpha v_1 + v$ be any element in V_1. Then, $\lambda_1(\alpha v_1 + v) = \alpha\beta + \lambda_0(v)$. If $\alpha = 0$, then $\lambda_1(\alpha v_1 + v) = \lambda_0(v) \leq p(v) = p(\alpha v_1 + v)$. If $\alpha > 0$, then

$$\begin{aligned} \beta \leq \inf\{p(v_1 + u) - \lambda_0(u) \,;\, u \in V_0\} \quad &\Rightarrow \quad \beta \leq p(v_1 + \frac{1}{\alpha}v) - \lambda_0(\frac{1}{\alpha}v) \\ &\qquad \leq \frac{1}{\alpha}p(\alpha v_1 + v) - \frac{1}{\alpha}\lambda_0(v) \\ &\Rightarrow \quad \alpha\beta \leq p(\alpha v_1 + v) - \lambda_0(v) \\ &\Rightarrow \quad \alpha\beta + \lambda_0(v) \leq p(\alpha v_1 + v) \\ &\Rightarrow \quad \lambda_1(\alpha v_1 + v) \leq p(\alpha v_1 + v). \end{aligned}$$

If $\alpha < 0$, then

$$\begin{aligned} \beta \geq \sup\{-p(-v_1 + u) + \lambda_0(u) \,:\, u \in V_0\} \quad &\Rightarrow \quad \beta \geq -p(-v_1 - \frac{1}{\alpha}v) + \lambda_0(-\frac{1}{\alpha}v) \\ &\qquad \geq \frac{1}{\alpha}p(\alpha v_1 + v) - \frac{1}{\alpha}\lambda_0(v) \\ &\Rightarrow \quad \alpha\beta \leq p(\alpha v_1 + v) - \lambda_0(v) \\ &\Rightarrow \quad \alpha\beta + \lambda_0(v) \leq p(\alpha v_1 + v) \end{aligned}$$

$$\Rightarrow \quad \lambda_1(\alpha v_1 + v) \leq p(\alpha v_1 + v).$$

Thus, $\lambda_1 \leq p$ on V_1. Hence, λ_1 is the desired of extension of λ_0 to V_1.

Now, consider the collection, $\mathcal{S}$, of pairs of the form (λ, W), where W is a linear subspace of V containing V_0 and λ is a linear functional on W that agrees with λ_0 on V_0 and satisfies $\lambda(w) \leq p(w)$ for all $w \in W$. By our hypothesis and our previous construction, $\mathcal{S}$ is nonempty. Define a relation $\prec$ on $\mathcal{S}$ by $(\lambda_1, W_1) \prec (\lambda_2, W_2)$ if $W_1 \subset W_2$ and $\lambda_2 = \lambda_1$ on W_1. The relation $\prec$ is clearly reflexive. If $(\lambda_1, W_1) \prec (\lambda_2, W_2)$ and $(\lambda_2, W_2) \prec (\lambda_1, W_1)$, then $W_1 = W_2$ and $\lambda_2|_{W_1} = \lambda_1$ and $\lambda_1|_{W_2} = \lambda_2$. Thus, $(\lambda_1, W_1) = (\lambda_2, W_2)$, and $\prec$ is antisymmetric. Moreover, if $(\lambda_1, W_1) \prec (\lambda_2, W_2) \prec (\lambda_3, W_3)$, then $W_1 \subset W_2 \subset W_3$ and $\lambda_2|_{W_1} = \lambda_1$ and $\lambda_3|_{W_2} = \lambda_2$. Thus, $\lambda_3|_{W_1} = \lambda_2|_{W_1} = \lambda_1$, and $(\lambda_1, W_1) \prec (\lambda_3, W_3)$. So, $\prec$ is transitive, and $\prec$ is a partial order on $\mathcal{S}$.

We next claim that each chain in $\mathcal{S}$ has an upper bound. Let $\mathcal{C} = \{(\lambda_\gamma, W_\gamma) : \gamma \in \Gamma\}$ be a chain in $\mathcal{S}$ indexed by some set Γ, so for any (λ_1, W_1) and (λ_2, W_2) in $\mathcal{C}$, we have either $(\lambda_1, W_1) \prec (\lambda_2, W_2)$ or $(\lambda_2, W_2) \prec (\lambda_1, W_1)$.

Let $\widetilde{W} = \bigcup_{\gamma \in \Gamma} W_\gamma$. Suppose $c_1, c_2 \in \Re$ and $w_1, w_2 \in \widetilde{W}$. Then $w_1 \in W_{\gamma_1}$ and $w_2 \in W_{\gamma_2}$ for some γ_1 and γ_2. Since $\mathcal{C}$ is a chain, $(\lambda_{\gamma_1}, W_{\gamma_1})$ and $(\lambda_{\gamma_2}, W_{\gamma_2})$ are comparable. We can assume without loss of generality that $(\lambda_{\gamma_1}, W_{\gamma_1}) \prec (\lambda_{\gamma_2}, W_{\gamma_2})$. Hence, $W_{\gamma_1} \subset W_{\gamma_2}$, which implies that $w_1, w_2 \in W_{\gamma_2}$. So, $c_1 w_1 + c_2 w_2 \in W_{\gamma_2} \Rightarrow c_1 w_1 + c_2 w_2 \in \widetilde{W}$, and $\widetilde{W}$ is a linear subspace of V, and $V_0 \subset W_\gamma \; \forall \gamma \in \Gamma$, implying that $V_0 \subset \widetilde{W}$.

Now, we define a mapping $\tilde{\lambda} : \widetilde{W} \to \Re$. For any $w \in \widetilde{W}$, there is some $\gamma \in \Gamma$ such that $w \in W_\gamma$. Define $\tilde{\lambda}(w) = \lambda_\gamma(w)$. To see that this is well-defined, suppose there exist $\gamma_1, \gamma_2 \in \Gamma$ such that $w \in W_{\gamma_1}$ and $w \in W_{\gamma_2}$. Since $\mathcal{C}$ is a chain, we can assume, as before, that $(\lambda_{\gamma_1}, W_{\gamma_1}) \prec (\lambda_{\gamma_2}, W_{\gamma_2})$. Hence, $W_{\gamma_1} \subset W_{\gamma_2}$, and λ_{γ_2} agrees with λ_{γ_1} on W_{γ_1}. Since $w \in W_{\gamma_1}$, it follows that $\lambda_{\gamma_2}(w) = \lambda_{\gamma_1}(w)$. Thus, the value of $\tilde{\lambda}(w)$ is independent of the element $(\lambda_\gamma, W_\gamma)$ we choose in $\mathcal{C}$ so that $w \in W_\gamma$.

To see that $\tilde{\lambda}$ is linear, let $c_1, c_2 \in \Re$ and $w_1, w_2 \in \widetilde{W}$. Again, we can assume that there exist $\gamma_1, \gamma_2 \in \Gamma$ such that $w_1 \in W_{\gamma_1}$ and $w_2 \in W_{\gamma_2}$ and $(\lambda_{\gamma_1}, W_{\gamma_1}) \prec (\lambda_{\gamma_2}, W_{\gamma_2})$. Hence, $w_1, w_2 \in W_{\gamma_2}$, which implies that $c_1 w_1 + c_2 w_2 \in W_{\gamma_2}$. Then,

$$\begin{aligned} \tilde{\lambda}(c_1 w_1 + c_2 w_2) &= \lambda_{\gamma_2}(c_1 w_1 + c_2 w_2) \\ &= c_1 \lambda_{\gamma_2}(w_1) + c_2 \lambda_{\gamma_2}(w_2) \\ &= c_1 \tilde{\lambda}(w_1) + c_2 \tilde{\lambda}(w_2) \end{aligned}$$

and we see that $\tilde{\lambda}$ is linear. Now, every $W_\gamma \in \mathcal{C}$ contains V_0, and $\lambda_\gamma = \lambda_0$ on V_0 for all $\gamma \in \Gamma$. So, if $v \in V_0$, then $v \in W_\gamma$ for some $\gamma \in \Gamma$. Hence, $\tilde{\lambda}(v) = \lambda_\gamma(v) = \lambda_0(v)$, and we see that $\tilde{\lambda}|_{V_0} = \lambda_0$.

Finally, let w be any element of $\widetilde{W}$, and let $\gamma \in \Gamma$ be such that $w \in W_\gamma$. Then $\tilde{\lambda}(w) = \lambda_\gamma(w) \leq p(w)$. Hence, $(\tilde{\lambda}, \widetilde{W})$ is an element of $\mathcal{S}$. Moreover, for any $(\lambda_\gamma, W_\gamma) \in \mathcal{C}$, we have $W_\gamma \subset \widetilde{W}$. In addition, if $w \in W_\gamma$, then $\tilde{\lambda}(w) = \lambda_\gamma(w)$. That is, $\tilde{\lambda}$ agrees with λ_γ on W_γ. Thus, $(\lambda_\gamma, W_\gamma) \prec (\tilde{\lambda}, \widetilde{W})$, and this holds for all $\gamma \in \Gamma$. So, $\mathcal{C}$ has an upper bound, namely $(\tilde{\lambda}, \widetilde{W})$.

Since $\mathcal{C}$ was arbitrary, we can conclude that every chain in $\mathcal{S}$ has an upper bound. Applying Zorn's Lemma, we conclude that there is a maximal element $(\tilde{\lambda}, \widetilde{W})$ in $\mathcal{S}$ with respect to $\prec$. That is, if $(\lambda, W) \in \mathcal{S}$ and $(\tilde{\lambda}, \widetilde{W}) \prec (\lambda, W)$, then $(\tilde{\lambda}, \widetilde{V}) = (\lambda, W)$.

We next claim that $\widetilde{V} = V$. Suppose the contrary, and let v' be in $\widetilde{V}^c$. Then we can apply the exact same construction used in the first part of the proof to extend $\tilde{\lambda}$ to a linear functional $\tilde{\lambda}'$ on $\mathbf{span}\{\,\widetilde{\mathbf{V}} \cup \{\mathbf{v}'\}\,\}$ that agrees with $\tilde{\lambda}$ on $\widetilde{V}$. It follows that $(\tilde{\lambda}', \mathbf{span}\{\,\widetilde{\mathbf{V}} \cup \{\mathbf{v}'\}\,\}) \in \mathcal{S}$ and $(\tilde{\lambda}, \widetilde{V}) \prec (\tilde{\lambda}', \mathbf{span}\{\,\widetilde{\mathbf{V}} \cup \{\mathbf{v}'\}\,\})$. But this implies that $\widetilde{V} = \widetilde{V} \cup \{v'\}$, because $(\tilde{\lambda}, \widetilde{V})$ is maximal. So, we have reached a contradiction. It follows that $\widetilde{V} = V$, and $\tilde{\lambda}$ is a linear functional on V. Since $(\tilde{\lambda}, \widetilde{V}) = (\tilde{\lambda}, V) \in \mathcal{S}$, we know that $\tilde{\lambda} = \lambda_0$ on V_0 and $\tilde{\lambda}(v) \leq p(v)$ for all $v \in V$. So, $\tilde{\lambda}$ is the desired functional. ■

Homework

Exercise 10.2.1 *$C([a,b])$ is a subspace of $B([a,b])$. Show $p(x) = (b-a)\|\boldsymbol{x}\|_\infty$ is a subadditive and positive homogeneous functional on $B([a,b])$. Prove the linear functional $f(\boldsymbol{x}) = \int_a^b x(t)dt$ on $C([a,b])$ has an extension $\widetilde{f}$ to $B([a,b])$ satisfying $\widetilde{f}|_{C([a,b])} = f$ with $\widetilde{f}(\boldsymbol{x}) \leq p(\boldsymbol{x})$ for all $\boldsymbol{x} \in B([a,b])$.*

Exercise 10.2.2 *$RI([a,b])$ is a subspace of $B([a,b])$. Show $p(x) = (b-a)\|\boldsymbol{x}\|_\infty$ is a subadditive and positive homogeneous functional on $B([a,b])$. Prove the linear functional $f(\boldsymbol{x}) = \int_a^b x(t)dt$ on $C([a,b])$ has an extension $\widetilde{f}$ to $B([a,b])$ satisfying $\widetilde{f}|_{C([a,b])} = f$ with $\widetilde{f}(\boldsymbol{x}) \leq p(\boldsymbol{x})$ for all $\boldsymbol{x} \in B([a,b])$.*

10.3 Consequences

Theorem 10.3.1 Generalized Hahn - Banach Theorem

Let V be a real or complex linear space, and suppose the subadditive functional $p : V \to \Re$ satisfies $p(\alpha v) = |\alpha| p(v)$ for all $\alpha \in F$, the scalar field, and $v \in V$. Suppose V_0 is a subspace of V, and $\lambda_0 : V_0 \to \Re$ is a linear functional on V_0 satisfying $|\lambda_0(\boldsymbol{x})| \leq p(\boldsymbol{x})$ for all $\boldsymbol{x} \in V_0$. Then there is a linear functional, λ, on V such that $\lambda|_{V_0} = \lambda_0$ and $|\lambda(\boldsymbol{x})| \leq p(\boldsymbol{x})$ for all $\boldsymbol{x} \in V$.

Proof 10.3.1
($F = \Re$):
Since $\lambda_0(\boldsymbol{x}) \leq |\lambda_0(\boldsymbol{x}) \leq p(\boldsymbol{x})$ for all $\boldsymbol{x} \in V_0$, the Hahn - Banach Theorem implies that there is a linear functional, λ, on V such that $\lambda|_{V_0} = \lambda_0$ and $\lambda(\boldsymbol{x}) \leq p(bsx)$ for all $\boldsymbol{x} \in V$. So, for any $\boldsymbol{x} \in V$, $-\lambda(\boldsymbol{x}) = \lambda(-\boldsymbol{x}) \leq p(-\boldsymbol{x}) = p(\boldsymbol{x})$, which implies that $\lambda(\boldsymbol{x}) > -p(\boldsymbol{x})$. Hence, we have $-p(\boldsymbol{x}) \leq \lambda(\boldsymbol{x}) \leq p(\boldsymbol{x})$ for all $\boldsymbol{x} \in V$. Thus, $|\lambda(\boldsymbol{x})| \leq p(\boldsymbol{x})$ for all $\boldsymbol{x} \in V$, and λ, therefore, is the desired linear functional.

($F = \mathbb{C}$):
Now Z is a complex vector space and f is complex-valued. Let's remind you of some basic things. If $f : V \to \mathbb{C}$, we can write $f(\boldsymbol{x}) = f_1(\boldsymbol{x}) + \boldsymbol{i} f_2(\boldsymbol{x})$ where f_1 and f_2 map V to $\Re$. The function f_1 is the real part of the function f and is denoted by $Re(f)$ and f_2 is called the imaginary part of f, denoted by $Im(f)$. Also, if $z \in \mathbb{C}$, we write $z = x + \boldsymbol{i}y$ for some x and y in $\Re$. We have $|z| = \sqrt{x^2 + y^2}$ which if you think of mapping z into a two dimensional vector $\boldsymbol{u_z} = \begin{bmatrix} x \\ y \end{bmatrix}$ is just $\|\boldsymbol{u}\|_2$, the usual Euclidean norm. Also, we can write $z = |z| e^{Arg(z)\boldsymbol{i}}$ where $Arg(z) = \tan^{-1}(\frac{y}{x})$ implying $\cos(Arg(z)) = x$ and $\sin(Argz) = y$. Of course, this is equivalent to the polar coordinate representation of the two dimensional vector $\boldsymbol{u_z}$. We often simply write $\theta = Arg(z)$, $r = |z|$ and $z = re^{i\theta}$.

Now $Z_{\Re} = \{\alpha \boldsymbol{x} \mid \alpha \in \Re, \boldsymbol{x} \in V_0\}$ is a subspace, over $\Re$, of the real vector space $V_{\Re} = \{\alpha \boldsymbol{x} \mid \alpha \in \Re, \boldsymbol{x} \in V\}$. We have $V_0 \subset Z_{\Re}$ and in fact $V_0 = Z_{\Re}$ and $V \subset X_{\Re}$ and indeed, $V = V_{\Re}$.

Now let's return to the proof at hand. Let $f_{1,\Re} = f_1|_{Z_{\Re}}$ and $f_{2,\Re} = f_2|_{Z_{\Re}}$. Then $|f(\boldsymbol{x})| = \sqrt{(f_1(\boldsymbol{x}))^2 + (f_2(\boldsymbol{x}))^2}$.

- *On $Z_{\Re}$, $f_{1,\Re}(\boldsymbol{x}) = f_1(\boldsymbol{x})$ and $f_{2,\Re}(\boldsymbol{x}) = f_2(\boldsymbol{x})$.*
- *$f_1(\boldsymbol{x}) \leq |f_1(\boldsymbol{x})| \leq |f(\boldsymbol{x})|$ and $f_2(\boldsymbol{x}) \leq |f_2(\boldsymbol{x})| \leq |f(\boldsymbol{x})|$.*
- *$f_{1,\Re}(\boldsymbol{x}) \leq |f(\boldsymbol{x})| \leq p(\boldsymbol{x})$ on $Z_{\Re} = V_0$ and $f_{2,\Re}(\boldsymbol{x}) \leq |f(\boldsymbol{x})| \leq p(\boldsymbol{x})$ on $Z_{\Re} = V_0$*

By the Hahn - Banach Theorem, there are linear extensions $\widetilde{f_{1,\Re}}$ and $\widetilde{f_{2,\Re}}$ on $V_{\Re}$ with $\widetilde{f_{1,\Re}}(\boldsymbol{x}) \leq p(\boldsymbol{x})$ and $\widetilde{f_{2,\Re}}(\boldsymbol{x}) \leq p(\boldsymbol{x})$ on $V_{\Re} = V$. In fact, there is a relationship between f_1 and f_2. If $\boldsymbol{x} \in V_0$,

$$\boldsymbol{i}f(\boldsymbol{x}) = \boldsymbol{i}(f_1(\boldsymbol{x}) + \boldsymbol{i}f_2(\boldsymbol{x})) = \boldsymbol{i}f_1(\boldsymbol{x}) + \boldsymbol{i}^2\, f_s(\boldsymbol{x}) = f_1(\boldsymbol{i}\boldsymbol{x}) + \boldsymbol{i}f_2(\boldsymbol{i}\boldsymbol{x})$$

We can also expand in another way

$$\boldsymbol{i}f(\boldsymbol{x}) = \boldsymbol{i}(f_1(\boldsymbol{x}) + \boldsymbol{i}f_2(\boldsymbol{x})) = \boldsymbol{i}f_1(\boldsymbol{x}) - f_2(\boldsymbol{x})$$

Comparing, we have $f_1(\boldsymbol{i}\boldsymbol{x}) + \boldsymbol{i}f_2(\boldsymbol{i}\boldsymbol{x}) = \boldsymbol{i}f_1(\boldsymbol{x}) - f_2(\boldsymbol{x})$. This implies $f_1(\boldsymbol{i}\boldsymbol{x}) + f_2(\boldsymbol{x}) + \boldsymbol{i}(f_2(\boldsymbol{i}\boldsymbol{x}) - f_1(\boldsymbol{x})) = 0$. Thus, the real and imaginary parts of the equation above must both be zero. This gives $f_1(\boldsymbol{i}\boldsymbol{x}) + f_2(\boldsymbol{x}) = 0$. So if $\boldsymbol{x} \in V_0$, $f_2(\boldsymbol{x}) = -f_1(\boldsymbol{i}\boldsymbol{x})$.

This suggests the complex-valued extension we seek should be $\widetilde{f}(\boldsymbol{x}) = \widetilde{f_{1,\Re}}(\boldsymbol{x}) + \boldsymbol{i}\widetilde{f_{2,\Re}}(\boldsymbol{x})$. However, we know on V_0, $f_2(\boldsymbol{x}) = -f_1(\boldsymbol{i}\boldsymbol{x})$, so we expect to be able to use $\widetilde{f}(\boldsymbol{x}) = \widetilde{f_{1,\Re}}(\boldsymbol{x}) - \boldsymbol{i}\widetilde{f_{1,\Re}}(\boldsymbol{i}\boldsymbol{x})$. Now we have to show the rest of the properties.

($\widetilde{f}|_{V_0} = f$):
If $\boldsymbol{x} \in V_0$, $\widetilde{f}|_{1,\Re}(\boldsymbol{x}) = f_1(\boldsymbol{x})$ and so $\widetilde{f}(\boldsymbol{x}) = f_1(\boldsymbol{x}) - \boldsymbol{i}f_1(\boldsymbol{i}\boldsymbol{x})$. But this is also $\widetilde{f}(\boldsymbol{x}) = f_1(\boldsymbol{x}) + \boldsymbol{i}f_2(\boldsymbol{x}) = f(\boldsymbol{x})$. This shows $\widetilde{f}|_{V_0} = f$.

($\widetilde{f}$ is linear on V): It is easy to see $\widetilde{f}(\boldsymbol{x_1} + \boldsymbol{x_2}) = \widetilde{f}(\boldsymbol{x_1}) + \widetilde{f}(\boldsymbol{x_2})$ for any choice of $\boldsymbol{x_1}$ and $\boldsymbol{x_2}$ in V. It suffices then to show $\widetilde{f}((a + \boldsymbol{i}b)\boldsymbol{x}) = (a + \boldsymbol{i}b)\widetilde{f}(\boldsymbol{x})$ for all complex numbers $a + \boldsymbol{i}b$ in $\mathbb{C}$. Now

$$\widetilde{f}((a + \boldsymbol{i}b)\boldsymbol{x}) = \widetilde{f_{1,\Re}}((a + \boldsymbol{i}b)\boldsymbol{x}) - \boldsymbol{i}\widetilde{f_{1,\Re}}(\boldsymbol{i}(a + \boldsymbol{i}b)\boldsymbol{x})$$

Since a and b are in $\Re$, $a\boldsymbol{x} \in V_{\Re}$ and $a(\boldsymbol{i}\boldsymbol{x}) \in V_{\Re}$. Further, $b(\boldsymbol{i}\boldsymbol{x}) \in V_{\Re}$ and $-b\boldsymbol{x} \in V_{\Re}$. Since $\widetilde{f_{1,\Re}}$ is linear on $V_{\Re}$, we have

$$\begin{aligned}
\widetilde{f}((a + \boldsymbol{i}b)\boldsymbol{x}) &= a\widetilde{f_{1,\Re}}(\boldsymbol{x}) + b\widetilde{f_{1,\Re}}(\boldsymbol{i}\boldsymbol{x}) - \boldsymbol{i}a\widetilde{f_{1,\Re}}(\boldsymbol{i}\boldsymbol{x}) + b\boldsymbol{i}\widetilde{f_{1,\Re}}(\boldsymbol{x}) \\
&= (a + \boldsymbol{i}b)\widetilde{f_{1,\Re}}(\boldsymbol{x}) - \boldsymbol{i}(a + \boldsymbol{i}b)\widetilde{f_{1,\Re}}(\boldsymbol{i}\boldsymbol{x}) \\
&= (a + \boldsymbol{i}b)(\widetilde{f_{1,\Re}}(\boldsymbol{x}) - \boldsymbol{i}\widetilde{f_{1,\Re}}(\boldsymbol{i}\boldsymbol{x})) = (a + \boldsymbol{i}b)\widetilde{f}(\boldsymbol{x})
\end{aligned}$$

$|\widetilde{f}(\boldsymbol{x})| \leq p(\boldsymbol{x})$ on V:
Note $p(\boldsymbol{0}) = p(0\,\boldsymbol{x}) = 0p(\boldsymbol{x}) = 0$. Thus, $0 = p(\boldsymbol{x} - \boldsymbol{x}) \leq p(\boldsymbol{x}) + p(-\boldsymbol{x}) = p(\boldsymbol{x}) + |-1|p(\boldsymbol{x})$. So we see $p(\boldsymbol{x}) \geq 0$ for all $\boldsymbol{x}$ in V and $p(\boldsymbol{0}) = 0$. If $\boldsymbol{x} \in ker(\widetilde{f})$, then $0 = |\widetilde{f}(\boldsymbol{x}) \leq p(\boldsymbol{x})$. On the other hand, if $\widetilde{f}(\boldsymbol{x}) \neq 0$, then we can write $\widetilde{f}(\boldsymbol{x}) = |\widetilde{f}(\boldsymbol{x})|e^{\boldsymbol{i}Arg(\boldsymbol{x})}$. Then,

$$\begin{aligned}
|\widetilde{f}(\boldsymbol{x})| &= \widetilde{f}(\boldsymbol{x})\, e^{-\boldsymbol{i}Arg(\boldsymbol{x})} = \widetilde{f}(e^{-\boldsymbol{i}Arg(\boldsymbol{x})}\,\boldsymbol{x}), \text{ as linear} \\
&= \widetilde{f_{1,\Re}}(e^{-\boldsymbol{i}Arg(\boldsymbol{x})}\,\boldsymbol{x}) - \boldsymbol{i}\widetilde{f_{1,\Re}}(\boldsymbol{i}e^{-\boldsymbol{i}Arg(\boldsymbol{x})}\,\boldsymbol{x})
\end{aligned}$$

Since $|\tilde{f}(\boldsymbol{x})|$ is real, we must have $|\tilde{f}(\boldsymbol{x})| = \tilde{f}_{1,\Re}(e^{-\boldsymbol{i}Arg(\boldsymbol{x})}\,\boldsymbol{x})$. Now $e^{-\boldsymbol{i}Arg(\boldsymbol{x})}\boldsymbol{x} \in V$ so it is in $V_\Re$ also. Then, since $\tilde{f}_{1,\Re}(\boldsymbol{u}) \leq p(\boldsymbol{u})$ for all $\boldsymbol{u} \in V_|Re$, we can say

$$|\tilde{f}(\boldsymbol{x})| = \tilde{f}_{1,\Re}(e^{-\boldsymbol{i}Arg(\boldsymbol{x})}\,\boldsymbol{x}) \quad \leq \quad p(e^{-\boldsymbol{i}Arg(\boldsymbol{x})}\boldsymbol{x}) = |e^{-\boldsymbol{i}Arg(\boldsymbol{x})}|p(\boldsymbol{x}) = p(\boldsymbol{x})$$

So $\tilde{f}$ is the extension we seek. ■

A good example of a functional which would work in the theorem above is a *seminorm* which is a mapping ρ that fails condition N2 for a norm but satisfies all the other conditions: i.e. $\rho(\boldsymbol{x}) = 0$ does not necessarily imply $\boldsymbol{x} = \boldsymbol{0}$. Note the seminorm $\|\cdot\|$ is an additive map (and so subadditive), is positively homogeneous and is non-negative.

Theorem 10.3.1 is referred to as the *Generalized Hahn - Banach Theorem* because it is typically more applicable than the Hahn - Banach Theorem, since the dominating function p is usually a seminorm, meaning that it will satisfy the conditions of the theorem. Clearly, this is a corollary of the Hahn - Banach Theorem. The next result specializes the Hahn - Banach Theorem to normed linear spaces.

Theorem 10.3.2 Hahn - Banach Theorem for Normed Linear Spaces

Let $(X, \|\cdot\|)$ be a normed linear space, and let $Z \subset X$ be a subspace. Suppose there is a linear functional, $f : Z \to F$, that is bounded. Then, there is a linear functional $\tilde{f} : X \to F$ such that $\tilde{f}|_Z = f$ and $\|\tilde{f}\|_X = \|f\|_Z$. That is, $\tilde{f}$ agrees with f on Z and the operator norm of $\tilde{f}$ is the same as that of f on Z.

Proof 10.3.2

If $Z = \{0\}$, then the result is trivial by letting $\tilde{f} = 0$. So, suppose $Z \neq \{0\}$. By definition,

$$\|f\|_Z = \sup_{x \in Z, x \neq 0} \frac{|f(x)|}{\|x\|} \Rightarrow \frac{|f(x)|}{\|x\|} \leq \|f\|_Z$$

for all $x \in Z$, $x \neq 0$. Define $p : X \to \Re$ by $p(x) = \|f\|_Z\|x\|$. Then $p(x + y) = \|f\|_Z\|x + y\| \leq \|f\|_Z(\|x\| + \|y\|)$. So, p is subadditive. Moreover, $p(\alpha x) = \|f\|_Z\|\alpha x\| = |\alpha|\|f\|_Z\|x\| = |\alpha|p(x)$. Also, $|f(x)| \leq \|f\|_Z\|x\|$ for all $x \in Z$, so $|f(x)| \leq p(x)$ for all $x \in Z$.

By Theorem 10.3.1, there exists $\tilde{f} : X \to \Re$ such that $\tilde{f}|_Z = f$ and $|\tilde{f}(x)| \leq p(x)$ for all $x \in X$. It follows that $|\tilde{f}(x)| \leq \|f\|_Z\|x\|$ for all $x \in X$. That is,

$$\frac{|\tilde{f}(x)|}{\|x\|} \leq \|f\|_Z \qquad \forall\, x \in X,\ x \neq 0.$$

So, it follows that

$$\sup_{x \in X, x \neq 0} \frac{|\tilde{f}(x)|}{\|x\|} \quad \leq \quad \|f\|_Z \Rightarrow \|\tilde{f}\|_X \leq \|f\|_Z.$$

But, since $\tilde{f}|_Z = f$, we have

$$\sup_{x \in X, x \neq 0} \frac{|\tilde{f}(x)|}{\|x\|} \quad \geq \quad \sup_{x \in Z, x \neq 0} \frac{|\tilde{f}(x)|}{\|x\|} = \sup_{x \in Z, x \neq 0} \frac{|f(x)|}{\|x\|} = \|f\|_Z.$$

Thus, $\|\tilde{f}\|_X \geq \|f\|_Z$, and $\|\tilde{f}\|_X = \|f\|_Z$. ■

Homework

Exercise 10.3.1 *Let $\boldsymbol{C}$ denote the vector subspace of ℓ_∞ that consists of all sequences that converge. Define the* **functional** $L : \boldsymbol{C} \rightarrow \Re$ *by* $L(x) = \lim x_j$ *where* $x = (x_j)$ *is any object in* $\boldsymbol{C}$*. Further, define* $\rho : \ell^\infty \rightarrow \Re$ *by* $\rho(x) = \overline{\lim}\, x_j$ *when* $x = (x_j)$ *is any object in* ℓ^∞*. Let's use the sup norm on both* $\boldsymbol{C}$ *and* ℓ_∞*.*

1. *Prove that L is a continuous linear functional.*
2. *Prove that ρ is sublinear and $L(x) \leq \rho(x)$ on $\boldsymbol{C}$.*
3. *Show there exists a linear extension of L, L^* to all of ℓ^∞ satisfying $L^*(x) \leq \rho(x)$ on ℓ_∞.*
4. *Show for all x in ℓ^∞, that*

$$\underline{\lim}\, x_j \quad \leq \quad L^*(x) \leq \overline{\lim}\, x_j$$

5. *Show that if $x \geq 0$, then $L^*(x) \geq 0$. This says L is a positive linear functional.*
6. *Show that L^* is a continuous linear functional with operator norm 1.*

Exercise 10.3.2 *Let $C([a, b])$ denote the vector subspace of $RI([a, b])$ that consists of all functions that are continuous. Define the* **functional** $L : C([a, b]) \rightarrow \Re$ *by* $L(x) = \int_a^b x(t)dt$ *where x is any function in $C([a, b])$. Further, define* $\rho : Ri([a, b]) \rightarrow \Re$ *by* $\rho(x) = \int_a^b |x(t)|dt$*. Let's use the* $\|\cdot\|_1$ *on* $C([a, b])$ *and* $RI([a, b])$*.*

1. *Prove that L is a continuous linear functional.*
2. *Prove that ρ is sublinear and $L(x) \leq \rho(x)$ on $\boldsymbol{C}$.*
3. *Show there exists a linear extension of L, L^* to all of $RI([a, b])$ satisfying $L^*(x) \leq \rho(x)$ on $RI([a, b])$.*
4. *Show that L is a positive linear functional.*
5. *Show that L^* is a continuous linear functional with operator norm 1.*

Exercise 10.3.3 *Let $PC([a, b])$ denote the vector subspace of $RI([a, b])$ that consists of all functions that are piecewise continuous; i.e. they are continuous at all but a finite number of points. Define the* **functional** $L : PC([a, b]) \rightarrow \Re$ *by* $L(x) = \int_a^b x(t)dt$ *where x is any function in $PC([a, b])$. Further, define* $\rho : RI([a, b]) \rightarrow \Re$ *by* $\rho(x) = \int_a^b |x(t)|dt$*. Let's use the* $\|\cdot\|_1$ *on* $PC([a, b])$ *and* $RI([a, b])$*.*

1. *Prove that L is a continuous linear functional.*
2. *Prove that ρ is sublinear and $L(x) \leq \rho(x)$ on $\boldsymbol{C}$.*
3. *Show there exists a linear extension of L, L^* to all of $RI([a, b])$ satisfying $L^*(x) \leq \rho(x)$ on $RI([a, b])$.*
4. *Show that L is a positive linear functional.*
5. *Show that L^* is a continuous linear functional with operator norm 1.*

The following theorem uses the previous results to show that there always exists a bounded linear functional having norm 1 on any normed linear space. For that reason, it is usually referred to as the *Bounded Linear Functional Theorem.*

Theorem 10.3.3 Bounded Linear Functional Theorem

Suppose $(X, \| * \cdot\|)$ is a normed linear space. Let $x_0 \in X$ be nonzero. Then there is a bounded linear functional $\tilde{f} : X \rightarrow \Re$ such that $\|\tilde{f}\|_{op} = 1$ and $\tilde{f}(x_0) = \|x_0\|$.

Proof 10.3.3

Let $Z = span\{x_0\}$. Define $f : Z \rightarrow \Re$ by $f(\alpha x_0) = \alpha\|x_0\|$. Then f is a bounded linear functional on Z, and, since

$$\frac{|f(\alpha x_0)|}{\|\alpha x_0\|} = \frac{|\alpha|\|x_0\|}{|\alpha|\|x_0\|} = 1 \quad for\, \alpha \neq 0$$

it follows that f is bounded. By Theorem 10.3.2, there is $\tilde{f} : X \rightarrow \Re$ such that $\tilde{f}$ is a bounded linear functional, $\tilde{f}|_Z = f$, and $\|\tilde{f}\|_X = \|f\|_Z = 1$. Moreover, $\tilde{f}(x_0) = f(x_0) = \|x_0\|$. ■

We can characterize the norm of an element in a normed linear space in terms of its linear functionals.

Theorem 10.3.4 Characterizing Norm in Terms of Linear Functionals

Suppose $(X, \| \cdot \|)$ is a normed linear space. Then

$$\|\boldsymbol{x}\| = \sup_{f \in X', f \neq 0} \frac{|f(\boldsymbol{x})|}{\|f\|_{op}}.$$

Proof 10.3.4

If $\boldsymbol{x_0} \in X$, then Theorem 10.3.3 implies that there exists a bounded linear functional $f_{\boldsymbol{x_0}} : X \rightarrow F$ such that $\|f_{\boldsymbol{x_0}}\| = 1$ and $f_{\boldsymbol{x_0}}(\boldsymbol{x_0}) = \|\boldsymbol{x_0}\|$. So, $f_{\boldsymbol{x_0}} \in X'$ and we have

$$\sup_{f \in X', x' \neq 0} \frac{|f(\boldsymbol{x_0})|}{\|f\|_{op}} \geq \frac{|f_{x_0}(x_0)|}{\|f_{x_0}\|_{op}} \Rightarrow \|x_0\| \leq \sup_{f \in X', f \neq 0} \frac{|f(x_0)|}{\|f\|_{op}}.$$

We also know that $|f(x_0)| \leq \|f\|_{op}\|\boldsymbol{x_0}\|$, so

$$\frac{|f(\boldsymbol{x_0})|}{\|f\|_{op}} \leq \|\boldsymbol{x_0}\| \Rightarrow \sup_{f \in X', f \neq 0} \frac{|f(\boldsymbol{x_0})|}{\|f\|_{op}} \leq \|\boldsymbol{x_0}\|$$

It follows that

$$\sup_{f \in X', f \neq 0} \frac{|f(\boldsymbol{x_0})|}{\|f\|_{op}} = \|\boldsymbol{x_0}\|.$$

■

As a corollary to this result, we can prove

Theorem 10.3.5 $f(\boldsymbol{x}) = 0$ for all $f \in X'$ implies $\boldsymbol{x} = 0$

Suppose $(X, \| \cdot \|)$ is a normed linear space. Then if $f(\boldsymbol{x}) = 0$ for all $f \in X'$, $\boldsymbol{x} = \boldsymbol{0}$.

Proof 10.3.5

We have $\|\boldsymbol{x}\| = \sup_{f \in X', f \neq 0} \frac{|f(\boldsymbol{x})|}{\|f\|_{op}} = 0$. Thus, $\boldsymbol{x} = \boldsymbol{0}$. ■

Homework

Exercise 10.3.4 *Characterize the norm of* $x \in \ell^2$ *in terms of its dual space.*

Exercise 10.3.5 *Characterize the norm of* $x \in \ell^1$ *in terms of its dual space.*

Exercise 10.3.6 *Characterize the norm of* $x \in \Re^n$ *in terms of its dual space.*

Exercise 10.3.7 *If* $x \in \Re^n$, *prove* $< y, x >= 0$ *for all* $y \in \Re^n$ *implies* $x = 0$.

Chapter 11

More About Dual Spaces

Let's look at the double dual again and explicitly find the dual space of some familiar spaces.

11.1 Reflexive Spaces

Let's go back to the double dual space.

Theorem 11.1.1 The Double Dual Canonical Embedding

Let $(X, \|\cdot\|)$ be a normed linear space. Define $J : X \rightarrow X''$ by $(J(\boldsymbol{x}))(\mathbf{f}) = \boldsymbol{f}(\boldsymbol{x})$ for all $\boldsymbol{f} \in X'$. Then J is linear, $1-1$ and norm preserving onto $J(X)$.

Proof 11.1.1
J **is linear**: *For any scalars α and β, any $\boldsymbol{x}$ and $\boldsymbol{y} \in X$ and any $\boldsymbol{f} \in X'$, by definition we have*

$$\begin{aligned}(J(\alpha\boldsymbol{x}+\beta\boldsymbol{y}))(\mathbf{f}) &= \boldsymbol{f}(\alpha\boldsymbol{x}+\beta\boldsymbol{y}) = \alpha\boldsymbol{f}(\boldsymbol{x})+\beta\boldsymbol{f}(\boldsymbol{y})\\ &= \alpha(J(\boldsymbol{x}))(\mathbf{f})+\beta(\mathbf{J}(\boldsymbol{y}))(\mathbf{f})\end{aligned}$$

But $\boldsymbol{f}$ is arbitrary, so we have shown $J(\alpha\boldsymbol{x}+\beta\boldsymbol{y}) = \alpha J(\boldsymbol{x}) + \beta J(\boldsymbol{y})$ telling us J is linear.

J preserves norm $\|J\|_{op} = \|\boldsymbol{x}\|$: *Given $\boldsymbol{x_0} \in X$, nonzero, there is a linear functional $f_{\boldsymbol{x_0}}$ so that $f_{\boldsymbol{x_0}}(\boldsymbol{x_0}) = \|\boldsymbol{x_0}\|$ and $\|f_{\boldsymbol{x_0}}\|_{op} = 1$. Thus,*

$$\|J(\boldsymbol{x_0})\|_{op} \geq \frac{|f_{\boldsymbol{x_0}}(\boldsymbol{x_0})|}{\|\boldsymbol{f}(\boldsymbol{x_0})\|} = \frac{\|\boldsymbol{x_0}\|}{1} = \boldsymbol{x_0}$$

Since $\boldsymbol{x_0}$ was arbitrary and nonzero, this implies $\|J(\boldsymbol{x})\|_{op} \geq \|\boldsymbol{x}\|$ for all $\boldsymbol{x}$.

To show the reverse inequality holds, note

$$\|J(\boldsymbol{x})\|_{op} = \sup_{\boldsymbol{f}\in X', \boldsymbol{f}\neq \boldsymbol{0}} \frac{|\boldsymbol{f}(\boldsymbol{x})|}{\|\boldsymbol{f}\|} \leq \sup_{\boldsymbol{f}\in X', \boldsymbol{f}\neq \boldsymbol{0}} \frac{\|\boldsymbol{f}\|\,\|\boldsymbol{x}\|}{\|\boldsymbol{f}\|} = \|\boldsymbol{x}\|$$

and so J preserves norms.

J **is** $1-1$*: If $J(\boldsymbol{x}) = J(\boldsymbol{y})$, then $\boldsymbol{f}(\boldsymbol{x}) = \boldsymbol{f}(\boldsymbol{y})$ for all $\boldsymbol{f} \in X'$. Hence, $\boldsymbol{f}(\boldsymbol{x}-\boldsymbol{y}) = 0$. However, we know*

$$\|\boldsymbol{x}-\boldsymbol{y}\| = \sup_{\boldsymbol{f}\in X', \boldsymbol{f}\neq \boldsymbol{0}} \frac{|\boldsymbol{f}(\boldsymbol{x}-\boldsymbol{y})|}{\|\boldsymbol{f}\|} = 0$$

implying $\boldsymbol{x} = \boldsymbol{y}$ and so J is $1-1$. Another way to do this is to use the norm preservation. If $J(\boldsymbol{x}) = J(\boldsymbol{y})$, then $J(\boldsymbol{x}-\boldsymbol{y}) = \boldsymbol{0}$. Thus $\|J(\boldsymbol{x}-\boldsymbol{y})\| = 0$. But because J preserves norm, this tells us $\|\boldsymbol{x}-\boldsymbol{y}\| = 0$ or $\boldsymbol{x} = \boldsymbol{y}$. Thus $J : X \to X''$ is what is called a congruence between X and $J(X) \subset X''$. ■

Comment 11.1.1 *Thus $J : X \to X''$ is what is called a congruence between X and $J(X) \subset X''$. A congruence is thus a linear map which is a norm preserving bijection. The symbol we use for congruence is $\cong$; so we would say $X \cong J(X)$.*

The above work leads to a useful idea: a reflexive space.

Definition 11.1.1 Reflexive Space

A normed linear space $(X, \|\cdot\|)$ is said to be **reflexive** *if $X \cong X''$. Recall this means there is a map $\alpha : X \to X''$ which is linear, $1-1$, onto and preserves norms.*

We can prove the following:

Theorem 11.1.2 The Sequence Space ℓ^p is reflexive for $1 < p < \infty$

ℓ^p is reflexive for $1 < p < \infty$.

Proof 11.1.2

We will do the proof for real sequences and leave the complex sequence case to you. We already know $(\ell^p)' \cong \ell^q$ for $1 < p < \infty$ with $\frac{1}{p} + \frac{1}{q} = 1$. The congruence is set up by $T : (\ell^p)' \to \ell^q$ defined by $T(\boldsymbol{f}) = <\cdot, \boldsymbol{z_f}>$ where $\boldsymbol{z_f} \in \ell^q$ is unique and $<\cdot,\cdot>_{pq}$ is the bilinear pairing on $\ell^p \times \ell^q$ defined by $<\boldsymbol{x}, \boldsymbol{y}>_{pq} = \sum_{n=1}^{\infty} x_i y_i$ for any $\boldsymbol{x} \in \ell^p$ and $\boldsymbol{y} \in \ell^q$. Recall the bilinear pairing is well-defined because of Hölder's Inequality. Further, $(\ell^q)' \cong \ell^p$ also. So there is a mapping $S : (\ell^q)' \to \ell^p$ defined by $S(\boldsymbol{g}) = <\cdot, \boldsymbol{z_g}>_{qp}$ where $\boldsymbol{z_g} \in \ell^p$ is unique and $<\cdot,\cdot>_{qp}$ is the bilinear pairing on $\ell^q \times \ell^p$ defined by $<\boldsymbol{y}, \boldsymbol{x}> = \sum_{n=1}^{\infty} y_i x_i$ for any $\boldsymbol{y} \in \ell^q$ and $\boldsymbol{x} \in \ell^p$. Since we have real sequences, we see $<\cdot,\cdot>_{pq} = <\cdot,\cdot>_{qp}$ although with complex-valued sequences these bilinear pairings would require complex conjugations. Hence, we will just use the symbol $<\cdot,\cdot>$ to indicate this common bilinear pairing.

Let $\boldsymbol{h} \in ((\ell^p)')'$, Then given $\boldsymbol{f} \in (\ell^p)'$, there is a unique $\boldsymbol{z_f} \in \ell^q$ with $\boldsymbol{f} = <\cdot, \boldsymbol{z_f}>$ and $\boldsymbol{f} = T^{-1}(\boldsymbol{z_f})$. Thus,

$$\boldsymbol{h}(\boldsymbol{f}) = \boldsymbol{h}(T^{-1}(\boldsymbol{z_f})) = \boldsymbol{h} \circ T^{-1}(\boldsymbol{z_f})$$

Now $\boldsymbol{h} \circ T^{-1}(\boldsymbol{z_f}) \in (\ell^q)'$. Hence, there is a unique $\boldsymbol{z_h} \in \ell^p$ so that $\boldsymbol{h} \circ T^{-1} = S^{-1}(\boldsymbol{z_h})$. So if $\boldsymbol{x} \in \ell^q$, $(\boldsymbol{h} \circ T^{-1})(\boldsymbol{x}) = <\boldsymbol{x}, \boldsymbol{z_h}>$. Thus,

$$\boldsymbol{h}(\boldsymbol{f}) = (\boldsymbol{h} \circ T^{-1})(\boldsymbol{z_f}) = (S^{-1}(\boldsymbol{z_h}))(\boldsymbol{z_f}) = <\boldsymbol{z_f}, \boldsymbol{z_h}>$$

Hence there are unique elements $\boldsymbol{z_f} \in \ell^p$ and $\boldsymbol{z_h} \in \ell^q$ so that $\boldsymbol{h}(\boldsymbol{f}) = <\boldsymbol{z_f}, \boldsymbol{z_h}>$.

Finally, the canonical embedding J for ℓ^p is defined by $J(\boldsymbol{z_h})(\boldsymbol{g}) = \boldsymbol{g}(\boldsymbol{z_h})$ for any $\boldsymbol{g} \in (\ell^p)'$. Any $\boldsymbol{g} \in (\ell^p)'$ has unique representation $\boldsymbol{g}(\boldsymbol{z_h}) =< \boldsymbol{z_h}, \boldsymbol{z_f} >=< \boldsymbol{z_f}, \boldsymbol{z_h} >$. So $J(\boldsymbol{z_h})(\boldsymbol{g}) =< \boldsymbol{z_f}, \boldsymbol{z_h} >$. Hence, $\boldsymbol{h} = J(\boldsymbol{z_h})$. Hence, the canonical mapping J is onto and $(\ell^p)'' \cong \ell^p$. ■

Riesz's Lemma 5.4.3 is a useful result which is similar to the result we prove next. Recall

Theorem Riesz's Lemma

Let $(X, \|\cdot\|)$ be a normed linear space and let Y and Z be subspaces of X with Y closed and $Y \subset Z$ a proper subset. Then given $\lambda \in (0,1)$, there is $\boldsymbol{z_\lambda} \in Z$, $\|\boldsymbol{z_\lambda}\| = 1$ so that $\inf_{\boldsymbol{y} \in Y} \|\boldsymbol{z_\lambda} - \boldsymbol{y}\| \geq \lambda$.

Now we want to phrase this using ideas from the dual space and the distance from a point to a subspace.

Lemma 11.1.3 A Relationship between the Distance from a Point to a Set and Linear Functionals

Let $(X, \|\cdot\|)$ be a normed linear space and let Y be a closed subspace that is a proper subset of X. Let $\boldsymbol{x_0} \in X \setminus Y$ and let $\delta = \inf_{\boldsymbol{y} \in Y} \|\boldsymbol{y} - \boldsymbol{x_0}\| = dist(\boldsymbol{x_0}, Y)$. Then there is $\boldsymbol{f} \in X'$ with $\|\boldsymbol{f}\|_{op} = 1$, $\boldsymbol{f} = 0$ on Y and $\boldsymbol{f}(\boldsymbol{x_0}) = \delta$.

Proof 11.1.3

Let $Z = span(\{\boldsymbol{x_0}\}, Y)$. By standard arguments, any $\boldsymbol{z} \in Z$ has a unique representation $\boldsymbol{z} = \boldsymbol{y} + a\boldsymbol{x_0}$ for some $\boldsymbol{y} \in Y$ and scalar a. Define $\boldsymbol{f} : Z \rightarrow F$ by $f(\boldsymbol{z}) = f(\boldsymbol{y} + a\boldsymbol{z_0}) = a\delta$ where F is the field used by X. It is easy to see $\boldsymbol{f}$ is linear. Also, $|\boldsymbol{f}(\boldsymbol{z})| = |a|\delta$ and $\|\boldsymbol{z}\| = \|\boldsymbol{y} + a\boldsymbol{x_0}\|$. Now if $a = 0$, $\boldsymbol{f}(\boldsymbol{z}) = \boldsymbol{f}(\boldsymbol{y} + 0 \cdot \boldsymbol{x_0}) = 0$ so f is zero on Y. If $a \neq 0$, then $\boldsymbol{f}(\boldsymbol{z}) = |a|\delta = |a| \inf_{\boldsymbol{y} \in Y} \|\boldsymbol{y} - \boldsymbol{x_0}\|$. In particular, this is true for the elements of Y of the form $-\frac{1}{a}\boldsymbol{y}$. From the definition of the infimum, we then have

$$|f(\boldsymbol{z})| \leq |a| \left\| -\frac{1}{a}\boldsymbol{y} - \boldsymbol{x_0} \right\| = \| - \boldsymbol{y} - a\boldsymbol{x_0}\| = \|\boldsymbol{y} + a\boldsymbol{x_0}\| = \|\boldsymbol{z}\|$$

Thus, if $a \neq 0$,

$$|f(\boldsymbol{z})| \leq \|\boldsymbol{z}\| \Longrightarrow \sup_{\boldsymbol{z} \in Z,\, \boldsymbol{z} \neq \boldsymbol{0}} \frac{|f(\boldsymbol{z})|}{\|\boldsymbol{z}\|} \leq 1$$

and so $\|\boldsymbol{f}\|_{op} \leq 1$ and $\boldsymbol{f}$ is a bounded linear functional on X.

We know from the above argument that $\boldsymbol{f} = 0$ on Y and since Y is a closed subspace we must have $\delta > 0$. Therefore if $a \neq 0$, $F(\boldsymbol{y} + a\boldsymbol{x_0}) = a\delta > 0$. We conclude $\boldsymbol{f} \neq 0$ on Z. Also, $\boldsymbol{f}(\boldsymbol{0} + 1 \cdot \boldsymbol{x_0}) = \delta$ or $\boldsymbol{f}(\boldsymbol{x_0}) = \delta$. Finally, from the definition of the infimum, there is a sequence $(\boldsymbol{y_n}) \subset Y$ so that $\|\boldsymbol{y_n} - \boldsymbol{x_0}\| \rightarrow \delta$. Let $\boldsymbol{z_n} = \boldsymbol{y_n} - \boldsymbol{x_0}$. Then $(\boldsymbol{z_n}) \subset Z$ and $\boldsymbol{f}(\boldsymbol{z_n}) = -\delta$. We see

$$\|\boldsymbol{f}\|_{op} = \sup_{\boldsymbol{z} \in Z,\, \boldsymbol{z} \neq \boldsymbol{0}} \frac{|\boldsymbol{f}(\boldsymbol{z})|}{\|\boldsymbol{z_n}\|} \geq \frac{|\boldsymbol{f}(\boldsymbol{z} - \boldsymbol{n})|}{\|\boldsymbol{z}\|} = \frac{\delta}{\|\boldsymbol{z_n}\|}$$

It follows immediately that as $n \rightarrow \infty$, we obtain $\|\boldsymbol{f}\|_{op} \geq \lim_{n \rightarrow \infty} \frac{\delta}{\|\boldsymbol{z_n}\|} = \frac{\delta}{\delta} = 1$. We previously showed $\|\boldsymbol{f}\|_{op} \leq 1$ so combining we have $\|\boldsymbol{f}\|_{op} = 1$.

Now invoke the Hahn - Banach Theorem for normed linear spaces to conclude there is $\widetilde{\boldsymbol{f}} \in X'$ so that $\widetilde{\boldsymbol{f}}|_Z = \boldsymbol{f}$ which implies $\widetilde{\boldsymbol{f}} = 0$ on Y. We also know $\|\widetilde{\boldsymbol{f}}\|_{op} = \|\boldsymbol{f}\|_{op} = 1$ and $\widetilde{\boldsymbol{f}}(\boldsymbol{x_0}) = \boldsymbol{f}(\boldsymbol{x_0}) = \delta$.

This is the bounded linear functional we seek. ■

We know $(\ell^1)' \cong \ell^\infty$ but we can show ℓ^1 is not reflexive as $(\ell^\infty)' \ncong \ell^1$!. To prove this we need a result about separability.

Theorem 11.1.4 If the Dual is Separable, so is the Original Space

Let $(X, \|\cdot\|)$ be a normed linear space. If X' is separable, then X is separable also.

Proof 11.1.4
If X' is separable, letting $U = \{\boldsymbol{f} \,\|\, : \|\boldsymbol{f}\|_{op} = 1\} \subset X'$ also contains a countable dense subset $(\boldsymbol{f_n})$ and for each n, $\|\boldsymbol{f_n}\|_{op} = \sup_{\|\boldsymbol{x}\|=1} |f_n(\boldsymbol{x})| = 1$. By the definition of supremum, this means there is a sequence $(\boldsymbol{x_n})$ in X of norm 1 so that $|f_n(\boldsymbol{x_n})| \geq \frac{1}{2}$.

Let $Y = \overline{span(\boldsymbol{x_n})}$ and let M be the collection of all finite linear combinations from $(\boldsymbol{x_n})$ with rational coefficients. Then M is a countable subset of Y which is dense. This tells us Y is separable.

If $Y \neq X$, there is $\boldsymbol{x_0} \in X \setminus Y$ which is nonzero. Apply Lemma 11.1.3 to conclude there is a $\widetilde{\boldsymbol{f}} \in X'$ with $\|\widetilde{\boldsymbol{f}}\|_{op} = 1$, $\widetilde{\boldsymbol{f}}$ is zero on Y and $\widetilde{\boldsymbol{f}}(\boldsymbol{x_0}) = \delta$ where $\delta = dist(\boldsymbol{x_0}, Y) > 0$. Since each $\boldsymbol{x_n} \in Y$, $\widetilde{\boldsymbol{f}}(\boldsymbol{x_n}) = 0$ and

$$\begin{aligned} \frac{1}{2} &\leq |\boldsymbol{f_n}(\boldsymbol{x_n})| = |\boldsymbol{f_n}(\boldsymbol{x_n}) - \widetilde{\boldsymbol{f}}(\boldsymbol{x_n})| = |(\boldsymbol{f_n} - \widetilde{\boldsymbol{f}})(\boldsymbol{x_n})| \\ &\leq \|\boldsymbol{f_n} - \widetilde{\boldsymbol{f}}\|_{op} \, \|\boldsymbol{x_n}\| = \|\boldsymbol{f_n} - \widetilde{\boldsymbol{f}}\|_{op} \end{aligned}$$

Hence $\|\boldsymbol{f_n} - \widetilde{\boldsymbol{f}}\|_{op} \geq \frac{1}{2}$ for all n. But $(\boldsymbol{f_n}) \subset U$ is dense, so there is an index N so that $\|\boldsymbol{f_N} - \widetilde{\boldsymbol{f}}\|_{op} < \frac{1}{4}$ because $\widetilde{\boldsymbol{f}} \in U$. This is a contradiction and so $Y = X$ implying X is separable. ■

From this it immediately follows that ℓ^1 cannot be reflexive.

Theorem 11.1.5 ℓ^1 is not Reflexive

ℓ^1 is not reflexive.

Proof 11.1.5
We know ℓ^1 is separable and $(\ell^1)' \cong \ell^\infty$. We also know ℓ^∞ is **not** *separable. If ℓ^1 was reflexive, then $\ell^1 \cong (\ell^1)'' = ((\ell^1)')'$. From the previous theorem this would tell us $(\ell^1)'$ must be separable since ℓ^1 is separable. But that would mean ℓ^∞ is separable which it is not. Hence, ℓ^1 is* **not** *reflexive.* ■

We finish with the statement that ℓ^2 is really special. First the bilinear pairing on $\ell^2 \times \ell^2$ is an inner product so that ℓ^2 becomes an inner product space. Then we see ℓ^2 and $(\ell^2)''$ are congruent. Hence, the canonical embedding $J(\ell^2)$ just gives ℓ^2 back.

Theorem 11.1.6 $\ell^2 \cong \ell^2$

$\ell^2 \cong \ell^2$

Proof 11.1.6
This is easy. $(\ell^2)'' = ((\ell^2)')'$. But $(\ell^2)' \cong \ell^2$, so $(\ell^2)'' \cong (\ell^2)' \cong \ell^2$. ■

Comment 11.1.2 *Since ℓ^2 is a complete inner product space, we call it a Hilbert Space. So ℓ^2 is a complete Hilbert Space which is reflexive.*

We can apply the idea of weak convergence to X''. If $(\boldsymbol{f_n}) \subset X'$ and $\boldsymbol{f} \in X'$, we can say $(\boldsymbol{f_n})$ converges weakly to $\boldsymbol{f}$ in X' if $\boldsymbol{g}(\boldsymbol{f_n}) \to \boldsymbol{g}(\boldsymbol{f})$ for all $\boldsymbol{g} \in X''$. We know $J(X) \subset X''$ so for all $\boldsymbol{g} \in J(X)$, we have $(J(\boldsymbol{x}))(\boldsymbol{\phi}) = \boldsymbol{\phi}(\boldsymbol{x})$ for all $\boldsymbol{x} \in X$. In this case, weak convergence in X' becomes $\boldsymbol{f_n}(\boldsymbol{x}) \to \boldsymbol{f}(\boldsymbol{x})$ for all $\boldsymbol{x} \in X$. If you stop and think about this a minute, what we have just described is what we call the pointwise convergence of the sequence $(\boldsymbol{f_n}) \subset X'$ to a functional $\boldsymbol{f}$. We define this type of convergence as **weak** $*$ convergence.

Definition 11.1.2 Weak * Convergence

Let $(X, \|\cdot\|)$ be a normed linear space. We say $(\boldsymbol{f_n}) \subset X'$ converges weak $*$ to $\boldsymbol{f} \in X'$ if $\boldsymbol{f_n}(\boldsymbol{x}) \to \boldsymbol{f}(\boldsymbol{x})$ for all $\boldsymbol{x} \in X$.

We can say a lot more about this kind of convergence and the kind of topology it implies if we study topological spaces in general. We discuss this in (Peterson (17) 2020).

11.1.1 Homework

Exercise 11.1.1 *Prove weak convergence implies weak $*$ convergence.*

Exercise 11.1.2 *There is no norm on $C([0,1])$ that defines pointwise convergence.*

1. *Let (f_n) be a sequence of functions in $C([0,1])$. Define what it means for this sequence to converge pointwise to a function f defined on $[0,1]$.*

2. *Assume $||.||$ is a norm on $C([0,1])$ such that f_n converges to f with respect to this norm iff f_n converges pointwise to f. We will show we can find a sequence of functions f_n which does not converge in norm to the zero function, 0, even though the sequence does converge to the zero function pointwise. Let's define for $n \geq 1$,*

$$g_n(t) = \begin{cases} 2^n\, t & 0 \leq t \leq 2^{-n} \\ 2 - 2^n\, t & 2^{-n} \leq t \leq 2^{1-n} \\ 0 & else \end{cases}$$

   Graph g_n for an arbitrary n and hence show g_n is never identically zero on $[0,1]$.

3. *Let $f_n = \frac{g_n}{||g_n||}$ and show f_n does not converge to the zero function in norm.*

4. *Show f_n does converge to the zero function pointwise.*

5. *Explain carefully why this gives us the contradiction we need.*

11.2 The Dual of the Continuous Functions

We already know the Riemann Integral of a continuous function f on $[a,b]$, $\int_a^b f dx$ is a particular example of a continuous linear functional on $C([a,b])$. Note if we denote by ϕ the linear functional defined by $\phi(f) = \int_a^b f dx$,

$$\|\phi\|_{op} = \sup_{f \in C([a,b]), \|f\|_\infty \neq 0} \frac{|\phi(f)|}{\|f\|_\infty} \leq \frac{\|f\|_\infty\,(b-a)}{\|f\|_\infty} = b - a$$

it is clearly bounded and so continuous. Hence, one bounded linear functional on $C([a,b])$ is the usual Riemann Integral. We now want to characterize all bounded linear functionals on $C([a,b])$ as what are called Riemann - Stieltjes Integrals. To construct the dual to $(C([a,b]), \|\cdot\|_\infty)$ we need to know about an extension to Riemann Integration called Riemann - Stieltjes Integration. We cover this in detail in (Peterson (16) 2019).

11.2.1 A Quick Look at Riemann - Stieltjes Integration

First, we look at monotone functions.

Definition 11.2.1 Monotone Functions

Let $f : [a, b] \to \Re$ be given. We say

- *f is* **increasing** *on $[a, b]$ if $x_1 < x_2$ implies $f(x_1) \leq f(x_2)$.*
- *f is* **strictly increasing** *on $[a, b]$ if $x_1 < x_2$ implies $f(x_1) < f(x_2)$.*
- *f is* **decreasing** *on $[a, b]$ if $x_1 < x_2$ implies $f(x_1) \geq f(x_2)$.*
- *f is* **strictly decreasing** *on $[a, b]$ if $x_1 < x_2$ implies $f(x_1) > f(x_2)$.*

Next, we can define what we mean by the extension of Riemann Integration to what is called Riemann - Stieltjes integration. Let f and g be any bounded functions on the finite interval $[a, b]$. If $\boldsymbol{\pi}$ is any partition of $[a, b]$ and $\boldsymbol{\sigma}$ is any evaluation set, we can extend the notion of the Riemann sum $S(f, \boldsymbol{\pi}, \boldsymbol{\sigma})$ to the more general **Riemann - Stieltjes** sum as follows:

Definition 11.2.2 The Riemann - Stieltjes Sum

Let $f, g \in B([a, b])$, $\boldsymbol{\pi} \in \Pi([a, b])$ and $\boldsymbol{\sigma} \subseteq \boldsymbol{\pi}$. Let the partition points in $\boldsymbol{\pi}$ be $\{x_0, x_1, \ldots, x_p\}$ and the evaluation points be $\{s_1, s_2, \ldots, s_p\}$ as usual. Define

$$\Delta g_j = g(x_j) - g(x_j - i),\ 1 \leq j \leq p.$$

and the Riemann - Stieltjes sum for **integrand** *f and* **integrator** *g for partition $\boldsymbol{\pi}$ and evaluation set $\boldsymbol{\pi}$ by*

$$S(f, g, \boldsymbol{\pi}, \boldsymbol{\sigma}) = \sum_{j \in \boldsymbol{\pi}} f(s_j)\, \Delta g_j$$

This is also called the **Riemann - Stieltjes sum for the function f with respect to the function** *g for partition $\boldsymbol{\pi}$ and evaluation set $\boldsymbol{\sigma}$.*

You should compare this definition to the one we use to define Riemann Integration in (Peterson (18) 2020). Note if the integrator is $g(x) = x$, this is just the usual Riemann sum. If the integrator was any monotonic function g on $[a, b]$, then the difference Δg_j is non-negative and we have extended the idea of a traditional Riemann sum to a monotonic integrator. We can also think of g as the difference of two monotonic functions as well. We can then define the new type of integral as follows:

Definition 11.2.3 The Riemann - Stieltjes Integral

Let $f, g \in B([a,b])$. If there is a real number I so that for all positive ϵ, there is a partition $\pi_0 \in \Pi([a,b])$ so that

$$\left| S(f, g, \pi, \sigma) - I \right| < \epsilon$$

for all partitions π that refine π_0 and evaluation sets σ from π, then we say f is Riemann - Stieltjes integrable with respect to g on $[a,b]$. We call the value I the Riemann - Stieltjes Integral of f with respect to g on $[a,b]$. We use the symbol

$$I = RS(f, g; a, b)$$

to denote this value. We call f the **integrand** *and g the* **integrator**.

It is also common to denote $RS(f, g; a, b)$ as $\int_a^b f dg$ or $\int_a^b f(t)dg(t)$ although the first looks better!

As usual, there is the question of what pairs of functions (f, g) will turn out to have a finite Riemann - Stieltjes Integral. The collection of the functions f from $B([a,b])$ that are Riemann - Stieltjes integrable with respect to a given integrator g from $B([a,b])$ is denoted by $RS([g, a, b])$.

Comment 11.2.1 *If $g(x) = x$ on $[a,b]$, then $RS([g, a, b]) = RI([a,b])$ and $RS(f, g; a, b) = \int_a^b f(x)dx$.*

Comment 11.2.2 *We will use the standard conventions: $RS(f, g; a, b) = -RS(f, g; b, a)$ and $RS(f, g; a; a) = 0$.*

Homework

Exercise 11.2.1 *Let $f(x) = 2x^4$ on $[1,4]$ and $g(x) = x^2$. Write down the Riemann - Stieltjes sums for various partitions* **pi** *and evaluation sets of* $[1,4]$.

Exercise 11.2.2 *Let $f(x) = 2 + 3x + 6x^2$ on $[1,2]$ and $g(x) = x^3$. Write down the Riemann - Stieltjes sums for various partitions* **pi** *and evaluation sets of* $[1,2]$.

Exercise 11.2.3 *Construct a MATLAB function to compute a Riemann - Stieltjes sum for a given f, g, π, σ and $[a,b]$.*

We can easily prove the usual properties that we expect an *integration* type mapping to have.

Theorem 11.2.1 The Linearity of the Riemann - Stieltjes Integral

If f_1 and f_2 are in $RS([g,a,b])$, then

(i)

$$c_1 f_1 + c_2 f_2 \in RS([g,a,b]), \ \forall c_1, c_2 \in \Re$$

(ii)

$$RS(c_1 f_1 + c_2 f_2, g; a, b) = c_1 RS(f_1, g; a, b) \ + \ c_2 RS(f_2, g; a, b)$$

If $f \in RS([g_1,a,b])$ and $f \in RS([g_2,a,b])$ then

(i)

$$f \in RS([c_1 g_1 + c_2 g_2, a, b]), \ \forall c_1, c_2 \in \Re$$

(ii)

$$RS(f, c_1 g_1 + c_2 g_2; a, b) = c_1 RS(f, g_1; a, b) \ + \ c_2 RS(f, g_2; a, b)$$

Proof 11.2.1
These are easy modifications of the proofs for the Riemann Integral. ■

Homework

Exercise 11.2.4 *Prove if If f_1 and f_2 are in $RS([g,a,b])$, so is $2f_1 - 5f_2$.*

Exercise 11.2.5 *Prove if f is in $RS([f,g_1,a,b])$ and f is in $RS([f,g_2,a,b])$, then f is also in $RS([f,g_1+g_2,a,b])$.*

Exercise 11.2.6 *Prove if f is in $RS([f,g_1,a,b])$ and f is in $RS([f,g_2,a,b])$, then f is also in $RS([f,3g_1+6g_2,a,b])$.*

To give you a feel for the kind of partition arguments we use for Riemann - Stieltjes proofs, we will go through the proof of the standard **Integration By Parts** formula in this context.

Theorem 11.2.2 Riemann - Stieltjes Integration By Parts

If $f \in RS([g,a,b])$, then $g \in RS([f,a,b])$ and

$$RS(g, f; a, b) = f(x)g(s)\Big|_a^b - RS(f, g; a, b)$$

Proof 11.2.2
Since $f \in RS([g,a,b])$, there is a number $I_f = RS(f,g;a,b)$ so that given a positive ϵ, there is a partition π_0 such that

$$\left| S(f,g,\boldsymbol{\pi},\boldsymbol{\sigma}) - I_f \right| \ < \ \epsilon, \ \boldsymbol{\pi}_0 \preceq \boldsymbol{\pi}, \ \boldsymbol{\sigma} \subseteq \boldsymbol{\pi}. \qquad (\alpha)$$

For such a partition $\boldsymbol{\pi}$ and evaluation set $\boldsymbol{\sigma} \subseteq \boldsymbol{\pi}$, we have

$$\begin{aligned} \boldsymbol{\pi} &= \{x_0, x_1, \ldots, x_p\}, \\ \boldsymbol{\sigma} &= \{s_1, \ldots, s_p\} \end{aligned}$$

and

$$S(g, f, \boldsymbol{\pi}, \boldsymbol{\sigma}) \quad = \quad \sum_{\boldsymbol{\pi}} g(s_j)\Delta f_j.$$

We can rewrite this as

$$S(g, f, \boldsymbol{\pi}, \boldsymbol{\sigma}) \quad = \quad \sum_{\boldsymbol{\pi}} g(s_j)f(x_j) \; - \; \sum_{\boldsymbol{\pi}} g(s_j)f(x_{j-1}) \qquad (\boldsymbol{\beta})$$

Also, we have the identity (it is a collapsing sum)

$$\sum_{\boldsymbol{\pi}} \left(f(x_j)g(x_j) - f(x_{j-1})g(x_{j-1}) \right) = f(b)g(b) - f(a)g(a). \qquad (\boldsymbol{\gamma})$$

Thus, using Equation $\boldsymbol{\beta}$ and Equation $\boldsymbol{\gamma}$, we have

$$\begin{aligned} f(b)g(b) - f(a)g(a) \; - \; S(g, f, \boldsymbol{\pi}, \boldsymbol{\sigma}) \quad &= \quad \sum_{\boldsymbol{\pi}} f(x_j)\left(g(x_j) - g(s_j) \right) \\ &+ \quad \sum_{\boldsymbol{\pi}} f(x_{j-1})\left(g(s_j) - g(x_{j-1}) \right) \end{aligned} \qquad (\boldsymbol{\xi})$$

Since $\boldsymbol{\sigma} \subseteq \boldsymbol{\pi}$, we have the ordering

$$a = x_0 \leq s_1 \leq x_1 \leq s_2 \leq x_2 \leq \ldots \leq x_{p-1} \leq s_p \leq x_p = b.$$

Hence, the points above are a refinement of $\boldsymbol{\pi}$ we will call $\boldsymbol{\pi}'$. Relabel the points of $\boldsymbol{\pi}'$ as

$$\boldsymbol{\pi}' = \{y_0, y_1, \ldots, y_q\}$$

and note that the original points of $\boldsymbol{\pi}$ now form an evaluation set $\boldsymbol{\sigma}'$ of $\boldsymbol{\pi}'$. We can therefore rewrite Equation $\boldsymbol{\xi}$ as

$$f(b)g(b) - f(a)g(a) \; - \; S(g, f, \boldsymbol{\pi}, \boldsymbol{\sigma}) = \sum_{\boldsymbol{\pi}} f(y_j)\Delta g_j \; = \; S(f, g, \boldsymbol{\pi}', \boldsymbol{\sigma}')$$

Let $I_g = f(b)g(b) - f(a)g(a) - I_f$. Then since $\boldsymbol{\pi}_0 \preceq \boldsymbol{\pi} \preceq \boldsymbol{\pi}'$, we can apply Equation $\boldsymbol{\alpha}$ to conclude

$$\begin{aligned} \epsilon \quad &> \quad \left| S(f, g, \boldsymbol{\pi}', \boldsymbol{\sigma}') - I_f \right| \\ &= \quad \left| f(b)g(b) - f(a)g(a) \; - \; S(g, f, \boldsymbol{\pi}, \boldsymbol{\sigma}) \; - \; I_f \right| \\ &= \quad \left| S(g, f, \boldsymbol{\pi}, \boldsymbol{\sigma}) \; - \; I_g \right| \end{aligned}$$

Since our choice of refinement $\boldsymbol{\pi}$ of $\boldsymbol{\pi}_0$ and evaluation set $\boldsymbol{\sigma}$ was arbitrary, we have shown that $g \in RS([f, a, b])$ with value

$$RS(g, f, a, b) = f(x)g(x)\Big|_a^b \; - RS(f, g, a, b).$$

■

The next important topic for us is to consider the class of functions of *bounded variation.*

Definition 11.2.4 Functions of Bounded Variation

Let $f : [a, b] \rightarrow \Re$ and let $\pi \in \Pi([a, b])$ be given by $\pi = \{x_0 = a, x_1, \ldots, x_p = b\}$. Define $\Delta f_j = f(x_j) - f(x_{j-1})$ for $1 \leq j \leq p$. Define $\ell_{pi}(f; a, b) = \sum_\pi |\Delta f_j|$. The **Total Variation** *of f on $[a, b]$ is defined by*

$$V(f; a, b) \quad = \quad \sup\{\ell_\pi(f; a, b) \mid \pi \in \Pi([a, b])\}$$

If $V(f; a, b) < \infty$, we say f is a function of **bounded variation** *on $[a, b]$. We let $BV([a, b])$ denote the set of all functions f of bounded variation on $[a, b]$.*

Comment 11.2.3

1. *Also, if f is of bounded variation on $[a, b]$, then, for any $x \in (a, b)$, the set $\{a, x, b\}$ is a partition of $[a, b]$. Hence, there exists $M > 0$ such that $|f(x) - f(a)| + |f(b) - f(x)| \leq M$. But this implies*

$$|f(x)| - |f(a)| \quad \leq \quad |f(x) - f(a)| + |f(b) - f(x)| \leq M$$

This tells us that $|f(x)| \leq |f(a)| \leq M$. Since our choice of x in $[a, b]$ was arbitrary, this shows that f is bounded, i.e. $\|f\|_\infty < \infty$.

2. *Thus $BV([a, b]) \subset B([a, b])$.*

Comment 11.2.4 *For any $f \in BV([a, b])$, we clearly have $V(f; a, b) = V(-f; a, b)$ and $V(f; a, b) \geq 0$. Moreover, we also see that $V(f; a, b) = 0$ if and only if f is constant on $[a, b]$.*

Comment 11.2.5 *We can show if $f \in BV([a, b])$, then*

- *$u(t) = V(f; a, t)$ and $v(t) = V(f : a, t) - f(t)$ are both monotonic increasing functions of bounded variation on $[a, b]$.*

- *$f = u - v$; i.e. any function of bounded variation can be written as the difference of two monotonic increasing functions.*

We discuss all of this carefully in (Peterson (16) 2019).

We can prove

Theorem 11.2.3 Riemann - Stieltjes Integrable Pairs (f, g)

- *If $f \in C([a, b])$ and $g \in BV([a, b])$, then $f \in RS([g; a, b])$.*
- *If $f \in BV([a, b])$ and $g \in C([a, b])$, then $f \in RS([g; a, b])$.*

Proof 11.2.3
This is proved in (Peterson (16) 2019). ■

Homework

Exercise 11.2.7 *Compute $\ell_\pi(f : a, b)$ for $f(x) = 2x^2 + 3x + 4$ on $[-1, 4]$ for various partitions π.*

Exercise 11.2.8 *Write a MATLAB function to compute $\ell_\pi(f : a, b)$ for a given f, interval $[a, b]$ and partition π.*

Exercise 11.2.9 *Write a uniform partition function to compute $\ell_\pi(f : a, b)$ for a given f, interval $[a, b]$ and a uniform partition of norm $\frac{b-a}{n}$.*

Exercise 11.2.10 *Write a MATLAB function to graph $\ell_\pi(f : a, b)$ for a given f, interval $[a, b]$ and a uniform partition of norm $\frac{b-a}{n}$.*

We will also need an standard approximation result for Riemann - Stieltjes Integrals. You should compare this to the Riemann Integration approximation result in (Peterson (18) 2020).

Theorem 11.2.4 Riemann - Stieltjes Approximation

If $f \in C([a, b])$, $g \in BV([a, b])$, then $f \in RS([g; a, b])$ and if $\boldsymbol{\pi_n}$ is a sequence of partitions of $[a, b]$ whose norm goes to zero, then for any choice of evaluation set $\boldsymbol{\sigma_n}$ for the partition $\boldsymbol{\pi_n}$, we have $S(f, g, \boldsymbol{\pi_n}, \boldsymbol{\sigma_n}) \to \int_a 6b f dg$.

Proof 11.2.4
This proof is also in (Peterson (16) 2019). ■

We need to think about the set of functions of bounded variation as a normed space too; this requires we find a useful norm.

Theorem 11.2.5 $BV([a, b])$ is a Normed Linear Space

$BV([a, b])$ is a normed linear space with norm $\|w\| = |w(a)| + V(w; a, b)$ for all w in $BV([a, b])$.

Proof 11.2.5
It is straightforward to show that linear combinations of functions of bounded variation are also of bounded variation, so we will leave that argument to you.

Let's show this is a norm.

(N1): *It is easy to see $\|w\| \geq 0$.*

(N2:) *Now if $\|w\| = 0$, this implies $w(a) = 0$ and $V(w; a, b) = 0$. Hence, $\sup_\pi \sum_\pi |\Delta w_j| = 0$. Thus, $\ell_\pi(w; a, b) = 0$ for all **pi**. If there was a t_0 so that $w(t_0) \neq 0$, then for the partition $\{a, t_0, b\}$, we have $|w(t_0) - w(a)| + |w(b) - w(t_0)| = 0$.*
If $t_0 = a$, this contracts the fact that $w(a) = 0$, Hence $a < t_0 \leq b$. If $t_0 < b$, then we have $|w(t_0)| + |w(b) - w(t_0)| = 0$. But $w(t_0) > 0$ by assumption so this expression is not possible. Finally, if $t_0 = b$, we can choose the partition $\{a, b\}$ and then we must have $|w(b) = w(a)| = |w(b)| = 0$ which contradicts the fact that here $t_0 = b$ and $w(t_0) \neq 0$.
We conclude our assumption $w(t_0) \neq 0$ is wrong for any $a \leq t_0 \leq b$. Hence, $w = 0$ on $[a, b]$. The converse is easy to see.

(N3): *For any α,*

$$\|\alpha w\| = |(\alpha w)(a)| + V(\alpha w; a, b) = |\alpha||w(a)| = |\alpha| V(w; a, b) = |\alpha|\|w\|$$

(N4):

$$\|w_1 + w_2\| = |(w_1 + w_2)(a)| + V(w_1 + w_2; a, b) \leq V(w_1; a, b) + V(w_2; a.b)$$

$$
\begin{aligned}
&\leq (|w_1(a)| + V(w_1; a, b)) + (|w_2(a)| + V(w_2; a, b)) \\
&= \|w_1\| + \|w_2\|
\end{aligned}
$$

Thus $\|\cdot\|$ is a norm on this space. ■

Homework

Exercise 11.2.11 *If f and g are two functions of bounded variation on $[a, b]$. prove any linear combination of them is also of bounded variation on $[a, b]$. This shows $BV([a, b])$ is a vector space over either the $\Re$ or $\mathbb{C}$.*

Exercise 11.2.12 *If f is continuously differentiable on $[a, b]$, prove $f \in BV([a, b])$.*

Exercise 11.2.13 *If f is Lipschitz on $[a, b]$, prove $f \in BV([a, b])$.*

Exercise 11.2.14 *If f has a bounded derivative on $[a, b]$, prove $f \in BV([a, b])$.*

Exercise 11.2.15 *Prove if f and g are both functions of bounded variation on $[a, b]$, so if fg.*

11.2.2 Characterizing the Dual of the Set of Continuous Functions

We are now ready to try to characterize the dual of $C([a, b])$.

Theorem 11.2.6 Riesz's Theorem for the Linear Functionals on $C([a, b])$.

Every $f \in (C([a, b]))'$ can be represented by a Riemann - Stieltjes Integral $f(\boldsymbol{x}) = \int_a^b x dw$ for some $w \in BV([a, b])$ which satisfies $V(w; a, b) = \|f\|_{op}$.

Proof 11.2.6
We note $C([a, b])$ is a normed linear space with the $\|\cdot\|_\infty$ norm and $B([a, b])$ is a normed space with that norm also. Further $C([a, b])$ is a vector subspace of $B([a, b])$. Apply the Hahn - Banach Theorem to extend $f \in (C([a, b]))'$ to $\widetilde{f} \in (B([a, b]))'$. We know the extensions satisfies

$$
\|\widetilde{f}\|_{op} = \sup_{\boldsymbol{x} \neq \boldsymbol{0} \in B([a,b])} \frac{|\widetilde{f}(\boldsymbol{x})|}{\|\boldsymbol{x}\|_\infty} = \|f\|_{op} = \sup_{\boldsymbol{x} \neq \boldsymbol{0} \in C([a,b])} \frac{|f(\boldsymbol{x})|}{\|\boldsymbol{x}\|_\infty}
$$

For a given $t \in [a, b]$, define $x_t : [a, b] \to \Re$ by

$$
x_t(s) = \begin{cases} 1, & a \leq s \leq t \\ 0, & t < s \leq b \end{cases}
$$

If you draw a picture of x_t, you'll see x_t is a step function which jumps down from 1 to 0 at the point t. Clearly, x_t is a bounded function. Define $w : [a, b] \to \Re$ by $w(a) = 0$ and $w(t)\widetilde{f}(x_t)$ for $a < t \leq b$. Note $\widetilde{f}$ is a functional on $B([a, b])$ so it assigns a real number to each x_t.

*(**Step 1**:)*
We show w is a function of bounded variation with $V(w; a, b) \leq \|\widetilde{f}\|_{op}$.

Let π be a partition of $[a,b]$; i.e. $\pi = \{a = t_0 < t_1 < \ldots < t_{p-1} < t_p = b\}$. Let $c_1 = sign(\widetilde{f}(x_{t_1}))$ and $c_j = sign(\widetilde{(f)}(x_{t_j}) - \widetilde{(f)}(x_{t_{j-1}}))$ for $2 \leq j \leq p$. Then,

$$\begin{aligned}\sum_{\pi} |w(t_j) - w(t_{j-1})| &= |w(t_1) - w(t_0)| + \sum_{j=2}^{p} |w(t_j) - w(t_{j-1})| \\ &= |\widetilde{f}(x_{t_1}) - w(a)| + \sum_{j=2}^{p} |\widetilde{f}(x_{t_j}) - \widetilde{f}(x_{t_{j-1}})|\end{aligned}$$

But $w(a) = 0$, so we have

$$\begin{aligned}\ell_{\pi}(w; a, b) = \sum_{\pi} |w(t_j) - w(t_{j-1})| &= |\widetilde{f}(x_{t_1})| + \sum_{j=2}^{p} |\widetilde{f}(x_{t_j}) - \widetilde{f}(x_{t_{j-1}})| \\ &= c_1\widetilde{f}(x_{t_1}) + \sum_{j=2}^{p} c_j(\widetilde{f}(x_{t_j}) - \widetilde{f}(x_{t_{j-1}}))\end{aligned}$$

But $\widetilde{f}$ is linear. Thus,

$$\ell_{\pi}(w; a, b) = \widetilde{f}\left(c_1 x_{t_1} + \sum_{j=2}^{p} c_j(x_{t_j} - x_{t_{j-1}})\right)$$

We see

$$\ell_{\pi}(w; a, b) \leq \|\widetilde{f}\|_{op} \left\|c_1 x_{t_1} + \sum_{j=2}^{p} c_j(x_{t_j} - x_{t_{j-1}})\right\|_{\infty}$$

For convenience, let

$$u_{\pi}(s) = (c_1 x_{t_1} + \sum_{j=2}^{p} c_j(x_{t_j} - x_{t_{j-1}}))(s)$$

If you look at the graphs of x_{t_j} and $x_{t_{j-1}}$, you'll see we find

$$x_{t_j}(s) - x_{t_{j-1}}(s) = \begin{cases} 0, & a < s < t_{j-1} \\ 1, & t_{j-1} \leq s \leq t_j \\ 0, & t_j < s \leq b \end{cases}$$

So, given $s \in [a,b]$,

- *if $s \in [a, t_1]$ only $x_{t_1}(s) = 1$ with all other terms zero.*
- *if $s \in [t_{j-1}, t_j]$ only $x_{t_j}(s) - x_{t_{j-1}}(s) = 1$ with all other terms zero.*

Hence, if $s \in [t_{k-1}, t_k]$,

$$u_{\pi}(s) = (c_1 x_{t_1} + \sum_{j=2}^{p} c_j(x_{t_j} - x_{t_{j-1}}))(s) = c_k$$

and so since $|c_k| = 1$ always,

$$\left\| c_1 x_{t_1} + \sum_{j=2}^{p} c_j(x_{t_j} - x_{t_{j-1}}) \right\|_\infty = \sup_{0 \leq s \leq 1} u_{\boldsymbol{\pi}}(s)| = 1$$

We conclude $\ell_{\boldsymbol{\pi}}(w; a, b) \leq \|\widetilde{f}\|_{op}$. Since the choice of $\boldsymbol{\pi}$ is arbitrary, this immediately implies $V(w; a, b) \leq \|\widetilde{f}\|_{op}$. We see w is a function of bounded variation.

*(**Step 2**:)*
We show w is a function of bounded variation with $\|\widetilde{f}\|_{op} \leq V(w; a, b)$.

For a given partition $\boldsymbol{\pi} = \{a = t_0 < t_1 < \ldots < t_{p-1} < t_p = b\}$, define $z(\boldsymbol{x}, \boldsymbol{\pi})$ by

$$z(\boldsymbol{\pi}, \boldsymbol{x}) = x(t_0)x_{t_1} + \sum_{j=2}^{p} x(t_{j-1}) \left(x_{t_j} - x_{t_{j-1}} \right)$$

for each continuous x on $[a, b]$. Since each x_{t_j} is bounded, we have $z(\boldsymbol{\pi}, \boldsymbol{x})$ is in $B([a, b])$. Then

$$\begin{aligned} \widetilde{f}(z(\boldsymbol{\pi}, \boldsymbol{x})) &= x(t_0)\widetilde{f}(x_{t_1}) + \sum_{j=2}^{p} x(t_{j-1}) \left(\widetilde{f}(x_{t_j}) - \widetilde{f}(x_{t_{j-1}}) \right) \\ &= x(t_0)w(t_1) + \sum_{j=2}^{p} x(t_{j-1}) \left(w(t_j) - w(t_{j-1}) \right) \end{aligned}$$

But $w(t_0) = 0$ and so

$$\begin{aligned} \widetilde{f}(z(\boldsymbol{\pi}, \boldsymbol{x})) &= x(t_0)(w(t_1) - w(t_0)) + \sum_{j=2}^{p} x(t_{j-1}) \left(w(t_j) - w(t_{j-1}) \right) \\ &= \sum_{j=1}^{p} x(t_{j-1}) \left(w(t_j) - w(t_{j-1}) \right) = \sum_{j=1}^{p} x(t_{j-1}) \, \Delta w_j \end{aligned}$$

Since $\boldsymbol{x}$ is continuous and w is of bounded variation, the Riemann - Stieltjes Integral, $\int_a^b \boldsymbol{x} d\boldsymbol{w}$ exists. Also, by the Riemann - Stieltjes approximation Theorem, for any $(\boldsymbol{\pi_n})$ with $\|\boldsymbol{\pi_n}\| \to 0$, we know $\sum_{\boldsymbol{\pi_n}} x(t_{j-1}^n)\Delta w_j \to \int_a^b \boldsymbol{x} d\boldsymbol{w}$ as the points $\boldsymbol{\sigma_n} = \{t_0^n, \ldots t_{p_n-1}^n\}$ give an Evaluation set for $\boldsymbol{\pi_n}$. Note our notation here is quite messy as these points come from the partition $\boldsymbol{\pi_n}$ and so are labeled with the extra subscript and superscript n!

Next, we show $\|\widetilde{f}(z(\boldsymbol{\pi}, \boldsymbol{x}) - \boldsymbol{x})\|_\infty \to 0$. We have so far

$$z(\boldsymbol{\pi_n}, \boldsymbol{x}) = x(a)x_{t_1^n} + \sum_{j=2}^{p} x(t_{j-1}^n) \left(x_{t_j^n} - x_{t_{j-1}^n} \right)$$

We know $x_{t_j^n} - x_{t_{j-1}^n}$ evaluated at a is zero for all $j \geq 2$, so $z(\boldsymbol{\pi_n}, \boldsymbol{x})(a) = x(a)x(t_1^n)(a) = x(a)$. Thus, $z(\boldsymbol{\pi_n}, \boldsymbol{x})(a) - x(a) = 0$.

Now pick $s \in (t_{k-1}, t_k]$. Then the only nonzero term in the sum that defines $z(\boldsymbol{\pi_n}, \boldsymbol{x})(s)$ is $x_{t_k}(s) - x_{t_{k-1}}(s) = 1$. We conclude

$$|z(\boldsymbol{\pi_n}, \boldsymbol{x})(s) - x(s)| = |x(t_{k-1}) - x(s)|$$

on $(t_{k-1}, t_k]$. Since x is continuous on $[a, b]$, x is uniformly continuous there. So given $\epsilon > 0$, there is a $\delta > 0$ so $|x(s') - x(s)| < \frac{\epsilon}{2}$ when $|s' - s| < \delta$. Since $\|\boldsymbol{\pi_n}\| \to 0$, there is N so that $n > N$ implies $\|\boldsymbol{\pi_n}\| < \delta$ which implies $|x(t_{k-1}) - x(s)| < \frac{\epsilon}{2}$.We can therefore say

$$|z(\boldsymbol{\pi_n}, \boldsymbol{x})(s) - x(s)| < \frac{\epsilon}{2}, \; s \in (t_{k-1}, t_k], \; 1 \leq k \leq p_n$$

This implies $|z(\boldsymbol{\pi_n}, \boldsymbol{x})(s) - x(s)| < \frac{\epsilon}{2}$ on $[a, b]$ and thus $\|z(\boldsymbol{\pi_n}, \boldsymbol{x}) - \boldsymbol{x}\|_\infty < \epsilon$ if $n > N$. We have thus shown $z(\boldsymbol{\pi_n}, \boldsymbol{x}) \to x$ in $\| \cdot \|_\infty$.

By the continuity of $\widetilde{f}$, it then follows that $\widetilde{(f)}(z(\boldsymbol{\pi_n}, \boldsymbol{x})) \to \widetilde{f}(\boldsymbol{x})$. But we also know

$$\widetilde{f}(z(\boldsymbol{\pi}, \boldsymbol{x})) = \sum_{\boldsymbol{\pi_n}} x(t^n_{j-1}) \Delta w_j \to \int_a^b x dw \to \int_a^b x dw$$

Since limits are unique, we must have $\widetilde{f}(\boldsymbol{x}) = \int_a^b x dw$.

The Riemann - Stieltjes sums here satisfy

$$\sum_{\boldsymbol{\pi_n}} x(t^n_{j-1}) \Delta w_j \leq \|\boldsymbol{x}\|_\infty \sum_{\boldsymbol{pi_n}} \Delta w_j \leq \|x\|_\infty V(w; a, b)$$

Thus taking the limit as $n \to \infty$,

$$\left| \int_a^b x dw \right| = \left| \lim_n \sum_{\boldsymbol{\pi_n}} x(t^n_{j-1}) \Delta w_j \right| = \lim_n \left| \sum_{\boldsymbol{\pi_n}} x(t^n_{j-1}) \Delta w_j \right| \leq \|x\|_\infty V(w; a, b)$$

Thus, $| \int_a^b x dw| \leq \|\boldsymbol{x}\|_\infty V(w; a, b)$ and we see $\widetilde{f}(\boldsymbol{x}) \leq \|\boldsymbol{x}\|_\infty V(w; a, b)$. Therefore

$$\|\widetilde{f}\|_{op} = \sup_{\boldsymbol{x} \neq \boldsymbol{0} \in C([a,b])} \frac{|\widetilde{f}|}{\|\boldsymbol{x}\|_\infty} \leq \frac{\|\boldsymbol{x}\|_\infty V(w; a, b)}{\|\boldsymbol{x}\|_\infty}$$

which implies $\|\widetilde{f}\|_{op} \leq V(w; a, b)$.

We have now shown both parts of the inequality we need and we can say $\|\boldsymbol{z_f}\|_{op} = V(w : a, b)$. This completes our proof. ■

11.2.3 A Norm Isometry for the Dual of the Continuous Functions

We now know every continuous linear functional on $C([a, b])$ has a characterization in terms of a Riemann - Stieltjes Integral. We can find a way to make this characterization unique. Hence, given $\boldsymbol{f} \in (C([a, b]))'$ there is a $\boldsymbol{w} \in BV([a, b])$ so that $\boldsymbol{f}(\boldsymbol{x}) = \int_a^b \boldsymbol{x} d\boldsymbol{w}$. Let's find a subspace $Z \subset (C([a, b]))'$so that Z is norm isometric to $(C([a, b]))'$.

Definition 11.2.5 Normalized Functions of Bounded Variation

Let $\boldsymbol{w} \in BV([a,b])$. We say $\boldsymbol{w}$ is a normalized function of bounded variation if $w(a) = 0$ and $w(t^+) = w(t)$ for all t. Hence, these functions $\boldsymbol{w}$ are continuous from the right on $[a,b]$.

It is easy to see the set of all normalized functions of bounded variation on $[a,b]$ is a subspace of $BV([a,b])$ which inherits the usual norm. Thus $\|\boldsymbol{w}\| = V(\boldsymbol{w}; a, b)$.

Definition 11.2.6 The Subspace of Normalized Functions of Bounded Variation $NBV([a,b])$

The subspace of all normalized functions of bounded variation on $[a,b]$ is denoted by $NBV([a,b])$.

To show the norm isometry we are looking for, we need a few key facts which we establish in the next two lemmas.

Lemma 11.2.7 Characterizing $\phi(c^+)$ for a Function of Bounded Variation ϕ

Let $\boldsymbol{\phi} \in BV([a,b])$. Then for $a \leq c < b$,

$$\lim_{h \to 0^+} \int_c^{c+h} \phi(t)\, dt = \phi(c^+)$$

Proof 11.2.7

There is a standard decomposition theorem for functions of bounded variation (see (Peterson (16) 2019)) which say $\boldsymbol{\phi} = \boldsymbol{f} - \boldsymbol{g}$ where $\boldsymbol{f}$ and $\boldsymbol{g}$ are both monotone increasing on $[a,b]$. Thus, given $c \in [a,b)$, since $\boldsymbol{f}$ is monotone increasing, it follows $f(c+h) \geq f(c)$ for all $c > 0$ with $c+h \in [a,b)$. Thus, if (h_n) is a sequence with $h_n \downarrow 0^+$, $(f(c+h_n))$ is a decreasing sequence which is bounded below. Hence, this sequence must have a limiting value. It is easy to show the value of this limit is independent of the choice of sequence (h_n). Thus $\lim_{h\to 0^+} f(c+h)$ exists.

In a similar way, we argue that $\lim_{h\to 0^+} g(c+h)$ exists and so $\phi(c^+) = \lim_{h\to 0^+}(f(c+h) - g(c+h))$ exists. Then,

$$\begin{aligned}
\left| \frac{1}{h} \int_c^{c+h} \phi(t)\, dt - \phi(c^+) \right| &= \left| \frac{1}{h} \int_c^{c+h} \phi(t)\, dt - \frac{1}{h} \int_c^{c+h} \phi(c^+) \right| \\
&= \left| \frac{1}{h} \int_c^{c+h} (\phi(t) - \phi(c^+))\, dt \right| \\
&\leq \frac{1}{h} \int_c^{c+h} |\phi(t) - \phi(c^+)|\, dt
\end{aligned}$$

Since $\lim_{h\to 0^+} \phi(c+h) = \phi(c^+)$, given the tolerance $\eta > 0$, there is $\delta > 0$ so that $|\phi(t) - \phi(c^+)| < \eta$ if $0 < |t - c| < \delta$. So if $|h| < \delta$, $|\phi(t) - \phi(c^+)| < \eta$. This implies

$$\left| \frac{1}{h} \int_c^{c+h} \phi(t)\, dt - \phi(c^+) \right| < \frac{1}{h}\, \eta\, h = \eta$$

when $|h| < \delta$. This shows if $c \in [a,b)$,

$$\lim_{h \to 0^+} \frac{1}{h} \int_c^{c+h} | : \phi(t)\, dt = \phi(c^+)$$

■

Lemma 11.2.8 Characterizing $\phi(c^-)$ for a Function of Bounded Variation ϕ

Let $\phi \in BV([a, b])$. Then for $a \leq c < b$,

$$\lim_{h \to 0^+} \int_{c-h}^{c} \phi(t)\, dt = \phi(c^-)$$

Proof 11.2.8
The proof is similar to the one we just did. ■

We are now ready to prove our main result.

Theorem 11.2.9 $NBV([a, b])$ is Norm Isometric to $(C([a, b]))'$

$NBV([a, b])$ is norm isometric to $(C([a, b]))'$; i.e. there is a mapping $T : NBV([a, b]) \rightarrow (C([a, b]))'$ which is $1-1$ and onto and $\|T(\boldsymbol{w})\|_{(C([a,b]))'} = \|\boldsymbol{w}\|_{NBV([a,b])}$ where the norm on $NBV([a, b])$ is the same as the norm on $BV([a, b])$.

Proof 11.2.9
Define the equivalence relationship on $BV([a, b])$ by $\boldsymbol{w_1} \sim \boldsymbol{w_2}$ if $\int_a^b \boldsymbol{x} d\boldsymbol{w_1} = \int_a^b \boldsymbol{x} d\boldsymbol{w_2}$ for all $\boldsymbol{x} \in C([a, b])$. It is easy to show $\sim$ is indeed an equivalence relationship. We need to look at the structure of the equivalence classes $[\boldsymbol{w}]$ generated by $\sim$.

Claim One: $\boldsymbol{w} \sim \boldsymbol{0} \Longleftrightarrow \left(w(a) = w(b), \quad w(c^{\pm}) = w(a), \; c \in (a, b) \right)$

($\Longrightarrow$):
Assume $\boldsymbol{w} \sim \boldsymbol{0}$. Then $0 = \int_a^b \boldsymbol{x} d\boldsymbol{w}$ for all $\boldsymbol{x} \in C([a, b])$. In particular for the constant function $\mathbf{1}$, $0 = \int_a^b 1\, d\boldsymbol{w}$. Standard results from Riemann - Stieltjes integration theory then tell us $0 = w(b) - w(a)$. To prove the rest, define $\boldsymbol{y} \in C([a, b])$ by

$$y(t) = \begin{cases} 1, & a \leq t \leq c \\ 1 - \frac{t-c}{h}, & c < t < c + h \\ 0, & c + h \leq t \leq b \end{cases}$$

We know $\int_a^b \boldsymbol{y} d\boldsymbol{w} = 0$ by assumption. We also know

$$0 = \int_a^c 1\, d\boldsymbol{w} + \int_c^{c+h} \left(1 - \frac{t-c}{h}\right) d\boldsymbol{w}$$

Now we will use the integration by parts formula we proved for Riemann - Stieltjes integration on the second integral.

$$\int_c^{c+h} \left(1 - \frac{t-c}{h}\right) d\boldsymbol{w} = \left(1 - \frac{t-c}{h}\right) \boldsymbol{w} \Big|_c^{c+h} - \int_c^{c+h} \boldsymbol{w} d\boldsymbol{u}$$

where $\boldsymbol{u}$ is the function of bounded variation defined by $u(t) = \left(1 - \frac{t-c}{h}\right)$. Evaluating, we obtain

$$\int_c^{c+h} \left(1 - \frac{t-c}{h}\right) d\boldsymbol{w} \quad = \quad -w(c) + \frac{1}{h}\int_c^{c+h} \boldsymbol{w}dt$$

Then, combining with the other result, we have

$$\begin{aligned} 0 &= \int_a^c 1\, d\boldsymbol{w} - w(c) + \frac{1}{h}\int_c^{c+h} \boldsymbol{w}dt = w(c) - w(a) - w(c) + \frac{1}{h}\int_c^{c_h} \boldsymbol{w}dt \\ &= -w(a) + \frac{1}{h}\int_c^{c_h} \boldsymbol{w}dt \end{aligned}$$

Since this holds for all $h > 0$, taking the limit, we find for any $c \in (a, b)$

$$0 \quad = \quad -w(a) + \lim_{h \to 0^+} \frac{1}{h}\int_c^{c_h} \boldsymbol{w}dt = -w(a) + w(c^+)$$

This shows part of the past piece we need to show. In a similar way, we can show $w(b) = w(c^-)$ for any $c \in (a, b)$. Since $w(a) = w(b)$ already, we have shown $w(a) = w(c^\pm)$ for $c \in (a, b)$.

($\Longleftarrow$):
Conversely, if $w(a) = w(b) = w(c^\pm)$ for $c \in (a, b)$, then $\lim_{c \to a} w(c) = w(a)$. We must show in this case $\boldsymbol{w} \sim \boldsymbol{0}$. Now if $\boldsymbol{w} \in BV([a, b])$, it is known that the set of points in $[a, b]$ where $\boldsymbol{w}$ is not continuous is a countable set. You can refer to (Peterson (16) 2019) for the proof. However, since we have taken it as true that $\boldsymbol{w}$ is the difference of two monotonic functions, we can also apply the Riemann - Lebesgue Lemma from (Peterson (18) 2020). This says since each part of the decomposition into monotonic functions is Riemann Integrable, it must be continuous a.e.; hence $\boldsymbol{w}$ is continuous a.e. Let S be the set of discontinuities of $\boldsymbol{w}$. We can also prove extensions of our standard theorems from Riemann Integration to Riemann - Stieltjes integration and prove that if $\boldsymbol{w_1}$ and $\boldsymbol{w_2}$ are in $BV([a, b])$ and $\boldsymbol{w_1} = \boldsymbol{w_2}$ except on a set of content zero, then $\int_a^b \boldsymbol{x}\, d\boldsymbol{w_1} = \int_a^b \boldsymbol{x}\, d\boldsymbol{w_2}$ for any $\boldsymbol{x} \in C([a, b])$. Although we do not prove these facts carefully here, we think these are very plausible results and we have sketched enough of the Riemann - Stieltjes Integration theory to see how this would work out.

We also know $\lim_{h \to 0} w(c + h) = w(a)$ on (a, b). This implies immediately that if $w(t) = w(a)$, $\boldsymbol{w}$ will have to be continuous at that t. Define $\widetilde{\boldsymbol{w}}$ by

$$\widetilde{w}(t) \quad = \quad \begin{cases} w(a), & t \in S^C \\ w(a), & t \in S \end{cases} = w(a)$$

Then $\widetilde{\boldsymbol{w}} \in BV([a, b])$ and $\boldsymbol{w} = \widetilde{\boldsymbol{w}}$ except on S which has content zero. Thus, $\int -a^b \boldsymbol{x}\, d\boldsymbol{w} = \int_a^b \boldsymbol{x}\, d\widetilde{\boldsymbol{w}}$ for all continuous $\boldsymbol{x}$. This tells us $\boldsymbol{w} \sim \widetilde{\boldsymbol{w}}$. But since $\widetilde{\boldsymbol{w}}$ is a constant, $\int_a^b \boldsymbol{x}\, d\widetilde{\boldsymbol{w}} = 0$. Thus, $\int_a^b \boldsymbol{x}\, d\boldsymbol{w} = 0$ too. This holds for $\boldsymbol{x} \in C([a, b])$ and so $\boldsymbol{w} \sim \widetilde{\boldsymbol{w}} \sim \boldsymbol{0}$.

Claim Two: *Let $\boldsymbol{w} \in BV([a, b])$. The $[\boldsymbol{w}]$ contains one and only one member $\widetilde{\boldsymbol{w}} \in NBV([a, b])$.*

Let $\boldsymbol{w} \in BV([a, b])$. Define $\widetilde{\boldsymbol{w}}$ as follows

$$\widetilde{w}(t) \quad = \quad \begin{cases} 0, & t = a \\ w(t^+) - w(a), & a < t < b \\ w(b) - w(a), & t = b \end{cases}$$

Note,

$$\lim_{h\to 0^+} \widetilde{w}(t+h) = \lim_{h\to 0^+} (w(t+h) - w(a)) = w(t^+) - w(a) = \widetilde{w}(t)$$

and so $\widetilde{w}(t^+) = \widetilde{w}(t)$ *for* $t \in (a,b)$. *It is also clear* $\widetilde{\boldsymbol{w}} \in BV([a,b])$ *and so* $\widetilde{\boldsymbol{w}} \in NBV([a,b])$.

As we said earlier, the set of discontinuities for $\boldsymbol{w}$ *is a set of content zero. Let this set be* S. *Then*

$$\widetilde{w}(t) = \begin{cases} 0, & t = a \\ w(t) - w(a), & t \in (a,b) \cap S^C \\ w(t^+) - w(a), & t \in (a,b) \cap S \\ w(b) - w(a), & t = b \end{cases}$$

which is the same as $\boldsymbol{w} - w(a)$ *except on a set of content zero. We have then*

$$\int_a^b \boldsymbol{x}\, d\widetilde{\boldsymbol{w}} = \int_a^b \boldsymbol{x}\, d\boldsymbol{w}$$

for all continuous $\boldsymbol{x}$. *We conclude* $\boldsymbol{w} \sim \widetilde{\boldsymbol{w}}$.

Next, if $\boldsymbol{w}^* \in [\boldsymbol{w}]$ *is also in* $NBV([a,b])$, *then* $\boldsymbol{w}^* - \widetilde{\boldsymbol{w}} \sim \boldsymbol{0}$. *Thus, by* **Claim One**,

$$w^*(a) - \widetilde{w}(a) = w^*(b) - \widetilde{w}(b) \tag{11.5}$$

$$w^*(c^-) - \widetilde{w}(c^-) = w^*(c^+) - \widetilde{w}(c^+) = w^*(a) - \widetilde{w}(a) \tag{11.6}$$

Since $\widetilde{\boldsymbol{w}}$ *is normalized,* $\widetilde{w}(a) = 0$ *and* $\widetilde{w}(c^+) = \widetilde{w}(c)$. *Also,* $\boldsymbol{w^a st}$ *is normalized and so* $w^*(a) = 0$ *and* $w^*(c^+) = w^*(c)$. *Further,* $\widetilde{w}(b) = w(b) - w(a)$ *and* $\widetilde{w}(c) = w(c^+) - w(a)$. *Now use this information in Equation 11.5 to find* $w^*(b) = w(b) - w(a) = \widetilde{w}(b)$.

From Equation 11.6, we have $0 = w^*(c^+) - \widetilde{w}(c^+) = w^*(c) - \widetilde{w}(c)$. *Therefore,* $w^*(c) = \widetilde{w}(c)$ *for* $c \in (a,b)$. *We already know* $\boldsymbol{w}^*$ *and* $\widetilde{\boldsymbol{w}}$ *match at* a *and* b, *so* $\boldsymbol{w}^* = \widetilde{\boldsymbol{w}}$ *and so* $\widetilde{\boldsymbol{w}} \in [\boldsymbol{w}]$ *is unique.*

Claim Three*: Let* $\boldsymbol{v}^*$ *be the unique element of* $NBV([a,b])$ *in* $[\boldsymbol{v}]$ *for* $\boldsymbol{v} \in BV([a,b])$. *Then* $V(\boldsymbol{v}^*; a, b) \leq V(\boldsymbol{v}; a, b)$.

From our discussions, we know

$$v^*(t) = \begin{cases} 0, & t = a \\ v(t^+) - v(a), & a < t < b \\ v(b) - v(a), & t = b \end{cases}$$

Let $P = \{t_0, t_1, \ldots, t_p\}$ *be any partition of* $[a,b]$. *We know* $v(t^+)$ *always exists as* $\boldsymbol{v}$ *is the difference of two monotone functions. So* $v(t_k^+)$ *exists at each point in the partition* P. *Note* $v^*(t_k) = v(t_k^+) - v(a)$ *for all partition points* t_k. *Let* $\epsilon > 0$ *be given. Then we can find a points*$_k \in (t_k, t_{k+1})$ *so that* $|v(t_k^+) - v(s_k)| < \frac{\epsilon}{2p}$. *Let* $s_0 = a$ *and* $s_p = b$. *Then* $Q = \{s_0, s_1, \ldots, s_p\}$ *is another partition of* $[a,b]$. *We see*

$$\begin{aligned} \sum_{k\in P} |v^*(t_k) - v^*(t_{k-1})| &= \sum_{k\in P} |v(t_k^+) - v(t_{k-1}^+)| \\ &\leq \sum_{k\in P} |v(t_k^+) - v(s_k)| + \sum_{k\in P} |v(s_k) - v(s_{k-1})| \end{aligned}$$

$$
\begin{aligned}
&+ \sum_{k \in P} |v(s_{k-1}) - v(t_{k-1}^{+})| \\
&< \sum_{k \in P} \frac{\epsilon}{2p} + \sum_{k \in P} |v(s_k) - v(s_{k-1})| + \sum_{k \in P} \frac{\epsilon}{2p} \\
&= \sum_{k \in P} |v(s_k) - v(s_{k-1})| + \epsilon
\end{aligned}
$$

By definition, $\sum_{k \in P} |v(s_k) - v(s_{k-1})| \leq V(\boldsymbol{v}; a, b)$*. So we have shown,* $\sum_{k \in P} |v^*(t_k) - v^*(t_{k-1})| \leq V(\boldsymbol{v}; a, b) + \epsilon$*. Taking the supremum over all partitions, we find* $V(\boldsymbol{v}^*; a, b) \leq V(\boldsymbol{v}; a, b) + \epsilon$*. But* $\epsilon > 0$ *is arbitrary and therefore we have established the claim:* $V(\boldsymbol{v}^*; a, b) \leq V(\boldsymbol{v}; a, b)$*.*

Claim Four*: Define* $T : NBV([a, b]) \to (C([a, b]))'$ *by* $(T(\boldsymbol{z}))(\boldsymbol{y}) = \int_a^b \boldsymbol{y} \, d\boldsymbol{z}$ *for all* $\boldsymbol{y} \in C([a, b])$*. Then* T *is a norm isometry.*

Now

$$
|(\,T(\boldsymbol{z})\,)(\boldsymbol{y})| \;=\; \left| \int_a^b \boldsymbol{y} \, d\boldsymbol{z} \right| \leq \|\boldsymbol{y}\|_\infty V(\boldsymbol{z}; a, b)
$$

using a standard estimate from Riemann - Stieltjes Integration Theory. This is the result $|\int_a^b d\boldsymbol{z}| \leq V(\boldsymbol{z}; a, b)$ *which is not surprising as the variation is defined in terms of a supremum over all partitions of* $[a, b]$*. This means for all* $\boldsymbol{y} \neq \boldsymbol{0}$*, we have*

$$
\frac{|(\,T(\boldsymbol{z})\,)(\boldsymbol{y})|}{\|\boldsymbol{y}\|_\infty} \;\leq\; V(\boldsymbol{z}; a, b)
$$

Taking the supremum over all $\boldsymbol{y} \neq \boldsymbol{0}$*, then gives an estimate of the operator norm of the linear functional* $T(\boldsymbol{z})$*;* $\|T(\boldsymbol{z})\| \leq V(\boldsymbol{z}; a, b) = \|z\|_{NBV}$ *as* $z(a) = 0$*.*

Our first representation theorem says there exists $\boldsymbol{u} \in BV([a, b])$ *with*

$$
(\,T(\boldsymbol{z})\,)(\boldsymbol{y}) \;=\; \int_a^b \boldsymbol{y} \, d\boldsymbol{u}, \quad \|T(\boldsymbol{z})\| = V(\boldsymbol{u}; a, b)
$$

for all continuous $\boldsymbol{y}$*. Let* $\boldsymbol{u}^*$ *be the unique element in* $[\boldsymbol{u}]$ *which is in* $NBV([a, b])$*. Then*

$$
(\,T(\boldsymbol{z})\,)(\boldsymbol{y})| \;=\; \int_a^b \boldsymbol{y} \, d\boldsymbol{u} = \int_a^b \boldsymbol{y} \, d\boldsymbol{u}^*
$$

Using the same argument we used before

$$
|(\,T(\boldsymbol{z})\,)(\boldsymbol{y})| \;=\; \left| \int_a^b \boldsymbol{y} \, d\boldsymbol{u}^* \right| \leq \|\boldsymbol{y}\|_\infty V(\boldsymbol{u}^*; a, b) \Longrightarrow \|T(\boldsymbol{z})\| \leq V(\boldsymbol{u}^*; a, b)
$$

Then by **Claim Three***, we have*

$$
\|T(\boldsymbol{z})\| \leq V(\boldsymbol{u}^*; a, b) \leq V(\boldsymbol{u}; a, b) = \|T(\boldsymbol{z})\|
$$

We also know $\int_a^b \boldsymbol{y} \, d\boldsymbol{z} = \int_a^b \boldsymbol{y} \, d\boldsymbol{u}^*$*, so* $\boldsymbol{z} \sim \boldsymbol{u}^*$*. Hence* $\boldsymbol{z}$ *and* $\boldsymbol{u}^*$ *are both in* $[\boldsymbol{u}]$*. But there is only one element of* $NBV([a, b])$ *in* $[\boldsymbol{u}]$*. So we must have* $\boldsymbol{z} = \boldsymbol{u}^*$*. Thus,*

$$
\|T(\boldsymbol{z})\| \leq V(\boldsymbol{z}; a, b) = \|\boldsymbol{z}\| \leq \|T(\boldsymbol{z})\|
$$

and so T is norm isometric.

Claim Five: *T is bijective.*

If $T(\boldsymbol{z}) = T(\boldsymbol{w})$, then for all continuous $\boldsymbol{y}$, $\int_a^b \boldsymbol{y}\, d\boldsymbol{z} = \int_a^b \boldsymbol{y}\, d\boldsymbol{w}$. This implies $\boldsymbol{z} - \boldsymbol{w} \sim \boldsymbol{0}$. But there is only one element of $NBV([a,b])$ in $[\boldsymbol{0}]$ which is $\boldsymbol{0}$. So $\boldsymbol{z} = \boldsymbol{w}$ which says T is $1-1$.

If $\boldsymbol{f} \in (C([a,b]))'$, there is $\boldsymbol{u} \in BV([a,b])$ with $f(\boldsymbol{y}) = \int_a^b \boldsymbol{y}\, d\boldsymbol{u} = \int_a^b \boldsymbol{y}\, d\boldsymbol{u}^*$ where $\boldsymbol{u}^*$ is the unique element of $NBV([a,b])$ in $[\boldsymbol{u}]$. Thus, $f(\boldsymbol{y}) = (T(\boldsymbol{u}^*))(\boldsymbol{y})$ for all continuous $\boldsymbol{y}$. We see $f = T(\boldsymbol{u}^*)$ and so T is onto. ■

This is a very nice application of all the material we have been learning. There is just a small bit of discomfort as we do not provide a full and careful discussion of all the ideas we need from Riemann - Stieltjes Integration theory, but we think we have sketched out most of what you need. And it is easy enough to look this stuff up in the full treatment in (Peterson (16) 2019).

11.3 Riesz's Characterization of the Hilbert Space Dual

Let's look closely at the dual of a Hilbert Space H. A linear functional on H is a map $f : H \to F$ where F is the underlying field.

Theorem 11.3.1 Riesz's Theorem: the Characterization of the Hilbert Space Dual

Let $(X, <\cdot,\cdot>)$ be a Hilbert Space. Let $f \in H'$. Then there is a unique $\boldsymbol{z_f} \in H$ so that $f(\boldsymbol{x}) = <\boldsymbol{x}, \boldsymbol{z_f}>$ with $\|\boldsymbol{z_f}\| = \|f\|_{op}$.

Proof 11.3.1
(Finding $\boldsymbol{z_f}$):
If $f = 0$, we can choose $\boldsymbol{z_f} = \boldsymbol{0}$.

If $f \neq 0$, note $\boldsymbol{z_f}$ cannot be zero as otherwise $<\boldsymbol{x}, \boldsymbol{z_f}> = 0$ for all $\boldsymbol{x}$. Since $f \neq 0$, the kernel of f is not all of H. We know $ker(f)$ is a subspace of f as f is linear. Also, if $(\boldsymbol{x_n})$ in $ker(f)$ converges to $\boldsymbol{x}$ then since f is continuous,

$$f(\boldsymbol{x}) = f(\lim_{n\to\infty} \boldsymbol{x_n}) = \lim_{n\to\infty} f(\boldsymbol{x_n}) = 0$$

Thus $ker(f)$ is a closed subspace and it follows $H = ker(f) \oplus (ker(f))^\perp$. Now the $\boldsymbol{z_f}$ we seek must satisfy $f(\boldsymbol{x}) = <\boldsymbol{x}, \boldsymbol{z_f}>$ for all $\boldsymbol{x}$. In particular, if $\boldsymbol{x} \in ker(f)$, we would have $0 = <\boldsymbol{x}, \boldsymbol{z_f}>$ also. We conclude $\boldsymbol{z_f} \in (ker(f))^\perp$.
Since $ker(f) \neq \boldsymbol{0}$, $(ker(f))^\perp \neq \boldsymbol{0}$ either. So there is a $\boldsymbol{z_0} \in (ker(f))^\perp$ which is nonzero. We will use this $\boldsymbol{z_0}$ to build $\boldsymbol{z_f}$. Let $\boldsymbol{x} \in H$ be arbitrary. Then, $\boldsymbol{u} = f(x)\boldsymbol{z_0} - f(\boldsymbol{z_0})\boldsymbol{x}$ satisfies $f(\boldsymbol{u}) = 0$ and hence $\boldsymbol{u} \in ker(f)$. This tells us $<\boldsymbol{z_0}, \boldsymbol{u}> = 0$ as $\boldsymbol{z_0} \in (ker(f))^\perp$. Hence,

$$0 = <\boldsymbol{z_0}, f(x)\boldsymbol{z_0} - f(\boldsymbol{z_0})\boldsymbol{x} = \overline{f(\boldsymbol{x})} <\boldsymbol{z_0}, \boldsymbol{z_0}> - \overline{f(\boldsymbol{z_0})} <\boldsymbol{z_0}, \boldsymbol{x}>$$

Since $\boldsymbol{z_0} \neq \boldsymbol{0}$, we can then say

$$\begin{aligned} \overline{f(\boldsymbol{x})} &= \frac{\overline{f(\boldsymbol{z_0})} <\boldsymbol{z_0}, \boldsymbol{x}>}{\|\boldsymbol{z_0}\|^2} \Longrightarrow f(\boldsymbol{x}) = \frac{f(\boldsymbol{z_0}) <\boldsymbol{x}, \boldsymbol{z_0}>}{\|\boldsymbol{z_0}\|^2} \\ &= \left\langle \boldsymbol{x}, \frac{\overline{f(\boldsymbol{z_0})}}{\|\boldsymbol{z_0}\|^2} \boldsymbol{z_0} \right\rangle = <\boldsymbol{x}, \boldsymbol{z_f}> \end{aligned}$$

where $\boldsymbol{z_f} = \frac{\overline{f(\boldsymbol{z_0})}}{\|\boldsymbol{z_0}\|^2}\,\boldsymbol{z_0}$*. Thus, we have found our* $\boldsymbol{z_f}$*.*

(Is $\boldsymbol{z_f}$ *unique?)*
If $f(\boldsymbol{x}) =< \boldsymbol{x}, \boldsymbol{z} >=< \boldsymbol{x}, \boldsymbol{z_f} >$*, then* $< \boldsymbol{x}, \boldsymbol{z} - \boldsymbol{z_f} >= 0$ *for all* $\boldsymbol{x}$*. In particular, for* $\boldsymbol{x} = \boldsymbol{z} - \boldsymbol{z_f}$*, we find* $\|\boldsymbol{z} - \boldsymbol{z_f}\|^2 = 0$ *which forces* $\boldsymbol{z} = \boldsymbol{z_f}$*.*

($\|\boldsymbol{z_f}\| = \|f\|_{op}$*):*
First, $f(\boldsymbol{x}) =< \boldsymbol{x}, \boldsymbol{z_f} >$ *for all choices of* $\boldsymbol{x}$*. Thus,* $f(\boldsymbol{z_f}) = \|\boldsymbol{z_f}\|^2$ *which implies* $\frac{f(\boldsymbol{z_f})}{\|\boldsymbol{z_f}\|} = \|\boldsymbol{z_f}\|$*. This tells us that* $\|f\|_{op} \geq \|\boldsymbol{z_f}\|$*.*

To get the reverse inequality, since

$$|f(\boldsymbol{x})| \;=\; |< \boldsymbol{x}, \boldsymbol{z_f} >| \leq \|\boldsymbol{x}\|\,\|\boldsymbol{z_f}\| \Longrightarrow \|f\|_{op} = \sup_{\boldsymbol{x}\neq\boldsymbol{0}} \frac{|f(\boldsymbol{x})|}{\|\boldsymbol{x}\|} \leq \|\boldsymbol{z_f}\|$$

Combining, we see $\|\boldsymbol{z_f}\| = \|f\|_{op}$*.* ■

Homework

Exercise 11.3.1 *Let* $\boldsymbol{f}$ *be in* $(\Re^n)'$ *which we know is a Hilbert Space and assume* $\boldsymbol{E}$ *is a basis of* $\Re^n$*. Let the vector* $\boldsymbol{z_f}$ *be defined by* $(z_f)_i = f(\boldsymbol{E_i})$*. Prove for any* $\boldsymbol{x} \in \Re^n$ $f(\boldsymbol{x}) =< \boldsymbol{x}, \boldsymbol{z_f} >= \sum_{i=1}^{n} x_i (z_f)_i$*. Also prove* z_f *is unique with* $\|\boldsymbol{z_f}\| = \|f\|_{op}$*. Note we can prove Riesz's Theorem easier here!*

- *Let* $T : (\Re^n)' \rightarrow \Re^n$ *be defined by* $T(\boldsymbol{f}) = \boldsymbol{z_f}$*. Prove* T *is* $1-1$*, onto and isometric.*
- *Prove* $\Re^n$ *is reflexive showing all the details of the mappings involved.*

Exercise 11.3.2 *Let* $\boldsymbol{f}$ *be in* $(\ell^2)'$ *we know is a Hilbert Space with the usual inner product. Let* $\boldsymbol{E}$ *be the standard Schauder basis for* ℓ^2*. Let the* $\boldsymbol{z_f} \in \ell^2$ *be defined by* $(z_f)_i = f(\boldsymbol{E_i})$*. Then for any* $\boldsymbol{x} \in \ell^2$ $< \boldsymbol{x}, \boldsymbol{z_f} >= \sum_{i=1}^{\infty} x_i (z_f)_i$*. Prove* z_f *is unique and* $f(\boldsymbol{x}) =< \boldsymbol{x}, \boldsymbol{z_f} >$ *with* $\|\boldsymbol{z_f}\| = \|f\|_{op}$*. Note we can prove Riesz's Theorem easier here.*

- *Let* $T : (\ell^2)' \rightarrow \ell^2$ *be defined by* $T(\boldsymbol{f}) = \boldsymbol{z_f}$*.* T *is* $1-1$*, onto and isometric.*
- *Prove* ℓ^2 *is reflexive showing all the details of the mappings involved.*

Exercise 11.3.3 *Let* f *be defined by* $f(\boldsymbol{y}) =< \boldsymbol{y}, \boldsymbol{x} >$ *for all* $\boldsymbol{y} \in \Re^n$ *for* $\boldsymbol{x} = \begin{bmatrix} 2 \\ 3 \\ -3 \end{bmatrix}$*. Prove* $f \in (\Re^n)'$ *and compute its operator bound.*

Exercise 11.3.4 *Let* A *be an* $n \times n$ *matrix and let* f *be defined by* $f(\boldsymbol{y}) =< A(\boldsymbol{x}), \boldsymbol{y} >$ *for all* $\boldsymbol{y} \in \Re^n$ *for* $\boldsymbol{x} = \begin{bmatrix} 12 \\ 13 \\ -23 \end{bmatrix}$*. Prove* $f \in (\Re^n)'$ *and compute its operator bound.*

We can then show $H' \equiv H$.

Theorem 11.3.2 $H' \equiv H$

Let $(X, < \cdot, \cdot >)$ *be a Hilbert Space. Then to* $f \in H'$ *there is a unique* $\boldsymbol{z_f} \in H$ *with* $f(\boldsymbol{x}) =< \boldsymbol{x}, \boldsymbol{z_f} >$ *and* $\|\boldsymbol{z_f}\| = \|f\|_{op}$*. The mapping* $T : H' \rightarrow H$ *defined by* $T(f) = \boldsymbol{z_f}$ *is a linear bijective norm isometry. Thus* $H' \equiv H$*.*

Proof 11.3.2
We already know the mapping T is well-defined and since $\|T(f)\| = \|\boldsymbol{z_f}\| = \|f\|_{op}$, it is a norm isometry. Therefore the map is $1-1$.

Now

$$f(\boldsymbol{x}) = <\boldsymbol{x}, \boldsymbol{z_f}> \Longrightarrow (\alpha f)(\boldsymbol{x}) = \alpha\,(f(\boldsymbol{x})) = \alpha <\boldsymbol{x}, \boldsymbol{z_f}> = <\boldsymbol{x}, \overline{\alpha}\boldsymbol{z_f}>$$

This says $T(\alpha f) = <\boldsymbol{x}, \overline{\alpha}\boldsymbol{z_f}>$ and so $\boldsymbol{z_{\alpha f}} = \overline{\alpha}\boldsymbol{z_f}$. We can argue this way for any linear functional. Also, it is easy to see since $(T(f+g))(\boldsymbol{x}) = <\boldsymbol{x}, \boldsymbol{z_f} + \boldsymbol{z_g}>$ that T is additive. Finally, we note

$$\begin{aligned}(T(\alpha f + \beta g))(\boldsymbol{x}) &= <\boldsymbol{x}, \boldsymbol{z_{\alpha f}} + \boldsymbol{z_{\beta g}}> = <\boldsymbol{x}, \overline{\alpha}\boldsymbol{z_f} + \overline{\beta}\boldsymbol{z_g}> \\ &= \alpha <\boldsymbol{x}, \boldsymbol{z_f}> + \beta <\boldsymbol{x}, \boldsymbol{z_g}> = (\alpha T(f) + \beta T(g))(\boldsymbol{x})\end{aligned}$$

This shows T is linear.

It is easy to see T is onto. Any $\boldsymbol{z} \in H$ defines the bounded linear functional $f(\boldsymbol{x}) = <\boldsymbol{x}, \boldsymbol{z}>$ and since the choice of element assigned to f must be unique, we must have $T(f) = \boldsymbol{z}$ here. ■

As a last introduction to these sorts of results, let's prove a powerful theorem about compactness. We know if the normed linear space is infinite dimensional, the unit ball about zero cannot be compact. However, we can prove any sequence in $\overline{B(\boldsymbol{0}, 1)} \subset H$ for a Hilbert Space H has a weakly convergent subsequence to an element of $\overline{B(\boldsymbol{0}, 1)}$. This is what we mean by weakly compact. This is a special case of a series of results which we will not prove here. However, we will state some of them and discuss a bit about their proofs before we prove the Hilbert Space result. We know about weak and weak * convergence and so we can discuss whether sequences are compact with respect to weak or weak* convergence. Such sequences would obviously be called sequentially weakly compact or sequentially weak * compact. We have not discussed the idea of topological compactness here and we actually discuss that in (Peterson (17) 2020). Roughly speaking, recall that maps are continuous if the inverse images of open sets are open. So in the context of linear functionals on a normed linear space X, if we put a topology on X and a topology on $\Re$ if the functionals are real-valued, we would want the inverse image of an open set in $\Re$ to be an open set in X. As you can imagine there are lots of ways to define the collections of open sets that gives a topology on X. The **weak** topology is the one where all the linear functionals on X are continuous and if we remove open sets from the collection, some of the linear functionals stop being continuous. We say it is the coarsest topology which guarantees the continuity we need.

We do a similar thing for linear functionals on X'. The **weak *** topology is the coarsest one which will make all the linear functionals continuous in the sense we have outlined above. Once we have topologies, we can then talk intelligently about topological compactness of sets using the standard ideas of finite subcovers of open covers like before. We don't have this machinery yet, so our discussions only involve sequential compactness ideas. The important theorem here is this:

Theorem 11.3.3 The Banach-Alaoglu Theorem: the Closed Unit Ball of the Dual Space of a Normed Linear Space is Compact in the weak* Topology

Let X be a normed linear space and $U = \overline{B(\boldsymbol{0}, 1)}$ be the closed unit ball. Then U is topologically compact in the weak topology: i.e. any open cover of U must have a finite subcover.*

Proof 11.3.3

The proof here is not constructive and uses concepts we don't yet know such as product topologies over uncountable products and Tychonoff's theorem. If you know X is separable, the proof can be more constructive as you see in our proof below for the Hilbert Space setting. ■

Theorem 11.3.4 The Closed Unit Ball in Hilbert Space is Weakly Compact

> *Let H be a Hilbert Space. Then $\overline{B(\mathbf{0},1)}$ is weakly sequentially compact.*

Proof 11.3.4

The proof of this result uses the same sort of argument we use in the proof of the Arzela - Ascoli theorem: the Cauchy Diagonalization Process.

Let $(\boldsymbol{u_n})$ be a sequence in $\overline{B(\mathbf{0},1)}$. We want to show there is a subsequence $(\boldsymbol{u_n^1})$ and an element $\boldsymbol{u}$ in $\overline{B(\mathbf{0},1)}$ so that $\boldsymbol{u_n^1} \overset{weak}{\rightarrow} \boldsymbol{u}$. This means $\boldsymbol{f}(\boldsymbol{u_n^1}) \rightarrow \boldsymbol{f}(\boldsymbol{u})$ for all $\boldsymbol{f} \in H'$.

We do not know if H is separable, but we do not need that here. Let $U = \overline{span(\boldsymbol{u_n})}$ be the closure of the span of sequence in H. Then $H = U \oplus U^\perp$. If $\boldsymbol{y} \in U$, then there is a sequence $(\boldsymbol{s_n})$ from U so that $\boldsymbol{s_n} \rightarrow \boldsymbol{y}$. Each $\boldsymbol{s_n}$ has a representation: $\boldsymbol{s_n} = \sum_{i=1}^{p_n} c_{n,i}\boldsymbol{u_{n,i}}$. Hence, given $\epsilon > 0$, there is an element s_p with $\|\boldsymbol{s_p} - \boldsymbol{y}\| < \epsilon$. If we restrict our attention to rational coefficients, it is straightforward to then show there is an element $\widetilde{\boldsymbol{s_p}}$ with rational coefficients in the representation which is within ϵ of $\boldsymbol{y}$ in norm. The set of all such finite representations with rational coefficients in countable and dense in U. Label the elements of this set as $(\boldsymbol{y_k})$ as it is countable. Use these elements to create an orthogonal sequence $\boldsymbol{x_k} = \frac{\boldsymbol{y_k}}{\|\boldsymbol{x_k}\|}$ Hence, $\|\boldsymbol{x_k}\| = 1$ for all indices. We now know U is separable.

The sequence $(< \boldsymbol{u_n}, \boldsymbol{x_1} >)$ is bounded and so there is a convergent subsequence $(\boldsymbol{u_n^1})$ and a scalar c_1 so that $< \boldsymbol{u_n^1}, \boldsymbol{x_1} > \rightarrow c_1$. Thus $|c_1| \leq 2$ using a standard argument. Next, since the sequence $(< \boldsymbol{u_n^1}, \boldsymbol{x_2} >)$ is bounded, there is a convergent subsequence $(\boldsymbol{u_n^2}) \subset (\boldsymbol{u_n^1})$ and a scalar c_2 so that $< \boldsymbol{u_n^2}, \boldsymbol{x_2} > \rightarrow c_2$. Again, it is easy to see $|c_2| \leq 2$. In fact, note we also know $< \boldsymbol{u_n^2}, \boldsymbol{x_1} > \rightarrow c_1$.

It is now clear how the process goes. After P steps, we have extracted a subsequence $(\boldsymbol{u_n^P}) \subset \ldots \subset (\boldsymbol{u_n^1})$ and a scalar c_P so that $< \boldsymbol{u_n^P}, \boldsymbol{x_i} > \rightarrow c_i$ for $1 \leq i \leq P$ with $|c_P| \leq 2$. Since (c_n) is a bounded sequence by the Bolzano - Weierstrass theorem there is a subsequence (c_n^1) and a number c so that $c_n^1 \rightarrow c$. Now keep only $\boldsymbol{u_n}$ elements whose indices correspond to the indices in the subsequence (c_n^1).

We now use the Cauchy diagonalization process on these elements. We look only at the **diagonal** *elements $\boldsymbol{u_p^p}$ in this process. For convenience of exposition, let's let $\boldsymbol{u_p'} = \boldsymbol{u_p^p}$. We then know $< \boldsymbol{u_p'}, \boldsymbol{x_k} > \rightarrow c_k$ for all p and k. Given any $\boldsymbol{y}$ in U, there is a sequence $(\boldsymbol{x_k^1})$ so that $\boldsymbol{x_k^1} \rightarrow \boldsymbol{y}$. Note by the continuity of the inner product*

$$< \boldsymbol{u_n'}, \boldsymbol{y} > \;=\; \lim_{k\rightarrow\infty} < \boldsymbol{u_n'}, \boldsymbol{x_k^1} >$$

Let $\epsilon > 0$ be given. then there is a K_1 so that $\|\boldsymbol{y} - \boldsymbol{x_k^1}\| < \frac{\epsilon}{3}$ if $k > K_1$. Also, there is a K_2 so that if $k > K_2$, then $|c - c_k| < \frac{\epsilon}{3}$. Hence, if $k > max\{K_1, K_2\}$, we have

$$\begin{aligned} |< \boldsymbol{u_n'}, \boldsymbol{y} > -c| &= |< \boldsymbol{u_n'}, \boldsymbol{y} - \boldsymbol{x_k^1} + \boldsymbol{x_k^1} > -c_k + c_k - c| \\ &\leq |< \boldsymbol{u_n'}, \boldsymbol{y} - \boldsymbol{x_k^1} > | + | < \boldsymbol{u_n'}, \boldsymbol{x_k^1} > -c_k| + |c_k - c| \\ &\leq \|\boldsymbol{y} - \boldsymbol{x_k^1}\| + | < \boldsymbol{u_n'}, \boldsymbol{x_k^1} > -c_k| + \frac{\epsilon}{3} \end{aligned}$$

$$< \frac{\epsilon}{3} + | < \boldsymbol{u}'_{\boldsymbol{n}}, \boldsymbol{x}^1_{\boldsymbol{k}} > -c_k| + \frac{\epsilon}{3} = 2\frac{\epsilon}{3} + | < \boldsymbol{u}'_{\boldsymbol{n}}, \boldsymbol{x}^1_{\boldsymbol{k}} > -c_k|$$

Finally, there is N so that if $n > N$, we have $| < \boldsymbol{u}'_{\boldsymbol{n}}, \boldsymbol{x}^1_{\boldsymbol{k}} > -c_k| < \frac{\epsilon}{3}$. Thus, if $n > N$, $| < \boldsymbol{u}'_{\boldsymbol{n}}, \boldsymbol{y} > -c| < \epsilon$. This shows the $\lim_{n\to\infty} < \boldsymbol{u}'_{\boldsymbol{n}}, \boldsymbol{y} >$ always exists. Hence, the map f defined on U by

$$f(\boldsymbol{y}) \quad = \quad \lim_{n\to\infty} < \boldsymbol{u}'_{\boldsymbol{n}}, \boldsymbol{y} >$$

with $f = 0$ on $U^\perp$ is well-defined. This is clearly a linear map. Note for all $\boldsymbol{y}$ not zero

$$\frac{f(\boldsymbol{y})}{\|\boldsymbol{y}\|} \quad \leq \quad \frac{\lim_{n\to\infty} \|\boldsymbol{u}'_{\boldsymbol{n}} \ \|\boldsymbol{y}\|\|}{\|\boldsymbol{y}\|} = 1$$

which implies f is a bounded linear functional in H'.

Now use the Riesz Representation theorem: there is a unique $\boldsymbol{z} \in H$ so that $f(\boldsymbol{y}) =< \boldsymbol{y}, \boldsymbol{z} >$ for all $\boldsymbol{y} \in H$ and $\|\boldsymbol{z}\| = \|f\| \leq 1$. This shows $\lim_{n\to\infty} < \boldsymbol{u}'_{\boldsymbol{n}}, \boldsymbol{y} >=< \boldsymbol{z}, \boldsymbol{y} >$ on H with $\boldsymbol{z} \in \overline{B(\boldsymbol{0}, 1)}$ and establishes the weak sequential compactness. ■

Homework

Exercise 11.3.5 *Let $\boldsymbol{H}$ be a Hilbert Space. Let $B = \{\boldsymbol{u} \in \boldsymbol{H} \mid \|\boldsymbol{x}\| \leq 1\}$. Let f be a continuous linear functional on $\boldsymbol{H}$. Consider the extremal problem $\min_{\boldsymbol{u}\in B} |f(\boldsymbol{u})|$. Let $(\boldsymbol{u_n})$ be a minimizing sequence for this problem. Prove there is a subsequence of $(\boldsymbol{u_n})$ which converges weakly to an element of B.*

Exercise 11.3.6 *Let $\boldsymbol{H}$ be the Hilbert Space $\Re^n$. Let $B = \{\boldsymbol{u} \in \boldsymbol{H} | \|\boldsymbol{x}\| \leq 1\}$. Let f be a continuous linear functional on $\boldsymbol{H}$. Consider the extremal problem $\min_{\boldsymbol{u}\in B} |f(\boldsymbol{u})|$. Let $(\boldsymbol{u_n})$ be a minimizing sequence for this problem. Prove there is a subsequence of $(\boldsymbol{u_n})$ which converges weakly to an element of B. Prove there is also a subsequence of $(\boldsymbol{u_n})$which converges in norm to an element of B. Can we assume these are the same sequence? What can we say in general?*

Exercise 11.3.7 *Let $\boldsymbol{H}$ be the Hilbert Space $\Re^n$. Let A be a $n \times n$ matrix and let $B = \{\boldsymbol{u} \in \boldsymbol{H} \mid \|\boldsymbol{x}\| \leq 1\}$. For any $\boldsymbol{x} \in \boldsymbol{H}$, define $f(\boldsymbol{x}) =< A(\boldsymbol{x}), \boldsymbol{x} >$. Consider the extremal problem $\min_{\boldsymbol{u}\in B} |f(\boldsymbol{u})|$. Let $(\boldsymbol{u_n})$ be a minimizing sequence for this problem. Prove there is a subsequence of $(\boldsymbol{u_n})$ which converges weakly to an element of B. Prove there is also a subsequence of $(\boldsymbol{u_n})$which converges in norm to an element of B. Can we assume these are the same sequence? What can we say in general?*

11.4 Sesquilinear Forms

The next results have to do with mappings similar to inner products called sesquilinear forms.

Definition 11.4.1 Sesquilinear Forms

Let X and Y be vector spaces over a field F. Let $h : X \times Y \to F$ satisfy for all inputs from X and Y

SF1: $h(\boldsymbol{x_1} + \boldsymbol{x_2}, \boldsymbol{y}) = h(\boldsymbol{x_1}, \boldsymbol{y}) + h(\boldsymbol{x_2}, \boldsymbol{y})$. *This is the linearity of h in slot one.*

SF2: $h(\boldsymbol{x}, \boldsymbol{y_1} + \boldsymbol{y_2}) = h(\boldsymbol{x}, \boldsymbol{y_1}) + h(\boldsymbol{x}, \boldsymbol{y_2})$. *This is linearity in slot two.*

SF3: $h(\alpha\boldsymbol{x}, \boldsymbol{y}) = \alpha\, h(\boldsymbol{x}, \boldsymbol{y})$. *This is a scaling law for slot one.*

SF4: $h(\boldsymbol{x}, \alpha\boldsymbol{y}) = \overline{\alpha}\, h(\boldsymbol{x}, \boldsymbol{y})$. *This is a complex conjugate scaling law for slot two.*

Note the usual property that you can interchange the order of $\boldsymbol{x}$ and $\boldsymbol{y}$ in h is not possible as they come from different normed linear spaces!

We are primarily interested in sesquilinear forms that are bounded in the following sense.

Definition 11.4.2 Bounded Sesquilinear Forms

Let X and Y be vector spaces over a field F and let $h : X \times Y \to F$ be a sesquilinear form. We say h is bounded if

$$S_B(h) = \sup_{\boldsymbol{x} \neq \boldsymbol{0} \in X} \sup_{\boldsymbol{y} \neq \boldsymbol{0} \in Y} \frac{|h(\boldsymbol{x}, \boldsymbol{y})|}{\|\boldsymbol{x}\|_X \, \|\boldsymbol{y}\|_Y} < \infty$$

So if h is a bounded sesquilinear form, we have the fundamental inequality $|h(\boldsymbol{x}, \boldsymbol{y})| \leq S_B(h)\, \|\boldsymbol{x}\|_X \, \|\boldsymbol{y}\|_Y$.

Here are some examples:

Any inner product: Let $(X, < \cdot, \cdot >)$ be any inner product space. Then the inner product is a sesquilinear form on $X \times X$.

The pairing between ℓ^1 and ℓ^∞: For $(\ell^1, \|\cdot\|_1)$ and $(\ell^\infty, \|\cdot\|_\infty)$ the bilinear pairing between these two spaces is computed as the sum $h(\boldsymbol{x}, \boldsymbol{y}) = \sum_{n=1}^{\infty} x_n \overline{y_n}$ for $\boldsymbol{x} = (x_n) \in \ell^1$ and $\boldsymbol{y} = (y_n) \in \ell^\infty$. Hölder's Inequality shows the calculation is well-defined. It is easy to show properties $SF1$ to $SF4$ are satisfied. We also have

$$\sup_{\boldsymbol{x} \neq \boldsymbol{0} \in X} \sup_{\boldsymbol{y} \neq \boldsymbol{0} \in Y} \frac{|h(\boldsymbol{x}, \boldsymbol{y})|}{\|\boldsymbol{x}\|_X \, \|\boldsymbol{y}\|_Y} \leq \frac{\|\boldsymbol{x}\|_1 \|\boldsymbol{y}\|_\infty}{\|\boldsymbol{x}\|_1 \|\boldsymbol{y}\|_\infty} \leq 1$$

So $S_B(h)$ is finite and this is a bounded sesquilinear form.

The pairing between ℓ^p and ℓ^q: This is similar to the previous example. If $1 < p < \infty$ and q is the number conjugate to p, for $(\ell^p, \|\cdot\|_p)$ and $(\ell^q, \|\cdot\|_q)$, the bilinear pairing between these two spaces is computed as the sum $h(\boldsymbol{x}, \boldsymbol{y}) = \sum_{n=1}^{\infty} x_n \overline{y_n}$ for $\boldsymbol{x} = (x_n) \in \ell^1$ and $\boldsymbol{y} = (y_n) \in \ell^\infty$. Hölder's Inequality shows the calculation is well-defined. It is easy to show properties $SF1$ to $SF4$ are satisfied. We also have

$$\sup_{\boldsymbol{x} \neq \boldsymbol{0} \in X} \sup_{\boldsymbol{y} \neq \boldsymbol{0} \in Y} \frac{|h(\boldsymbol{x}, \boldsymbol{y})|}{\|\boldsymbol{x}\|_X \, \|\boldsymbol{y}\|_Y} \leq \frac{\|\boldsymbol{x}\|_p \|\boldsymbol{y}\|_q}{\|\boldsymbol{x}\|_p \|\boldsymbol{y}\|_\infty} \leq 1$$

So $S_B(h)$ is finite and this is a bounded sesquilinear form.

Homework

Exercise 11.4.1 *Let $(X_1, \|\cdot\|_1)$ and $(X_2, \|\cdot\|_2)$ be normed spaces over $\Re$ and let $h : X_1 \times X_2 \to F$ be a bounded sesquilinear form. Prove if $(\boldsymbol{x_n})$ and $(\boldsymbol{y_n})$ are convergent sequences in X_1 and x_2 respectively, then $(h(\boldsymbol{x_n}, \boldsymbol{y_n}))$ also converges.*

Exercise 11.4.2 *Let f be a vector function $f : \Re^n \to \Re^n$ which satisfies $\|f(\boldsymbol{y})\| \leq B\|\boldsymbol{y}\|$ for some $B > 0$ for all $\boldsymbol{y} \in \Re^n$. This is a global bound condition. Define $h : \Re^n \times \Re^n \to \Re$ by $h(\boldsymbol{x}, \boldsymbol{y}) =< \boldsymbol{x}, f(\boldsymbol{y}) >$.*

- *Prove that $S_B(h)$ is bounded.*
- *Is h a sesquilinear form for any f?*

Exercise 11.4.3 *Let A be an $n \times n$ matrix and define $h : \Re^n \times \Re^n \to \Re$ by $h(\boldsymbol{x}, \boldsymbol{y}) =< \boldsymbol{x}, A(\boldsymbol{y}) >$.*

- *Prove that $S_B(h)$ is bounded.*
- *Prove h is a sesquilinear form for any A.*
- *Prove $\boldsymbol{x}^T A(\boldsymbol{x}) \leq S_B(h) \leq \|A\|_{op}$ for all $\boldsymbol{x} \in \Re^n$.*
- *Can you think of a condition to impose on A to ensure $S_B(h) = \|A\|_{op}$?*

We can now prove a standard representation theorem for bounded sesquilinear forms.

Theorem 11.4.1 Riesz Representation Theorem for Bounded Sesquilinear Forms

Let $(H_1, < \cdot, \cdot >_1)$ and $(H_2, < \cdot, \cdot >_2)$ be two Hilbert Spaces and assume $h : H_1 \times H_2 \to F$ is a bounded sesquilinear form. Then there is a unique bounded linear operator $S : H_1 \to H_2$ so that $\|S\|_{op} = S_B(h)$ and $h(\boldsymbol{x}, \boldsymbol{y}) =< S(\boldsymbol{x}), y >_2$ for all $\boldsymbol{x} \in H_1$ and $\boldsymbol{y} \in H_2$.

Proof 11.4.1
Fix $\boldsymbol{x}$ in H_1 and define $\phi_{\boldsymbol{x}} : H_2 \to H_1$ by $\phi_{\boldsymbol{x}}(\boldsymbol{y}) = \overline{h(\boldsymbol{x}, \boldsymbol{y})}$.

($\phi_{\boldsymbol{x}}$ is a bounded linear functional on H_2):
Note

$$\begin{aligned}
\phi_{\boldsymbol{x}}(\alpha \boldsymbol{y_1} + \beta \boldsymbol{y_2}) &= \overline{h(\boldsymbol{x}, \alpha \boldsymbol{y_1} + \beta \boldsymbol{y_2})} = \overline{h(\boldsymbol{x}, \alpha \boldsymbol{y_1}) + h(\boldsymbol{x}, \beta \boldsymbol{y_2})} \\
&= \overline{\overline{\alpha}\, h(\boldsymbol{x}, \boldsymbol{y_1}) + \overline{\beta}\, h(\boldsymbol{x}, \boldsymbol{y_2})} = \alpha\, \overline{h(\boldsymbol{x}, \boldsymbol{y_1})} + \beta\, \overline{h(\boldsymbol{x}, \boldsymbol{y_2})} \\
&= \alpha\, \phi_{\boldsymbol{x}}(\boldsymbol{y_1}) + \beta\, \phi_{\boldsymbol{x}}(\boldsymbol{y_2})
\end{aligned}$$

Hence $\phi_{\boldsymbol{x}}$ is linear.

We also know

$$|\phi_{\boldsymbol{x}}(\boldsymbol{y})| = |\overline{h(\boldsymbol{x}, \boldsymbol{y})}| \leq S_b h\, \|\boldsymbol{x}\|_1\, \|\boldsymbol{y}\|_2 \Longrightarrow \sup_{\boldsymbol{y} \neq \boldsymbol{0} \in H_2} \frac{|\phi_{\boldsymbol{x}}(\boldsymbol{y})|}{\|\boldsymbol{y}\|_2} \leq S_B(h)\|\boldsymbol{x}\|_1$$

and so $\phi_{\boldsymbol{x}}$ is bounded.
(Each $\phi_{\boldsymbol{x}}$ corresponds to $\boldsymbol{z_x} \in H_2$ with $\|\boldsymbol{z_x}\|_2 = \|\phi_{\boldsymbol{x}}\|_{op}$ and $\phi_{\boldsymbol{x}}(\boldsymbol{y}) =< \boldsymbol{y}, \boldsymbol{z_x} >_2$):
Apply the Riesz Representation Theorem to $\phi_{\boldsymbol{x}}$.
(The mapping $S : H_1 \to H_2$ defined by $S(\boldsymbol{x}) = \boldsymbol{z_x}$ is a bounded linear operator):
First, we show S is linear.

$$< S(\alpha \boldsymbol{x_1} + \beta \boldsymbol{x_2}), \boldsymbol{y} > = h(\alpha \boldsymbol{x_1} + \beta \boldsymbol{x_2}), \boldsymbol{y} >= \alpha\, h(\boldsymbol{x_1}, \boldsymbol{y}) + \beta h(\boldsymbol{x_2}, \boldsymbol{y}) >$$

$$= \quad \alpha < S(\boldsymbol{x_1}), \boldsymbol{y} > +\beta < S(\boldsymbol{x_2}), \boldsymbol{y} >$$

This implies for all $\boldsymbol{y}$ *in* H_2

$$< S(\alpha \boldsymbol{x_1} + \beta \boldsymbol{x_2}) - (\alpha S(\boldsymbol{x_1}) + \beta S(\boldsymbol{x_2})), \boldsymbol{y} >_2 = 0$$

Letting $\boldsymbol{y_0} = S(\alpha \boldsymbol{x_1} + \beta \boldsymbol{x_2}) - \alpha S(\boldsymbol{x_1}) - \beta S(\boldsymbol{x_2})$, *we see* $< \boldsymbol{y_0}, \boldsymbol{y_0} >_2 = 0$. *This tell us* $\boldsymbol{y_0} = 0$ *or* $S(\alpha \boldsymbol{x_1} + \beta \boldsymbol{x_2}) = \alpha S(\boldsymbol{x_1}) + \beta S(\boldsymbol{x_2})$ *which shows* S *is linear.*

($\|S\|_{op} = S_B(h)$*):*
We have

$$\begin{aligned} S_B(h) &= \sup_{(\boldsymbol{x},\boldsymbol{y}) \neq (\boldsymbol{0},\boldsymbol{0}) \in H_1 \times H_2} \frac{|h(\boldsymbol{x},\boldsymbol{y})|}{\|\boldsymbol{x}\|_1 \, \|\boldsymbol{y}\|_2} = \sup_{(\boldsymbol{x},\boldsymbol{y}) \neq (\boldsymbol{0},\boldsymbol{0}) \in H_1 \times H_2} \frac{| < S(\boldsymbol{x}), \boldsymbol{y} >_2 |}{\|\boldsymbol{x}\|_1 \, \|\boldsymbol{y}\|_2} \\ &\geq \sup_{\boldsymbol{x} \neq \boldsymbol{0} \in H_1, S(\boldsymbol{x}) \neq 0} \frac{| < S(\boldsymbol{x}), S(\boldsymbol{x}) >_2 |}{\|\boldsymbol{x}\|_1 \, \|\boldsymbol{S(x)}\|_2} = \sup_{\boldsymbol{x} \neq \boldsymbol{0} \in H_1} \frac{\|S(\boldsymbol{x})\|_2}{\|\boldsymbol{x}\|_1} = \|S\|_{op} \end{aligned}$$

This immediately implies S *is bounded with* $\|S\|_{op} \leq S_B(h)$.

To show the reverse inequality, we have

$$\begin{aligned} S_B(h) &= \sup_{(\boldsymbol{x},\boldsymbol{y}) \neq (\boldsymbol{0},\boldsymbol{0}) \in H_1 \times H_2} \frac{|h(\boldsymbol{x},\boldsymbol{y})|}{\|\boldsymbol{x}\|_1 \, \|\boldsymbol{y}\|_2} = \sup_{(\boldsymbol{x},\boldsymbol{y}) \neq (\boldsymbol{0},\boldsymbol{0}) \in H_1 \times H_2} \frac{| < S(\boldsymbol{x}), \boldsymbol{y} >_2 |}{\|\boldsymbol{x}\|_1 \, \|\boldsymbol{y}\|_2} \\ &\leq \sup_{(\boldsymbol{x},\boldsymbol{y}) \neq (\boldsymbol{0},\boldsymbol{0}) \in H_1 \times H_2} \frac{\|S(\boldsymbol{x})\|_2 \, \|\boldsymbol{y}\|_2}{\|\boldsymbol{x}\|_1 \, \|\boldsymbol{y}\|_2} = \sup_{\boldsymbol{x} \neq \boldsymbol{0} \in H_1} \frac{\|S(\boldsymbol{x})\|_2}{\|\boldsymbol{x}\|_1} = \|S\|_{op} \end{aligned}$$

This establishes the reverse inequality and we now know $\|S\|_{op} = S_B(h)$.

(S *is unique):*
If we assume there is a mapping $T : H_1 \to H_2$ *with* $h(\boldsymbol{x}, \boldsymbol{y}) =< T(\boldsymbol{x}), \boldsymbol{y} >_2 =< S(\boldsymbol{x}), \boldsymbol{y} >_2$, *then* $< T(\boldsymbol{x}) - S(\boldsymbol{x}), \boldsymbol{y} >_2 = 0$ *for all choices of* $\boldsymbol{y}$. *In particular, choose* $\boldsymbol{y_0} = T(\boldsymbol{x}) - S(\boldsymbol{x})$. *Then* $\|\boldsymbol{y_0}\|_2 = 0$ *which implies* $T(\boldsymbol{x}) = S(\boldsymbol{x})$. *We can make this argument at each* $\boldsymbol{x} \in H_1$. *Thus* $T = S$ *and we conclude* S *is unique.* ■

Homework

Exercise 11.4.4 *Let* A *and* B *be* $n \times n$ *matrices and define* $h : \Re^n \times \Re^n \to \Re$ *by* $h(\boldsymbol{x}, \boldsymbol{y}) =< A(\boldsymbol{x}), B(\boldsymbol{y}) >$. *Find the unique operator* S *for which* $S_B(h) = \|S\|_{op}$ *with* $h(\boldsymbol{x}, \boldsymbol{y}) =< S(\boldsymbol{x}), y >$.

Exercise 11.4.5 *Let* A *be an invertible* $n \times n$ *matrix and define* $h : \Re^n \times \Re^n \to \Re$ *by* $h(\boldsymbol{x}, \boldsymbol{y}) =< A^{-1}(\boldsymbol{x}), A(\boldsymbol{y}) >$. *Find the unique operator* S *for which* $S_B(h) = \|S\|_{op}$ *with* $h(\boldsymbol{x}, \boldsymbol{y}) =< S(\boldsymbol{x}), y >$.

Exercise 11.4.6 *Let* A *be an invertible symmetric* $n \times n$ *matrix and define* $h : \Re^n \times \Re^n \to \Re$ *by* $h(\boldsymbol{x}, \boldsymbol{y}) =< A^{-1}(\boldsymbol{x}), A(\boldsymbol{y}) >$. *Find the unique operator* S *for which* $S_B(h) = \|S\|_{op}$ *with* $h(\boldsymbol{x}, \boldsymbol{y}) =< S(\boldsymbol{x}), y >$.

We can now discuss the adjoints of linear operators.

Definition 11.4.3 The Hilbert Adjoint Operator

Let $(H_1, < \cdot, \cdot >_1)$ and $(H_2, < \cdot, \cdot >_2)$ be two Hilbert Spaces and let $T : H_1 \to H_2$ be a linear operator. Then the operator $T^* : H_2 \to H_1$ by $< T(\boldsymbol{x}), \boldsymbol{y} >_2 = < \boldsymbol{x}, T^*(\boldsymbol{y}) >_1$ is called the Hilbert Adjoint if it exists.

The Hilbert adjoint exists for a bounded linear operator.

Theorem 11.4.2 The Hilbert Adjoint Exists for a Bounded Linear Operator

Let $(H_1, < \cdot, \cdot >_1)$ and $(H_2, < \cdot, \cdot >_2)$ be two Hilbert Spaces and let $T : H_1 \to H_2$ be a bounded linear operator. Then the operator $T^* : H_2 \to H_1$ by $< T(\boldsymbol{x}), \boldsymbol{y} >_2 = < \boldsymbol{x}, T^*(\boldsymbol{y}) >_1$ exists, is unique and is a bounded linear operator also.

Proof 11.4.2
Define $h : H_2 \times H_1 \to F$ by $h(\boldsymbol{y}, \boldsymbol{x}) = < \boldsymbol{y}, T(\boldsymbol{x}) >_2$.

(h is a sesquilinear form):
($SF1$): $h(\boldsymbol{y_1} + \boldsymbol{y_2}, \boldsymbol{x}) = < \boldsymbol{y_1} + \boldsymbol{y_2}, T(\boldsymbol{x}) >_2 = < \boldsymbol{y_1}, T(\boldsymbol{x}) >_2 + < \boldsymbol{y_2}, T(\boldsymbol{x}) >_2$. But then $h(\boldsymbol{y_1} + \boldsymbol{y_2}, \boldsymbol{x}) = H(\boldsymbol{y_1}, \boldsymbol{x}) + H(\boldsymbol{y_2}, \boldsymbol{x})$.

($SF2$): $h(\alpha \boldsymbol{y}, \boldsymbol{x}) = < \alpha \boldsymbol{y_1}, T(\boldsymbol{x}) >_2 = \alpha < \boldsymbol{y_1}, T(\boldsymbol{x}) >_2 = \alpha h(\boldsymbol{y}, \boldsymbol{x})$.

($SF3$): $h(\boldsymbol{y_1}, \alpha \boldsymbol{x}) = < \boldsymbol{y_1}, T(\alpha \boldsymbol{x}) >_2 = \overline{\alpha} < \boldsymbol{y_1}, T(\boldsymbol{x}) >_2 = \overline{\alpha} h(\boldsymbol{y_1}, \boldsymbol{x})$.

($SF4$): $h(\boldsymbol{y}, \boldsymbol{x_1} + \boldsymbol{x_2}) = < \boldsymbol{y}, T(\boldsymbol{x_1} + \boldsymbol{x_2}) >_2 = < \boldsymbol{y}, T(\boldsymbol{x_1}) + T(\boldsymbol{x_2}) >_2$. But this implies $h(\boldsymbol{y_1}, \boldsymbol{x_1} + \boldsymbol{x_2}) = h(\boldsymbol{x}, \boldsymbol{y_1}) + h(\boldsymbol{x}, \boldsymbol{y_2})$.

(h is bounded):
We note

$$\begin{aligned} |h(\boldsymbol{y}, \boldsymbol{x})| &= |< \boldsymbol{y}, T(\boldsymbol{x}) >_2 \leq \|\boldsymbol{y}\|_2 \, \|T\|_{op} \, \|\boldsymbol{x}\|_1 \\ &\Longrightarrow S_B(h) = \sup_{(\boldsymbol{x}, \boldsymbol{y}) \neq (\boldsymbol{0}, \boldsymbol{0}) \in H_1 \times H_2} \frac{|h(\boldsymbol{y}, \boldsymbol{x})|}{\|\boldsymbol{x}\|_1 \, \|\boldsymbol{y}\|_2} \leq \|T\|_{op} \end{aligned}$$

So h is bounded.

(There is a unique bounded linear operator $T^* : H_2 \to H_1$ with $h(\boldsymbol{y}, \boldsymbol{x}) = < T^*(\boldsymbol{y}), \boldsymbol{x} >_1$ and $\|T^*\|_{op} = S_B(h)$):
We apply the Riesz Representation Theorem for bounded sesquilinear operators to h. Then, for all choices of $\boldsymbol{x} \in H_1$ and $\boldsymbol{y} \in H_2$, there is a unique operator $T^* : H_2 \to H_1$ so that

$$h(\boldsymbol{y}, \boldsymbol{x}) = < \boldsymbol{y}, T(\boldsymbol{x}) >_2 = < T^*(\boldsymbol{y}), \boldsymbol{x} >_1, \quad \|T^*\|_{op} = S_B(h)$$

So,

$$\overline{< \boldsymbol{y}, T(\boldsymbol{x}) >_2} = \overline{< T^*(\boldsymbol{y}), \boldsymbol{x} >_1} \Longrightarrow < T(\boldsymbol{x}), \boldsymbol{y} >_2 = < \boldsymbol{x}, T^*(\boldsymbol{y}) >_1$$

($\|T^*\|_{op} = \|T\|_{op}$):
To see this, we note

$$S_B(h) = \sup_{(\boldsymbol{x}, \boldsymbol{y}) \neq (\boldsymbol{0}, \boldsymbol{0}) \in H_1 \times H_2} \frac{|h(\boldsymbol{y}, \boldsymbol{x})|}{\|\boldsymbol{x}\|_1 \, \|\boldsymbol{y}\|_2} \geq \sup_{\boldsymbol{x} \neq \boldsymbol{0} \in H_1, T(\boldsymbol{x}) \neq \boldsymbol{0} \in H_2} \frac{|h(\boldsymbol{T}(\boldsymbol{x}), \boldsymbol{x})|}{\|\boldsymbol{x}\|_1 \, \|\boldsymbol{T}(\boldsymbol{x})\|_2}$$

$$= \sup_{\boldsymbol{x} \neq \boldsymbol{0} \in H_1, T(\boldsymbol{x}) \neq \boldsymbol{0} \in H_2} \frac{| < T(\boldsymbol{x}), T(\boldsymbol{x}) >_2 |}{\|\boldsymbol{x}\|_1 \, \|\boldsymbol{T}(\boldsymbol{x})\|_2}$$

$$= \sup_{\boldsymbol{x} \neq \boldsymbol{0} \in H_1} \frac{|\|T(\boldsymbol{x})\|_2|}{\|\boldsymbol{x}\|_1} = \|T\|_{op}$$

Thus, $S_B(h) \geq \|T\|_{op}$. *We also know* $S_B(h) \leq \|T\|_{op}$. *Hence,* $\|T\|_{op} = S_B(h) = \|T^*\|_{op}$. ■

Homework

Exercise 11.4.7 *Let* A *be an* $n \times n$ *matrix. Find the Hilbert Adjoint of* A.

Exercise 11.4.8 *Let* A *be an* $n \times n$ *symmetric matrix. Find the Hilbert Adjoint of* A. *Under what conditions is such a matrix self-adjoint?*

11.5 Adjoints on Normed Linear Spaces

Definition 11.5.1 Adjoints on Normed Linear Spaces

Let $(X, \|\cdot\|_X)$ *and* $(Y, \|\cdot\|_Y)$ *be normed linear spaces and* X' *and* Y' *their respective dual spaces. Given the linear operator* $T : X \rightarrow Y$ *define the adjoint of* T *to be* $T^* : Y' \rightarrow X'$ *by*

$$T^*(f) : X \rightarrow F \Longrightarrow (T^*(f))(\boldsymbol{x}) = f(T(\boldsymbol{x}))$$

Let's construct the adjoint of a linear operator $T : X \rightarrow Y$ where both X and Y are finite dimensional of dimension n and m respectively. We can assume the operator $T : \Re^n \rightarrow \Re^m$ for our calculations as once we choose a basis for the two normed linear spaces, we will find a matrix representation of T to work with. The spaces $\Re^n$ and $\Re^m$ are Hilbert Spaces under the $\|\cdot\|_2$ norm so we have a lot of structure here to work with. Since linear operators on finite dimensional normed spaces are bounded, By Theorem 11.4.2, there is a unique bounded linear operator T^* with $\|T\|_{op} = \|T^*\|_{op}$ and $< T(\boldsymbol{x}), \boldsymbol{y} >=< \boldsymbol{x}, T^*(\boldsymbol{y}) >$ for all choices of $\boldsymbol{x} \in \Re^n$ and $\boldsymbol{y} \in \Re^m$. Let $\boldsymbol{E}$ be a basis for $\Re^n$ and $\boldsymbol{F}$, a basis for $\Re^m$. Then $[T]_{\boldsymbol{EF}}$ is a representation of T. Let $[T^*]_{\boldsymbol{FE}}$ be a representation of T^*. Remember $T^* : \Re^m \rightarrow \Re^n$. Then

$$[T]_{\boldsymbol{EF}} = \begin{bmatrix} T_{11} & \ldots & T_{1n} \\ \vdots & \vdots & \vdots \\ T_{m1} & \ldots & T_{mn} \end{bmatrix}, \quad [T^*]_{\boldsymbol{FE}} = \begin{bmatrix} T^*_{11} & \ldots & T^*_{1m} \\ \vdots & \vdots & \vdots \\ T^*_{n1} & \ldots & T^*_{nm} \end{bmatrix}$$

and

$$[\boldsymbol{x}]_{\boldsymbol{E}} = \begin{bmatrix} x_1 \\ \vdots \\ x_n \end{bmatrix}, \quad [\boldsymbol{y}]_{\boldsymbol{F}} = \begin{bmatrix} y_1 \\ \vdots \\ y_m \end{bmatrix}$$

The adjoint condition gives

$$< T(\boldsymbol{x}), \boldsymbol{y} > = \sum_{i=1}^{m} (T(\boldsymbol{x}))_i y_i = \sum_{j=1}^{n} x_j (T^*(\boldsymbol{y}))_j =< \boldsymbol{x}, T^*(\boldsymbol{y}) >$$

We can expand these expression out as

$$(T(\boldsymbol{x}))_i = \sum_{j=1}^{n} T_{ij}x_j, \quad (T^*(\boldsymbol{y}))_j = \sum_{i=1}^{m} T^*_{ji}y_i$$

to give

$$\sum_{i=1}^{m}\sum_{j=1}^{n} T_{ij}x_jy_i = \sum_{j=1}^{n}\sum_{i=1}^{m} T^*_{ji}x_jy_i \Longrightarrow \sum_{i=1}^{m}\sum_{j=1}^{n}(T_{ij} - T^*_{ji})x_iy_j = 0$$

This is true for all choices of $\boldsymbol{x}$ and $\boldsymbol{y}$. In particular for $\boldsymbol{x} = \boldsymbol{E_p}$ and $\boldsymbol{y} = \boldsymbol{F_q}$, we find

$$\sum_{i=1}^{m}\sum_{j=1}^{n}(T_{ij} - T^*_{ji})\delta_{ip}\delta_{jq} = 0 \Longrightarrow T_{pq} = T^*qp$$

We see if the underlying field F is $\Re$, the adjoint of T is the transpose of T; i.e. $T^* = T^T$.

If the underlying field $F = \mathbb{C}$, then calculations similar to what was done above give us $\overline{T_{qp}} = T^*_{pq}$. Hence if the scalar field is complex, T^* is the conjugate transpose of T.

Comment 11.5.1 *If the bounded linear operator satisfies $T = T^*$, T is said to be self-adjoint. We have discussed this possibility in detail in our study of Stürm - Liouville differential equations.*

Note we don't really know how to calculate the adjoint of the differential operator $L(x) = x'$ as to do so we would need to characterize the dual space of $C^1([a, b])$.

Homework

Exercise 11.5.1 *Show all the details in calculating the adjoint of*

$$A = \begin{bmatrix} 3 & 2 & -1 \\ -4 & 5 & 1 \\ 6 & 7 & -2 \end{bmatrix}$$

Exercise 11.5.2 *Show all the details in calculating the adjoint of*

$$A = \begin{bmatrix} 3 & 2 & -1 & 5 \\ -4 & 5 & 1 & 1 \\ 6 & 7 & -2 & 9 \\ -1 & 2 & -1 & 4 \end{bmatrix}$$

Chapter 12

Some Classical Results

There are three classical results in linear functional analysis which are very useful.

- The Uniform Boundedness Theorem
- The Open Mapping Theorem
- The Closed Graph Theorem

Their proofs will help your grow in your ability to look for underlying structure in mathematical objects and understand how to use that previously unseen information.

12.1 First and Second Category Metric Spaces

We first need to return to the notion of **nowhere dense** sets we introduced earlier in Chapter 5. Recall a subset of a normed linear space is nowhere dense if its closure has empty interior. This is defined in Definition 5.4.3. In (Peterson (18) 2020), a standard project is to construct Cantor sets and we prove a Cantor set cannot contain any interval. Hence it is a nowhere dense subset of $[0, 1]$. We also prove that compact subsets of infinite dimensional normed linear spaces are nowhere dense in Theorem 5.4.5. We now want to study these sets more carefully.

Definition 12.1.1 First Category Sets of a metric space

Let (X, d) be a metric space. We say X is of **first category** *if there is a sequence (A_n) of subsets of X which are all nowhere dense so that $X = \cup_{n=1}^{\infty} A_n$. If X is not of first category, we say X is of* **second category**.

A most important result is then

Theorem 12.1.1 The Baire Category Theorem

Let (X, d) be a nonempty complete metric space. Then X is of **second category**.

Proof 12.1.1
We assume X is of first category and derive a contradiction. This is a somewhat long proof so be patient! Since X is first category, we can write $X = \cup_{n=1}^{\infty} M_n$, where each M_n is of first category.

*(***Step 1***:)*
Since M_1 is nowhere dense, $\overline{M_1}$ has no interior points; i.e. $\overline{M_1}$ does not contains a nonempty open

set. But X does contain nonempty open sets – trivially X itself so we cannot have $\overline{M_1} = X$. Hence $(\overline{M_1})^C = X \setminus M_1$ is nonempty and open as it is the complement of a closed set.

Choose $p_1 \in (\overline{M_1})^C$. Since this set is open, there is a radius $\epsilon_1 > 0$ so that $B(p_1; \epsilon_1) \subset (\overline{M_1})^C$. We can assume $\epsilon_1 < \frac{1}{2}$.

(Step 2:)
By an argument similar to that in Step 1, we know $(\overline{M_2})^C$ is both open and nonempty. Also $\overline{M_2}$ cannot contain an open set so $B(p_1; \frac{\epsilon_1}{2}) \subset (\overline{M_2})^C$. This means $(\overline{M_2})^C \cap B(p_1, \frac{\epsilon}{2})$ is a nonempty open set. Therefore there is p_2 and associated radius ϵ_2 which we can choose so that $\epsilon_2 < \frac{\epsilon_1}{2} < \frac{1}{2^2}$ so that

$$p_2 \in (\overline{M_2})^C \cap B(p_1; \frac{\epsilon}{2}), \quad B(p_2; \epsilon_2) \subset (\overline{M_2})^C \cap B(p_1; \frac{\epsilon}{2})$$

(Step 3:)
Now continue by induction. We obtain a sequence (p_n) and a sequence (ϵ_n) satisfying

$$\begin{aligned} p_n &\in (\overline{M_n})^C \cap B(p_{n-1}; \frac{\epsilon_{n-1}}{2}), \quad \epsilon_n < \frac{\epsilon_{n-1}}{2} < \frac{1}{2^n} \\ B(p_n; \epsilon_n) &\subset (\overline{M_n})^C \cap B(p_{n-1}; \frac{\epsilon_{n-1}}{2}) \end{aligned}$$

Thus, for $n = 2, 3, \ldots$

$$B(p_n; \epsilon_n) \subset B(p_{n-1}; \frac{\epsilon_{n-1}}{2}) \subset B(p_{n-1}, \epsilon_{n-1})$$

Pick a tolerance $\eta > 0$. There is N so that $\frac{1}{2^N} < \eta$. We see then that if $n > N$, $p_n \in B(p_N; \epsilon_N) \subset B(p_N; \frac{1}{2^N})$. It follows that if $m, n > N$,

$$d(p_n, p_m) \leq d(p_n, p_N) + d(p_m, p_N) < \frac{1}{2^N} + \frac{1}{2^N} < \eta$$

This shows (p_n) is a Cauchy sequence in X which is complete. Therefore there is a $p \in X$ so that $p_n \to d$. Hence, given the tolerance $\xi > 0$, there is M so $d(p_n, p) < \xi$ if $n > M$. However, we also know $\epsilon_n < \frac{1}{2^n}$. This implies $p \in B(p_n; \epsilon_n)$ if $\frac{1}{2^n} < \xi$. Since ξ is arbitrary, we see $p \in B(p_n; \epsilon_n) \subset (\overline{M_n})^C$ for all n. But that means

$$p \in \cap_{n=1}^{\infty} (\overline{M_n})^C = \left(\cup_{n-1}^{\infty} \overline{M_n}\right)^C$$

by DeMorgan's Laws. Thus, $p \in X^C$ since we assume $X = \cup_{n=1}^{\infty} \overline{M_n}$. But this is not possible. Thus our assumption X is first category is wrong and we conclude X is second category. ■

Comment 12.1.1 *If (X, d) is a nonempty complete space, if we can write $X = \cup_{n=1}^{\infty} M_n$, then at least one set M_p cannot be nowhere dense. So we know $(\overline{M_p})^C$ has a nonempty interior. Hence there is $P \in \overline{M_p}$ and a radius $r > 0$ so that $B(p; r) \subset \overline{M_p}$.*

12.1.1 The Uniform Boundedness Theorem

The next important result is called the **Uniform Boundedness Theorem**. It uses the fact that complete metric spaces are of second category.

Theorem 12.1.2 The Uniform Boundedness Theorem

Let $(X, \|\cdot\|_X)$ and $(Y, \|\cdot\|_Y)$ be nonempty normed linear spaces. Let $(T_n) \subset B(X, Y)$ be a sequence of bounded linear operators from X to Y. Further assume X is complete. Assume given $\boldsymbol{x} \in X$, there is a constant $c_{\boldsymbol{x}}$ so that $\|T_n(\boldsymbol{x})\|_Y \leq c_{\boldsymbol{x}}$ for all n. Then there is a constant $c > 0$ satisfying $\|T_n\|_{op} \leq c$ for all n. In this case, we say the sequence of norms $(\|T_n\|_{op})$ is uniformly bounded.

Proof 12.1.2

For each k, define

$$A_k = \{\boldsymbol{x} \in X \mid \|T_n(\boldsymbol{x})\|_Y \leq k, \forall n\}$$

First, note the set A_k is closed as if $\boldsymbol{x} \in \overline{A_k}$. Then there is a sequenced $(\boldsymbol{x_j})$ in A_k so that $\boldsymbol{x_j} \to \boldsymbol{x}$. Since each $\boldsymbol{x_j} \in A_k$, we then know $\|T_n(\boldsymbol{x_j})\|_Y \leq k$ for all j. But $\|\cdot\|_Y$ is continuous in Y and so

$$\lim_{j\to\infty} \|T_n(\boldsymbol{x_j})\|_Y \leq \lim_{j\to\infty} k = k$$

Further,

$$\lim_{j\to\infty} \|T_n(\boldsymbol{x_j})\|_Y = \|\lim_{j\to\infty} T_n(\boldsymbol{x_j})\| = \|T_n(\boldsymbol{x})\|_Y$$

So $\|T_n(\boldsymbol{x})\|_Y \leq k$ and hence $\boldsymbol{x} \in A_k$. This shows A_k is closed.

Second, we show $X = \cup_{k=1}^{\infty} A_k$. It is clear $\cup_{k=1}^{\infty} A_k \subset X$. To show the converse, let $\boldsymbol{x_0} \in X$. Then there is a $c_{\boldsymbol{x_0}}$ so that $\|T_n(\boldsymbol{x_0})\|_Y \leq c_{\boldsymbol{x_0}}$ for all n. Pick any positive integer k_0 satisfying $c_{\boldsymbol{x_0}} \leq k_0$. Then $\|T_n(\boldsymbol{x_0})\|_Y \leq k_0$ for all n showing us $\boldsymbol{x_0} \in A_{k_0}$. This shows $X \subset \cup_{k=1}^{\infty} A_k$. Combining, we have $X = \cup_{k=1}^{\infty} A_k$.

Then since X is nonempty and complete, X is of second category. There is then an index k_0 so that $\overline{A_{k_0}}$ contains an open ball. Also, since A_{k_0} is already closed this means there is $\mathbf{p} \in A_{k_0}$ and $r > 0$ with $B(\mathbf{p}; r) \subset A_{k_0}$. Now if $\boldsymbol{x} \in X$, $\boldsymbol{x} \neq \mathbf{0}$, let $\gamma_{\boldsymbol{x}} = \frac{r}{2\|\boldsymbol{x}\|_X}$. and $\boldsymbol{z} = \mathbf{p} + \gamma_{\boldsymbol{x}}$. Then

$$\|\boldsymbol{z} - \mathbf{p}\|_X = \|\gamma_{\boldsymbol{x}}\|_X = \frac{r}{2\|\boldsymbol{x}\|_X} \|\boldsymbol{x}\|_X < r$$

and so $\boldsymbol{z} \in B(\mathbf{p}; r) \subset A_{k_0}$. From this, we have $\|T_n(\boldsymbol{z})\|_Y \leq k_0$ for all n. Since $\mathbf{p} \in A_{k_0}$ too, we also have $\|T_n(\mathbf{p})\|_Y \leq k_0$ for all n. Therefore

$$\begin{aligned} \|T_n(\boldsymbol{x})\|_Y &= \left\| T_n\left(\frac{\boldsymbol{z} - \mathbf{p}}{\gamma_{\boldsymbol{x}}}\right)\right\|_Y \leq \frac{1}{\gamma_{\boldsymbol{x}}} \left(\|T_n(\boldsymbol{z})\|_Y + \|T_n(\mathbf{p})\|_Y\right) \\ &\leq \frac{2k_0}{\gamma_{\boldsymbol{x}}} = \frac{4k_0}{r} \|\boldsymbol{x}\|_X \end{aligned}$$

This implies that for $\boldsymbol{x} \neq \mathbf{0}$ and for all n,

$$\frac{\|T_n(\boldsymbol{x})\|_Y}{\|\boldsymbol{x}\|_X} \leq \frac{4k_0}{r} \Longrightarrow \|T_n\|_{op} \leq \frac{4k_0}{r}$$

Hence, choosing $c = \frac{4k_0}{r}$, we establish the claim. ■

12.1.2 Some Fourier Series Do Not Converge Pointwise

Let X be the set of all continuous functions on $\Re$ which are 2π periodic where 2π periodic means $x(t+2\pi) = x(t)$ for all $t \in \Re$. Then X is a normed linear space that is complete with respect to $\|\cdot\|_\infty$. For any $\boldsymbol{x} \in X$, define the scalar sequences $(a_m)_{m=0}^{\infty}$ and $(b_m)_{m=1}^{\infty}$ by

$$\begin{aligned} a_0 &= \frac{1}{\pi}\int_0^{2\pi} x(t)\,dt \\ a_m &= \frac{1}{\pi}\int_0^{2\pi} x(t)\,\cos(mt)\,dt, \quad m \geq 1 \\ b_m &= \frac{1}{\pi}\int_0^{2\pi} x(t)\,\sin(mt)\,dt, \quad m \geq 1 \end{aligned}$$

The **Fourier Series** associated with $\boldsymbol{x} \in X$ is the formal series defined by

$$(S(\boldsymbol{x}))(t) = \frac{a_0}{2} + \sum_{m=1}^{\infty} (\,a_m\,\cos(mt) + b_m\,\sin(mt)\,)$$

which may diverge at some points. We have worked through the basic theoretical background to discuss pointwise convergence of this series in (Peterson (18) 2020). There the discussion was on the interval $[0, 2L]$ and we worked with functions on $[0, L]$ and extended them in odd and even ways to obtain periodic extensions that are in X. In this text, we discuss how this series is a particular case of the kind of expansions we obtain from sequences of functions arising from Stürm - Liouville problems. To obtain pointwise convergence we had to impose various conditions on the differentiability of a given $\boldsymbol{x}$. Here, we will show how there is at least one $\boldsymbol{x}$ which has a Fourier series which fails to converge at at least one point. To do this we will use the **Uniform Boundedness Theorem** we have just proved. So this is a great way to see how that idea is used for something.

Theorem 12.1.3 Some Fourier Series Do Not Converge Pointwise

Let X be the set of all continuous functions on $\Re$ which are 2π periodic Then there are functions $\boldsymbol{x} \in X$ whose Fourier Series fail to converge at some points.

Proof 12.1.3
Step One: *We rewrite the partial sums S_n in terms of auxiliary functions $\tilde{q}_n$.*

Look at the nth partial sum. At a fixed t_0, we find

$$(S_n(\boldsymbol{x}))(t_0) = \frac{a_0}{2} + \sum_{m=1}^{n} (\,a_m\,\cos(mt_0) + b_m\,\sin(mt_0)\,)$$

Now, $\sin(u)\sin(v) + \cos(u)\cos(v) = \cos(u-v)$ *and hence we can rewrite the above as follows:*

$$\begin{aligned} S_n(x) &= \frac{1}{2\pi}\int_0^{2\pi} x(t)\,dt + \sum_{m=1}^{n} \frac{1}{\pi}\int_0^{2\pi} x(t)\,(\cos(mt)\cos(mt_0) + \sin(mt)\sin(mt_0)) \\ &= \frac{1}{2\pi}\int_0^{2\pi} x(t)\,dt + \sum_{m=1}^{n} \frac{1}{\pi}\int_0^{2\pi} x(t)\cos(m(t-t_0)) \end{aligned}$$

We also know from trigonometry that

$$\cos(my)\sin\left(\frac{y}{2}\right) = \frac{1}{2}\left(\sin\left(\left(m+\frac{1}{2}\right)y\right) - \sin\left(\left(m-\frac{1}{2}\right)y\right)\right)$$

and so

$$\begin{aligned}
&\left(\frac{1}{2} + \sum_{m=1}^{n}\cos(m((t-x))\right)\sin\left(\frac{1}{2}(t-x)\right) \\
&= \frac{1}{2}\sin\left(\frac{1}{2}(t-x)\right) + \sum_{m=1}^{n}\left(\sin\left(\left(m+\frac{1}{2}\right)(t-x)\right) - \sin\left(\left(m-\frac{1}{2}\right)(t-x)\right)\right) \\
&= \frac{1}{2}\sin\left(\frac{1}{2}(t-x)\right) + \frac{1}{2}\sin\left(\frac{3}{2}(t-x)\right) - \frac{1}{2}\sin\left(\frac{1}{2}(t-x)\right) \\
&\quad + \frac{1}{2}\sin\left(\frac{5}{2}(t-x)\right) - \frac{1}{2}\sin\left(\frac{3}{2}(t-x)\right) \\
&\quad \ldots \\
&\quad + \frac{1}{2}\sin\left(\frac{(2n+1)}{2}(t-x)\right) - \frac{1}{2}\sin\left(\frac{(2n-1)}{2}(t-x)\right) \\
&= \frac{1}{2}\sin\left(\left(n+\frac{1}{2}\right)(t-x)\right)
\end{aligned}$$

We have found for $t \neq x$

$$\frac{1}{2} + \sum_{m=1}^{n}\cos(m(t-x)) = \frac{\frac{1}{2}\sin\left(\left(n+\frac{1}{2}\right)(t-x)\right)}{\sin\left(\frac{1}{2}(t-x)\right)}$$

and in particular at $x = 0$*, we have*

$$\frac{1}{2} + \sum_{m=1}^{n}\cos(mt) = \frac{\frac{1}{2}\sin\left(\left(n+\frac{1}{2}\right)t\right)}{\sin(\frac{1}{2}t)}$$

Now let

$$q_n(t) = \frac{\sin\left(\left(n+\frac{1}{2}\right)t\right)}{\sin(\frac{1}{2}t)}$$

The function $\boldsymbol{q_n}$ *has removeable discontinuities at* $t = 2j\pi$*. Note*

$$\begin{aligned}
\lim_{t\to 2j\pi} q_n(t) &= \lim_{t\to 2j\pi}\frac{\sin\left(\left(n+\frac{1}{2}\right)t\right)}{\sin(\frac{1}{2}t)} = \lim_{t\to 2j\pi}\frac{\cos\left(\left(n+\frac{1}{2}\right)t\right)\left(n+\frac{1}{2}\right)}{\cos(\frac{1}{2}t)\,\frac{1}{2}} \\
&= \frac{(-1)^j\left(n+\frac{1}{2}\right)}{(-1)^j\,\frac{1}{2}} = 2n+1
\end{aligned}$$

Hence if we define $\tilde{\boldsymbol{q}}_n$ by

$$\tilde{q}_n(t) = \begin{cases} q_n(t), & t \neq 2j\pi \\ 2n+1, & t = 2j\pi \end{cases}$$

then $\tilde{\boldsymbol{q}}_n$ is continuous and it is easy to see it is 2π periodic on $\Re$. Note we can then say

$$\begin{aligned} (S_n(\boldsymbol{x}))(0) &= \frac{1}{\pi}\int_0^{2\pi} x(t)\left(\frac{1}{2} + \sum_{m=1}^{n} \cos(mt)\right) dt = \frac{1}{\pi}\int_0^{2\pi} x(t)\, \frac{\frac{1}{2}\sin\left(\left(n+\frac{1}{2}\right)t\right)}{\sin(\frac{1}{2}t)} \\ &= \frac{1}{2\pi}\int_0^{2\pi} x(t)\, q_n(t)dt = \frac{1}{2\pi}\int_0^{2\pi} x(t)\, \tilde{q}_n(t)dt \end{aligned}$$

Step Two*: Defining the bounded linear functionals T_n.*

Now define the functionals $T_n : X \to \Re$ by

$$T_n(\boldsymbol{x}) = S_n(\boldsymbol{x})(0) = \frac{1}{2\pi}\int_0^{2\pi} x(t)\, \tilde{q}_n(t)dt$$

We see T_n is clearly linear on X. Next, we show T_n is bounded. First, remember $\boldsymbol{x}$ is 2π periodic, so $\|\boldsymbol{x}\|_\infty = \max_{0 \leq t \leq 2\pi} |x(t)$ is finite.

$$|T_n(\boldsymbol{x})| = \left|\frac{1}{2\pi}\int_0^{2\pi} x(t)\, \tilde{q}_n(t)dt\right| \leq \frac{1}{2\pi}\int_0^{2\pi} |x(t)|\, |\tilde{q}_n(t)|dt \leq \frac{1}{2\pi}\, \|\boldsymbol{x}\|_\infty \int_0^{2\pi} |\tilde{q}_n(t)|dt$$

Since each $\tilde{\boldsymbol{q}}$ is continuous on $[0, 2\pi]$, we can further estimate this by

$$|T_n(\boldsymbol{x})| \leq \frac{1}{2\pi}\|\boldsymbol{x}\|_\infty\, \|\tilde{\boldsymbol{q}}_n\|_\infty\, (2\pi) = \|\boldsymbol{x}\|_\infty\, \|\tilde{\boldsymbol{q}}_n\|_\infty$$

where $\|\tilde{\boldsymbol{q}}_n\|_\infty$ is calculated on $[0, 2\pi]$. This inequality immediately implies the operator norm $\|T_n\|$ is finite and

$$\|T_n\| \leq \frac{1}{2\pi}\int_0^{2\pi} |\tilde{q}_n(t)|dt \leq \|\tilde{\boldsymbol{q}}_n\|_\infty$$

Step Three*: The structure of $\tilde{\boldsymbol{q}}_n$.*

We can do more. Define $\boldsymbol{y}_n$ by

$$y_n(t) = \begin{cases} 1, & \tilde{q}_n(t) \geq 0 \\ -1, & \tilde{q}_n < 0 \end{cases}$$

Then we have $|\tilde{q}_n(t)| = y_n(t)\tilde{q}_n(t)$. We can now analyze the structure of $\boldsymbol{y}_n$.

Since $\tilde{\boldsymbol{q}}_n$ is 2π periodic we only have to figure out what $\boldsymbol{y}_n$ does on $[0, 2\pi]$. Now we know $\tilde{q}_n(t) = 0$ when $\sin((n + 1/2)t) = 0$ or when $t = \frac{2k\pi}{2n+1}$. Now in $[0, 2\pi]$, we must have $0 \leq \frac{2k\pi}{2n+1} \leq 2\pi$ which implies $0 \leq k \leq 2n+1$. These are the zeros of $\tilde{\boldsymbol{q}}_n$ in $[0, 2\pi]$. We know $\tilde{q}_n(0) = \tilde{q}_n(2\pi) = 2n+1$. Thus the zeros are at $2k\pi/(2n+1)$ for $1 \leq k \leq 2n$ giving the set of zeroes U

$$U = \left\{\frac{2\pi}{2n+1}, \ldots, \frac{4n\pi}{2n+1}\right\} = \{t_{1,n}, \ldots, t_{2n,n}\}$$

Now $\sin(s) > 0$ *on* $(0, \pi)$ *so* $\sin((n+1/2)t) > 0$ *when* $0 < (n+1/2)t < \pi$ *or* $0 < t < \frac{2\pi}{2n+1} = t_{1,n}$. *Also,* $\sin(t/2) > 0$ *when* $0 < t < \pi$ *or* $0 < t < 2\pi$. *So the ratio* $\widetilde{q}_n(t) > 0$ *on* $(0, t_{1,n})$. *The ratio is zero at the endpoints, so the sign is* $+1$ *there, Hence* $y_n(t) = 1$ *on* $[0, t_{1,n}]$.

We know $\sin(s) < 0$ *on* $(\pi, 2\pi)$. *Hence, we want* $\pi < (n+1/2)t < 2\pi$ *or* $\frac{2\pi}{2n+1} < t < \frac{4\pi}{2n+1}$ *or* $t_{1,n} < t < t_{2,n}$. *The denominator* $\sin(t/2)$ *is still positive here, so the ratio* $\widetilde{q}_n(t) < 0$ *on* $(t_{1,n}, t_{2,n})$. *This means* $y_n(t) = -1$ *on the open interval* $(t_{1,n}, t_{2,n})$.

It is easy to see $\boldsymbol{y}_n$ *switches sign on successive* $(t_{j,n}, t_{(j+1)\pi})$ *intervals. At the last subinterval* $[\frac{4n\pi}{2n+1}, 2\pi]$, *we know* $\widetilde{q}_n(2\pi) = 2n+1$ *so* $y_n(2\pi) = 1$. *The denominator of* $\widetilde{q}_n(t)$ *is positive here so the sign is determined by the numerator. However,* $\widetilde{q}_n(t) = 0$ *only at the points* $t_{1,n}$. *Hence we can determine the sign of* $\widetilde{q}_n$ *on this interval by checking what happens at one point. At the midpoint of this interval,* $\frac{2n\pi}{2n+1} + \pi$, *we have*

$$\sin\left(n + \frac{1}{2}\left(\frac{2n\pi}{2n+1} + \pi\right)\right) = \sin\left(n\pi + \left(n + \frac{1}{2}\right)\pi\right) = \sin\left(\frac{\pi}{2}\right) = 1$$

We see $\widetilde{q}_n(t) > 0$ *on* $[\frac{4n\pi}{2n+1}, 2\pi]$ *implying* $y_n(t) = 1$ *there. To see the structure of* $\boldsymbol{y}_n$ *is it easiest to show some explicitly.*

$$y_1(t) = \begin{cases} 1, & 0 \le t \le t_{1,1} \\ -1, & t_{1,2} < t < t_{2,1} \\ 1, & t_{2,1} \le t \le 2\pi \end{cases} = \begin{cases} 1, & 0 \le t \le \frac{2\pi}{3} \\ -1, & \frac{2\pi}{3} < t < \frac{4\pi}{3} \\ 1, & \frac{4\pi}{3} \le t \le 2\pi \end{cases}$$

$$y_2(t) = \begin{cases} 1, & 0 \le t \le t_{1,2} \\ -1, & t_{1,2} < t < t_{2,2} \\ 1, & t_{2,2} \le t \le t_{3,2} \\ -1, & t_{3,2} < t < t_{4.2} \\ 1, & t_{4,2} \le t \le 2\pi \end{cases} = \begin{cases} 1, & 0 \le t \le \frac{2\pi}{5} \\ -1, & \frac{2\pi}{5} < t < \frac{4\pi}{5} \\ 1, & \frac{4\pi}{3} \le t \frac{6\pi}{5} \\ -1, & \frac{6\pi}{5} < t < \frac{8\pi}{5} \\ 1, & \frac{8\pi}{5} \le t \le 2\pi \end{cases}$$

and so on. Each $\boldsymbol{y}_n$ *starts at* 1 *and ends at* 1.

Step Four: *Smoothing* $\widetilde{\boldsymbol{q}}_n$.

Now replace each $\boldsymbol{y}_n$ *by a smoothed version* $\boldsymbol{x}_{\delta,n}$. *We will explicitly show* $\boldsymbol{x}_{\delta,1}$ *and* $\boldsymbol{x}_{\delta,2}$ *before we define the general version.*

$$x_{\delta,1}(t) = \begin{cases} 1, & 0 \le t \le t_{1,1} - \delta \\ L(t, (t_{1,1} - \delta, 1), (t_{1,1} + \delta, -1)), & t_{1,1} - \delta \le t \le t_{1,1} + \delta \\ -1, & t_{1,1} + \delta < t < t_{2,1} - \delta \\ L(t, (t_{2,1} - \delta, -1), (t_{2,1} + \delta, 1)), & t_{2,1} - \delta \le t \le t_{2.1} + \delta \\ 1, & t_{2,1} + \delta \le t \le 2\pi \end{cases}$$

$$x_{\delta,2}(t) = \begin{cases} 1, & 0 \le t \le t_{1,2} - \delta \\ L(t, (t_{1,2} - \delta, 1), (t_{1,2} + \delta, -1)), & t_{1,2} - \delta \le t \le t_{1,2} + \delta \\ -1, & t_{1,2} + \delta < t < t_{2,2} - \delta \\ L(t, (t_{2,2} - \delta, -1), (t_{2,2} + \delta, 1)), & t_{2,2} - \delta \le t \le t_{2.2} + \delta \\ 1, & t_{2,2} + \delta \le t_{3,2} - \delta \\ L(t, (t_{3,2} - \delta, 1), (t_{3,2} + \delta, -1)), & t_{3,2} - \delta \le t \le t_{3,2} + \delta \\ -1, & t_{3,2} + \delta < t < t_{4,2} - \delta \\ L(t, (t_{4,2} - \delta, -1), (t_{4,2} + \delta, 1)), & t_{4,2} - \delta \le t \le t_{4.2} + \delta \\ 1, & t_{4,2} + \delta \le 2\pi \end{cases}$$

where the functions denoted by L are the straight lines that transition between the constant value of $\boldsymbol{y}_n$. *These formulae are indeed messy and the functional forms are not so important. But to be clear in a few examples*

- $L(t, (t_{1,2} - \delta, 1), (t_{1,2} + \delta, -1))$ *is the straight line connecting the point* $(t_{1,2} - \delta, 1)$ *to the point* $(t_{1,2} + \delta, -1)$. *This has a negative slope as we are connecting from* $+1$ *to* -1. *Note* δ *has to be sufficiently small so that we do not spill over into another* $+1$ *interval.*

- $L(t, (t_{2,2} - \delta, -1), (t_{2,2} + \delta, 1))$ *is the straight line connecting the point* $(t_{2,2} - \delta, -1)$ *to the point* $(t_{2,2} + \delta, 1)$. *This has a positive slope as we are connecting from* -1 *to* 1. *Note* δ *has to be sufficiently small for this to work.*

It is not hard to see how this works, so we encourage you to make some sketches to make sure you understand the idea. This replacement of a piecewise constant function with a smooth function of this sort is pretty standard. To make this work (draw a picture!), we must choose $\delta < \frac{t_{1,n}}{4} = \frac{2\pi}{2n+1}\frac{1}{4}$.

Step Five: *Computing* $\|T_n\|$.

From all this, we get an estimate

$$\begin{aligned} |(x_{\delta,n}(t) - y_n(t))\widetilde{q}_n(t)| &= \begin{cases} 0, & t \in (t_{j,n} - \delta, t_{j,n} + \delta)^C \\ |x_{\delta,n}(t) - y(t)| : |\widetilde{q}_n(t)|, & t \in (t_{j,n} - \delta, t_{j,n} + \delta) \end{cases} \\ &\leq \max_{t \in (t_{j,n} - \delta, t_{j,n} + \delta)} |x_{\delta,n}(t) - y(t)| \; |\widetilde{q}_n(t)| \leq 2|\widetilde{q}_n(t)| \end{aligned}$$

Thus,

$$\begin{aligned} \left| \frac{1}{2\pi} \int_0^{2\pi} (x_{\delta,n}(t) - y_n(t))\widetilde{q}_n(t) : dt \right| &\leq \frac{1}{2\pi} \int_0^{2\pi} |x_{\delta,n}(t) - y_n(t)| \, |\widetilde{q}_n(t)| \, dt \\ &= \frac{1}{2\pi} \sum_{j=1}^{2n} \int_{t_{j,n} - \delta}^{t_{j,n} + \delta} |x_{\delta,n}(t) - y_n(t)| \, |\widetilde{q}_n(t)| \, dt \\ &\leq \frac{1}{2\pi} \sum_{j=1}^{2n} \int_{t_{j,n} - \delta}^{t_{j,n} + \delta} 2 \, |\widetilde{q}_n(t)| \, dt \end{aligned}$$

From the definition of $\widetilde{q}_n(t)$, *we see*

$$|\widetilde{q}_n(t)| \leq \left| \frac{1}{2} + \sum_{j=1}^{n} \cos(jt) \right| \leq \frac{1}{2} + n$$

Thus, we have

$$\begin{aligned} \left| \frac{1}{2\pi} \int_0^{2\pi} (x_{\delta,n}(t) - y_n(t))\widetilde{q}_n(t) : dt \right| &\leq \frac{1}{\pi} \sum_{j=1}^{2n} \int_{t_{j,n} - \delta}^{t_{j,n} + \delta} (\frac{1}{2} + n) \, dt \\ &= \frac{1}{\pi}(2n)(2\delta)(\frac{1}{2} + n) \leq \frac{6n^2\delta}{\pi} \end{aligned}$$

Now pick $\epsilon > 0$ *and fix* n. *Choose* $\delta_n < \min\{\frac{\pi}{2(2n+1)}, \frac{6n^2\delta}{\pi} \frac{\epsilon}{2}\}$. *Then, if* $\delta < \delta_n$,

$$\left| \frac{1}{2\pi} \int_0^{2\pi} (x_{\delta,n}(t) - y_n(t))\widetilde{q}_n(t) \, dt \right| = \left| \frac{1}{2\pi} \int_0^{2\pi} x_{\delta,n}(t)\widetilde{q}_n(t) \, dt - \frac{1}{2\pi} \int_0^{2\pi} y_n(t)\widetilde{q}_n(t) \, dt \right|$$

$$= \left| T_n(x_{\delta,n}) - \frac{1}{2\pi} \int_0^{2\pi} y_n(t))\widetilde{q}_n(t)\, dt \right| \leq \frac{\epsilon}{2}$$

as $T_n(x_{\delta,n}) = \frac{1}{2\pi} \int_0^{2\pi} x_{\delta,n}(t)\widetilde{q}_n(t)\, dt$*. Also, noting* $y_n(t)\widetilde{q}_n(t) = |\widetilde{q}_n(t)|$ *and using the reverse triangle inequality, we then find*

$$|T_n(x_{\delta,n})| \geq \frac{1}{2\pi} \int_0^{2\pi} |\widetilde{q}_n(t)|\, dt - \frac{\epsilon}{2}$$

Since $\|\boldsymbol{x}_{\delta,n}\|_\infty = 1$*, the definition of* $\|T_n\|$ *immediately tells us*

$$\|T_n\| \geq \frac{1}{2\pi} \int_0^{2\pi} |\widetilde{q}_n(t)|\, dt - \frac{\epsilon}{2}$$

However, the choice of $\epsilon > 0$ *was arbitrary, so this shows*

$$\|T_n\| \geq \frac{1}{2\pi} \int_0^{2\pi} |\widetilde{q}_n(t)|\, dt$$

In **Step Two***, we showed the reverse inequality, so we can conclude*

$$\|T_n\| = \frac{1}{2\pi} \int_0^{2\pi} |\widetilde{q}_n(t)|\, dt$$

Step Six*: Showing there is an* $\boldsymbol{x} \in X$ *whose Fourier Series diverges at some points.*

Finally, we see

$$\|T_n\| = \frac{1}{2\pi} \int_0^{2\pi} |\widetilde{q}_n(t)|\, dt = \frac{1}{2\pi} \int_0^{2\pi} \left| \frac{\sin\left(\left(n + \frac{1}{2}\right)t\right)}{\sin(\frac{1}{2}t)} \right| dt$$

On $[0, 2\pi]$*,* $sin(t/2) < t/2$*, thus*

$$\|T_n\| > \frac{1}{2\pi} \int_0^{2\pi} \left| \frac{\sin\left(\left(n + \frac{1}{2}\right)t\right)}{\frac{t}{2}} \right| dt$$

Letting $v = (n + 1/2)t$*, we have*

$$\|T_n\| > \frac{1}{2\pi} \int_0^{(2n+1)\pi} \frac{|\sin(v)|}{\frac{v}{2n+1}} \frac{1}{n + 1/2}\, dv = \frac{1}{\pi} \sum_{k=0}^{2n} \int_{k\pi}^{(k+1)\pi} \frac{|\sin(v)|}{v}\, dv$$

On $[k\pi, (k + 1)\pi]$*,* $v \leq (k + 1)\pi$ *implies* $\frac{1}{v} \geq \frac{1}{(k+1)\pi}$*. Thus,*

$$\|T_n\| > \frac{1}{\pi} \sum_{k=0}^{2n} \frac{1}{(k + 1)\pi} \int_{k\pi}^{(k+1)\pi} |\sin(v)|\, dv$$

On $[k\pi, (k+1)\pi]$, the area under the curve $|\sin(v)|$ is exactly the same as the area under the curve $\sin(v)$ on $[0, \pi]$ which is $\int_0^{\pi} \sin(v) dv = -\cos(\pi) + \cos(0) = 2$. We conclude

$$\|T_n\| \quad > \quad \frac{1}{\pi} \sum_{k=0}^{2n} \frac{2}{(k+1)\pi}$$

Since $\|T_n\|$ is the partial sum of a divergent series, we have $\|T_n\| \to \infty$. Now we apply the **Uniform Boundedness Theorem***: if at each $\boldsymbol{x} \in X$, there was a constant $c_{\boldsymbol{x}}$ so that $\|T_n(\boldsymbol{x})\| \leq c_{\boldsymbol{x}}$ for all n, then there would be another constant c so that $\|T_n\| \leq c$ for all n. But we don't have that situation here. So there has to be some $\boldsymbol{x_0} \in X$ for which there is no finite constant $c_{\boldsymbol{x_0}}$ with $\|T_n(\boldsymbol{x_0})\| \leq c_{\boldsymbol{x_0}}$ for all n. For this $\boldsymbol{x_0}$, we must have $\|T_n(\boldsymbol{x_0})\| \to \infty$.*

Finally,

$$\begin{aligned} T_n(\boldsymbol{x_0}) &= (S_n(\boldsymbol{x_0}))(0) = \frac{1}{2\pi} \int_0^{2\pi} x_0(t) \widetilde{q}_n(t) dt = \frac{1}{2\pi} \int_0^{2\pi} x_0(t) \left(\frac{1}{2} + \sum_{m=1}^{n} \cos(mt) \right) dt \\ &= \frac{a_0}{2} + a_1 + \ldots + a_m \end{aligned}$$

Since $\|T_n(\boldsymbol{x_0})\| \to \infty$, we must have $|\frac{a_0}{2} + a_1 + \ldots + a_m| \to \infty$ which tells us the series $(S_n(\boldsymbol{x_0}))(0)$ diverges and so $S_n(\boldsymbol{x_0})$ does not converge for all t. Hence, we have shown there is a continuous 2π periodic function whose Fourier series fails to converge at at least one point. ■

Comment 12.1.2 *Again, recall our discussions about Fourier Series back in (Peterson (18) 2020). We had to impose additional derivative conditions on $\boldsymbol{x} \in X$ to ensure convergence of the Fourier Series.*

12.2 The Open Mapping Theorem

This is another really important theorem of great use. We start with a definition.

Definition 12.2.1 Open Mappings

Let (X, d_X) and (Y, d_Y) be metric spaces and : $T : D(T) \subset X \to Y$ be a mapping. We say T is an **open mapping** *if $T(U)$ is an open set in Y for all open sets U in X.*

Here, an open set is defined like usual in a metric space: each point p in an open set must be an interior point. We want to prove that bounded linear operators from X to Y must be open mappings. Of course, the requirement that T is linear means we must let X and Y be normed linear spaces whose metrics are induced by a norm. Note if T is also $1-1$, the inverse T^{-1} exists and $T^{-1} : T(dom(T)) \subset Y \to X$ where $T(dom(T))$ is the range of T. Let $U \subset dom(T)$ be an open set. Then for $g = t^{-1}$ to be continuous, $g^{-1}(U)$ must be open in Y. But $g^{-1} = (T^{-1})^{-1} = T$. So if T is an open mapping, it will make T^{-1} a continuous mapping.

To prove the result we want, which is called the **Open Mapping Theorem**, we need a preliminary fact. We will need the Baire Category Theorem here.

Lemma 12.2.1 The Open Unit Ball Lemma

Let $T : X \to Y$ be a mapping between two complete normed linear spaces where $\|\cdot\|_X$ is the norm on X and $\|\cdot\|_Y$ on Y. Let T be a bounded, linear and onto operator. Let $B_0 = B(\mathbf{0};1) \subset X$. Then $T(B_0)$ contains an open ball $B(\mathbf{0};r) \subset Y$ for some $r > 0$.

Proof 12.2.1

Consider $B(\mathbf{0};\frac{1}{8}) \subset X$. Let $\boldsymbol{x_0} \in X$ be nonzero. Then $\|\boldsymbol{x_0}\|_X < 0$ and so there is a positive integer k_0 with $\|\boldsymbol{x_0}\|_X < \frac{k_0}{8}$ which tells us $\boldsymbol{x_0} \in k_0\, B(\mathbf{0};\frac{1}{8})$. We can do this for all nonzero $\boldsymbol{x} \in X$. And all these balls contain $\mathbf{0}$. Thus, $X = \cup_{k=1}^{\infty} k\, B(0;\frac{1}{8})$.

Claim One: $Y = \cup_{k=1}^{\infty} k\, \overline{T(B(0;\frac{1}{8}))}$.

If $\boldsymbol{y} \in T(X)$, there is an $\boldsymbol{x} \in X$ with $\boldsymbol{y} = T(\boldsymbol{x})$ as we have assumed T is onto. But since $\boldsymbol{x} \in X$, there is some k so that $\boldsymbol{x} \in kB(\mathbf{0};\frac{1}{8})$. Hence, $\underline{s}y = T(\boldsymbol{x}) \in T(kB(\mathbf{0};\frac{1}{8}))$. But T is linear, so if $\boldsymbol{u} \in kB(\mathbf{0};\frac{1}{8})$, there is an $\boldsymbol{u_0} \in B(\mathbf{0};\frac{1}{8})$ with $\boldsymbol{u} = k\boldsymbol{u_0}$ and $T(\boldsymbol{u}) = kT(\boldsymbol{u_0})$. We conclude $T(kB(\mathbf{0};\frac{1}{8})) = k\, T(B(\mathbf{0};\frac{1}{8}))$ and so

$$T(\boldsymbol{x}) \in k\,T\Big(B(\mathbf{0};\frac{1}{8})\Big) \subset \cup_{k=1}^{\infty} k\,T\Big(B(\mathbf{0};\frac{1}{8})\Big) \subset \cup_{k=1}^{\infty} \overline{k\,T\Big(B(\mathbf{0};\frac{1}{8})\Big)}$$

But it is easy to see the linearity of T implies

$$\overline{k\,T\Big(B(\mathbf{0};\frac{1}{8})\Big)} = k\,\overline{T\Big(B(\mathbf{0};\frac{1}{8})\Big)}$$

Since the choice of $\boldsymbol{y} = T(\boldsymbol{x})$ was arbitrary, we have shown $Y = T(X) \subset \cup_{k=1}^{\infty} k\,\overline{T(B(0;\frac{1}{8}))}$. The reverse containment is obvious, so we have shown the claim is true.

Now since T is complete, Y must be **category two**. *So there is $k_0 > 0$ so that $k_0\overline{T(B(0;\frac{1}{8}))}$ contains an open ball $B(\boldsymbol{z};r)$ for some $\boldsymbol{z} \in Y$ and some $r > 0$. Hence,*

$$B(\boldsymbol{z};r) \subset k_0\overline{T(B(0;\frac{1}{8}))} \Longrightarrow B\Big(\frac{\boldsymbol{z}}{k_0};\frac{r}{k_0}\Big) \subset \overline{T(B(0;\frac{1}{8}))} \tag{12.1}$$

$$\Longrightarrow B\Big(\mathbf{0};\frac{r}{k_0}\Big) \subset B\Big(\frac{\boldsymbol{z}}{k_0};\frac{r}{k_0}\Big) - \Big\{\frac{\boldsymbol{z}}{k_0}\Big\} \subset \overline{T(B(0;\frac{1}{8}))} - \Big\{\frac{\boldsymbol{z}}{k_0}\Big\} \tag{12.2}$$

Claim Two: $\overline{T(B(0;\frac{1}{8}))} - \{\frac{\boldsymbol{z}}{k_0}\} \subset \overline{T(B(0;\frac{1}{4}))}$.

Let $\boldsymbol{y} \in \overline{T(B(0;\frac{1}{8}))} - \{\frac{\boldsymbol{z}}{k_0}\}$. Then $\boldsymbol{y} + \{\frac{\boldsymbol{z}}{k_0}\} \in \overline{T(B(0;\frac{1}{8}))}$. We know from Equation 12.1 that $\{\frac{\boldsymbol{z}}{k_0}\} \in \overline{T(B(0;\frac{1}{8}))}$. Hence there are sequences

$$(\boldsymbol{v}_n),\quad \boldsymbol{v}_n = T(\boldsymbol{u}_n^0),\quad T(\boldsymbol{u}_n^0) \in T(B(0;\frac{1}{8})),\quad \boldsymbol{v}_n \to\to \frac{\boldsymbol{z}}{k_0}$$

$$(\boldsymbol{w}_n),\quad \boldsymbol{v}_n = T(\boldsymbol{u}_n^1),\quad T(\boldsymbol{u}_n^1) \in T(B(0;\frac{1}{8})),\quad \boldsymbol{w}_n \to\to \boldsymbol{y} + \frac{\boldsymbol{z}}{k_0}$$

Note,

$$\|\boldsymbol{u}_n^0 - \boldsymbol{u}_n^1\|_X \leq \|\boldsymbol{u}_n^0\|_X + \|\boldsymbol{u}_n^1\|_X < \frac{1}{8} + \frac{1}{8} = \frac{1}{4}$$

We see $\boldsymbol{u}_n^0 - \boldsymbol{u}_n^1 \in B(\mathbf{0}; \frac{1}{4})$. *Then*

$$T(\boldsymbol{u}_n^1 - \boldsymbol{u}_n^0) = \boldsymbol{w_n} - \boldsymbol{v_n} \to -\frac{\boldsymbol{z}}{k_0} + \boldsymbol{y} + \frac{\boldsymbol{z}}{k_0} = \boldsymbol{y}$$

This implies $\boldsymbol{y} \in \overline{T(B(\mathbf{0}; \frac{1}{4}))}$. *Since the choice of* $\boldsymbol{y}$ *was arbitrary, we have shown* **Claim Two** *is true.*

Claim Three: $B(\mathbf{0}; \frac{r}{k_0}) \subset \overline{T(B(\mathbf{0}; \frac{1}{4}))}$.

By Equation 12.2, we have $B(\mathbf{0}, \frac{r}{k_0}) \subset \overline{T(B(0; \frac{1}{8}))} - \left\{\frac{\boldsymbol{z}}{k_0}\right\}$ *and by* **Claim Two**, *we have* $\overline{T(B(0; \frac{1}{8}))} - \{\frac{\boldsymbol{z}}{k_0}\} \subset \overline{T(B(0; \frac{1}{4}))}$. *This shows the claim is true.*

Claim Four: *Let* $V_n = B(\mathbf{0}; \frac{\epsilon_0}{2^n})$. *Then* $V_n \subset \overline{T(B_{n+2})}$.

Let $B_n = B(\mathbf{0}; \frac{1}{2^n})$ *for* $n \geq 1$. *Since T is linear,* $\overline{T(B_n)} = \frac{1}{2^n}\overline{T(B_0)}$. *Now if* $\epsilon_0 = \frac{r}{k_0}$, *we have*

$$B\left(\mathbf{0}; \epsilon_0 \frac{1}{2^n}\right) = \frac{1}{2^n} B\left(\mathbf{0}; \epsilon_0\right) \subset \frac{1}{2^n}\, \overline{T(B(\mathbf{0}; \frac{1}{4}))}, \quad \textit{(by } \textbf{Claim Three}\textit{)}$$

Then by the linearity of T, we have

$$B\left(\mathbf{0}; \epsilon_0 \frac{1}{2^n}\right) \subset \overline{T(B(\mathbf{0}; \frac{1}{2^{n+2}}))} = \overline{T(B_{n+2})}$$

This establishes the claim.

Claim Five: *Given* $\boldsymbol{y} \in B(\mathbf{0}; \frac{\epsilon_0}{2}) = V_1$, *there is* $\boldsymbol{z} \in B(\mathbf{0}; 1)$ *with* $T(\boldsymbol{z}) = \boldsymbol{y}$.

The case $n = 1$:
Let $\boldsymbol{y} \in V_1$. *By* **Claim Four**, $V_1 \subset \overline{T(B_3)} = \overline{T(B(\mathbf{0}; \frac{1}{8}))}$. *So there is a sequence* $\boldsymbol{\xi}_n^1 \subset T(B(\mathbf{0}; \frac{1}{8}))$ *with* $\boldsymbol{\xi}_n^1 \to \boldsymbol{y}$. *Thus, given* $\tau = \frac{\epsilon_0}{32}$, *there is N with*

$$\ell > N \implies \|\boldsymbol{y} - \boldsymbol{\xi}_n^1\|_Y < \tau$$

Choose $\ell > N$ *and label it* ℓ^1. *Then* $\|\boldsymbol{y} - \boldsymbol{\xi}_n^1\|_Y < \frac{\epsilon_0}{32}$ *and* $\boldsymbol{\xi}_{\ell^1}^1 \in T(B(\mathbf{0}; \frac{1}{8}))$. *Let* $\boldsymbol{v_1} = \boldsymbol{\xi}_{\ell^1}^1$ *which implies there is* $\boldsymbol{u_1} \in B(\mathbf{0}; \frac{1}{8})$ *with* $\boldsymbol{v_1} = T(\boldsymbol{u_1})$ *and* $\|\boldsymbol{y} - T(\boldsymbol{u_1})\|_X < \frac{\epsilon_0}{32}$. *Let* $\boldsymbol{z_1} = \boldsymbol{u_1}$.

Thus,

$$\boldsymbol{y} - T(\boldsymbol{u_1}) \in B\left(\mathbf{0}; \frac{\epsilon_0}{32}\right) = B\left(\mathbf{0}, \frac{\epsilon_0}{2^{4+1}}\right) \subset B\left(\mathbf{0}, \frac{\epsilon_0}{2^2}\right) = V_2$$

The case $n = 2$:
Let $\boldsymbol{y} \in V_2$. *By* **Claim Four**, $V_2 \subset \overline{T(B_4)} = \overline{T(B(\mathbf{0}; \frac{1}{16}))}$ *and we know* $\boldsymbol{y} - T(\boldsymbol{u_1}) \in V_2$. *So there is a sequence* $\boldsymbol{\xi}_n^2 \subset T(B(\mathbf{0}; \frac{1}{16}))$ *with* $\boldsymbol{\xi}_n^2 \to \boldsymbol{y} - T(\boldsymbol{u_1})$. *By an argument similar to that of the case* $n = 1$, *there is an index* $\ell^2 > \ell^1$ *so that* $\|\boldsymbol{y} - T(\boldsymbol{u_1}) - \boldsymbol{\xi}_{\ell^2}^2\|_Y < \frac{\epsilon_0}{64}$ *and* $\boldsymbol{\xi}_{\ell^1}^2 \in T(B(\mathbf{0}; \frac{1}{16}))$. *Let* $\boldsymbol{v_2} = \boldsymbol{\xi}_{\ell^2}^2$, *which implies there is* $\boldsymbol{u_2} \in B(\mathbf{0}; \frac{1}{16})$ *with* $\boldsymbol{v_2} = T(\boldsymbol{u_2})$ *Thus,* $\boldsymbol{u_1} \in B(\mathbf{0}; \frac{1}{8})$, $\boldsymbol{u_2} \in B(\mathbf{0}; \frac{1}{16})$ *and*

$$\|\boldsymbol{y} - T(\boldsymbol{u_1}) - T(\boldsymbol{u_2})\|_Y < \frac{\epsilon_0}{64}$$

and

$$\boldsymbol{y} - T(\boldsymbol{u_1} + \boldsymbol{u_2}) \in B\left(\boldsymbol{0}; \frac{\epsilon_0}{64}\right) = B\left(\boldsymbol{0}, \frac{\epsilon_0}{2^{4+2}}\right) \subset B\left(\boldsymbol{0}, \frac{\epsilon_0}{2^3}\right) = V_3$$

Let $\boldsymbol{z_2} = \boldsymbol{u_1} + \boldsymbol{u_2}$.

The case $n = 3$*:*
Let $\boldsymbol{y} \in V_3$. *By* **Claim Four,** $V_3 \subset \overline{T(B_{2^3})} = \overline{T(B(\boldsymbol{0}; \frac{1}{32}))}$ *and we know* $\boldsymbol{y} - T(\boldsymbol{u_1}) - T(\boldsymbol{u_2}) \in V_3$. *So there is a sequence* $\boldsymbol{\xi}_n^3 \subset T(B(\boldsymbol{0}; \frac{1}{32}))$ *with* $\boldsymbol{\xi}_n^3 \to \boldsymbol{y} - T(\boldsymbol{u_1}) - T(\boldsymbol{u_2})$. *By an argument similar to that of the case* $n = 2$, *there is an index* $\ell^3 > \ell^2$ *so that* $\|\boldsymbol{y} - T(\boldsymbol{u_1})T(\boldsymbol{u_2}) - \boldsymbol{\xi}_{\ell^3}^3\|_Y < \frac{\epsilon_0}{128}$ *and* $\boldsymbol{\xi}_{\ell^2}^3 \in T(B(\boldsymbol{0}; \frac{1}{32}))$. *Let* $\boldsymbol{v_3} = \boldsymbol{\xi}_{\ell^3}^3$. *which implies there is* $\boldsymbol{u_3} \in B(\boldsymbol{0}; \frac{1}{32})$ *with* $\boldsymbol{v_3} = T(\boldsymbol{u_3})$ *Thus,* $\boldsymbol{u_1} \in B(\boldsymbol{0}; \frac{1}{8})$, $\boldsymbol{u_2} \in B(\boldsymbol{0}; \frac{1}{16})$, $\boldsymbol{u_3} \in B(\boldsymbol{0}; \frac{1}{32})$

$$\|\boldsymbol{y} - T(\boldsymbol{u_1}) - T(\boldsymbol{u_2}) - T(\boldsymbol{u_3})\|_Y < \frac{\epsilon_0}{128}$$

and

$$\boldsymbol{y} - T(\boldsymbol{u_1} + \boldsymbol{u_2} + \boldsymbol{u_3}) \in B\left(\boldsymbol{0}; \frac{\epsilon_0}{128}\right) = B\left(\boldsymbol{0}, \frac{\epsilon_0}{2^{4+3}}\right) \subset B\left(\boldsymbol{0}, \frac{\epsilon_0}{2^4}\right) = V_4$$

Let $\boldsymbol{z_3} = \boldsymbol{u_1} + \boldsymbol{u_2} + \boldsymbol{u_3}$.

The cases $n = Q$*:*
After Q steps, we obtain $\boldsymbol{u_Q} \in B(\boldsymbol{0}, \frac{1}{2^{Q+2}})$, $\boldsymbol{z_Q} = \boldsymbol{u_1} + \boldsymbol{u_2} + \ldots + \boldsymbol{u_Q}$ *with*

$$\begin{aligned} \|\boldsymbol{z_Q}\|_X &\leq \frac{1}{2^{2+1}} + \frac{1}{2^{2+2}} + \ldots + \frac{1}{2^{2+Q}} \\ &= \frac{1}{8}\left(1 + \frac{1}{2} + \ldots + \frac{1}{2^{Q-1}}\right) < \frac{1}{8}\,\frac{1 - 2^{-Q}}{1 - 1/2} < \frac{1}{4} \end{aligned}$$

and

$$\|\boldsymbol{y} - T(\boldsymbol{u_1} + \ldots + \boldsymbol{u_Q})\|_Y = \|\boldsymbol{y} - T(\boldsymbol{z_Q})\|_Y < \frac{\epsilon_0}{2^{4+Q}}$$

We can now finish the proof. Consider the sequence $(\boldsymbol{z_k})$. *If* $p > q$, *we have*

$$\begin{aligned} \|\boldsymbol{z_p} - \boldsymbol{z_q}\|_X &= \|\boldsymbol{u_{q+1}} + \ldots + \boldsymbol{u_p}\|_X \leq \sum_{k=q+1}^{p} \|\boldsymbol{u_k}\| < \sum_{k=q+1}^{p} \frac{1}{2^{k+2}} \\ &= \frac{1}{2^2}\,\frac{1}{2^{q+1}}\left(1 + \ldots + \frac{1}{2^{p-q-1}}\right) = \frac{1}{2^2}\,\frac{1}{2^{q+1}}\,\frac{1 - 2^{p-q}}{1 - 1/2} \\ &< \frac{1}{2^{q+1}}(1 - 2^{p-q}) < \frac{1}{2^{q+2}} \end{aligned}$$

This shows $(\boldsymbol{z_k})$ *is a Cauchy sequence in the complete normed linear space* X. *Hence, there is* $\boldsymbol{z} \in X$ *with* $\boldsymbol{z_k} \to \boldsymbol{z}$. *By the continuity of* T, *we then have* $T(\boldsymbol{z_k}) \to T(\boldsymbol{z})$. *However, we also know* $\|\boldsymbol{y} - T(\boldsymbol{z_k})\|_Y < \frac{\epsilon_0}{2^{4+k}}$ *which implies* $T(\boldsymbol{z_k}) \to \boldsymbol{y}$. *Since limits are unique, we have* $T(\boldsymbol{z}) = \boldsymbol{y}$. *We also had* $\|\boldsymbol{z_k}\|_X < \frac{1}{4}$ *for all k, so* $\|\boldsymbol{z}\|_X \leq \frac{1}{4} < 1$. *Therefore* $\boldsymbol{z} \in B(\boldsymbol{0}; 1)$ *with* $T(\boldsymbol{z}) = \boldsymbol{y}$.

Since our choice of $\boldsymbol{y} \in B(\boldsymbol{0}, \frac{\epsilon_0}{2})$ *was arbitrary, this tells us* $B(\boldsymbol{0}; \frac{\epsilon_0}{2}) \subset T(B(\boldsymbol{0}; 1))$. *This completes the proof of The Open Unit Ball Lemma.* ■

We can now prove the Open Mapping Theorem (OMT).

Theorem 12.2.2 The Open Mapping Theorem

Let $T : X \rightarrow Y$ be a mapping between two complete normed linear spaces where $\|\cdot\|_X$ is the norm on X and $\|\cdot\|_Y$ on Y. Let T be a bounded, linear and onto operator. Let $U \subset X$ be open. Then $T(U)$ is open in Y.

Proof 12.2.2
Let $U \subset X$ be open. Choose $\boldsymbol{y} \in T(U)$. Then $\boldsymbol{y} = T(\boldsymbol{u})$ for some $\boldsymbol{u} \in U$. We will show $\boldsymbol{y}$ is an interior point of $T(U)$. Now U is open and $\boldsymbol{u} \in U$. Hence, there is $\epsilon > 0$ so that $B(\boldsymbol{u}; \epsilon) \subset U$. Thus,

$$\begin{aligned} B(\mathbf{0}; \epsilon) &= B(\boldsymbol{u}; \epsilon) - \{\boldsymbol{u}\} \subset U - \{\boldsymbol{u}\} \\ &\Longrightarrow B(\mathbf{0}; 1) = \frac{1}{\epsilon} B(\mathbf{0}; \epsilon) = \frac{1}{\epsilon}\Big(B(\boldsymbol{u}; \epsilon) - \{\boldsymbol{u}\}\Big) \subset \frac{1}{\epsilon}\Big(U - \{\boldsymbol{u}\}\Big) \end{aligned}$$

By the Open Unit Ball Lemma, we know $T(B(\mathbf{0}; 1))$ contains an open ball about $\mathbf{0}$. Hence, there is $r_0 > 0$ so that

$$B(\mathbf{0}; r_0) \subset T(B(\mathbf{0}; 1)) \subset T\left(\frac{1}{\epsilon}\Big(U - \{\boldsymbol{u}\}\Big)\right) = \frac{1}{\epsilon}(T(U) - \{T(\boldsymbol{u})\})$$

by the linearity of T. We can rewrite this as

$$\epsilon\, B(\mathbf{0}; r_0) + \{T(\boldsymbol{u})\} \subset T(U) \Longrightarrow \epsilon\, B(T(\boldsymbol{u}); r_0) \subset T(U) \Longrightarrow B(T(\boldsymbol{u}); \epsilon\, r_0) \subset T(U)$$

But this says $B(\boldsymbol{y}; \epsilon\, r_0) \subset T(U)$ which shows $\boldsymbol{y}$.is an interior point of $T(U)$. Since the choice of $\boldsymbol{y}$ was arbitrary, this shows $T(U)$ is open. ■

12.2.1 Homework

Exercise 12.2.1 *Prove*

$$\overline{k\, T\left(B(\mathbf{0}; \frac{1}{8})\right)} = k\, T\left(\overline{B(\mathbf{0}; \frac{1}{8})}\right)$$

Exercise 12.2.2 *Prove if $\alpha > 0$, then*

$$B(\alpha \boldsymbol{z}; \alpha r) = \alpha B(\boldsymbol{z}; r)$$

Exercise 12.2.3 *Let X and Y be Banach Spaces and let $T : X \rightarrow Y$ be a bounded linear operator. Prove that either T is surjective or that $T(X)$ is of first category.*

12.3 The Closed Graph Theorem

This result is another famous one which should be in everyone's toolkit. First, we need to define what a **closed linear operator** means.

Definition 12.3.1 Closed Linear Operators

Let X and Y be normed linear spaces with $\|\cdot\|_X$ the norm for X and $\|\cdot\|_Y$, the norm for Y. Let $T : dom(T) \subset X \rightarrow Y$ be linear. We say T is a closed linear operator if $G(T) = \{(\boldsymbol{x}, \boldsymbol{y}) \mid \boldsymbol{x} \in D(T), \boldsymbol{y} = T(\boldsymbol{x})\}$ is closed as a subset of $X \times Y$. Here closure is interpreted in terms of the norm $\|\cdots\|$ on $X \times Y$ defined by $\|(\boldsymbol{x}, \boldsymbol{y})\| = \|\boldsymbol{x}\|_X + \|\boldsymbol{y}\|_Y$. We call $G(T)$ the graph of T.

Let's look at some examples before we prove the main result.

Example 12.3.1 *Let $X = (C([0,1]), \|\cdot\|_\infty)$ and*

$$Y = C^1([0,1) = \{\boldsymbol{x} : [0,1] \rightarrow \Re : \boldsymbol{x}' \text{ exists and } \boldsymbol{x}' \in C([0,1])\}$$

and the norm on Y is also $\|\cdot\|_\infty$. Define $T : dom(T) = Y \subset X \rightarrow X$ by $T(\boldsymbol{x}) = \boldsymbol{x}'$. Note T is not bounded as for $x_n(t) = t^n$,

$$\frac{\|T(\boldsymbol{x_n})\|_\infty}{\|\boldsymbol{x_n}\|_\infty} = n \rightarrow \infty$$

and so $\|T\| = \infty$. We see T is a linear operator which is not bounded. Consider the set

$$G(T) = \{(\boldsymbol{x}, \boldsymbol{x}') : \boldsymbol{x} \in C^1([0,1]) \subset C^1([0,1]) \times C([0,1])\}$$

Assume the sequence $(\boldsymbol{x_n}, \boldsymbol{x_n}') \subset G(T)$ and there is a pair $(\boldsymbol{x}, \boldsymbol{y})$ so that $((\boldsymbol{x_n}, \boldsymbol{x_n}')) \rightarrow (\boldsymbol{x}, \boldsymbol{y})$ in the $X \times Y$ norm. Hence, given $\epsilon > 0$, there is N so that

$$n > N \longrightarrow \|\boldsymbol{x_n} - \boldsymbol{x}\|_\infty + \|\boldsymbol{x_n}' - \boldsymbol{y}\|_\infty < \epsilon$$

This implies $\boldsymbol{x_n} \overset{unif}{\longrightarrow} \boldsymbol{x}$ and $\boldsymbol{x_n}' \overset{unif}{\longrightarrow} \boldsymbol{y}$. Since each $\boldsymbol{x_n}$ is continuous, it follows $\boldsymbol{x}$ is continuous. Also, since each $\boldsymbol{x_n}'$ is continuous, it follows $\boldsymbol{y}$ is also continuous. From the fundamental theorem of calculus, we have

$$x_n(t) = x_n(0) + \int_0^t x_n'(s)$$

Since $\boldsymbol{x_n}' \overset{unif}{\longrightarrow} \boldsymbol{y}$, it follows that

$$\int_0^t x_n'(s)ds \rightarrow \int_0^t y(s)ds \Longrightarrow x_n(t) \rightarrow x(0) + \int_0^t y(s)ds$$

Thus, $x(t) = x(0) + \int_0^t y(s)ds$. From the fundamental theorem of calculus, we then know $x'(t) = y(t)$. This tells us $\boldsymbol{x} \in C^1([0,1])$, with $T(\boldsymbol{x}) = \boldsymbol{y}$ and so $((\boldsymbol{x_n}, \boldsymbol{x_n}')) \rightarrow (\boldsymbol{x}, T(\boldsymbol{x}))$. This shows $G(T)$ is closed in what is called the cross product topology of $C^1([0,1]) \times C([0,1])$.

Example 12.3.2 *Consider $C^1([0,1]) \subset C([0,1])$ as described in the previous example with the sup - norm. Is $C^1([0,1])$ a closed subset of $C([0,1])$? This means if $(\boldsymbol{x_n}) \subset C^1([0,1])$ with $\boldsymbol{x_n} \overset{unif}{\longrightarrow} \boldsymbol{x}$, is it always true $\boldsymbol{x} \in C^1([0,1])$?*

The answer is **No**! *Here is an example. Let*

$$x_n(t) = \begin{cases} \frac{1}{2} - t, & 0 \le t \le \frac{1}{2} - \frac{1}{n} \\ \frac{n}{2}\left(t - \frac{1}{2}\right)^2 + \frac{1}{2n}, & \frac{1}{2} - \frac{1}{n} < t < \frac{1}{2} + \frac{1}{n} \\ -\frac{1}{2} + t, & \frac{1}{2} + \frac{1}{n} \le t \le 1 \end{cases}$$

Then,

$$\begin{aligned} t \to (\frac{1}{2} - \frac{1}{n})^- &\implies x_n(t) \to \frac{1}{2} - (\frac{1}{2} - \frac{1}{n}) = \frac{1}{n} \\ t \to (\frac{1}{2} - \frac{1}{n})^+ &\implies x_n(t) \to \frac{n}{2}\left(\left(\frac{1}{2} - \frac{1}{n}\right) - \frac{1}{2}\right)^2 + \frac{1}{2n} = \frac{1}{n} \\ t \to (\frac{1}{2} + \frac{1}{n})^- &\implies x_n(t) \to \frac{1}{n} \\ t \to (\frac{1}{2} + \frac{1}{n})^+ &\implies x_n(t) \to \frac{1}{n} \end{aligned}$$

and so $\boldsymbol{x_n} \in C([0,1])$. *Also note*

$$x_n'(t) = \begin{cases} -1 & 0 \le t < \frac{1}{2} - \frac{1}{n} \\ n\left(t - \frac{1}{2}\right), & \frac{1}{2} - \frac{1}{n} < t < \frac{1}{2} + \frac{1}{n} \\ 1, & \frac{1}{2} + \frac{1}{n} < t \le 1 \end{cases}$$

Further,

$$\begin{aligned} t \to (\frac{1}{2} - \frac{1}{n})^- &\implies x'\left(\frac{1}{2} - \frac{1}{n}\right)^- = -1 \\ t \to (\frac{1}{2} - \frac{1}{n})^+ &\implies x'\left(\frac{1}{2} - \frac{1}{n}\right)^+ = -1 \\ t \to (\frac{1}{2} + \frac{1}{n})^- &\implies x'\left(\frac{1}{2} - \frac{1}{n}\right)^- = 1 \\ t \to (\frac{1}{2} + \frac{1}{n})^+ &\implies x'\left(\frac{1}{2} - \frac{1}{n}\right)^+ = 1 \end{aligned}$$

So $\boldsymbol{x'} \in C([0,1])$.

Claim: $\boldsymbol{x_n} \overset{unif}{\longrightarrow} \boldsymbol{x}$ *where* $x(t) = |t - 1/2|$.

$$|x_n(t) - x(t)| = \begin{cases} 0 & 0 \le t \le \frac{1}{2} - \frac{1}{n} \\ \left|\frac{n}{2}\left(t - \frac{1}{2}\right)^2 + \frac{1}{2n} - |t - \frac{1}{2}|\right|, & \frac{1}{2} - \frac{1}{n} < t < \frac{1}{2} + \frac{1}{n} \\ 0, & \frac{1}{2} + \frac{1}{n} \le t \le 1 \end{cases}$$

$$\leq \frac{n}{2}\frac{1}{n^2} + \frac{1}{2n} + \frac{1}{n} = \frac{2}{n}$$

This inequality holds on $[0,1]$, *so we see* $\boldsymbol{x_n} \overset{unif}{\longrightarrow} \boldsymbol{x}$ *but it is clear the limit function* $x(t)$ *is not in* $C^1([0,1])$. *So* $C^1([0,1])$ *is* **not** *a closed subset (in fact a subspace)of* $C([0,1])$ *using the sup - norm.*

Now we are ready to tackle the theorem. First, a preliminary fact.

Theorem 12.3.1 Completeness of the Product Norm

Let X and Y be complete normed linear spaces with norms $\|\cdot\|_X$ *and* $\|\cdot\|_Y$, *respectively. The* $X \times Y$ *is complete with respect to the norm* $\|(\cdot,\cdot)\| = \cdot\|_X + \|\cdot\|_Y$.

Proof 12.3.1
The product topology is the topology induced by the product norm $\|\boldsymbol{x}\|_X + \|\boldsymbol{y}\|_Y$ *on* $(\boldsymbol{x},\boldsymbol{y}) \in X \times Y$. *It is straightforward to show this is indeed a norm and we let you do that in an exercise. Let* $(\boldsymbol{x_n}, \boldsymbol{y_n}) \subset X \times Y$ *be a Cauchy sequence. Then given* $\epsilon > 0$, *there is* N *so that*

$$n, m > N \implies \|(\boldsymbol{x_n}, \boldsymbol{y_n}) - (\boldsymbol{x_n}, \boldsymbol{y_n})\| < \epsilon$$

Hence,

$$n, m > N \implies \|\boldsymbol{x_n} - \boldsymbol{x_m}\|_X + \|\boldsymbol{y_n} - \boldsymbol{y_m}\|_Y < \epsilon$$

This immediately tells us $(\boldsymbol{x_n})$ *is a Cauchy sequence in* X *and* $(\boldsymbol{y_n})$ *is a Cauchy sequence in* Y. *Thus, there is a pair* $(\boldsymbol{x},\boldsymbol{y})$ *so that* $\boldsymbol{x_n} \to \boldsymbol{x}$ *in* $\|\cdot\|_X$ *and* $\boldsymbol{y_n} \to \boldsymbol{y}$ *in* $\|\cdot\|_Y$; *ii.e.* $(\boldsymbol{x_n}, \boldsymbol{y_n}) \to (\boldsymbol{x},\boldsymbol{y})$ *in* $\|(\cdot,\cdot)\| = \cdot\|_X + \|\cdot\|_Y$ *norm on* $X \times Y$. *This shows* $X \times Y$ *is complete with respect to this norm.*
■

Now we can prove the main result.

Theorem 12.3.2 The Closed Graph Theorem

Let X and Y be complete normed linear spaces with norms $\|\cdot\|_X$ *and* $\|\cdot\|_Y$, *respectively. Let* $T : dom(T) \subset X \to Y$ *be a closed linear operator. Then if* $dom(T)$ *is closed,* T *is bounded.*

Proof 12.3.2
Since $D(T)$ *is assumed closed with respect to convergence in* $\|\cdot\|_X$, *and it is therefore a closed subspace of a complete normed linear space, we know* $D(T)$ *is complete with respect to its norm too.*

Now let's look at $G(T)$. *Since* T *is a closed linear operator, we know* $G(T)$ *is closed with respect to convergence in* $\|(\cdot,\cdot)\|$. *We have proven that* $X \times Y$ *is complete in this norm and closed subsets of a complete normed linear space are complete.*

To show T *is bounded, we define* $\rho : G(T) \to D(T)$ *by* $\rho(\boldsymbol{x}, T(\boldsymbol{x})) = \boldsymbol{x}$. *Then* ρ *is a projection mapping which is clearly linear. Note*

$$\|\rho(\boldsymbol{x}, T(\boldsymbol{x}))\|_X = \|\boldsymbol{x}\|_X \leq \|\boldsymbol{x}\|_X + \|T(\boldsymbol{x})\|_Y = \|(\boldsymbol{x}, T(\boldsymbol{x}))\|$$

It follows that

$$\|\rho\| = \sup_{(\boldsymbol{x},T(\boldsymbol{x}))\neq(\boldsymbol{0},\boldsymbol{0})} \frac{\|\rho(\boldsymbol{x}, T(\boldsymbol{x}))\|_X}{\|(\boldsymbol{x}, T(\boldsymbol{x}))\|} \leq 1$$

and so ρ is a bounded linear operator. It is easy to see ρ is onto. Also, $\rho(\boldsymbol{x}, T(\boldsymbol{x})) = \rho(\boldsymbol{y}, T(\boldsymbol{y}))$ implies $\boldsymbol{x} = \boldsymbol{y}$ and so $(\boldsymbol{x}, T(\boldsymbol{x})) = (\boldsymbol{y}, T(\boldsymbol{y}))$ implying ρ is $1-1$. From the Open Mapping Theorem, it follows that ρ is an open mapping. But we also know ρ^{-1} exists. By the comment after the statement of the Open Mapping Theorem, we see ρ^{-1} must be continuous. Since $\rho^{-1}(\boldsymbol{x}) = (\boldsymbol{x}, T(\boldsymbol{x}))$, we note ρ^{-1} is also linear. We conclude ρ^{-1} is bounded. We have

$$\begin{aligned} \|\rho^{-1}(\boldsymbol{x})\| &\leq \|\rho^{-1}\| \, \|\boldsymbol{x}\|_X \Longrightarrow \|(\boldsymbol{x}, T(\boldsymbol{x}))\| \leq \|\rho^{-1}\| \, \|\boldsymbol{x}\|_X \\ &\Longrightarrow \|\boldsymbol{x}\|_X + \|T(\boldsymbol{x})\|_Y \leq \|\rho^{-1}\| \, \|\boldsymbol{x}\|_X \end{aligned}$$

We then see $\|T(\boldsymbol{x})\|_Y \leq \|\rho^{-1}\| \, \|\boldsymbol{x}\|_X$ and so $\|T\| \leq \|\rho^{-1}\|$. ■

In Example 12.3.1 we saw an operator which is closed and unbounded. In Example 12.3.2 tells us the domain of this operator $T(\boldsymbol{x}) = \boldsymbol{x}'$ is not closed in $C([0,1])$. The contrapositive form of the Closed Graph Theorem tells this also and requires less work!

We can characterize closed linear operators too.

Theorem 12.3.3 Characterization of Closed Linear Operators

Let X and Y be normed linear spaces with norms $|$ cdot$\|_X$ and $\| \cdot \|_Y$ respectively. Let $T : D(t) \subset X \rightarrow Y$ be linear. Then

$$T \text{ is closed} \iff \left\{ \begin{array}{l} (\boldsymbol{x_n}) \subset D(T),\ (T(\boldsymbol{x_n})) \subset Y \\ \text{with } \boldsymbol{x_n} \rightarrow \boldsymbol{x},\ T(\boldsymbol{x_n}) \rightarrow \boldsymbol{y} \\ \Longrightarrow \boldsymbol{x} \in D(T),\ T(\boldsymbol{x}) = \boldsymbol{y} \end{array} \right\}$$

Proof 12.3.3

($\Longrightarrow$):
If T is closed, $G(T)$ is closed and so if $(\boldsymbol{x_n}, T(\boldsymbol{x_n})) \subset G(T)$ with $\boldsymbol{x_n} \rightarrow \boldsymbol{x}$, $T(\boldsymbol{x_n}) \rightarrow \boldsymbol{y}$, we must have $\boldsymbol{x} \in D(T)$, $T(\boldsymbol{x}) = \boldsymbol{y}$.

($\Longleftarrow$):
If we assume the right-hand side, then $(\boldsymbol{x_n}, T(\boldsymbol{x_n})) \subset G(T)$ and $(\boldsymbol{x_n}, T(\boldsymbol{x_n})) \rightarrow (\boldsymbol{x}, \boldsymbol{y} = T(\boldsymbol{x})) \in G(T)$. Thus, $G(T)$ is closed and so T is a closed operator. ■

Let's finish with a convenient lemma.

Lemma 12.3.4 The Connections between a Closed Domain and a Closed Operator

Let X and Y be normed linear spaces with norms $|$ cdot$\|_X$ and $\| \cdot \|_Y$ respectively. Let $T : D(t) \subset X \rightarrow Y$ be linear and bounded. Then

1. *If $D(T)$ is closed, T is a closed operator.*
2. *If T is closed and Y complete, then $D(T)$ is closed.*

Proof 12.3.4
One: *If $(\boldsymbol{x_n}) \subset D(T)$ with $\boldsymbol{x_n} \rightarrow \boldsymbol{x}$ and $T(\boldsymbol{x_n}) \rightarrow \boldsymbol{y}$, then $\boldsymbol{x} \in \overline{D(T)}$. But $D(T)$ is closed, so $\boldsymbol{x} \in D(T)$. We also know T is continuous as it is bounded. Thus,*

$$\boldsymbol{y} = \lim_{n \rightarrow \infty} T(\boldsymbol{x_n}) = T(\boldsymbol{x})$$

implying $(\boldsymbol{x}, \boldsymbol{y} = T(\boldsymbol{x})) \in G(T)$. *By Theorem 12.3.3, it follows* T *is closed.*

Two:
If $\boldsymbol{x} \in \overline{D(T)}$, *there is a sequence* $(\boldsymbol{x_n}) \subset D(T)$ *so that* $\boldsymbol{x_n} \to \boldsymbol{x}$. *Since* T *is bounded*

$$\|T(\boldsymbol{x_n}) - T(\boldsymbol{x_m})\|_Y \leq \|T\| \, \|\boldsymbol{x_n} - \boldsymbol{x_m}\|_X$$

This tells us $(T(\boldsymbol{x_n}))$ *is a Cauchy sequence in* Y *which is complete. Hence, there is* $\boldsymbol{y} \in Y$ *so that* $T(\boldsymbol{x_n}) \to \boldsymbol{y}$. *Thus,* $(\boldsymbol{x_n}, T(\boldsymbol{x_n})) \in G(T)$ *with* $\boldsymbol{x_n} \to \boldsymbol{x}$, $T(\boldsymbol{x_n}) \to \boldsymbol{y}$. *We assume* T *is closed, we must have* $T(\boldsymbol{x}) = \boldsymbol{y}$ *and so* $\boldsymbol{x} \in D(T)$. *This shows* $D(T)$ *is closed.* ■

12.3.1 Homework

Exercise 12.3.1 *Let* X *and* Y *be normed linear spaces with* $\|\cdot\|_X$ *the norm for* X *and* $\|\cdot\|_Y$, *the norm for* Y. *Prove the mapping* $\|\cdots\|$ *on* $X \times Y$ *defined by* $\|(\boldsymbol{x}, \boldsymbol{y})\| = \|\boldsymbol{x}\|_X + \|\boldsymbol{y}\|_Y$ *is a norm.*

Exercise 12.3.2 *Show* $T(\boldsymbol{x}) = 3\boldsymbol{x}' + 2\boldsymbol{x}$ *is a closed linear operator on appropriate spaces.*

Exercise 12.3.3 *Show* $T(\boldsymbol{x}) = 3\boldsymbol{x}'' + 4\boldsymbol{x}' + 2\boldsymbol{x}$ *is a closed linear operator on appropriate spaces.*

Exercise 12.3.4 *Let* p *be a finite number greater than one. Let* $y = (a_j)$ *be a sequence of numbers and let* $x = (x_j)$ *be any object in* ℓ^p. *Assume that*

$$\sum_{j=1}^{\infty} a_j \, x_j$$

converges for all such x. *You will prove that the* y *belongs to* ℓ^q *where* p *and* q *are conjugate indices.*

1. *For each* x *and index* i, *let*

$$z_i = \sum_{j=1}^{i} a_j \, x_j$$

   This defines a sequence $z = (z_i)$. *Show that this sequence is in* ℓ_∞.

2. *Define the mapping* T *from* ℓ^p *to* ℓ^∞ *using the above equation; i.e.* $Tx = z$. *Prove that* T *is linear and a closed operator.*

3. *Prove that* T *is continuous and hence*

$$\left| \sum_{j=1}^{i} a_j \, x_j \right| \leq ||T|| \left(\sum_{j=1}^{\infty} |x_j|^p \right)^{\frac{1}{p}}$$

4. *Now since our choice of* x *is arbitrary, let's choose* x *like this:*

$$x_j = a_j \, |a_j|^{q-2}$$

   for $1 \leq j \leq i$ *and* x_k *nonzero. Otherwise, set* x_k *to be zero. Using the results above, you can then show*

$$\sum_{j=1}^{i} |a_j|^q \leq ||T|| \left(\sum_{j=1}^{i} |a_j|^q \right)^{\frac{1}{p}}$$

for all i.

5. *Show this implies y is in ℓ^q.*

Exercise 12.3.5 *Now let p and q be two numbers in $[1, \infty]$. Let A be the set of numbers (a_{ij}) for all integer indices i and j at least one. You can think of this as an* infinite *matrix. Let $x = (x_j)$ be any object in ℓ^p. Assume that*

$$y_i = \sum_{j=1}^{\infty} a_{ij}\, x_j$$

converges for each index i and all such x. Further, if we define $y = (y_i)$ to be the sequence defined by the above, we assume y is always in ℓ^q.

1. *The above defines a linear operator T from ℓ^p to ℓ^q by $Tx = y$.*

2. *Prove that x'_i defined by*

$$x'_i(x) = \sum_{j=1}^{\infty} a_{ij}\, x_j$$

defines a continuous linear functional on ℓ^p.

   (a) *If p is bigger than one, you can use the previous problem by realizing that since the first index i in the doubly subscripted a_{ij} is fixed, all the results of that problem apply to the sequence $b_i = (a_{ij})$.*

   (b) *If p is one, the result follows if you prove that if you know that*

$$\sum_{j=1}^{\infty} c_j\, d_j$$

   converges whenever $\sum |d_j|$ converges, then $||(c_j)||_\infty$ is finite.

   (c) *If p is infinity, prove that if you know that*

$$\sum_{j=1}^{\infty} c_j\, d_j$$

   converges whenever $\lim d_j = 0$ converges, then $\sum |c_j|$ converges.

3. *Prove T is closed.*
To do this, if $x_n \rightarrow x$ and $Tx_n \rightarrow w$, we know $x'_i(x_n) \rightarrow x'_i(x)$. Now by our construction, we know $Tx_n = (x'_i(x_n))$ and since $Tx_n \rightarrow w$, we see that $x'_i(x_n) \rightarrow w_i$. Show this then implies that $x'_i(x) = w_i$. This shows that $Tx = w$ and so T is closed.

4. *Prove T is continuous.*

Part V

Operators

Chapter 13

Stürm - Liouville Operators

Let's look at some differential equation models more abstractly. We will now discuss a few theoretical results surrounding certain types of second order nonlinear systems of ordinary differential equations.

13.1 ODE Background

Consider the following system for some $L > 0$ and some Θ.

$$\begin{aligned} u'' + \Theta u &= f, \; 0 \leq x \leq L, \\ \ell_1(u(0), u'(0)) &= \alpha_1 \, u(0) + \alpha_2 \, u'(0) = 0 \\ \ell_2(u(L), u'(L)) &= \beta_1 \, u(L) + \beta_2 \, u'(L) = 0 \end{aligned}$$

We will start by assuming the data or driving force f is continuous on $[0, L]$. We can relax this later. We also assume the boundary conditions are not degenerate by noting that α_1, α_2, β_1 and β_2 satisfy

$$\alpha_1^2 + \alpha_2^2 > 0, \quad \beta_1^2 + \beta_2^2 > 0.$$

The model above can be rewritten in a more abstract form by defining the operator $\mathscr{L}$ acting on a suitable domain of functions. We let $\mathscr{D}(\mathscr{L})$ denote the domain of $\mathscr{L}$ which for us will be

$$\mathscr{D}(L) = C^2([0, L]) \cap \{x \in C^2([0, L]) \text{ with } \ell_1(u(0), u'(0)) = 0; \ell_2(u(L), u'(L)) = 0\}$$

where $C^2([0, L])$ is the set of functions that are twice differentiable on $[0, L]$. We only assume the second derivative exists, but of course, if f is continuous and u is also continuous, since $u'' = f - \Theta u$, we actually know the second derivative is continuous also. However, if we relax the smoothness of f, we will only find the second derivative exists. For convenience, we will denote this collection of functions by $C^2([0, L]) \cap \{BC\}$. The operator $\mathscr{L}$ is then defined on this domain to be

$$\mathscr{L}(u) = u''.$$

There are two standard types of boundary value problems which specify zero values for the derivative on the boundary and which specify zero values for the solution itself. The two different models based on the operator $\mathscr{L}$ are then

$$\begin{aligned} u'' + \Theta u &= f, \; 0 \leq x \leq L \\ u'(0) &= 0, \quad u'(L) = 0, \end{aligned}$$

and

$$\begin{aligned} u'' + \Theta u &= f, \; 0 \leq x \leq L \\ u(0) &= 0, \quad u(L) = 0, \end{aligned}$$

These boundary values can be written very succinctly as

$$L(u) + \Theta u = f$$

and note the boundary conditions are hidden in the domain of $\mathscr{L}$. It is easy to see that we can solve this model if and only if the operator $\mathscr{L} + \Theta$ has a zero kernel. If the nullspace of $\mathscr{L} + \Theta$ is zero, then $\mathscr{L} + \Theta$ is invertible and we could write

$$u = (\mathscr{L} + \Theta)^{-1}(f)$$

for any continuous f on $[0, L]$. Of course, it is then clear that the inverse $(\mathscr{L} + \Theta)^{-1}$ has a much larger domain than just continuous functions. For example, we know that a function continuous a.e. is still Riemann Integrable. Thus, the data f could have many discontinuities and still be Riemann Integrable. The inverse operator handles that fine, but it is pretty easy to see that the resulting u is going to have problems in its second derivative at the points of discontinuity of f. These statements require proof, of course, but going into those issues in full detail requires advanced notions from abstract integration theory which is discussed in (Peterson (16) 2019). We have discussed carefully the completion of $C([a, b])$ to $\mathbb{L}_2([a, b])$ in this text and so we will be able to look at the inverse operator acting on the full $\mathbb{L}_2([a, b])$ but you should know we can say even more with the tools of measure theory at our disposable.

Homework

Exercise 13.1.1

$$\begin{aligned} u'' + 3u &= f, \; 0 \leq x \leq 6, \\ \ell_1(u(0), u'(0)) &= \alpha_1 \, u(0) + \alpha_2 \, u'(0) = 0 \\ \ell_2(u(L), u'(L)) &= \beta_1 \, u(L) + \beta_2 \, u'(L) = 0 \end{aligned}$$

Assume the data f is a step function such as

$$f(x) = \begin{cases} 0, & 0 \leq x < 2 \\ 10, & 2 \leq x \leq 4 \\ 0, & 4 < x \leq 6 \end{cases}$$

Explain what is happening in this differential equation at the points $x = 2$ and $x = 4$. How would you solve this problem?

Exercise 13.1.2

$$\begin{aligned} u'' + 9u &= f, \; 0 \leq x \leq 6, \\ \ell_1(u(0), u'(0)) &= \alpha_1 \, u(0) + \alpha_2 \, u'(0) = 0 \\ \ell_2(u(L), u'(L)) &= \beta_1 \, u(L) + \beta_2 \, u'(L) = 0 \end{aligned}$$

Assume the data f is a step function such as

$$f(x) = \begin{cases} 10, & 0 \le x < 2 \\ 2, & 2 \le x \le 4 \\ 20, & 4 < x \le 6 \end{cases}$$

Explain what is happening in this differential equation at the points $x = 2$ and $x = 4$. How would you solve this problem?

Exercise 13.1.3 *If $u' = f$ where f is continuous, what does the Fundamental Theorem of Calculus say about u?*

Exercise 13.1.4 *If $u' = f$ where f is a step function what does the Fundamental Theorem of Calculus say about u?*

13.2 The Stürm - Liouville Models

We are now going to discuss an important class of operators that are based on Stürm - Liouville problems. These come in a variety of forms, so we need some notation. We will do everything now on the interval $[a, b]$ instead of $[0, L]$. Our models earlier will be special cases of this class of differential operators. Let

$$\begin{aligned} C([a,b]) &= \{\boldsymbol{x} : [a,b] \to \Re \mid \boldsymbol{x} \text{ is continuous on } [a,b]\} \\ C^1([a,b]) &= \{\boldsymbol{x} : [a,b] \to \Re \mid \boldsymbol{x}' \text{ exists and is continuous on } [a,b]\} \\ C^2([a,b]) &= \{\boldsymbol{x} : [a,b] \to \Re \mid \boldsymbol{x}'' \text{ exists and is continuous on } [a,b]\} \end{aligned}$$

The differential equation models we want to solve will involve different types of boundary conditions.

$$\begin{aligned} \boldsymbol{BC_1} &= \{\boldsymbol{x} \in C^1([a,b]) \mid \alpha_a x(a) + \beta_a x'(a) = 0 \\ &\qquad \alpha_b x(b) + \beta_b x'(b) = 0, \ (\alpha_a, \beta_a) \text{ and } (\alpha_b, \beta_b) \ne (0,0)\} \\ \boldsymbol{BC_2} &= \{\boldsymbol{x} \in C^1([a,b]) \mid \boldsymbol{x} \text{ or } \boldsymbol{x}' \text{ is bounded at } t = a \\ &\qquad \text{i.e. } \lim_{t \to a^+} x(t) \text{ exists or the } \lim_{t \to a^+} x'(t) \text{ exists for some value} \\ &\qquad \alpha_b x(b) + \beta_b x'(b) = 0, \ (\alpha_b, \beta_b) \ne (0,0)\} \\ \boldsymbol{BC_3} &= \{\boldsymbol{x} \in C^1([a,b]) \mid \boldsymbol{x} \text{ and } \boldsymbol{x}' \text{ are bounded at } t = a, \ t = b\} \\ \boldsymbol{BC_4} &= \{\boldsymbol{x} \in C^1([a,b]) \mid x(a) = x(b), \ x'(a) = x'(b)\} \end{aligned}$$

Boundary condition $\boldsymbol{BC_2}$ can also be flipped with bounded functions at b^- for either the function or its derivative for some specific value with the usual other condition at $t = a$. We will leave that to you to work out. The differential equation to solve subject to the boundary conditions above is

$$(\boldsymbol{p}\boldsymbol{u}')' - (\lambda \boldsymbol{r} + \boldsymbol{q})\boldsymbol{u} = \boldsymbol{0} \tag{13.1}$$

where $\boldsymbol{0}$ is the function which is always 0 on $[a, b]$. For $\boldsymbol{p} \in C^1([a, b])$ with $p(t) > 0$ on (a, b), we add the following restrictions to $\boldsymbol{p}$

$$\begin{aligned} \boldsymbol{u} \text{ satisfies } \boldsymbol{BC_1} &\Longrightarrow p(a), p(b) > 0, \quad \boldsymbol{u} \text{ satisfies } \boldsymbol{BC_2} \Longrightarrow p(a) = 0 \\ \boldsymbol{u} \text{ satisfies } \boldsymbol{BC_3} &\Longrightarrow p(a) = p(b) = 0, \quad \boldsymbol{u} \text{ satisfies } \boldsymbol{BC_4} \Longrightarrow p(a) = p(b) \end{aligned}$$

We also assume

$$\boldsymbol{r} \in C([a,b]) \quad , r(t) > 0, \ a \le t \le b, \quad \boldsymbol{q} \in C([a,b]).$$

unless we are using $\boldsymbol{BC_2}$ in which case we relax continuity of the functions $\boldsymbol{q}$ to the interval (a, b) possibly. We can rewrite Equation 13.1 then as

$$(\boldsymbol{p}\boldsymbol{u}')' - \boldsymbol{q}\boldsymbol{u} = \lambda \boldsymbol{r}\boldsymbol{u}, \quad \boldsymbol{u} \text{ satisfies } \boldsymbol{BC_i} \tag{13.2}$$

Since $\boldsymbol{r}$ is positive on $[a, b]$, we can rewrite again as

$$\frac{1}{\boldsymbol{r}}\left(\, (\boldsymbol{p}\boldsymbol{u}')' - \boldsymbol{q}\boldsymbol{u}\,\right) = \lambda \boldsymbol{r}\boldsymbol{u}, \quad \boldsymbol{u} \text{ satisfies } \boldsymbol{BC_i} \tag{13.3}$$

where it is understood this equation holds at each t in $[a, b]$. Define the differential operator $\boldsymbol{L} : C^2([a, b]) \rightarrow C([a, b])$ by

$$\boldsymbol{L}(\boldsymbol{u}) = \frac{1}{\boldsymbol{r}}\left(\, (\boldsymbol{p}\boldsymbol{u}')' - \boldsymbol{q}\boldsymbol{u}\,\right)$$

This is well-defined because $\boldsymbol{p}'\boldsymbol{u}' + \boldsymbol{p}\boldsymbol{u}'' - \boldsymbol{q}\boldsymbol{u}$ is a continuous function on $[a, b]$ for the class of functions $\boldsymbol{p}$, $\boldsymbol{r}$ and $\boldsymbol{q}$ we have assumed whenever $\boldsymbol{u} \in C^2([a, b])$. Let the domain of $\boldsymbol{L}$ be $dom(\boldsymbol{l}) = C^2([a, b])$. Then the differential equation plus boundary conditions we wish to solve is equivalent to

$$\boldsymbol{L}(\boldsymbol{u}) = \lambda \boldsymbol{u}, \quad \boldsymbol{u} \text{ satisfies } \boldsymbol{BC_i} \tag{13.4}$$

Let

$$\boldsymbol{\Omega_i} = dom(\boldsymbol{L}) \cap \boldsymbol{BC_i} = C^2([a, b]) \cap \{\boldsymbol{x} \in C^1([a, b]) \ni \boldsymbol{BC_i} \text{ holds }\}$$

Then the differential equation to solve becomes an operator equation to solve: $\boldsymbol{L}(\boldsymbol{u}) = \lambda\, \boldsymbol{u}$ for $\boldsymbol{L} : \boldsymbol{\Omega_i} \subset C^2([a, b]) \rightarrow C([a, b])$. Clearly $\boldsymbol{L}$ is a linear operator and it is easy to check $\boldsymbol{\Omega_i}$ is a vector subspace of $C^2([a, b])$. We could endow the vector space $C([a, b])$ with usual inner product: $< \boldsymbol{x}, \boldsymbol{y} >= \int_a^b x(t)y(t)\, dt$. However, it is more useful to define a new inner product:

$$\omega(\boldsymbol{x}, \boldsymbol{y}) = \int_a^b r(t)\, x(t)y(t)\, dt.$$

We can easily show this is an inner product.

IP1:

$$\omega(\alpha\boldsymbol{x}, \boldsymbol{y}) = \alpha \int_a^b r(t)\, x(t)y(t)\, dt = \int_a^b r(t)\, x(t)\alpha y(t)\, dt = \alpha\omega(\boldsymbol{x}, \boldsymbol{y})$$

IP2:

$$\begin{aligned}
\omega(\boldsymbol{x} + \boldsymbol{y}, \boldsymbol{z}) &= \int_a^b r(t)\, (x(t) + y(t))\, z(t)\, dt \\
&= \int_a^b r(t)\, x(t)\, z(t)\, dt + \int_a^b r(t)\, y(t)\, z(t)\, dt \\
&= \omega(\boldsymbol{x}, \boldsymbol{z}) + \omega(\boldsymbol{y}, \boldsymbol{z})
\end{aligned}$$

IP3: It is clear this function is symmetric in its arguments.

IP4:

$$\omega(\boldsymbol{x}, \boldsymbol{x}) = \int_a^b r(t)\, (x(t))^2\, dt \geq 0$$

Moreover

$$\omega(\boldsymbol{x}, \boldsymbol{x}) \;=\; \int_a^b r(t)\,(x(t))^2\,dt = 0$$

implies the integrand is identically zero by a standard argument. Thus $\boldsymbol{x} = \boldsymbol{0}$.

Thus ω is an inner product on $C([a,b])$ and so is an inner product on $\boldsymbol{\Omega_i}$ too. From now on, in this discussion, we will simply use $< \cdot, \cdot >$ to denote the norm $\boldsymbol{\omega}$.

Homework

Exercise 13.2.1 *Is this a Stürm - Liouville problem on the interval* $[-1,2]$*?* $(1+t^2)u''(t)+2tu'(t) - t^3u(t) = 3(1+t^4)u(t)$ *for appropriate boundary conditions? If so, state* $\mathbf{p}$*,* $\boldsymbol{q}$ *and* $\boldsymbol{r}$*.*

Exercise 13.2.2 *For the Stürm - Liouville problem on* $[-1,1]$ $(2+4t^2)u''(t) + 8tu'(t) - 3tu(t) = 3(1+2t^2)u(t)$ *use power series methods to find the recurrence relation for the coefficients of the solution for the boundary conditions* $u(-1)+u'(-1) = 0$ *and* $u(1)+u'(1) = 0$*. Generate plots for solutions as well.*

Exercise 13.2.3 *For the Stürm - Liouville problem on* $[-1,1]$ $(2+4t^2)u''(t) + 8tu'(t) - 3tu(t) = 3(1+2t^2)u(t)$ *what is the appropriate inner product?*

Exercise 13.2.4 *Prove the differential operator* $L(u) = u''$ *is not bounded using* $\|\cdot\|_\infty$*.*

Exercise 13.2.5 *Prove the differential operator* $L(u) = u''$ *is not bounded using* $\|\cdot\|_2$*.*

Here are some examples of Stürm - Liouville differential equations.

Example 13.2.1 Bessel's Equation

$$t^2\boldsymbol{u}'' + t\boldsymbol{u}' - (\lambda t^2 + n^2)\boldsymbol{u} = 0, \quad , a \le t \le b$$

$$\begin{aligned} \boldsymbol{u}'' + \frac{1}{t}\boldsymbol{u}' - \left(\lambda + \frac{n^2}{t^2}\right)\boldsymbol{u} &= 0 \\ t\boldsymbol{u}'' + \boldsymbol{u}' - \left(\lambda\, t + \frac{n^2}{t}\right)\boldsymbol{u} &= 0 \\ (t\boldsymbol{u}')' - \left(\lambda t - \frac{n^2}{t}\right)\boldsymbol{u} &= 0 \end{aligned}$$

Here $\mathbf{p} = t$*,* $\boldsymbol{r} = t$ *and* $\boldsymbol{q} = \frac{n^2}{t}$*. Note this is rewritten as*

$$(t\boldsymbol{u}')' - \frac{n^2}{t}\boldsymbol{u} \;=\; \lambda\, t\, \boldsymbol{u}$$

The differential operator is

$$\boldsymbol{L}(\boldsymbol{u}) \;=\; \frac{1}{\boldsymbol{r}}\,(\,(\boldsymbol{p}\boldsymbol{u}')' - \boldsymbol{q}\boldsymbol{u}\,) \Longrightarrow \boldsymbol{L}(\boldsymbol{u}) = \frac{1}{t}\,((t\boldsymbol{u}')' - \frac{n^2}{t}\boldsymbol{u}\,)$$

and we use $\boldsymbol{BC_2}$*. The inner product is*

$$< \boldsymbol{x}, \boldsymbol{y} > \;=\; \int_a^b t\, x(t) y(t)\, dt.$$

Example 13.2.2 Legendre Functions

$$\begin{aligned} (1-t^2)\boldsymbol{u}'' - 2t\boldsymbol{u}' + n(n+1)\boldsymbol{u} &= 0, \quad -1 \le t \le 1 \\ ((1-t^2)\boldsymbol{u}')' - n(n+1)\boldsymbol{u} = 0 \end{aligned}$$

Here, $\boldsymbol{p} = 1 - t^2$, $\boldsymbol{q} = 0$, $\boldsymbol{r} = 1$ *and* $\lambda = n(n+1)$. *The differential operator is*

$$\boldsymbol{L}(\boldsymbol{u}) = \frac{1}{\boldsymbol{r}}\,(\,(\boldsymbol{p}\boldsymbol{u}')' - \boldsymbol{q}\boldsymbol{u}\,) \Longrightarrow \boldsymbol{L}(\boldsymbol{u}) = (\,((1-t^2)\boldsymbol{u}')'\,)$$

and we use $\boldsymbol{BC_3}$. *The inner product is*

$$< \boldsymbol{x}, \boldsymbol{y} > = \int_a^b x(t)y(t)\,dt.$$

Example 13.2.3 Hermite Polynomials

$$\begin{aligned} \boldsymbol{u}'' - 2t\boldsymbol{u}' - 2n\boldsymbol{u} &= 0, \quad -\infty < t < \infty \\ e^{-t^2}\boldsymbol{u}'' - 2te^{-t^2}\boldsymbol{u}' - 2ne^{-t^2}\boldsymbol{u} &= 0 \\ (e^{-t^2}\boldsymbol{u}')' - 2ne^{-t^2}\boldsymbol{u} &= 0 \end{aligned}$$

Here, $\boldsymbol{p} = e^{-t^2}$, $\boldsymbol{q} = 0$, $\boldsymbol{r} = e^{-t^2}$ *and* $\lambda = 2n$. *The differential operator is*

$$\boldsymbol{L}(\boldsymbol{u}) = \frac{1}{\boldsymbol{r}}\,(\,(\boldsymbol{p}\boldsymbol{u}')' - \boldsymbol{q}\boldsymbol{u}\,) \Longrightarrow \boldsymbol{L}(\boldsymbol{u}) = e^{t^2}(\,(e^{-t^2}\boldsymbol{u}')'\,) = \boldsymbol{u}'' - 2t\boldsymbol{u}'$$

and we use $\boldsymbol{BC_4}$. *The inner product is*

$$< \boldsymbol{x}, \boldsymbol{y} > = \int_{-\infty}^{\infty} e^{-t^2} x(t)y(t)\,dt.$$

Example 13.2.4 Laguerre Polynomials

$$\begin{aligned} t\boldsymbol{u}'' + (1-t)\boldsymbol{u}' + n\boldsymbol{u} &= 0, \quad a \le t < \infty \\ te^{-t}\boldsymbol{u}'' + (1-t)e^{-t}\boldsymbol{u}' + ne^{-t}\boldsymbol{u} &= 0 \\ (te^{-t}\boldsymbol{u}')' + ne^{-t}\boldsymbol{u} &= 0 \end{aligned}$$

Here, $\boldsymbol{p} = te^{-t}$, $\boldsymbol{q} = 0$, $\boldsymbol{r} = e^{-t}$ *and* $\lambda = -n$. *The differential operator is*

$$\boldsymbol{L}(\boldsymbol{u}) = \frac{1}{\boldsymbol{r}}\,(\,(\boldsymbol{p}\boldsymbol{u}')' - \boldsymbol{q}\boldsymbol{u}\,) \Longrightarrow \boldsymbol{L}(\boldsymbol{u}) = e^{t}(\,(te^{-t}\boldsymbol{u}')'\,) = t\boldsymbol{u}'' + (1-t)\boldsymbol{u}'$$

and we use like $\boldsymbol{BC_2}$ *except the solutions are bounded at* ∞. *The inner product is*

$$< \boldsymbol{x}, \boldsymbol{y} > = \int_0^{\infty} e^{-t} x(t)y(t)\,dt.$$

Example 13.2.5 Standard ODE from Separation of Variables applied to Linear PDE

$$\boldsymbol{u}'' - \lambda\boldsymbol{u} = 0, \quad a \le t \le b$$

Here, $\boldsymbol{p} = 1$*,* $\boldsymbol{q} = 0$*,* $\boldsymbol{r} = 1$*. The differential operator is*

$$\boldsymbol{L}(\boldsymbol{u}) \quad = \quad \frac{1}{\boldsymbol{r}}\,(\,(\boldsymbol{p}\boldsymbol{u}')' - \boldsymbol{q}\boldsymbol{u}\,) \Longrightarrow \boldsymbol{L}(\boldsymbol{u}) = \boldsymbol{u}''$$

and we use BC_1 *or* BC_4*. The inner product is*

$$< \boldsymbol{x}, \boldsymbol{y} > \quad = \quad \int_a^b x(t)y(t)\,dt.$$

Homework

Exercise 13.2.6 *Use MATLAB to plot a variety of Bessel functions.*

Exercise 13.2.7 *Use MATLAB to plot a variety of Legendre functions.*

Exercise 13.2.8 *Use MATLAB to plot a variety of Laguerre polynomials.*

Exercise 13.2.9 *Use MATLAB to plot a variety of Hermite polynomials.*

Stürm - Liouville models have many properties that we can use to understand their behavior.

13.3 Properties

We begin with a standard result that is at the heart of the usefulness of these operators.

Lemma 13.3.1 $< L(x), \boldsymbol{y} >=< \boldsymbol{x}, L(\boldsymbol{y}) >$

The differential operator $\boldsymbol{L}(\boldsymbol{u}) = \frac{1}{\boldsymbol{r}}((\boldsymbol{p}\boldsymbol{u}')' - \boldsymbol{q}\boldsymbol{u})$ *from* $\boldsymbol{\Omega_i} = dom(\boldsymbol{L}) \cap \boldsymbol{BC_i}$ *to* $C([a, b])$ *is self-adjoint: ff* $\boldsymbol{x}, \boldsymbol{y} \in dom(L) \cap \boldsymbol{BC_i}$*, then* $< \boldsymbol{L}(\boldsymbol{x}), \boldsymbol{y} >=< \boldsymbol{x}, \boldsymbol{L}(\boldsymbol{y}) >$*.*

Proof 13.3.1

$$\begin{aligned}
< \boldsymbol{L}(\boldsymbol{x}), \boldsymbol{y} > &= \int_a^b r(t)\left(\frac{(\,p(t)x'(t)\,)' - q(t)x(t)}{r(t)}\right) y(t)\,dt \\
&= \int_a^b (\,(p(t)x'(t)\,)' - q(t)x(t)\,)\,y(t)\,dt \\
&= \int_a^b (p(t)x'(t))'\,y(t)dt - \int_a^b q(t)\,x(t)y(t)\,dt \\
&= \left(p(t)x'(t)y(t)\right)_a^b - \int_a^b p(t)\,x'(t)\,y'(t)dt - \int_a^b q(t)\,x(t)y(t)\,dt
\end{aligned}$$

For BC_1**:**

$$\left(p(t)x'(t)y(t)\right)_a^b \quad = \quad p(b)x'(b)y(b) - p(a)x'(a)y(a) = \xi_1$$

For BC_2:

We know $\lim_{t\to a^+} x'(t) = f_{12}$ *and* $\lim_{t\to a^+} y(t) = g_{11}$

$$\left(p(t)x'(t)y(t)\right)_a^b = p(b)x'(b)y(b) - p(a)f_{12}g_{11}$$

But $p(a) = 0$ *here so*

$$\left(p(t)x'(t)y(t)\right)_a^b = p(b)x'(b)y(b) - 0(f_{12}g_{11}) = p(b)x'(b)y(b) = \xi_2$$

For BC_3:

We assume $\lim_{t\to a^+} x'(t) = f_{12}$, $\lim_{t\to b^-} x'(t) = f_{22}$ $\lim_{t\to a^+} y(t) = g_{11}$ *and* $\lim_{t\to b^-} y(t) = g_{21}$

$$\left(p(t)x'(t)y(t)\right)_a^b = p(b)f_{22}g_{21} - p(a)f_{12}g_{11} = 0 = \xi_3$$

because $p(a) = p(b) = 0$.

For BC_4:

Here $p(a) = p(b)$ *and*

$$\begin{aligned}\left(p(t)x'(t)y(t)\right)_a^b &= p(b)x'(b)y(b) - p(a)x'(a)y(a)\\ &= p(a)x'(a)y(a) - p(a)x'(a)y(a) = 0 = \xi_4\end{aligned}$$

We conclude

$$\begin{aligned}< \boldsymbol{L}(\boldsymbol{x}), \boldsymbol{y} > &= \xi_i - \int_a^b p(t)\, x'(t)\, y'(t) dt - \int_a^b q(t)\, x(t)y(t)\, dt\\ &= \xi_i - \int_a^b (p(t)\, y'(t))\, x'(t) dt - \int_a^b x(t)y(t)\, q(t)\, dt\\ &= \xi_i - \left\{\left(x(t)(p(t)y'(t))\right)_a^b - \int_a^b x(t)\, (p(t)y'(t))' dt\right\} - \int_a^b x(t)y(t)\, q(t)\, dt\\ &= \xi_i - \left(x(t)(p(t)y'(t))\right)_a^b + \int_a^b ((p(t)y'(t))' - q(t))x(t) dt\\ &= \xi_i - \left(x(t)(p(t)y'(t))\right)_a^b + < x, \boldsymbol{L}(\boldsymbol{y}) >\end{aligned}$$

For BC_1:

$$\left(p(t)x(t)y'(t)\right)_a^b = p(b)x(b)y'(b) - p(a)x(a)y'(a) = \zeta_1$$

For BC_2:

We know $\lim_{t\to a^+} x(t) = f_{11}$ *and* $\lim_{t\to a^+} y'(t) = g_{12}$

$$\left(p(t)x(t)y'(t)\right)_a^b = p(b)x(b)y'(b) - p(a)f_{11}g_{12}$$

But $p(a) = 0$ *here so*

$$\left(p(t)x'(t)y(t)\right)_a^b = p(b)x(b)y'(b) = \zeta_2$$

For BC_3:

We assume $\lim_{t\to a^+} x(t) = f_{11}$, $\lim_{t\to b^-} x(t) = f_{21}$ $\lim_{t\to a^+} y'(t) = g_{12}$ *and* $\lim_{t\to b^-} y'(t) = g_{22}$

$$\left(p(t)x'(t)y(t)\right)_a^b = p(b)f_{21}g_{22} - p(a)f_{11}g_{12} = 0 = \zeta_3$$

because $p(a) = p(b) = 0$.

For BC_4:

Here $p(a) = p(b)$ *and*

$$\begin{aligned}\left(p(t)x'(t)y(t)\right)_a^b &= p(b)x'(b)y(b) - p(a)x'(a)y(a) \\ &= p(a)x'(a)y(a) - p(a)x'(a)y(a) = 0 = \zeta_4\end{aligned}$$

Combining, we find

$$< \boldsymbol{L}(\boldsymbol{x}), \boldsymbol{y} > = \xi_i - \zeta_i + < x, \boldsymbol{L}(\boldsymbol{y}) >$$

Substituting,

$$\begin{aligned}\xi_1 - \zeta_1 &= (p(b)x'(b)y(b) - p(a)x'(a)y(a) - p(b)x(b)y'(b) + p(a)x(a)y'(a)) \\ &= p(b)(x'(b)y(b) - x(b)y'(b)) - p(a)(x'(a)y(a) \quad x(a)y'(a))\end{aligned}$$

Any functions $\boldsymbol{u}$ *satisfying* $\boldsymbol{BC_1}$, *satisfy* $\alpha_a u(a) + \beta_a u'(a) = 0$ *and* $\alpha_b u(b) + \beta_b u'(b) = 0$. *Further, we know* (α_a, β_a) *and* (α_b, β_b) *are not* $(0, 0)$. *Let's look at just the term* $x'(b)y(b) - x(b)y'(b)$. *If only* $\alpha_b \neq 0$, *then* $x(b) = y(b) = 0$ *and* $\xi_1 - \zeta_1 = 0$. *If only* β_b *is not zero, a similar argument shows* $\xi_1 - \zeta_1 = 0$. *If both are not zero, then* $x(b) = -\frac{-\beta_b}{\alpha_b} x'(b)$ *and* $y(b) = -\frac{-\beta_b}{\alpha_b} y'(b)$. *Then*

$$x'(b)y(b) - x(b)y'(b) = -x'(b)\,\frac{\beta_b}{\alpha_b} y'(b) + \frac{\beta_b}{\alpha_b} x'(b)y'(b) = 0$$

A similar argument shows the second term is zero as well. So we conclude $\xi_1 - \zeta_1 = 0$. *In addition, this sort of argument shows* $\xi_2 - \zeta_2 - 0$. *We already know* $\xi_3 - \zeta_3 = \xi_4 - \zeta_4 = 0$. *We have therefore shown* $< \boldsymbol{L}(\boldsymbol{x}), \boldsymbol{y} > = < \boldsymbol{x}, \boldsymbol{L}(\boldsymbol{y}) >$. ■

Homework

Exercise 13.3.1 *Verify* $< \boldsymbol{L}(\boldsymbol{x}), \boldsymbol{y} >=< \boldsymbol{x}, \boldsymbol{L}(\boldsymbol{y}) >$ *for the* $\boldsymbol{L}$ *defined by this model on* $[-1, 1]$ $(1 + t^2)u''(t) + 2tu'(t) - t^3 u = 3(1 + t^4)u(t)$ *with the boundary conditions* $u(-1) + u'(-1) = 0$ *and* $u(1) + u'(1) = 0$.

Exercise 13.3.2 *Verify* $< \boldsymbol{L}(\boldsymbol{x}), \boldsymbol{y} >=< \boldsymbol{x}, \boldsymbol{L}(\boldsymbol{y}) >$ *for the* $\boldsymbol{L}$ *defined by this model on* $[-2, 2]$ $(2+4t^2)u''(t)+8tu'(t)-3tu = 3(1+2t^2)u(t)$ *for the boundary conditions* $2u(-2)+3u'(-2) = 0$ *and* $4u(2) + 6u'(2) = 0$.

Exercise 13.3.3 *Verify* $< \boldsymbol{L}(\boldsymbol{x}), \boldsymbol{y} >=< \boldsymbol{x}, \boldsymbol{L}(\boldsymbol{y}) >$ *for the* $\boldsymbol{L}$ *defined by this model on* $[0, 5]$ $(2+3t+4t^2)u''(t) + (3 + 8t)u'(t) + 9tu(t) = 3(1 + t^2)u(t)$ *for the boundary conditions* $u(0) - 2u'(0) = 0$ *and* $-2u(5) + u'(5) = 0$.

We can extend these ideas to complex-valued functions $\boldsymbol{u}$ which we write as $\boldsymbol{x} = \boldsymbol{u} + \boldsymbol{iy}$ where $\boldsymbol{x}$ and $\boldsymbol{y}$ are in our original domain $\boldsymbol{\Omega_i}$. The space of functions we use then is extended to

$$\begin{aligned} C^*([a,b]) &= \{\boldsymbol{x} : [a,b] \to \mathbb{C} \mid \boldsymbol{x} \text{ is continuous on } [a,b]\} \\ C^{*,1}([a,b]) &= \{\boldsymbol{x} : [a,b] \to \mathbb{C} \mid \boldsymbol{x}' \text{ exists and is continuous on } [a,b]\} \\ C^{*,2}([a,b]) &= \{\boldsymbol{x} : [a,b] \to \mathbb{C} \mid \boldsymbol{x}'' \text{ exists and is continuous on } [a,b]\} \end{aligned}$$

Continuity and differentiability of $\boldsymbol{x}$ is then defined in terms of the continuity and differentiability of the components $\boldsymbol{u}$ and $\boldsymbol{v}$. The function $\boldsymbol{u}$ is called the real part of $\boldsymbol{x}$ and the function $\boldsymbol{v}$ is the imaginary part of $\boldsymbol{x}$. Now the functions $\boldsymbol{p}$, $\boldsymbol{q}$ and $\boldsymbol{r}$ that comprise L are specifically real-valued. Since our function spaces involve complex values, we need to use a complex inner product now which must satisfy properties $IPC1$, $IPC2$, $IPC3$ and $IPC4$. We also define Riemann Integration here as $\int_a^b x(t)dt = \int_a^b u(t)dt + \boldsymbol{i} \int_a^b v(t)dt$ and this will satisfy all the usual properties for the Riemann Integral of real-valued functions. Hence, we now use the inner product

$$\omega(\boldsymbol{x}, \boldsymbol{y}) = \int_a^b r(t)\, x(t)\overline{y(t)}\, dt.$$

where $\overline{y(t)}$ is the complex conjugate of $\boldsymbol{y} = \boldsymbol{f} + \boldsymbol{ig} = \boldsymbol{f} - \boldsymbol{ig}$. We can easily show this is a complex inner product.

IPC1: Let $\boldsymbol{x} = \boldsymbol{u} + \boldsymbol{iv}$ and $\boldsymbol{y} = \boldsymbol{f} + \boldsymbol{ig}$

$$\begin{aligned} \omega(\boldsymbol{x}, \boldsymbol{y}) &= \int_a^b r(t)\, x(t)\overline{y(t)}\, dt = \int_a^b r(t)\, (u(t) + \boldsymbol{i}v(t))(f(t) - \boldsymbol{i}g(t))\, dt \\ &= \int_a^b r(t) \Big((u(t)f(t) + v(t)g(t)) \; + \boldsymbol{i}(v(t)f(t) - u(t)g(t)) \Big) dt \end{aligned}$$

and

$$\begin{aligned} \omega(\boldsymbol{y}, \boldsymbol{x}) &= \int_a^b r(t)\, y(t)\overline{x(t)}\, dt = \int_a^b r(t)\, (f(t) + \boldsymbol{i}g(t))(u(t) - \boldsymbol{i}v(t))\, dt \\ &= \int_a^b r(t) \Big((u(t)f(t) + v(t)g(t)) \; - \boldsymbol{i}(v(t)f(t) - u(t)g(t)) \Big) dt \\ &= \overline{\omega(\boldsymbol{x}, \boldsymbol{y})} \end{aligned}$$

IPC2: let $\alpha = c + \boldsymbol{i}d$. Then

$$\omega(\alpha\boldsymbol{x}, \boldsymbol{y}) = \int_a^b r(t)\, \alpha \Big((u(t)f(t) - v(t)g(t)) \; + \boldsymbol{i}(v(t)f(t) - u(t)g(t)) \Big) dt$$

Let $U(t) + \boldsymbol{i}V(t) = (u(t)f(t) - v(t)g(t)) + \boldsymbol{i}(v(t)f(t) - u(t)g(t))$. Then

$$\begin{aligned}\omega(\alpha \boldsymbol{x}, \boldsymbol{y}) &= \int_a^b r(t)\,(c + \boldsymbol{i}d)(U(t) + \boldsymbol{i}V(t))dt \\ &= \int_a^b r(t)\left((cU(t) - dV(t)) + \boldsymbol{i}(cV(t) + dU(t))\right)dt\end{aligned}$$

Now,

$$\begin{aligned}\alpha\omega(\boldsymbol{x}, \boldsymbol{y}) &= (c + \boldsymbol{i}d)\int_a^b r(t)\,(U(t) + \boldsymbol{i}V(t))dt \\ &= c\int_a^b r(t)U(t)dt - d\int_a^b V(t)dt + \boldsymbol{i}d\int_a^b r(t)U(t)dt + \boldsymbol{i}c\int_a^b V(t)dt \\ &= \int_a^b r(t)(cU(t) - dV(t))dt + \boldsymbol{i}\int_a^b r(t)(cV(t) + dU(t))dt\end{aligned}$$

and these two quantities are the same.

IPC3:

$$\begin{aligned}\omega(\boldsymbol{x} + \boldsymbol{y}, \boldsymbol{z}) &= \int_a^b r(t)\,(x(t) + y(t))\,\overline{z(t)}\,dt \\ &= \int_a^b r(t)\,x(t)\,\overline{z(t)}\,dt + \int_a^b r(t)\,y(t)\,\overline{z(t)}\,dt \\ &= \omega(\boldsymbol{x}, \boldsymbol{z}) + \omega(\boldsymbol{y}, \boldsymbol{z})\end{aligned}$$

IPC4:

$$\omega(\boldsymbol{x}, \boldsymbol{x}) = \int_a^b r(t)\,x(t)\overline{x(t)}\,dt = \int_a^b r(t)\,((u(t))^2 + (v(t))^2)\,dt \neq 0.$$

Moreover

$$\omega(\boldsymbol{x}, \boldsymbol{x}) = \int_a^b r(t)\,((u(t))^2 + (v(t))^2)\,dt = 0$$

implies the $u(t)$ and $v(t)$ is identically zero by a standard argument. Thus $\boldsymbol{x} = \boldsymbol{0}$.

We will let $\boldsymbol{\Omega_i}$ continue to denote the domain of the operator here; it will be understood we are now looking at the complex-valued spaces. Thus ω is an inner product on $C^*([a, b])$ and so is an inner product on $\boldsymbol{\Omega_i}$ too. From now on, in this discussion, we will simply use $< \cdot, \cdot >$ to denote the complex-valued norm $\boldsymbol{\omega}$.

Homework

Exercise 13.3.4 *Let $f(t) = t + \boldsymbol{i}t^2$ and $g(t) = -2t + \boldsymbol{i}5t^2$. Let $r(t) = (1 + t^2)$. Compute $< \boldsymbol{f}, \boldsymbol{g} >$.*

Exercise 13.3.5 *If you had defined the complex inner product by $< \boldsymbol{f}, \boldsymbol{g} >= \int_a^b \boldsymbol{r}\boldsymbol{f}\boldsymbol{g}$ instead of using the complex conjugate of $\boldsymbol{g}$, what happens to $< \boldsymbol{x}, \boldsymbol{x} >$ in that situation?*

Exercise 13.3.6 *Let $f(t) = R_1 e^{i\alpha t}$ and assume $r(t) = e^{-\sigma r^2 t}$. The inner product is $< f, g >= \int_{\Re} r(t) f(t) \overline{g(t)} dt$. Compute $\|f\|_2$. Can you interpret this as a probability if this value is suitably normalized?*

Exercise 13.3.7 *Find the general complex solution to $u'' - 4u' + 13u = 0$. Use the weighting function $r = 1$. Find the two linearly independent complex solutions and compute their norm.*

We can now show the eigenvalues of L are always real. To do this, let's temporarily think of L as a differential operator that acts on complex-valued functions. Thus

Lemma 13.3.2 The eigenvalues of L are Real

> *The differential operator $L(u) = \frac{1}{r}\,(\,(pu')' - qu\,)$ from $\Omega_i = dom(L) \cap BC_i$ to $C^*([a,b])$ has real eigenvalues.*

Proof 13.3.2
If an eigenvalue was complex, we would have $L(x) = (a + bi)x$. Let $x = u + iv$. It is an easy calculation to then show $L(\overline{x}) = (a - bi)\overline{x}$. Then, for $y = f + ig$, we have

$$\begin{aligned} < L(x), y > &= < L(u + iv), f + ig >=< L(u) + iL(v), f + ig > \\ &= < L(u), f + ig > + < iL(v), f + ig > \\ &= < L(u), f > + < L(u), ig > + < iL(v), f > + < iL(v), ig > \\ &= < L(u), f > -i < L(u), g > +i < L(v), f > +i\,\bar{i}\;< L(v), g > \end{aligned}$$

Now use the fact that for real-valued functions x and y, $< L(x), y >=< x, L(y) >$. The complex inner product when used on real-valued functions is the same as our original inner product, so we can now say

$$\begin{aligned} < L(x), y > &= < u, L(f) > -i < u, L(g) > +i < v, L(f) > + < v, L(g) > \\ &= < u + iv, L(f) > + < -iu + v, L(g) > \\ &= < u + iv, L(f) > -i\; < u + iv, L(g) > \\ &= < u + iv, L(f) > + < u + iv, i\,L(g) > \\ &= < u + iv, L(f) + iL(g) >=< x, L(y) > \end{aligned}$$

We can now use this to show eigenvalues must be real. We have

$$< L(x), x > \;=\; < (a + ib)x, x >= (a + ib) < x, x >$$

But we also know

$$< L(x), x > \;=\; < x, L(x) >=< x, (a + ib)x >= (a - ib) < x, x >$$

Since these two calculations must match, we have $(a + ib) < x, x >= (a - ib) < x, x >$. Since $a + ib$ is an eigenvalue, we must have $< x, x > > 0$ as x is an eigenvector for $a + ib$. We conclude $a - ib = a + ib$ implying $b = 0$. So the eigenvalues must be real. ■

We can also prove

Lemma 13.3.3 Eigenvectors of L for Distinct Eigenvalues are Orthogonal

The eigenvectors corresponding to distinct eigenvalues of the differential operator $\boldsymbol{L}(\boldsymbol{u}) = \frac{1}{r}\left((\boldsymbol{pu'})' - \boldsymbol{qu} \right)$ from $\boldsymbol{\Omega_i} = dom(\boldsymbol{L}) \cap \boldsymbol{BC_i}$ to $C^*([a,b])$ are orthogonal.

Proof 13.3.3
Let λ and μ be two distinct eigenvalues with eigenvectors $\boldsymbol{x}$ and $\boldsymbol{y}$ respectively.. Then

$$\begin{aligned} < \boldsymbol{L}(\boldsymbol{x}), \boldsymbol{y} > &= < \lambda \boldsymbol{x}, \boldsymbol{x} >= \lambda < \boldsymbol{x}, \boldsymbol{y} > \\ &= < \boldsymbol{x}, \boldsymbol{L}(\boldsymbol{y}) >=< \boldsymbol{x}, \mu \boldsymbol{y} >= \mu < \boldsymbol{x}, \boldsymbol{y} > \end{aligned}$$

because μ is real. We see these two calculations must be the same. Thus $(\lambda - \mu) < \boldsymbol{x}, \boldsymbol{y} >= 0$. Since $\lambda \neq \mu$, we must have $< \boldsymbol{x}, \boldsymbol{y} >= 0$ and the eigenvectors are orthogonal. ■

Homework

Exercise 13.3.8 *Prove explicitly that the eigenvectors for distinct eigenvalues of Bessel's equation with appropriate boundary conditions are orthogonal.*

Exercise 13.3.9 *Prove explicitly that the eigenvectors for distinct eigenvalues of Laguerre's equation with appropriate boundary conditions are orthogonal.*

Exercise 13.3.10 *Prove explicitly that the eigenvectors for distinct eigenvalues of Legendre's equation with appropriate boundary conditions are orthogonal.*

Exercise 13.3.11 *Prove explicitly that the eigenvectors for distinct eigenvalues of Hermite's equation with appropriate boundary conditions are orthogonal.*

13.4 Linear Independence of the Solutions

Consider the ordinary differential equation on the interval $[a, b]$:

$$L(\boldsymbol{u}) = \boldsymbol{u}'' + \frac{\boldsymbol{a_1}}{\boldsymbol{a_0}}\boldsymbol{u}' + \frac{\boldsymbol{a_2}}{\boldsymbol{a_0}}\boldsymbol{u} = \boldsymbol{0}$$

where $\boldsymbol{a_0}$ is nonzero on $[a, b]$ We assume the domain of L is $C^2([a,b])$ and $\boldsymbol{a_0}$, $\boldsymbol{a_1}$ and $\boldsymbol{a_2}$ are at least in $C([a,b])$. Consider these separate problems

$$L(\boldsymbol{u}) = \boldsymbol{0}, \quad u(a) = 1, \; u'(a) = 0 \tag{13.5}$$

$$L(\boldsymbol{u}) = \boldsymbol{0}, \quad u(a) = 0, \; u'(a) = 1 \tag{13.6}$$

If $\boldsymbol{u_1}$ solves Equation 13.5 and $\boldsymbol{u_2}$ solves Equation 13.6, the solutions are linearly independent if

$$\alpha \boldsymbol{u_1} + \beta \boldsymbol{u_2} = \boldsymbol{0} \Longrightarrow \alpha = \beta = 0$$

We want a tool that checks for independence. Let the **Wronskian**, $\boldsymbol{W}$ be defined by

$$W(t) = \det \begin{bmatrix} u_1(t) & u_2(t) \\ u_1'(t) & u_2'(t) \end{bmatrix} = u_1(t)u_2'(t) - u_1'(t)u_2(t)$$

Lemma 13.4.1 Characterizing the Wronskian

$$W(t) = W(c)\exp\left\{\int_c^t -\frac{a_1(s)}{a_0(s)}\,ds\right\}$$

for any c in $[a,b]$.

Proof 13.4.1
We know

$$\begin{aligned}W(t) &= u_1(t)u_2'(t) - u_1'(t)u_2(t)\\ W'(t) &= u_1'(t)u_2'(t) + u_1(t)u_2''(t) - u_1'(t)u_2'(t) - u_2(t)u_1''(t)\\ &= u_1(t)u_2''(t) - u_2(t)u_1''(t)\end{aligned}$$

but

$$\begin{aligned}\boldsymbol{u_1}'' &= -\frac{\boldsymbol{a_1}}{\boldsymbol{a_0}}\boldsymbol{u_1}' - \frac{\boldsymbol{a_2}}{\boldsymbol{a_0}}\boldsymbol{u_1}\\ \boldsymbol{u_2}'' &= -\frac{\boldsymbol{a_1}}{\boldsymbol{a_0}}\boldsymbol{u_2}' - \frac{\boldsymbol{a_2}}{\boldsymbol{a_0}}\boldsymbol{u_2}\end{aligned}$$

Substituting, we have

$$\begin{aligned}W'(t) &= u_1(t)\left(-\frac{a_1(t)}{a_0(t)}u_2(t)' - \frac{a_2(t)}{a_0(t)}u_2(t)\right) - u_2(t)\left(-\frac{a_1(t)}{a_0(t)}u_1(t)' - \frac{a_2(t)}{a_0(t)}u_1(t)\right)\\ &= -\frac{a_1(t)}{a_0(t)}u_1(t)u_2(t)' - \frac{a_2(t)}{a_0(t)}u_1(t)u_2(t) + \frac{a_1(t)}{a_0(t)}u_2(t)u_1(t)' + \frac{a_2(t)}{a_0(t)}u_1(t)u_2(t)\\ &= \frac{a_1(t)}{a_0(t)}(u_2(t)u_1(t)' - u_1(t)u_2(t)') = -\frac{a_1(t)}{a_0(t)}W(t)\end{aligned}$$

This immediately implies the family of solutions

$$W(t) = A\exp\left\{\int_c^t -\frac{a_1(s)}{a_0(s)}\,ds\right\}$$

where A is arbitrary. At $c \in [a,b]$, we note $A = W(c)$ leading to the result

$$W(t) = W(c)\exp\left\{\int_c^t -\frac{a_1(s)}{a_0(s)}\,ds\right\}$$

■

Note in particular, if $c = a$,

$$W(a) = \det\begin{bmatrix}u_1(a) & u_2(a)\\ u_1'(a) & u_2'(a)\end{bmatrix} = \begin{bmatrix}1 & 0\\ 0 & 1\end{bmatrix} = 1$$

So we can say

$$W(t) = \exp\left\{\int_a^t -\frac{a_1(s)}{a_0(s)}\,ds\right\}$$

We can use the Wronskian to determine linearly independence of solutions. In general, the Wronskian can be applied to any two functions on $[a,b]$. So to be clear, we often use the notation $\boldsymbol{W(f,g)}$

when we want to talk about the Wronskian of the functions $\boldsymbol{f}$ and $\boldsymbol{g}$.

Lemma 13.4.2 Two solutions are Linearly Independent if and only if the Wronskian of the Two Solutions is not Zero on $[a, b]$.

The two continuously differentiable functions $\boldsymbol{f}$ and $\boldsymbol{g}$ are linearly independent if and only if $\boldsymbol{W}(\boldsymbol{f}, \boldsymbol{g})$ is not zero on $[a, b]$.

Proof 13.4.2

The functions $\boldsymbol{f}$ and $\boldsymbol{g}$ are linearly independent on $[a, b]$ if

$$\alpha \boldsymbol{f} + \beta \boldsymbol{g} = \boldsymbol{0} \Longrightarrow \alpha = \beta = 0$$

Now whether or not we have independence, this equation implies

$$\begin{aligned} \alpha \boldsymbol{f} + \beta \boldsymbol{g} &= \boldsymbol{0} \\ \alpha \boldsymbol{f}' + \beta \boldsymbol{g}' &= \boldsymbol{0} \Longrightarrow \begin{bmatrix} f(t) & f'(t) \\ g(t) & g'(t) \end{bmatrix} \begin{bmatrix} \alpha \\ \beta \end{bmatrix} = \begin{bmatrix} 0 \\ 0 \end{bmatrix}, \quad a \leq t \leq b \end{aligned}$$

Since $\alpha = \beta = 0$, this implies

$$\det \begin{bmatrix} f(t) & f'(t) \\ g(t) & g'(t) \end{bmatrix} \neq 0, \quad a \leq t \leq b \implies \boldsymbol{W}(\boldsymbol{f}, \boldsymbol{g}) \neq 0$$

Conversely is $\boldsymbol{W}(\boldsymbol{f}, \boldsymbol{g}) \neq 0$, then the only solution to

$$\begin{bmatrix} f(t) & f'(t) \\ g(t) & g'(t) \end{bmatrix} \begin{bmatrix} \alpha \\ \beta \end{bmatrix} = \begin{bmatrix} 0 \\ 0 \end{bmatrix}, \quad a \leq t \leq b$$

is $\alpha = \beta = 0$ which tells us $\boldsymbol{f}$ and $\boldsymbol{g}$ are linearly independent. ■

Homework

Exercise 13.4.1 *For the Stürm - Liouville problem on $[-1, 1]$ $(2 + 4t^2)u''(t) + 8tu'(t) - 3tu(t) = 3(1 + 2t^2)u(t)$ use power series methods to find the recurrence relation for the coefficients of the solution for the boundary conditions $u(-1) + u'(-1) = 0$ and $u(1) + u'(1) = 0$ and from that the two linearly independent solutions.*

- *Find the Wronskian using the determinant form.*
- *Find the Wronskian using the integral characterization.*

Exercise 13.4.2 *For the Stürm - Liouville problem on $[-1, 1]$ $(2 + 4t^2)u''(t) + 8tu'(t) - 3tu(t) = 3(1 + 2t^2)u(t)$ use power series methods to find the recurrence relation for the coefficients of the solution for the boundary conditions $u'(-1) = 0$ and $u'(1) = 0$ for the coefficients of the solution for the boundary and from that the two linearly independent solutions.*

- *Find the Wronskian using the determinant form.*
- *Find the Wronskian using the integral characterization.*

Exercise 13.4.3 *For the Stürm - Liouville problem on $[0, 10]$ $u''(t) + 5u(t) = 2u(t)$ for the boundary conditions $u'(0) = 0$ and $u'(10) = 0$ find the two linearly independent solutions.*

- *Find the Wronskian using the determinant form.*

- *Find the Wronskian using the integral characterization.*

Exercise 13.4.4 *For the Stürm - Liouville problem on* $[0, 2]$ $u''(t) + 12u(t) = 2u(t)$ *for the boundary conditions* $u(0) = 0$ *and* $u(2) = 0$ *find the two linearly independent solutions.*

- *Find the Wronskian using the determinant form.*
- *Find the Wronskian using the integral characterization.*

Let's apply this result to the ordinary differential equation model.

Lemma 13.4.3 If $L(u) = 0$ has Two Solutions, They are Linearly Independent if the Wronskian is Nonzero at One Point

If $\boldsymbol{u_1}$ *and* $\boldsymbol{u_2}$ *are two solutions to* $\boldsymbol{L(u) = u'' + \frac{a_1}{a_0}u' + \frac{a_2}{a_0}u = 0}$, *then* $\boldsymbol{u_1}$ *and* $\boldsymbol{u_2}$ *are linearly independent on* $[a, b]$ *if* $\boldsymbol{W(u_1, u_2)}$ *is not zero for at least one point in* $[a, b]$.

Proof 13.4.3
For $\boldsymbol{W} = \boldsymbol{W(u_1, u_2)}$, *we have*

$$W(t) \quad = \quad W(c) \exp\left\{ \int_c^t -\frac{a_1(s)}{a_0(s)}\, ds \right\}$$

for any c *in* $[a, b]$. *Thus, if* $W(c) \neq 0$, $W(t) \neq 0$ *on* $[a, b]$ *and* $\boldsymbol{u_1}$ *and* $\boldsymbol{u_2}$ *are linearly independent.* ■

There are some other facts about the Stürm - Liouville models that are very useful.

Lemma 13.4.4 Lagrange's Identity for Stürm - Liouville Models

If $\boldsymbol{u}$ *and* $\boldsymbol{v}$ *are in* $\boldsymbol{\Omega_i}$ *then*

$$\begin{aligned} \boldsymbol{uL(v) - vL(u)} &= \frac{1}{\boldsymbol{r}}\Big(\boldsymbol{p\,(uv' - vu')}\Big)' \\ &= \frac{1}{\boldsymbol{r}}\Big(\boldsymbol{p\,W(u, v)}\Big)' \end{aligned}$$

Proof 13.4.4
$\boldsymbol{L(u) = \frac{(pu')' - qu}{r}}$, *so*

$$\begin{aligned} \boldsymbol{uL(v) - vL(u)} &= \boldsymbol{u}\left(\frac{\boldsymbol{(pv')' - qv}}{\boldsymbol{r}}\right) - \boldsymbol{v}\left(\frac{\boldsymbol{(pu')' - qu}}{\boldsymbol{r}}\right) \\ &= \frac{\boldsymbol{upv'' + up'v' - uqv - vpu'' - vp'u' + vqu}}{\boldsymbol{r}} \\ &= \frac{\boldsymbol{upv'' + up'v' - vpu'' - vp'u'}}{\boldsymbol{r}} \\ &= \frac{\boldsymbol{p'}}{\boldsymbol{r}}\,\boldsymbol{(uv' - vu')} + \frac{\boldsymbol{p}}{\boldsymbol{r}}\,\boldsymbol{(uv'' - vu'')} \end{aligned}$$

Next, we can calculate

$$\frac{1}{\boldsymbol{r}}\Big(\boldsymbol{pW(u_1, u_2)}\Big)' \quad = \quad \frac{1}{\boldsymbol{r}}\Big(\boldsymbol{p'\,(uv' - vu') + p(u'v' + uv'' - v'u' - vu'')}\Big)$$

$$
\begin{aligned}
&= \frac{1}{r}\Big(p'\,(uv' - vu') + p(uv'' - vu'')\Big) \\
&= \frac{p'}{r}(uv' - vu') + \frac{p}{r}(uv'' - vu'')
\end{aligned}
$$

These calculations are the same and so the result is verified. ■

and

Lemma 13.4.5 Abel's Identity for Stürm - Liouville Models

If $\boldsymbol{u}$ and $\boldsymbol{v}$ are in $\boldsymbol{\Omega_i}$ and satisfy $\boldsymbol{L(u)} = \lambda \boldsymbol{u}$ and $\boldsymbol{L(v)} = \lambda \boldsymbol{v}$; i.e. $\boldsymbol{u}$ and $\boldsymbol{v}$ are both eigenvectors for the eigenvalue λ, then

$$
\boldsymbol{pW(u,v)} = \boldsymbol{p(uv' - vu')} = c
$$

for some constant c.

Proof 13.4.5

Since $\mathbf{u}$ and $\mathbf{v}$ are both eigenvectors for the eigenvalue λ, we have

$$
\frac{(\boldsymbol{pu'})' - \boldsymbol{qu}}{\boldsymbol{r}} - \lambda \boldsymbol{u} = \mathbf{0}, \quad \frac{(\boldsymbol{pv'})' - \boldsymbol{qv}}{\boldsymbol{r}} - \lambda \boldsymbol{v} = \mathbf{0}
$$

Hence,

$$
0 = \boldsymbol{v}\Big(\frac{(\boldsymbol{pu'})' - \boldsymbol{qu}}{\boldsymbol{r}} - \lambda \boldsymbol{u}\Big) - \boldsymbol{u}\Big(\frac{(\boldsymbol{pv'})' - \boldsymbol{qv}}{\boldsymbol{r}} - \lambda \boldsymbol{v}\Big) = \frac{\boldsymbol{v}(\boldsymbol{pu'})' - \boldsymbol{u}(\boldsymbol{pv'})'}{\boldsymbol{r}}
$$

But $\boldsymbol{r}$ is positive, so we conclude

$$
\boldsymbol{v}(\boldsymbol{pu'})' - \boldsymbol{u}(\boldsymbol{pv'})' = \mathbf{0}
$$

But this implies

$$
\begin{aligned}
0 &= \int_a^t \Big(v(s)(p(s)u(s)')' - u(s)(p(s)v(s)')'\Big)\,ds \\
&= \Big(v(s)(p(s)u(s)')\Big)_a^t - \int_a^t v'(s)(p(s)u(s)')\,ds \\
&\quad -\Big(u(s)(p(s)v(s)')\Big)_a^t + \int_a^t u'(s)(p(s)v(s)')\,ds \\
&= \Big(v(s)(p(s)u(s)')\Big)_a^t - \Big(u(s)(p(s)v(s)')\Big)_a^t \\
&= v(t)p(t)u'(t) - v(a)p(a)u'(a) - u(t)p(t)v'(t) + u(a)p(a)v'(a)
\end{aligned}
$$

or for all $t \in [a,b]$,

$$
v(t)p(t)u'(t) - u(t)p(t)v'(t) = v(a)p(a)u'(a) - u(a)p(a)v'(a)
$$

This says

$$
p(t)(v(t)u'(t) - u(t)v'(t)) = p(a)(v(a)u'(a) - u(a)v'(a))
$$

This implies $p(t)(u(t)v'(t) - v(t)u'(t))$ is a constant c on $[a, b]$. ■

Homework

Exercise 13.4.5 *Verify explicitly that the eigenvectors Abel's and Lagrange's Identity for Bessel's equation.*

Exercise 13.4.6 *Verify explicitly that the eigenvectors Abel's and Lagrange's Identity for Laguerre's equation.*

Exercise 13.4.7 *Verify explicitly that the eigenvectors Abel's and Lagrange's Identity for Legendre's equation.*

Exercise 13.4.8 *Verify explicitly that the eigenvectors Abel's and Lagrange's Identity for Hermite's equation.*

We can now prove the eigenspace for an eigenvalue of a Stürm - Liouville system is one dimensional.

Lemma 13.4.6 The Eigenspace for Eigenvalue λ for a Stürm - Liouville System on Domain Ω_i is One Dimensional

If $\boldsymbol{u}$ and $\boldsymbol{v}$ are in $\boldsymbol{\Omega_i}$ for $i = 1, 2, 3$ and satisfy $\boldsymbol{L}(\boldsymbol{u}) = \lambda \boldsymbol{u}$ and $\boldsymbol{L}(\boldsymbol{v}) = \lambda \boldsymbol{v}$; i.e. $\boldsymbol{u}$ and $\boldsymbol{v}$ are both eigenvectors for the eigenvalue λ, then $\boldsymbol{u}$ and $\boldsymbol{v}$ are multiples; i.e. the eigenspace for eigenvalue λ is one dimensional.

Proof 13.4.6
By Abel's Identity, on $[a, b]$.

$$p(t) \begin{bmatrix} u(t) & v(t) \\ u'(t) & v'(t) \end{bmatrix} = c$$

$(\boldsymbol{BC_1})$:
Here $p(a) > 0$, $\alpha_a u(a) + \beta_a u'(a) = 0$ and $\alpha_b v(a) + \beta_a v'(a) = 0$ with $(\alpha_a, \beta_a) \neq (0, 0)$. Thus, the system

$$\begin{bmatrix} u(a) & u'(a) \\ v(a) & v'(a) \end{bmatrix} \begin{bmatrix} \alpha_a \\ \beta_a \end{bmatrix} = \begin{bmatrix} 0 \\ 0 \end{bmatrix}$$

must have a zero determinant. This tells us the Wronskian is zero at a. Thus, $p(a)(u(a)v'(a) - v(a)u'(a)) = 0 = c$. A similar argument shows the Wronskian is zero at b. Since $p(t) > 0$ on (a, b), we see the Wronskian is zero on (a, b) also. Thus the Wronskian is zero on $[a, b]$ implying $\boldsymbol{u}$ and $\boldsymbol{v}$ are linearly dependent.

$(\boldsymbol{BC_2})$:
Here $p(a) = 0$ but we can argue just like the case for $\boldsymbol{BC_1}$ using the point b to show $\boldsymbol{u}$ and $\boldsymbol{v}$ are linearly dependent.

$(\boldsymbol{BC_3})$:
Here $p(a) = p(b) = 0$ and so $p(a)(u(a)v'(a) - v(a)u'(a)) = 0 = c$. The same is true at b. So $p(a)(u(a)v'(a) - v(a)u'(a)) = 0 = c$. Thus, since $p(t) > 0$ on (a, b), the Wronskian is zero on $[a, b]$ and we see $\boldsymbol{u}$ and $\boldsymbol{v}$ are linearly dependent. ■

This is not true for $\boldsymbol{\Omega}_4$ in general. Consider the Stürm - Liouville system

$$\begin{aligned} u'' + \mu^2 u(t) &= 0 \\ u(-L) &= u(L), \quad u'(-L) = u'(L) \end{aligned}$$

on $[-L, L]$ for some $L > 0$. The general solution is $u(t) = A\cos(\mu t) + B\sin(\mu t)$ and the boundary conditions give

$$\begin{aligned} A\cos(-\mu L) + B\sin(-\mu L) &= A\cos(\mu L) + B\sin(\mu L) \\ -\mu A\sin(-\mu L) + \mu B\cos(-\mu L) &= -\mu A\sin(\mu L) + \mu B\cos(\mu L) \end{aligned}$$

or

$$\begin{aligned} A\cos(\mu L) - B\sin(\mu L) &= A\cos(\mu L) + B\sin(\mu L) \\ \mu A\sin(\mu L) + \mu B\cos(\mu L) &= -\mu A\sin(\mu L) + \mu B\cos(\mu L) \end{aligned}$$

We conclude $2B\sin(\mu L) = 0$ and $2A\sin(\mu L) = 0$. For nonzero solutions, we must have $\mu = \frac{n\pi}{L}$ for $n = 0, 1, 2, \ldots$. We see the eigenvectors for the eigenvalues $-\frac{n^2\pi^2}{L^2}$ form a two parameter family $A_n\cos(\frac{n\pi}{L}) + B_n\sin(\frac{n\pi}{L})$. Thus, the eigenspaces are two dimensional, not one.

Homework

Exercise 13.4.9 *Verify explicitly that the eigenspaces for Bessel's equation with appropriate boundary conditions are one dimensional.*

Exercise 13.4.10 *Verify explicitly that the eigenspaces for Laguerre's equation with appropriate boundary conditions are one dimensional.*

Exercise 13.4.11 *Verify explicitly that the eigenspaces for Hermite's equation with appropriate boundary conditions are one dimensional.*

Exercise 13.4.12 *Verify explicitly that the eigenspaces for Legendre's equation with appropriate boundary conditions are one dimensional.*

13.5 Eigenvalue Behavior

In general, it is not so hard to show there is a solution to an initial value problem when the coefficient functions in a differential operator are reasonably smooth. We cannot make this discussion completely self contained though, so we will just assume this. You really need to read a good book on differential equation theory! But we digress: for us then, we will take it as known the problems $\boldsymbol{L}(\boldsymbol{u_1}) = \lambda \boldsymbol{u_1}$ will have two nonzero linearly independent solutions on the interval $[a, b]$, $\boldsymbol{u_1^\lambda}$ and $\boldsymbol{u_2^\lambda}$. If we could find a linear combination of $\boldsymbol{u_1^\lambda}$ and $\boldsymbol{u_2^\lambda}$ that satisfied the boundary conditions BC_1, we would have

$$\begin{aligned} \alpha_a(c_1 u_1^\lambda(a) + c_2 u_2^\lambda(a)) + \beta_a(c_1 (u_1^\lambda)'(a) + c_2(u_2^\lambda)(a))' &= 0 \\ \alpha_b(c_1 u_1^\lambda(b) + c_2 u_2^\lambda(b)) + \beta_b(c_1 (u_1^\lambda)'(b) + c_2(u_2^\lambda)(b))' &= 0 \end{aligned}$$

or

$$\begin{bmatrix} \alpha_a u_1^\lambda(a) + \beta_a (u_1^\lambda)'(a) & \alpha_a u_2^\lambda(a) + \beta_a (u_2^\lambda)'(a) \\ \alpha_b u_1^\lambda(b) + \beta_b (u_1^\lambda)'(b) & \alpha_b u_2^\lambda(b) + \beta_b (u_2^\lambda)'(b) \end{bmatrix} \begin{bmatrix} c_1 \\ c_2 \end{bmatrix} = \begin{bmatrix} 0 \\ 0 \end{bmatrix}$$

This can be rewritten again as

$$\begin{bmatrix}\alpha_a & \beta_a\end{bmatrix}\begin{bmatrix}u_1^\lambda(a) & u_2^\lambda(a)\\(u_1^\lambda)'(a) & (u_2^\lambda)'(a)\end{bmatrix}\begin{bmatrix}c_1\\c_2\end{bmatrix} = 0 \quad (13.7)$$

$$\begin{bmatrix}\alpha_b & \beta_b\end{bmatrix}\begin{bmatrix}u_1^\lambda(b) & u_2^\lambda(b)\\(u_1^\lambda)'(b) & (u_2^\lambda)'(b)\end{bmatrix}\begin{bmatrix}c_1\\c_2\end{bmatrix} = 0 \quad (13.8)$$

We know also there is a constant c so that on $[a,b]$

$$p(t)\begin{bmatrix}u_1^\lambda(a) & u_2^\lambda(a)\\(u_1^\lambda)'(a) & (u_2^\lambda)'(a)\end{bmatrix} = c$$

Since $p(a) > 0$, for an eigenvalue, the Wronskian of $\boldsymbol{u_1^\lambda}$ and $\boldsymbol{u_2^\lambda}$ is zero on all $[a,b]$. In this case, we would find the kernel of

$$A^\lambda(a) = \begin{bmatrix}u_1^\lambda(a) & u_2^\lambda(a)\\(u_1^\lambda)'(a) & (u_2^\lambda)'(a)\end{bmatrix}$$

is nonzero and so there is a choice c_1 and c_2 not $(0,0)$ so that

$$\begin{bmatrix}u_1^\lambda(a) & u_2^\lambda(a)\\(u_1^\lambda)'(a) & (u_2^\lambda)'(a)\end{bmatrix}\begin{bmatrix}c_1\\c_2\end{bmatrix} = \begin{bmatrix}0\\0\end{bmatrix}$$

which would then satisfy Equation 13.7 and also Equation 13.8. On the other hand, if λ is not an eigenvalue, the only solution possible is the zero solution. Thus $c_1\boldsymbol{u_1^\lambda} + c_2\boldsymbol{u_2^\lambda} = \boldsymbol{0}$. Since the Wronskian of $\boldsymbol{u_1^\lambda}$ and $\boldsymbol{u_2^\lambda}$ is not zero on all $[a,b]$, this gives

$$\begin{bmatrix}u_1^\lambda(a) & u_2^\lambda(a)\\(u_1^\lambda)'(a) & (u_2^\lambda)'(a)\end{bmatrix}\begin{bmatrix}c_1\\c_2\end{bmatrix} = \begin{bmatrix}0\\0\end{bmatrix}$$

which implies $c_1 = c_2 = 0$.

We have now determined some important things about the eigenvalues of this system. Let f be a complex-valued function defined by

$$\begin{aligned} f(\lambda) &= \det\begin{bmatrix}\alpha_a u_1^\lambda(a) + \beta_a (u_1^\lambda)'(a) & \alpha_a u_2^\lambda(a) + \beta_a (u_2^\lambda)'(a)\\ \alpha_b u_1^\lambda(b) + \beta_b (u_1^\lambda)'(b) & \alpha_b u_2^\lambda(b) + \beta_b (u_2^\lambda)'(b)\end{bmatrix}\\ &= \begin{cases} 0, & \lambda \text{ is an eigenvalue; note it is then real}\\ \neq 0, & \lambda \text{ is not an eigenvalue}\end{cases}\end{aligned}$$

Homework

Exercise 13.5.1 *For the Stürm - Liouville problem on* $[0,10]$ $u''(t) + 5u(t) = \lambda u(t)$ *for the two generic linearly independent solutions* u_1^λ *and* u_2^λ *and find explicitly for boundary condition* **BC1**

$$\begin{aligned} f(\lambda) &= \det\begin{bmatrix}\alpha_a u_1^\lambda(a) + \beta_a (u_1^\lambda)'(a) & \alpha_a u_2^\lambda(a) + \beta_a (u_2^\lambda)'(a)\\ \alpha_b u_1^\lambda(b) + \beta_b (u_1^\lambda)'(b) & \alpha_b u_2^\lambda(b) + \beta_b (u_2^\lambda)'(b)\end{bmatrix}\\ &= \begin{cases} 0, & \lambda \textit{ is an eigenvalue; note it is then real}\\ \neq 0, & \lambda \textit{ is not an eigenvalue}\end{cases}\end{aligned}$$

Exercise 13.5.2 *For the Stürm - Liouville problem on* $[0,2]$ $u''(t) + 12u(t) = \lambda u(t)$ *for the two generic linearly independent solutions* u_1^λ *and* u_2^λ *and find explicitly for boundary condition* **BC1**

$$\begin{aligned} f(\lambda) &= \det \begin{bmatrix} \alpha_a u_1^\lambda(a) + \beta_a (u_1^\lambda)'(a) & \alpha_a u_2^\lambda(a) + \beta_a (u_2^\lambda)'(a) \\ \alpha_b u_1^\lambda(b) + \beta_b (u_1^\lambda)'(b) & \alpha_b u_2^\lambda(b) + \beta_b (u_2^\lambda)'(b) \end{bmatrix} \\ &= \begin{cases} 0, & \lambda \textit{ is an eigenvalue; note it is then real} \\ \neq 0, & \lambda \textit{ is not an eigenvalue} \end{cases} \end{aligned}$$

Exercise 13.5.3 *For the Stürm - Liouville problem on* $[0,1]$ $u''(t) + 7u(t) = \lambda u(t)$ *for the two generic linearly independent solutions* u_1^λ *and* u_2^λ *and find explicitly for boundary condition* **BC1**

$$\begin{aligned} f(\lambda) &= \det \begin{bmatrix} \alpha_a u_1^\lambda(a) + \beta_a (u_1^\lambda)'(a) & \alpha_a u_2^\lambda(a) + \beta_a (u_2^\lambda)'(a) \\ \alpha_b u_1^\lambda(b) + \beta_b (u_1^\lambda)'(b) & \alpha_b u_2^\lambda(b) + \beta_b (u_2^\lambda)'(b) \end{bmatrix} \\ &= \begin{cases} 0, & \lambda \textit{ is an eigenvalue; note it is then real} \\ \neq 0, & \lambda \textit{ is not an eigenvalue} \end{cases} \end{aligned}$$

Exercise 13.5.4 *For the Stürm - Liouville problem on* $[0,20]$ $u''(t) + 12u(t) = \lambda u(t)$ *for the two generic linearly independent solutions* u_1^λ *and* u_2^λ *and find explicitly for boundary condition* **BC1**

$$\begin{aligned} f(\lambda) &= \det \begin{bmatrix} \alpha_a u_1^\lambda(a) + \beta_a (u_1^\lambda)'(a) & \alpha_a u_2^\lambda(a) + \beta_a (u_2^\lambda)'(a) \\ \alpha_b u_1^\lambda(b) + \beta_b (u_1^\lambda)'(b) & \alpha_b u_2^\lambda(b) + \beta_b (u_2^\lambda)'(b) \end{bmatrix} \\ &= \begin{cases} 0, & \lambda \textit{ is an eigenvalue; note it is then real} \\ \neq 0, & \lambda \textit{ is not an eigenvalue} \end{cases} \end{aligned}$$

Now we need to use facts we can only state:

- f depends in a differentiable way on the complex number λ. This is a result from theory of ordinary differential equations.

- f is therefore complex differentiable and only has zeros that are real numbers. Hence f is a complex-valued function which is not identically zero which has a complex derivative at all complex numbers λ. Such a function is called an **entire** function.

- In a complex variable class, we prove the zeros of an entire function are isolated; i.e. if $\boldsymbol{p}$ is a zero of f, then there is a radius $r > 0$ so that f is not zero in $B(\boldsymbol{p}, r)$.

The last fact has an amazing consequence. If $(\boldsymbol{p_n})$ was a sequence of zeros of f which converged to $\boldsymbol{p}$, then if $\boldsymbol{p}$ was also a zero of f, then there would be a radius $r > 0$ so that f is not zero in $B(\boldsymbol{p}, r)$. This is not possible as $\boldsymbol{p_n} \rightarrow \boldsymbol{p}$. Hence, all of the zeros of an entire function are isolated and there can be no cluster points. This tells us immediately the zeros (eigenvalues) of our Stürm - Liouville system are either countably finite or countably infinite. If there are infinitely many eigenvalues λ_n we must have $|\lambda_n| \rightarrow \infty$ as otherwise there would be a cluster point! Booyah!!!

13.6 The Inverse of the Stürm - Liouville Differential Operator

We start by finding two linearly independent solutions to the model.

Lemma 13.6.1 The Standard Linearly Independent Solutions to the Stürm - Liouville Models

Let $\boldsymbol{u_1}$ solve $\boldsymbol{L(u_1)} = \lambda \boldsymbol{u_1}$ with $\alpha_a u_1(a) + \beta_a u_1'(a) = 0$ and $\alpha_b u_1(b) + \beta_b u_1'(b) = 1$

Let $\boldsymbol{u_2}$ solve $\boldsymbol{L(u_2)} = \lambda \boldsymbol{u_2}$ with $\alpha_a u_2(a) + \beta_a u_2'(a) = 1$ and $\alpha_b u_2(b) + \beta_b u_2'(b) = 0$

for (α_a, β_a) and (α_b, β_b) not $(0, 0)$. Then $\boldsymbol{u_1}$ and $\boldsymbol{u_2}$ are linearly independent.

Proof 13.6.1

Note $\boldsymbol{u_1}$ and $\boldsymbol{u_2}$ are not in $\boldsymbol{BC_1}$. We let W denote the Wronskian here. Then, $W(a) = u_1(a)u_2'(a) - u_1'(a)u_2(a)$. Then

$$\begin{aligned}
\alpha_a W(a) &= \alpha_a(u_1(a)u_2'(a) - u_1'(a)u_2(a)) \\
&= \alpha_a u_1(a)u_2'(a) - \alpha_a u_1'(a)u_2(a) \\
&= -\beta_a u_1'(a)u_2'(a) - \alpha_a u_1'(a)u_2(a) \\
&= -u_1'(a)(\beta_a u_2'(a) + \alpha_a u_2(a)) = -u_1'(a)(\,1\,) = -u_1'(a)
\end{aligned}$$

Thus $\alpha_a W(a) = -\boldsymbol{u}_1'(a)$. A similar calculation shows

$$\begin{aligned}
\beta_a W(a) &= \beta_a(u_1(a)u_2'(a) - u_1'(a)u_2(a)) \\
&= \beta_a u_1(a)u_2'(a) - \beta_a u_1'(a)u_2(a) \\
&= \beta_a u_1(a)u_2'(a) + \alpha_a u_1(a)u_2(a) \\
&= u_1(a)(\beta_a u_2'(a) + \alpha_a u_2(a)) = u_1(a)(\,1\,) = u_1(a)
\end{aligned}$$

and so $\beta_a W(a) = u_1(a)$.
If both $u_1(a) = u'(a) = 0$, then since at least one of α_a and β_a must be nonzero, our work above would force $W(a) = 0$. The linearly independence equation is

$$c_1\boldsymbol{u_1} + c_2\boldsymbol{u_2} = \boldsymbol{0}$$

Thus at a, we find

$$\begin{bmatrix} u(a) & u'(a) \\ v(a) & v'(a) \end{bmatrix} \begin{bmatrix} c_1 \\ c_2 \end{bmatrix} = \begin{bmatrix} 0 \\ 0 \end{bmatrix}$$

has a nonzero solution since $W(a) = 0$. This would mean $\boldsymbol{u_1}$ and $\boldsymbol{u_2}$ were linearly dependent and $\boldsymbol{u_2} = \gamma \boldsymbol{u_1}$ for some nonzero constant γ. But this forces $u_2(a) = \gamma u_1(a) = 0$ and $u_2'(a) = \gamma u_1'(a) = 0$. Thus, $\boldsymbol{u_2}$ does not satisfy its boundary conditions. We conclude at least one of $u_1(a)$ and $u'(a)$ is not zero. If both α_a and β_a are not zero, this gives $W(a) \neq 0$ from one of our equations. If only α_a is not zero, this means $u_1(a) = 0$ and so $u'(a) \neq 0$ giving $W(a) \neq 0$ in that case. We can argue in a similar fashion in all cases. But if $W(a) \neq 0$, then the only way we can satisfy the linear independence equation is with $c_1 = c_2 = 0$. Hence, the solutions are linearly independent. ■

Homework

Exercise 13.6.1 *For the Stürm - Liouville problem on $[0, 10]$ $u''(t) + 5u(t) = \lambda u(t)$ find the two standard linearly independent solutions u_1^λ and u_2^λ.*

Exercise 13.6.2 *For the Stürm - Liouville problem on $[0, 2]$ $u''(t) + 12u(t) = \lambda u(t)$ find the two standard linearly independent solutions u_1^λ and u_2^λ.*

Exercise 13.6.3 *For the Stürm - Liouville problem on $[0, 1]$ $u''(t) + 7u(t) = \lambda u(t)$ find the two standard linearly independent solutions u_1^λ and u_2^λ.*

Exercise 13.6.4 *For the Stürm - Liouville problem on* $[0, 20]$ $u''(t) + 12u(t) = \lambda u(t)$ *find the two standard linearly independent solutions* u_1^λ *and* u_2^λ.

13.6.1 The Actual Inversion

We now have enough tools to invert the differential operator. Let's assume λ is not an eigenvalue. We will use the two linearly independent solutions from Lemma 13.6.1 to do this.

Now remember, our operator on $\boldsymbol{\Omega_1}$ is $\boldsymbol{L}(\boldsymbol{u}) = \frac{1}{\boldsymbol{r}}\,(\,(\boldsymbol{p}\boldsymbol{u}')' - \boldsymbol{q}\boldsymbol{u}\,)$ and the associated eigenvalue problem is to find the values of λ and associated nonzero functions $\boldsymbol{u}$ so that $\boldsymbol{L}(\boldsymbol{u}) = \lambda \boldsymbol{u}$. Expanding, this becomes

$$\begin{aligned} \frac{1}{\boldsymbol{r}}\,(\,(\boldsymbol{p}\boldsymbol{u}')' - \boldsymbol{q}\boldsymbol{u}) &= \lambda \boldsymbol{u} \\ &\Longrightarrow\ (\boldsymbol{p}\boldsymbol{u}')' - (\lambda \boldsymbol{r} + \boldsymbol{q})\boldsymbol{u} = 0 \end{aligned}$$

Any value of λ which is not an eigenvalue, means the only solution in $\boldsymbol{\Omega_1}$ of $(\boldsymbol{p}\boldsymbol{u}')' - (\lambda\boldsymbol{r} + \boldsymbol{q})\boldsymbol{u} = 0$ is $\boldsymbol{u} = \boldsymbol{0}$. Of course, the differential operator here is a bit different from $\boldsymbol{L}$ so let's define a new one: $\boldsymbol{L_\lambda}$ by

$$\boldsymbol{L_\lambda}(\boldsymbol{u}) = (\boldsymbol{p}\boldsymbol{u}')' - (\lambda \boldsymbol{r} + \boldsymbol{q})\boldsymbol{u}$$

Consider the non homogeneous problem

$$\boldsymbol{L}(\boldsymbol{u}) = \boldsymbol{f} \Longrightarrow (\boldsymbol{p}\boldsymbol{u}')' - (\lambda \boldsymbol{r} + \boldsymbol{q})\boldsymbol{u} = \boldsymbol{r}\boldsymbol{f}$$

for a continuous f on $[a, b]$. We know

$$\begin{aligned} (p(t)u_1'(t))' - (\lambda r(t) + q(t))u_1(t) &= 0,\ a \le t \le b \\ \alpha_a u_1(a) + \beta_a u_1'(a) &= 0, \quad \alpha_b u_1(b) + \beta_b u_1'(b) = 1 \\ (p(t)u_2'(t))' - (\lambda r(t) + q(t))u_2(t) &= 0,\ a \le t \le b \\ \alpha_a u_2(a) + \beta_a u_2'(a) &= 1, \quad \alpha_b u_2(b) + \beta_b u_2'(b) = 0 \end{aligned}$$

We use Variation of Parameters to find a particular solution. Let $u_p(t) = A(t)u_1(t) + B(t)u_2(t)$. The equation to solve is

$$\boldsymbol{u}'' + \frac{\boldsymbol{p}'}{\boldsymbol{p}}\boldsymbol{u}' - \frac{\lambda \boldsymbol{r} + \boldsymbol{q}}{\boldsymbol{p}}\,\boldsymbol{u} = \frac{\boldsymbol{r}\boldsymbol{f}}{\boldsymbol{p}}$$

Then, letting

$$\boldsymbol{W} = \boldsymbol{W}(\boldsymbol{u_1}, \boldsymbol{u_2}) = \begin{bmatrix} \boldsymbol{u_1} & \boldsymbol{u_2} \\ \boldsymbol{u_1}' & \boldsymbol{u_2}' \end{bmatrix}$$

we have $\boldsymbol{p}\boldsymbol{W} = c$ on $[a, b]$ for a nonzero constant c and

$$\begin{aligned} A'(t) &= \frac{1}{W(t)} \begin{bmatrix} 0 & u_2(t) \\ \frac{r(t)f(t)}{p(t)} & u_2'(t) \end{bmatrix} = -\frac{u_2(t)r(t)f(t)}{p(t)W(t)} \\ B'(t) &= \frac{1}{W(t)} \begin{bmatrix} u_1 t & 0 \\ u'(t) & \frac{r(t)f(t)}{p(t)} \end{bmatrix} = \frac{u_1(t)r(t)f(t)}{p(t)W(t)} \end{aligned}$$

This implies

$$A(t) = -\int_a^t \frac{u_2(s)r(s)f(s)}{p(s)W(s)}\,ds, \quad B(t) = \int_a^t \frac{u_1(s)r(s)f(s)}{p(s)W(s)}\,ds$$

and

$$u_p(t) = -\int_a^t \frac{u_1(t)u_2(s)r(s)f(s)}{p(s)W(s)}\,ds + \int_a^t \frac{u_1(s)u_2(t)r(s)f(s)}{p(s)W(s)}\,ds$$

Thus, the general solution is

$$\begin{aligned}
u(t) &= C_1u_1(t) + C_2u_2(t) + \int_a^t \frac{u_1(s)u_2(t) - u_1(t)u_2(s)}{p(s)W(s)}\,r(s)f(s)\,ds \\
u'(t) &= C_1u_1'(t) + C_2u_2'(t) + \frac{u_1(t)u_2(t) - u_1(t)u_2(t)}{p(t)W(t)}\,f(t) \\
&\quad + \int_a^t \frac{u_1(s)u_2'(t) - u_1'(t)u_2(s)}{p(s)W(s)}\,r(s)f(s)\,ds \\
&= C_1u_1'(t) + C_2u_2'(t) + \int_a^t \frac{u_1(s)u_2'(t) - u_1'(t)u_2(s)}{p(s)W(s)}\,r(s)f(s)\,ds \\
u(a) &= C_1u_1(a) + C_2u_2(a) \\
u(b) &= C_1u_1(b) + C_2u_2(b) + \int_a^b \frac{u_1(s)u_2(b) - u_1(b)u_2(s)}{p(s)W(s)}\,r(s)f(s)\,ds \\
u'(a) &= C_1u'(a) + C_2u'(a) \\
u'(b) &= C_1u_1'(b) + C_2u_2'(b) + \int_a^b \frac{u_1(s)u_2'(b) - u_1'(b)u_2(s)}{p(s)W(s)}\,r(s)f(s)\,ds
\end{aligned}$$

Now apply the boundary conditions:

$$\begin{aligned}
0 &= \alpha_a u(a) + \beta_a u'(a) = \alpha_a(C_1u_1(a) + C_2u_2(a)) + \beta_a(C_1u'(a) + C_2u'(a)) \\
&= C_1(\alpha_a u_1(a) + \beta_a u'(a)) + C_2(\alpha_a u_2(a) + \beta_a u_2'(a)) \\
&= C_1(0) + C_2(1) = C_2
\end{aligned}$$

To find the value of C_1 we use the other boundary condition (since $C_2 = 0$ we can drop that term):

$$\begin{aligned}
0 = \alpha_b u(b) + \beta_b u'(b) &= \alpha_b\left(C_1u_1(b) + \int_a^b \frac{u_1(s)u_2(b) - u_1(b)u_2(s)}{p(s)W(s)}\,r(s)f(s)\,ds\right) \\
&+ \beta_b\left(C_1u_1'(b) + \int_a^b \frac{u_1(s)u_2'(b) - u_1'(b)u_2(s)}{p(s)W(s)}\,r(s)f(s)\,ds\right) \\
&= C_1(\alpha_b u_1(b) + \beta_b u_1'(b)) \\
&+ \int_a^b \frac{u_1(s)(\alpha_b u_2(b) + \beta_b u_2'(b))}{p(s)W(s)}\,r(s)f(s)\,ds \\
&- \int_a^b \frac{u_2(s)(\alpha_b u_1(b) + \beta_b u_1'(b))}{p(s)W(s)}\,r(s)f(s)\,ds \\
&= C_1(1) + \int_a^b \frac{u_1(s)(0)}{p(s)W(s)}\,r(s)f(s)\,ds - \int_a^b \frac{u_2(s)(1)}{p(s)W(s)}\,r(s)f(s)\,ds
\end{aligned}$$

$$= \quad C_1 - \int_a^b \frac{u_2(s)}{p(s)W(s)}\, r(s)f(s)\, ds$$

We conclude $C_1 = \int_a^b \frac{u_2(s)}{p(s)W(s)}\, r(s)f(s)\, ds$. Thus the solution is

$$\begin{aligned} u(t) &= \int_a^b \frac{u_1(t)u_2(s)}{p(s)W(s)}\, r(s)f(s)\, ds + \int_a^t \frac{u_1(s)u_2(t) - u_1(t)u_2(s)}{p(s)W(s)}\, r(s)f(s)\, ds \\ &= \int_a^t \left(\frac{u_1(t)u_2(s) + u_1(s)u_2(t) - u_1(t)u_2(s)}{p(s)W(s)}\, r(s)f(s)\, ds \right) \\ &+ \int_t^b \frac{u_1(t)u_2(s)}{p(s)W(s)}\, r(s)f(s)\, ds \\ &= \int_a^t \frac{u_1(s)u_2(t)}{p(s)W(s)}\, r(s)f(s)\, ds + \int_t^b \frac{u_1(t)u_2(s)}{p(s)W(s)}\, r(s)f(s)\, ds \end{aligned}$$

Define the **kernel** function $k_\lambda(s,t)$ by

$$k_\lambda(s,t) \quad = \quad \begin{cases} \frac{u_1(s)u_2(t)}{p(s)W(s)}, & a \le s \le t \\ \frac{u_1(t)u_2(s)}{p(s)W(s)}, & t \le s \le b \end{cases}$$

Note $k_\lambda(s,t)$ is continuous on the compact set $[a,b] \times [a,b]$ and so it is uniformly continuous and bounded on this domain. Define the linear operator $\boldsymbol{J_\lambda}$ on $C([a,b])$ by $(\boldsymbol{J_\lambda}(\boldsymbol{f}))(t) = \int_a^b k_\lambda(s,t)r(s)$ $f(s)ds$. We have inverted the differential operator $\boldsymbol{L_\lambda}$ and for $\boldsymbol{u} \in \boldsymbol{\Omega_1}$

$$\boldsymbol{L}(\boldsymbol{u}) - \lambda \boldsymbol{u} \quad = \quad \boldsymbol{f} \Longrightarrow (\boldsymbol{p}\boldsymbol{u}')' - (\lambda \boldsymbol{r} + \boldsymbol{q})\boldsymbol{u} = \boldsymbol{r}\boldsymbol{f} \Longrightarrow \boldsymbol{u} = \boldsymbol{J_\lambda}(\boldsymbol{f})$$

Hence, ${\boldsymbol{L_\lambda}}^{-1} = \boldsymbol{J_\lambda}(\boldsymbol{f})$. Right now, it seems that the range of $\boldsymbol{J_\lambda}$ should be $C^2([a,b]) \cap \boldsymbol{\Omega_1}$ due to our construction process, but we need to verify this.

Homework

Exercise 13.6.5 *For the Stürm - Liouville problem on* $[0,10]$ $u''(t) + 5u(t) = \lambda u(t)$ *if* λ *is not an eigenvalue, find the inverse of this operator.*

Exercise 13.6.6 *For the Stürm - Liouville problem on* $[0,2]$ $u''(t) + 12u(t) = \lambda u(t)$ *if* λ *is not an eigenvalue, find the inverse of this operator.*

Exercise 13.6.7 *For the Stürm - Liouville problem on* $[0,1]$ $u''(t) + 7u(t) = \lambda u(t)$ *if* λ *is not an eigenvalue, find the inverse of this operator.*

Exercise 13.6.8 *For the Stürm - Liouville problem on* $[0,20]$ $u''(t) + 12u(t) = \lambda u(t)$ *if* λ *is not an eigenvalue, find the inverse of this operator.*

13.6.2 Verifying the Solution

We have

$$k_\lambda(s,t) \quad = \quad \begin{cases} \frac{u_1(s)u_2(t)}{p(s)W(s)}, & a \le s \le t \\ \frac{u_1(t)u_2(s)}{p(s)W(s)}, & t \le s \le b \end{cases}$$

Thus, for $s < t$,

$$\frac{\partial k_\lambda(s,t)}{\partial t} = \begin{cases} \frac{u_1(s)u_2'(t)}{p(s)W(s)}, & a \leq s < t \\ \frac{u_1'(t)u_2(s)}{p(s)W(s)}, & t < s \leq b \end{cases}$$

We note

$$\lim_{s \to t^-} \frac{\partial k_\lambda(s,t)}{\partial t} = \frac{u_1(t^-)u_2'(t)}{p(t^-)W(t^-)} = \frac{u_1(t)u_2'(t)}{p(t)W(t)}$$

because $\boldsymbol{u_1}$ is continuous and $\boldsymbol{pW} = c$. Similarly, for $s > t$,

$$\lim_{s \to t^+} \frac{\partial k_\lambda(s,t)}{\partial t} = \frac{u_1'(t)u_2(t^+)}{p(t^+)W(t^+)} = \frac{u_1'(t)u_2'(t)}{p(t)W(t)}$$

because $\boldsymbol{u_2}$ is continuous and $\boldsymbol{pW} = c$. Therefore.

$$\frac{\partial k_\lambda(t^+,t)}{\partial t} - \frac{\partial k_\lambda(t^-,t)}{\partial t} = \frac{u_1'(t)u_2'(t)}{p(t)W(t)} - \frac{u_1(t)u_2'(t)}{p(t)W(t)} = -\frac{W(t)}{p(t)W(t)} = -\frac{1}{p(t)}$$

Next, since

$$u(t) = \int_a^t \frac{u_1(s)u_2(t)}{p(s)W(s)}\, r(s)f(s)\, ds + \int_t^b \frac{u_1(t)u_2(s)}{p(s)W(s)}\, r(s)f(s)\, ds$$

$$\begin{aligned} u'(t) &= \frac{u_1(t)u_2(t)}{p(t)W(t)}\, r(t)f(t) + \int_a^t \frac{u_1(s)u_2'(t)}{p(s)W(s)}\, r(s)f(s)\, ds \\ &- \frac{u_1(t)u_2(t)}{p(t)W(t)}\, r(t)f(t) + \int_t^b \frac{u_1'(t)u_2(s)}{p(s)W(s)}\, r(s)f(s)\, ds \\ &= \int_a^t \frac{u_1(s)u_2'(t)}{p(s)W(s)}\, r(s)f(s)\, ds + \int_t^b \frac{u_1'(t)u_2(s)}{p(s)W(s)}\, r(s)f(s)\, ds \end{aligned}$$

and

$$\begin{aligned} u''(t) &= \frac{u_1(t)u_2'(t) - u_1'(t)u_2(t)}{p(t)W(t)} r(t)f(t) + \int_a^t \frac{u_1(s)u_2''(t)}{p(s)W(s)}\, r(s)f(s)\, ds \\ &+ \int_t^b \frac{u_1''(t)u_2(s)}{p(s)W(s)}\, r(s)f(s)\, ds \end{aligned}$$

Now we also know for $i = 1, 2$,

$$p'(t)u_i'(t) + p(t)u_i''(t) - q(t) - \lambda r(t)u_i(t) = r(t)f(t)$$

Thus

$$\begin{aligned} & p'(t)u'(t) + p(t)u''(t) - (\lambda r(t) + q(t))u(t) \\ = \; & p'(t)\left\{\int_a^t \frac{u_1(s)u_2'(t)}{p(s)W(s)}\, r(s)f(s)\, ds + \int_t^b \frac{u_1'(t)u_2(s)}{p(s)W(s)}\, r(s)f(s)\, ds\right\} \\ & + p(t)\left\{\frac{u_1(t)u_2'(t) - u_1'(t)u_2(t)}{p(t)W(t)} r(t)f(t) + \int_a^t \frac{u_1(s)u_2''(t)}{p(s)W(s)}\, r(s)f(s)\, ds\right. \end{aligned}$$

$$+\int_t^b \frac{u_1''(t)u_2(s)}{p(s)W(s)}\, r(s)f(s)\, ds\Bigg\}$$

$$-(\lambda r(t)+q(t))\left\{\int_a^t \frac{u_1(s)u_2(t)}{p(s)W(s)}\, r(s)f(s)\, ds + \int_t^b \frac{u_1(t)u_2(s)}{p(s)W(s)}\, r(s)f(s)\, ds\right\}$$

$$= \frac{p(t)W(t)}{p(t)W(t)} r(t)f(t) + \int_a^t \frac{p'(t)u_2'(t)+p(t)u_2''(t)-(q(t)+\lambda r(t))u_2(t)}{p(s)W(s)}\, u_1(s)\, r(s)f(s)\, ds$$

$$+\int_t^b \frac{p'(t)u_1'(t)+p(t)u_1''(t)-(q(t)+\lambda r(t))u_1(t)}{p(s)W(s)}\, u_2(s)\, r(s)f(s)\, ds$$

$$= f(t) + \int_a^t \frac{0}{p(s)W(s)}\, u_1(s)\, r(s)f(s)\, ds + \int_t^b \frac{0}{p(s)W(s)}\, u_2(s)\, r(s)f(s)\, ds = r(t)f(t)$$

Next, we check the boundary conditions.

$$\alpha_a u(a) + \beta_a u'(a) = \alpha_a \int_a^b \frac{u_1(a)u_2(s)}{p(s)W(s)}\, r(s)f(s)\, ds + \beta_a \int_a^b \frac{u_1'(a)u_2(s)}{p(s)W(s)}\, r(s)f(s)\, ds$$

$$= \int_a^b \frac{(\alpha_a u_1(a)+\beta_1 u_1'(a))u_2(s)}{p(s)W(s)}\, r(s)f(s)\, ds = \int_a^b \frac{(0)u_2(s)}{p(s)W(s)}\, r(s)f(s)\, ds = 0$$

A similar calculation shows $u(t)$ satisfies the other boundary condition. Thus, $u(t) = (\boldsymbol{J_\lambda}(\boldsymbol{f}))(t) = \int_a^b k_\lambda(s,t)\, r(s)f(s)\, ds$ satisfies the differential equation $\boldsymbol{rL}(\boldsymbol{u}) - \lambda \boldsymbol{ru} = \boldsymbol{rf}$ and the boundary conditions. In particular we see function $\boldsymbol{J_\lambda}(\boldsymbol{f})$ is in $C^2([a,b]) \cap \boldsymbol{\Omega_1}$.

13.6.3 More on the Eigenvalues

We have the solution to $\boldsymbol{Lu} - \lambda_0 \boldsymbol{u} = \boldsymbol{f}$ plus boundary conditions can be written as $\boldsymbol{u} = \mathscr{J}_{\lambda_0}(\boldsymbol{f})$ where λ_0 is not an eigenvalue of $\boldsymbol{L}$. Note

$$\boldsymbol{L}(\boldsymbol{u_n}) = \lambda_n \boldsymbol{u_n}$$

for the eigenvalue and eigenfunction pairs of $\boldsymbol{L}$. Thus,

$$\boldsymbol{L}(\boldsymbol{u_n}) - \lambda_0 \boldsymbol{u_n} = (\lambda_n - \lambda_0)\boldsymbol{u_n} \Longrightarrow (\boldsymbol{L} - \lambda_o \boldsymbol{I})(\boldsymbol{u_n}) = (\lambda_n - \lambda_0)\boldsymbol{u_n}$$

and so the sequence $(\lambda_0 - \lambda_n)$ is the eigenvalue sequence for the operator $\boldsymbol{L} - \lambda_o \boldsymbol{I}$ with associated eigenfunction sequence $(\boldsymbol{u_n})$. Further, this tells us

$$\mathscr{J}_{\lambda_0}((\lambda_n - \lambda_0)\boldsymbol{u_n}) = \boldsymbol{u_n} \Longrightarrow \mathscr{J}_{\lambda_0}(\boldsymbol{u_n}) = \frac{1}{\lambda_n - \lambda_0}\boldsymbol{u_n}$$

and so $(\frac{1}{\lambda_n - \lambda_0})$ is the eigenvalue sequence for $\mathscr{J}_{\lambda_0}$ and the eigenfunction sequence is the same as the one for $\boldsymbol{L} - \lambda_0 \boldsymbol{I}$. Since we know $|\lambda_n| \to \infty$ we know $\frac{1}{\lambda_n - \lambda_0} \to 0$.

Homework

Exercise 13.6.9 *For the Stürm - Liouville problem on* $[0, 10]$ $u''(t) + 5u(t) = \lambda u(t)$ *if* λ *is not an eigenvalue, find the inverse of this operator and verify the solution.*

Exercise 13.6.10 *For the Stürm - Liouville problem on* $[0, 2]$ $u''(t) + 12u(t) = \lambda u(t)$ *if* λ *is not an eigenvalue, find the inverse of this operator and verify the solution.*

Exercise 13.6.11 *For the Stürm - Liouville problem on* $[0, 1]$ $u''(t) + 7u(t) = \lambda u(t)$ *if* λ *is not an eigenvalue, find the inverse of this operator and verify the solution.*

Exercise 13.6.12 *For the Stürm - Liouville problem on* $[0, 20]$ $u''(t) + 12u(t) = \lambda u(t)$ *if* λ *is not an eigenvalue, find the inverse of this operator and verify the solution.*

13.7 A Bessel's Equation Example

Let's look at a particular Bessel's Equation: (our λ here is the negative of the one we used earlier in our definition, but that is not important).

$$t^2\boldsymbol{x}'' + t\boldsymbol{x}' + (\lambda t^2 - k^2)\boldsymbol{x} = 0$$

on $[0, L]$ for an $L > 0$ for a positive integer k. In self-adjoint form this is

$$\frac{(t\boldsymbol{x}')' - \frac{k^2}{t}}{t} = -\lambda\boldsymbol{x}$$

and we will use $\boldsymbol{BC_2}$ here with the solution bounded as we approach 0^+ with $x(0) = 0$ and $x(L) = 0$. Note $p(t) = t$ which is 0 at $t = 0$, $q(t) = \frac{k^2}{t}$ and $r(t) = t$. The inner product is therefore given by $< f, g >= \int_0^L tf(t)g(t)dt$. Since $p(0) = 0$, this ODE has what is called a **regular singular point** at $t = 0$ and we use the method of Frobenius to find appropriate solutions. You should look up these ideas for yourself. We discussed power series solutions to ODEs in (Peterson (18) 2020) and referred you to (E. Ince (1) 1956) for the background theory. The proofs needed to show there is a power series solution with a nonzero radius of convergence require the use of complex variables and so we will not cover that here. But more reading for you! Hence, we assume a solution of the form

$$\begin{aligned} x(t) &= t^r \sum_{n=0}^{\infty} a_n t^n = \sum_{n=0}^{\infty} a_n t^{n+r} \\ t\,x'(t) &= \sum_{n=0}^{\infty} (n+r)\, a_n t^{n+r} \\ t^2\, x''(t) &= \sum_{n=0}^{\infty} (n+r)\,(n+r-1)\, a_n t^{n+r} \end{aligned}$$

The ODE then becomes

$$\left(r\,(r-1)a_0 + ra_0 - k^2a_0\right)t^r + \left((r+1)\,(r)a_1 + (r+1)a_1 - k^2a_1\right)t^{r+1}$$
$$+\left(\sum_{n=2}^{\infty}(r+n)\,(r+n-1)a_n + (r+n)a_1 - k^2a_n - \lambda a_{n-2}\right)t^{r+n} = 0.$$

Now the solution looks like $a_0t^r + a_1t^{r+1} \ldots$, so to get a bounded solution, the integer r must satisfy $r \geq 0$. This series above implies all the coefficients must be zero:

$$\begin{aligned} (r^2 - k^2)a_0 &= 0 \\ ((r+1)^2 - k^2)a_1 &= 0 \\ ((n+r)^2 - k^2)a_n - \lambda a_{n-2} &= 0 \end{aligned}$$

Now if $r < k$, we would have $a_0 = 0$, $a_1 = 0$ and by recursion $a_n = 0$ for all other n. But this is the zero solution and so not an eigenfunction. So we must have $a_0 \neq 0$, $r = k$. We will choose $a_0 = 1$. This implies

$$\begin{aligned}(r^2 - k^2)a_0 &= 0 \Longrightarrow r = \pm k \\ ((k+1)^2 - k^2)a_1 &= 0 \Longrightarrow a_1 = 0 \\ ((n+k)^2 - k^2)a_n - \lambda a_{n-2} &= 0 \Longrightarrow a_n = \frac{\lambda}{(n+k)^2 - k^2} a_{n-2}\end{aligned}$$

Since $a_1 = 0$, all the odd coefficient's must be 0 also and so only the even coefficients matter. Therefore, the solution has the form

$$x(t) = t^1 \left(1 + \sum_{n=1}^{\infty} a_{2n} t^{2n}\right)$$

The coefficients are solved recursively:

$$\begin{aligned}a_0^\lambda &= 1 \\ a_2^\lambda &= \frac{\lambda}{(2+k)^2 - k^2} a_0 = \frac{\lambda}{(2+k)^2 - k^2} \\ a_2^\lambda &= \frac{\lambda}{(4+k)^2 - k^2} a_2 = \frac{\lambda^2}{((4+k)^2 - k^2)((2+k)^2 - k^2)} \\ \vdots &= \vdots \\ a_{2n}^\lambda &= \frac{\lambda}{(2n+k)^2 - k^2} a_{2n-2} = \frac{\lambda^n}{((2n+k)^2 - k^2) \cdots ((2+k)^2 - k^2)}\end{aligned}$$

Let

$$\Lambda(n,k) = \begin{cases} 1, & n = 0 \\ ((2n+k)^2 - k^2) \cdots ((2+k)^2 - k^2), & n > 0 \end{cases}$$

Then, the solution we seek is

$$x(t) = t^1 \sum_{n=0}^{\infty} a_{2n}^\lambda t^{2n} = t^1 \left(\sum_{n=0}^{\infty} \frac{\lambda^n}{\Lambda(k,n)} t^{2n}\right)$$

To satisfy the last boundary condition $x(L) = 0$, we must have

$$0 = L^1 \left(\sum_{n=0}^{\infty} \frac{\lambda^n}{\Lambda(k,n)} L^{2n}\right)$$

Since all the terms in this sum are positive if λ is positive, we must assume $\lambda = -\omega^2$ for some $\omega > 0$ in order to satisfy this last condition. This gives the equation

$$0 = L^1 \left(\sum_{n=0}^{\infty} (-1)^n \frac{\omega^{2n}}{\Lambda(k,n)} L^{2n}\right)$$

Defining for convenience $a_{2n}^{-1} = \frac{(-1)^n}{\Lambda(k,n)}$, to satisfy the boundary condition, we must have

$$\sum_{n=0}^{\infty} a_{2n}^{-1}(L\omega)^{2n} = 0$$

To find the values of ω that work, we will find the zeroes of the function

$$y(\omega) = \sum_{n=0}^{\infty} a_{2n}^{-1}(L\omega)^{2n}$$

using MATLAB code. Working with power series is prone to numerical problems, so although we know for a given k there is an infinite set of eigenvalues and eigenfunctions we will only be able to find a few. There are many other more theoretical ways to find the eigenvalues and eigenfunctions for the Bessel's Equation and we encourage you to look them up. Lots of complex analysis!

13.7.1 The Bessel Function Code Implementation

Here is the code.

Listing 13.1: **Our Bessel Function Template**

```
function [c,p,q,Z,E,W] = mybessel(k,n,m,u,L)
%
% solution is x(t) = t^r( 1 + sum_{n=1}^\infty a_{2n} t^{2n} )
% k = power
% n = number of terms in expansion of the Bessel function
% m = number of points in the plot
% L = the interval is [0,L]
% Z contains the zeros of the Bessel function
% p is the Bessel function partial sum of n+1 terms
% E contains the eigenvalues
% phi contains the eigenfunctions
%
end
```

The coefficients we need are stored in a MATLAB vector `c` which has a starting index of 1 and the coefficients we compute from the recursion equation are labeled a_{2n} so first we have to make sure we map the coefficients correctly. This is the first part of the code.

Listing 13.2: **Index Mapping**

```
c = zeros(1,n+1);
% this is a_0
% a_0 = 1 ==> c(1) = a_{i-1}
c(1) = 1;
for i=2:n+1
  % a_{2i} = -a_{2i-2}/[ (2i+k)^2 - k^2] = -a_{2i-2}/[4i^2 + 4iK]
```

```
%          = -a_{2i-2}/[4i(i+k)]
% c_2 = a_2  = a_{2(2-1)}
% c_3 = a_4 = a_{2(3-1)}
% c_4 = a_6 = a_{2(4-1)}
% c_{p} = a_{2(p-1)}
% c_{p+1} = a_{2(p+1-1)} = a_{2p}
m1 = 4*(i-1)*(i-1+k);
% a_{2i} = -1 a_{2i-2}/[(2i+k)^2-k^2]
% c_{i+1} = -1 c_{i}/[(2i+k)^2 - k^2]
% c_i = -1 c_{i-1}/[(2(i-1)+k)^2 - k^2] = -1 c_{i-1}/[4(i-1)(i-1+k)]
c(i) = -c(i-1)/m1;
end
b = c(1,2:n+1);
% this computes c(2) + c(3) x + ...+ c(n+1) x^n or
% a(2) + a(4) x + ... + a(2n) x^n
```

The coefficients in `c` give coefficients for a polynomial of degree n of form $c(1)+c(2)x+c(2)x^2+\ldots+c(n+1)x^n$. The line `b = c(1,2:n+1)` strips off the first coefficient and gives $b = \begin{bmatrix} c_2 & c_3 & \ldots & c_{n+1} \end{bmatrix}$. We then use standard Horner's Method code to compute the function which represents this polynomial.

Listing 13.3: **The Polynomial Calculation**

```
% this computes
% c(2) + c(3) x + ...+ c(n+1) x^n
% a(2) + a(4) x + ... + a(2n) x^n
q = MyHorner(b');
```

Here we use `MyHorner.m` which returns our needed polynomial as a function handle.

Listing 13.4: **The Horner's Algorithm**

```
function p = MyHorner(a)
% a is a vector of coefficients
% we evaluate the polynomial
% p(x) = a(1) + a(2)*x + ... +
% a(n)*x^{n-1} + a(n+1)*x^n
% using Horner's method
% p is returned as a function
%
% Example:
% y = a(1) + a(2) x + a(3)*x^2
%   = a(1) + x * ( a(2) + a(3) * x )
% q = a(2) + a(3) * x
% y = a(1) + x * q
%
% So set up as a loop
[n,m] = size(a);
if ( n >= 3 )
```

```
    p = @(x) a(n-1) + a(n)*x;
    for i = n-2:-1:1
      p = @(x) ( a(i)+ x.*p(x) );
    end
  elseif ( n == 2 )
    p = @(x) a(1) + a(2)*x;
  else
    p = @(x) a(1);
  end
  end
```

We then evaluate this polynomial at $(LZ)^2$ and compute the function we need to find the zeroes.

Listing 13.5: **The Function Whose Zeroes We Seek**

```
% this computes a(2) + a(4)t^2 + a(6) t^4 + ... + a(2n) t^{2n-2}
y = @(z) power(L*z,2);
p = @(z) c(1) + y(z).*q(y(z));
```

We then set up a linspace and find where this function has zeros by finding zero crossings. The zeros are stored in `W`. We don't use the whole interval $[0, L]$ for the `linspace` as we will have trouble with explosive round off error. So we choose to use $[0, u]$ instead with some $u < L$. We usually have to experiment with u to get useful plot.

Listing 13.6: **Zero Crossing Code**

```
X = linspace(0,u,m);
for i = 1:m
  Y(i) = p(X(i));
end
plot(X,Y);
% find zeros
Z = [];
E = [];
W = [];
for i = 1:m-1
  j = 1;
  if ( (Y(i) < 0) && (Y(i+1) > 0) ) || ( (Y(i) > 0) && (Y(i+1) < 0) )
    M = ( X(i) + X(i+1) )/2.0;
    Z = [Z,L*M];
    E = [E,-power(M,2) ];
    W = [W,M];
    p(M)
  end
end
```

The full code with comments removed is thus

Listing 13.7: **Full Code**

```
function [c,p,q,Z,E,W] = mybessel(k,n,m,u,L)
%
c = zeros(1,n+1);
c(1) = 1;
for i=2:n+1
  m1 = 4*(i-1)*(i-1+k);
  c(i) = -c(i-1)/m1;
end
b = c(1,2:n+1);
q = MyHorner(b');
y = @(z) power(L*z,2);
p = @(z) c(1) + y(z).*q(y(z));
X = linspace(0,u,m);
for i = 1:m
  Y(i) = p(X(i));
end
plot(X,Y);
% find zeros
Z = [];
E = [];
W = [];
for i = 1:m-1
  j = 1;
  if ( (Y(i) < 0) && (Y(i+1) > 0) ) || ( (Y(i) > 0) && (Y(i+1) < 0) )
    M = ( X(i) + X(i+1) )/2.0;
    Z = [Z,L*M];
    E = [E,-power(M,2) ];
    W = [W,M];
    p(M)
  end
end

end
```

Here is an example: we use the interval $[0,5]$ so $L = 5$ and we use partial sums of size 25 so $n = 25$. The linspace is set up on the full interval $[0,3.9]$ so $u = 3.9$. We use $u = 3.9$ so the values of x don't grow out of control.

Listing 13.8: **Finding the Eigenvalues**

```
>> [c,p,q,Z,E,W] = mybessel(1,25,481,4.2,5);
>> W
W =

    0.76562    1.40438    2.03438    2.66438    3.29438    3.93313    4.16062
```

The plot is nice here as by choosing $u = 4.2$ we avoid overflow calculations. This generates the plot we see in Figure 13.1.

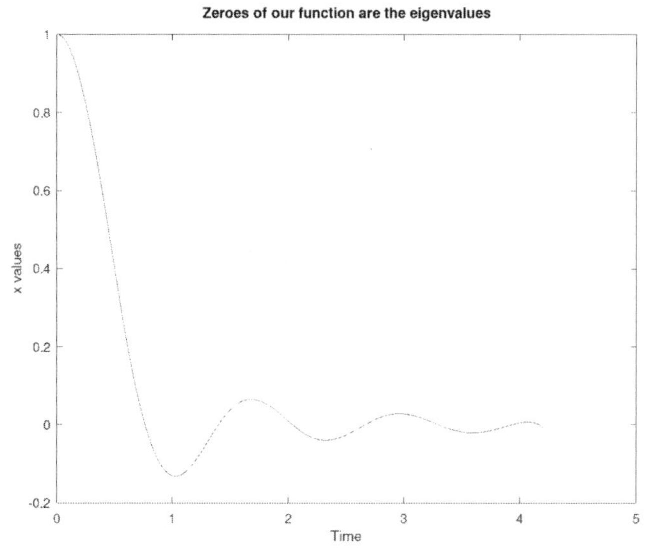

Figure 13.1: The zeros of this function are the eigenvalues.

We then find the corresponding eigenfunctions using the code below. Recall, the eigenfunctions will be

$$\phi_i(t) \quad = \quad t^k \left(\sum_{n=0}^{\infty} a_{2n}^{-1} (t\omega_i)^{2n} \right)$$

Once we have found the eigenfunctions, we can check to see if they are mutually orthogonal and compute the first few terms of the complete orthonormal sequence. We also print out the matrix D which is $(< \phi_i, \phi_j >)$. Here is the code we use to do this.

Listing 13.9: **Inner Product Blocks**

```
function [ip ,norm, phi] = FindInnerProducts(k,c,W,L,m, pts ,mycol)
%
%Let's find eigenfunctions.
%
% c is the coefficient vector from the previous approximation of the
% coefficients a_{2n}
% k is the positive integer chosen
% W is a vector of zeros
% m is the number of points to use in the plots
% pts is the number of points to use in the Riemann sum approximations
% col is the number of eigenfunctions to try to plot.
% Numerical problems make it difficult to get them all, so
% we can choose to plot fewer.
%
[row, col] = size(W);
q2 = MyHorner(c');
```

```
p2 = @(w,z) power(w*z,2);
f= @(w,t) power(t,k).*q2(p2(w,t));
phi = @(i,t) f(W(i),t);
EF = zeros(col,pts);
TT = linspace(0,L,pts);
for i = 1:col
  for j = 1:pts
    EF(i,j) = phi(i,TT(j));
  end
end
norm = zeros(1,col);
for i = 1:col
  for j = 1: pts
    norm(i) = norm(i) +  EF(i,j) * EF(i,j) *TT(j) *(L/pts);
  end
  norm(i) = sqrt(norm(i));
end
ip = zeros(mycol,mycol);
for i = 1:mycol
  for j = 1:mycol
    for k = 1: pts
      ip(i,j)  = ip(i,j) + EF(i,k) * EF(j,k) *TT(k) *(L/pts);
    end
    ip(i,j) = ip(i,j)/(norm(i)*norm(j));
  end
end
clf;
hold on;
for i = 1:mycol
  for j = 1:pts
    W(j) = EF(i,j);
  end
  plot(TT,W);
end
hold off;
end
```

We see the eigenfunctions are reasonably orthogonal given our round off errors.

Listing 13.10: **The Eigenfunction Orthonormality Check**

```
>> [ip,norm,phi] = FindInnerProducts(1,c,W,5,181,500,5);
>> ip
ip =

   1.000000000  -0.002840515   0.000682541  -0.000500264   0.000496489
  -0.002840515   1.000000000   0.002776112  -0.001506875   0.000854676
   0.000682541   0.002776112   1.000000000  -0.000074068   0.000463502
  -0.000500264  -0.001506875  -0.000074068   1.000000000  -0.000991149
   0.000496489   0.000854676   0.000463502  -0.000991149   1.000000000
```

This generates the plot we see if Figure 13.2.

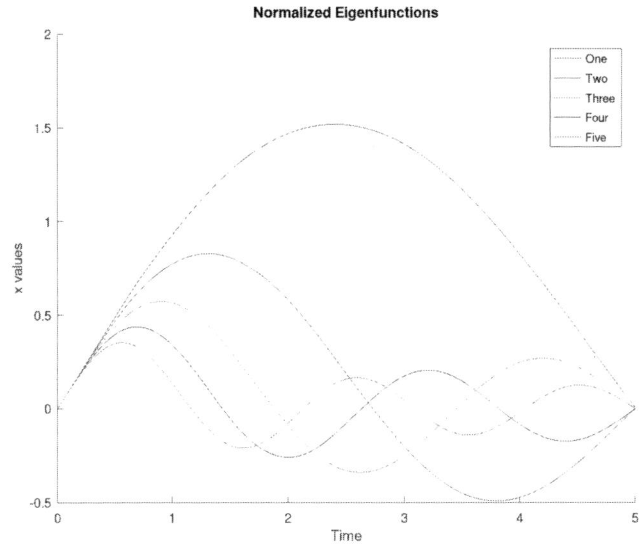

Figure 13.2: Normalized functions.

We can do these sorts of calculations for all of the traditional Stürm - Liouville problems although, as we mentioned, there are better ways to generate the orthonormal sequences.

Homework

Exercise 13.7.1 *Find eigenfunctions and eigenvalues for the Bessel's Equation as discussed above for* $k = 2$.

Exercise 13.7.2 *Find eigenfunctions and eigenvalues for the Bessel's Equation as discussed above for* $k = 3$.

Exercise 13.7.3 *Find eigenfunctions and eigenvalues for the Bessel's Equation as discussed above for* $k = 1$ *but use* $x'(0) = 0$.

Exercise 13.7.4 *Find eigenfunctions and eigenvalues for the Bessel's Equation as discussed above for* $k = 2$ *but use* $x'(0) = 0$.

Exercise 13.7.5 *Find eigenfunctions and eigenvalues for the Bessel's Equation as discussed above for* $k = 3$ *but use* $x'(0) = 0$.

13.7.2 Approximation with the Bessel Functions

If we are given a function f on $[0, L]$, we can now find the best $\|\cdot\|_2$ approximation to the subspace $\{\hat{\phi}_1, \ldots, \hat{\phi}_N\}$ for a given N. We calculated the normalized eigenfunctions and returned them in `FindInnerProduct` as `phihat`. To do the desired approximation, you can use this code:

Listing 13.11: **Full Best Approximation**

```
function [c,fstar] = FindBestApprox(check,n,N,pts,L,phihat,f)
%
% n is the number of normalized eigenfunctions
% L is the interval length
% phihat is the function of normalized eigenfunctions
% phihat(i,t) where i is the eigenfunction number
%             and t is the time
% f is the function to be approximated
% pts is the number of points in the Riemann Sum Approximations
% N is the number of points to use in the [0,L] linspace command
% check when 1 verifies the phihat's generate a reasonable
% ip matrix
% check
if (check == 1)
ip = zeros(n,n);
for i = 1:n
  for j = 1:n
    g1 = @(t) phihat(i,t);
    g2 = @(t) phihat(j,t);
    ip(i,j) = innerproduct(g1,g2,0,L,pts);
  end
end
ip
endif
%
c = zeros(n,1);
for i = 1:n
  g = @(t) phihat(i,t);
  c(i) = innerproduct(f,g,0,L,pts);
end
c
fstar = @(t) 0;
for i=1:n
  g = @(t) phihat(i,t);
  fstar = @(t) ( fstar(t) + c(i)*g(t) );
end

T = linspace(0,L,N);
for i = 1:N
  Y(i) = fstar(T(i));
end
plot(T,Y,T,f(T));

end
```

Let $f(x) = 2 * \cos(2 * x)e^{0.3*x}$. Using 5 of the Bessel normalized eigenfunctions we have found, we can find the approximation f^* as follows:

Listing 13.12: **The Best Approximation**

```
f = @(x) 2*cos(2*x).*exp(0.3*x);
[c,p,q,Z,E,W] = mybessel(1,25,481,4.2,5);
[ip,norm,phihat] = FindInnerProducts(1,c,W,5,101,301,5);
[c,fstar] = FindBestApprox(1,5,101,301,5,phihat,f);
```

This generates the plot of Figure 13.3. Better if we had more eigenfunctions!

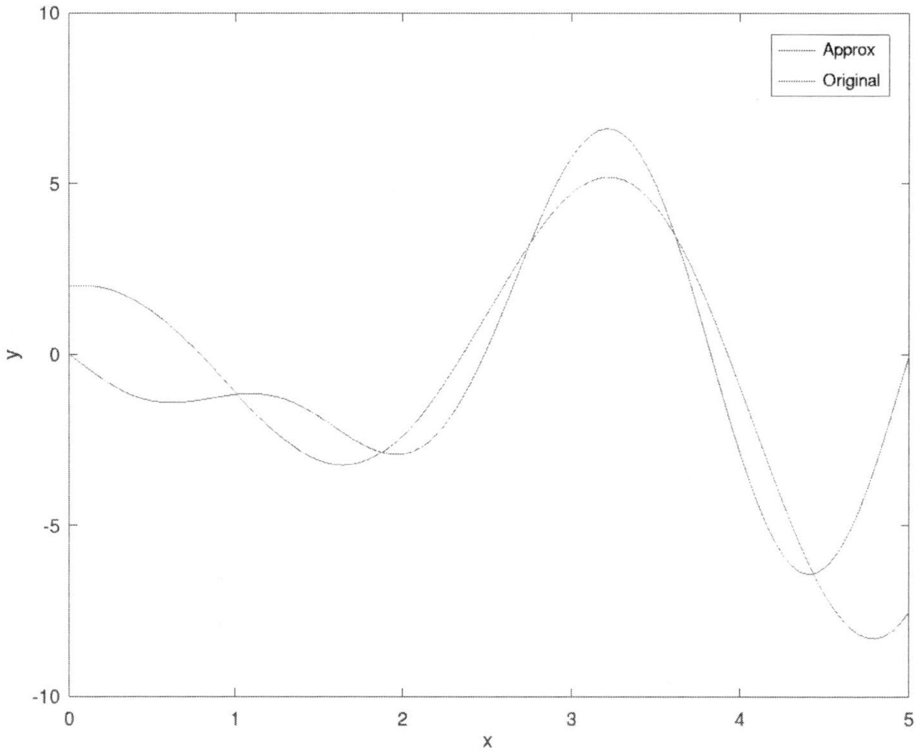

Figure 13.3: Best approximation to $f(x) = 2 * \cos(2 * x)e^{0.3*x}$; 5 eigenfunctions.

Homework

Exercise 13.7.6 *Find eigenfunctions and eigenvalues for the Bessel's Equation as discussed above on* $[0, 7]$ *for* $k = 2$ *and use them to find the best approximation to* $f(x) = 3\sin(4x)e^{0.7x}$.

Exercise 13.7.7 *Find eigenfunctions and eigenvalues for the Bessel's Equation as discussed above on* $[0, 3]$ *for* $k = 2$ *and use them to find the best approximation to* $f(x) = 3\sin(4x)e^{-0.05x}$.

Exercise 13.7.8 *Find eigenfunctions and eigenvalues for the Bessel's Equation on* $[0, 2]$ *as discussed above for* $k = 3$ *and use them to find the best approximation to* $f(x) = 3\cos(4x)e^{1.7x}$.

Exercise 13.7.9 *Find eigenfunctions and eigenvalues for the Bessel's Equation on* $[0, 4]$ *as discussed above for* $k = 3$ *and use them to find the best approximation to* $f(x) = 5\cos(4x^2)e^{-0.7x}$.

Chapter 14

Self-Adjoint Operators

In the last chapter, we inverted the Stürm - Liouville differential operator to obtain the inverse in the case where λ was not an eigenvalue. We can now determine the eigenfunctions of this operator form a complete orthonormal set in the Hilbert Space $\mathbb{L}_2([a, b])$.

First, let's think about integration in $\mathbb{L}_2([a, b])$. $RI([a, b])$ is a subspace of $\mathbb{L}_2([a, b])$. The mapping $\phi(\boldsymbol{f}) = \int_a^b f(t)dt$ for all $\boldsymbol{f} \in RI([a, b])$ is a bounded linear functional which satisfies $\left|\int_a^b f(t)dt\right| \leq \|\boldsymbol{f}\|_2\sqrt{b-a} = p(\boldsymbol{f})$ where p satisfies $p(\alpha v) = |\alpha| p(v)$ for all $\alpha \in \Re$ and $v \in \mathbb{L}_2([a, b])$. Hence, by the Generalized Hahn - Banach Theorem, there is a linear functional $\widetilde{\phi}$ which satisfies $|\widetilde{\phi}(\boldsymbol{v})| \leq \|\boldsymbol{f}\|_2(b-a)$ for all $\boldsymbol{v} \in \mathbb{L}_2([a, b])$. and $\widetilde{\phi} = \phi$ on the subspace $RI([a, b])$. From the way we completed the Riemann Integrable functions to construct $\mathbb{L}_2([a, b])$, we have

$$\widetilde{\phi}(\mathscr{F}) = \lim_{n \to \infty} \phi(\boldsymbol{x_n}) = \lim_{n \to \infty} \int_a^b x_n(t)dt$$

where $\boldsymbol{x_n}$ is any representative of the equivalence class $([\boldsymbol{f_n}])$ of Cauchy sequences which converge to the equivalence class associated with $\boldsymbol{f}$ in $\mathbb{L}_2([a, b])$ which we call $\mathscr{F}$.

We can construct this extension of Riemann Integration in other ways. Most notably, we can develop Lebesgue Integration as we do in (Peterson (16) 2019). In our future discussions, we will refer to $\widetilde{\phi}(\mathscr{F})$ as simply $\int_a^b f(t)dt$ where it will be understood this is the extension of the standard Riemann Integral to $\mathbb{L}_2([a, b])$.

Homework

Exercise 14.0.1 *Think about all of this carefully and define the function spaces* $\mathbb{L}_p([a, b])$ *for* $1 \leq p < \infty$.

Exercise 14.0.2 *Think about all of this carefully and define the function spaces* $\mathbb{L}_\infty([a, b])$.

Exercise 14.0.3 *Prove Hölder's Inequality for these spaces. Look carefully at how we did this for the Riemann Integrable functions.*

Exercise 14.0.4 *Prove Minkowski's Inequality for these spaces. Look carefully at how we did this for the Riemann Integrable functions.*

14.1 Integral Operators

Let's look at the inverse forms of our models. We have an operator $\mathscr{J}_\Theta$ whose domain is at least $C([0, L])$ defined by

$$\mathscr{J}_\Theta(f) \quad = \quad \int_0^L k_\Theta(x, s) f(s)\, ds$$

where $k_\Theta(x, s)$ is a symmetric function of the form

$$k_\Theta(x, s) \quad = \quad \frac{1}{W(x_1, x_2)} \begin{cases} x_1(s)\, x_2(x) & 0 \leq s \leq x \\ x_1(x)\, x_2(s) & x < s \leq L \end{cases}$$

where x_1 and x_2 are the two standard linearly independent solutions obtained as follows: x_1 solves

$$\begin{aligned} \mathscr{L}(u) + \Theta u &= 0, \ 0 \leq x \leq L \\ \ell_1(u(0), u'(0)) &= \alpha_1\, u(0) + \alpha_2\, u'(0) = 0 \end{aligned}$$

and x_2 satisfies the homogeneous model with the second boundary condition

$$\begin{aligned} \mathscr{L}(u) + \Theta u &= 0, \ 0 \leq x \leq L \\ \ell_2(u(L), u'(L)) &= \beta_1\, u(L) + \beta_2\, u'(L) = 0 \end{aligned}$$

In all of these problems, it is assumed $-\Theta$ is not an eigenvalue of $\mathscr{L}$. We already know the eigenvalues of $\mathscr{L}$ for various boundary conditions form an infinite sequence ω_n which goes to $-\infty$ as $n \to \infty$. In some cases 0 is an eigenvalue and in others, it is not. Let BC denote the subset of functions which satisfy the boundary conditions. We have already shown for any u and v in the $C^2([0, L]) \cap \{BC\}$, $< \mathscr{L}(u), v >=< u, \mathscr{L}(v) >$ using the $\boldsymbol{r}$ weighted inner product. Operators with this kind of behavior are called self-adjoint operators. Since $\mathscr{L}$ is defined on the subspace $C^2([0, L]) \cap \{BC\}$, we note the following important things.

- The subspace $C^2([0, L]) \cap \{BC\}$ is dense in $\mathbb{L}_2([a, b])$.
- Our notion of self-adjoint can thus be applied to the situation of densely defined linear operators. Since $\mathscr{J}_\Theta$ is a linear operator defined via integration of a kernel function, these are called **integral operators**.

Definition 14.1.1 The Densely Defined Hilbert Adjoint Operator

Let $(H, < \cdot, \cdot >)$ be a Hilbert Space and let $T : dom(T) \subset H \to H$ be a linear operator whose domain is dense in H. Then the operator $T^* : dom(T) \subset H \to H$ by $< T(\boldsymbol{x}), \boldsymbol{y} >=< \boldsymbol{x}, T^*(\boldsymbol{y}) >$ for all $\boldsymbol{x}$ and $\boldsymbol{y}$ in H. is called the Hilbert Adjoint if it exists.

If $T^* = T$, we say T is **self-adjoint**.

We can also check the self-adjointness of the corresponding operator $\mathscr{J}_\Theta$ when Θ is not an eigenvalue. For convenience, let $\mathscr{J}_\Theta$ and k_Θ simply be denoted by $\mathscr{J}$ and k to avoid the subscripting issues.

14.1.1 Integral Operator with Symmetric and Hermitian Kernels

Let's focus on integral operators with symmetric or Hermitian kernels. Thus, the operator $\mathscr{J}$ does not have to be the inverse of a differential operator on a densely defined subspace of $\mathbb{L}_2([a, b])$. Consider an operator $\mathscr{J}$ whose domain is at least $C([0, L])$ defined by

$$\mathscr{J}(f) = \int_0^L k(x, s) f(s)\, ds$$

where $k(x, s)$ is a symmetric function continuous on $[0, L] \times [0, L]$

We have

$$< \mathscr{J}(f), g > = \int_a^b \left(\int_a^b k(x, s) f(s) ds \right) \overline{g(x)}\, dx$$

Note this integral is well-defined as the product $k(x, \cdot) f(\cdot)$ defines a function in $\mathbb{L}_2([a, b])$. We do need the fact that the function $\int_a^b k(x, s) f(s) ds$ defines a function in $\mathbb{L}_2([a, b])$ also which is not too hard to see. We also need a version of Fubini's theorem to do the next step. But since the Riemann Integral satisfies this and the extension integral is constructed via a limit process, this is also not hard to verify. Now interchange the order of integration to get

$$< \mathscr{J}(f), g > = \int_a^b \left(\int_a^b k(x, s) \overline{g(x)} dx \right) f(s)\, ds$$

If the kernel is symmetric, use the symmetry of the kernel to find

$$< \mathscr{J}(f), g > = \int_a^b \left(\int_a^b k(s, x) \overline{g(x)} dx \right) f(s)\, ds$$

But, $\int_a^b k(s, x) \overline{g(x)} dx = \mathscr{J}(\overline{g})$ and so

$$< \mathscr{J}(f), g > = \int_a^b \mathscr{J}(\overline{g})\, f(s)\, ds = \int_a^b \overline{\mathscr{J}(g)}\, f(s)\, ds =< f, \mathscr{J}(g) > .$$

On the other hand, if the kernel is Hermitian, we find

$$< \mathscr{J}(f), g > \;=\; \int_a^b \left(\int_a^b \overline{k(s,x)g(x)dx} \right) f(s)\, ds$$

But, $\int_a^b \overline{k(s,x)g(x)}dx = \overline{\mathscr{J}(g)}$ and so

$$< \mathscr{J}(f), g > \;=\; \int_a^b \overline{\mathscr{J}(g)}\, f(s)\, ds =< f, \mathscr{J}(g) > .$$

We have shown

Theorem 14.1.1 Integral Operators with Symmetric and Hermitian kernels are Self-adjoint

If $\boldsymbol{k}$ *is symmetric and Hermitian and continuous on* $[a,b] \times [a,b]$*, then* $\mathscr{J}$ *defined on* $\mathbb{L}_2([a,b])$ *by* $\mathscr{J}(\boldsymbol{f}) = \int_a^b k(x,s)f(s)ds$ *is self-adjoint on* $\mathbb{L}_2([a,b])$*.*

Proof 14.1.1

We have just presented this argument. ■

Thus, symmetric and Hermitian kernel integral operators $\mathscr{J}$ is also self-adjoint on the domain $C([0,L])$.

Homework

Exercise 14.1.1 *Define the* kernel *function* k_ω, $\omega L \neq n\pi$, *on* $[0,L] \times [0,L]$ *by*

$$k_\omega(x,s) \;=\; \frac{1}{\omega \sin(L\omega)} \begin{cases} \cos(\omega s)\, \cos(\omega(L-x)) & 0 \le s \le x \\ \cos(\omega x)\, \cos(\omega(L-s)) & x < s \le L \end{cases}$$

and define the operator J *by* $J(f) = \int_0^L k_\omega(x,s)f(s)ds$ *for any* $f \in C([0,L])$.

- *Prove* k_ω *is continuous on* $[0,L] \times [0,L]$.
- *Prove* k_ω *is symmetric.*
- *Prove* J *is a linear bounded operator.*
- *Prove* J *is a continuous mapping.*
- *Prove* J *is self-adjoint using the usual inner product.*

Exercise 14.1.2 *Define the* kernel *function* k_ω, $\omega L \neq n\pi$, *on* $[0,L] \times [0,L]$ *by*

$$k_\omega(x,s) \;=\; \frac{1}{\omega \sin(L\omega)} \begin{cases} \sin(\omega s)\, \sin(\omega(L-x)) & 0 \le s \le x \\ \sin(\omega x)\, \sin(\omega(L-s)) & x < s \le L \end{cases}$$

and define the operator J *by* $J(f) = \int_0^L k_\omega(x,s)f(s)ds$ *for any* $f \in C([0,L])$.

- *Prove* k_ω *is continuous on* $[0, L] \times [0, L]$.
- *Prove* k_ω *is symmetric.*
- *Prove* J *is a linear bounded operator.*
- *Prove* J *is a continuous mapping.*
- *Prove* J *is self-adjoint using the usual inner product.*

Exercise 14.1.3 *Define the* kernel *function* k_ω *on* $[0, L] \times [0, L]$ *by*

$$k_\omega(x, s) = \frac{1}{\omega \sin(L\omega)} \begin{cases} s(L - x) & 0 \leq s \leq x \\ x(L - s) & x < s \leq L \end{cases}$$

and define the operator J *by* $J(f) = \int_0^L k_\omega(x, s) f(s) ds$ *for any* $f \in C([0, L])$.

- *Prove* k_ω *is continuous on* $[0, L] \times [0, L]$.
- *Prove* k_ω *is symmetric.*
- *Prove* J *is a linear bounded operator.*
- *Prove* J *is a continuous mapping.*
- *Prove* J *is self-adjoint using the usual inner product.*

14.1.2 Properties of the Integral Operator

The integral operator $\mathscr{I}_\Theta$ has many properties. Again, for ease of exposition, we will now just call the operator $\mathscr{I}$ and simply denote the kernel by k, thereby avoiding messy subscripting.

14.1.3 $\mathscr{I}$ is Well-Defined

First, let's verify that $\mathscr{I}(f)$ is actually well-defined by showing $\mathscr{I}(f)$ is a continuous function for any f in $\mathbb{L}_2([a, b])$. To do this analysis, we have to use the extension of Riemann Integration $\widetilde{\phi}$ to $\mathbb{L}_2([a, b])$. We haven't proven any of the standard integral like properties of this extension but we will use them anyway. Again, to do this properly, we need to develop Lebesgue Integration. But assuming our extensions behaves in similar ways to what we expect from Riemann Integration, the argument for this is as follows. We see for any t_0 and t in $[a, b]$, we have

$$|(\mathscr{I}(f))(t) - (\mathscr{I}(f))(t_0)| \leq \int_a^b |k(s, t) - k(s, t_0)| \; |f(s)| \, ds$$

But k is continuous on $[a, b] \times [a, b]$, so given $\epsilon > 0$, there is a $\delta > 0$ so that

$$|k(s, t) - k(s, t_0)| < \frac{1}{\sqrt{b - a} \, (\|f\|_2 + 1)} \epsilon$$

where $\|f\|_2$ is the usual $\mathscr{L}^2$ norm. We then see

$$|(\mathscr{J}(f))(t) - (\mathscr{J}(f))(t_0)| \leq \frac{1}{\sqrt{b-a}\,(\|f\|_2 + 1)}\,\epsilon \int_a^b |f(s)|\,ds$$

if $|t - t_0| < \delta$ (we make obvious adjustments to this argument if t_0 is a or b of course). By the Cauchy - Schwartz Inequality (the extension of Hölder's Inequality for $\widetilde{\phi}$ is in the problems), we also know

$$\begin{aligned} \int_a^b |f(s)|\,ds &\leq \sqrt{\int_a^b |f(s)|^2\,ds}\,\sqrt{\int_a^b \mathbf{1}\,ds} \\ &= \|f\|_2\,\sqrt{b-a}. \end{aligned}$$

Thus,

$$|(\mathscr{J}(f))(t) - (\mathscr{J}(f))(t_0)| \leq \frac{1}{\sqrt{b-a}\,(\|f\|_2 + 1)}\,\epsilon\,\|f\|_2\,\sqrt{b-a} < \epsilon$$

This shows that $\mathscr{J}(f)$ is a continuous function. Note this argument does not require that f be continuous; only that it is in $\mathbb{L}_2([a,b])$.

Homework

Exercise 14.1.4 *Define the* kernel *function* k_ω, $\omega L \neq n\pi$, *on* $[0,L] \times [0,L]$ *by*

$$k_\omega(x,s) = \frac{1}{\omega \sin(L\omega)} \begin{cases} \cos(\omega s)\,\cos(\omega(L-x)) & 0 \leq s \leq x \\ \cos(\omega x)\,\cos(\omega(L-s)) & x < s \leq L \end{cases}$$

and define the operator J *by* $J(f) = \int_0^L k_\omega(x,s)f(s)ds$. *Prove* $J(f)$ *is continuous for any* f *in* $\mathbb{L}_2([a,b])$.

Exercise 14.1.5 *Define the* kernel *function* k_ω, $\omega L \neq n\pi$, *on* $[0,L] \times [0,L]$ *by*

$$k_\omega(x,s) = \frac{1}{\omega \sin(L\omega)} \begin{cases} \sin(\omega s)\,\sin(\omega(L-x)) & 0 \leq s \leq x \\ \sin(\omega x)\,\sin(\omega(L-s)) & x < s \leq L \end{cases}$$

and define the operator J *by* $J(f) = \int_0^L k_\omega(x,s)f(s)ds$. *Prove* $J(f)$ *is continuous for any* f *in* $\mathbb{L}_2([a,b])$.

Exercise 14.1.6 *Define the* kernel *function* k_ω *on* $[0,L] \times [0,L]$ *by*

$$k_\omega(x,s) = \frac{1}{\omega \sin(L\omega)} \begin{cases} s(L-x) & 0 \leq s \leq x \\ x(L-s) & x < s \leq L \end{cases}$$

and define the operator J *by* $J(f) = \int_0^L k_\omega(x,s)f(s)ds$. *Prove* $J(f)$ *is continuous for any* f *in* $\mathbb{L}_2([a,b])$.

14.1.4 Equicontinuous Families Determined by the Kernel

We can say more. Recall, if $\mathscr{F}$ is a family of functions which is uniformly bounded in $\mathbb{L}_2([a,b])$ by the constant $B > 0$, this means that $\|f\|_2 < B$ for all f in the family.

Theorem 14.1.2 $\mathscr{J}(\overline{B(\mathbf{0},1)})$ **is an Equicontinuous Family**

Let $\Omega = \{\mathscr{J}(\boldsymbol{u}) \mid \boldsymbol{u} \in C([a,b]),\ \|\boldsymbol{u}\|_\infty \leq 1\}$. Then Ω is a bounded set of equicontinuous functions in $(C([a,b]), \|\cdot\|_\infty)$.

Proof 14.1.2

Recall a set of functions $\mathcal{F}$ on $[a,b]$ is equicontinuous on $[a,b]$ means for all $\epsilon > 0$, there is a $\delta > 0$ (independent of $f \in \mathcal{F}$), so that

$$|t - s| < \delta \quad \Longrightarrow \quad |f(t) - f(s)| < \epsilon,\ t, s \in [a,b],\ f \in \mathcal{F}$$

We know k is uniformly continuous on $[a,b] \times [a,b]$ and so fixing $s \in [a,b]$, for all $\epsilon > 0$, there is a $\delta > 0$ such that

$$|k(s,t_1) - k(s,t_2)| \quad < \quad \frac{\epsilon}{b-a},\ t_1, t_2 \in [a,b],\ |t_1 - t_2| < \delta$$

So if $\boldsymbol{u} \in C([a,b])$ and $|t_1 - t_2| < \delta$, $t_1, t_2 \in [a,b]$, we have

$$\begin{aligned} |\mathscr{J}(u(t_1)) - \mathscr{J}(u(t_2))| &= \left| \int_a^b k(s,t_1)u(s)\,ds - \int_a^b k(s,t_2)u(s)\,ds \right| \\ &\leq \int_a^b |k(s,t_1) - k(s,t_2)|\,|u(s)|ds \\ &\leq \int_a^b \frac{\epsilon}{b-a}\,\|\boldsymbol{u}\|_\infty = \frac{\epsilon}{b-a}\,(1)\,(b-a) = \epsilon \end{aligned}$$

We conclude Ω is an equicontinuous family.

Finally, since k is uniformly continuous on $[a,b] \times [a,b]$, there is $b > 0$ so that $|k(s,t)| \leq B$ for all $(s,t) \in [a,b] \times [a,b]$. Thus,

$$|\mathscr{J}(u(t))| \quad \leq \quad \int_a^b k(s,t)u(s)\,ds \leq B\,\|\boldsymbol{u}\|_\infty (b-a)$$

This shows Ω is bounded. ■

Clearly $\mathscr{J}$ can act on functions that are not continuous.

Theorem 14.1.3 $\mathscr{J}(\overline{B(\mathbf{0},1)}), \overline{B(\mathbf{0},1)} \subset \mathbb{L}_2([a,b])$ **is an Equicontinuous Family**

Let $\Omega = \{\mathscr{J}(\boldsymbol{u}) \mid \boldsymbol{u} \in \mathbb{L}_2([a,b]).\ \|\boldsymbol{u}\|_2 \leq 1\}$. The Ω is a bounded set of equicontinuous functions in $(\mathbb{L}_2([a,b]), \|\cdot\|_2)$.

Proof 14.1.3

We know k is uniformly continuous on $[a,b] \times [a,b]$ and so fixing $s \in [a,b]$, for all $\epsilon > 0$, there is a $\delta > 0$ such that

$$|k(s,t_1) - k(s,t_2)| \quad < \quad \frac{\epsilon}{\sqrt{b-a}}, \; t_1, t_2 \in [a,b], \; |t_1 - t_2| < \delta$$

So if $\boldsymbol{u} \in \mathbb{L}_2([a,b])$ and $|t_1 - t_2| < \delta$, $t_1, t_2 \in [a,b]$, we have

$$\begin{aligned} |\mathscr{J}(u(t_1)) - \mathscr{J}(u(t_2))| &= \left| \int_a^b k(s,t_1)u(s)\, ds - \int_a^b k(s,t_2)u(s)\, ds \right| \\ &\leq \int_a^b |k(s,t_1) - k(s,t_2)| \, |u(s)| ds \\ &\leq \int_a^b \frac{\epsilon}{\sqrt{b-a}} \|\boldsymbol{u}\|_2 = \frac{\epsilon}{\sqrt{b-a}} (1) \sqrt{b-a} = \epsilon \end{aligned}$$

using Hölder's Inequality. We conclude Ω is an equicontinuous family.

Finally, since k is uniformly continuous on $[a,b] \times [a,b]$, there is $b > 0$ so that $|k(s,t)| \leq B$ for all $(s,t) \in [a,b] \times [a,b]$. Thus,

$$|\mathscr{J}(u(t))| \quad \leq \quad \int_a^b k(s,t)u(s)\, ds \leq B \, \|\boldsymbol{u}\|_2 \sqrt{b-a}$$

This shows Ω is bounded. ■

In either case, Ascoli's Theorem then applies to such families and tells us that $\mathcal{F}$ has a sequence (f_n) so that $f_n \overset{\text{sup}}{\to} f$ for some $f \in C([a,b])$.

These operators belong to the class of **compact** operators.

Definition 14.1.2 Compact Linear Operators

Let $(X, \|\cdot\|_X)$ and $(Y, \|\cdot\|_Y)$ be two normed linear spaces and let $T : X \to Y$ be a bounded linear operator which satisfies $\overline{T(\Omega)}$ is sequentially compact when $\Omega \subset X$ is bounded.

The integral operator with symmetric or Hermitian kernel k which is continuous on $[a,b] \times [a,b]$ satisfies this definition. If $\Omega = B(\boldsymbol{0}; 1) \subset \mathbb{L}_2([a,b])$, the arguments in the proof of Theorem 14.1.3 show us $\mathscr{J}(\Omega) = \{\mathscr{J}(\boldsymbol{u}) | \boldsymbol{u} \in \Omega\}$ is a bounded set of equicontinuous functions in $(\mathbb{L}_2([a,b]), \|\cdot\|_2)$. Ascoli's Theorem then applies to such families and tells us that $\mathcal{F}$ has a sequence (f_n) so that $f_n \overset{\text{sup}}{\to} f$ for some $f \in C([a,b])$ which is in the closure of $\mathscr{J}(\Omega)$. Hence, these integral operators are compact operators.

Homework

Exercise 14.1.7 *Define the* kernel *function* k_ω, $\omega L \neq n\pi$, *on* $[0, L] \times [0, L]$ *by*

$$k_\omega(x, s) = \frac{1}{\omega \sin(L\omega)} \begin{cases} \cos(\omega s)\,\cos(\omega(L - x)) & 0 \leq s \leq x \\ \cos(\omega x)\,\cos(\omega(L - s)) & x < s \leq L \end{cases}$$

and define the operator J *by* $J(f) = \int_0^L k_\omega(x, s) f(s) ds$. *Let* $B = \{f \in C([0, L]), \mid \|f\|_\infty \leq 1\}$. *Prove* $J(B)$ *is an equicontinuous family in* $C([0, L])$.

Exercise 14.1.8 *Define the* kernel *function* k_ω, $\omega L \neq n\pi$, *on* $[0, L] \times [0, L]$ *by*

$$k_\omega(x, s) = \frac{1}{\omega \sin(L\omega)} \begin{cases} \cos(\omega s)\,\cos(\omega(L - x)) & 0 \leq s \leq x \\ \cos(\omega x)\,\cos(\omega(L - s)) & x < s \leq L \end{cases}$$

and define the operator J *by* $J(f) = \int_0^L k_\omega(x, s) f(s) ds$. *Let* $B = \{f \in \mathbb{L}_2([a, b]) \mid \|f\|_2 \leq 1\}$. *Prove* $J(B)$ *is an equicontinuous family in* $C([0, L])$.

Exercise 14.1.9 *Define the* kernel *function* k_ω, $\omega L \neq n\pi$, *on* $[0, L] \times [0, L]$ *by*

$$k_\omega(x, s) = \frac{1}{\omega \sin(L\omega)} \begin{cases} \sin(\omega s)\,\sin(\omega(L - x)) & 0 \leq s \leq x \\ \sin(\omega x)\,\sin(\omega(L - s)) & x < s \leq L \end{cases}$$

and define the operator J *by* $J(f) = \int_0^L k_\omega(x, s) f(s) ds$. *Let* $B = \{f \in C([0, L]), \mid \|f\|_\infty \leq 1\}$. *Prove* $J(B)$ *is an equicontinuous family in* $C([0, L])$.

Exercise 14.1.10 *Define the* kernel *function* k_ω, $\omega L \neq n\pi$, *on* $[0, L] \times [0, L]$ *by*

$$k_\omega(x, s) = \frac{1}{\omega \sin(L\omega)} \begin{cases} \sin(\omega s)\,\sin(\omega(L - x)) & 0 \leq s \leq x \\ \sin(\omega x)\,\sin(\omega(L - s)) & x < s \leq L \end{cases}$$

and define the operator J *by* $J(f) = \int_0^L k_\omega(x, s) f(s) ds$. *Let* $B = \{f \in \mathbb{L}_2([a, b]) \mid \|f\|_2 \leq 1\}$. *Prove* $J(B)$ *is an equicontinuous family in* $C([0, L])$.

Exercise 14.1.11 *Define the* kernel *function* k_ω *on* $[0, L] \times [0, L]$ *by*

$$k_\omega(x, s) = \frac{1}{\omega \sin(L\omega)} \begin{cases} s(L - x) & 0 \leq s \leq x \\ x(L - s) & x < s \leq L \end{cases}$$

and define the operator J *by* $J(f) = \int_0^L k_\omega(x, s) f(s) ds$. *Let* $B = \{f \in C([0, L]), \mid \|f\|_\infty \leq 1\}$. *Prove* $J(B)$ *is an equicontinuous family in* $C([0, L])$.

Exercise 14.1.12 *Define the* kernel *function* k_ω *on* $[0, L] \times [0, L]$ *by*

$$k_\omega(x, s) = \frac{1}{\omega \sin(L\omega)} \begin{cases} s(L - x) & 0 \leq s \leq x \\ x(L - s) & x < s \leq L \end{cases}$$

and define the operator J by $J(f) = \int_0^L k_\omega(x,s)f(s)ds$. Let $B = \{f \in \mathbb{L}_2([a,b]) \mid \|f\|_2 \leq 1\}$. Prove $J(B)$ is an equicontinuous family in $C([0,L])$.

14.1.5 Characterization of the Self-Adjoint Operator Norm

The self-adjoint operators we have here have many special properties. We'll now prove some results.

Theorem 14.1.4 Characterization of the Self-Adjoint Operator Norm

Let $\mathscr{J}$ be defined as above. Then $\mathscr{J}$ is a linear operator from $\mathbb{L}_2([a,b])$ to $C([a,b]) \subset \mathbb{L}_2([a,b])$ and

$$\|\mathscr{J}\|_{op} = \sup_{\|u\|_2 = 1} |< \mathscr{J}(u), u >|.$$

Proof 14.1.4

It is easy to see that this supremum is bounded as if $\|\boldsymbol{u}\|_2 = 1$, then

$$\begin{aligned} |< \mathscr{J}(\boldsymbol{u}), \boldsymbol{u} >| &\leq \|\mathscr{J}(\boldsymbol{u})\|_2 \, \|\boldsymbol{u}\|_2 \leq \|\mathscr{J}(\boldsymbol{u})\|_2 \\ &\leq \|\mathscr{J}\|_{op} \, \|\boldsymbol{u}\|_2 \leq \|\mathscr{J}\|_{op}. \end{aligned}$$

We conclude

$$\sup_{\|\boldsymbol{u}\|_2 = 1} |< \mathscr{J}(\boldsymbol{u}), \boldsymbol{u} >| \leq \|\mathscr{J}\|_{op}.$$

For convenience, let $B = \sup_{\|\boldsymbol{u}\|_2 = 1} |< \mathscr{J}(\boldsymbol{u}), \boldsymbol{u} >|$. Then, we know for any $\boldsymbol{u} \neq \boldsymbol{0}$

$$\left|\left\langle \mathscr{J}\left(\frac{\boldsymbol{u}}{\|\boldsymbol{u}\|_2}\right), \frac{\boldsymbol{u}}{\|\boldsymbol{u}\|_2}\right\rangle\right| \leq B$$

which implies immediately for all $\boldsymbol{x}$,

$$|< \mathscr{J}(\boldsymbol{x}), \boldsymbol{x} >| \leq B \, \|\boldsymbol{x}\|_2^2.$$

Next, we have

$$\begin{aligned} < \mathscr{J}(\boldsymbol{x}+\boldsymbol{y}), \boldsymbol{x}+\boldsymbol{y} > &= < \mathscr{J}(\boldsymbol{x}), \boldsymbol{x} > + < \mathscr{J}(\boldsymbol{y}), \boldsymbol{y} > +2 < \mathscr{J}(\boldsymbol{x}), \boldsymbol{y} > \\ &\leq B < x+y, x+y > = B \, \|x+y\|^2. \\ < \mathscr{J}(\boldsymbol{x}-\boldsymbol{y}), \boldsymbol{x}-\boldsymbol{y} > &= < \mathscr{J}(\boldsymbol{x}), \boldsymbol{x} > + < \mathscr{J}(\boldsymbol{y}), \boldsymbol{y} > -2 < \mathscr{J}(\boldsymbol{x}), \boldsymbol{y} > \\ &\geq -B < \boldsymbol{x}-\boldsymbol{y}, \boldsymbol{x}-\boldsymbol{y} > = -B \, \|\boldsymbol{x}-\boldsymbol{y}\|^2. \end{aligned}$$

Subtracting, we have for $\boldsymbol{x}$ and $\boldsymbol{y}$ of norm 1,

$$\begin{aligned} 4 < \mathscr{J}(\boldsymbol{x}), \boldsymbol{y} > &\leq B\Big(\|\boldsymbol{x}+\boldsymbol{y}\|^2 + \|\boldsymbol{x}-\boldsymbol{y}\|^2 \Big) \\ &= 2B\left(\|\boldsymbol{x}\|_2^2 + \|\boldsymbol{y}\|_2^2 \right) = 4B. \end{aligned}$$

Thus, for all $\boldsymbol{x}$ and $\boldsymbol{y}$ of norm 1, $< \mathscr{J}(\boldsymbol{x}), \boldsymbol{y} > \leq B$. Finally, let $\boldsymbol{y} = \mathscr{J}(\boldsymbol{x})$. If $\boldsymbol{y}$ is $\boldsymbol{0}$, we trivially have $\| \mathscr{J}(\boldsymbol{x})\|^2 \leq B$. Otherwise, we have for all $\boldsymbol{x}$ of norm 1,

$$\left\langle \mathscr{J}(\boldsymbol{x}), \frac{\mathscr{J}(\boldsymbol{x})}{\|\mathscr{J}(\boldsymbol{x})\|_2} \right\rangle \leq B$$

which tells us

$$\| \mathscr{J}(\boldsymbol{x})\|_2 \leq B$$

We conclude, from both cases, that

$$\| \mathscr{J} \|_{op} = \sup_{\|\boldsymbol{x}\|_2 = 1} \| \mathscr{J}(\boldsymbol{x})\|_2 \leq B$$

This establishes the reverse inequality and the result is proved. ■

Note our argument here is pretty much the same as the ones we gave for symmetric and Hermitian matrices. The infinite dimensional nature of the underlying spaces did not make a difference.

Homework

Exercise 14.1.13 *Define the* kernel *function k_ω, $\omega L \neq n\pi$, on $[0, L] \times [0, L]$ by*

$$k_\omega(x, s) = \frac{1}{\omega \sin(L\omega)} \begin{cases} \cos(\omega s)\, \cos(\omega(L-x)) & 0 < s < x \\ \cos(\omega x)\, \cos(\omega(L-s)) & x < s \leq L \end{cases}$$

and define the operator J by $J(f) = \int_0^L k_\omega(x,s) f(s) ds$. Estimate the operator norm of J on $C([0, L])$ and express $< J(f), f >$ as an integral.

Exercise 14.1.14 *Define the* kernel *function k_ω, $\omega L \neq n\pi$, on $[0, L] \times [0, L]$ by*

$$k_\omega(x, s) = \frac{1}{\omega \sin(L\omega)} \begin{cases} \cos(\omega s)\, \cos(\omega(L-x)) & 0 \leq s \leq x \\ \cos(\omega x)\, \cos(\omega(L-s)) & x < s \leq L \end{cases}$$

and define the operator J by $J(f) = \int_0^L k_\omega(x,s) f(s) ds$. Estimate the operator norm of J on $\mathbb{L}_2([a, b])$ and express $< J(f), f >$ as an integral.

Exercise 14.1.15 *Define the* kernel *function* k_ω, $\omega L \neq n\pi$, *on* $[0, L] \times [0, L]$ *by*

$$k_\omega(x, s) = \frac{1}{\omega \sin(L\omega)} \begin{cases} \sin(\omega s)\,\sin(\omega(L - x)) & 0 \leq s \leq x \\ \sin(\omega x)\,\sin(\omega(L - s)) & x < s \leq L \end{cases}$$

and define the operator J *by* $J(f) = \int_0^L k_\omega(x, s) f(s) ds$. *Estimate the operator norm of* J *on* $C([0, L])$ *and express* $< J(f), f >$ *as an integral.*

Exercise 14.1.16 *Define the* kernel *function* k_ω, $\omega L \neq n\pi$, *on* $[0, L] \times [0, L]$ *by*

$$k_\omega(x, s) = \frac{1}{\omega \sin(L\omega)} \begin{cases} \sin(\omega s)\,\sin(\omega(L - x)) & 0 \leq s \leq x \\ \sin(\omega x)\,\sin(\omega(L - s)) & x < s \leq L \end{cases}$$

and define the operator J *by* $J(f) = \int_0^L k_\omega(x, s) f(s) ds$. *Estimate the operator norm of* J *on* $\mathbb{L}_2([a, b])$ *and express* $< J(f), f >$ *as an integral.*

Exercise 14.1.17 *Define the* kernel *function* k_ω *on* $[0, L] \times [0, L]$ *by*

$$k_\omega(x, s) = \frac{1}{\omega \sin(L\omega)} \begin{cases} s(L - x) & 0 \leq s \leq x \\ x(L - s) & x < s \leq L \end{cases}$$

and define the operator J *by* $J(f) = \int_0^L k_\omega(x, s) f(s) ds$. *Estimate the operator norm of* J *on* $C([0, L])$ *and express* $< J(f), f >$ *as an integral.*

Exercise 14.1.18 *Define the* kernel *function* k_ω *on* $[0, L] \times [0, L]$ *by*

$$k_\omega(x, s) = \frac{1}{\omega \sin(L\omega)} \begin{cases} s(L - x) & 0 \leq s \leq x \\ x(L - s) & x < s \leq L \end{cases}$$

and define the operator J *by* $J(f) = \int_0^L k_\omega(x, s) f(s) ds$. *Estimate the operator norm of* J *on* $\mathbb{L}_2([a, b])$ *and express* $< J(f), f >$ *as an integral.*

14.2 Eigenvalues of Self-Adjoint Operators

Next we want to look at the eigenvalues of self-adjoint operators which is quite a journey. We can do this using an iterative construction process using the same procedure we used for the symmetric and Hermitian matrices. This also gives us useful estimates for later use.

Theorem 14.2.1 The First Eigenvalue

Either $\|\mathscr{J}\|_{op}$ *or* $-\|\mathscr{J}\|_{op}$ *is an eigenvalue for* $\mathscr{J}$. *Letting the eigenvalue be* x_1, *then* $|x_1| = \|\mathscr{J}\|_{op}$ *with an associated eigenfunction of norm* 1, $\boldsymbol{u_1}$.

Proof 14.2.1

Let $(\boldsymbol{u_n})$ *be an extremizing sequence for* $\sup_{\|\boldsymbol{u}\|_2=1} | < \mathscr{J}(\boldsymbol{u}), \boldsymbol{u} > |$. *Then, it is easy to see that the corresponding sequence* $(\mathscr{J}(\boldsymbol{u}) : \|\boldsymbol{u}\|_2 = 1)$ *is a sequence of continuous functions which is uniformly bounded in* $\mathbb{L}_2([a,b])$ *norm. By the Arzela - Ascoli theorem, there is a subsequence* $(\boldsymbol{u_n}^1)$ *so that* $\mathscr{J}(\boldsymbol{u_n}^1)$ *converges uniformly to a continuous function* $\boldsymbol{v}$ *on* $[a,b]$ *in the usual sup-norm on* $C([a,b])$*. Since* $(\boldsymbol{u_n}^1)$ *is bounded in the Hilbert Space* $\mathbb{L}_2([a,b])$*, we know it has a weakly convergent subsequence* $(\boldsymbol{u_n}^2)$ *with*

$$< \boldsymbol{u_n}^2, g > \quad \rightarrow \quad < \boldsymbol{u}, \boldsymbol{g} >$$

for all $\boldsymbol{g}$ *in* $\mathbb{L}_2([a,b])$*. We claim* $< \mathscr{J}(\boldsymbol{u_n}^2), \boldsymbol{u_n}^2 > \rightarrow < \mathscr{J}(\boldsymbol{u}), \boldsymbol{u} >$*. To see this note*

$$\mathscr{J}(\boldsymbol{u_n}^2) \quad = \quad \int_a^b k_t(s)\boldsymbol{u_n}^2(s)\, ds \; = < k_t, \boldsymbol{u_n}^2 >$$

where $k_t(s) = k(s,t)$ *for fixed* t *is a continuous function and so is in* $\mathbb{L}_2([a,b])$*. Then because* $\boldsymbol{u_n}^2$ *converges weakly to* u*, we have*

$$\mathscr{J}(\boldsymbol{u_n}^2) \quad = \quad < k_t, \boldsymbol{u_n}^2 > \rightarrow < k_t, u > = \; \mathscr{J}(\boldsymbol{u}).$$

Next, note

$$\begin{aligned} | < \mathscr{J}(\boldsymbol{u_n}^2), \boldsymbol{u_n}^2 > - < \mathscr{J}(\boldsymbol{u}), \boldsymbol{u} > | \quad &= \quad | < \mathscr{J}(\boldsymbol{u_n}^2) \; - \; \mathscr{J}(\boldsymbol{u}) \; + \; \mathscr{J}(\boldsymbol{u}), \boldsymbol{u_n}^2 > \\ & \qquad - < \mathscr{J}(\boldsymbol{u}), \boldsymbol{u} > | \\ &\leq \quad | < \mathscr{J}(\boldsymbol{u_n}^2) \; - \; \mathscr{J}(\boldsymbol{u}), \boldsymbol{u_n}^2 > | \\ & \qquad + | < \mathscr{J}(\boldsymbol{u}), \boldsymbol{u_n}^2 - \boldsymbol{u} > | \\ &\leq \quad \| \mathscr{J}(\boldsymbol{u_n}^2) \; - \; \mathscr{J}(\boldsymbol{u}) \| \; \| \boldsymbol{u_n}^2 \| \\ & \qquad + | < \mathscr{J}(\boldsymbol{u}), \boldsymbol{u_n}^2 - \boldsymbol{u} > |. \end{aligned}$$

Since $(\boldsymbol{u_n}^2)$ *is a subsequence of* $(\boldsymbol{u_n}^1)$ *and so* $\mathscr{J}(\boldsymbol{u_n}^2) \rightarrow \mathscr{J}(\boldsymbol{u})$ *in sup norm, we see the first term vanishes as* n *goes to* ∞*. Further, because* $\mathscr{J}(\boldsymbol{u})$ *is continuous and so is in* $\mathbb{L}_2([a,b])$*, the second term vanishes in the limit also because* $\boldsymbol{u_n}^2$ *converges weakly to* u*. Hence, the result is true and we have*

$$\lim_n \; | < \mathscr{J}(\boldsymbol{u_n}^2), \boldsymbol{u_n}^2 > | \quad = \quad | < \mathscr{J}(\boldsymbol{u}), \boldsymbol{u} > |.$$

Let $\alpha_n = < \mathscr{J}(\boldsymbol{u_n}^2), \boldsymbol{u_n}^2 >$ *and let* $\alpha = \|\mathscr{J}\|_{op}$*. Then* $|\alpha_n| \rightarrow \alpha$ *and so by passing to a subsequence if necessary, we can assume without loss of generality that* $\alpha_n \rightarrow \|\mathscr{J}\|_{op}$ *or* $\alpha_n \rightarrow -\|\mathscr{J}\|_{op}$*.*

Case I*: We assume* $\alpha_n \rightarrow -\|\mathscr{J}\|_{op} = -\alpha$*. This implies* $< \mathscr{J}(\boldsymbol{u}), \boldsymbol{u} > = -\alpha$*. Then,*

$$\begin{aligned} < \mathscr{J}(\boldsymbol{u}) - (-\alpha)\boldsymbol{u}, \mathscr{J}(\boldsymbol{u}) - (-\alpha)\boldsymbol{u} > &= < \mathscr{J}(\boldsymbol{u}) + \alpha u, \mathscr{J}(\boldsymbol{u}) + \alpha \boldsymbol{u} > \\ &= < \mathscr{J}(\boldsymbol{u}), \mathscr{J}(\boldsymbol{u}) > + 2\alpha < \mathscr{J}(\boldsymbol{u}), \boldsymbol{u} > + \alpha^2 < \boldsymbol{u}, \boldsymbol{u} > \\ &= \; \|\mathscr{J}(\boldsymbol{u})\|_2^2 + 2\alpha < \mathscr{J}(\boldsymbol{u}), \boldsymbol{u} > + \alpha^2 \end{aligned}$$

$$= < \mathscr{J}(\boldsymbol{u}), \mathscr{J}(\boldsymbol{u}) > -2\alpha^2 + \alpha^2$$
$$= \|\mathscr{J}(\boldsymbol{u})\|_2^2 - \alpha^2$$

since $\|\boldsymbol{u}\|_2 = 1$. Then, overestimating, we have

$$< \mathscr{J}(\boldsymbol{u}) - (-\alpha)\boldsymbol{u}, \mathscr{J}(\boldsymbol{u}) - (-\alpha)\boldsymbol{u} > \;\leq\; \|\mathscr{J}\|_{op}^2 \, \|\boldsymbol{u}\|_2^2 - \alpha^2.$$

But $\alpha = \|\mathscr{J}\|_{op}$, so we have

$$< \mathscr{J}(\boldsymbol{u}) - (-\alpha)\boldsymbol{u}, \mathscr{J}(\boldsymbol{u}) - (-\alpha)\boldsymbol{u} > \;\leq\; \alpha^2 - \alpha^2 \,=\, 0.$$

We conclude $\|\mathscr{J}(\boldsymbol{u}) - (-\alpha)u\|_2 = 0$ which tells us $\mathscr{J}(\boldsymbol{u}) = -\alpha\mathbf{u}$. Hence, $-\alpha$ is an eigenvalue of $\mathscr{J}$ and we can pick $\mathbf{u}$ as the associated eigenvector of norm 1.

Case II: *We assume $\alpha_n \to \|\mathscr{J}\|_{op} = \alpha$. This implies $< \mathscr{J}(\boldsymbol{u}), \boldsymbol{u} >= \alpha$. The argument is then quite similar. We conclude $\|\mathscr{J}(\boldsymbol{u}) - \alpha\mathbf{u}\|_2 = 0$ which tells us $\mathscr{J}(\boldsymbol{u}) = \alpha\mathbf{u}$. Hence, α is an eigenvalue of $\mathscr{J}$ and we can pick $\mathbf{u}$ as the associated eigenvector of norm 1.* ■

Compare the argument here to the ones we used for the symmetric and Hermitian matrices. In the matrix case, we knew we had a maximum value which gave us the eigenvector right away. If we had followed the proof we do here, we would have used the fact that there is a maximizing sequence which converges to the maximum value. In the matrix case, we then would have a bounded sequence of vectors and we could apply the Bolzano - Weierstrass Theorem to find a convergent subsequence. Thus, we had $(\boldsymbol{u_n}) \subset \Re^n$ which had a subsequence $(\boldsymbol{u_n^1})$ converging to $\boldsymbol{u}$ of norm 1 so that $A(\boldsymbol{u_n^1}) \to A(\boldsymbol{u})$. We would then show $\boldsymbol{u}$ is the eigenvector we seek.

Here, we do not have this compactness. We find the extremizing sequence and then use the fact that the integral operator is compact to extract a subsequence so that $\mathscr{J}(\boldsymbol{u_n^1}) \to \boldsymbol{v}$. However, we do not know the subsequence $(\boldsymbol{u_n^1})$ converges in norm as since the spaces here are infinite dimensional the unit ball is not compact. However, we do know the unit ball is weakly compact and we use that to extract the subsequence of $(\boldsymbol{u_n^1})$ we need. Hence, weak compactness plays an important role here.

Homework

Exercise 14.2.1 *Define the* kernel *function k_ω, $\omega L \neq n\pi$, on $[0, L] \times [0, L]$ by*

$$k_\omega(x, s) \;=\; \frac{1}{\omega \sin(L\omega)} \begin{cases} \cos(\omega s)\, \cos(\omega(L - x)) & 0 \leq s \leq x \\ \cos(\omega x)\, \cos(\omega(L - s)) & x < s \leq L \end{cases}$$

Write down explicitly the optimization problem for the first eigenvalue.

Exercise 14.2.2 *Define the* kernel *function k_ω, $\omega L \neq n\pi$, on $[0, L] \times [0, L]$ by*

$$k_\omega(x, s) = \frac{1}{\omega \sin(L\omega)} \begin{cases} \sin(\omega s)\, \sin(\omega(L - x)) & 0 \leq s \leq x \\ \sin(\omega x)\, \sin(\omega(L - s)) & x < s \leq L \end{cases}$$

and define the operator J by $J(f) = \int_0^L k_\omega(x, s) f(s) ds$. Write down explicitly the optimization problem for the first eigenvalue.

Exercise 14.2.3 *Define the* kernel *function k_ω on $[0, L] \times [0, L]$ by*

$$k_\omega(x, s) = \frac{1}{\omega \sin(L\omega)} \begin{cases} s(L - x) & 0 \leq s \leq x \\ x(L - s) & x < s \leq L \end{cases}$$

and define the operator J by $J(f) = \int_0^L k_\omega(x, s) f(s) ds$. Write down explicitly the optimization problem for the first eigenvalue.

We can now find the next eigenvalue using a similar argument.

Theorem 14.2.2 The Second Eigenvalue

There is a second eigenvalue x_2 which satisfies $|x_2| \leq |x_1|$. Let the new kernel k_2 be defined by

$$k_2(s, t) = k(s, t) - x_1\, u_1(s) u_1(t),$$

and the new operator $\mathscr{J}_2$ be defined by

$$\begin{aligned} (\mathscr{J}_2(\boldsymbol{x}))(t) &= \int_a^b k_2(s, t)\, x(s)\, ds \\ &= \int_a^b k(s, t)\, x(s)\, ds - x_1 \int_a^b u_1(s) u_1(t) x(s)\, ds \\ &= (\mathscr{J}(\boldsymbol{x}))(t) - x_1 \int_a^b u_1(s) u_1(t) x(s)\, ds \\ &= (\mathscr{J}(\boldsymbol{x}))(t) - x_1\, u_1(t) < \boldsymbol{u_1}, \boldsymbol{x} > . \end{aligned}$$

Then, $|x_2| = \|\mathscr{J}_2\|_{op}$ and the associated eigenfunction $\boldsymbol{u_2}$ is orthogonal to $\boldsymbol{u_1}$; that is $< \boldsymbol{u_1}, \boldsymbol{u_2} >= 0$.

Proof 14.2.2

The reasoning here is virtually identical to what we did for Theorem 14.2.1. First, note, if $\boldsymbol{x}$ is a multiple of $\boldsymbol{u_1}$, we have $\boldsymbol{x} = \mu \boldsymbol{u_1}$ for some μ giving

$$\begin{aligned} (\mathscr{J}_2(\boldsymbol{x}))(t) &= (\mathscr{J}(\mu \boldsymbol{u_1}))(t) - x_1 \mu u_1(t) < \boldsymbol{u_1}, \boldsymbol{u_1} > \\ &= \mu x_1 u_1(t) - \mu x_1 u_1(t) = 0. \end{aligned}$$

since $\boldsymbol{u_1}$ is an eigenfunction with eigenvalue x_1. Hence, it is easy to see that

$$\sup_{\|\boldsymbol{u}\|_2=1} |< \mathscr{J}_2(\boldsymbol{u}), \boldsymbol{u} >| = \sup_{\|\boldsymbol{u}\|_2=1, \boldsymbol{u}\in W_1} |< \mathscr{J}_2(\boldsymbol{u}), \boldsymbol{u} >|$$

where W_1 is the set of functions in $\mathbb{L}_2([a,b])$ that are orthogonal to eigenfunction $\boldsymbol{u_1}$. Since this is a smaller set of functions, the supremum we obtain will, of course, be possibly smaller. Hence, we know $\sup_{\|\boldsymbol{u}\|_2=1} |< \mathscr{J}(\boldsymbol{u}), \boldsymbol{u} >| \geq \sup_{\|\boldsymbol{u}\|_2=1,\ \boldsymbol{u}\in W_1} |< \mathscr{J}_2(\boldsymbol{u}), \boldsymbol{u} >|$. The rest of the argument, applied to the new operator $\mathscr{J}_2$ is identical. Thus, we find

$$|x_2| = \|\mathscr{J}_2\| = \sup_{\|\boldsymbol{u}\|_2=1, \boldsymbol{u}\in W_1} |< \mathscr{J}_2(\boldsymbol{u}), \boldsymbol{u} >|.$$

Further, $|x_2| \leq |x_1|$ and eigenfunction $\boldsymbol{u_2}$ is orthogonal to $\boldsymbol{u_1}$. ∎

Homework

Exercise 14.2.4 *Define the* kernel *function k_ω, $\omega L \neq n\pi$, on $[0,L]\times[0,L]$ by*

$$k_\omega(x,s) = \frac{1}{\omega\sin(L\omega)} \begin{cases} \cos(\omega s)\ \cos(\omega(L-x)) & 0 \leq s \leq x \\ \cos(\omega x)\ \cos(\omega(L-s)) & x < s \leq L \end{cases}$$

Write down explicitly the optimization problem for the second eigenvalue.

Exercise 14.2.5 *Define the* kernel *function k_ω, $\omega L \neq n\pi$, on $[0,L]\times[0,L]$ by*

$$k_\omega(x,s) = \frac{1}{\omega\sin(L\omega)} \begin{cases} \sin(\omega s)\ \sin(\omega(L-x)) & 0 \leq s \leq x \\ \sin(\omega x)\ \sin(\omega(L-s)) & x < s \leq L \end{cases}$$

and define the operator J by $J(f) = \int_0^L k_\omega(x,s)f(s)ds$. Write down explicitly the optimization problem for the second eigenvalue.

Exercise 14.2.6 *Define the* kernel *function k_ω on $[0,L]\times[0,L]$ by*

$$k_\omega(x,s) = \frac{1}{\omega\sin(L\omega)} \begin{cases} s(L-x) & 0 \leq s \leq x \\ x(L-s) & x < s \leq L \end{cases}$$

and define the operator J by $J(f) = \int_0^L k_\omega(x,s)f(s)ds$. Write down explicitly the optimization problem for the second eigenvalue.

We can now find the next eigenvalue using the same argument. So let's do one more to see the pattern!

Theorem 14.2.3 The Third Eigenvalue

There is a third eigenvalue x_3 which satisfies $|x_3| \leq |x_2| \leq |x_1|$. Let the new kernel k_3 be defined by

$$k_3(s,t) \quad = \quad k(s,t) - \sum_{i=1}^{2} x_i\, u_i(s) u_i(t),$$

and the new operator $\mathscr{J}_3$ be defined by

$$\begin{aligned} (\mathscr{J}_3(\boldsymbol{x}))(t) &= \int_0^L k_3(s,t)\, x(s)\, ds \\ &= \int_a^b k(s,t)\, x(s)\, ds - \sum_{i=1}^{2} \int_a^b x_i\, u_i(s) u_i(t) x(s)\, ds \\ &= (\mathscr{J}(\boldsymbol{x}))(t) - \sum_{i=1}^{2} x_i u_i(t) \int_a^b u_i(s) x(s)\, ds \\ &= (\mathscr{J}(\boldsymbol{x}))(t) - \sum_{i=1}^{2} x_i u_i(t) < \boldsymbol{u_i}, \boldsymbol{x} > . \end{aligned}$$

Then, $|x_3| = \|\mathscr{J}_3\|_{op}$ and the associated eigenfunction $\boldsymbol{u_3}$ is orthogonal to $\boldsymbol{u_2}$ and $\boldsymbol{u_1}$., i.e., $< \boldsymbol{u_1}, \boldsymbol{u_3} >= 0$ and $< \boldsymbol{u_2}, \boldsymbol{u_3} >= 0$.

Proof 14.2.3

First, note, if $\boldsymbol{x}$ is in the subspace generated by $\boldsymbol{u_1}$ and $\boldsymbol{u_2}$, then $\boldsymbol{x} = \mu_1\, \boldsymbol{u_1} + \mu_2\, \boldsymbol{u_2}$ for some μ_1 and μ_2 giving

$$\begin{aligned} (\mathscr{J}_3(\boldsymbol{x}))(t) &= (\mathscr{J}(\boldsymbol{x}))(t) - \sum_{i=1}^{2} x_i u_i(t) < u_i, x > \\ &= \mu_1 x_1 u_1(t) + \mu_2 x_2 u_2(t) - \sum_{i=1}^{2} x_i u_i(t) < \boldsymbol{u_i}, \mu_1 \boldsymbol{u_1} + \mu_2 \boldsymbol{u_2} > \\ &= \mu_1 x_1 u_1(t) + \mu_2 x_2 u_2(t) - \sum_{i=1}^{2} \mu_i x_i u_i(t) = 0. \end{aligned}$$

since $\boldsymbol{u_1}$ and $\boldsymbol{u_2}$ are eigenfunctions with eigenvalues x_1 and x_2. Hence, it is easy to see that

$$\sup_{\|\boldsymbol{u}\|_2=1} | < \mathscr{J}_3(\boldsymbol{u}), \boldsymbol{u} > | \quad = \quad \sup_{\|\boldsymbol{u}\|_2=1, \boldsymbol{u} \in W_2} | < \mathscr{J}_3(\boldsymbol{u}), \boldsymbol{u} > |$$

where W_2 is the set of functions in $\mathbb{L}_2([a,b])$ that are orthogonal to eigenfunctions $\boldsymbol{u_1}$ and $\boldsymbol{u_2}$. Since this is a smaller set of functions, the supremum we obtain will, of course, be possibly smaller. Hence, we know

$$\sup_{\|\boldsymbol{u}\|_2=1} | < \mathscr{J}(\boldsymbol{u}), \boldsymbol{u} > | \quad \geq \quad \sup_{\|\boldsymbol{u}\|_2=1, \boldsymbol{u} \in W_1} | < \mathscr{J}_2(\boldsymbol{u}), \boldsymbol{u} > |$$

$$\geq \sup_{\|\boldsymbol{u}\|_2=1, \boldsymbol{u}\in W_2} | < \mathscr{J}_3(\boldsymbol{u}), \boldsymbol{u} > |$$

The rest of the argument, applied to the new operator $\mathscr{J}_3$ is again identical. Thus, we find

$$|x_3| = \|\mathscr{J}_3\| = \sup_{\|\boldsymbol{u}\|_2=1, \boldsymbol{u}\in W_2} | < \mathscr{J}_3(\boldsymbol{u}), \boldsymbol{u} > |.$$

Further, $|x_3| \leq |x_2| \leq |x_1|$ and eigenfunction $\boldsymbol{u_3}$ is orthogonal to both $\boldsymbol{u_1}$ and $\boldsymbol{u_2}$. ■

Homework

Exercise 14.2.7 *Define the* kernel *function k_ω, $\omega L \neq n\pi$, on $[0, L] \times [0, L]$ by*

$$k_\omega(x, s) = \frac{1}{\omega \sin(L\omega)} \begin{cases} \cos(\omega s)\, \cos(\omega(L - x)) & 0 \leq s \leq x \\ \cos(\omega x)\, \cos(\omega(L - s)) & x < s \leq L \end{cases}$$

Write down explicitly the optimization problem for the third eigenvalue.

Exercise 14.2.8 *Define the* kernel *function k_ω, $\omega L \neq n\pi$, on $[0, L] \times [0, L]$ by*

$$k_\omega(x, s) = \frac{1}{\omega \sin(L\omega)} \begin{cases} \sin(\omega s)\, \sin(\omega(L - x)) & 0 \leq s \leq x \\ \sin(\omega x)\, \sin(\omega(L - s)) & x < s \leq L \end{cases}$$

and define the operator J by $J(f) = \int_0^L k_\omega(x, s) f(s) ds$. Write down explicitly the optimization problem for the third eigenvalue.

Exercise 14.2.9 *Define the* kernel *function k_ω on $[0, L] \times [0, L]$ by*

$$k_\omega(x, s) = \frac{1}{\omega \sin(L\omega)} \begin{cases} s(L - x) & 0 \leq s \leq x \\ x(L - s) & x < s \leq L \end{cases}$$

and define the operator J by $J(f) = \int_0^L k_\omega(x, s) f(s) ds$. Write down explicitly the optimization problem for the third eigenvalue.

We can clearly continue this process. Since the underlying space is infinite dimensional, we can terminate with only a finite number of eigenvalues but if so the last eigenvalue will have an infinite dimensional eigenspace.

Theorem 14.2.4 All Eigenvalues

There is a sequence of eigenvalues x_i which satisfies $|x_{n+1}| \leq |x_n|$ for all positive integers n. Let the new kernel k_{n+1} be defined by

$$k_{n+1}(s,t) = k(s,t) - \sum_{i=1}^{n} x_i\, u_i(s)u_i(t),$$

and the new operator $\mathscr{I}_{n+1}$ be defined by

$$\begin{aligned}(\mathscr{I}_{n+1}(\boldsymbol{x}))(t) &= \int_a^b k_{n+1}(s,t)\, x(s)\, ds \\ &= \int_a^b k(s,t)\, x(s)\, ds - \sum_{i=1}^{n} \int_a^b x_i\, u_i(s)u_i(t)x(s)\, ds \\ &= (\mathscr{I}(\boldsymbol{x}))(t) - \sum_{i=1}^{n} x_i u_i(t) \int_a^b u_i(s)x(s)\, ds \\ &= (\mathscr{I}(\boldsymbol{x}))(t) - \sum_{i=1}^{n} x_i u_i(t) < \boldsymbol{u_i}, \boldsymbol{x} > .\end{aligned}$$

Then, $|x_{n+1}| = \|\mathscr{I}_{n+1}\|_{op}$ and the associated eigenfunction $\boldsymbol{u_{n+1}}$ is orthogonal to $\boldsymbol{u_i}$ for all $i \leq n$. Note it is possible for there to be a last distinct eigenvalue, say k_N so that for all $n \geq N$, $x_n = x_N$.

Proof 14.2.4

We have already sketched this induction out. ■

Let's summarize what we have learned. If the integral operator has a symmetric or Hermitian kernel, we have found a sequence of descending absolute eigenvalues whose eigenvectors are mutually orthogonal. We figured this out already for the Stürm - Liouville problems but if you examine the proof we used only the self-adjoint nature of the differential operator $\boldsymbol{L}$ to show the eigenvalues were real and the associated eigenvectors were mutually disjoint for eigenvalues that are distinct. In the Stürm - Liouville model, we showed each eigenspace was one dimensional using specific properties of these models such as Abel's identity and so forth.

Of course, if all we had was a symmetric or Hermitian kernel, we could have a finite number of distinct eigenvalues. Also, there is a nice connection between the eigenvalues of the Stürm - Liouville differential operator and its inverse.

Note the construction process works on linear bounded self-adjoint operators. Hence, it works on the integral operator obtained by inverting the Stürm - Liouville differential operator which is linear but not bounded. The inversion process helps a lot as it gives us the boundedness we need.

14.3 Back to the Stürm - Liouville Operator

In the context of the Stürm - Liouville Operator, we know much more than we do for a general integral operator with symmetric or Hermitian kernel. The eigenvalue construction theorem needs to be modified a bit as we use the inner product with weight $\boldsymbol{r}$, $<\boldsymbol{f},\boldsymbol{g}>=\int_a^b \boldsymbol{r}\boldsymbol{f}\boldsymbol{g}$. First, there is the characterization of the self-adjoint norm. Let's recall what we know. The differential operator is $\boldsymbol{L}: C^2([a,b]) \rightarrow C([a,b])$ defined by

$$\boldsymbol{L}(\boldsymbol{u}) = \frac{1}{\boldsymbol{r}}\left((\boldsymbol{p}\boldsymbol{u}')' - \boldsymbol{q}\boldsymbol{u} \right)$$

We solve the problem

$$\boldsymbol{L}(\boldsymbol{u}) - \lambda_0 \boldsymbol{u} = \boldsymbol{f}$$

for a continuous f on $[a,b]$ for any λ which is not an eigenvalue of $\boldsymbol{L}$. For this ODE problem, it is straightforward to use the boundary conditions and the two standard linearly independent solutions to the ODE to find an infinite number of eigenvalues and we have shown you how to do this in a few examples. We choose the two linearly independent solutions $\boldsymbol{x_1}$ and $\boldsymbol{x_2}$ where $\boldsymbol{x_1}$ solves $\boldsymbol{L}(\boldsymbol{x_1}) = \lambda \boldsymbol{x_1}$ with $\alpha_a x_1(a) + \beta_a x_1'(a) = 0$ and $\alpha_b x_1(b) + \beta_b x_1'(b) = 1$ and $\boldsymbol{x_2}$ solves $\boldsymbol{L}(\boldsymbol{x_2}) = \lambda \boldsymbol{x_2}$ with $\alpha_a x_2(a) + \beta_a x_2'(a) = 1$ and $\alpha_b x_2(b) + \beta_b x_2'(b) = 0$ for (α_a, β_a) and (α_b, β_b) not $(0,0)$. Then $\boldsymbol{x_1}$ and $\boldsymbol{x_2}$ are linearly independent. We then solve the problem using Variation of Parameters obtaining the **kernel** function $k_\lambda(s,t)$ defined by

$$k_\lambda(s,t) = \begin{cases} \frac{x_1(s)x_2(t)}{p(s)W(s)}, & a \leq s \leq t \\ \frac{x_1(t)x_2(s)}{p(s)W(s)}, & t \leq s \leq b \end{cases}$$

Then $k_\lambda(s,t)$ is uniformly continuous and so bounded on the compact set $[a,b] \times [a,b]$ We define the linear operator $\boldsymbol{J_\lambda}$ on $C([a,b])$ by $(\mathscr{J}_\lambda(\boldsymbol{f}))(t) = \int_a^b k_\lambda(s,t) r(s) f(s) ds$. We thus have inverted the differential operator

$$(\boldsymbol{L} - \lambda \boldsymbol{I})(\boldsymbol{u}) = \boldsymbol{f} \Longrightarrow \boldsymbol{u} = \mathscr{J}_\lambda(\boldsymbol{f})$$

Hence, $(\boldsymbol{L} - \lambda \boldsymbol{I})^{-1}(\boldsymbol{f}) = \mathscr{J}_\lambda(\boldsymbol{f})$.
We then prove, for this new inner product:

Theorem 14.3.1 Characterization of the Stürm - Liouville operator norm

Let $\mathscr{J}_\lambda$ be defined as above. Then $\mathscr{J}_\lambda$ is a linear operator from $\mathbb{L}_2([a,b])$ to $C([a,b]) \subset \mathbb{L}_2([a,b])$ and

$$\|\mathscr{J}_\lambda\|_{op} = \sup_{\|u\|_2 = 1} |<\mathscr{J}_\lambda(u), u>|$$

where recall the inner product here is weighted by $\boldsymbol{r}$.

Proof 14.3.1

The argument is exactly the same as the one for the symmetric and Hermitian kernel. ■

Next, we construct the eigenvalues. The only difference in the argument is we have to use the weighting function $\boldsymbol{r}$ appropriately.

Theorem 14.3.2 Eigenvalue Construction for the Stürm - Liouville Operator

There is a sequence of eigenvalues x_i which satisfies $|x_{n+1}| \leq |x_n|$ for all positive integers n. Let the new kernel k_{n+1} be defined by

$$k_{n+1}(s,t) = k(s,t) - \sum_{i=1}^{n} x_i\, u_i(s)u_i(t),$$

and the new operator $\mathscr{I}_{n+1,\lambda}$ be defined by

$$\begin{aligned}
(\mathscr{I}_{n+1,\lambda}(\boldsymbol{x}))(t) &= \int_a^b k_{n+1}(s,t)\, r(s)x(s)\, ds \\
&= \int_a^b k(s,t)\, r(s)x(s)\, ds - \sum_{i=1}^{n} \int_a^b x_i\, r(s)u_i(s)u_i(t)x(s)\, ds \\
&= (\mathscr{I}_\lambda(\boldsymbol{x}))(t) - \sum_{i=1}^{n} x_i u_i(t) \int_a^b r(s)u_i(s)x(s)\, ds \\
&= (\mathscr{I}_\lambda(\boldsymbol{x}))(t) - \sum_{i=1}^{n} x_i u_i(t) < \boldsymbol{u_i}, \boldsymbol{x} >
\end{aligned}$$

Then, $|x_{n+1}| = \|\mathscr{I}_{n+1,\lambda}\|_{op}$ and the associated eigenfunction $\boldsymbol{u_{n+1}}$ is orthogonal to $\boldsymbol{u_i}$ for all $i \leq n$.

Proof 14.3.2

This proof is essentially the same as for the symmetric or Hermitian self-adjoint operator even though we use the weighted norm. ■

We know much more here about the eigenvalues than what is said in the theorem above. If λ is not an eigenvalue of $\boldsymbol{L}$, then the eigenvalues of $\boldsymbol{L} - \lambda \boldsymbol{I}$ are $\lambda_n - \lambda$ with $|\lambda_n| \rightarrow \infty$ and the sequence (λ_n) does not have any cluster points. The eigenvalues for the inverse $(\boldsymbol{L} - \lambda \boldsymbol{I})^{-1} = \mathscr{I}_\lambda$ are the reciprocals, $\frac{1}{\lambda_n - \lambda}$ and so they converge to 0. Both $\boldsymbol{L} - \lambda \boldsymbol{I}$ and $\mathscr{I}_\lambda$ have the same sequence of eigenfunctions $(\boldsymbol{u_n})$. Further, all the eigenspaces are one dimensional.

Let's look at a very simple Stürm - Liouville differential operator to make this concrete. The model we are solving

$$\begin{aligned}
u'' &= f,\ 0 \leq x \leq L, \\
\ell_1(u(0), u'(0)) &= \alpha_1\, u(0) + \alpha_2\, u'(0) = 0 \\
\ell_2(u(L), u'(L)) &= \beta_1\, u(L) + \beta_2\, u'(L) = 0
\end{aligned}$$

has infinitely many eigenvalues β_n with $|\beta_n| \rightarrow \infty$ and all of the associated eigenfunctions $\boldsymbol{u_n}$ are mutually orthogonal. If Θ is not an eigenvalue, and if β_n is an eigenvalue of the differential equation model, then

$$\boldsymbol{u_n} = (\Theta + \beta_n) \mathscr{J}_\Theta(\boldsymbol{u_n})$$

so that $\frac{1}{\Theta+\beta_n}$ is an eigenvalue of $\mathscr{J}_\Theta$ with associated eigenfunction $\boldsymbol{u_n}$. Further, if 0 is not an eigenvalue, as discussed, we have

$$\boldsymbol{u_n} = (\beta_n) \mathscr{J}_0(\boldsymbol{u_n})$$

so that $\frac{1}{\beta_n}$ are the eigenvalues of $\mathscr{J}_0$.

For this model, we know the eigenfunctions associated with different eigenvalues must be orthogonal and we know the eigenspaces are one dimensional. The simple Stürm - Liouville operator $u'' = 0$ plus boundary conditions was easy to solve and gave us several orthonormal sequences.

14.3.1 Derivative Boundary Conditions

The nonhomogeneous model to solve is

$$\begin{aligned} u'' + u &= f \\ u'(0) &= 0, \\ u'(L) &= 0. \end{aligned}$$

We find the eigenvalues for this model are $\beta_0 = 0$ with eigenfunction $u_0(x) = 1$ and $\beta_n = -\omega_n^2 = -\frac{n^2\pi^2}{L^2}$ ($\omega_n = \frac{n\pi}{L}$) for any nonzero integer n with associated eigenfunctions $\boldsymbol{u_n}(x) = \cos(\omega_n\, x)$. Hence, there are an infinite number of eigenvalues satisfying $|\beta_n| \rightarrow \infty$. Further, these eigenfunctions are mutually orthogonal. The model

$$\begin{aligned} u'' + u &= f \\ u'(0) &= 0, \quad u'(L) = 0. \end{aligned}$$

is therefore invertible. We can construct the kernel function as follows. Let

$$\boldsymbol{x_1}(x) = \cos(x), \quad x_2(x) = \cos(L - x).$$

and $W(\boldsymbol{x_1}, x_2)$ be the determinant

$$W(\boldsymbol{x_1}, x_2) = \begin{vmatrix} \boldsymbol{x_1} & x_2 \\ \boldsymbol{x_1}' & x_2' \end{vmatrix} = \omega \sin(L).$$

The kernel function is then

$$k(x,s) = \frac{1}{W(\boldsymbol{x_1},x_2)} \begin{cases} \boldsymbol{x_1}(s)\,x_2(x) & 0 \le s \le x \\ \boldsymbol{x_1}(x)\,x_2(s) & x < s \le L \end{cases}$$

and any solution u satisfies

$$u = \mathscr{I}(\boldsymbol{f}) = \int_0^L k(x,s)\,f(s)\,ds$$

with the kernel k a symmetric function. We can then find the eigenvalues and eigenfunctions for the resulting symmetric integral operator which inverts the differential operator. The integral operators $\mathscr{I}_n$ we defined in that process can then be used to find the eigenvalues which we already know.

Homework

Our general kernel function is

$$k(s,t) = \begin{cases} \frac{x_1(s)x_2(t)}{p(s)W(s)}, & a \le s \le t \\ \frac{x_1(t)x_2(s)}{p(s)W(s)}, & t \le s \le b \end{cases}$$

where $\boldsymbol{x_1}$ and $\boldsymbol{x_2}$ solve

$$\boldsymbol{L(u)} = \boldsymbol{0} \Longrightarrow (\boldsymbol{pu'})' - (\boldsymbol{\lambda r} + \boldsymbol{q})\boldsymbol{u} = \boldsymbol{0}$$

and

$$\begin{aligned} (p(t)u_1'(t))' - (\lambda r(t) + q(t))u_1(t) &= 0,\ a \le t \le b \\ \alpha_a u_1(a) + \beta_a u_1'(a) &= 0, \quad \alpha_b u_1(b) + \beta_b u_1'(b) = 1 \\ (p(t)u_2'(t))' - (\lambda r(t) + q(t))u_2(t) &= 0,\ a \le t \le b \\ \alpha_a u_2(a) + \beta_a u_2'(a) &= 1, \quad \alpha_b u_2(b) + \beta_b u_2'(b) = 0 \end{aligned}$$

Exercise 14.3.1 *Prove*

$$\frac{\partial k}{\partial t}(t^+,t) - \frac{\partial k}{\partial t}(t^-,t) = \frac{x_1'(t)x_2(t) - x_1(t)x_2'(t)}{pW(\boldsymbol{x_1},\boldsymbol{x_2})}$$

Exercise 14.3.2 *Prove*

$$\frac{\partial k}{\partial t}(t^+,t) - \frac{\partial k}{\partial t}(t^-,t) = -\frac{1}{p(t)}$$

Exercise 14.3.3 *Define* $\boldsymbol{\xi}$ *on* $[a,b]$ *by* $\xi(s_0) = k(s_0,t)$ *for* $s \ne s_0$ *in* $[a,b]$*. Find* $\boldsymbol{\xi}'$ *and* $\boldsymbol{\xi}''$*.*

Exercise 14.3.4 *Define* $\boldsymbol{\xi}$ *on* $[a,b]$ *by* $\xi(t) = k(s_0,t)$ *for* $t \ne s_0$ *in* $[a,b]$*. Find* $\mathbf{p}'\boldsymbol{\xi}' + \mathbf{p}\boldsymbol{\xi}'' - (\lambda \boldsymbol{q} + \boldsymbol{r})\boldsymbol{\xi}$ *and since* $\boldsymbol{x_1}$ *and* $\boldsymbol{x_2}$ *satisfy the homogeneous Stürum Liouville problem with their boundary* *conditions, we find* $\mathbf{p}'\boldsymbol{\xi}' + \mathbf{p}\boldsymbol{\xi}'' - (\lambda \boldsymbol{q} + \boldsymbol{r})\boldsymbol{\xi} = \mathbf{0}$ *for all* $t \neq s_0$, $s_0 \in [a, b]$ *and* $t \in [a, b]$.

What is the data f is not continuous?

Let's look at the solution $\boldsymbol{u}$ a bit more closely. We have done this sort of calculation before, but it is good to look at details like this multiple times. Let's assume Θ is not an eigenvalue of our models and $\boldsymbol{u} = \mathscr{J}_\Theta(\boldsymbol{z})$. Again, for convenience of notation, let's just use $\mathscr{J}$ and k in the arguments that follows. From the definition of the kernel k, we have for any $\boldsymbol{z}$ in $\mathbb{L}_2([0, L])$,

$$u(x) = \frac{1}{W(\boldsymbol{x_1}, x_2)} \int_0^x \boldsymbol{x_1}(s)\, x_2(x)\, z(s)\, ds + \frac{1}{W(\boldsymbol{x_1}, x_2)} \int_x^L \boldsymbol{x_1}(x)\, x_2(s)\, z(s)\, ds$$

If the data $\boldsymbol{z}$ is continuous, we can repeatedly use Leibnitz's Theorem to find the derivatives. We have

$$\begin{aligned} u'(x) &= \left(\frac{1}{W(\boldsymbol{x_1}, x_2)} \int_0^x \boldsymbol{x_1}(s)\, x_2(x)\, z(s)\, ds\right)' \\ &\quad + \left(\frac{1}{W(\boldsymbol{x_1}, x_2)} \int_x^L \boldsymbol{x_1}(x)\, x_2(s)\, z(s)\, ds\right)' \\ &= \frac{1}{W(\boldsymbol{x_1}, x_2)} \int_0^x \boldsymbol{x_1}(s)\, x_2'(x)\, z(s)\, ds + \frac{1}{W(\boldsymbol{x_1}, x_2)} \boldsymbol{x_1}(x)\, x_2(x)\, z(x) \\ &\quad + \frac{1}{W(\boldsymbol{x_1}, x_2)} \int_x^L \boldsymbol{x_1}'(x)\, x_2(s)\, z(s)\, ds - \frac{1}{W(\boldsymbol{x_1}, x_2)} \boldsymbol{x_1}(x)\, x_2(x)\, z(x) \\ &= \frac{1}{W(\boldsymbol{x_1}, x_2)} \int_0^x \boldsymbol{x_1}(s)\, x_2'(x)\, z(s)\, ds + \frac{1}{W(\boldsymbol{x_1}, x_2)} \int_x^L \boldsymbol{x_1}'(x)\, x_2(s)\, z(s)\, ds \end{aligned}$$

Next, we find another derivative again using Leibnitz's Theorem:

$$\begin{aligned} u''(x) &= \left(\frac{1}{W(\boldsymbol{x_1}, \boldsymbol{x_2})} \int_0^x \boldsymbol{x_1}(s)\, \boldsymbol{x_2}'(x)\, z(s)\, ds\right)' \\ &\quad + \left(\frac{1}{W(\boldsymbol{x_1}, x_2)} \int_x^L \boldsymbol{x_1}'(x)\, \boldsymbol{x_2}(s)\, z(s)\, ds\right)' \\ &= \frac{1}{W(\boldsymbol{x_1}, \boldsymbol{x_2})} \int_0^x \boldsymbol{x_1}(s)\, \boldsymbol{x_2}''(x)\, z(s)\, ds + \frac{1}{W(\boldsymbol{x_1}, x_2)} \boldsymbol{x_1}(x)\, x_2'(x)\, z(x) \\ &\quad \frac{1}{W(\boldsymbol{x_1}, \boldsymbol{x_2})} \int_x^L \boldsymbol{x_1}''(x)\, \boldsymbol{x_2}(s)\, z(s)\, ds - \frac{1}{W(\boldsymbol{x_1}, x_2)} \boldsymbol{x_1}'(x)\, x_2(x)\, z(x). \end{aligned}$$

Now, recall that $\boldsymbol{x_1}$ and $\boldsymbol{x_2}$ satisfy $\boldsymbol{x_1}'' = -\Theta\, \boldsymbol{x_1}$ and $\boldsymbol{x_2}'' = -\Theta\, \boldsymbol{x_2}$ with appropriate boundary conditions. Hence, we have

$$\begin{aligned} u''(x) &= -\Theta \frac{1}{W(\boldsymbol{x_1}, \boldsymbol{x_2})} \left(\int_0^x \boldsymbol{x_1}(s)\, \boldsymbol{x_2}(x)\, z(s)\, ds + \int_x^L \boldsymbol{x_1}(x)\, \boldsymbol{x_2}(s)\, z(s)\, ds\right) \\ &\quad + \frac{1}{W(\boldsymbol{x_1}, \boldsymbol{x_2})} \Big(\boldsymbol{x_1}(x)\, \boldsymbol{x_2}'(x) - \boldsymbol{x_1}'(x)\, \boldsymbol{x_2}(x)\Big) z(x). \end{aligned}$$

Using the definition of $\mathscr{J}$ and $W(\boldsymbol{x_1}, \boldsymbol{x_2})$, this simplifies to

$$u''(x) = -\Theta\Big(\mathscr{J}(z)\Big)(x) + z(x).$$

Finally, $\mathscr{J}(z) = u$ and so we have

$$u''(x) = -\Theta u(x) + z(x).$$

or $u'' + \Theta u = z$. So if the data z is continuous, the inversion process via $\mathscr{J}$ leads us to a classically understood differential equation.

What is the data $\boldsymbol{z}$ lacks continuity? In this case, we know $\boldsymbol{z}$ is in $\mathbb{L}_2([0, L])$ and so there is a sequence $(\boldsymbol{g_n})$ of functions in $C([0, L])$ with $\boldsymbol{g_n} \to \boldsymbol{z}$ in $\mathbb{L}_2([0, L])$ norm. This is so because $C([0, L])$ is dense in $\mathbb{L}_2([0, L])$. Let $\boldsymbol{u_n}$ be defined by $\boldsymbol{u_n} = \mathscr{J}(\boldsymbol{g_n})$. Thus, $\boldsymbol{u_n}'' + \Theta \boldsymbol{u_n} = \boldsymbol{g_n}$. Since $\mathscr{J}$ is continuous, we know $\boldsymbol{u_n} = \mathscr{J}(\boldsymbol{g_n}) \to \mathscr{J}(\boldsymbol{z}) = \boldsymbol{u}$. Hence, the function defined by

$$\boldsymbol{u_n}'' = \boldsymbol{v_n} = -\Theta \boldsymbol{u_n} + \boldsymbol{g_n} = -\Theta \mathscr{J}(\boldsymbol{g_n}) + \boldsymbol{g_n}$$

converges in the $\mathbb{L}_2([0, L])$ norm to $-\Theta \boldsymbol{u} + \boldsymbol{z}$. This defines the function v in $\mathbb{L}_2([0, L])$ by $\boldsymbol{v} = -\Theta \boldsymbol{u} + \boldsymbol{z}$. Hence, the limiting differential equation, in some sense, is $\boldsymbol{v} + \Theta \boldsymbol{u} = \boldsymbol{z}$ and $\boldsymbol{v}$ is some sort of generalized second derivative.

What does this mean? Well, we can no longer say the inversion process gives us a function which is in $C^2([0, L])$. All we can say is that the result is square integrable and is in $\mathbb{L}_2([a, b])$. Hence, we have generalized the idea of a solution of a differential equation model to the case where the data z is very non classical. The full consequences of this are profound, but we will not dwell on them further here.

Homework

Exercise 14.3.5 *For the problem*

$$\begin{aligned} u'' + 9u &= f \\ u'(0) &= 0, \\ u'(16) &= 0. \end{aligned}$$

find the sequence of eigenvalues and eigenfunctions.

Exercise 14.3.6 *For the problem*

$$\begin{aligned} u'' + 8u &= f \\ u'(0) &= 0, \\ u'(31) &= 0. \end{aligned}$$

find the sequence of eigenvalues and eigenfunctions.

14.3.1.1 Proving Completeness

The eigenvalues of the model

$$\begin{aligned} u'' + u &= f \\ u'(0) &= 0, \quad u'(L) = 0. \end{aligned}$$

are $1 - \beta_n^2$ for integers $n \geq 0$. We know we can write

$$x = \sum_{n=0}^{\infty} < x, \hat{u}_n > \hat{u}_n + z.$$

Assume z is not zero on $[0, L]$ and $< z, \boldsymbol{u_n} >= 0$ for all n. Consider $\mathscr{J}(z)$. Using the construction $\mathscr{J}_n$ from the construction process for the eigenvalues of a self-adjoint operator, we can write

$$\begin{aligned} (\mathscr{J}(z))(t) &= (\mathscr{J}_n(z))(t) + \sum_{i=0}^{n} \frac{1}{1 - \beta_i^2} \hat{u}_i(t) < \hat{u}_i, z > \\ &= (\mathscr{J}_n(z))(t) \end{aligned}$$

because $< z, \hat{u}_n >= 0$ by assumption. Hence, for all n,

$$\|\mathscr{J}(z)\| = \|\mathscr{J}_n(z)\| \leq \|\mathscr{J}_n\| \|z\|.$$

However, we know $\|\mathscr{J}_n(z)\| = \frac{1}{|1-\beta_n^2|} \to 0$. Hence, $\|\mathscr{J}(z)\| = 0$ implying $\mathscr{J}(z) = 0$.

Now for any $\boldsymbol{z}$, $\mathscr{J}(\boldsymbol{z})$ is a continuous function. If we assume $\boldsymbol{u} = \mathscr{J}(\boldsymbol{z}) = \boldsymbol{0}$, we see $\boldsymbol{u}$ is actually twice differentiable and satisfies the boundary conditions. Hence, if $\boldsymbol{L} + \Theta$ denotes the differential operator

$$(\boldsymbol{L} + \Theta)(\mathscr{J}(\boldsymbol{z})) = \boldsymbol{z}$$

But $(\boldsymbol{L} + \Theta)(\boldsymbol{u}) = \boldsymbol{0}$ and so we must have $\boldsymbol{z} = \boldsymbol{0}$.
Hence the sequence of eigenfunctions $(\cos(\omega_n(L - x)))$ for non-negative integers n is complete once normalized.

Homework

Exercise 14.3.7 *For the problem*

$$\begin{aligned} u'' + 9u &= f \\ u'(0) &= 0, \end{aligned}$$

$$u'(16) = 0.$$

find the sequence of eigenfunctions and prove it is complete.

Exercise 14.3.8 *For the problem*

$$\begin{aligned} u'' + 8u &= f \\ u'(0) &= 0, \\ u'(31) &= 0. \end{aligned}$$

find the sequence of eigenfunctions and prove it is complete.

14.3.2 State Boundary Conditions

The nonhomogeneous model to solve is

$$\begin{aligned} u'' + u &= f \\ u(0) &= 0, \quad u(L) = 0. \end{aligned}$$

The eigenvalues for this model are $\beta_n = -\omega_n^2 = -\frac{n^2\pi^2}{L^2}$, $\omega_n = \frac{n\pi}{L}$, for any nonzero integer n with associated eigenfunctions $\boldsymbol{u_n}(x) = \sin(\omega_n\, x)$. Hence, there are an infinite number of eigenvalues satisfying $|\beta_n| \to \infty$. Further, these eigenfunctions are mutually orthogonal. The model

$$\begin{aligned} u'' + u &= f \\ u(0) &= 0, \quad u(L) = 0. \end{aligned}$$

is therefore invertible. Let

$$\begin{aligned} \boldsymbol{x_1}(x) &= \sin(x) \\ x_2(x) &= \sin(L - x). \end{aligned}$$

and $W(\boldsymbol{x_1}, x_2)$ be the determinant

$$W(\boldsymbol{x_1}, x_2) = \begin{vmatrix} \boldsymbol{x_1} & x_2 \\ \boldsymbol{x_1}' & x_2' \end{vmatrix} = -\omega \sin(L).$$

The kernel function is then

$$k(x,s) = \frac{1}{W(\boldsymbol{x_1}, x_2)} \begin{cases} \boldsymbol{x_1}(s)\, x_2(x) & 0 \le s \le x \\ \boldsymbol{x_1}(x)\, x_2(s) & x < s \le L \end{cases}$$

and any solution u satisfies

$$u = \mathscr{J}(\boldsymbol{u}) = \int_0^L k(x,s)\, u(s)\, ds$$

with the kernel k a symmetric function. The mutually orthogonal functions $\boldsymbol{u_n}$ can be normalized to give the sequence $\hat{u}_n$ which all have norm 1. Let $\bar{P}$ be defined as the $\mathscr{L}^2$ norm closure of P where P is the union of P_n which are defined as usual. We have, in this case, P_n is the set of all linear combinations of the functions $\hat{u}_i$ for $0 \leq i \leq n$ where each $\boldsymbol{u_n}(x) = \sin(\omega_n(L - x))$ this time. Then $\bar{P}$ is a closed subspace and there is another closed subspace Q so that

$$\mathbb{L}_2([0, L]) = \bar{P} \oplus Q.$$

Hence, any x can be written uniquely as $x = y + z$ for an $y \in \bar{P}$ and a $z \in Q$. Now, as before, to any x there is an associated $y \in \bar{P}$ with

$$y = \sum_{n=0}^{\infty} < x, \hat{u}_n > \hat{u}_n$$

and so completeness of $(\hat{u}_n)$ would tell us any $x = y$, i.e. x has a unique expansion

$$x = \sum_{n=0}^{\infty} < x, \hat{u}_n > \hat{u}_n$$

with convergence of the infinite series interpreted in the sense of $\mathscr{L}^2$ norm convergence, which is often called least squares convergence or convergence in the mean.

Homework

Exercise 14.3.9 *For the problem*

$$\begin{aligned} u'' + 5u &= f \\ u(0) &= 0, \\ u(6) &= 0. \end{aligned}$$

find the sequence of eigenvalues and eigenfunctions.

Exercise 14.3.10 *For the problem*

$$\begin{aligned} u'' + 8u &= f \\ u(0) &= 0, \\ u(31) &= 0. \end{aligned}$$

find the sequence of eigenvalues and eigenfunctions.

14.3.2.1 Proving Completeness

It is easy to see that the eigenvalues of the model

$$\begin{aligned} u'' + u &= f \\ u(0) &= 0, \\ u(L) &= 0. \end{aligned}$$

are $1 - \beta_n^2$ for integers $n \geq 1$. We know we can write

$$x = \sum_{n=0}^{\infty} < x, \hat{u}_n > \hat{u}_n + z.$$

Again, assume z is not zero on $[0, L]$ and $< z, \boldsymbol{u_n} >= 0$ for all n. Consider $\mathscr{J}(z)$. Using our construction $\mathscr{J}_n$, we can write

$$\begin{aligned} (\mathscr{J}(z))(t) &= (\mathscr{J}_n(z))(t) + \sum_{i=1}^{n} \frac{1}{1 - \beta_i^2|} \hat{u}_i(t) < \hat{u}_i, z > \\ &= (\mathscr{J}_n(z))(t) \end{aligned}$$

because $< z, \hat{u}_n >= 0$ by assumption. Hence, for all n,

$$\|\mathscr{J}(z)\| = \|\mathscr{J}_n(z)\| \leq \|\mathscr{J}_n\|\|z\|.$$

However, we know $\|\mathscr{J}_n(z)\| = \frac{1}{|1-\beta_n^2|} \rightarrow 0$. Hence, $\|\mathscr{J}(z)\| = 0$ implying $\mathscr{J}(z) = 0$.

Now for any $\boldsymbol{z}$, $\mathscr{J}(\boldsymbol{z})$ is a continuous function. If we assume $\boldsymbol{u} = \mathscr{J}(\boldsymbol{z}) = \boldsymbol{0}$, we see $\boldsymbol{u}$ is twice differentiable and satisfies the boundary conditions. Hence, if $\boldsymbol{L}+\Theta$ denotes the differential operator

$$(\boldsymbol{L} + \Theta)(\mathscr{J}(\boldsymbol{z})) = \boldsymbol{z}$$

But $(\boldsymbol{L} + \Theta)(\boldsymbol{u}) = \boldsymbol{0}$ and so we must have $\boldsymbol{z} = \boldsymbol{0}$.

Hence, the sequence $(\sin(\omega_n \, x))$ for positive integers n is complete once normalized.

Comment 14.3.1 *Note how even if an integral with symmetric or Hermitian kernel has infinitely many eigenvalues, proving the completeness of the sequence of eigenfunctions is not so easy. We do not have the extra piece which is that the solutions satisfy a differential equation. It is that fact that helps us prove completeness.*

Homework

Exercise 14.3.11 *For the problem*

$$\begin{aligned} u'' + 4u &= f \\ u(0) &= 0, \\ u(3) &= 0. \end{aligned}$$

find the sequence of eigenfunctions and prove it is complete.

Exercise 14.3.12 *For the problem*

$$\begin{aligned} u'' + 7u &= f \\ u(0) &= 0, \\ u(10) &= 0. \end{aligned}$$

find the sequence of eigenfunctions and prove it is complete.

Exercise 14.3.13 *For the problem*

$$\begin{aligned} u'' + 4u &= f \\ u'(0) &= 0, \\ u(3) &= 0. \end{aligned}$$

find the sequence of eigenfunctions and prove it is complete.

Exercise 14.3.14 *For the problem*

$$\begin{aligned} u'' + 7u &= f \\ u(0) &= 0, \\ u'(10) &= 0. \end{aligned}$$

find the sequence of eigenfunctions and prove it is complete.

14.3.3 Completeness for the Stürm - Liouville Eigenfunction Sequences

The inverse operator here has a symmetric kernel which is continuous on $[a, b] \times [a, b]$ and the operator is compact. We also know each eigenvalues λ to the Stürm - Liouville ODE mode has a one dimensional eigenspaces. We will assume none of the eigenvalues are zero; otherwise we simply shift our arguments by adding a $-\theta$ to the operator as we have done before. We know that $\frac{1}{\lambda}$ is an eigenvalue of the corresponding inverse operator $\mathscr{J}$. From our construction process for the eigenvalues $\mathscr{J}$, since the domain is infinite dimensional there must be an infinite number of eigenvalues which we can therefore label as λ_n. Thus, we know the eigenvalues cannot have a cluster point. Hence we know $(|\lambda_n)|)$ is not bounded and so $\frac{1}{|\lambda_n|} \to 0$. Let the eigenfunction orthonormal sequence be $\boldsymbol{u_n}$. We know $\overline{span((\boldsymbol{u_n}))} = \mathbb{L}_2([a,b])$ if and only if $< \boldsymbol{u_n}, \boldsymbol{f} >= 0$ for all n implies $\boldsymbol{f} = \boldsymbol{0}$; that is, $(\ span((\boldsymbol{u_n}))\ )^{\perp} = \boldsymbol{0}$.

Theorem 14.3.3 Completeness Conditions

The following conditions are equivalent:

- $\mathscr{J}(\boldsymbol{f}) = \boldsymbol{0}$
- $< \boldsymbol{u_n}, \boldsymbol{f} >= 0$ *for all* n.

Proof 14.3.3

First, let's recall for the general Stürm - Liouville problem, the inner product we use is $< \boldsymbol{f}, \boldsymbol{g} >= \int_a^b r(t) f(t) \overline{g(t)} dt$. *This will modify some of the approximations below.*

If we assume $\mathscr{J}(\boldsymbol{f}) = \boldsymbol{0}$, *then* $< \boldsymbol{u_n}, \mathscr{J}(\boldsymbol{f}) >= 0$ *for all* n. *Let the eigenvalues of* $\mathscr{J}$ *be denoted by* λ_j. *But* $< \boldsymbol{u_n}, \mathscr{J}(\boldsymbol{f}) >=< \mathscr{J}(\boldsymbol{u_n}), \boldsymbol{f} >$. *Thus,* $\lambda_n < \boldsymbol{u_n}, \boldsymbol{f} >= 0$ *for all* n. *This tells us* $< \boldsymbol{u_n}, \boldsymbol{f} >= 0$ *for all* n.

Conversely, if $< \boldsymbol{u_n}, \boldsymbol{f} >= 0$ *for all* n, *recall we built auxiliary kernels in the construction of the eigenvalues for the symmetric integral operator. These constructions are modified by adding* $\boldsymbol{r}$ *into the* $\boldsymbol{k_{n+1}}$ *kernels. Note where the* $\boldsymbol{r}$ *appears. Thus, we have*

$$k_{n+1}(s,t) = k(s,t) - \sum_{j=1}^{n} \lambda_j r(s) u_j(s) u_j(t)$$

and so

$$\begin{aligned} \int_a^b k_{n+1}(s,t) f(s) ds &= \mathscr{J}(\boldsymbol{f}) - \sum_{j=1}^{n} \lambda_j u_j(t) < \boldsymbol{u_j}, \boldsymbol{f} > \\ &= \mathscr{J}(\boldsymbol{f}) \end{aligned}$$

However, this also tells us $\| \mathscr{J}(\boldsymbol{f}) \|_2 = \| \mathscr{J}_{n+1}(\boldsymbol{f}) \|_2 = |\lambda_{n+1}|$. *Since* $|\lambda_{n+1} \to 0$, *we see* $\| \mathscr{J}(\boldsymbol{f}) \|_2 = 0$. *Thus,* $\mathscr{J}(\boldsymbol{f}) = \boldsymbol{0}$. ■

Thus, we can show the orthonormal sequence is complete and therefore is a Schauder basis for $\mathbb{L}_2([a,b])$ if we can show either of these conditions is satisfied. So let's look at $\mathscr{J}(\boldsymbol{z}) = \boldsymbol{0}$.

Now for any $\boldsymbol{z}$, $\mathscr{J}(\boldsymbol{z})$ is a continuous function. If we assume $\boldsymbol{u} = \mathscr{J}(\boldsymbol{z}) = \boldsymbol{0}$, we see $\boldsymbol{u}$ is actually twice differentiable and satisfies the boundary conditions. Hence, for the differential operator $\boldsymbol{L_{\lambda_0}}$, we have

$$\boldsymbol{L_{\lambda_0}}(\mathscr{J}(\boldsymbol{z})) = \boldsymbol{z}$$

But $\boldsymbol{L}_{\boldsymbol{\lambda}_0}(\boldsymbol{u}) = \boldsymbol{0}$ and so we must have $\boldsymbol{z} = \boldsymbol{0}$.

Hence the eigenfunctions for a Stürm - Liouville model give an complete orthonormal sequence which is therefore a Schauder basis for $\mathbb{L}_2([a,b])$.

Theorem 14.3.4 Generalized Fourier Expansions from Stürm - Liouville Models

Given a Stürm - Liouville model $\boldsymbol{L}_{\boldsymbol{\lambda}_0}(\boldsymbol{u}) = \boldsymbol{f}$ with λ_0 not an eigenvalue of $\mathbf{L}$, then there is a complete orthonormal sequence $(\boldsymbol{u}_{\boldsymbol{n}})$ and associated eigenvalues (λ_n) with $|\lambda_n| \to \infty$ so that $\boldsymbol{L}_{\boldsymbol{\lambda}_0}$ has inverse $\mathscr{J}$ with eigenvalues $\frac{1}{\lambda_j}$ and any $\boldsymbol{x} \in \mathbb{L}_2([a,b])$ can be written $\boldsymbol{x} = \sum_{n=1}^{\infty} < \boldsymbol{x}, \boldsymbol{u}_{\boldsymbol{n}} > \boldsymbol{u}_{\boldsymbol{n}}$. The coefficients $< \boldsymbol{x}, \boldsymbol{u}_{\boldsymbol{n}} >$ are the generalized Fourier coefficients of $\boldsymbol{x}$.

Proof 14.3.4
We have proven all these things in our discussions. ■

Comment 14.3.2 *Note also using Bessel's Inequality, we also know $< \boldsymbol{u}_{\boldsymbol{n}}, \boldsymbol{x} > \to 0$.*

Let's go back and think about the eigenvalue behavior of the Stürm - Liouville problem. Earlier, we showed these eigenvalues formed a sequence of real numbers and using ideas from complex variable theory, we found this sequence of eigenvalues could not have a cluster points and hence all the eigenvalues λ_n had to be isolated and hence $|\lambda_n| \to \infty$. Here is another way to see that using Bessel's Inequality and therefore bypassing the ideas from complex variables.

For fixed t, let $h(s) = \sqrt{r(t)}k(s,t)$ which is ok as $r(t) > 0$. We then have

$$(< \boldsymbol{h}, \boldsymbol{u}_{\boldsymbol{n}} >)(t) = \sqrt{r(t)} \int_a^b k(s,t)r(s)u_n(s)ds = \sqrt{r(t)}\,(\mathscr{J}(\boldsymbol{u}_{\boldsymbol{n}}))(t) = \mu_n \sqrt{r(t)}\, u_n(t)$$

where we let μ_n be the eigenvalues of $\mathscr{J}$. Of course $\frac{1}{\mu_n} = \lambda_n$ are the eigenvalues of the ODE model. As usual, we assume no eigenvalues are zero. Then by Bessel's Inequality

$$\sum_{j=0}^{\infty} | < \boldsymbol{h}, \boldsymbol{u}_{\boldsymbol{j}} > |^2 = \sum_{j=0}^{\infty} \mu_j^2 r(t) u_j(t)^2 \leq \int_a^b |h(s)|^2\, ds$$

Hence,

$$\sum_{j=0}^{\infty} \mu_j^2\, r(t) u_j(t)^2 = \leq \int_a^b |k(s,t)|^2\, ds$$

$$\Longrightarrow \sum_{j=0}^{\infty} \mu_j^2 \int_a^b r(t)|u_j(t)|^2 dt \leq \int_a^b \int_a^b |k(s,t)|^2\, ds\, dt$$

However the kernel is continuous on $[a, b] \times [a, b]$ and $< \boldsymbol{u_j}, \boldsymbol{u_j} >= 1$, so we can say

$$\sum_{j=0}^{\infty} \mu_j^2 \leq \|\boldsymbol{k}\|_\infty (b-a)^2$$

where $\|\boldsymbol{k}\|_\infty$ is the obvious extension of the sup norm to the square:

$$\|\boldsymbol{k}\|_\infty \quad = \quad \sup_{(s,t) \in [a,b] \times [a,b]} |k(s,t)|$$

This clearly tells us the series converges and so by the nth term test we must have $\mu_n \rightarrow 0$. The construction process to find the eigenvalues μ_n also tells us that the values $|\mu_n|$ are monotonically decreasing. That then implies $\lambda_n \rightarrow \infty$.

Theorem 14.3.5 Convergence of Approximations to the Stürm - Liouville Model

Let $\boldsymbol{f} \in C([a, b])$ and $\boldsymbol{u} = \mathscr{J}_{\lambda_0}(\boldsymbol{f})$. Then the partial sums $\sum_{j=0}^{n} \mu_j \boldsymbol{u_j} < \boldsymbol{u_j}, \boldsymbol{f} >$ converge to the continuous function $\boldsymbol{u} = \mathscr{J}_{\lambda_0}(\boldsymbol{f})$ both uniformly and in $\mathbb{L}_2([a, b])$ norm.

Proof 14.3.5

Let $\boldsymbol{f} \in C([a, b])$ and let $\boldsymbol{u} = \mathscr{J}(\boldsymbol{f})$. The eigenvalues of $\boldsymbol{L} - \lambda_0 \boldsymbol{I}$ are $\lambda_n - \lambda_0$ and $\lambda_n \rightarrow \infty$. The eigenvalues of $\mathscr{J}_{\lambda_0}$ are the reciprocals $\frac{1}{\lambda_n - \lambda_0}$ which go to 0. To help us with our typing here, we will rename all of these for our convenience. We will let $\boldsymbol{L} - \lambda_0 \boldsymbol{I} \equiv \boldsymbol{L}$ and $\mathscr{J}_{\lambda_0} \equiv \mathscr{J}$. The eigenvalues of $\boldsymbol{L} - \lambda_0 \boldsymbol{I} \equiv \boldsymbol{L}$ will just be λ_n and the reciprocal eigenvalues will be μ_n. Hence, $\mathscr{J}(\boldsymbol{u_n}) = \mu_n \boldsymbol{u_n}$. The operator $\mathscr{J}$ is then defined pointwise by $(\mathscr{J}(\boldsymbol{f})) = \int_a^b k(s,t) r(s) f(s) ds$ where $\boldsymbol{r}$ is the weighting function for this Stürm - Liouville problem. Let's also assume the eigenvalues are numbered from 0 on. Also, for convenience, we will do these calculations in a real vector space setting as the use of complex conjugations is straightforward.

We know $\mu_n \rightarrow 0$. In the construction process for the eigenvalues we used the kernels and approximate inverses

$$\begin{aligned} k_{n+1}(s,t) &= k(s,t) - \sum_{i=1}^{n} \mu_i \, u_i(s) u_i(t) \\ (\mathscr{J}_{n+1}(\boldsymbol{x}))(t) &= \int_a^b k_{n+1}(s,t) \, r(s) x(s) \, ds = (\mathscr{J}(\boldsymbol{x}))(t) - \sum_{i=1}^{n} \mu_i u_i(t) \ < \boldsymbol{u_i}, \boldsymbol{x} > . \end{aligned}$$

Then

$$(\mathscr{J}_{n+1}(\boldsymbol{f}))(t) \quad = \quad (\mathscr{J}(\boldsymbol{f}))(t) - \sum_{i=1}^{n} \mu_i u_i(t) \ < \boldsymbol{u_i}, \boldsymbol{f} >$$

We also know

$$\|\mathscr{J}_{n+1}(\boldsymbol{f})\|_2 \ \leq \ \|\mathscr{J}_{n+1}\|_{op} \, \|\boldsymbol{f}\|_2, \quad |\mu_{n+1}| = \sup_{\|\boldsymbol{u}\|_2 = 1} \sqrt{< \mathscr{J}_{n+1}(\boldsymbol{u}), \boldsymbol{u} >}$$

Hence,

$$\| \mathscr{J}(\boldsymbol{f}) - \sum_{i=1}^{n} \mu_i \boldsymbol{u_i} < \boldsymbol{u_i}, \boldsymbol{f} > \|_2 \quad \leq \quad |\mu_{n+1}| \, \|\boldsymbol{f}\|_2$$

Since $\mu_n \to 0$, *we therefore know* $\| \mathscr{J}(\boldsymbol{f}) - \sum_{i=1}^{n} \mu_i \boldsymbol{u_i} \ < \boldsymbol{u_i}, \boldsymbol{f} > \|_2 \to 0$. *We can conclude then that* $\mathscr{J}(\boldsymbol{f}) = \sum_{j=0}^{\infty} \mu_j \boldsymbol{u_j} \ < \boldsymbol{u_j}, \boldsymbol{f} >$ *in* $\mathbb{L}_2([a,b])$ *norm. Note since* $\boldsymbol{f}$ *is continuous, we also know* $\mathscr{J}(\boldsymbol{f})$ *is continuous.*

Now let $m > n$. *Then*

$$\mathscr{J}\left(\sum_{j=n+1}^{m} < \boldsymbol{u_j}, \boldsymbol{f} > \boldsymbol{u_j} \right) \quad = \quad \sum_{j=n+1}^{m} < \boldsymbol{u_j}, \boldsymbol{f} > \ \mathscr{J}(\boldsymbol{u_j}) = \sum_{j=n+1}^{m} \mu_j < \boldsymbol{u_j}, \boldsymbol{f} > \boldsymbol{u_j}$$

and for $\boldsymbol{v} \in C([a,b])$, *we have*

$$\begin{aligned} |(\mathscr{J}(\boldsymbol{v}))(t)| &= \left| \int_a^b k(s,t) r(s) v(s) ds \right| \leq \int_a^b |k(s,t)||r(s)v(s)| ds \\ &\leq \sqrt{\int_a^b \left| \sqrt{r(s)} |k(s,t)\right|^2 ds} \sqrt{\int_a^b \left| \sqrt{r(s)} v(s) \right|^2 ds} \\ &\leq \|\boldsymbol{r}\|_\infty \|\boldsymbol{k}\|_\infty \, (b-a) \, \|\boldsymbol{v}\|_2 \end{aligned}$$

Now apply this calculation to $\boldsymbol{v} = \sum_{j=n+1}^{m} < \boldsymbol{u_j}, \boldsymbol{f} > \boldsymbol{u_j}$ *for* $m > n$. *Thus,*

$$\left\| \mathscr{J}\left(\sum_{j=n+1}^{m} < \boldsymbol{u_j}, \boldsymbol{f} > \boldsymbol{u_j} \right) \right\|_\infty \quad \leq \quad \|\boldsymbol{r}\|_\infty \|\boldsymbol{k}\|_\infty \, (b-a) \left\| \sum_{j=n+1}^{m} < \boldsymbol{u_j}, \boldsymbol{f} > \boldsymbol{u_j} \right\|_2$$

The orthonormality of $(\boldsymbol{u_n})$ *then allows us to say*

$$\left\| \sum_{j=n+1}^{m} < \boldsymbol{u_j}, \boldsymbol{f} > \boldsymbol{u_j} \right\|_2 \quad = \quad \sqrt{\sum_{j=n+1}^{m} | < \boldsymbol{u_j}, \boldsymbol{f} > |^2}$$

Then using Bessel's Inequality we have

$$\sum_{j=0}^{\infty} | < \boldsymbol{u_j}, \boldsymbol{f} > |^2 \quad \leq \quad \|f\|_2^2$$

which tells us the sequence $| < \boldsymbol{u_j}, \boldsymbol{f} > |^2$ *is a Cauchy sequence. Thus, given* $\epsilon > 0$, *there is* N *so that if* $m > n > N$,

$$\sum_{j=n+1}^{m} | < \boldsymbol{u_j}, \boldsymbol{f} > |^2 \quad < \quad \frac{\epsilon^2}{\|\boldsymbol{r}\|_\infty^2 \|\boldsymbol{k}\|_\infty^2 \, (b-a)^2}$$

It then follows immediately the sequence of partial sums $\sum_{j=0}^{n} \mu_j < \boldsymbol{u_j}, \boldsymbol{f} > \boldsymbol{u_j}$ *is a Cauchy sequence in* $C([a,b])$ *in the* sup *norm and since* $(C([a,b]), \|\cdot\|_\infty)$ *is complete, there is a continuous function* Ψ *in* $C([a,b])$ *so that*

$$\sum_{j=0}^{n} \mu_j < \boldsymbol{u_j}, \boldsymbol{f} > \boldsymbol{u_j} \xrightarrow{u} \Psi$$

We know now that $\sum_{j=0}^{n} \mu_j < \boldsymbol{u_j}, \boldsymbol{f} > \boldsymbol{u_j}$ *converges uniformly to a continuous function* Ψ *and in* $\|\cdot\|_2$ *to a continuous function* $\mathscr{J}(\boldsymbol{f})$.

It is straightforward to show the convergence to Ψ *is also in the* $\|\cdot\|_2$ *norm. You are asked to do this in the following exercise.*

From the above it is also easy to see Ψ *and* $\mathscr{J}(\boldsymbol{f})$ *are in the same equivalence class in* $\mathbb{L}_2([a,b])$. *Thus* $\Psi - \mathscr{J}(\boldsymbol{f}) = \boldsymbol{0}$ *in* $\mathbb{L}_2([a,b])$. *Since both of these functions are continuous on* $[a,b]$, *the Riemann Integral of* $\Psi - \mathscr{J}(\boldsymbol{f})$ *exists and so* $g(t) = \Psi(t) - \mathscr{J}(\boldsymbol{f})(t) = 0$ *except on a set of content zero. In (Peterson (18) 2020), we said* $\boldsymbol{g} = \boldsymbol{0}$ *a.e. Let's refresh your memory here.*

Definition 14.3.1 Sets of Content Zero

A subset S *of* $\Re$ *is said to have content zero if and only if given any positive* ϵ *we can find a sequence of bounded open intervals* $\{J_n^\epsilon = (a_n, b_n)\}$ *either finite in number or infinite so that*

$$S \subseteq \cup J_n,$$

with the total length

$$\sum (b_n - a_n) < \epsilon$$

If the sequence only has a finite number of intervals, the union and sum are written from 1 *to* N *where* N *is the number of intervals and if there are infinitely many intervals, the sum and union are written from* 1 *to* ∞.

If a set of content zero contained an interval of finite length, say $r > 0$, *for* $0 < \epsilon < r$, *we would violate the covering condition giving a total summed length less than* r. *Hence a set of content zero cannot contain an interval of positive length. We know* $\boldsymbol{g}$ *is zero except on a set* E *of content zero. Thus, if* $t_0 \in E$, $g(t) \neq 0$. *For concreteness, let's assume* $g(t_0) > 0$; *the other argument is similar. Then by the continuity of* g, *for* $\epsilon = \frac{g(t_0)}{2}$, *there is* $\delta > 0$ *so that*

$$|t - t_0| < \delta, \ t_0, t \in [a,b] \implies |g(t) - g(t_0)| < \frac{g(t_0)}{2}$$

which implies

$$|t - t_0| < \delta, \ t_0, t \in [a,b] \implies 0 < \frac{g(t_0)}{2} < g(t) < 3\frac{g(t_0)}{2}$$

This tells us $g(t) \neq 0$ *on the set* $(t_0 - \delta, t_0 + \delta) \cap [a, b] \cap E^C$. *If* $(t_0 - \delta, t_0 + \delta) \cap [a, b] \cap E^C = \emptyset$, *we would know* $(t_0 - \delta, t_0 + \delta) \cap [a, b] \subset E$ *telling us* E *contains an interval of finite length which is impossible. So* $(t_0 - \delta, t_0 + \delta) \cap [a, b] \cap E^C$ *is not empty. But on* E^C, $g(t)$ *is zero. This is a contradiction and so we cannot find a point* $t_0 \in E$ *with* $g(t_0) > 0$. *As we said, the argument is similar for the case* $g(t_0) < 0$. *Hence,* $g(t) = 0$ *on all of* $[a, b]$ *and* $\Phi = \mathscr{J}(\boldsymbol{f})$ *on* $[a, b]$.

We have shown $\mathscr{J}(\boldsymbol{f})$ *is continuous on* $[a, b]$ *and* $\sum_{j=0}^n \mu_j < \boldsymbol{u_j}, \boldsymbol{f} > \boldsymbol{u_j} \xrightarrow{u} \mathscr{J}(\boldsymbol{f})$ *and* $\sum_{j=0}^n \mu_j < \boldsymbol{u_j}, \boldsymbol{f} > \boldsymbol{u_j} \xrightarrow{\|\cdot\|_2} \mathscr{J}(\boldsymbol{f})$. ■

Comment 14.3.3 *Given a linear partial differential equation with the right boundary conditions, if we use the method of separation of variables, we will often get a Stürm - Liouville problem in the space variable* x. *For the applied data* $\boldsymbol{f}$ *on* $[a, b]$, *we then solve* $(\boldsymbol{L} - \lambda_0 \boldsymbol{I})(\boldsymbol{u}) = \boldsymbol{f}$ *for the appropriate Stürm - Liouville boundary conditions, where* λ_0 *is any useful choice that is not an eigenvalue. Following our inversion procedure, we can then approximate the solution* $\boldsymbol{u}$ *in both uniform and* $\mathbb{L}_2([a, b])$ *norm with* $\sum_{j=0}^n \mu_j < \boldsymbol{u_j}, \boldsymbol{f} > \boldsymbol{u_j}$ *using the eigenfunctions* $\boldsymbol{u_n}$ *of* $\boldsymbol{L} - \lambda_0 \boldsymbol{I}$ *and the reciprocal eigenvalues* $\mu_j = \frac{1}{\lambda_n - \lambda_0}$.

Note this is what we do in separation of variables. We write the solution as an expansion in terms of the eigenfunctions of $\boldsymbol{L} - \lambda_0 \boldsymbol{I}$, *compute the generalized Fourier coefficients of the data function* f *with respect to this complete orthonormal set and then equate coefficients.*

Comment 14.3.4 *We always know for any* $\boldsymbol{f} \in \mathbb{L}_2([a, b])$, *we have* $\sum_{j=0}^n < \boldsymbol{u_j}, \boldsymbol{f} > \boldsymbol{u_j}$ *converges in* $\mathbb{L}_2([a, b])$ *because the eigenfunctions are a complete orthonormal sequence. In particular, this holds for continuous* $\boldsymbol{f}$.

We can also find an expansion for an inverse operator of the form $\mathscr{J}$ in terms of projections.

Theorem 14.3.6 Expanding $\mathscr{J}$ in terms of projections

Let $(\boldsymbol{u_n})$ *be the complete orthonormal sequence associated with the Stürm - Liouville operator* $\boldsymbol{L} - \lambda_0 \boldsymbol{I}$ *for a non eigenvalue* λ_0. *Let* (μ_n) *be the eigenvalue sequence for* $\mathscr{J}_{\lambda_0} \equiv \mathscr{J}$. *Let* $E_n = span(\{\boldsymbol{u_n}\})$ *and let* P_j *be the projection operator from* $\mathbb{L}_2([a, b])$ *to* E_n *defined by* $P_n(\boldsymbol{f}) =< \boldsymbol{u_n}, \boldsymbol{f} > \boldsymbol{u_n}$. *Then* $\sum_{j=0}^n \mu_j P_j$ *converges to* $\mathscr{J}$ *in operator norm; i.e.* $\mathscr{J} = \sum_{j=0}^\infty \mu_j P_j$ *where the convergence of this infinite series is interpreted in the operator norm.*

Proof 14.3.6

We know $\|\mathscr{J}_{n+1}\|_{op} = |\mu_{n+1}| \rightarrow 0$. *Hence, given* $\epsilon > 0$, *there is* N *so* $\|\mathscr{J}_{n+1}\|_{op} < \epsilon$ *if* $n > N$. *Note we also know*

$$\mathscr{J}_{n+1}(\boldsymbol{f}) = \mathscr{J}(\boldsymbol{f}) - \sum_{j=0}^n \mu_j < \boldsymbol{u_j}, \boldsymbol{f} > \boldsymbol{u_j}$$

Let $E_n = span(\{\boldsymbol{u_n}\})$ and let P_j be the projection operator from $\mathbb{L}_2([a,b])$ to E_n defined by $P_n(\boldsymbol{f}) =< \boldsymbol{u_n}, \boldsymbol{f} > \boldsymbol{u_n}$. Then

$$\sum_{j=0}^{n} \mu_j < \boldsymbol{u_j}, \boldsymbol{f} > \boldsymbol{u_j} = \sum_{j=0}^{n} \mu_j P_j(\boldsymbol{f})$$

Thus, we can say

$$\mathscr{I}_{n+1} = \mathscr{I} - \sum_{j=0}^{n} \mu_j P_j$$

and so if $n > N$, $\|\mathscr{I} - \sum_{j=0}^{n} \mu_j P_j\|_{op} < \epsilon$.

This tells us $\sum_{j=0}^{n} \mu_j P_j$ converges to $\mathscr{I}$ in operator norm; i.e. $\mathscr{I} = \sum_{j=0}^{\infty} \mu_j P_j$ where the convergence of this infinite series is interpreted in the operator norm. ■

Homework

Exercise 14.3.15 *In our above discussions, we use*

$$\sum_{j=0}^{\infty} \mu_j^2 \, |u_n(t)|^2 = \sum_{j=0}^{\infty} \mu_j^2 \int_a^b |u_n(t)|^2 dt$$

Prove this statement.

Exercise 14.3.16 *Since we know now that $\sum_{j=0}^{n} \mu_j < \boldsymbol{u_j}, \boldsymbol{f} > \boldsymbol{u_j}$ converges uniformly to a continuous function Ψ and in $\|\cdot\|_2$ to a continuous function $\mathscr{I}(\boldsymbol{f})$, prove the convergence to Ψ is also in the $\|\cdot\|_2$ norm.*

Exercise 14.3.17 *Prove $\|\boldsymbol{\Psi} - \mathscr{I}(\boldsymbol{f})\|_2 = 0$.*

Exercise 14.3.18 *Let's extend these results to data with a jump. The data is now*

$$f(t) = \begin{cases} 0, & a \leq t \leq c \\ H, & c < t \leq b \end{cases}$$

for some $a < c < b$. Let $f_1(t) = 0$ on $[a, c]$ and $f_2(t) = H$ on $[c, b]$. Solve $(\boldsymbol{L} - \lambda_0 \boldsymbol{I})(\boldsymbol{u_1}) = \boldsymbol{f_1}$, $(\boldsymbol{L} - \lambda_0 \boldsymbol{I})(\boldsymbol{u_2}) = \boldsymbol{f_2}$ and discuss how these solutions compare to the solution $(\boldsymbol{L} - \lambda_0 \boldsymbol{I})(\boldsymbol{u}) = \boldsymbol{f}$. Explain how we obtain a solution which satisfies $(\boldsymbol{L} - \lambda_0 \boldsymbol{I})(\boldsymbol{u}) = \boldsymbol{f}$ as a classical differential equation on $[a, b]$ except at c; i.e. $(\boldsymbol{L} - \lambda_0 \boldsymbol{I})(\boldsymbol{u}) = \boldsymbol{f}$ a.e.

Exercise 14.3.19 *Let's extend these results to data with two jumps. The data is now*

$$f(t) = \begin{cases} H_1, & a \leq t \leq c_1 \\ H_2, & c_1 < t \leq c_2 \\ H_3, & c_2 < t \leq b \end{cases}$$

for some $a < c_1 < c_2 < b$. Define f_1, f_2 and f_3 appropriately similar to what is done in the previous exercise. Solve $(\boldsymbol{L} - \lambda_0 \boldsymbol{I})(\boldsymbol{u_i}) = \boldsymbol{f_i}$ for $i = 1, 2, 3$ and discuss how these solutions compare to the solution $(\boldsymbol{L} - \lambda_0 \boldsymbol{I})(\boldsymbol{u}) = \boldsymbol{f}$. Explain how we obtain a solution which satisfies $(\boldsymbol{L} - \lambda_0 \boldsymbol{I})(\boldsymbol{u}) = \boldsymbol{f}$ as a classical differential equation on $[a, b]$ except at c_1 and c_2; i.e. $(\boldsymbol{L} - \lambda_0 \boldsymbol{I})(\boldsymbol{u}) = \boldsymbol{f}$ a.e.

Exercise 14.3.20 *Explain how to extend the previous two results to data which has a finite number of discontinuities.*

Exercise 14.3.21 *Prove the following statements:*

- *From our approximation theorem, we also know*

$$T_N^0 = \sum_{j=0}^{n} \frac{1}{\Theta - \frac{n^2\pi^2}{L^2}} < \hat{u}_j, f > \hat{u}_j$$

  is the best approximation to the solution to $x'' + \Theta u = f, u'(0) = u'(L) = 0$ for $f \in \mathbb{L}_2([0, L])$ data to the subspace spanned by $\{\hat{u}_n\}_{n=0}^N$.

- $$T_N^1 = \sum_{j=1}^{n} \frac{1}{\Theta - \frac{n^2\pi^2}{L^2}} < \hat{v}_j, f > \hat{v}_j$$

  is the best approximation to the solution to $u'' + \Theta u = f, u(0) = u(L) = 0$ for $f \in \mathbb{L}_2([0, L])$ data to the subspace spanned by $\{\hat{v}_n\}_{n=1}^N$.

14.4 The Ball and Stick Model

Next let's look at a differential operator which is not from a Stürm - Liouville problem but whose solution and conversion to an integral equation is similar in spirit and illustrates some complications. This is another linear differential operator arising from using the technique of separation of variables on a standard linear partial differential equation.

However, the boundary conditions are quite different and our knowledge of self-adjoint operators does not help us! The mode we focus on is based on a simple model of information processing in the neuron. However, the exact same model arises in other situations such as understanding the emission of volatile compounds from new carpet after it has been laid down in a room. You can look up (Little et al. (13) 1994) for the details. But here, since we like neurological things, we will look at a neuron model.

Let's review the basics of information processing in a typical neuron. There are many first sources for this material; some of them are *Introduction to Neurobiology* (Hall (7) 1992), *Ionic Channels of Excitable Membranes* (Hille (9) 1992), *Foundations of Cellular Neurophysiology* (Johnston and Wu (10) 1995), *Rall's review of cable theory in the 1977 Handbook of Physiology* (Rall (21) 1977) and *Cellular Biophysics: Transport and Electrical Properties* (Weiss (25) 1996) and (Weiss (26) 1996).

Our basic model consists of the following structural elements: A neuron which consists of a *dendritic tree* (which collects sensory stimuli and sums this information in a temporally and spatially dependent way), a cell body (called the *soma*) and an output fiber (called the *axon*). Individual dendrites of the dendritic tree and the axon are all modeled as cylinders of some radius a whose length ℓ is very long compared to this radius and whose walls are made of a bilipid membrane. The inside of each cylinder consists of an intracellular fluid and we think of the cylinder as lying in a bath of extracellular fluid. So for many practical reasons, we can model a dendritic or axonal fiber as two concentric cylinders; an inner one of radius a (this is the actual dendrite or axon) and an outer one with the extracellular fluid contained in the space between the inner and outer membranes.

The potential difference across the inner membrane is essentially due to a balance between the electromotive force generated by charge imbalance, the driving force generated by charge concentration differences in various ions and osmotic pressures that arise from concentration differences in water molecules on either side of the membrane. Roughly speaking, the ions of importance in our simplified model are the potassium K^+, sodium Na^+ and chloride Cl^- ions. The equilibrium potential across the inner membrane is about -70 millivolts and when the membrane potential is driven above this rest value, we say the membrane is *depolarized* and when it is driven below the rest potential, we say the membrane is *hyperpolarized*. The axon of one neuron interacts with the dendrite of another neuron via a site called a *synapse*. The synapse is physically separated into two parts: the *presynaptic* side (the side the axon is on) and the *postsynaptic* side (the side the dendrite is on). There is an actual physical gap, the *synaptic cleft*, between the two parts of the synapse. This cleft is filled with extracellular fluid.

If there is a rapid depolarization of the presynaptic site, a chain of events is initialized which culminates in the release of specialized molecules called *neurotransmitters* into the synaptic cleft. There are pores embedded in the postsynaptic membrane whose opening and closing are dependent on the potential across the membrane that are called *voltage-dependent gates*. In addition, the gates generally allow the passage of a specific ion; so for example, there are sodium, potassium and chloride gates. The released neurotransmitters bind with the sites specific for the Na^+ ion. Such sites are called *receptors*. Once bound, Na^+ ions begin to flow across the membrane into the fiber at a greater rate than before. This influx of positive ions begins to drive the membrane potential above the rest value; that is, the membrane begins to depolarize. The flow of ions across the membrane is measured in gross terms by what are called *conductances*. Conductance has the units of reciprocal ohms; hence, high conductance implies high current flow per unit voltage. Thus the conductance of a gate is a good way to measure its flow. We can say that as the membrane begins to depolarize, the sodium conductance, g_{Na}, begins to increase. This further depolarizes the membrane. However, the depolarization is self-limited as the depolarization of the membrane also triggers the activation of voltage-dependent gates for the potassium ion, K^+, which allow potassium ions to flow through the membrane out of the cell. So the increase in the sodium conductance, g_{Na} triggers a delayed increase in potassium conductance, g_K (there are also conductance effects due to chloride ions which we will not mention here). The net effect of these opposite driving forces is the generation of a potential pulse that is fairly localized in both time and space. It is generated at the site of the synaptic contact and then begins to propagate down the dendritic fiber toward the soma. As it propagates, it attenuates in both time and space. We call these voltage pulses *Post Synaptic Pulses* or PSPs.

We model the soma itself as a small isopotential sphere, small in surface area compared to the surface area of the dendritic system. The possibly attenuated values of the PSPs generated in the dendritic system at various times and places are assumed to propagate without change from any point on the soma body to the initial segment of the axon which is called the *axon hillock*. This is a specialized piece of membrane which generates a large output voltage pulse in the axon by a coordinated rapid increase in g_{Na} and g_K once the axon hillock membrane depolarizes above a critical trigger value. The axon itself is constructed in such a way that this output pulse, called the *action potential*, travels without change throughout the entire axonal fiber. Hence, the initial depolarizing voltage impulse that arrives at a given presynaptic site is due to the action potential generated in the presynaptic neuron by its own dendritic system.

The salient features of our model are thus:

- Axonal and dendritic fibers are modeled as two concentric membrane cylinders.
- The axon carries action potentials which propagate without change along the fiber once they are generated. Thus if an axon makes 100 synaptic contacts, we assume that the depolarizations of each presynaptic membrane are the same.
- Each synaptic contact on the dendritic tree generates a time and space localized depolarization of the postsynaptic membrane which is attenuated in space as the pulse travels along the fiber from the injection site and which decrease in magnitude the longer the time is since the pulse was generated.
- The effect of a synaptic contact is very dependent on the position along the dendritic fiber that the contact is made–in particular, how far was the contact from the axon hillock (i.e., in our model, how far from the soma)? Contacts made in essentially the same space locality have a high probability of reinforcing each other and thereby possibly generating a depolarization high enough to trigger an action potential.
- The effect of a synaptic contact is very dependent on the time at which the contact is made. Contacts made in essentially the same time frame have a high probability of reinforcing each other and thereby possibly generating a depolarization high enough to trigger an action potential.

A simple dendritic cable model is called the *ball and stick* neuron model. This consists of an isopotential sphere to model the cell body or soma coupled to a single dendritic fiber input line. We will model the soma as a simple parallel resistance/ capacitance network and the dendrite as a finite length cable as previously discussed (see Figure 14.1). In Figure 14.1, you see the terms I_0, the input current at the soma/ dendrite junction starting at $\tau = 0$; I_D, the portion of the input current that enters the dendrite (effectively determined by the input conductance to the finite cable, G_D); I_S, the portion of the input current that enters the soma (effectively determined by the soma conductance G_S); and C_S,

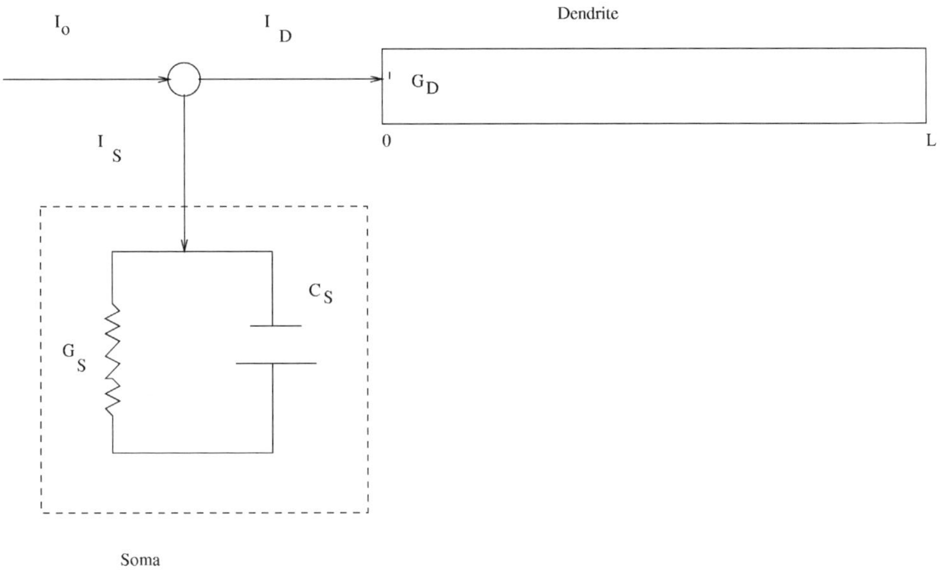

Figure 14.1: The ball and stick model.

the soma membrane capacitance. We assume that the electrical properties of the soma and dendrite membrane are the same; this implies that the fundamental time and space constants of the soma and dendrite are given by the same constant (we will use our standard notation τ_M and λ_C as usual). It takes a bit of work, but it is possible to show that with a reasonable zero-rate left end cap condition the appropriate boundary condition at $\lambda = 0$ is given by

$$\rho \frac{\partial \hat{v}_m}{\partial \lambda}(0,\tau) = \tanh(L)\left[\hat{v}_m(0,\tau) + \frac{\partial \hat{v}_m}{\partial \tau}(0,\tau)\right], \tag{14.1}$$

where we introduce the fundamental ratio $\rho = \frac{G_D}{G_S}$, the ratio of the dendritic conductance to soma conductance. For more discussion of the ball and stick model boundary conditions we use here, you can look at the treatment in (Rall (21, Chapter 7, Section 2) 1977). The full system to solve is therefore:

$$\frac{\partial^2 \hat{v}_m}{\partial \lambda^2} = \hat{v}_m + \frac{\partial \hat{v}_m}{\partial \tau}, \quad 0 \le \lambda \le L, \ \tau \ge 0. \tag{14.2}$$

$$\frac{\partial \hat{v}_m}{\partial \lambda}(L,\tau) = 0, \tag{14.3}$$

$$\rho \frac{\partial \hat{v}_m}{\partial \lambda}(0,\tau) = \tanh(L)\left[\hat{v}_m(0,\tau) + \frac{\partial \hat{v}_m}{\partial \tau}(0,\tau)\right]. \tag{14.4}$$

Applying the technique of separation of variables, $\hat{v}_m(\lambda,\tau) = u(\lambda)w(\tau)$, leads to the system:

$$\begin{aligned} u''(\lambda)w(\tau) &= u(\lambda)w(\tau) + u(\lambda)w'(\tau) \\ \rho u'(0)w(\tau) &= \tanh(L)\,(u(0)w(\tau) + u(0)w'(\tau)) \\ u'(L)w(\tau) &= 0 \end{aligned}$$

This leads again to the ratio equation

$$\frac{u''(\lambda) - u(\lambda)}{u(\lambda)} = \frac{w'(\tau)}{w(\tau)}.$$

Since these ratios hold for all τ and λ, they must equal a common constant Θ. Thus, we have

$$\frac{d^2u}{d\lambda^2} = (1+\Theta)u, \; 0 \leq \lambda \leq L, \tag{14.5}$$

$$\frac{dw}{d\tau} = \Theta w, \; \tau \geq 0. \tag{14.6}$$

The boundary conditions then become

$$\begin{aligned} u'(L) &= 0 \\ \rho u'(0) &= (1 + \Theta) \tanh(L)u(0). \end{aligned}$$

There are several cases to consider.

The case where $1 + \Theta = \omega^2$, where $\omega \neq 0$.

This gives

$$\begin{aligned} u'' - \omega^2 u &= 0 \leq \lambda \leq L, \\ u'(L) &= 0 \\ \rho u'(0) - \omega^2 \tanh(L)u(0) &= 0. \end{aligned}$$

The general solution is then

$$\begin{aligned} u(\lambda) &= A\cosh(\omega\lambda) + B\sinh(\omega\lambda) \\ u'(\lambda) &= \omega A\sinh(\omega\lambda) + \omega B\cosh(\omega\lambda). \end{aligned}$$

Applying the first boundary condition, we have

$$0 = u'(L) = \omega A\sinh(\omega L) + \omega B\cosh(\omega L)$$

Thus,

$$B = -A\,\frac{\sinh(\omega L)}{\cosh(\omega L)}.$$

leading to

$$\begin{aligned}
u(\lambda) &= A\cosh(\omega\lambda) - A\,\frac{\sinh(\omega L)}{}\cosh(\omega L)\sinh(\omega\lambda) \\
&= \frac{A}{\cosh(\omega L)}\left(\cosh(\omega L)\cosh(\omega\lambda) - \sinh(\omega L)\sinh(\omega\lambda)\right) \\
&= \frac{A}{\cosh(\omega L)}\cosh(\omega(L-\lambda)).
\end{aligned}$$

The second boundary condition then gives

$$-\rho\,\frac{\omega\,A}{\cosh(\omega L)}\sinh(\omega L) - A\omega^2 \;=\; 0.$$

Rearranging a bit, we find

$$\omega\,A\,(-\rho\,\tanh(\omega L) - \omega) \;=\; 0.$$

Hence, the solution is trivial unless

$$\tanh(\omega L) \;=\; \frac{1}{\rho\,L}\,(\omega L)$$

Letting $y = \omega L$, we see we have a trivial solution unless $\tanh(y) = y/(L\rho)$. The tangent line to $\tanh(y)$ at $y = 0$ is $T(y) = y$ which has slope 1. The line $y/(L\rho)$ has slope $1/(L\rho)$. For the ball stick model, rho is about 10 and L for most dendrites is 4 or 5. So the slope of the second line is quite small; perhaps .02. Hence, we will not have any intersections as we can see in Figure 14.2. Thus, we only have a trivial solution here and so we reject this case.

The case where $1 + \Theta = 0$.

This is the easiest.

$$\begin{aligned}
u'' &= 0 \leq \lambda \leq L, \\
u'(L) &= 0 \\
\rho u'(0) - \omega^2\,\tanh(L)u(0) &= 0.
\end{aligned}$$

Hence, $u(\lambda) = A + B\lambda$ and the first boundary condition tells us $B = 0$. Hence, $u(\lambda) = A$ and this solution satisfies the second boundary condition. The corresponding w solution is then $w(t) = Be^{-t}$. We see our first building block for the ball stick solution is $u_0(\lambda) = 1$ and $w_0(t) = e^{-t}$.

The case where $1 + \Theta = -\omega^2$ where $\omega \neq 0$.

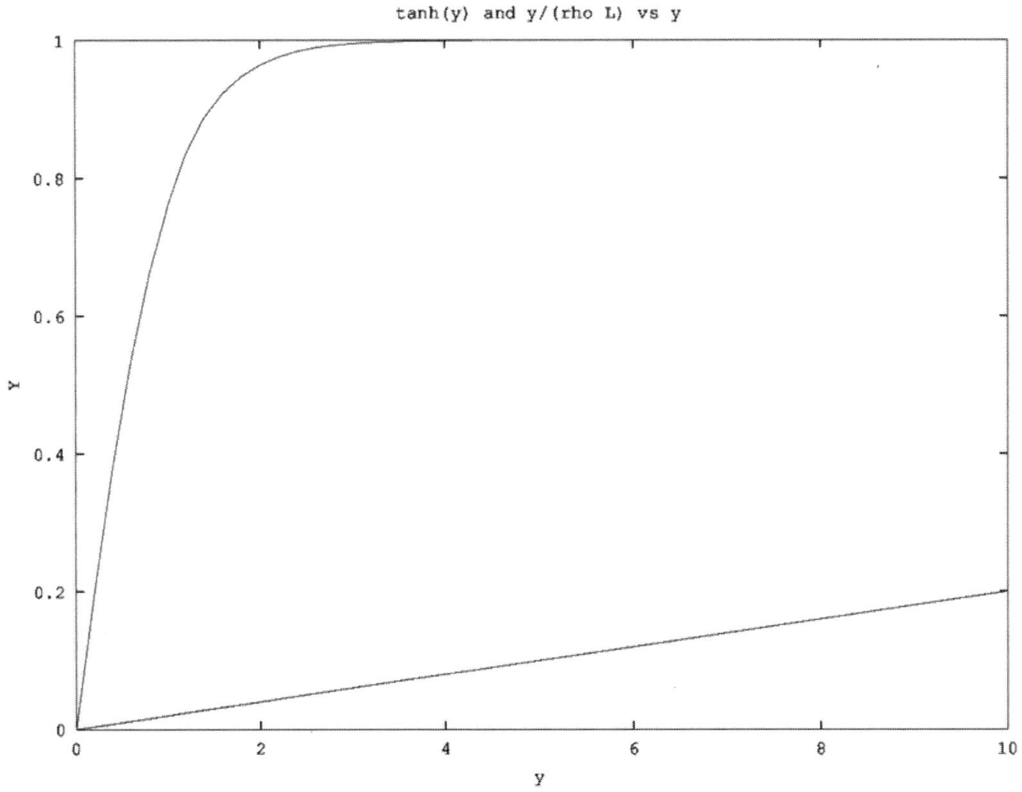

Figure 14.2: $tanh(y)$ vs. $y/(\rho L)$ for $rhoL = 50$.

This case is, as usual, the most interesting. The model to solve is now

$$\begin{aligned} u'' + \omega^2 u &= 0 \leq \lambda \leq L, \\ u'(L) &= 0 \\ \rho u'(0) - \omega^2 \tanh(L) u(0) &= 0. \end{aligned}$$

The details of finding the solution here are essentially the same as what we did in the first case. We write the general solution is $u(x) = A\cos(\omega(L-\lambda)) + B\sin(\omega(L-\lambda))$. Since

$$u'(\lambda) = \omega A \sin(\omega(L-\lambda)) - \omega B \cos(\omega(L-\lambda))$$

we see

$$u'(L) = -\omega B = 0,$$

and so $B = 0$. Then,

$$u(0) = A\cos(\omega L)$$

$$u'(0) \quad = \quad \omega A \sin(\omega L)$$

to satisfy the last boundary conditions, we find, since $1 + \Theta = -\omega^2$,

$$\omega \; A\Big(\rho \sin(\omega L) \; + \; \omega \tanh(L) \cos(\omega L)\Big)$$

A non trivial solution for A requires ω must satisfy the transcendental equation

$$\tan(\alpha L) \quad = \quad -\alpha \, \frac{\tanh(L)}{\rho} \; = \; -\kappa(\alpha L), \tag{14.7}$$

where $\kappa = \frac{\tanh(L)}{\rho L}$. The values of α that satisfy Equation 14.7 give us the values for our original problem, $\Theta = -1 - \alpha^2$. Hence, these values can be determined by the solution of the transcendental equation 14.7. This is easy to do graphically as you can see in Figure 14.3. It can easily be shown that these numbers form a monotonically increasing sequence starting with $\alpha_0 = 0$ and with the values α_n approaching asymptotically the values $\frac{2n-1}{2} \, \pi$. But you shouldn't call the values eigenvalues as having the ω^2 in the boundary condition gives us a problem that is **not** an eigenvalue problem! Hence, there is a countable number of values $\Theta_n = -1 - \alpha_n^2$ leading to a general solution of the form

$$\hat{v}_m^n(\lambda, \tau) \quad = \quad A_n \, \cos(\alpha_n \lambda) \, e^{-(1+\alpha_n^2)\tau}. \tag{14.8}$$

Hence, this system uses the pairs α_n (the solution to the transcendental equation 14.7) and $\cos[\alpha_n(L-\lambda)]$. In fact, we can show by direct integration that for $n \neq m$,

$$\int_0^L \cos(\alpha_n(L-\lambda)) \, \cos(\alpha_m(L-\lambda)) \, d\lambda \quad = \quad \frac{\sin((\alpha_n + \alpha_m)L)}{2(\alpha_n + \alpha_m)} + \frac{\sin((\alpha_n - \alpha_m)L)}{2(\alpha_n - \alpha_m)} \neq 0.$$

Since $\lim \alpha_n = \frac{(2n-1)\pi}{2}$, we see there is an integer Q so that

$$\int_0^L \cos(\alpha_n(L-\lambda)) \, \cos(\alpha_m(L-\lambda)) \quad \approx \quad 0$$

if n and m exceed Q.

Homework

Exercise 14.4.1 *We can construct the solutions in the case* $1 + \Theta - -\omega^2$ *a different way. The model to solve is*

$$\begin{aligned} \boldsymbol{u}'' + \omega^2 \boldsymbol{u} &= \boldsymbol{0}, \quad 0 \leq \lambda \leq L \\ u'(L) &= 0, \quad \rho u'(0) - \omega^2 \tanh L u(0) = 0 \end{aligned}$$

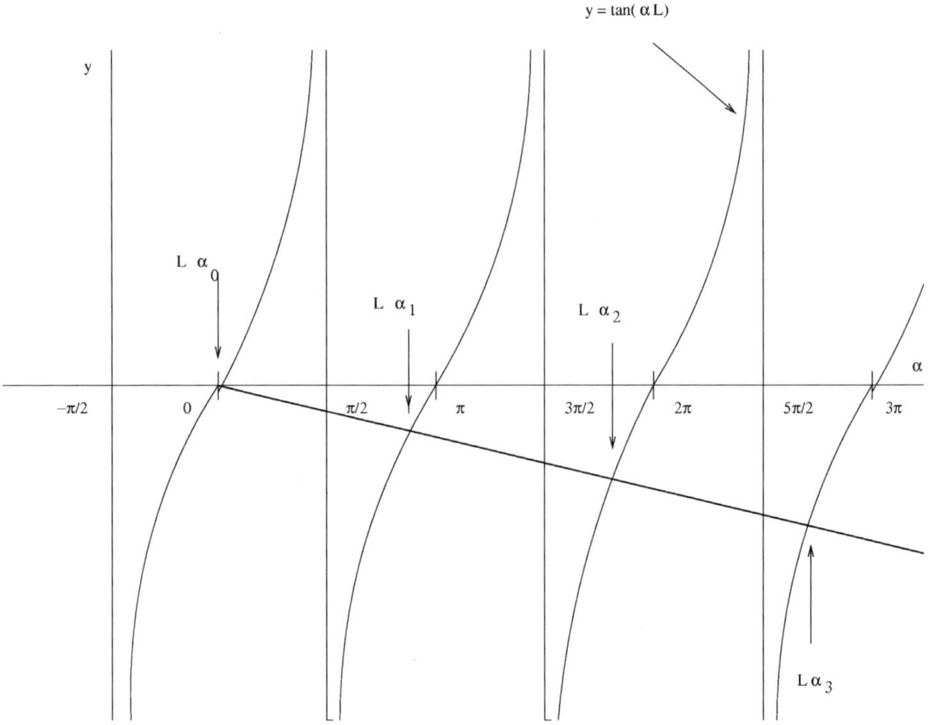

Figure 14.3: The ball and stick eigenvalue problem.

Let $\boldsymbol{u_1}$ solve this model with just $u'(L) = 0$ and $\boldsymbol{u_2}$ solve this model with $\rho u'(0) - \omega^2 \tanh Lu(0) = 0$. Show these two solutions are linearly independent as long as ω does not solve $\rho u'(0) - \omega^2 \tanh Lu(0) = 0$.

Exercise 14.4.2 *This refers to the previous problem. For any value ω_n that solves $\rho u'(0) - \omega^2 \tanh Lu(0) = 0$, prove $u_n(t) = \cos(\omega_n t) + \tanh(L) \sin(\omega_n t)$ solves*

$$\begin{aligned} \boldsymbol{u_n}'' + \omega_n^2 \boldsymbol{u_n} &= \boldsymbol{0}, \quad 0 \leq \lambda \leq L \\ u_n'(L) &= 0, \quad \rho u_n'(0) - \omega_n^2 \tanh L u_n(0) = 0 \end{aligned}$$

Exercise 14.4.3 *If $\boldsymbol{u_n}$ and $\boldsymbol{u_m}$ solve*

$$\begin{aligned} \boldsymbol{u_n}'' + \omega_n^2 \boldsymbol{u_n} &= \boldsymbol{0}, \quad 0 \leq \lambda \leq L \\ u_n'(L) &= 0, \quad \rho u_n'(0) - \omega_n^2 \tanh L u_n(0) = 0 \end{aligned}$$

and

$$\begin{aligned} \boldsymbol{u_m}'' + \omega_m^2 \boldsymbol{u_m} &= \boldsymbol{0}, \quad 0 \leq \lambda \leq L \\ u_m'(L) &= 0, \quad \rho u_m'(0) - \omega_m^2 \tanh L u_m(0) = 0 \end{aligned}$$

Use the differential equations directly and integration by parts to show $< \boldsymbol{u_n}, \boldsymbol{u_m} >= -\frac{\tanh(L)}{\rho} \neq 0$ *so that these functions are not orthogonal.*

14.4.1 The Eigenfunctions Are Complete

The homogeneous model to solve is

$$\begin{aligned} u'' - (1 + \Theta)u &= 0 \\ \rho u'(0) - \tanh(L)(1 + \Theta)u(0) &= 0, \\ u'(L) &= 0. \end{aligned}$$

We have already found the nonzero solutions for this model use $1 + \theta_n = -\alpha_n^2$ where $\rho \sin(\alpha_n L) = -\alpha_n \tanh(L) \cos(\alpha_n L)$ and we have seen these values form an infinite sequence with $|\alpha_n| \to \infty$. The solutions then are $\phi_n(\lambda) = \cos(\alpha_n(L - \lambda))$ and they are not mutually orthogonal. Note, this is not a typical boundary value problem as we have discussed since the critical values α_n also occur in the second boundary condition. We can rewrite the model slightly to show this better. Our model is equivalent to

$$\begin{aligned} u'' + \alpha^2 u &= 0 \\ \rho u'(0) - \tanh(L)u''(0) &= 0, \\ u'(L) &= 0. \end{aligned}$$

which clearly shows how different its structure is. With these nonstandard boundary conditions, we see the values $-\alpha_n^2$ are the eigenvalues of the model with associated eigenfunctions $\phi(\lambda)$. However, if you try to construct the kernel, you will not be successful because the boundary condition given in terms of the second derivative makes the operator here not self-adjoint. Still, the separation of variables technique gives rise to these solutions and we will be building general solutions of the form

$$\hat{v}_m(\lambda, \tau) = A_0\, e^{-\tau} + \sum_{n=1}^{\infty} A_n \cos(\alpha_n(L - \lambda))e^{-(1+\alpha_n^2)\tau}.$$

When $\tau = 0$, this means we want to know if the series

$$\hat{v}_m(\lambda, 0) = A_0 + \sum_{n=1}^{\infty} A_n \cos(\alpha_n(L - \lambda))$$

converges. Further, if we apply a known voltage pulse V to the cable we will want to expand V in terms of the eigenfunctions $\phi(\lambda)$ and equate coefficients. That is, we would write

$$V(\lambda) = V_0 + \sum_{n=1}^{\infty} V_n \cos(\alpha_n(L - \lambda))$$

and match coefficients from the expansion of $\hat{v}_m(\lambda, 0)$ given above. However, to do that we need to know if any V can be written like this and to know that the expansion of $\hat{v}_m$ at $\tau = 0$ is valid. Note for $\tau > 0$, the series expansions are eventually so strongly damped by the decay exponentials, that convergence is guaranteed. We will solve this problem a bit obliquely.

14.4.1.1 Auxiliary Model One

Consider the following homogeneous model

$$\begin{aligned} u'' + \omega^2 u &= 0 \\ \rho u(0) - \tanh(L)\, u'(0) &= 0 \\ u(L) &= 0. \end{aligned}$$

The general solution is

$$u(\lambda) = A \cos(\omega\, \lambda) + B \sin(\omega\, \lambda).$$

with derivative

$$u'(\lambda) = -\omega\, A \sin(\omega\, \lambda) + B\, \omega \cos(\omega\, \lambda).$$

The boundary conditions give

$$\begin{aligned} \rho A - \tanh(L)\, \omega B &= 0 \\ A \cos(\omega\, L) + B \sin(\omega\, L) &= 0. \end{aligned}$$

This implies $A = -B \tan(\omega L)$ and hence, using the other boundary condition, we have

$$-B\Big(\rho \tan(\omega L) + \tanh(L)\, \omega\Big) = 0.$$

Simplifying, we find the eigenvalues of this model occur when

$$\tan(\omega L) = -\frac{\tanh(L)}{\rho\, L}\, (\omega L)$$

which is the same equation as Equation 14.7. Hence, the eigenvalues are still α_n^2 starting at 0 with $\alpha_n \to \infty$. The corresponding eigenfunctions are

$$w_n(\lambda) = \sin(\alpha_n(L - \lambda)).$$

Hence, the nonhomogeneous model

$$\begin{aligned} u'' + u &= f \\ \rho u(0) - \tanh(L)\, u'(0) &= 0 \end{aligned}$$

$$u(L) \;=\; 0.$$

is invertible with corresponding self-adjoint operator $\mathscr{J}_1$ and u solves the model if and only if $\mathscr{J}_1(f) = u$. The eigenvalues of $\mathscr{J}_1$ are $\frac{1}{1-\alpha_n^2}$. These eigenfunctions are orthogonal and using our standard arguments, if f is a nonzero function which is orthogonal to each w_n, then it follows $\mathscr{J}_1(f) = 0$. This then implies that $u = 0$ is a solution to a model with a nonzero forcing function which is not possible. Hence, the sequence of eigenfunctions (w_n), once normalized by dividing by their length, becomes an orthonormal sequence $(\hat{w}_n)$ and any f in $\mathbb{L}_2([0, L])$ has an expansion

$$f(\lambda) \;=\; \sum_{n=0}^{\infty} < \hat{w}_n, f > \hat{w}_n(\lambda).$$

14.4.1.2 Auxiliary Model Two

Consider the following homogeneous model

$$\begin{aligned} u'' + \omega^2 u &= 0 \\ u(0) &= 0 \\ \rho u(L) + \tanh(L)\, u'(L) &= 0 \end{aligned}$$

The general solution is

$$u(\lambda) \;=\; A\,\cos(\omega\,\lambda) + B\,\sin(\omega\,\lambda).$$

with derivative

$$u'(\lambda) \;=\; -\omega\, A \sin(\omega\,\lambda) + B\,\omega\,\cos(\omega\,\lambda).$$

The boundary condition $u(0) = 0$ means $A = 0$ and so we have know $u(\lambda) = B\,\sin(\omega\lambda)$. The other boundary condition gives

$$B\Big(\rho\,\sin(\omega L) + \tanh(L)\,\omega\,\cos(\omega L)\Big) \;=\; 0$$

Simplifying, we find the eigenvalues of this model occur when

$$\tan(\omega L) \;=\; -\frac{\tanh(L)}{\rho\, L}\,(\omega L)$$

which is again the same equation as Equation 14.7. Hence, the eigenvalues are still α_n^2 starting at 0 with $\alpha_n \to \infty$. The corresponding eigenfunctions are now

$$z_n(\lambda) \;=\; \sin(\alpha_n\,\lambda).$$

Hence, the nonhomogeneous model

$$\begin{aligned} u'' + u &= f \\ u(0) &= 0 \\ \rho u(L) + \tanh(L)\, u'(L) &= 0 \end{aligned}$$

is invertible with corresponding self-adjoint operator $\mathscr{J}_2$ and u solves the model if and only if $\mathscr{J}_2(f) = u$. The eigenvalues of $\mathscr{J}_2$ are also $\frac{1}{1-\alpha_n^2}$. These eigenfunctions are orthogonal and using our standard arguments, if f is a nonzero function which is orthogonal to each z_n, then it follows $\mathscr{J}(f) = 0$. This then implies that $u = 0$ is a solution to a model with a nonzero forcing function which is not possible. Hence, the sequence of eigenfunctions (z_n) can also be made into an orthonormal sequence by dividing by their lengths, $(\hat{z}_n)$. Any f in $\mathbb{L}_2([0, L])$ then has an expansion

$$f(\lambda) = \sum_{n=0}^{\infty} < \hat{z}_n, f > \hat{z}_n(\lambda).$$

14.4.1.3 Completeness of (ϕ_n)

First, note that

$$\begin{aligned} \sin(\alpha_n L)\phi_n(\lambda) - \cos(\alpha_n L)w_n(\lambda) &= \sin(\alpha_n L)\Big(\cos(\alpha_n L)\cos(\alpha_n \lambda) + \sin(\alpha_n L)\sin(\alpha_n \lambda)\Big) \\ &\quad - \cos(\alpha_n L)\Big(\sin(\alpha_n L)\cos(\alpha_n \lambda) - \cos(\alpha_n L)\sin(\alpha_n \lambda)\Big) \\ &= \Big(\sin^2(\alpha_n L) + \cos^2(\alpha_n L)\Big)\ \sin(\alpha_n \lambda) \\ &= z_n(\lambda). \end{aligned}$$

Hence,

$$\sin(\alpha_n L)\phi_n(\lambda) = \cos(\alpha_n L)w_n(\lambda) + z_n(\lambda).$$

Now, since $\int_0^L \sin^2(\alpha_n(L-\lambda))\, d\lambda = \int_0^L \sin^2(\alpha_n \lambda)\, d\lambda$, we see $\|z_n\| = \|w_n\|$. Thus, dividing by $\|z_n\|$, we have

$$\frac{\sin(\alpha_n L)}{\|z_n\|}\, \phi_n(\lambda) = \cos(\alpha_n L)\hat{w}_n(\lambda) + \hat{z}_n(\lambda).$$

Now assume f is a function in $\mathbb{L}_2([0, L])$ which satisfies $< \phi_n, f >= 0$ for all n. Then, it follows that

$$\cos(\alpha_n L) < \hat{w}_n(\lambda), f > + < \hat{z}_n(\lambda), f > = 0.$$

Then summing up to Q, we have

$$\sum_{n=0}^{Q} \cos(\alpha_n L) < \hat{w}_n, f > \hat{z}_n(\lambda) + \sum_{n=0}^{Q} < \hat{z}_n, f > \hat{z}_n(\lambda) = 0.$$

Then, take the inner product with f to obtain

$$\sum_{n=0}^{Q} \cos(\alpha_n L) < \hat{w}_n, f >< \hat{z}_n, f > + \sum_{n=0}^{Q} < \hat{z}_n, f >< \hat{z}_n, f > = 0.$$

Letting $Q \to \infty$, we have

$$\sum_{n=0}^{\infty} \cos(\alpha_n L) < \hat{w}_n, f >< \hat{z}_n, f > + \sum_{n=0}^{\infty} < \hat{z}_n, f >< \hat{z}_n, f > .$$

The first partial sum clearly converges in $\mathbb{L}_2([0, L])$ norm as the terms $| \cos(\alpha_n L) < w_n, f > |$ are bounded by $| < w_n, f > |$ which is in ℓ^2 since (w_n) is an orthonormal sequence. The second series also converges since (z_n) is complete. However, since $(\hat{w}_n)$ and $(\hat{z}_n)$ are both complete, we know $\sum(< \hat{w}_n, f >)^2 = \sum(< \hat{z}_n, f >)^2$. Hence,

$$\sum_{n=0}^{\infty} \left(\cos(\alpha_n L) < \hat{w}_n, f >< \hat{z}_n, f > + < \hat{w}_n, f >< \hat{w}_n, f > \right) = 0.$$

or

$$\sum_{n=0}^{\infty} < \hat{w}_n, f > \left(\cos(\alpha_n L) < \hat{z}_n, f > + < \hat{w}_n, f > \right) = 0.$$

Finally, recall that $\cos(\alpha_n L) < \hat{w}_n, f > + < \hat{z}_n(\lambda), f >= 0$ and so substituting, we have

$$\sum_{n=0}^{\infty} < \hat{w}_n, f > \left(- \cos^2(\alpha_n L) < \hat{w}_n, f > + < \hat{w}_n, f > \right) = 0.$$

This simplifies to

$$\sum_{n=0}^{\infty} \sin^2(\alpha_n L) \, (< \hat{w}_n, f >)^2 = 0.$$

Since each term is positive, this tells us $< \hat{w}_n, f >= 0$ for all n. Since $(\hat{w}_n)$ is complete, this tells us $f = 0$. This shows that (ϕ_n) is complete in the sense that any function f in $\mathbb{L}_2([0, L])$ can be written as the series $f(\lambda) = \sum_{n=0}^{\infty} < \phi_n, f > \phi_n(\lambda)$. Thus, as usual, we expect the most general solution is given by

$$\hat{v}_m(\lambda, \tau) = A_0 \, e^{-\tau} + \sum_{n=1}^{Q} A_n \, \cos(\alpha_n (L - \lambda)) e^{-(1+\alpha_n^2)\tau} \tag{14.9}$$

$$+ \sum_{n=Q+1}^{\infty} A_n \cos(\alpha_n(L-\lambda))e^{-(1+\alpha_n^2)\tau} \tag{14.10}$$

Since the spatial functions are approximately orthogonal, the computation of the coefficients A_n for $n > Q$ can be handled with a straightforward inner product calculation. The calculation of the first Q coefficients must be handled as a linear algebra problem.

Part VI

Topics in Applied Modeling

Chapter 15

Fields and Charges on a Set

We are now going on an extended journey into how we could use ℓ^1 sequences in economics. Along the way, we will need to look at everything a bit differently so that is the purpose. Looking at stuff you think you know from a different point of view helps us grow and learn how to attack new problems never seen before by thinking outside of our box. This is background material for our brief introduction to a certain type of game. Our game theory introduction is inspired by the much more complete discussion in (Kunnai (12, Chapter 12) 1992).

15.1 Rings and Fields of Subsets

We need to set up a fair amount of notation, so let's get started. These will be set in a given set Ω and concern collections of subset of Ω that satisfy various properties. We need to define

- A **semi-ring** of subsets.
- A **semi-field** of subsets.
- A **ring** of subsets.
- A **field** of subsets.
- An **additive class** of subsets
- A σ **ring** of subsets.
- A σ **field** of subsets.
- A σ **class** of subsets.

and then mappings μ on field $\mathcal{F}$ of subsets of Ω with various properties also. In general, we have a map $\mu : \mathcal{F} \to [-\infty, \infty]$ which will be a **charge**. Then, we add properties to the charge such as it is **bounded**, a **positive charge**, a charge with values in an interval $[0, a]$ and so forth.

We start with the idea of a **semi-ring**.

Definition 15.1.1 A Semi-ring and Semi-fields of Subsets

Let Ω be a set and $\mathcal{F}$ a collection of subsets of Ω. We say $\mathcal{F}$ is a **semi-ring** *on Ω if*

1. *$\emptyset \in \mathcal{F}$.*
2. *If $A, B \in \mathcal{F}$, then $A \cap B \in \mathcal{F}$.*
3. *If $A, B \in \mathcal{F}$ with $A \subset B$, then there exists $A_0, A_1, \ldots A_n \in \mathcal{F}$ for some n so that $A = A_0 \subset A_1 \subset A_2 \subset \ldots \subset A_n = B$ and $A_i \setminus A_{i-1} \in \mathcal{F}$ for $1 \leq i \leq n$.*

If $\Omega \in \mathcal{F}$ and $\mathcal{F}$ is a semi-ring then we say $\mathcal{F}$ is a **semi-field**.

Now we add the definitions rings and fields.

Definition 15.1.2 Rings and Fields of Subsets

We say $\mathcal{F}$ is a **ring** *on Ω if*

1. *$\emptyset \in \mathcal{F}$.*
2. *If $A, B \in \mathcal{F}$, then $A \cup B \in \mathcal{F}$.*
3. *If $A, B \in \mathcal{F}$, then $A \setminus B \in \mathcal{F}$.*

If Ω is also in $\mathcal{F}$ and $\mathcal{F}$ is a ring, we say $\mathcal{F}$ is a **field**.

Homework

Exercise 15.1.1 *Prove if $\mathcal{F}$ is a ring, it is also a semi-ring.*

Exercise 15.1.2 *Prove if $\mathcal{F}$ is a field of subsets, it is closed under intersection and complements.*

The above exercises show we could have defined a field like so:

Definition 15.1.3 A Field of Subsets

Let Ω be a set and $\mathcal{F}$ a collection of subsets of Ω which contains the empty set $\emptyset$. Then $\mathcal{F}$ is a field if

- *If A and B are in $\mathcal{F}$, so is $A \cup B$.*
- *If A and B are in $\mathcal{F}$, so is $A \cap B$.*
- *If A is in $\mathcal{F}$ so is A^C.*

it is easy to see $\mathcal{F}$ is closed under finite unions, intersections and complements.

Homework

Exercise 15.1.3 *Prove the field $\mathcal{F}$ is closed under finite unions, intersections and complements.*

Exercise 15.1.4 *Let $\Omega = \Re$. Let $\mathcal{F}$ be the collection of all intervals of $\Re$ and the empty set. This means intervals which either contain their endpoints or not and which can be finite or infinite as well as $\emptyset$. Is $\mathcal{F}$ a semi-ring? Is it a ring? Is it a semi-field? Is it a field?*

Now we add the idea of an **additive class**. First, we focus on collections of subsets which are closed under disjoint unions.

Definition 15.1.4 An Additive Class of Subsets

We say $\mathcal{F}$ is an **additive class** *on Ω if*

1. $\emptyset \in \mathcal{F}$.
2. *If $A, B \in \mathcal{F}$ and $A \cap B = \emptyset$, then $A \cup B \in \mathcal{F}$.*
3. *If $A, B \in \mathcal{F}$, then $A \setminus B \in \mathcal{F}$.*

The notions of **σ ring** and **σ field** extend the notion of an additive class to countable unions of sets in the collection. Here we do not need to assume these are unions of disjoint sets.

Definition 15.1.5 σ Rings and Fields

$\mathcal{F}$ is a **σ ring** *in Ω if*

1. $\emptyset \in \mathcal{F}$.
2. *If $(A_n)_{n \geq 1} \in \mathcal{F}$, then $\cup_{n \geq 1} A_n \in \mathcal{F}$.*
3. *If $A, B \in \mathcal{F}$, then $A \setminus B \in \mathcal{F}$.*

We then say $\mathcal{F}$ is a **σ field** *if it is a σ ring which contains Ω.*

The last type of collection is called a **σ class**. This is different from a σ - field: look at the definition carefully!

Definition 15.1.6 A σ Class of Subsets

1. $\emptyset \in \mathcal{F}$.
2. *If $(A_n)_{n \geq 1} \in \mathcal{F}$ is a sequence of pairwise disjoint sets in $\mathcal{F}$, then $\cup_{n \geq 1} A_n \in \mathcal{F}$.*
3. *If $A \in \mathcal{F}$, then $A^C \in \mathcal{F}$.*

15.2 Charges

Now we are ready to look at mappings from collections of subsets into the extended reals. The minimal structure we need on the collection of subsets $\mathcal{F}$ is that it is a field. Hence, we are interested in mapping of the type $\mu : \mathcal{F} \rightarrow [-\infty, \infty]$, where $\mathcal{F}$ is a field of subsets. We begin with the idea of a **charge**.

Definition 15.2.1 A Charge

Let $\mathcal{F}$ be a field of subsets of the set Ω. The map $\mu : \mathcal{F} \rightarrow [-\infty, \infty]$ is a **charge** *if*

1. $\mu(\emptyset) = 0$.
2. *If $A, B \in \mathcal{F}$ and $A \cap B = \emptyset$, then $\mu(A \cup B) = \mu(A) + \mu(B)$.*

There are a number of flavors of charges:

1. *We say μ is a real charge if $\mu(F)$ is finite for all $F \in \mathcal{F}$.*
2. *We say μ is a bounded charge if $\sup_{F \in \mathcal{F}} |\mu(F)|$ is finite.*
3. *We say μ is a positive charge if $\mu(F) \geq 0$ for all $F \in \mathcal{F}$.*
4. *We say μ is a positive bounded charge if μ is both positive and bounded.*
5. *We say μ is $\mathbf{0} - \boldsymbol{a}$ valued for some $\boldsymbol{a} \neq 0$ if $\mu(F) = 0$ or $\mu(F) = a$ for all $F \in \mathcal{F}$ and there is at least one F in $\mathcal{F}$ which satisfies $\mu(F) = a$.*
6. *We say μ is a probability charge if μ is positive and $\mu(\Omega) = 1$.*

Homework

Exercise 15.2.1 *Let $\Omega = [a, b]$, a finite interval in $\Re$. Let $\mathcal{F}$ be the collection of all subsets of Ω which are intervals. Is $\mathcal{F}$ a field? Let μ be defined by the usual length of an interval. Is μ a charge?*

Exercise 15.2.2 *Let $\Omega = \Re$. Let $\mathcal{F}$ be the collection of all subsets of Ω which are intervals. Is $\mathcal{F}$ a field? Let μ be defined by the usual length of an interval. Is μ a charge?*

Exercise 15.2.3 *Let $\mathcal{F}$ be a field of subsets of Ω. If $\mathcal{F}$ is a field, note the analogue of $\oplus$ in this ring is set inclusion and the analogue of $\odot$ is set intersection. Define an* **ideal** *and* **filter** *in the field $\mathcal{F}$ like this:*

- *$\mathcal{I}$ is an* **ideal** *in $\mathcal{F}$ if*

  1. *$\Omega \notin \mathcal{I}$; i.e. does not contain the whole set.*
  2. *$A, B \in \mathcal{I}$ imply $A \cup B \in \mathcal{I}$, i.e. closed under unions. Thus, $A \oplus B \in \mathcal{I}$ when A, B are in $\mathcal{I}$ – i.e. it is a subgroup for $\oplus$.*
  3. *$A \in \mathcal{I}$, $B \in \mathcal{F}$ implies $B \cap A \in \mathcal{I}$. Thus, $A \odot B \in \mathcal{I}$ for all $B \in \mathcal{F}$ when $A \in \mathcal{I}$.*

  We say $\mathcal{I}$ is a **maximal ideal** *if no other ideal in $\mathcal{F}$ properly contains $\mathcal{I}$ and a $\mathcal{J}$ is a* **filter** *in $\mathcal{F}$ if*

  1. *$\emptyset \notin \mathcal{J}$; i.e. does not contain the empty set.*
  2. *$A, B \in \mathcal{J}$ imply $A \cap B \in \mathcal{J}$, i.e. closed under intersections; that is $A \odot B \in \mathcal{I}$ when A, B are in $\mathcal{I}$ – i.e. it is a subgroup for $\odot$.*
  3. *$A \in \mathcal{J}$, $B \in \mathcal{F}$ implies $A \cup B \in \mathcal{J}$. Thus, $A \oplus B \in \mathcal{I}$ for all $B \in \mathcal{F}$ when $A \in \mathcal{I}$.*

We say $\mathcal{J}$ is a **maximal filter** *if no other filter in $\mathcal{F}$ properly contains $\mathcal{J}$. Now to the exercise itself. Let $\mathcal{I}$ in $\mathcal{F}$ be a maximal ideal. Define μ by*

$$\mu(F) = \begin{cases} 0, & F \in \mathcal{I} \\ 1, & F \notin \mathcal{I} \end{cases}$$

Prove μ is a charge.

Conversely, if μ is a $0-1$ charge on $\mathcal{F}$, then $\mathcal{I} = \{F \in \mathcal{F} \mid \mu(F) = 0\}$ is a maximal ideal in $\mathcal{F}$ and $\mathcal{J} = \{F \in \mathcal{F} \mid \mu(F) = 1\}$ is a maximal filter in $\mathcal{F}$.

15.3 Ordered Vector Spaces and Lattices

We need further background. We know we are descending into a blizzard of definitions, but we must be careful. So be patient with all this for a bit longer.

Definition 15.3.1 Ordered Vector Spaces

A Vector Space V over $\Re$ is said to be ordered if there is a partial ordering $\preceq$ which is compatible with the vector space operations $\oplus$ and $\odot$ (the scalar multiplication operation). This means

1. $x \preceq y \longrightarrow x \oplus z \preceq y \oplus z$ *for all* $x, y, z \in V$.
2. $x \preceq y \longrightarrow c \odot x \preceq c \odot y$ *for all* $c \neq 0$ V *and all* $x, y in V$.

We denote this ordered vector space by $(V, \preceq)$.

Homework

Exercise 15.3.1 *Prove $C([0,1])$ with the partial ordering $\boldsymbol{f} \preceq \boldsymbol{g} \Longleftrightarrow f(x) \leq g(x)$ for all $x \in [0,1]$ is an ordered vector space.*

Exercise 15.3.2 *Prove $\Re^2$ with $(a,b) \preceq (c,d)$ if and only if $a < c$ or $a = c$ and $b \leq d$ is an ordered vector space.*

In an ordered vector space we can find the infimum and supremum of two elements of the vector space.

Definition 15.3.2 The Infimum and Supremum of Elements in an Ordered Vector Space

- *If $x, y \in (V, \preceq)$ and if there is $z \in (V, \preceq)$ so that $x \preceq z$ and $y \preceq z$ and if $x \preceq z'$ and $y \preceq z'$ implies $z \preceq z'$, we call z the supremum of x and y and denote it by $\sup\{x, y\}$ or $x \vee y$. Of course, z may not exist!*

- *If $x, y \in (V, \preceq)$ and if there is $w \in (V, \preceq)$ so that $w \preceq x$ and $w \preceq y$ and if $w' \preceq x$ and $w' \preceq y$ implies $w' \preceq w$, we call w the infimum of x and y and denote it by $\inf\{x, y\}$ or $x \wedge y$. Such an element w need not exist of course.*

if Γ is an index set and if $\{x_\alpha \,|\, \alpha \in \Gamma\}$ is a subset of $(V, \preceq)$, then the definition above easily extends to $\sup(x_\alpha) = \vee_{\alpha \in \Gamma} x_\alpha$ and $\inf(x_\alpha) = \wedge_{\alpha \in \Gamma} x_\alpha$ as follows:

- *If there is $z \in (V, \preceq)$ so that $x_\alpha \preceq z$ for all α and if $x_\alpha \preceq z'$ for all α implies $z \preceq z'$, we call z the supremum of (x_α)*

- *If there is $w \in (V, \preceq)$ so that $w \preceq x_\alpha$ and if $w' \preceq x_\alpha$ for all α implies $w' \preceq w$, we call w the infimum of (x_α).*

We can now define a vector lattice.

Definition 15.3.3 A Vector Lattice

We say an ordered vector space $(V, \preceq)$ is a Vector Lattice if $x \vee y$ and $x \wedge y$ exist for all $x, y \in (V, \preceq)$.

Further, in a vector lattice, we define the positive part of x to be $x^+ = x \vee 0$ and the negative part of x to be $x^- = x \wedge 0$. The modulus or size of x is then $|x| = x^+ + x^-$. Finally, we can say x and y are orthogonal if $|x| \wedge |y| = 0$ and then we write $x \perp y$.

Homework

Exercise 15.3.3 *In $C([0,1])$ using the usual preference relation $\preceq$, prove $\boldsymbol{f} \vee \boldsymbol{g}$ is the function defined by $h(x) = \max\{f(x), g(x)\}$ and $k(x) = \min\{f(x), g(x)\}$. You must prove $\boldsymbol{h}$ and $\boldsymbol{k}$ are continuous, of course.*

Exercise 15.3.4 *In $C([0,1])$, explain what $\boldsymbol{f}^+$ and $\boldsymbol{f}^-$ look like for any $\boldsymbol{f} \in C([0,1])$.*

Exercise 15.3.5 *In $C([0,1])$ explain what $|\boldsymbol{f}| \wedge |\boldsymbol{g}| = 0$ means for any $\boldsymbol{f}, \boldsymbol{g} \in C([0,1])$.*

Exercise 15.3.6 *How does this definition of orthogonality of two continuous functions compare to our previous one involving the inner product?*

We often want to consider sublattices.

Definition 15.3.4 Sublattices

A subset $W \subset (V, \preceq)$ is called a sublattice if W is a vector subspace of V and

1. *$x \vee y \in W$ for all $x, y \in W$.*
2. *$x \wedge y \in W$ for all $x, yinW$.*

A vector sublattice can also be considered **normal** which is more involved to define.

Definition 15.3.5 Normal sublattices

A vector sublattice W of $(V, \preceq)$ is **normal** *if*

1. *If $x \in W$, $0 \leq |y| \leq |x|$ implies $y \in W$.*
2. *Let (x_α) be a nonempty collection of elements of W indexed by the set Γ. If $\sup_{\alpha \in \Gamma}$ exists, then $\sup_{\alpha \in \Gamma} \in W$.*

We have known for a long time what the orthogonal complement of a subset S of an inner product space is: $S^\perp = \{v \in V \mid < u, v >= 0, \forall u \in S\}$. We have enough structure here to define this same concept in a vector lattice setting.

Definition 15.3.6 The Orthogonal Complement of a Subset of a Vector Lattice

The orthogonal complement of $S \subset (V, \preceq)$ is denoted by $S^\perp$ and defined by $S^\perp = \{v \in V \mid v \perp u, \forall u \in S\}$.

Homework

Exercise 15.3.7 *What is the orthogonal complement of the unit ball in $C([0, 1])$? Here we use the* sup*-norm on $C([0, 1])$ and also think of it as an ordered vector space in the usual way.*

Exercise 15.3.8 *Consider $\Re^2$ as an ordered vector space as we have defined earlier. Let $\Re^2$ also be endowed with the usual Euclidean norm. What is the orthogonal complement of the unit ball?*

We are now ready to prove our first result.

Theorem 15.3.1 The Vector Lattice Theorem One

Let $(V, \preceq)$ be a vector lattice. **We will start abbreviating this as VL for convenience.** *Let (x_α) for α in an index set Γ be nonempty. Then the following are true:*

1. *If $x \in V$ and one of $\sup_\alpha\{x_\alpha\}$ and $\sup_\alpha\{x + x_\alpha\}$ exists, then $x + \sup_\alpha\{x_\alpha\} = \sup_\alpha\{x + x_\alpha\}$.*
2. *If $x \in V$ and one of $\inf_\alpha\{x_\alpha\}$ and $\inf_\alpha\{x + x_\alpha\}$ exists, then $x + \inf_\alpha\{x_\alpha\} = \inf_\alpha\{x + x_\alpha\}$.*
3. *If one of $\sup_\alpha\{x_\alpha\}$ and $\inf_\alpha\{x_\alpha\}$ exists, then the other one also exists and $\inf_\alpha\{x_\alpha\} = -\sup_\alpha\{-x_\alpha\}$.*
4. *For $\beta > 0$, if one of $\sup_\alpha\{x_\alpha\}$ and $\sup_\alpha\{\beta x_\alpha\}$ exists, then the other exists also and $\beta \sup_\alpha\{x_\alpha\} = \sup_\alpha\{\beta x_\alpha\}$.*
5. *For $\beta > 0$, if one of $\inf_\alpha\{x_\alpha\}$ and $\inf_\alpha\{\beta x_\alpha\}$ exists, then the other exists also and $\beta \inf_\alpha\{x_\alpha\} = \inf_\alpha\{\beta x_\alpha\}$.*

Proof 15.3.1
You will prove these assertions in the exercise below. ■

Homework

Exercise 15.3.9 *Prove Theorem 15.3.1:* **1**.

Exercise 15.3.10 *Prove Theorem 15.3.1:* **2**.

Exercise 15.3.11 *Prove Theorem 15.3.1:* **3**.

Exercise 15.3.12 *Prove Theorem 15.3.1:* **4**.

Exercise 15.3.13 *Prove Theorem 15.3.1:* **5**.

Theorem 15.3.2 The Vector Lattice Theorem Two

Let $(V, \preceq)$ be a VL. Let (x_α) for α in an index set Γ be nonempty. Then the following are true for all $x, y \in V$:

1. $(x \vee y) + (x \wedge y) = x + y$.
2. $x = x^+ - x^-$; *the standard decomposition result.*
3. $x^+ \wedge x^- = 0$.
4. $|x| = x^+ \vee x^- = x \vee (-x) = x^+ + x^-$.

Proof 15.3.2
1:

$$\begin{aligned} (-x) \vee (-y) &= -(x \wedge y), \text{ by Theorem 15.3.1: } \mathbf{3} \implies \\ (x+y) - (x \wedge y) &= (x+y) + ((-x) \vee (-y)) \\ &= ((x+y) - x) \vee ((x_y) - y), \text{ by Theorem 15.3.1: } \mathbf{1} \\ &= y \vee x = x \vee y \end{aligned}$$

2:
Note $x^+ - x^- = (x \vee 0) - ((-x) \vee 0)$. *Also by Theorem 15.3.1:* **3**, *we know* $-(x \wedge 0) = ((-x) \vee 0)$. *Thus, by Theorem 15.3.1:* **6** $x^+ - x^- = (x \vee 0) + (x \wedge 0) = x + 0 = x$

3:
We know from Theorem 15.3.1: **6** *that* $(x^+ \vee x^-) + (x^+ \wedge x^-) = x^+ + x^-$. *We will show* $x^+ + x^- = (x^+) \vee (x^-)$ *which implies* $(x^+) \wedge (x^-) = 0$. *Since,* $x = x^+ - x^-$, $x^+ + x^- = x + 2x^-$. *But* $x^- = (-x) \vee 0$ *and so*

$$x^- + x^- = ((-x) \vee 0) + ((-x) \vee 0) = 2((-x) \vee 0)$$

So

$$\begin{aligned} x + x^- + x^- &= x + 2((-x) \vee 0) = x + ((-2x) \vee 0), \text{ by Theorem 15.3.1: } \mathbf{4} \\ &= (x - 2x) \vee x, \text{ by Theorem 15.3.1: } \mathbf{1} \end{aligned}$$

$$\begin{aligned} &= (-x) \vee (x) + 0 \\ &= ((-x) \vee 0) \vee (x \vee 0) \text{ by Theorem 15.3.1: } \mathbf{4} \\ &= x^- \vee x^+ \end{aligned}$$

4:
In the proof of the last assertion, we found $x^+ + x^- = x^+ \vee x^- = |x|$ by definition. ■

Theorem 15.3.3 The Vector Lattice Theorem Three

Let $(V, \preceq)$ be a VL. Let (x_α) for α in an index set Γ be nonempty. Then the following are true for all $x, y, z \in V$:

1. *If $x = y - z$, $y \geq 0$ and $z \geq 0$, then $y \geq x^+$ and $Z \geq x^-$.*
2. *If $x = y - z$ and $y \wedge z = 0$, then $y = x^+$ and $z = x^-$.*

Proof 15.3.3
1:
If $x = y - z$, then $y - x = z \geq 0$ by assumption. Hence, $(y - x) + x \geq 0 + x$ as $\preceq$ is compatible with the vector space operations. Thus, $y \geq x$. We also know $y \geq 0$ by assumption and so by the definition of the supremum here $x \vee 0 \leq y$ implying $x^+ \leq y$.

Also, $y - x = z \geq 0$ and so $y = z + x$. Since $y \geq 0$ by assumption, we have $(z + x)0x \geq 0 + (-x)$ or $z \geq -x$. Since $z \geq 0$ and $z \geq -x$, the definition of supremum tells us $(-x) \vee 0 = x^- \leq z$.
2:
$y \wedge (z) = 0$ implies by definition that $y \geq y \wedge z = 0$ and $z \geq y \wedge z = 0$. By the previous assertion, we then have $y \geq x^+$ and $z \geq x^-$. Thus, $y - x^+ \geq 0$ and $z - x^- \geq 0$.

Now $x^+ = x \vee 0$ implies $x^+ \geq 0$ and $-x^+ \leq 0$ or $y - x^+ \leq y$. In a similar fashion, we can also show $z = x^- \leq z$. Combining these results, we have $0 \leq y = x^+ \leq y$ and $0 \leq z - x^- \leq z$.

Now we assume $y \wedge z = 0$ and so by Theorem 15.3.1: **2** *$(y = x^+) \wedge (z - x^+) = 0$ and since $x = y - z$ by assumption*

$$x = x^+ - x^- \Longrightarrow y - z = x^+ - x^- \Longrightarrow y = x^+ = z - x^-$$

Let $u = y - x^+ = z - x^-$ and so $u \wedge u = 0$ implying $0 \leq u$. But we already know $u \geq 0$ and so combining $u = 0$. This immediately implies $y = x^+$ and $z = x^-$. ■

Theorem 15.3.4 The Vector Lattice Theorem Four

Let $(V, \preceq)$ be a VL. Let (x_α) for α in an index set Γ be nonempty. Then the following are true for all $x \in V$ and scalar c

1. *$|x| = 0$ if and only if $x = 0$.*
2. *$|cx| = |c|\,|x|$.*
3. *$|x + y| \leq |x| + |y|$.*

Proof 15.3.4

1:
If $|x| = 0$*, then* $x^+ \vee x^- = 0$ *and* $x \vee (-x) = 0$ *by Theorem 15.3.2:* **4**. *Thus,* $x \leq 0$ *and* $(-x) \leq 0$ *implying* $x = 0$*. Conversely, if* $x = 0$*, then* $|x| = 0$*.*

3:
$x^+ = x \vee 0$ *implies* $x^+ \geq 0$ *and* $x^+ \geq x$ *and also* $y^+ = y \vee 0$ *implies* $y^+ \geq 0$ *and* $y^+ \geq y$ *or* $x^+ y^+ \geq x + y^+ \geq x + y$*.*

Also

$$\begin{aligned} x^- &= (-x) \vee 0 \Longrightarrow x^- \geq 0,\ x^- \geq -x \\ y^- &= (-y) \vee 0 \Longrightarrow y^- \geq 0,\ y^- \geq -y \end{aligned}$$

Hence, $x^- + y^- \geq -x + y^- \geq -x - y$*. Also, since* $x^+ + y^+ \geq 0$ *and* $x^- + y^- \geq 0$*. Combining,*

$$\begin{aligned} x^+ + y^+ &\geq 0, \quad x^+ + y^+ \geq x + y \Longrightarrow x^+ + y^+ \geq (x + y) \vee 0 = (x + y)^+ \\ x^- + y^- &\geq 0, \quad x^- + y^- \geq -x - y \Longrightarrow x^- + y^- \geq (-x - y) \vee 0 = (-x - y)^+ = (x + y)^- \end{aligned}$$

Thus,

$$|x| + |y| = (x^+ + y^+) + (x^- + y^-) \quad \geq \quad (x_y)^+ (x + y)^- = |x + y|$$

2:
If $c > 0$*,*

$$(cx)^- \quad = \quad (-cx) \vee 0 = -((cx) \wedge 0) \Longrightarrow -(cx) = (cx) \wedge 0 = c(x \wedge 0)$$

Now, $x^- = (-x) \vee 0 = -(x \wedge 0)$ *implies* $-x^- = x \wedge 0$*. Thus,* $-(cx)^- = c(x \wedge 0) = c(-x^-)$*; that is,* $-(cx)^- = -cx^-$ *or* $(cx^-) = cx^-$*. Also, since* $c > 0$*,* $(cx^+) - (cx) \vee 0 = c(x \vee 0) = cx^+$*. This tells us*

$$(cx)^+ + (cx)^- \quad = \quad cx^+ + cx^- = c|x| \Longrightarrow |cx| = c|x|$$

when $c > 0$*.*

Now, if $c < 0$*,* $-c > 0$ *and* $c = -|c|$*. Thus,* $|cx| = (cx)^+ + (cx)^- = ((-|c|x)^+) + ((-|c|x)^-)$*. It follows*

$$\begin{aligned} (-|c|x)^- &= (|c|x) \vee 0 = |c|(x \vee 0) = |c|x^+ \\ (-|c|x)^+ &= (-||c|x) \vee 0 = -(|c|x) \wedge 0 \end{aligned}$$

which implies

$$-(-|c|x)^+ \quad = \quad (|c|x) \wedge 0 = |c|(x \wedge 0) = |c|(-x^-) = -|c|x^-$$

We conclude $(-|c|x)^+ = |c|x^-$*. Combining, we have* $|cx| = |c|x^+ + |c|x^- = |c|(x^+ + x^-) = |c||x|$*.*
■

15.4 The Structure of Bounded Charges

The set of all bounded charges comes up a lot. We need a definition.

Definition 15.4.1 The Set of Bounded Charges

Let $\mathcal{F}$ be a field of subsets of the set Ω. We let

$$ba(\Omega, \mathcal{F}) = \{\mu : \mathcal{F} \rightarrow \Re \mid \mu \text{ is a bounded charge}\}$$

Define a partial order $\preceq$ on $ba(\Omega, \mathcal{F})$ as follows: $\mu \leq \nu$ if $\mu(F) \leq \nu(F)$ for all $F \in \mathcal{F}$. Then, it is easy to see $\preceq$ is reflexive, antisymmetric and transitive and so is a partial order. We leave the proof of this to you in the exercise below.

Homework

Exercise 15.4.1 *Prove $\preceq$ is a partial order on $ba(\Omega, \mathcal{F})$.*

We need some more notation.

Definition 15.4.2 Boundedly Complete Ordered Vector Spaces

$(V, \preceq)$ is a boundedly complete VL if for all sequences (x_α) for α in an index set Γ that are bounded above (i.e. there is an $x \in V$ so that $x_\alpha \leq x$ for all $\alpha \in \Gamma$), then $\sup_{\alpha \in \Gamma} x_\alpha$ exists in V.

We can add a norm to the VL:

Definition 15.4.3 Normed VL

$(V, \preceq)$ is a normed VL if there is a norm $\|\cdot\|_V$ so that $\|x\|_V \leq \|y\|_V$ for all $x, y \in V$ with $|x| \leq |y|$.

We can also assume the normed VL is complete with respect to this norm giving us a Banach Space.

Definition 15.4.4 Complete Normed VL

$(V, \preceq)$ is a complete normed VL V is a normed VL and V is complete with respect to this norm; i.e. $(V, \preceq, \|\cdot\|_V)$ is a complete normed linear space.

Now we will prove quite a few properties about the bounded charges. First, we show $ba(V, \preceq)$ is an ordered vector space.

Theorem 15.4.1 The Bounded Charge Theorem One

1. $\mu, \nu \in ba(\Omega, \mathcal{F})$ *and* $c, d \in \Re$ *imply* $c\mu + d\nu \in ba(\Omega, \mathcal{F})$; *closed under linear combinations.*
2. $ba(\Omega, \mathcal{F})$ *is a vector space.*
3. $ba(\Omega, \mathcal{F})$ *is an ordered vector space using the partial order* $\preceq$ *defined above which means*
   - $\mu, \nu \in ba(\Omega, \mathcal{F})$ *with* $\mu \preceq \nu$ *implies* $\mu + \gamma \leq \nu + \gamma$ *for all* $\gamma \in ba(\Omega, \mathcal{F})$.
   - $\mu, \nu \in ba(\Omega, \mathcal{F})$ *with* $\mu \preceq \nu$ *implies* $c\mu \leq c\nu$ *for all* $c \geq 0$.

Proof 15.4.1
We will leave these arguments to you. ■

Homework

Exercise 15.4.2 *Prove Theorem 15.4.1:* **1**.

Exercise 15.4.3 *Prove Theorem 15.4.1:* **2**.

Exercise 15.4.4 *Prove Theorem 15.4.1:* **3**.

Next, we characterize $\mu \vee \nu$ and $\mu \wedge \nu$ for any $\mu, \nu \in ba(\Omega, \mathcal{F})$.

Theorem 15.4.2 Characterize $\mu \vee \nu$ and $\mu \wedge \nu$ in $ba(\Omega, \mathcal{F})$

1. *For* $\mu, \nu \in ba(\Omega, \mathcal{F})$, *define* $\lambda : \mathcal{F} \to [-\infty, \infty]$ *by*
$$\lambda(F) = \sup\{\mu(E) + \nu(F \setminus E) \mid E \subset F, \ E \in \mathcal{F}\}$$
2. $\mu, \nu \in ba(\Omega, \mathcal{F})$ *implies* $\mu \vee \nu \in ba(\Omega, \mathcal{F})$ *and* $\lambda = \mu \vee \nu$.
3. *Let* $\mu, \nu \in ba(\Omega, \mathcal{F})$ *and define* $\tau : \mathcal{F} \to [-\infty, \infty]$ *by*
$$\tau(F) = \inf\{\mu(E) + \nu(F \setminus E) \mid E \subset F, \ E \in \mathcal{F}\}$$
Then $\tau \in ba(\Omega, \mathcal{F})$.
4. $\mu, \nu \in ba(\Omega, \mathcal{F})$ *implies* $\mu \wedge \nu \in ba(\Omega, \mathcal{F})$ *and* $\tau = \mu \wedge \nu$ *where* τ *is defined as above.*
5. $(ba(\Omega, \mathcal{F}), \preceq)$ *is a vector lattice.*

Proof 15.4.2
We leave these arguments to you also. ■

Homework

Exercise 15.4.5 *Prove Theorem 15.4.2:* **1.**

Exercise 15.4.6 *Prove Theorem 15.4.2:* **2.**

Exercise 15.4.7 *Prove Theorem 15.4.2:* **3.**

Exercise 15.4.8 *Prove Theorem 15.4.2:* **4.**

Now we can prove $ba(\Omega, \mathcal{F})$ is a vector lattice as we can now characterize the infimum and supremum of two elements.

Theorem 15.4.3 $ba(\Omega, \mathcal{F})$ **is a Vector Lattice**

The vector space $(ba(\Omega, \mathcal{F}), \preceq)$ *is a vector lattice.*

Proof 15.4.3
This proof is easy as we have shown the infimum and supremum of two elements exist. ■

Next, we show the vector lattice $(ba(\Omega, \mathcal{F}), \preceq)$ is boundedly complete.

Theorem 15.4.4 $(ba(\Omega, \mathcal{F}), \preceq)$ **is a Boundedly Complete Vector Lattice**

The vector lattice $(ba(\Omega, \mathcal{F}), \preceq)$ *is a boundedly complete.*

Proof 15.4.4
Let (μ_α) *be a collection in* $ba(V, \preceq)$ *for the index set* Γ *be bounded above so there is* $\mu \in ba(V, \preceq)$ *so that* $\mu_\alpha \leq \mu$ *for all* $\alpha \in \Gamma$. *We will show* $\sup_{\alpha \in \Gamma} \mu_\alpha$ *exists.*

Let $\mathcal{A}$ *be the collection of finite subsets of* Γ. *Let a finite subset of* Γ *be denoted by* γ *and let* $\mu_\gamma = \sup_{c \in \gamma} \mu_c$ *be the supremum over the finite index set. Now consider the collection* $\{\mu_\gamma \,|\, \gamma \in \mathcal{A}\}$. *Note if* $\gamma \subset \xi$, *then* $\mu_\gamma \leq \mu_\xi$. *Also, by assumption* $\mu_\gamma \leq \mu$ *for all* $\gamma \in \mathcal{A}$.

Define τ *as* $\tau(F) = \sup_{\gamma \in \mathcal{A}} \mu_\gamma(F)$. *Since* $\mu_\gamma(F) \leq \mu(F)$ *for all choices of* γ, *we see the supremum exists. Thus* τ *is a charge.*

Also, since $\tau \leq \mu$ *as* μ *is an upper bound, we see* τ *is a bounded charge. Next, since* $\{\alpha\}$ *is a singleton set for all* α, *it is a valid choice for* $\mathcal{A}$. *For ease of notation, let* $\mu_\alpha = \mu_{\{\alpha\}}$. *Then,* $\mu_\alpha \leq \tau$. *Hence,* $\sup_{\alpha \in \Gamma} \mu_\alpha \leq \tau$.

Also, $\mu_\gamma \leq \sup_{\alpha \in \Gamma} \mu_\alpha$ *for all* γ. *This implies* $\tau = \sup_{\gamma \in \Gamma} \mu_\gamma \leq \sup_{\alpha \in \Gamma} \mu_\alpha \leq \tau$ *or* $\tau = \sup_{\alpha \in \Gamma} \mu_\alpha$ *which completes the proof.*

■

Next, we show the vector lattice $(ba(\Omega, \mathcal{F}), \preceq)$ is a Banach Space.

Theorem 15.4.5 $(ba(\Omega, \mathcal{F}), \preceq)$ **is a Banach Space**

1. *For $\mu \in ba(\Omega, \mathcal{F})$, let $\|\mu\| = |\mu|(\Omega)$ where recall, $|\mu| = \mu^+ + \mu^-$. The $\|\cdot\|$ is a norm on $(ba(\Omega, \mathcal{F}), \preceq)$.*
2. *$(ba(\Omega, \mathcal{F}), \preceq, \|\cdot\|)$ is a Banach Lattice.*

Proof 15.4.5

1:

If $\mu \in ba(\Omega, \mathcal{F})$, let $|\mu| = \mu^+ + \mu^-$. If $\mu = 0$, then $|\mu| = 0$ which implies $|\mu|(\Omega) = 0$ and so $|\mu| = 0$. Conversely, if $\|\mu\| = 0$, then $|mu|(\Omega) = 0$. Since $|\mu|$ is a positive charge on $\mathcal{F}$, this tells us $|\mu| = 0$. Then by Theorem 15.3.4: **1**, *it follows that $\mu = 0$.*

Now if $c \in \Re$ and $\mu \in ba(\Omega, \mathcal{F})$, we know $\|c\mu\| = |c\mu|(\Omega) = |c||\mu|(\Omega) = |c|\|\mu\|$ by Theorem 15.3.4: **2**.

If $\mu, \nu \in ba(\Omega, \mathcal{F})$, then $\|\mu + \nu\| = |\mu + \nu|(\Omega) \leq |\mu|(\Omega) + |\nu|(\Omega)$ by Theorem 15.3.4: **3**. *This says $\|\mu + \nu\| \leq \|\mu\| + \|\nu\|$*

If $\mu, \nu \in ba(\Omega, \mathcal{F})$ and $\mu \leq \nu$, then $|\mu|(\Omega) \leq |\nu|(\Omega)$ implying $\|\mu\| \leq \|\nu\|$. Hence the norm is compatible with the ordering $\preceq$ and $ba(\Omega, \mathcal{F})$ is a normed VL.

It follows $\|\cdot\|$ is a norm on $ba(\Omega, \mathcal{F})$.

2:

Let (μ_n) be a Cauchy sequence in $ba(\Omega, \mathcal{F})$. then given $\epsilon > 0$ there is N so that

$$n, m > N \implies \|\mu_m - \mu_n\| < \frac{\epsilon}{2}$$

Also, for any $\xi \in ba(\Omega, \mathcal{F})$, we know $|\xi| \geq \xi$ and $|\xi| \geq \xi^-$. Thus,

$$\mu_m - \mu_n \leq |\mu_m - \mu_n|, \quad -(\mu_m - \mu_n) \leq |\mu_m - \mu_n|$$

We conclude for any $F \in \mathcal{F}$,

$$|\mu_m(F) - \mu_n(F)| = |\mu_m - \mu_n|(F) \leq |\mu_m - \mu_n|(\Omega) = \|\mu_m - \mu_n\| < \frac{\epsilon}{2}$$

if $n, m > N$. Hence, for all $F \in \mathcal{F}$, $(\ \mu_n(F)\ )$ is a Cauchy sequence in $\Re$ and so converges to some number. Define $\mu : \mathcal{F} \to \Re$ by $\mu(F) = \lim_n \mu_n(F)$.

It is clear μ is well-defined on $\mathcal{F}$ is a charge. Also note for $n > N$,

$$\lim_m |\mu_m(F) - \mu_n(F)| \leq \frac{\epsilon}{2} < \epsilon \Longrightarrow |\mu(F) - \mu_n(F)| < \epsilon$$

It is easy to see the $|\cdot|$ satisfies a reverse triangle inequality, so this implies for $\hat{n} > N$

$$|\mu(F) \leq |\mu_{\hat{n}}(F)| + \epsilon \leq |\mu_{\hat{n}}(\Omega)| + \epsilon$$

This shows μ is a bounded charge as $\mu_{\hat{n}}$ is bounded. Hence, $\mu - \mu_{\hat{n}} \in ba(\Omega, \mathcal{F})$.

Now $\mu_n(F) \rightarrow \mu(F)$ implies $\mu_n^+(F) \rightarrow \mu^+(F)$ and $\mu_n^-(F) \rightarrow \mu^-(F)$. Thus, in particular, $\mu_n^+(\Omega) \rightarrow \mu^+(\Omega)$ and $\mu_n^-(\Omega) \rightarrow \mu^-(F\Omega)$. It follows,

$$|\mu_n(\Omega)| \quad = \quad \mu_n^+(\Omega) + \mu_n^-(\Omega) \rightarrow \mu^+(\Omega) + \mu^-(\Omega) = |\mu(\Omega)|$$

Thus, μ_n converges to μ is norm. ■

Let's summarize what we know into a single theorem.

Theorem 15.4.6 Properties of $ba(\Omega, \mathcal{F})$

Let $\mathcal{F}$ be a field of subsets of the set Ω. If $\mu \in ba(\Omega, \mathcal{F})$, then

1. $\mu^+(F) = \sup\{\mu(E) \mid E \subset F, E \in \mathcal{F}\}$, $\forall F \in \mathcal{F}$.
2. $\mu^-(F) = \inf\{\mu(E) \mid E \subset F, E \in \mathcal{F}\}$, $\forall F \in \mathcal{F}$.
3. $|\mu|(F) = \sup\{\mu(E) + (-\mu)(F \setminus E) \mid E \subset F, E \in \mathcal{F}\}$, $\forall F \in \mathcal{F}$.
4. *If $\mu = \lambda - \nu$ for $\lambda, \nu \in ba(\Omega, \mathcal{F})$ with $\lambda \geq 0$ and $\nu \geq 0$, then $\lambda \geq \mu^+$ and $\nu \geq \mu^-$.*
5. *If $\mu = \lambda - \nu$ for $\lambda, \nu \in ba(\Omega, \mathcal{F})$ with $\lambda \wedge \nu = 0$, then $\lambda = \mu^+$ and $\nu = \mu^-$.*
6. $\mu^+ = (1/2)(|\mu| + \mu)$, $\quad \mu^- = (1/2)(|\mu| - \mu)$
7. *If $\nu \in ba(\Omega, \mathcal{F})$ with $\nu \wedge \mu = 0$, then $(\mu + \nu)^+ = |\mu| + |\nu|$.*

Proof 15.4.6

1:
*By Theorem 15.4.2: **1** and **2**, since $\mu^+ \; \mu \vee 0$, for $F \in \mathcal{F}$,*

$$\mu^+(F) \quad = \quad \sup\{\mu(E) + \mathbf{0}(fF \setminus E) \mid E \subset F, \; E \in \mathcal{F}\}$$

Thus, $\mu^+(F) = \sup\{\mu(E) \mid E \subset F, E \in \mathcal{F}\}$, $\forall F \in \mathcal{F}$.

2:
*This is essentially the same proof as the one above except we use Theorem 15.4.2: **3** and **4**.*

3:
*Since $\mu| = \mu^+ + \mu^-$ and $|\mu| = \mu \vee (-\mu)$ by Theorem 15.3.2: **4**, by Theorem 15.4.2: **1** and **2** or $F \in \mathcal{F}$,*

$$|\mu|(F) \quad = \quad \sup\{\mu(E) + (-\mu)(fF \setminus E) \mid E \subset F, \; E \in \mathcal{F}\}$$

4 - 7:
We leave these arguments to you. ■

Comment 15.4.1 *μ^+ is called the **positive variation** of μ and μ^- is the **negative variation** of μ. Hence, $|\mu|$ is the **absolute variation** of μ.*

Homework

Exercise 15.4.9 *If $\mu_n(F) \rightarrow \mu(F)$, then $\mu_n^+(F) \rightarrow \mu^+(F)$ and $\mu_n^-(F) \rightarrow \mu^-(F)$.*

Exercise 15.4.10 *Prove Theorem 15.4.6:* **4**.

Exercise 15.4.11 *Prove Theorem 15.4.6:* **5**.

Exercise 15.4.12 *Prove Theorem 15.4.6:* **6**.

Exercise 15.4.13 *Prove Theorem 15.4.6:* **7**.

Comment 15.4.2 *μ^+ is called the* **positive variation** *of μ and μ^- is the* **negative variation** *of μ. Hence, $|\mu|$ is the* **absolute variation** *of μ.*

Comment 15.4.3 *The theorem above tells us that is $\mu \in ba(\Omega, \mathcal{F})$, then there is a decomposition $\mu = \mu^+ - \mu^-$ with $\mu^+ \wedge \mu^- = 0$. This is the* **Jordan Decomposition of** μ. *Further, $|\mu| = \mu^+ + \mu^- = \mu^+ \vee \mu^- = \mu \vee (-\mu)$.*

We now prove another characterization of the total variation of μ.

Theorem 15.4.7 Characterizing the Total Variation of μ in Terms of Partitions

A partition P of the set $F \in \mathcal{F}$ is a finite collection $\{F_1, \ldots, F_n\}$ for some n with $F_i \cap F_j = \emptyset$ if $i \neq j$ (i.e. the sets are pairwise disjoint) with $F = \cup_{i=1}^n F_i$.

Then $|\mu|(F) = \sup_P \sum_{i=1}^n |\mu(F_i)|$. We usually use the standard abuse of notation for this kind of sum (see our earlier discussion of Riemann and Riemann - Stieltjes Integration) and write $|\mu|(F) = \sup_P \sum_{i \in P} |\mu(F_i)|$.

Further, $|\mu|(\Omega) \leq 2 \sup_{F \in \mathcal{F}} |\mu(F)|$.

Proof 15.4.7
Let $\{F_1, \ldots, F_n\}$ be a partition of F and let I be the set of indices where $\mu(F_i) \geq 0$ And J, the set of indices where $\mu(F_i) < 0$. Then

$$\sum_{i \in P} |\mu(F_i)| \quad = \quad \sum_{i \in I} \mu(F_i) + \sum_{j \in J} |\mu(F_j)|$$

Now for $j \in J$, $|\mu(F_j)| = \mu^+(F_j) + \mu^-(F_j)$ where $\mu^+(F_j) = \mu(F_j) \vee 0 = 0$ and $\mu^-(F_j) = (-\mu(F_j)) \vee 0 = \mu(F_j)$ as $\mu(F_j) < 0$. So, $|\mu(F_j)| = -\mu(F_j)$. We conclude

$$\begin{aligned} \sum_{i \in P} |\mu(F_i)| &= \sum_{i \in I} \mu(F_i) - \sum_{j \in J} \mu(F_j) = \mu\Big(\cup_{i \in I} F_i\Big) - \mu\Big(\cup_{j \in J} F_{i}j\Big) \\ &= \mu\Big(\cup_{i \in I} F_i\Big) - \mu\Big(F \setminus \cup_{i \in I} F_i\Big) \end{aligned}$$

From Theorem 15.4.6: **3**, *we know $|\mu|(F) = \sup\{\mu(E) + (-\mu)(F \setminus E) \mid E \subset F, E \in \mathcal{F}\}$. This is true for all partitions of F. It follows immediately from this that*

$$|\mu|(F) \quad \geq \quad \mu\Big(\cup_{i \in I} F_i\Big) - \mu\Big(F \setminus \cup_{i \in I} F_i\Big) = \sum_{i \in P} |\mu(F_i)|$$

Conversely, let $B \in \mathcal{F}$ with $B \subset F$. Then $\{B, F \setminus B\}$ is a partition of F and so

$$\mu(B) - \mu(F \setminus B) \;\; \leq \;\; |\mu(B)| + |\mu(F \setminus B)| \leq \sup_P \sum_{i \in P} |\mu(F_i)|$$

where P is any partition of F. Then again by Theorem 15.4.6: **3**

$$|\mu|(F) \;\; = \;\; \sup_B \{\mu(B) - \mu(F \setminus B)\} \leq \sup_P \sum_{i \in P} |\mu(F_i)|$$

Combining, we have the desired result.

Also, $\{F, \Omega \setminus F\}$ is a partition of Ω so $|\mu|(\Omega) = \sup_F\{\mu(F) - \mu(\Omega \setminus F)\}$. This tells us

$$\mu(F) - \mu(\Omega \setminus F) \;\; \leq \;\; |\mu(F)| + |\mu(\Omega \setminus F)| \leq \sup_{G \in \mathcal{F}} |\mu(G)| + \sup_{G \in \mathcal{F}} |\mu(G)| = 2 \sup_{G \in \mathcal{F}} |\mu(G)|$$

which proves the second assertion. ■

Homework

Exercise 15.4.14 *Look back at Riemann Integration. For a given bounded function $\boldsymbol{f}$ which is Riemann Integrable on $[a, b]$, we know $\int_a^b \boldsymbol{f} = \lim_n \sum_{i \in P_n} m_i \ell(I_i)$ where $\ell(I_i)$ is the length of the ith subinterval determined by the uniform partition P_n of $[a, b]$. Show how any of these Riemann sums give a charge in $ba([a, b], \mathcal{F})$ where $\mathcal{F}$ is the collection of all open, closed and half open subintervals of $[a, b]$. There are some subtleties here. Each partition of $[a, b]$ in the usual Riemann Integral way divides up the interval $[a, b]$ into disjoint open subintervals. We can use those to create the pairwise disjoint sets F_i in our partition scheme in Theorem 15.4.7. Of course, how we do this is not unique. However, the infimum α and supremum β of any open subinterval of $[a, b]$ are uniquely defined and we can define $\ell(F_i) = \beta - \alpha$ whether the endpoints of the interval are in F_i or not. Hence, we can define a charge μ_f by $\mu_f(F) = \sup_P \sum_{i \in P} m_i \ell(F_i)$ for any subinterval F of $[a, b]$. Again, there are all these subtleties of endpoint stuff here so be careful to think this through. Does it seem reasonable to you that Riemann Integration gives us charges?*

Exercise 15.4.15 *Do the same as what you did in the above exercise but for Riemann - Stieltjes Integration.*

15.5 Measures

We can add one more idea: that of a measure on the field $\mathcal{F}$ of subsets of Ω. Let $\mu : \mathcal{F} \to [-\infty, \infty]$ be a charge. We already know a charge is finitely additive: i.e. $\mu(\cup_{n=1}^N A_n) = \sum_{n=1}^\infty \mu(A_n)$ for all finite sequences of sets (A_n) that are pairwise disjoint. Since $\mathcal{F}$ is a field it is a ring containing Ω and therefore it is closed under countable unions. However, we don't know in general if a charge satisfies $\mu(\cup_{n=1}^\infty A_n) = \sum_{n=1}^\infty mu(E_n)$ for all sequences of sets (A_n) of pairwise disjoint $A_n \in \mathcal{F}$ with $\cup_{n=1}^\infty A_n \in \mathcal{F}$. If we assume the charge satisfies that condition, we get a **measure** on $\mathcal{F}$.

Definition 15.5.1 Measures

Let μ be a charge on $\mathcal{F}$. Assume $\mu(\cup_{n=1}^{\infty} A_n) = \sum_{n=1}^{\infty} \mu(E_n)$ for all sequences of sets (A_n) of pairwise disjoint $A_n \in \mathcal{F}$ which satisfy $\cup_{n=1}^{\infty} A_n \in \mathcal{F}$. Then we say μ is a measure.

1. *If μ is bounded, it is a* **bounded** *measure.*
2. *If μ is non - negative, it is a* **positive** *measure.*

Note if $\mathcal{F}$ is a σ - field, it is closed under countable unions and so the condition $\cup_{n=1}^{\infty} A_n \in \mathcal{F}$ is always true.

Comment 15.5.1 *Since any charge is finitely additive, any charge can be called a* **finitely additive measure**.

There is a very important subset of $ba(\Omega, \mathcal{F})$ which is very useful in practice.

Definition 15.5.2 The Set of All Bounded Measures

The set of all bounded measures on a field $\mathcal{F}$ is denoted by $ca(\Omega, \mathcal{F})$.

It is easy to see the following properties hold for bounded measures.

Theorem 15.5.1 Properties of Bounded Measures

If $\mu, \nu \in ca(\Omega, \mathcal{F})$, then μ^+, μ^-, $|\mu|$, $\mu \vee \nu$ and $\mu \wedge \nu$ are also in $ca(\Omega, \mathcal{F})$.

Proof 15.5.1
We will let you prove these assertions. ■

Also, we can show $ca(\Omega, \mathcal{F})$ is a normal vector sublattice of $ba(\Omega, \mathcal{F})$.

Theorem 15.5.2 The set of Bounded Measures is a Normal Sublattice in the Set of Bounded Charges

The set $ca(\Omega, \mathcal{F})$ is a normal vector sublattice of $ba(\Omega, \mathcal{F})$. In particular, this implies $ca(\Omega, \mathcal{F})$ is a closed subspace of $ba(\Omega, \mathcal{F})$.

Proof 15.5.2
We will let you prove this one also. ■

Homework

Exercise 15.5.1 *Prove the assertions of Theorem 15.5.1.*

Exercise 15.5.2 *Prove $ca(\Omega, \mathcal{F})$ is a normal vector sublattice of $ba(\Omega, \mathcal{F})$.*

Exercise 15.5.3 *Prove $ca(\Omega, \mathcal{F})$ is a closed subspace of $ba(\Omega, \mathcal{F})$.*

Chapter 16

Games

A nice application to use all of our material on charges, measures and the sequence space ℓ^1 and dual spaces is something called **game theory**. A good theoretical discussion of this at a graduate level in analysis can be found in (Kunnai (12, Chapter 12) 1992). In Chapter 15, we gave you the background material on charges you will need. Now we can talk about simple games and connect the discussion to the sequence space $\mathbf{c_0}$, its dual ℓ^1 and finally the sequence space ℓ^∞ and its dual $ba(\Omega, \mathcal{F})$. Then we stop there as to do more is far beyond our brief here.

16.1 Finite Numbers of Players

Consider a set of players which is finite. This set of players is immersed in a type of interaction from which they receive some benefit – we call this benefit a **payoff**. We discuss a very simple game in the chapter on altruism in (Peterson (15) 2016) but here we will have more players. A fundamental concept that helps us understand how players which interact (i.e. **cooperate**) can choose interaction strategies which benefit all players as much as possible is the **CORE**. This type of player interaction is called a **cooperative game** and we want to find tools to analyze such a game.

The **CORE** is the set of all **feasible outcomes** (payoffs) that no player (i.e. participant) or group of players (a coalition) can improve upon by acting for themselves. That is, once an agreement in the core has been reached, no individual or group could gain by changing actions or creating new coalitions. Hence, in a free market economy, free market outcomes should be in the core; i.e. economic activities should be advantageous to all. But for many games, feasible outcomes which can't be improved upon may not exist.

We can now discuss the theory of cores of games of transferable utility. Let Ω be a set of players. Let $\mathcal{F}$ be a collection of coalitions of players that are permitted. We have no structure on $\mathcal{F}$ yet. Then any function $\nu : \mathcal{F} \to \Re$ assigns a number to each permissible coalition. Clearly, the worth of the empty coalition should be 0 and it seems reasonable to allow all players in the set to assemble into the largest coalition possible. We can call this the **grand** coalition if you want. Hence, it is reasonable to assume $\Omega \in \mathcal{F}$. Hence our map $\nu : \mathcal{F} \to \Re$ satisfies $\nu(\emptyset) = 0$. The value $\nu(F)$ for any coalition in $\mathcal{F}$ therefore assigns a value to a coalition. Assuming each member $\omega \in \Omega$ is a permissible singleton coalition, each member ω can obtain a payoff which is a real number x_ω. We can therefore talk about the total payoff given to a coalition $F \in \mathcal{F}$. If Ω is a finite set, we can label the elements of Ω as $\{y_1, \ldots, y_N\}$ for some N. Then a coalition F is defined by $F = \{y_{n_1}, \ldots, y_{n_k}\}$ where the indices n_j are increasing and each $n_j \in \{1, \ldots, N\}$. and the payoff for a coalition F would be $\sum_{i=1}^k x_{n_i}$. Using our standard abuse of notion for Riemann Sums and letting $p(F)$ denote the payoff for F, we can write

$$p(F) = \sum_{y_i \in F} x_{y_i} = \sum_{\omega \in F} x_\omega$$

which is more compact and easier to understand. Of course, if Ω was not countable but was countably infinite, these payoffs would become infinite series which may or may not converge. And if Ω was infinite but not countably infinite, these sums would have to be defined another way which we won't pursue here.

Our first condition is we only want to consider games where $p(\Omega) = \sum_{\omega \in \Omega} x_\omega \leq \nu(\Omega)$. These are the games that are called **games of transferable utility**.
Homework

Exercise 16.1.1 *Consider a game of 5 players; i.e. $\Omega = \{y_1, y_2, y_3, y_4, y_5\}$ and let $\mathcal{F}$ be the collection of all subsets of Ω. These are the coalitions of players that are permitted. Let $\Theta = \{x_1, x_2, x_3, x_4, x_5\}$ be a set of five numbers. Define $\nu : \mathcal{F} \rightarrow \Re$ by $\nu(F) = \sum_{j \in S} x_{y_j}$. Work out all the possible payoffs for $\Theta = \{0.1, 0.4, 0.8, 0.3, 0.7\}$. Does this game satisfy $\sum_{j \in \Omega} x_{y_j} \leq \nu(\Omega)$?*

Exercise 16.1.2 *Consider a game of 5 players; i.e. $\Omega = \{y_1, y_2, y_3, y_4, y_5\}$ and let $\mathcal{F}$ be the collection of all subsets of Ω. These are the coalitions of players that are permitted. Let $\Theta = \{x_1, x_2, x_3, x_4, x_5\}$ be a set of five numbers. Define $\nu : \mathcal{F} \rightarrow \Re$ by $\nu(F) = \sum_{j \in S} x_j$. Work out all the possible payoffs for $\Theta = \{-0.1, 0.4, -0.8, 0.3, -0.7\}$. Does this game satisfy $\sum_{j \in \Omega} x_{y_j} \leq \nu(\Omega)$?*

Exercise 16.1.3 *Do the same as in the previous exercises but for a game of 3 players. Write a program which stores the payoffs for a given choice of Θ in a matrix using an intelligent way to map each coalition to a unique positive integer. Note the storage requirements become excessive quickly so for large N this is not the way to go. Can you think of a good way to plot these results instead?*

Exercise 16.1.4 *Do the same as in the previous exercises but for a game of 10 players. Write a program which computes the payoff of a coalition for a given choice of Θ.*

Exercise 16.1.5 *Consider a game of 3 players; i.e. $\Omega = \{y_1, y_2, y_3\}$ and let $\mathcal{F}$ be the collection of all subsets of Ω. These are the coalitions of players that are permitted. Let $\Theta = \{x_1, x_2, x_3\}$ be a set of three numbers. Define $\nu : \mathcal{F} \rightarrow \Re$ by $\nu(F) = \sum_{j \in S} x_j$ if $F \neq \Omega$ and $\nu(F) = 1$. Work out all the possible payoffs for $\Theta = \{0.1, 0.4, 0.8, 0.3, 0.7\}$. Does this game satisfy $\sum_{j \in \Omega} x_{y_j} \leq \nu(\Omega)$?*

16.2 Games of Transferable Utility

We need to be precise about all of this, so let's start with a finite set of players we label using the positive integers $X = \{1, \ldots, N\}$.

Definition 16.2.1 Coalitions of Finite Sets of Players

A subset of $X = \{1, \ldots, N\}$ is called a coalition.

For such a finite set of players, we need a **characteristic** or **worth** function which assigns a score to each coalition.

Definition 16.2.2 Characteristic or Worth Functions

A real-valued function ν defined on coalitions, i.e. $\nu(F) \in \Re$ for all coalitions $F \subset X = \{1, \ldots, N\}$ with $\nu(\emptyset) = 0$ is called a characteristic or worth function.

The players interact with one another in what we call a game based on the worth function ν. So we often call ν a **game**. Each interaction assigns a **payoff** value to each player.

Definition 16.2.3 Payoffs, Pareto Optimality and Rationality

For a given worth function ν, each game of interaction has an outcome which assigns a vector in $\Re^n$ given by

$$\boldsymbol{x} = \begin{bmatrix} x_1 \\ \vdots \\ x_N \end{bmatrix}$$

which is interpreted as the ith player gets x_i. This vector is called a **payoff** *vector. We require that a payoff vector satisfy*

$$\sum_{i=1}^{N} x_i = \nu(X) \tag{16.1}$$

Equation 16.1 is called the **feasibility** *condition and also called the* **Pareto optimality condition**. *We also require that*

$$x_i \geq \nu(\{i\}), \; 1 \leq i \leq N \tag{16.2}$$

Equation 16.2 is called the **Individual Rationality Condition**.

Equation 16.1 combines the requirement that the members of the grand coalition $X = \{1, \ldots, N\}$ can actually achieve the outcome $\boldsymbol{x}$, $\sum_{i=1}^{N} x_i \leq \nu(X)$, which is **feasibility** with the recognition that the grand coalition cannot achieve more than the payoff, $\sum_{i=1}^{N} x_i \geq \nu(X)$, which is the idea of **Pareto Optimality**. Note Equation 16.2 combined with Equation 16.1 implies

$$\sum_{i=1}^{N} \nu(\{i\}) \leq \nu(N) \tag{16.3}$$

We will assume $\{\boldsymbol{x} \in \Re^N\}$ such that Equation 16.2 and Equation 16.1 is **not empty**. If Equation 16.3 becomes an equality, $\sum_{i=1}^{N} \nu(\{i\}) = \nu(X)$, then the $\boldsymbol{x}$ that works is

$$\boldsymbol{x} = \begin{bmatrix} \nu(\{1\}) \\ \vdots \\ \nu(\{N\}) \end{bmatrix}$$

which is the trivial case so to avoid it, we always assume $\sum_{i=1}^{N} \nu(\{i\}) < \nu(X)$.

Now if F is a coalition so that $\sum_{\omega \in F} x_\omega < \nu(F)$, then the members of F can improve their payoffs. This leads to the idea of the **Core** of the game.

Definition 16.2.4 The Core

The **core** *is the set of all feasible payoffs upon which* **no** *individual and* **no** *group can improve. That is the core is a subset of $\Re^n$, C_ν with*

$$C_\nu = \left\{ \boldsymbol{x} \in \Re^n \mid \forall\, F \subset X, \sum_{\omega \in F} x_\omega \geq \nu(F) \text{ and } \sum_{i=1}^{n} x_i \leq \nu(X) \right\} \quad (16.4)$$

Note this implies $\sum_{i=1}^{n} x_i = \nu(X)$.

Note, $\{E_1, \ldots, E_k\}$ is a partition of X, then by feasibility, $\sum_{i=1}^{N} x_i = \sum_{i=1}^{k} \sum_{j \in E_i} x_j \leq \nu(X)$ and by Equation 16.4, $\sum_{j \in E_i} x_j \geq \nu(FE_i)$. Thus, for the core to be nonempty, we must have $\sum_{i=1}^{k} \nu(E_i) \leq \nu(X)$ for all partitions of X.

Example 16.2.1 *Let $N = 3$ and ν be defined by*

$$\nu(F) = \begin{cases} 1, & \textit{if a coalition } F \textit{ has } 2 \textit{ or } 3 \textit{ members} \\ 0, & \textit{if } F = \{1\}, \{2\} \textit{ or } \{3\} \end{cases}$$

Here, $X = \{1, 2, 3\}$. Any partition of X can have only one coalition with 2 or 3 members. Hence $\sum_{i=1}^{k} \nu(S_i) \leq \nu X$ for all partitions of X.

Let's look explicitly at all possible coalitions with 2 players. To be in the core $\sum_{i \in F} x_i \geq \nu(F)$.

$$\begin{aligned} F &= \{1,2\} \Longrightarrow x_1 + x_2 \geq 1 \\ F &= \{1,3\} \Longrightarrow x_1 + x_3 \geq 1 \\ F &= \{2,3\} \Longrightarrow x_2 + x_3 \geq 1 \end{aligned}$$

Combining, we see $2x_1 + 2x_2 + 2x_3 \geq 3$ or $2(\sum_{i=1}^{3} x_i) \geq 3$ implying $\sum_{i=1}^{3} x_i \geq 1.5$. But feasibility requires that $\sum_{i=1}^{2} x_i \leq \nu(X) = 1$. These two inequalities are incompatible, so the core here is empty: $C_\nu = \emptyset$.

We need another condition to help us know what to do.

Definition 16.2.5 Balanced Collections and Games

A balanced collection of X means for the set of coalitions $\{S_1, \ldots, S_k\}$ of X, $\cup_{i=1}^{n} S_i = X$ and there are positive numbers $\{\lambda_i, \ldots, \lambda_k\}$ such that $\sum_{\{j \mid i \in S_j\}} \lambda_j = 1$ for all $i \in X$. The numbers $\{\lambda_i, \ldots, \lambda_k\}$ are called **balancing weights**.

A game ν satisfying $\sum_{i=1}^{k} \lambda_i \nu(S_i) \leq \nu(X)$ for all balanced collections is called a **balanced game**.

Note all partitions are balanced collections. Let $\{F_1, \ldots, F_k\}$ be a partition of X and $\{\lambda_1, \ldots, \lambda_k\}$ be balancing weights with $\lambda_i = 1$ for all i. Then given $i \in X$, there is only one index j_0 with $\imath \in F_{j+-}$. Hence, $\sum_{\{j \mid i \in S_j\}} \lambda_j = \lambda_{j_0} = 1$ and so the partition is a balanced collection.

Also, note if we let $F_j = X \setminus \{j\}$ and let $\lambda_j = \frac{1}{N-1}$, then given i, i is in every F_j except F_i and so $\sum_{\{j \mid i \in F_j\}} \lambda_j = \sum_{j \neq i} \lambda_j = \sum_{j \neq i} \frac{1}{N-1} = 1$. So we have found two kinds of balanced collections so far.

Comment 16.2.1 *Now, for an arbitrary $\{S_1, \ldots, S_k\}$ collection of subsets of X, define $I_{S_j}(i) = 1$ if $i \in S_j$ and 0 if $i \notin S_j$. Then for any choice $\{\lambda_1, \ldots, \lambda_k\}$ with $\lambda_j \in \Re$, we have $\sum_{j=1}^{k} \lambda_j I_{S_j}(i) =$*

$\sum_{\{j \mid i \in S_j\}} \lambda_j$. *So the* **balanced collection condition** *can also be written as* $\sum_{j=1}^{k} \lambda_j I_{S_j}(i) = 1 = I_X(\{i\})$.

We are now ready to prove a necessary and sufficient condition on when the core is nonempty.

Theorem 16.2.1 The Necessary Condition that the Core of a Game is Nonempty

The **core** *of a game* ν *is nonempty implies for all balanced collections* $\{S_1, \ldots, S_k\}$ *with balancing weights* $\{\lambda_1, \ldots, \lambda_k\}$*, we have*

$$\sum_{j=1}^{k} \lambda_j \nu(S_j) \quad \leq \quad \nu(X) \tag{16.5}$$

Proof 16.2.1

Let $\{S_1, \ldots, S_k\}$ *be a balanced collection with balancing weights* $\{\lambda_1, \ldots, \lambda_k\}$*. Since* C_ν *is nonempty, let* $\boldsymbol{x} \in C_\nu$*. Then by the definition of the core, Equation 16.4 holds and so* $\sum_{i \in S_j} x_i \geq \nu(S_j)$ *for* $1 \leq j \leq k$*. Hence,*

$$\lambda_j \sum_{i \in S_j} x_i \quad \geq \quad \lambda_j \, \nu(S_j), \; 1 \leq j \leq k \Longrightarrow \sum_{j=1}^{k} \lambda_j \sum_{i \in S_j} x_i \geq \sum_{j=1}^{k} \lambda_j \, \nu(S_j)$$

But we assume the collection is balanced and so $\left(\sum_{\{j \mid i \in S_j\}} \lambda_j\right) x_i = 1 \cdot x_i = x_i$.

Now since the collection is balanced we also have

$$\begin{aligned} \sum_{j=1}^{k} \lambda_j \left(\sum_{i \in S_j} x_i\right) &= \lambda_1 \left(\sum_{i \in S_1} x_i\right) + \ldots + \lambda_k \left(\sum_{i \in S_k} x_i\right) \\ &= \left(\sum_{\{j \mid 1 \in S_j\}} \lambda_j\right) x_1 + \ldots + \left(\sum_{\{j \mid N \in S_j\}} \lambda_j\right) x_N \\ &= x_1 \, + \, \ldots \, + \, x_N \end{aligned}$$

Thus, as $\boldsymbol{x}$ *is feasible,* $\sum_{i=1}^{N} x_i \leq \nu(X)$ *and since* $\boldsymbol{x}$ *is in the core,* $\sum_{i=1}^{N} x_i \geq \nu(X)$*. We conclude from the above that* $\nu(X) \geq \sum_{j=1}^{k} \lambda_j \, \nu(S_j)$*. Thus Equation 16.5 holds.* ∎

To go the other way, note we must assume $\nu(X) \geq \sum_{j=1}^{k} \lambda_j \; \nu(S_j)$ for all balanced collections $\{S_1, \ldots, S_K\}$ with balancing weights $\{\lambda_1, \ldots, \lambda_k\}$. To understand how this implies the core is not empty, we need to step back and think about linear programming.

Consider the problem

$$\min \begin{bmatrix} 1 \\ 1 \\ \vdots \\ 1 \end{bmatrix} \cdot \begin{bmatrix} x_1 \\ x_2 \\ \vdots \\ x_N \end{bmatrix} \qquad \text{subject to} \quad \sum_{i=1}^{N} I_S(i) x_i \geq \nu(S), \;\; \forall \, S \subset X$$

or letting $\boldsymbol{c} = \left[1, \ldots, 1\right]^T$,

$$\min \boldsymbol{c}^T \cdot \boldsymbol{x} \quad \text{subject to} \quad \sum_{i=1}^{N} I_S(i) x_i \geq \nu(S), \ \forall\, S \subset X$$

Since $X = \{1, \ldots, N\}$, the power set 2^X here is finite. Let $P = |2^X|$. Then the possible subsets are $\{S_1, \ldots, S_P\}$. Hence, we can rewrite our minimization problem as

$$\min \boldsymbol{c}^T \cdot \boldsymbol{x} \quad \text{subject to} \quad \sum_{i=1}^{N} I_{S_j}(i) x_i \geq \nu(S_j), \ 1 \leq j \leq P$$

This can be rewritten again as

$$\min \boldsymbol{c}^T \cdot \boldsymbol{x} \quad \text{subject to} \quad \sum_{i \in S_j} x_i \geq \nu(S_j), \ 1 \leq j \leq P$$

This is called the **Primal Problem**.

The sets $\{S_1, \ldots, S_P\}$ create a matrix. For example, for $X = \{1, 2, 3, 4\}$, we have, ignoring the empty set,

$$\begin{aligned}
S_1 &= \{1\}, \quad , S_2 = \{2\}, \quad , S_3 = \{3\}, \quad , S_4 = \{4\} \\
S_5 &= \{1,2\}, \quad , S_6 = \{1,3\}, \quad , S_7 = \{1,4\}, \quad , S_8 = \{2,3\} \\
S_9 &= \{2,4\}, \quad , S_{10} = \{3,4\} \\
S_{11} &= \{1,2,3\}, \quad , S_{12} = \{2,3,4\}, \quad , S_{13} = \{1,2,4\} \\
S_{14} &= \{1,3.4\}. \quad S_{15} = \{1,2,3,4\}
\end{aligned}$$

This gives

$$\begin{bmatrix} \begin{bmatrix} 1 & 0 & 0 & 0 \\ 0 & 1 & 0 & 0 \\ 0 & 0 & 1 & 0 \\ 0 & 0 & 0 & 1 \end{bmatrix} \\ \\ \begin{bmatrix} 1 & 1 & 0 & 0 \\ 1 & 0 & 1 & 0 \\ 1 & 0 & 0 & 1 \\ 0 & 1 & 1 & 0 \\ 0 & 1 & 0 & 1 \\ 0 & 0 & 1 & 1 \end{bmatrix} \\ \\ \begin{bmatrix} 1 & 1 & 1 & 0 \\ 0 & 1 & 1 & 1 \\ 1 & 1 & 0 & 1 \\ 0 & 0 & 0 & 1 \\ 1 & 0 & 1 & 1 \end{bmatrix} \\ \\ \begin{bmatrix} 1 & 1 & 1 & 1 \end{bmatrix} \end{bmatrix} \begin{bmatrix} x_1 \\ x_2 \\ x_3 \\ x_4 \end{bmatrix} \geq \begin{bmatrix} \begin{bmatrix} \nu(S_1) \\ \nu(S_2) \\ \nu(S_3) \\ \nu(S_4) \end{bmatrix} \\ \\ \begin{bmatrix} \nu(S_5) \\ \nu(S_6) \\ \nu(S_7) \\ \nu(S_8) \\ \nu(S_9) \\ \nu(S_{10}) \end{bmatrix} \\ \\ \begin{bmatrix} \nu(S_{11}) \\ \nu(S_{12}) \\ \nu(S_{13}) \\ \nu(S_{14}) \end{bmatrix} \\ \\ \begin{bmatrix} \nu(S_{15}) \end{bmatrix} \end{bmatrix}$$

This gives a 15×4 matrix we will call $\boldsymbol{A}$ and a 15×1 data vector we will call $\boldsymbol{d}$; i.e. $\boldsymbol{d} \in \Re^{15}$. Hence, our minimization problem can be written as $\min \boldsymbol{c}^T\boldsymbol{x}$ subject to $\boldsymbol{A}\boldsymbol{x} \geq \boldsymbol{d}$. In general, for these sorts of problems, $\boldsymbol{A}\boldsymbol{x} \geq \boldsymbol{d}$ has the structure

$$\begin{bmatrix} \text{one element blocks} \\ \text{two element blocks} \\ \vdots \\ \text{N-1 element blocks} \\ \text{N element block} \end{bmatrix} \begin{bmatrix} x_1 \\ \vdots \\ x_N \end{bmatrix} \geq \begin{bmatrix} \nu(S_1) \\ \vdots \\ \nu(S_P) \end{bmatrix}$$

So the primal problem

$$\min\ x_1 + \ldots x_N \quad \text{subject to} \quad \sum_{i \in S_j} x_i \geq \nu(S_j),\ \ 1 \leq j \leq P$$

can be rewritten as

$$\min\ \boldsymbol{c}^T\, \boldsymbol{x} \quad \text{subject to} \quad \boldsymbol{A}\, \boldsymbol{x} \geq \boldsymbol{d}$$

From the theory of linear programming (no, we won't go into that here!), we know the **Dual** problem is

$$\max\ \boldsymbol{d}^T\, \boldsymbol{y} \quad \text{subject to} \quad \boldsymbol{A}^T\, \boldsymbol{y} = \boldsymbol{c},\ \ \boldsymbol{y} \geq 0$$

We can rewrite this again by looking at the structure of $\boldsymbol{A}$. We know

$$\text{column i of } \boldsymbol{A} \;=\; \begin{bmatrix} \text{1 in column i and row p} \implies i \in S_p \\ \\ \\ \text{1 in column i and in row q} \implies i \in S_q \end{bmatrix}$$

Thus,

$$\begin{aligned} \text{row i of } \boldsymbol{A}^T\, \boldsymbol{y} = c_i \quad &\implies \quad \text{column i of } \boldsymbol{A}\, \boldsymbol{y} = c_i = 1 \\ &\implies \quad \sum_{\{j \mid i \in S_j\}} y_j = 1,\ \ 1 \leq i \leq N \end{aligned}$$

For a balanced collection $\{E_1, \ldots, E_k\}$ with balancing weights $\{y_1, \ldots, y_k\}$, we know we can identify each E_i with some S_{j_i}. The usual balanced condition is then $y_j \geq 0$ and

$$\begin{aligned} \sum_{\{j \mid i \in S_j\}} y_j = 1,\ \ 1 \leq i \leq N \quad &\implies \quad \sum_{j=1}^{P} I_{S_{j_i}}(i) y_{j_i} = 1,\ \ 1 \leq i \leq N \\ &\implies \quad \sum_{\{j \mid i \in S_j\}} y_j = 1,\ \ 1 \leq i \leq N \end{aligned}$$

From our discussion, this is the same as $\boldsymbol{A}^T\boldsymbol{y} = \boldsymbol{c}$ with $\boldsymbol{y} \geq 0$. So, we can rewrite the dual problem as

$$\max \sum_{j=1}^{P} \nu(S_j) y_j \quad \text{subject to} \quad \sum_{\{j \mid i \in S_j\}} y_j = 1,\ \ 1 \leq i \leq N,\ \ \boldsymbol{y} \geq 0$$

We are now ready to prove sufficiency.

Theorem 16.2.2 The Sufficient Condition that the Core of a Game is Nonempty

If for all balanced collections $\{S_1, \ldots, S_k\}$ with balancing weights $\{\lambda_1, \ldots, \lambda_k\}$, we have

$$\sum_{j=1}^{k} \lambda_j \nu(S_j) \;\le\; \nu(X) \tag{16.6}$$

then the **core** *of a game ν is nonempty.*

Proof 16.2.2
We assume $\nu(X) \ge \sum_{j=1}^{k} \lambda_j\ \nu(S_j)$ for all balanced collections $\{S_1, \ldots, S_K\}$ with balancing weights $\{\lambda_1, \ldots, \lambda_k\}$. From our linear programming discussions, we know this is equivalent to saying

$$\sum_{j=1}^{P} y_j\ \nu(S_j) \;\le\; \nu(X), \quad \boldsymbol{A}^T \boldsymbol{y} = \mathbf{c},\ \boldsymbol{y} \ge 0$$

Thus, if ω^* solves the dual problem, we have $\sum_{j=1}^{P} y_j\ \nu(S_j) \le \omega^*$ for all $\boldsymbol{y}$ such that $\boldsymbol{A}^T\boldsymbol{y} = \mathbf{c}$ with $\boldsymbol{y} \ge 0$. In particular, for $\{X\}$ with balancing coefficient $\{1\}$, we have $\nu(X) \cdot (1) \le \omega^*$. However, since $\nu(X) \ge \sum_{j=1}^{k} \lambda_j\ \nu(S_j)$ by assumption, this tells us $\omega^* \le \nu(X)$.

Combining, we have $\nu(X) = \omega^*$. The theory of linear programming then tells us the primal problem has optimal value ω^* also. Hence, there is a $\boldsymbol{x}^*$ so that $\sum_{i\in S_j} x_i^* \ge \nu(S_j)$ for $1 \le j \le P$ with $\sum_{i=1}^{N} x_i^* = \nu(X)$. Now $\boldsymbol{x}^*$ is in the core C_ν if $\sum_{i\in S_j} x_i^* \ge \nu(S_j)$ for $1 \le j \le P$ and $\boldsymbol{x}^*$ is feasible; i.e. $\sum_{i=1}^{N} x_i^* \le \nu(X)$. So it is true $\boldsymbol{x}^* \in C_\nu$ and C_ν is nonempty. ■

We combine this into a powerful necessary and sufficient condition for the core of a game to be nonempty.

Theorem 16.2.3 The Necessary and Sufficient Condition that the Core of a game is Nonempty

The **core** *of a game ν is nonempty if and only if for all balanced collections $\{S_1, \ldots, S_k\}$ with balancing weights $\{\lambda_1, \ldots, \lambda_k\}$, we have*

$$\sum_{j=1}^{k} \lambda_j \nu(S_j) \le \nu(X)$$

Proof 16.2.3
We have just proven this in separate arguments. ■

Homework

Exercise 16.2.1 *Consider a game of 5 players; i.e. $\Omega = \{y_1, y_2, y_3, y_4, y_5\}$ and let $\mathcal{F}$ be the collection of all subsets of Ω. These are the coalitions of players that are permitted. Let $\Theta = \{0.1, 0.4, 0.8, 0.3, 0.7\}$. Define $\nu : \mathcal{F} \to \Re$ by $\nu(F)\ \sum_{j\in S} 0.3\ (x_{y_j})$ for all $f \ne \omega$ and $\nu(\Omega) =$*

$\sum_{j=1}^{5} 7 : x_j$. *Work out all the possible payoffs. Does this game satisfy the Pareto Rationality Condition? Does this game satisfy the Individual Rationality Condition?*

Exercise 16.2.2 *Consider a game of* 5 *players; i.e.* $\Omega = \{y_1, y_2, y_3, y_4, y_5\}$ *and let* $\mathcal{F}$ *be the collection of all subsets of* Ω. *These are the coalitions of players that are permitted. Let* $\Theta = \{-0.1, -0.4, 0.8, 0.3, 0.7\}$. *Define* $\nu : \mathcal{F} \to \Re$ *by* $\nu(F) \sum_{j\in S} 0.3\ (x_{y_j})$ *for all* $f \neq \omega$ *and* $\nu(\Omega) = \sum_{j=1}^{5} 7 : x_j$. *Work out all the possible payoffs. Does this game satisfy the Pareto Rationality Condition? Does this game satisfy the Individual Rationality Condition?*

Exercise 16.2.3 *Let* $N = 3$ *and* ν *be defined by*

$$\nu(F) \quad = \quad \begin{cases} 2, & \text{if a coalition } F \text{ has } 2 \text{ or } 3 \text{ members} \\ 0, & \text{if } F = \{1\}, \{2\} \text{ or } \{3\} \end{cases}$$

Determine if the core is nonempty.

Exercise 16.2.4 *Let* $N = 3$ *and* ν *be defined by*

$$\nu(F) \quad = \quad \begin{cases} 2, & \text{if a coalition } F \text{ has } 1 \text{ or } 3 \text{ members} \\ 0, & \text{if } F = \{1\}, \{2\} \text{ or } \{3\} \end{cases}$$

Determine if the core is nonempty.

Exercise 16.2.5 *This is about the sufficient condition for the core to be nonempty. The sufficiency condition states if for all balanced collections* $\{S_1, \ldots, S_k\}$ *with balancing weights* $\{\lambda_1, \ldots, \lambda_k\}$, *we have*

$$\sum_{j=1}^{k} \lambda_j \nu(S_j) \quad \leq \quad \nu(X) \tag{16.7}$$

then the **core** *of a game* ν *is nonempty. The argument was based on primal and dual problems. In this exercise, you will work out the details of how this argument goes using* $X = \{1, 2\}$ *following the example in the text.*

Exercise 16.2.6 *This is again about the sufficient condition for the core to be nonempty. The sufficiency condition states if for all balanced collections* $\{S_1, \ldots, S_k\}$ *with balancing weights* $\{\lambda_1, \ldots, \lambda_k\}$, *we have*

$$\sum_{j=1}^{k} \lambda_j \nu(S_j) \quad \leq \quad \nu(X) \tag{16,8}$$

then the **core** *of a game* ν *is nonempty. The argument was based on primal and dual problems. In this exercise, you will work out the details of how this argument goes using* $X = \{1, 2, 3, 4\}$ *following the example in the text.*

16.3 Payoff Vectors as Charges

Recall $ba(\Omega, \mathcal{F}, \preceq, \|\cdot\|)$ is a complete normed linear space which is a lattice and it consists of all charges which are bounded on the field $\mathcal{F}$. Also recall μ is a charge if $\mu(\emptyset) = 0$ and $\mu(A \cup B) = \mu(A) + \mu(B)$ when A and B are disjoint. Hence, if $\mu \in ba(\Omega, \mathcal{F}, \preceq, \|\cdot\|)$ is a bounded finitely additive charge.

Let $\Omega = \{\omega_1, \ldots, \omega_N\}$ be a collection of N distinct elements. Since μ is finitely additive

$$\sum_{i=1}^{N} \mu(\{\omega_i\}) = \mu\left(\cup_{i=1}^{N} \omega_i\right) = \mu(\Omega)$$

Now define a mapping $T : ba(\Omega, \mathcal{F}, \preceq, \|\cdot\|) \rightarrow \Re^N$ where $\mathcal{F} = 2^\Omega$ by

$$T(\mu) = \begin{bmatrix} \mu(\{\omega_1\}) \\ \vdots \\ \mu(\{\omega_N\}) \end{bmatrix}$$

Note if $\boldsymbol{x} \in \Re^N$ is chosen, the map $\mu_{\boldsymbol{x}} : \mathcal{F} = 2^\Omega \rightarrow \Re$ defined by $\mu_{\boldsymbol{x}}(\emptyset) = 0$ and otherwise

$$\mu_{\boldsymbol{x}}(F) = \sum_{\omega_i \in F} x_i$$

defines a charge. Hence T is onto.

Now if $T(\mu_1) = T(\mu_2)$, we have

$$T(\mu_1) = \begin{bmatrix} \mu_1(\{\omega_1\}) \\ \vdots \\ \mu_1(\{\omega_N\}) \end{bmatrix} = T(\mu_2) = \begin{bmatrix} \mu_2(\{\omega_1\}) \\ \vdots \\ \mu_2(\{\omega_N\}) \end{bmatrix}$$

which tells us $\mu_1(\{\omega_i\}) = \mu_2(\{\omega_i\})$ for all i. Hence, $\mu_1(F) = \mu_2(F)$ for all $f \in 2^\Omega$ and so T is $1-1$.

Then given $\mu \in (ba(\Omega, \mathcal{F}), \preceq, \|\cdot\|)$,

$$\begin{aligned} \|\mu\| &= |\mu|(\Omega) = \sum_{i \in I} \mu(\{\omega_i\}) + \sum_{j \in J} (-\mu)(\{\omega_j\}) \\ &= \sum_{i=1}^{N} |\mu|(\{\omega_i\})| = \|T(\mu)\|_1 \end{aligned}$$

where I is the set of indices where μ is non-negative and J the set of indices where μ is negative. Hence, $\|\mu\| = \|T(\mu)\|_1$ and so we have shown T is a norm preserving bijection from $(ba(\Omega, \mathcal{F}), \preceq, \|\cdot\|)$ to $(\Re^n, \|\cdot\|_1)$. Now we know

$$C_\mu = \left\{ \boldsymbol{x} \in \Re^N \mid \sum_{i \in S_j} x_i \geq \nu(S_j), \ \forall\, S_j \in 2^\Omega, \ \sum_{1=1}^{N} x_i \leq \nu(\Omega) \right\}$$

Each $\boldsymbol{x} \in \Re^N$ defines the charge $\mu_{\boldsymbol{x}}$ and so we can rewrite the core condition as

$$C_\mu = \{ \boldsymbol{x} \in \Re^N \mid \mu_{\boldsymbol{x}}(S_j) \geq \nu(S_j), \ \forall\, S_j \in 2^\Omega, \ \mu_{\boldsymbol{x}}(\Omega) \leq \nu(\Omega) \}$$

Finally, since we can identify $(ba(\Omega, \mathcal{F}), \preceq, \|\cdot\|)$ and $(\Re^n, \|\cdot\|_1)$ via the norm preserving bijection T, we can state the core condition as

$$C_\mu = \{ \mu \in ba(\Omega, 2^\Omega) \mid \mu(S_j) \geq \nu(S_j), \ \forall\, S_j \in 2^\Omega, \ \mu(\Omega) \leq \nu(\Omega) \}$$

There is another way we can look at this. Each set S_j corresponds to a vector in $\Re^N$ with only 0's and 1's. Hence, each S_j corresponds to a vector of 0's and 1's in $\Re^N$ and the finitely additive charge μ above can be thought of as acting on a vector in $\Re^N$ and assigning it a real number. Since a vector in $\Re^N$ can be embedded in ℓ^∞ as a sequence with all 0's after slot N, we can think of μ as acting on a element in ℓ^∞ in a linear bounded way and assigning a real number. Thus, μ is a bounded linear functional acting on ℓ^∞ and so the core C_ν consists of elements of the dual of ℓ^∞ that satisfy certain properties.

16.3.1 Countably Infinite Numbers of Players

Now let's focus on extending the set of players from cardinality N to the cardinality of $\mathbb{N}$. The set of players is now $X = \mathbb{N}$ or $X = \{i\}_{i=1}^\infty$. We let $\mathcal{F} = 2^{\mathbb{N}}$ as usual and we look at mappings $\nu : \mathcal{F} \rightarrow \Re \cup \{-\infty, \infty\}$. We seek payoff vectors $\boldsymbol{x} = (x_i)_{i=1}^\infty$ so that we have individual rationality

$$
\begin{aligned}
x_i &\geq \nu(\{i\}), \; \forall\, i, \quad \textbf{individual rationality} && (16.9)\\
\sum_{i=1}^\infty x_i &\leq \nu(X), \quad \textbf{feasibility} && (16.10)\\
\sum_{i=1}^\infty x_i &\geq \nu(X), \quad \textbf{Pareto optimality} && (16.11)
\end{aligned}
$$

For now, let's specialize and assume $\nu(F) \geq 0$ for all $F \in \mathcal{F}$. Then Equation 16.9 implies each $x_i \geq 0$. Also, this result plus Equation 16.10 implies the series $\sum_{i=1}^\infty x_i < \infty$ and so $(x_i)_{i=1}^\infty \in \ell^1$. In this new context, we define the **core** as

Definition 16.3.1 The Core in the Countably Infinite Players Case

The **core** *is the set of all feasible payoffs upon which* **no** *individual and* **no** *group can improve. That is the core is a subset of ℓ^1, C_ν with*

$$
C_\nu = \left\{ \boldsymbol{x} \in \ell^1 \,|\, \forall\, F \subset X, \sum_{\omega \in F} x_\omega \geq \nu(F) \text{ and } \sum_{i=1}^\infty x_i \leq \nu(X) \right\} \quad (16.12)
$$

Note this implies $\sum_{i=1}^\infty x_i = \nu(X)$. Hence, we could define the core as

$$
C_\nu = \left\{ \boldsymbol{x} \in \ell^1 \,|\, \forall\, F \subset X, \sum_{\omega \in F} x_\omega \geq \nu(F) \text{ and } \sum_{i=1}^\infty x_i = \nu(X) \right\} \quad (16.13)
$$

Note a subset F of $\mathbb{N}$ could be finite or infinite. Each sequence $\boldsymbol{x} \in \ell^1$ defines a finitely additive charge μ and $\sum_{\omega \in F} x_\omega \geq \nu(F)$ can therefore be interpreted as $\mu(\boldsymbol{x})$ for some $\boldsymbol{x} \in \ell^\infty$. Hence, the ℓ^1 sequences in the core are particular choices of bounded linear functionals acting on ℓ^∞ which are therefore a member of $(\ell^\infty)'$. We can show $(\ell^\infty)'$ is equivalent to $ba(\mathbb{N}, 2^{\mathbb{N}})$ and so when we try to understand the structure of the balanced core $C(\nu)$ for a given $\nu : \mathcal{F} \rightarrow \Re$, we are led to studying $ba(\Omega, \mathcal{F})$ for appropriate Ω and $\mathcal{F}$.

To figure all this out, we need several steps:

1. The dual space to the set of all convergence sequences $\boldsymbol{c}$ and the dual space to the set of all sequences that converge to 0, $\boldsymbol{c_0}$.

2. Some understanding of measure theory which is covered fully in (Peterson (16) 2019). We will give a short introduction to the salient facts and ideas here. We need to define $B(\Omega, \mathcal{F})$ which is the set of functions $f : \Omega \rightarrow \Re$ which are uniform limits of finite linear combinations of specialized functions called characteristic functions which are quite similar to charges. This will take some time so be patient. We can show $B(\mathbb{N}, 2^{\mathbb{N}}) \equiv \ell^\infty$ also.

3. We show the dual of $B(\Omega, \mathcal{F})$ is $ba(\Omega, \mathcal{F})$.

Note we then now have a chain of correspondences:

$$(\boldsymbol{c_0})' \cong \ell^1, \quad (\ell^1) \cong \ell^\infty \Longrightarrow (\ell^\infty)' = (\; B(\mathbb{N}, 2^{\mathbb{N}})\;)' = ba(\mathbb{N}, 2^{\mathbb{N}})$$

So we have finally found the dual of ℓ^∞! So let's get to these equivalences. Note we are again exploring the duals of certain sequence spaces and then bootstrapping from that to understanding spaces of charge.

Homework

Exercise 16.3.1 *Let $X = \mathbb{N}$ and let $x_i = (-1)^i/i$. Let ν be defined by $\nu(\{i\}) = (-1)^i/i$. What are the Individual Rationality, feasibility and Pareto optimality conditions here? Are they satisfied?*

Exercise 16.3.2 *Let $X = \mathbb{N}$ and let $x_i = 1/i^2$. Let ν be defined by $\nu(\{i\}) = 1/i^2$. What are the Individual Rationality, feasibility and Pareto optimality conditions here?*

Exercise 16.3.3 *Let $X = \mathbb{N}$ and let ν be defined by $\nu(\{i\}) = 1/i^2$. What is the defining condition for the core C_ν?*

Exercise 16.3.4 *Let $X = \mathbb{N}$ and let ν be defined by $\nu(\{i\}) = 1/i^4$. What is the defining condition for the core C_ν?*

Exercise 16.3.5 *Let $X = \mathbb{N}$ and let ν be defined by $\nu(\{i\}) = 1/i^6$. What is the defining condition for the core C_ν?*

16.4 Some Additional Dual Spaces

We need to find the dual space of the set of sequences that converge and the set of sequences that converge to 0.

16.4.1 The Dual of Sequences That Converge

Here is the theorem we can prove:

Theorem 16.4.1 The Dual of Sequences that Converge

Let $\boldsymbol{c}$ be the set of all sequences that converge with the sup norm $\|\cdot\|_\infty$. Then $\boldsymbol{c}' \cong \ell^1$.

Proof 16.4.1

In Section 5.2.1, we found a Schauder Basis for $\boldsymbol{c}$: $E = (\boldsymbol{e_n})_{n=0}^\infty$ where

$$\begin{aligned} \boldsymbol{e_0} &= (1, 1, 1, \ldots), \text{ all ones} \\ \boldsymbol{e_n} &= (0, \ldots, 0, \underbrace{1}_{\text{slot } n}, 0, \ldots), n \geq 1 \end{aligned}$$

and there is a unique representation $\boldsymbol{x} = x_\infty \boldsymbol{e_0} + \sum_{j=1}^\infty (x_n - x_\infty)\boldsymbol{e_n}$ where x_∞ is the limit value of the sequence and the partial sums $S_n \rightarrow \boldsymbol{x}$ in $\|\cdot\|_\infty$.

We will show there is a mapping $T : \mathbf{c}' \rightarrow \ell^1$ so that

1. *T is $1-1$ and onto.*
2. *T is linear.*
3. *$\|T(\boldsymbol{f})\|_1 = \|\boldsymbol{f}\|_{op}$ for all $\boldsymbol{f} \in \mathbf{c}'$.*

Pick any $\boldsymbol{f} \in (\mathbf{c})'$. Let $\gamma_0 = f(\boldsymbol{e_0})$ and $\gamma_j = f(bse_j)$ for all $j \geq 1$. We will show $(\gamma_j)_{j=0}^{\infty} \in \ell^1$. Define $x_n \in \mathbf{c}$ as follows

$$(x_n)_i = \begin{cases} sign(\gamma_i), & 1 \leq i \leq n \\ 0, & i > n \end{cases}$$

Then $(x_n)_i \rightarrow 0$ and so the sequence x_n has canonical representation

$$x_n = x_\infty \boldsymbol{e_0} + \sum_{j=1}^{\infty}((x_n)_j - x_\infty)\boldsymbol{e_j} = \sum_{j=1}^{n}(x_n)_j \boldsymbol{e_j} = \sum_{j=1}^{n} sign(\gamma_j)\, \boldsymbol{e_j}$$

Since $\boldsymbol{f}$ is linear,

$$f(x_n) = \sum_{j=1}^{n}(sign(\gamma_j))\gamma_j = \sum_{j=1}^{n} |\gamma_j|$$

We know $\|x_n\|_\infty = 1$ and since in this case $|f(x_n)| \leq \|\boldsymbol{f}\|_{op} < \infty$, we have $\sum_{j=1}^{n} |\gamma_j| \leq \|\boldsymbol{f}\|_{op} < \infty$ for all n. This tells us the sequence of partial sums $\sum_{j=1}^{n} |\gamma_j|$ converges and so the sequence $(\gamma_j) \in \ell^1$.

Let $b_1 = \gamma_0 - \sum_{j=1}^{\infty} \gamma_j$ and $b_j = \gamma_{j+1}$ for $j > 1$. For any $x \in \mathbf{c}$, we know $x = x_\infty \boldsymbol{e_0} + \sum_{j=1}^{\infty}(x_j - x_\infty)\boldsymbol{e_j}$ implying $f(x) = x_\infty \gamma_0 + \sum_{j=1}^{\infty}(x_j - x_\infty)\gamma_j$. Now $(x_j - x_\infty)_{j \geq 1}$ is a bounded sequence and $(\gamma_j) \in \ell^1$ so it is easy to see $\sum_{j=1}^{\infty} x_j\gamma_j$, $\sum_{j=1}^{\infty} x_\infty\gamma_j$ and $\sum_{j=1}^{\infty}(x_j - x_\infty)\gamma_j$ all converge. In the exercise below, you can prove that $\sum_{j=1}^{\infty}(x_j - x_\infty)\gamma_j = \sum_{j=1}^{\infty} x_j\gamma_j - \sum_{j=1}^{\infty} x_\infty\gamma_j$. Thus,

$$f(x) = x_\infty(\gamma_0 - \sum_{j=1}^{\infty}\gamma_j) + \sum_{j=1}^{\infty} x_j\gamma_j = x_\infty b_1 + \sum_{j=1}^{\infty} b_{j+1}x_j$$

Since, $(\gamma_j)_{j=0}^{\infty} \in \ell^1$, it is clear $(b_j)_{j=1}^{\infty} \in \ell^1$ also.

Define $T : (\mathbf{c})' \rightarrow \ell^1$ by $T(\boldsymbol{f}) = (b_j)_{j=1}^{\infty}$.

- *T is linear.*
  Let $\boldsymbol{f}$ and $\boldsymbol{g}$ be in $(\mathbf{c})'$ and α and β be in $\Re$. Then $T(\alpha\boldsymbol{f} + \beta\boldsymbol{g}) = (b_j^{\alpha\boldsymbol{f}+\beta\boldsymbol{g}})$, $T(\boldsymbol{f}) = (b_j^{\boldsymbol{f}})$ and $T(\boldsymbol{g}) = (b_j^{\boldsymbol{g}})$. Now by linearity

$$\begin{aligned} b_0^{\alpha\boldsymbol{f}+\beta\boldsymbol{g}} &= \gamma_0^{\alpha\boldsymbol{f}+\beta\boldsymbol{g}} - \sum_{j=1}^{\infty}\gamma_j^{\alpha\boldsymbol{f}+\beta\boldsymbol{g}} = (\alpha\boldsymbol{f} + \beta\boldsymbol{g})(\boldsymbol{e_0}) - \sum_{j=1}^{\infty}(\alpha\boldsymbol{f} + \beta\boldsymbol{g})(\boldsymbol{e_j}) \\ &= \alpha\boldsymbol{f}(\boldsymbol{e_0}) + \beta\boldsymbol{g}(\boldsymbol{e_0}) - \sum_{j=1}^{\infty}(\alpha\boldsymbol{f}(\boldsymbol{e_j}) + \beta\boldsymbol{g}(\boldsymbol{e_j})) \\ &= \alpha\gamma_0^{\boldsymbol{f}} + \beta\gamma_0^{\boldsymbol{g}} - \sum_{j=1}^{\infty}(\alpha\gamma_j^{\boldsymbol{f}} + \beta\gamma_j^{\boldsymbol{g}}) = \alpha\gamma_0^{\boldsymbol{f}} + \beta\gamma_0^{\boldsymbol{g}} - \sum_{j=1}^{\infty}(\alpha\gamma_j^{\boldsymbol{f}} + \beta\gamma_j^{\boldsymbol{g}}) \end{aligned}$$

$$\begin{aligned} &= \alpha\gamma_0^{\boldsymbol{f}} - \sum_{j=1}^{\infty} \alpha\gamma_j^{\boldsymbol{f}} + \beta\gamma_0^{\boldsymbol{g}} - \sum_{j=1}^{\infty} \beta\gamma_j^{\boldsymbol{g}} = \alpha(\gamma_0^{\boldsymbol{f}} - \sum_{j=1}^{\infty} \gamma_j^{\boldsymbol{f}}) + \beta(\gamma_0^{\boldsymbol{g}} - \sum_{j=1}^{\infty} \gamma_j^{\boldsymbol{g}}) \\ &= \alpha b_0^{\boldsymbol{f}} + \beta_0^{\boldsymbol{g}} \end{aligned}$$

A similar calculation shows $b_j^{\alpha\boldsymbol{f}+\beta\boldsymbol{g}} = \alpha b_j^{\boldsymbol{f}} + \beta_j^{\boldsymbol{g}}$. *Thus,* T *is linear.*

- $\|T(\boldsymbol{f})\|_1 = \|\boldsymbol{f}\|_{op}$.

First note if $x \in \mathbf{c}$*, then* $x_n \to x_\infty$ *for some number* x_∞*. Hence, for* $\epsilon > 0$*, there is* N *so* $n > N$ *implies* $|x_n - x_\infty| < \epsilon$*. Thus* $|x_\infty| < \|x\|_\infty + \epsilon$*. Since* $\epsilon > 0$ *is arbitrary, this tells us* $|x_\infty| \leq \|x\|_\infty$*. From the representation of* x *we have determined, we also know*

$$\begin{aligned} |f(x) &= |b_1 x_\infty + \sum_{j=1}^{\infty} b_{j+1} x_j| \leq |b_1|\,|x_\infty| + \sum_{J=1}^{\infty} |b_{j+1}|\,|x_j| \\ &\leq |b_1|\|x\|_\infty + \sum_{j=1}^{\infty} |b_{j+1}|\,\|x\|_\infty = (|b_1| + \sum_{j=1}^{\infty} |b_{j+1}|)\|x\|_\infty \\ &= \|x\|_\infty \sum_{j=1}^{\infty} |b_j| = \|x\|_\infty\, \|(b_j)\|_1 \end{aligned}$$

Thus,

$$\|\boldsymbol{f}\|_{op} = \sup_{x \in \mathbf{c}, \|x\|_\infty \neq 0} \frac{|\boldsymbol{f}(x)|}{\|x\|_\infty} \leq \|(b_j)\|_1 = \|T(\boldsymbol{f})\|_1$$

It remains to prove the reverse inequality. To show that, define $x_n \in \mathbf{c}$ *by*

$$(x_n)_j = \begin{cases} sign(b_2), & slot\ 1 \\ \vdots & \\ sign(b_{n+1}), & slot\ n \\ sign(b_1), & slot\ \geq n+1 \end{cases}$$

Then $x_n \to sign(b_1)$ *for all* n *and so* $x_n \in \mathbf{c}$*. Further,*

$$\boldsymbol{f}(x_n) = b_1\, sign(b_1) + \sum_{j=1}^{n} (sign(b_{j+1}))b_{j+1} + \sum_{j=n+1}^{\infty} (sign b_1) b_{j+1}$$

We know $\sum_{j=n+1}^{\infty} b_{j+1}$ *converges because* $(b) \in \ell^1$*. Hence, we can write*

$$\boldsymbol{f}(x_n) = \sum_{j=1}^{n+1} |b_j| + (sign(b_1)) \sum_{j=n+1}^{\infty} b_{j+1}$$

Also, $\|x_n\|_\infty = 1$*, so* $|\boldsymbol{f}(x_n)| \leq \|\boldsymbol{f}\|_{op}$*; i.e.*

$$-(sign(b_1)) \sum_{j=n+1}^{\infty} b_{j+1}| - \|\boldsymbol{f}\|_{op} \leq \sum_{j=1}^{n+1} |b_j| \leq (sign(b_1)) \sum_{j=n+1}^{\infty} b_{j+1}| + \|\boldsymbol{f}\|_{op}$$

Since $(b_j) \in \ell^1$, for $\epsilon > 0$, there is N so that $\sum_{j=n+1}^{\infty} b_{j+1} < \epsilon$ if $n > N$. We see for $n > N$

$$-\|\boldsymbol{f}\|_{op} - \epsilon \quad \leq \quad \sum_{j=1}^{n+1} |b_j| \leq \|\boldsymbol{f}\|_{op} + \epsilon$$

Thus,

$$-\|\boldsymbol{f}\|_{op} - \epsilon \quad \leq \quad \|(b_j)\|_1 \leq \|f\|_{op} + \epsilon$$

implying $\|(b_j)\|_1 \leq \|\boldsymbol{f}\|_{op} + \epsilon$ for all ϵ. It follows that $\|(b_j)\|_1 - \epsilon \leq \|\boldsymbol{f}\|_{op}$. Combining with the other estimate, we see $\|(b_j)\|_1 = \|\boldsymbol{f}\|_{op}$.

- *T is one to one and onto.*

If $T(\boldsymbol{f}) = T(\boldsymbol{g})$, then $T(\boldsymbol{f} - \boldsymbol{g}) = 0$. Thus, $0 = \|T(\boldsymbol{f} - \boldsymbol{g})\|_1 = \|\boldsymbol{f} - \boldsymbol{g}\|_{op}$. Hence, $\boldsymbol{f} = \boldsymbol{g}$ and we have shown T is one to one.

Now to show T is onto, let $(w_j)_{j=1}^{\infty} \in \ell^1$. Define scalars γ_j for $j \geq 0$ by $\gamma_1 = w_2$, $\gamma_2 = w_3$ and so on. So $\gamma_j = w_{j+1}$ for $j \geq 1$. Hence, $\sum_{j=1}^{\infty} \gamma_j = \sum_{j=2}^{\infty} w_j$ is well-defined as $(w_j) \in \ell^1$. The scalar $\gamma_0 = w_1 + \sum_{j=1}^{\infty} \gamma_j = w_1 + \sum_{j=2}^{\infty} w_j = \sum_{j=1}^{\infty} w_j$ is also well-defined.

This gives us $\gamma_0 - \sum_{j=1}^{\infty} = w_1$ and $\gamma_j = w_{j+1}$ if $j \geq 1$. Define $\boldsymbol{f}$ on the usual Schauder Basis by $\boldsymbol{f}(\boldsymbol{e_j}) = \gamma_j$ and then $\boldsymbol{f}$'s action of any $x \in \ell^1$ is given by $\boldsymbol{f}(x) = x_\infty \gamma_0 + \sum_{j=1}^{\infty}(x_j - x_\infty)\gamma_j$. Then $\boldsymbol{f}$ is linear with $T(\boldsymbol{f}) = (w_j)_{J=1}^{\infty}$.

It is easy to see $\boldsymbol{f}$ is bounded on $\mathbf{c}$ and hence is continuous. This shows T is onto.

■

16.4.2 The Dual of Sequences That Converge to 0

The argument here is similar to that for $\mathbf{c}'$ except a bit simpler.

Theorem 16.4.2 The Dual of Sequences that Converge to Zero

> *Let $\mathbf{c_0}$ be the set of all sequences that converge to zero with the sup norm $\|\cdot\|_\infty$. Then $\mathbf{c_0}' \cong \ell^1$.*

Proof 16.4.2
In Section 5.2.1, we found a Schauder Basis for $\mathbf{c}$: $E = (\boldsymbol{e_n})_{n=1}^{\infty}$ where

$$\boldsymbol{e_n} \quad = \quad (0, \ldots, 0, \underset{\textit{slot } n}{1}, 0, \ldots), n \geq 1$$

and, for each $\boldsymbol{x}$, there is a unique representation $\boldsymbol{x} = \sum_{j=1}^{\infty} x_n \boldsymbol{e_n}$ where the partial sums $S_n \to \boldsymbol{x}$ in $\|\cdot\|_\infty$.

We can show there is a mapping $T : \mathbf{c_0}' \to \ell^1$ so that

1. *T is $1-1$ and onto.*
2. *T is linear.*

3. $\|T(\boldsymbol{f})\|_1 = \|\boldsymbol{f}\|_{op}$ *for all* $\boldsymbol{f} \in \boldsymbol{c_0}'$.

We will leave this to you as an exercise. You can mimic what we did in the proof of Theorem 16.4.1.
■

Homework

Exercise 16.4.1 *Proving rearrangement theorems is messy, but for our purposes here, use a standard* $\epsilon - N$ *argument to show* $S_1 - S_2 = S_3$ *where* $S_1 = \sum_{j=1}^{\infty} x_j \gamma_j$, $S_2 = \sum_{j=1}^{\infty} x_\infty \gamma_j$ *and* $S_3 = \sum_{j=1}^{\infty}(x_j - x_\infty)\gamma_j$.

Exercise 16.4.2 *Prove* $(\boldsymbol{c_0})' \cong \ell^1$.

16.5 A Digression to Integration Theory

Let Ω be a nonempty set and $\mathcal{F}$ be a field of subsets. If $\mathcal{F}$ is closed under countable unions, $\mathcal{F}$ is a $\boldsymbol{\sigma}$ - field. A good example of a $\boldsymbol{\sigma}$ - field is what is called the **Borel** field which is the collection of subsets which can be formed from open sets using the operations of countable union, countable intersection and complement. Hence, the Borel field $\mathcal{B}$ is usually defined to be the smallest $\boldsymbol{\sigma}$ - field generated by open intervals (a, b) where a, b are real numbers. More precisely, if $\mathcal{A} = \{(a, b) \mid a, b \in \Re\}$, then $\mathcal{B} = \cap_{\mathcal{F}}$ where $\mathcal{F}$ is a $\boldsymbol{\sigma}$ - field that contains $\mathcal{A}$.
Now let's define a minimum quality of niceness for a function to have: measurability.

Definition 16.5.1 $\mathcal{F}$ **Measurability**

If $f : \Omega \to \Re$, *we say* f *is* $\mathcal{F}$ *- measurable if for all* $B \in \mathcal{B}$, $f^{-1}(B) \in \mathcal{F}$.

Recall if $\mu \in ba(\Omega, \mathcal{F})$, then μ is finitely additive, μ is bounded with $|\mu| = \mu^+ + \mu^-$, $\mu^+ \ \mu \vee 0$ and $\mu^- = (-\mu) \vee 0$ and $\|\mu\| = |\mu|(\Omega)$. Further, by Theorem 15.4.7

$$\|\mu\| = |\mu|(\Omega) = \sup_{(E_i) \subset \mathcal{E}} \sum_i |\mu(E_i)|$$

where $\mathcal{E}$ is the set of all disjoint finite collections $\{E_1, \ldots, E_n\}$ for some n satisfying $\Omega \subset \cup_{i=1}^n E_i$. Recall such a collection of subsets is called a partition of Ω. We can extend the action of μ to all the subsets of Ω by constructing what is called the outer measure μ^* built from μ.

Definition 16.5.2 Outer Measure μ^*

Let $\mu : \mathcal{F} \to [0, \infty]$ *be finitely additive on the field of subsets* $\mathcal{F}$ *of the set* Ω. *Hence,* μ *is non - negative. Define* $\mu^* : 2^\Omega \to \Re \cup \{+\infty\}$ *by*

$$\mu^*(E) = \inf\{\mu(F) \mid E \subset F, F \in \mathcal{F}\}$$

We can prove many properties for the outer measure.

Lemma 16.5.1 Properties of μ^*

The outer measure μ^* satisfies

1. $\mu*(E) = \mu(E)$ *if* $E \in \mathcal{F}$.
2. $\mu^*(A \cup B) \leq \mu^*(A) + \mu^*(B)$ *for all* $A, B \subset \Omega$.
3. $\mu^*(A) \leq \mu^*(B)$ *for* $A \subset B \subset \Omega$.

Proof 16.5.1
1:
If $E \subset F \in \mathcal{F}$, *then* $\mu(F) = \mu(E) + \mu(F \setminus E)$ *since* μ *is additive. Hence,* $\mu(F) \geq \mu(E)$ *which implies* $\mu(F) \geq \mu(E) \geq \inf_{E \subset F} \mu(F)$ *as* $E \in \mathcal{F}$ *is an allowable choice in the infimum calculation* $\mu^* E$. *Thus* $\mu(E) \geq \mu^*(E)$. *But* $\mu(F) \geq \mu(E)$ *for all* F *that contain* E *implies* $\mu(E)$ *is a lower bound for* $\{\mu(F) \mid E \subset F, F \in \mathcal{F}\}$. *This immediately implies* $\mu^*(E) \geq \mu(E)$. *Combining* $\mu(E) = \mu^*(E)$.

2:
Let $\epsilon > 0$ *be given. By the infimum tolerance lemma, there are sets* A_1 *and* B_1 *in* $\mathcal{F}$ *so that* $A \subset A_1$, $B \subset B_1$ *and* $\mu(A_1) \leq \mu^*(A) + \frac{\epsilon}{2}$ *and* $\mu(B_1) \leq \mu^*(B) + \frac{\epsilon}{2}$. *Now* $A_1 \cup B_1 \subset A \cup B$ *and so* $\mu^*(A \cup B) \leq \mu(A_1 \cup B_1) = \mu(A_1) + \mu(B_1 - A_1)$. *Since* $B_1 \setminus A_1 \subset B_1$, *we then have* $\mu(B_1) = \mu(B_1 \cap A_1) + \mu(B_1 \setminus A_1)$. *Thus,* $\mu(B_1) \geq \mu(B_1 \setminus A_1)$.

Therefore $\mu^*(A \cup B) \leq \mu(A_1) + \mu(B_1) \leq \mu^*(A) + \mu^*(B) + \epsilon$. *But* $\epsilon > 0$ *is arbitrary, so the claim is true.*

3:
$\mu^*(A) \leq \mu^*(B)$ *if* $A \subset B \subset \mathcal{F}$ *by definition.* ■

The subsets E of Ω where $\mu^*(E) = 0$ are of great importance. At this point, go back and look at the definition of sets of content zero in (Peterson (18) 2020) and (Peterson (20) 2020) and compare what we say in that context to what we do below.

Definition 16.5.3 Null Sets

Let μ *be an additive set function on the field* $\mathcal{F}$. *Let* $\nu = |\mu|$, *the total variation of* μ *and* ν^* *be the outer measure built from* ν. *Then a set* E *in* Ω *is said to be a* $\boldsymbol{\mu}$ *- null set if* $\nu^*(E) = 0$.

Comment 16.5.1 *From Lemma 16.5.1, we see a subset of a* $\boldsymbol{\mu}$ *- null set is also a* $\boldsymbol{mu}$ *- null set.*

A new way of measuring how big a function is on Ω is called the **essential sup** of the function.

Definition 16.5.4 The Essential Supremum of a Function

Let $f : \Omega \rightarrow \Re \cup \{+\infty\}$. *If there is a* $\boldsymbol{mu}$ *- null set* $E \subset \Omega$ *so that* $f|_{\Omega \setminus E}$ *is bounded, we say* f *is* $\boldsymbol{\mu}$ *- essentially bounded. In this case, it follows*

$$\inf_{E \subset \Omega, E \mu \text{ null}} \quad \sup_{t \in \Omega \setminus E} |f(t)| \quad < \quad \infty$$

We define this number to be the **essential supremum** *or* **ess sup** *of* f *on* Ω. *It is denoted by* $\|f\|_\infty = esssup_{t \in \Omega} |f(t)|$.

Homework

Exercise 16.5.1 *Prove f is $\mathcal{F}$ - measurable if and only if f^+ and f^- are $\mathcal{F}$ - measurable.*

Exercise 16.5.2 *Let $\mathcal{F}$ be a field of subsets in $\Omega = \Re$. Is the singleton set $\{c\}$ in $\mathcal{F}$ for any $cin\Re$? Can you calculate $\nu * (\{c\})$ for such a set for $\nu = |\mu$ and ν^* the outer measure associated with ν?*

Exercise 16.5.3 *if f is continuous on $[0, 1]$, is the essential sup of f the same as the usual maximum? Note we have to define the field of subsets $\mathcal{F}$ here.*

16.5.1 A Riemann Integral Extension

We are now ready to extend Riemann Integration using finitely additive set functions.

Definition 16.5.5 Characteristic Functions

Let $|omega$ be a nonempty set and $\mathcal{F}$ a field of subsets. Let μ be a finitely additive set function on $\mathcal{F}$. if $S \in \mathcal{F}$, the characteristic function of S is denoted I_S and is defined to be $I_S(t) = 1$ if $t \in S$ and 0 otherwise.

Comment 16.5.2 *We have used this idea before when we discussed Riemann Integration in $\Re^2$ in (Peterson (20) 2020). So you might want to go back and look at that now.*

In general, we are interested in the equality of two $\mathcal{F}$ measurable functions. It is enough that the functions are equal except on a set which is $\boldsymbol{\mu}$ null.

Definition 16.5.6 Functions Equal μ Almost Everywhere

We say the $\mathcal{F}$ measurable functions f and g are equal μ almost everywhere if there is a $\boldsymbol{\mu}$ null set E with $f = g$ on $\Omega \setminus E$. We denote equal μ almost everywhere by the abbreviation μ a.e.

Integration in this setting is developed quite a bit different than the way we are used to in Riemann Integration which is done using Riemann Sums or Darboux upper and lower sums. However, in both a Riemann sum and a Darboux sum, a partition P of $[a, b]$ determines subintervals $[t_i, t_{i+1}]$ and we can use these to determine a disjoint collection of sets whose union is $[a, b]$. For example, we could let $E_1 = [t_0, t_1)$, $E_2 = [t_1, t_2)$ and so forth finishing with the last set $E_n = [t_{n+1}, t_n]$. These sets are disjoint. The length $t_{i+1} - t_i$ we can think of as the result of applying a finitely additive set function μ to E_i to get the usual length. Then in both the Riemann sum and the Darboux sum cases, we assign a number to each subinterval and compute a sum: $\sum_{i \in P} x_i \mu(E_i)$ where $\mu(E_i) = t_{i+1} - t_i$. There are lots of subtle details here such as what the field $\mathcal{F}$ is and how our μ is defined on $\mathcal{F}$ when we only know how to apply it to an interval. But let's just ignore that for the moment. We can define a function here by $f(x) = \sum_{i \in P} c_i I_{E_i}$ and then **define** the integral of f with respect to μ to be $\int_{[a,b]} f d\mu = \sum_{i \in P} x_i \mu(E_i)$. Call these types of function **simple** functions which are finite linear combinations of characteristic functions. This is a powerful idea which we can exploit. Note this means we are identifying Riemann and Darboux sums with the integrals of simple functions. Then, we use some sort of limiting process to define the integral of functions f which are appropriate limits of simple functions. So let's get started.

Definition 16.5.7 Simple Functions

Let $\mathcal{F}$ be a field of subsets of Ω and let $f : \Omega \rightarrow \Re$ have a range consisting of only a finite set of distinct values $\{x_1, \ldots, x_n\}$ with $E_i = f^{-1}(x_i) = \{t \in \Omega \mid f(t) = x_i\} \in \mathcal{F}$. In this case, we say f is a simple function. Clearly, a simple function is a finite linear combination of characteristic functions.

If g is a function equal to a simple function μ a.e., we say g is a μ - simple function.

We can then define the integral of a μ - simple function.

Definition 16.5.8 The μ integral of a μ - simple Function

A μ - simple function g is said to be μ-integrable if the only inverse image $E_i = f^{-1}(x_i)$ which has $|\mu|(E_i)$ infinite is the value $x_i = 0$. That is, since $g = f$ μ a.e. for some simple function $f = \sum_{i=1}^n x_i I_{E_i}$ and $|\mu|(E_i) < \infty$ except possibly the set $f^{-1}(0)$. The μ integral of g is then $\int_\Omega g d\mu = \sum_{i=1}^n x_i \mu(E_i)$ where we interpret the case $0 \cdot \mu(f^{-1}(0)) = 0 \cdot \infty = 0$. Note if $F \in \mathcal{F}$, we define $\int_F g \, d\mu = \sum_{i=1}^n x_i \mu(E_i \cap F)$.

Comment 16.5.3 *Let's go over the details here. Let $f = \sum_{i=1}^n x_i \, I_{E_i}$ where $\cup_{i=1}^n E_i = \Omega$. If $g = f$ μ a.e., there is a set $V \in \mathcal{F}$ with $\mu(V) = 0$ where $g \neq f$. The set $V = \cup_{j=1}^m F_j$ where $\mu(F_j) = 0$ and*

$$g(t) = \begin{cases} x_1, & t \in E_1 \setminus V \\ x_2, & t \in E_2 \setminus V \\ \vdots & \\ x_n, & t \in E_n \setminus V \\ y_j, & t \in F_j \end{cases}$$

Hence, $g = \sum_{i=1}^n x_i \mu(E_i \setminus V) + \sum_{j=1}^m y_j \mu(F_j)$. Note $\mu(E_i) = \mu(E_i \cap V) + \mu(E_i \cap V^C) = 0 + \mu(E_i \setminus V)$. So

$$\int_\Omega g d\mu = \sum_{i=1}^n x_i \mu(E_i \setminus V) + \sum_{j=1}^m y_j \mu(F_j) = \sum_{i=1}^n x_i \mu(E_i) = \int_\Omega f d\mu$$

To define the μ-integrability of a $\mathcal{F}$ measurable function, we proceed in stages.

Definition 16.5.9 The $\mu \geq 0$ integrability of Non-negative $\mathcal{F}$ Measurable Functions

Assume $\mu \geq 0$ and f is a non-negative $\mathcal{F}$ measurable function. We define the μ integral of f to be

$$\int_\Omega f \, d\mu = \sup_{\mu-\text{ simple } g} \left\{ \int_\Omega g \, d\mu \mid g \leq f \mu a.e. \text{ on } \Omega \right\}$$

Note $\int_\Omega f d\mu$ could be $+\infty$. Then, we extend to real-valued f.

Definition 16.5.10 The $\mu \geq 0$ integrability of Real-valued $\mathcal{F}$ Measurable Functions

Assume $\mu \geq 0$ and f is a real-valued $\mathcal{F}$ measurable function. We define $\int_\Omega f^+ d\mu$ and $\int_\Omega f^+ d\mu$ as in Definition 16.5.9. Then as long as both are finite, we define $\int_\Omega f \, d\mu = \int_\Omega f^+ \, d\mu - \int_\Omega f^- \, d\mu$. In this case, $\int_\Omega f \, d\mu$ is well-defined as we never are in the un resolvable case of $\infty - \infty$.

Then, if μ is real-valued, we go to the next step.

Definition 16.5.11 The μ-integrability of Real-valued $\mathcal{F}$ Measurable Functions

In this case $\mu = \mu^+ - \mu^-$. We define $\int_\Omega f\, d\mu = \int_\Omega f\, d\mu^+ - \int_\Omega f\, d\mu^-$.

Now a simple function is a finite linear combination of characteristic functions and it is straightforward to prove any finite linear combination of simple functions can be rewritten and relabeled to become a simple function over a suitable finite set of distinct range values. We will let you work this out in a few exercises.

Homework

Exercise 16.5.4 *Show the sum of two simple functions can be rewritten as one simple function for appropriate range values.*

Exercise 16.5.5 *Show the linear combination of two simple functions can be rewritten as one simple function for appropriate range values.*

Exercise 16.5.6 *Show the linear combination of three simple functions can be rewritten as one simple function for appropriate range values.*

Exercise 16.5.7 *Prove $\int_\Omega (f+g) d\mu = \int_\Omega f d\mu + \int_\Omega g d\mu$. Do this in stages: first for f and g non-negative and $\mu \geq 0$ and then the more general case.*

Exercise 16.5.8 *Prove $\int_\Omega (\alpha f) d\mu = \alpha \int_\Omega f d\mu$. Do this in stages: first for f and g non-negative and $\mu \geq 0$ and then the more general case.*

Next, let (f_n) be a sequence of finite linear combinations of characteristic functions. Given $F \in \mathcal{F}$, we can define the convergence of (f_n) to a function f as follows.

Definition 16.5.12 Convergence of Simple Function Combinations

Let (f_n) be a sequence of finite linear combinations of characteristic functions. Given $F \in \mathcal{F}$, we say $f_n \rightarrow f$ on F uniformly for a function $f : \Omega \rightarrow \Re$ if

$$\forall \epsilon > 0, \exists N, \ni n > N \implies \sup_{t \in F} |f_n(t) - f(t)| < \epsilon$$

Note for $F = \Omega$, this is simply uniform convergence of the sequence (f_n) to f on Ω.

We can now define two important collections of functions. We let

$$B(\Omega) = \left\{ f : \Omega \rightarrow \Re \mid \sup_{t \in \Omega} |f(t)| < \infty \right\}$$

and let the set of characteristic function on Ω be denoted by $BC(\Omega)$. Then, define

$$B(\Omega, \mathcal{F}) = \{ f : \Omega \rightarrow \Re \mid \exists\ (f_n) \subset span(\ BC(\Omega)\ ), f_n \rightarrow f \text{ uniformly } \}$$

We can show all bounded $\mathcal{F}$ - measurable functions are the uniform limit of a sequence of simple functions. Hence, all bounded $\mathcal{F}$ - measurable functions are in $B(\Omega, \mathcal{F})$.

Let's state some things we can prove, but we don't want to get too bogged down in all these details. Much of this careful discussion is best done more slowly and carefully as we do in (Peterson (16) 2019) so we will ask you to jump ahead to look at these kinds of arguments there. We will sketch here what these proofs entail.

1. $B(\Omega)$ is a complete normed linear space with respect to the sup norm. To prove this, we have to show that if (f_n) is a Cauchy sequence of bounded functions in the sup - norm, then there is a bounded function f with $f_n \rightarrow f$ on Ω uniformly. The usual argument works: since the sequence is Cauchy, at each $t \in \Omega$, the completeness of $\Re$ tells us there is a pointwise limit function f. It is then easy to see the pointwise limit function is bounded.

2. $B(\Omega, \mathcal{F})$ is a complete normed linear space with respect to the sup - norm. If (f_n) is a sequence of finite linear combinations of characteristic functions on Ω that converges uniformly to f, then it is easy to see each f_n is a bounded function and since $B(\Omega)$ is complete with respect to this norm, f is bounded too. The limit function f is by definition in $B(\Omega, \mathcal{F})$ so $B(\Omega, \mathcal{F})$ is a closed subspace of $B(\Omega)$ is therefore complete.

You should work out these details yourself.

Homework

Exercise 16.5.9 *$B(\Omega)$ is a complete normed linear space with respect to the sup norm.*

Exercise 16.5.10 *$B(\Omega, \mathcal{F})$ is a complete normed linear space with respect to the sup - norm.*

16.6 The Dual of Bounded Measurable Functions

We are now ready to characterize the dual space of $B(\Omega, \mathcal{F})$.

Theorem 16.6.1 The Dual Space of $B(\Omega, \mathcal{F})$

$(B(\Omega, \mathcal{F}))' \cong ba(\Omega, \mathcal{F})$.

Proof 16.6.1

We know if f is a μ integrable function, then $\int_\Omega f\, d\mu < \infty$. From the definition of μ integrability, it is easy to see $| \int_\Omega f d\mu| \leq \int_\Omega \|f\|_\infty d\mu = \|f\|_\infty |\mu|(\Omega) < \infty$ as each charge is bounded. Now define a mapping $\Theta : ba(\Omega, \mathcal{F}) \rightarrow \Re$ by $\Theta(f) = \int_\Omega f\, d\mu$.
From our discussions about the linearity of $\int_\Omega f d\mu$ (which you proved in an exercise!), we see Θ is a linear functional of $B(\Omega, \mathcal{F})$. Is it bounded? Consider

$$\sup_{f \neq 0, f \in B(\Omega,\mathcal{F})} \frac{|\Theta(f)|}{\|f\|_\infty} \leq \frac{\|f\|_\infty |\mu|(\Omega)}{\|f\|_\infty} = |\mu|(\Omega)$$

Thus, $\|\Theta\|_{op} \leq |\mu|(\Omega)$ and so $\Theta \in (B(\Omega, \mathcal{F}))'$.

Let's prove the reverse inequality. Let $\epsilon > 0$ be given. Let $\{E_1, \ldots, E_n\}$ be a partition of Ω into disjoint sets so that $|\mu|(\Omega) - \epsilon < \sum_{i=1}^n |\mu(E_i)|$. This follows from Theorem 15.4.7 and the supremum tolerance lemma. Define $\hat{f} : \Omega \rightarrow \Re$ by $\hat{f} = \sum_{i=1}^n sign(\mu(E_i)) I_{E_i}$. Then $\hat{f} \in B(\Omega, \mathcal{F})$

with $\|\hat{f}\|_\infty = 1$. Further, we have

$$\|\Theta\|_{op} \geq \int_\Omega \hat{f} d\mu = \sum_{i=1}^{n} |\mu(E_i)| > |\mu|(\Omega) - \epsilon$$

Thus, $\|\Theta\|_{op} \geq |\mu|(\Omega) - \epsilon$ for all $\epsilon > 0$. Hence, $\|\Theta\|_{op} \geq |\mu|(\Omega)$ which is the other half of the inequality we need. We conclude $\|\Theta\|_{op} = |\mu|(\Omega) = \|\mu\|$.

We have shown for all $\mu \in ba(\Omega, \mathcal{F})$ there is a map $\Theta_\mu \in (\ B(\Omega, \mathcal{F})\ )'$ defined by $\Theta_\mu(f) = \int_\Omega f d\mu$ with $\|\Theta_\mu\|_{op} = \|\mu\|$. Define $T : ba(\Omega, \mathcal{F}) \to (\ B(\Omega, \mathcal{F})\ )'$ by $T(\mu) = \Theta_\mu$. We have already shown T is a linear norm isometry which also tells us immediately T is $1 - 1$.

Next, if $g \in (\ B(\Omega, \mathcal{F})\ )'$, define the charge μ on Ω by $\mu(E) = g(I_E)$ for all $E \in \mathcal{F}$. It is easy to see μ is finitely additive and

$$|\mu(E)| = |g(I_E)| \leq \|g\|_{op} \|I_E\|_\infty = \|g\|_{op} < \infty$$

and so μ is bounded. Now let f be a linear combination of characteristic functions. Then $f = \sum_{i=1}^n I_{E_i}$ for some n and sets $E_i \in \mathcal{F}$. Thus,

$$\int_\Omega f d\mu = \sum_{i=1}^{n} \mu(E_i) = \sum_{i=1}^{n} g(I_{E_i}) = g(\sum_{i=1}^{n} I_{E_i}) = g(f)$$

We conclude $g(f) = \inf_\Omega f d\mu$ for all f that are finite linear combinations of characteristic functions.

Now let f be the uniform limit of a sequence of finite linear combinations of characteristic functions f_n; i.e. $f_n \to f$ uniformly. We know $g(f_n) = \int_\Omega f_n d\mu$ from the above and since g is continuous

$$g(f) = \lim_{n\to\infty} g(f_n) = \lim_{n\to\infty} \int_\Omega f_n d\mu$$

It is straightforward to show $\int_\Omega f_n d\mu \to \int_\Omega f d\mu$ since $f_n \to f$ uniformly on Ω. Thus, $g(f) = \int_\Omega f d\mu$. Our choice of f in $B(\Omega, \mathcal{F})$ is arbitrary and so we have shown given g in $(\ B(\Omega, \mathcal{F})\ )'$, there is a $\mu \in ba(\Omega, \mathcal{F})$ so that $g(f) = \int_\Omega f d\mu = \Theta_\mu(f) = T(\mu)$. Hence T is onto.

This competes the proof. ■

Homework

Exercise 16.6.1 *Prove if $f \in B(\Omega, \mathcal{F})$, then $|\int_\Omega f d\mu| \leq \int_\Omega \|f\|_\infty d\mu = \|f\|_\infty |\mu|(\Omega)$.*

Exercise 16.6.2 *Prove $B(\mathbb{N}, 2^{\mathbb{N}}) = \ell^\infty$.*

16.7 Connections to Game Theory

It is time to connect all of our theory to games again. We are going to look at this much more abstractly now using some powerful results known as Ky Fan's theorems. You can look up the original papers in (Fan (3) 1956). We are quoting them as they appear in (Kunnai (12) 1992) for convenience. The 1956 paper is a blast; look at how it is typeset before the wide use of LaTeX!

Theorem 16.7.1 Linear Inequalities in a Normed Linear Space

Let $(\boldsymbol{x}_\nu)$ be a family of elements in the normed linear space $(X, \|\cdot\|)$ and let $(\boldsymbol{\alpha}_\nu)$ be an associated family of real numbers. Then, for all $\rho \geq 0$, the following two conditions are equivalent:

1. *There is a continuous linear functional $f \in X'$ with $\|f\|_{op} \leq \rho$ so that $f(\boldsymbol{x}_\nu) \geq \boldsymbol{\alpha}_\nu$.*
2. *For any finite number n of indices $\{\nu_1, \ldots, \nu_n\}$ and any n positive numbers $\{\lambda_1, \ldots, \lambda_n\}$ the inequality*

$$\rho \left\| \sum_{i=1}^{n} \lambda_i \, \boldsymbol{x}_{\nu_i} \right\| \geq \sum_{i=1}^{n} \lambda_i \, \boldsymbol{\alpha}_{\nu_i}$$

holds.

Proof 16.7.1

See (Fan (3) 1956) for this. We usually like to be self contained but we just want to give the flavor of how our study of dual spaces turns out to be useful. Budding economic game theorists will have some more reading though! ■

Note the second conditions is the same as saying

$$\beta = \sup \left(\sum_{i=1}^{n} \lambda_i \, \boldsymbol{\alpha}_{\nu_i} \; : \; \left\| \sum_{i=1}^{n} \lambda_i \, \boldsymbol{x}_{\nu_i} \right\| = 1 \right)$$

exists, is finite and satisfies $\beta \leq \rho$. So if $\beta > 0$ we could choose $\beta = \rho$ and get a new version of this theorem.

Theorem 16.7.2 The Supremum Version of Linear Inequalities in a Normed Linear Space

Let $(\boldsymbol{x}_\nu)$ be a family of elements in the normed linear space $(X, \|\cdot\|)$ and let $(\boldsymbol{\alpha}_\nu)$ be an associated family of real numbers. Let σ be defined by

$$\sigma = \sup \left(\sum_{i=1}^{n} \lambda_i \, \boldsymbol{\alpha}_{\nu_i} \; : \; \left\| \sum_{i=1}^{n} \lambda_i \, \boldsymbol{x}_{\nu_i} \right\| = 1, \lambda_i > 0 \right)$$

Then if σ is finite there is $f \in X'$ with $\|f\|_{op} \leq \sigma$ so that $f(\boldsymbol{x}_\nu) \geq \boldsymbol{\alpha}_\nu$.

Proof 16.7.2

This is just a restatement of the previous theorem. ■

A useful version of this theorem is a slight restatement again.

Theorem 16.7.3 The Existence of a Solution to a System of Linear Inequalities in a Normed Linear Space

Let $(\boldsymbol{x_\nu})$ be a family of elements in the normed linear space $(X, \|\cdot\|)$ and let $(\boldsymbol{\alpha_\nu})$ be an associated family of real numbers. Let σ be defined by

$$\sigma = \sup\left(\sum_{i=1}^{n} \lambda_i \boldsymbol{\alpha_{\nu_i}} : \left\|\sum_{i=1}^{n} \lambda_i \boldsymbol{x_{\nu_i}}\right\| = 1, \lambda_i > 0\right)$$

Then

1. *There is $f \in X'$ with $\|f\|_{op} \leq \sigma$ so that $f(\boldsymbol{x_\nu}) \geq \boldsymbol{\alpha_\nu}$ if and only if σ is finite.*
2. *If there is a nonzero solution to $f(\boldsymbol{x_\nu}) \geq \boldsymbol{\alpha_\nu}$ then*

$$\sigma = \inf\left(\|f\|_{op} : f \text{ solves } f(\boldsymbol{x_\nu}) \geq \boldsymbol{\alpha_\nu}\right)$$

Note this implies $\sigma \leq \|f\|_{op}$.

Proof 16.7.3
Again, we refer you to (Fan (3) 1956). Bon appetit! ■

Now specialize to the set $\mathbb{N}$. Assume the game ν satisfies $\nu(S) = 0$ if $S \neq \mathbb{N}$ is an infinite subset of $\mathbb{N}$ with $\nu(\mathbb{N})$ some positive number. For any finite S, the indicator function I_S is 1 for any index $i \in S$ and zero otherwise. Hence, we can identify I_S with the sequence $(x) \in \boldsymbol{c}_0$ given by

$$x_i = \begin{cases} 1, & i \in S \\ 0, & i \notin S \end{cases}$$

Since S is finite, there is a largest index N beyond which $x_i = 0$. Hence, each I_S can be identified with a sequence y_i in $\boldsymbol{c}_0$. Here $\boldsymbol{x_{\nu_i}} \in \boldsymbol{c}_0$ is better written as $\boldsymbol{I_{S_i}}$. So the $\boldsymbol{c}_0$ sequence $\sum_{i=1}^{n} \lambda_i \boldsymbol{I_{S_i}}$ is the same as the $\boldsymbol{c}_0$ sequence z whose terms are

$$z_i = \begin{cases} \sum_{\{j: i \in S_j \,|\, 1 \leq j \leq n\}} \lambda_j, & \\ 0, & \text{else} \end{cases}$$

Clearly, z_i is zero after a finite number of terms and so is in $\boldsymbol{c}_0$. Let N be the maximum index where z_i is nonzero. Now consider the requirement that $\sup|\sum_{i=1}^{n} \lambda_i \boldsymbol{I_{S_i}}| = 1$ where we abuse notation by writing the $\boldsymbol{c}_0$ sequence as a linear combination of characteristic functions for convenience. The maximum value of the resulting z will be achieved at the index p_0 and so $\sum_{\{j: p_0 \in S_j \,|\, 1 \leq j \leq n\}} \lambda_j = 1$. Let

$$\begin{aligned} \xi_1 &= \sum_{\{j: 1 \in S_j\}} \lambda_j \\ \xi_2 &= \sum_{\{j: 2 \in S_j\}} \lambda_j \\ &\vdots \\ \xi_N &= \sum_{\{j: N \in S_j\}} \lambda_j \end{aligned}$$

and define new sets T_i by $T_i = \{i\}$ for all $i \neq p_0$ from $\{1, \ldots, N\}$. with associated scalars $\zeta_i = 1 - \xi_i$. Finally let $Z = \{N+1, \ldots\} \cup W$ where W is all the subsets of $\{1, \ldots, N\}$ not in the union of the S_j's. with scalar $\theta = 1$. Then $\mathbb{N} = \cup_{i=1}^{n} S_i \cup \cup_{i=1}^{N} T_i \cup Z$ and the collection of sets

$\{S_1, \ldots, S_n, (T_i)_{i \neq p_0, 1 \leq i \leq N}, Z\}$ with scalars $\{\lambda_1, \ldots, \lambda_n, (\zeta_i)_{i \neq p_0}, \theta\}$ satisfies

$$\begin{aligned}
&\sum_{j=1}^{n} \lambda_j I_{S_j} + \sum_{j \neq p_0, 1 \leq j \leq N} I_{T_j} + \theta I_Z \\
&= \{ \sum_{\{j:1 \in S_j \mid 1 \leq j \leq n\}} \lambda_j + 1 - \sum_{\{j:1 \in S_j \mid 1 \leq j \leq n\}} \lambda_j, \\
&\quad \ldots, \\
&\quad \sum_{\{j:p_0 \in S_j \mid 1 \leq j \leq n\}} \lambda_j, \ldots, \sum_{\{j:N \in S_j \mid 1 \leq j \leq n\}} \lambda_j + 1 - \sum_{\{j:N \in S_j \mid 1 \leq j \leq n\}} \lambda_j, \\
&\quad 1, \ldots\} \\
&= \{1, 1, 1, 1, \ldots\}
\end{aligned}$$

and so it is balanced. Thus, if ν **is a balanced game**, we have

$$\sum_{j=1}^{n} \lambda_j \nu(S_j) + \sum_{j \neq p_0, 1 \leq j \leq N} \nu(T_j) + \theta(Z) \leq \nu(\mathbb{N})$$

But $\nu(Z) = 0$ because it is an infinite set and so

$$\sum_{j=1}^{n} \lambda_j \nu(S_j) \leq \sum_{j=1}^{n} \lambda_j \nu(S_j) + \sum_{j \neq p_0, 1 \leq j \leq N} \nu(T_j) \leq \nu(\mathbb{N})$$

We can do this for any collection of sets satisfying $\sup |\sum_{i=1}^{n} \lambda_j \boldsymbol{I}_{\boldsymbol{S_i}}| = 1$. We conclude for all such collections $\sum_{i=1}^{k} \lambda_i \nu(S_i) \leq \nu(\mathbb{N})$ which immediately implies $\sigma \leq \nu(\mathbb{N})$.

From this discussion, we see for any collection of sets $\{S_1, \ldots, S_k\}$, we can augment them to get a new balanced collection which satisfies $\| \sum_{\ell=1}^{L} I_{W_\ell} \| = 1$ and always have $\sum_{i=1}^{k} \lambda_i \nu(S_i) \leq \nu(\mathbb{N})$. This tells us $\sigma = \sup(\sum_{i=1}^{k} \lambda_i \nu(S_i)) \leq \nu(\mathbb{N})$.

We have shown under this assumptions on ν that σ is finite. Since we are only interested in subsets of $\mathbb{N}$ which are finite, this collection is countable and so can be enumerated as $(S_i)_{i=1}^{\infty}$. We let the sequence of scalars $\boldsymbol{\alpha_i}$ be $\nu(S_i)$. We can then apply Theorem 16.7.3 and find $f \in \ell^1 \cong (\boldsymbol{c}_0)'$ so that $f(y_i) \geq \nu(S_i)$ where y_i is the $\boldsymbol{c}_0$ sequence associated with I_{S_i}. We know f has a representation in terms of the standard Schauder basis for ℓ^1. $f = \sum_{i=1}^{\infty} f(e_i) e_i$. Thus, $f(y_i) = \sum_{\{j:j \in S_i\}} f(e_j)$. Let the sequence $x \in \ell^1$ be defined by $x_i = f(e_i)$. Then we have $f(y_i) = \sum_{\{j:j \in S_i\}} x_j \geq \nu(S_i)$ for all i. So we can rewrite Theorem 16.7.3 in this context as

Theorem 16.7.4 The Existence of a Solution to a System of Linear Inequalities in $\boldsymbol{c}_0$

Let ν be a balanced game on $\mathbb{N}$ with $\nu(S) = 0$ unless S is finite or $S = \mathbb{N}$. Then the set of such subsets of $\mathbb{N}$ is countably infinite and can be denoted by $(S_i)_{i=1}^{\infty}$. Each characteristic function I_{S_i} then corresponds to an element y_i of $\boldsymbol{c}_0$. Then the family $(y_i \cong I_{S_i}) \subset \boldsymbol{c}_0$. Let $(\nu(S_i))$ be the associated family of real numbers. Let σ be defined by

$$\sigma = \sup\left(\sum_{i=1}^{n} \lambda_i\, \nu(S_i) \;:\; \sup\left\|\sum_{i=1}^{n} \lambda_i\, y_i\right\| = 1, \lambda_i > 0\right)$$

Then since σ is finite,

1. *There is $(x) \in \ell^1$ with $\|x\|_1 \leq \sigma$ so that $\sum_{\{j:j\in S_i\}} x_j \geq \nu(S_i)$ for all i. Note this implies $\|x\|_1 \geq \nu(\mathbb{N})$.*
2. *If there is a nonzero solution $z \in \ell^1$ to $\sum_{\{j:j\in S_i\}} z_i \geq \nu(S_i)$ for all i then*

$$\sigma = \inf\left(\|z\|_1\right)$$

Note this implies $\sigma \leq \|x\|_1$. Hence, combining, $\sigma = \|x\|_1$. Finally, $\|x\|_1 = \nu(\mathbb{N})$.

Proof 16.7.4
This is the argument we have just gone through. To show $\|x\|_1 = \nu(\mathbb{N})$, note since $\sigma = \|x\|_1$, we have $\|x\|_1 \leq \nu(\mathbb{N})$. This tells us $\sum_{i=1}^{\infty} x_i \leq \nu(\mathbb{N})$. Also, since $\sum_{\{j:j\in S_i\}} x_j \geq \nu(S_i)$ for all S_i, we have $\sum_{i=1}^{\infty} x_i \geq \nu(\mathbb{N})$. Hence, $\sum_{i=1}^{\infty} = \nu(\mathbb{N})$. ■

In this context, the **core** for the balanced game ν where $\nu(S) = 0$ if S is infinite unless $S = \mathbb{N}$ is

$$C_\nu = \{x \in \ell^1 \mid \sum_{i=1}^{\infty} x_i = \nu(\mathbb{N}),\; \sum_{\{j:j\in S\}} x_j \geq \nu(S),\; \forall\, S \subset \mathbb{N}\}$$

Our remarks above show clearly we have proved the following theorem.

Theorem 16.7.5 The Core to a c_0 Balanced Game ν is Nonempty

Let ν be a balanced game on $\mathbb{N}$ with $\nu(S) = 0$ unless S is finite or $S = \mathbb{N}$. There there is $x \in \ell^1$ so that $x \in C_\nu$.

Proof 16.7.5
This applies Theorem 16.7.4 and the comment right after its proof. ■

What do we do about the situation where the coalitions can be infinite in size? The identification we used between I_S and a sequence in $\boldsymbol{c}_0$ now fails as we see the limits of the sequences we construct need not be zero. However, these sequences are still bounded, so the appropriate sequence space is now ℓ^∞. The dual of ℓ^∞ is $ba(\mathbb{N}, 2^{\mathbb{N}})$ which is the space of all finitely additive charges or measures. Here our charges are to be non-negative. The core now becomes

$$C_\nu = \{\mu \in ba(\mathbb{N}, 2^{\mathbb{N}}) \mid \mu(\mathbb{N}) = \nu(\mathbb{N}),\; \mu(S) \geq \nu(S),\; \forall\, S \subset \mathbb{N}\}$$

where the concept of a payoff has been generalized to the term $\mu(S)$ for the finitely additive charge or measure μ. We can prove similar theorems about the non emptiness of the core in this context too, but we will leave our discussions here. We hope you will explore this in more detail on your own.

Homework

Exercise 16.7.1 *Consider the sets* $S_1 = \{1\}$, $S_2 = \{1,2\}$, $S_3 = \{1,2,3\}$, $S_4 = \{2\}$, $S_5 = \{2,3\}$ *and* $S_6 = \{3\}$. *Calculate* $\sum_{i=1}^{n} \lambda_i \boldsymbol{I}_{\boldsymbol{S_i}}$ *by hand and show the resulting* $\mathbf{c_0}$ *sequence you get is the same as the one defined by* (z_i) *in the text.*

Exercise 16.7.2 *Consider the sets* $S_1 = \{1\}$, $S_2 = \{1,4\}$, $S_3 = \{2,4,5\}$, $S_4 = \{5\}$, $S_5 = \{1,2,7\}$ *and* $S_6 = \{8\}$. *Calculate* $\sum_{i=1}^{n} \lambda_i \boldsymbol{I}_{\boldsymbol{S_i}}$ *by hand and show the resulting* $\mathbf{c_0}$ *sequence you get is the same as the one defined by* (z_i) *in the text.*

Exercise 16.7.3 *Consider the sets* $S_1 = \{1\}$, $S_2 = \{1,2\}$, $S_3 = \{1,2,3\}$, $S_4 = \{2\}$, $S_5 = \{2,3\}$ *and* $S_6 = \{3\}$. *Work out the sets* T_i *and the scalars* ζ_i *and show explicitly* $\sum_{j=1}^{n} \lambda_j I_{S_j} + \sum_{j \neq p_0, 1 \leq j \leq N} I_{T_j} + \theta I_Z$ *is the constant sequence of* 1*'s.*

Exercise 16.7.4 *Consider the sets* $S_1 = \{1\}$, $S_2 = \{1,4\}$, $S_3 = \{2,4,5\}$, $S_4 = \{5\}$, $S_5 = \{1,2,7\}$ *and* $S_6 = \{8\}$. *Work out the sets* T_i *and the scalars* ζ_i *and show explicitly* $\sum_{j=1}^{n} \lambda_j I_{S_j} + \sum_{j \neq p_0, 1 \leq j \leq N} I_{T_j} + \theta I_Z$ *is the constant sequence of* 1*'s.*

Part VII

Summing It All Up

Chapter 17

Summing It All Up

We have now come to the end of these notes which means you have covered the material in the first three courses. Let's summarize what we have done in the first three volumes. What we have been trying to do here is to balance the discussions of theory and abstraction with clear explanation and argument so students who are from many areas can follow the texts and use them profitably for self study even if they cannot take this material as a course. Many professionals need to add this sort of training to their toolkit later and they will do that if the text is accessible.

- **Basic Analysis One**: This is a primary text for a typical junior - senior year course in basic analysis. It can also be used as a supplementary text for anyone whose work requires that they begin to assimilate more abstract mathematical concepts as part of their professional growth after graduation or as a supplement to deficiencies in undergraduate preparation that leave them unprepared for the jump to the first graduate level analysis course. Students in other disciplines, such as biology and physics, also need a more theoretical discussion of these ideas and the writing is designed to help such students use this book for self study as their individual degree programs do not have enough leeway in them to allow them to take this course. which covers enough for a two semester sequence.

  This covers the fundamental ideas of calculus on the real line. Hence, sequences, function limits, continuity and differentiation, compactness and all the usual consequences as well as Riemann integration, sequences and series of functions and so forth. It is important to add pointers to extensions of these ideas to more general things regularly even though they are not gone over in detail. The problem with analysis on the real line is that the real line is everything: it is a metric space, a vector space, a normed linear space and an inner product space and so on. So many ideas that are actually separate things are conflated because of the special nature of $\Re$. **Basic Analysis One** is a two semester sequence which is different from the other texts to be discussed.

- **Basic Analysis Two**: A proper study of calculus in $\Re^n$ is no longer covered in most undergraduate and graduate curricula. Most students are learning this on their own or not learning it properly at all even though they get a master's or Ph.D. in mathematics. This course covers differentiation in $\Re^n$, integration in $\Re^2$ and $\Re^3$, the inverse and implicit function theorem and connections to these things to optimization.

- **Basic Analysis Three**: This covers the basics of three new kinds of spaces: metric spaces, normed linear spaces and inner product spaces. Since the students do not know measure theory at this point, many of the examples come from the sequence spaces ℓ^p. However, one can introduce many of the needed ideas on operators, compactness and completeness even with that restriction. This includes what is called **Linear Functional Analysis**. In general, the standards are discussed: the Hahn - Banach Theorem, the open and closed mapping theorems and some spectral theory for linear operators of various kinds. We try hard to push the reader to new levels of abstraction. We include a lot of discussion of dual spaces, some of their representations and so forth so you can have a lot of practical experience at working with the tools.

The last two volumes cover what is called measure theory and new ideas in topology and analysis. Briefly, these volumes are concerned with the following material:

- **Basic Analysis Four**: This covers the extension of what is meant by the length of an interval to more general ideas on the *length* or *measure* of a set. This is done abstractly first and then specialized to what is called **Lebesgue Measure**. In addition, more general notions are usually covered along with many ideas on the convergence of functions with respect to various notions.
- **Basic Analysis Five**: Here, we connect topology and analysis more clearly. There is a full discussion of topological and linear topological spaces, differential geometry and other topics.

You can see this sequence laid out in Figure 17.1. Since you have learned the first three volumes, the arrangement of this figure should make more sense. Since what we discuss in the first four volumes is still essentially what is a *primer* for the start of learning even more analysis and mathematics at this level, in Figure 17.1 we have explicitly referred to our texts using that label. The first volume is the one in the figure we call **A Primer on Analysis**, the second is **Primer Two: Escaping The Real Line**, the third (this text) is **Primer Three: Basic Abstract Spaces**, the fourth is **Primer Four: Measure Theory** and the fifth is **Primer Five: Functional Analysis**. Keep in mind these new labels as they show up in Figure 17.3 also. The typical way a graduate student takes this sequence is shown in Figure 17.2. There are lots of problems with how a mathematically interested student learns this material. An undergraduate major here would be required to take a two semester sequence in core real analysis, however, given the competition for graduate students in a mathematical sciences graduate program, it is often true that incoming master's degree students are admitted with deficiencies. One often has people come into the program with just one semester of core real analysis. Such students should take the second semester when they come into the program so that they can be exposed to Riemann integration theory and sequences of functions among other important things, but not all want to as it means taking a deficiency course. However, from Figure 17.2, you can see that jumping straight to **Basic Analysis Three** on Linear Analysis without proper training will almost always lead to inadequate understanding. Such students will have a poorly developed core set of mathematics which is not what is wanted. Hence, one needs to do as much as one can to train all undergraduate students in two full semesters of core real analysis. You, however, are now nicely trained and fit well into this sequence. Even if students take the full two semesters of core real analysis at the graduate level, the undergraduate curriculum today does not in general offer courses in:

- Basic Ordinary Differential Equation (ODE) Theory: existence of solutions, continuous dependence of the solution on the data and the structure of the solutions to linear ODEs. Since most students do not take **Basic Analysis Two**, they have not been exposed to differential calculus in $\Re^n$, the inverse and implicit function theorem and so forth. Many universities have replaced this type of course with a course in Dynamical Systems which cannot assume a rigorous background in the theory of ODE and so the necessary background must be discussed in a brief fashion. Again a text such as **Basic Analysis Two** designed for both self study and as a regular class is an important tool for the mathematical science major and other interested students to have available to them.
- Basic Theoretical Numerical Analysis: the discussion of the theory behind the numerical solution of ODEs, linear algebra factorizations such as QR and optimization algorithms. This is the course where one used to discuss operator norms for matrices as part of trying to understand our code implementations.

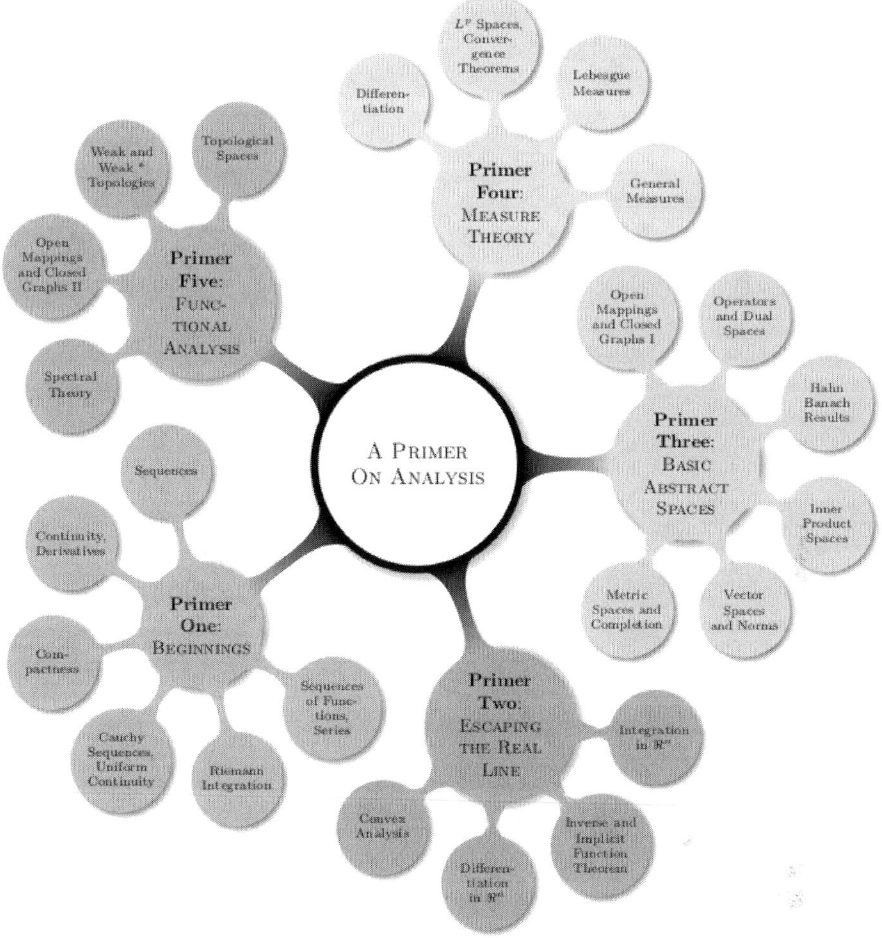

Figure 17.1: The general structure of the five core basic analysis courses.

- Basic Manifold Theory: the discussion of the extension of $\Re^n$ topology and analysis to the more general case of sets of objects locally like $\Re^n$. This material is pertinent to many courses in physics and to numerical Partial Differential Equations (PDE) as manifolds with a boundary are essential. Note this type of course needs **Basic Analysis Two** material, more topology and a more general theory of integration on chains.

All of the above areas of study are much enhanced by knowledge of the material in this third text. We have carefully developed the space $\mathbb{L}_2([a, b])$ so that we can talk intelligently about Hilbert Spaces of functions without needing full training in measure theory, but all of the above areas are greatly enhanced if you do have the training in Basic Analysis Four, (Peterson (16) 2019). We need to extend the idea of differential equations to non classical situations where the solution is not differentiable in the traditional sense. To do that we need more knowledge of analysis, topology and algebra. Ideas from algebraic topology can also be used to model signals into a complex biological or engineering system and both the mathematical sciences and physical sciences students have not been exposed to those ideas. Again, it is clear one must inculcate into the students the ability to read and think crit-

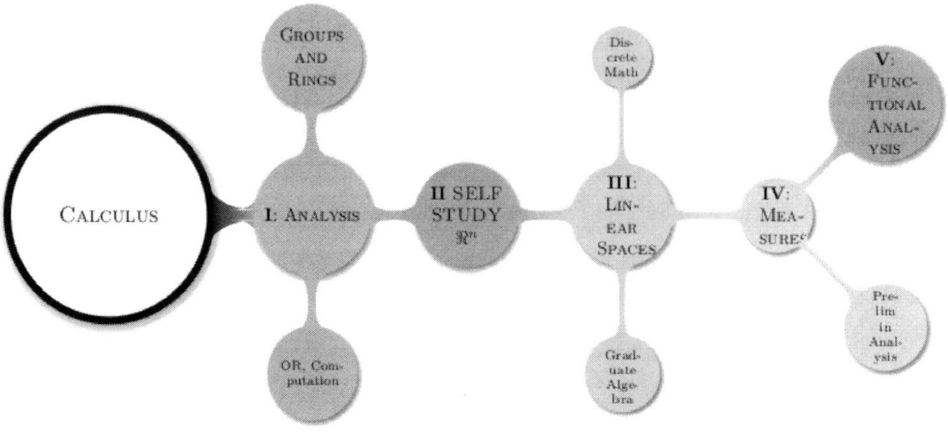

Figure 17.2: The analysis education pathway.

ically on their own, *outside of class and instructors* so they can become the scientists of the future. The way these proposed analysis courses fit into the grand scheme of things is shown in Figure 17.3. Now that you have read through the first three texts on basic analysis, you can see much better the many pathways you can take from this point on.

So here's to the future! Now that you have a much needed familiarity with basic analysis in $\Re^n$, you are ready to move into deeper levels of abstraction. The next journey starts in the next book (Peterson (16) 2019) which focuses on the study of abstract integration schemes using measures. You had a taste of this in this text when we talked about game theory and its connections to charges and the dual space of ℓ^∞. Also, in most graduate level mathematical sciences departments, you are now much closer to the preliminary examinations in analysis. These typically cover the material in books one, three and four of our series and you have already mastered the first three. So only one to go!

As always, remember to master the material in any courses you take at your university. Make sure you supplement as needed from books such as ours and learn to think carefully and deeply. You are preparing yourself for a life of research and the better you learn this material the easier it will be for you to read journal papers and think about solving problems that have not been solved before. Study all the proofs carefully as they tell you how to approach new problems. In the proof of each theorem, ask yourself what role each assumption plays in the proof so that you understand why the proof fails if you don't make that assumption. And remember, a new conjecture you have later in your own work always follows this pattern: it is true and you just haven't found the proof yet or it is false and there is a counterexample. This is a tricky thing to navigate and the best advice we can give you is to really, really learn the background material such as is in the books we have written and know the examples well. The final piece of advice is to learn to redefine your problem: when you are stuck a lot of times your failure leads to a better formulation of the problem you want to solve. It is the equivalent of walking around a wall instead of just butting into it over and over!
So enjoy your journey and get started!

Figure 17.3: Analysis connections.

Part VIII

References

References

[1] E. Ince. *Ordinary Differential Equations*. Dover Books on Mathematics, 1956.

[2] John W. Eaton, David Bateman, Søren Hauberg, and Rik Wehbring. *GNU Octave version 5.2.0 manual: a high-level interactive language for numerical computations*, 2020. URL https://www.gnu.org/software/octave/doc/v5.2.0/.

[3] Ky Fan. *On Systems of Linear Inequalities*, pages 99–155. Princeton University Press, 1956.

[4] Free Software Foundation. *GNU General Public License Version 3*, 2020. URL http://www.gnu.org/licenses/gpl.html.

[5] W. Fulks. *Advanced Calculus: An Introduction to Analysis*. John Wiley & Sons, third edition, 1978.

[6] B. Gelbaum and J. Olmstead. *Counter Examples in Analysis*. Dover Books, 2003.

[7] Z. Hall. *An Introduction to Molecular Neurobiology*. Sinauer Associates Inc., Sunderland, MA, 1992.

[8] E. Hewitt and K. Stromberg. *Real and Abstract Analysis*. Springer Verlag, 1975.

[9] B. Hille. *Ionic Channels of Excitable Membranes*. Sinauer Associates Inc., 1992.

[10] D. Johnston and S. Miao-Sin Wu. *Foundations of Cellular Neurophysiology*. MIT Press, 1995.

[11] E. Kreyszig. *Introduction to Functional Analysis and Applications*. Wiley, 1989.

[12] Y. Kunnai. *The Core and Balancedness*, pages 355–395. Elsevier Science Publishers, 1992.

[13] J. Little, A. Hodgson, and A. Gadgil. Modeling emissions of Volative Organic Compounds from New Carpets. *Atmospheric Environment*, 28:227–234, 1994.

[14] MATLAB. *Version Various (R2010a) - (R2019b)*, 2018 - 2020. URL https://www.mathworks.com/products/matlab.html.

[15] J. Peterson. *Calculus for Cognitive Scientists: Derivatives, Integration and Modeling*. Springer Series on Cognitive Science and Technology, Springer Science+Business Media Singapore Pte Ltd., 2016. URL http://dx.doi.org/10.1007/978-981-287-874-8.

[16] J. Peterson. *Basic Analysis IV: Measure Theory and Integration*. CRC Press, Boca Raton, Florida 33487, 2019.

[17] J. Peterson. *Basic Analysis V: Functional Analysis and Topology*. CRC Press, Boca Raton, Florida 33487, 2020.

[18] J. Peterson. *Basic Analysis I: Functions of a Real Variable*. CRC Press, Boca Raton, Florida 33487, 2020.

[19] J. Peterson. *Basic Analysis III: Mappings on Infinite Dimensional Spaces.* CRC Press, Boca Raton, Florida 33487, 2020.

[20] J. Peterson. *Basic Analysis II: A Modern Calculus in Many Variables.* CRC Press, Boca Raton, Florida 33487, 2020.

[21] W. Rall. *Core Conductor Theory and Cable Properties of Neurons*, chapter 3, pages 39 – 67. American Physiological Society, 1977.

[22] H. Royden. *Real Analysis (Classic Version) (4th Edition).* Pearson Modern Classics for Advanced Mathematics Series, 2017.

[23] G. Simmons. *Introduction to Topology and Modern Analysis.* McGraw-Hill Book Company, 1963.

[24] A. Taylor. *General Theory of Functions and Integration.* Dover Publications, Inc., 1985.

[25] T. Weiss. *Cellular Biophysics: Volume 1, Transport.* MIT Press, 1996.

[26] T. Weiss. *Cellular Biophysics: Volume 2, Electrical Properties.* MIT Press, 1996.

Part IX

Detailed Index

Index